Soil Microbiome in Green Technology Sustainability

Sesan Abiodun Aransiola • Habiba Iliyasu Atta
Naga Raju Maddela
Editors

Soil Microbiome in Green Technology Sustainability

Springer

Editors
Sesan Abiodun Aransiola
Department of Microbiology
Faculty of Science,
University of Abuja
Abuja, Nigeria

Habiba Iliyasu Atta
Department of Microbiology,
Faculty of Life Sciences
Ahmadu Bello University
Zaria, Nigeria

Naga Raju Maddela
Departamento de Ciencias Biológicas
Facultad de Ciencias de la Salud,
Universidad Técnica de Manabí
Portoviejo, Ecuador

ISBN 978-3-031-71843-4 ISBN 978-3-031-71844-1 (eBook)
https://doi.org/10.1007/978-3-031-71844-1

This Springer imprint is published by the registered company Springer Nature Switzerland AG
The registered company address is: Gewerbestrasse 11, 6330 Cham, Switzerland

Foreword

The book delves into the dynamics between the soil ecosystem and microorganisms; their interrelationships; and man's role in the soil health and quality. The book adequately captures the myriad of biological, chemical and physical activities and its vibrant nutrient-rich environment for green technological sustainability. The relevance of the application of the soil microbiome in sustainable solutions to global issues ranging from pollution, soil fertility, climate change, biocontrol of pathogens and nanotechnology is comprehensively addressed in the book. Also, the harmful effects of the pathogenic soil microorganisms including the toxigenic filamentous fungi are discussed in detail. The book provides a deep understanding of the metabolic and enzymatic capabilities of soil-dwelling bacterial and fungal species and their applications across various industries and sectors. The beneficial applications of soil microorganisms covered in this book highlight their potential as major players in nature-based solutions (NBS) and provide a shift in the ideology of microorganisms being harmful.

The authors of each chapter are experienced academics in the field of Life/Natural sciences with research centred on various aspects of the soil; their background in the field is reflected in the quality of the output. The authors have published scholarly works on the subject, and have also been teaching about the soil as an ecosystem. Their coverage on the various roles of microorganisms in the soil will be very helpful to undergraduate and postgraduate students as well as academics in the fields (but not limited to) of Microbiology, Soil Science, Biology, Botany, Biotechnology, Biochemistry and Environmental Engineering. The authors of the book are affiliated with tertiary institutions from various geographic regions of the world, thus providing a rich and diverse perspective on the subject.

The interdisciplinary approach in which the chapters were discussed by the authors provides an interphase for academics considering cross-cutting research or transitioning to a career in the industry. One of the appeals of the book is its coverage of recent technologies which also forms a knowledge base for fresh ideas for multidisciplinary research. The book also illuminates the promising roles microorganisms play in different professions including the broad fields of Agriculture and

Engineering, thus forming a potential template for projects with strong economic impact.

The book, "Soil Microbiome in Green Technology Sustainability", is an interesting read for not only professionals involved in teaching and research in tertiary institutions, but it could also be utilized as a source of ideas for industry experts seeking to apply beneficial soil microbes in green technology. It is a book for all stakeholders interested in achieving the sustainable development goals relating to the environment, health and terrestrial ecosystems.

Department of Pure and Applied Biology, Ladoke Akintola University of Technology
Ogbomoso, Oyo State, Nigeria
20th June 2024

Abiodun Ayanfemi Ayandele

Preface

The concept of the environment has been defined as the sum total of all surroundings of a living organism, including natural forces and other living things, which provide conditions for development and growth as well as of danger and damage. The deteriorating planetary condition, along with a deepening scientific understanding of and support for regenerative agriculture, is the ecological context for this book. Soils not only have a key role in fulfilling specific nutritional needs of plants. Soil also performs many ecosystem services, including carbon sequestration; water purification and soil contaminant reduction; climate regulation; nutrient cycling; providing a habitat for organisms; flood regulation; producing genetic resources and pharmaceuticals; and providing food, fuel and fibre. Most developing countries have already established laws and formal governmental structures to address their pertinent environmental problems, but few of them have been successful in alleviating those problems. The objective of this book is to stimulate discussion about the conservation of the environment and sustainability along regeneration of soil.

The first part of this book with seven chapters places heavy emphasis on the general state of the environment and revitalization of soil. It focuses on conservative agriculture, fauna and water for sustainable environment. Nevertheless, the administrative aspect of environmental laws has been a major issue of environmental conservation and protection policies, and it is now recognized that administration in many countries of the world is a neglected factor and may be tremendously inadequate. The book presents laws and policies for environmental conservation. Also, the analysis of these key issues in the book will draw the attention of readers and help them understand issues such as landscape and climate change perspective in conserving the environment, decision-making, rural and urban development, the community's role in the protection of the environment and realization of sustainability, recycling resources of soil and agroecosystem. The second part of the book has eight chapters and is designed to tackle different issues emanating from the "soil microbiomes and agricultural sustainability" through soil formation; soil health, desertification and soil biodiversity; topsoil regeneration and biosequestration; and soil erosion, mineral depletion and regeneration. The impact of pollutants in soil including organic, inorganic and other xenobiotics on soil degeneration is adequately discussed. The effect of anthropogenic activities on agricultural soils and the natural occurrence of soil dilapidation are also broadly discussed. The third section "soil microbiomes and roles in green technology" of this book with nine chapters lays emphasis on the factors influencing soil regeneration. Issues on regenerative agriculture for food security are addressed comprehensively, as well as soil regeneration and microbial community activities on terrestrial food chain. Finally, the future direction of environmental conservation and soil regeneration are examined.

The chapters were contributed by 92 Academicians/Scientists/Researchers from 14 different countries (Algeria, China, Ecuador, India, Nigeria, South Africa, the USA, the UK, Sweden, Rwanda, Thailand, Malaysia, Tanzania and Venezuela) across the world.

Abuja, Nigeria Sesan Abiodun Aransiola

Kaduna State, Nigeria Habiba Iliyasu Atta

Portoviejo, Ecuador Naga Raju Maddela

Acknowledgements

The Editorial team extends warm gratitude to all the contributors of this book project; their collective cooperation and prompt responses through the project eased the path to the successful completion of the book. Their expertise in the subject matter enhanced the quality of the chapters, and their dedication to meeting deadlines is highly commendable. We would also like to thank the reviewers for their valuable input and efforts in enhancing the overall quality of the chapters. We are grateful for the support and fulfilling experience gained from working with the Springer Editorial and Production unit. Finally, we would like to relay our deep appreciation to the Management of our respective institutions, University of Abuja, Abuja (Nigeria), Ahmadu Bello University, Zaria (Nigeria), and Universidad Técnica de Manabí (Ecuador) for their firm support throughout all the stages of the book preparation process, from the proposal to the completion of the project.

Sesan Abiodun Aransiola
Habiba Iliyasu Atta
Naga Raju Maddela

Contents

Contributors

A. K. Abdullahi Department of Biology, Faculty of Life Sciences, Ahmadu Bello University, Zaria, Nigeria

S. Abdullahi Department of Botany, Faculty of Life Sciences, Ahmadu Bello University, Zaria, Nigeria

Umar Abdulrahim School of Basic Medical Sciences, Department of Medical Microbiology, Central South University, Changsha, Hunan Province, China

Mustapha Abdulsalam Department of Microbiology, School of Sciences and Information Technology, Skyline University, Kano, Nigeria

Olabisi Peter Abioye Department of Microbiology, Federal University of Technology Minna, Niger State, Nigeria

Adeyemi Ojutalayo Adeeyo Institute of Nanotechnology and Water Sustainability, College of Science, Engineering and Technology (CSET), University of South Africa, The Science Campus, Roodeport, Johannesburg, South Africa

Olalekan Adekola Department of Geography, York St. John's University, York, UK

Ebenezer Adewuyi Ademola Department of Biological Sciences and Biotechnology, Caleb University, Imota, Lagos, Nigeria

Tarimoboere Agbalalah National Biotechnology Research and Development Agency, Abuja, Nigeria

Omosalewa Kikelomo Ajibola Department of Chemistry, Federal University of Technology, Akure, Nigeria

A. B. Alhassan Department of Biology, Faculty of Life Sciences, Ahmadu Bello University, Zaria, Nigeria

Hadiza Aliyu Bioresource Development Center National Biotechnology Research and Development Agency, Lugbe, Abuja, Nigeria

H. G. Amir School of Biological Sciences, Universiti Sains Malaysia, Pulau Pinang, Malaysia

Jean Carlo Andrade Universidad Tecnológica Indoamérica, Quito, Ecuador

Christian Bemgba Anjembe Department of Soil Science, Joseph Sarwuan Tarka University, Makurdi, Benue State, Nigeria

Idowu Blessing Apara Department of Chemistry, Federal University of Technology, Akure, Nigeria

Mounia Arab Laboratory of Applied Microbiology in Food, Biomedical and Environment, Department of Biology, Faculty of Nature Sciences and Life, University of Tlemcen, Tlemcen, Algeria
University of Sciences and Technology Houari Boumediene, Faculty of Biological Sciences, Algiers, Algeria

Sesan Abiodun Aransiola Department of Microbiology, Faculty of Science, University of Abuja, Abuja, Nigeria

Emily Avila Independent Researcher, Guayaquil, Ecuador

Babafemi Raphael Babaniyi Bioresources Development Centre, National Biotechnology Research and Development Agency, Abuja, Nigeria

Sunday Zeal Bala Department of Education, University of Maryland Eastern Shore, Princess Anne, MD, USA

Manjunatha Bangeppagari Department of Cell Biology and Molecular Genetics, Sri Devaraj Urs Academy of Higher Education and Research, Kolar, Karnataka, India

Suresh Babu Bastipati University College of Engineering Science and Technology Hyderabad, Jawaharlal Nehru Technological University Kukatpally, Hyderabad, Telangana, India

U. Bayero Department of Biology, Faculty of Life Sciences, Ahmadu Bello University, Zaria, Nigeria

B. H. Boje Gowd University College of Engineering Science and Technology Hyderabad JNTUH Kukatpally, Hyderabad, Telangana, India

Ademola Bisi-Omotosho Department of Sustainability Management and Innovation, London, UK

Aisha Bisola Bello Department of Biological Sciences, School of Applied and Natural Sciences, The Federal Polytechnic Bida, Niger State, Nigeria
Department of Pharmaceutical Microbiology, Faculty of Pharmaceutical Sciences, Ahmadu Bello University Zaria, Kaduna State, Nigeria

Arnab Chakraborty School of Pharmaceutical Technology, School of Health and Medical Sciences, Adamas University, Kolkata, India

Zubairu Umar Darma Department of Microbiology, Faculty of Natural and Applied Sciences, Umaru Musa 'Yar'adua University, Katsina, Nigeria

Bryan Xavier Daza Daza Departamento de Ciencias Biológicas, Facultad de Ciencias de la Salud, Universidad Técnica de Manabí, Portoviejo, Ecuador

Shehu-Alimi Elelu Department of Chemistry, Howard University, Washington, DC, USA

Kingsley Erhons Enerijiofi Department of Biological Sciences, College of Basic and Applied Sciences, Glorious Vision University, Ogwa, Edo State, Nigeria

P Gnana Deepu Center for Biotechnology, University College of Engineering Science and Technology JNTUH, Hyderabad, Telangana, India

Jogipeta Harihara Center for Biotechnology, University College of Engineering Science and Technology, JNTUH, Hyderabad, Telangana, India

Kathuroju Harikrishna Center for Biotechnology, University College of Engineering Science and Technology Hyderabad, Jawaharlal Nehru Technological University Kukatpally, Hyderabad, Telangana, India

Madinat Hassan Foresight Institute of Research and Translation (FIRAT), Kigali, Rwanda

José Herándes Universidad Tecnológica ECOTEC, Guayaquil, Ecuador Universidad Central de Venezuela, Caracas, Venezuela

Ibrahim Mohammed Hussaini Department of Microbiology, Faculty of Life Sciences, Ahmadu Bello University, Zaria, Kaduna State, Nigeria

Ganiyat Omotayo Ibrahim Department of Chemistry, Nottingham Trent University, Nottingham, UK

S. I. Ibrahim Department of Botany, Faculty of Life Sciences, Ahmadu Bello University, Zaria, Nigeria

Abdullahi Dabban Idris Department of Biological Sciences, School of Applied and Natural Sciences, The Federal Polytechnic Bida, Niger State, Nigeria
Department of Microbiology, Federal University of Technology Minna, Niger State, Nigeria

Musa Ojeba Innocent Department of Microbiology, School of Sciences and Information Technology, Skyline University Nigeria, Kano, Nigeria

Sirajo Salisu Jibia Department of Plant and Soil Sciences, Faculty of Agriculture, Chiang Mai University, Chiang Mai, Thailand

Srinivasan Kameswaran Department of Botany, Vikrama Simhapuri University College, Kavali, Andhra Pradesh, India

Priscilla Kini Department of Toxicology, University of Maryland Eastern Shore, Princess Anne, MD, USA

Nihel Klouche Laboratory of Applied Microbiology in Food, Biomedical and Environment, Department of Biology, Faculty of Nature Sciences and Life, University of Tlemcen, Tlemcen, Algeria

Patnam Prashanth Kumar Center for Biotechnology, University College of Engineering Science and Technology, JNTUH, Hyderabad, Telangana, India

Emmanuel Abayomi Ladapo School of Earth, Environment and Society, Bowling Green, OH, USA

Salami Olaitan Lateefat Department of Chemistry and Molecular Biology, University of Gothenburg, Gothenburg, Sweden

S. Stella Leh-Togi Zobeashia Naional Biotechnology Research and Development Agency, Abuja, Nigeria

Miracle Uwa Livinus Department of Microbiology, School of Sciences and Information Technology, Skyline University Nigeria, Kano, Nigeria

Naga Raju Maddela Departamento de Ciencias Biológicas, Facultad de Ciencias de la Salud, Universidad Técnica de Manabí, Portoviejo, Ecuador

Santiago Mafla Universidad Católica del Ecuador Ibarra, Ibarra, Ecuador

Wafa Malki Laboratory of Applied Microbiology in Food, Biomedical and Environment, Department of Biology, Faculty of Nature Sciences and Life, University of Tlemcen, Tlemcen, Algeria

Annie Julisa Delgado Mendoza Departamento de Ciencias Biológicas, Facultad de Ciencias de la Salud, Universidad Técnica de Manabí, Portoviejo, Ecuador

Katherine Georgina Menon Department of Microbiology, School of Allied Health Sciences, Malla Reddy University, Hyderabad, Telangana, India

Sandip Mondal Amity Institute of Pharmacy, Amity University, Kolkata, West Bengal, India

Nadjette Moumene Laboratory of Applied Microbiology in Food, Biomedical and Environment, Department of Biology, Faculty of Nature Sciences and Life, University of Tlemcen, Tlemcen, Algeria

Auwal Sagir Muhammad School of Informatics, Xiamen University, Xiamen, China

A. O. Musa Department of Botany, Faculty of Life Sciences, Ahmadu Bello University, Zaria, Nigeria

Anjaneyulu Musini Center for Biotechnology, University College of Engineering Science and Technology, JNTUH, Hyderabad, Telangana, India

Adamu Mustapha Department of Microbiology, Federal University of Technology, Minna, Niger State, Nigeria

Ignatius Mzungu Department of Microbiology, Federal University Dutsin-Ma, Katsina State, Nigeria

Surisetty Navyatha Center for Biotechnology, University College of Engineering Science and Technology, JNTUH, Hyderabad, Telangana, India

Pariki Nithin Department of Clinical Nutrition Dietetics and Food Science, School of Allied Health Sciences, Malla Reddy University, Hyderabad, Telangana, India

Amoge C. Ogu National Biotechnology Research and Development Agency, Abuja, Nigeria

Olusola David Ogundele Department of Chemical Sciences, Achievers University, Owo, Ondo State, Nigeria

Taiwo Stephen Okanlawon Department of Biological Sciences, College of Basic and Applied Sciences, Glorious Vision University, Ogwa, Edo State, Nigeria

Taiwo Hamidat Olaide Department of Chemistry, Federal University of Technology, Akure, Nigeria

Daramola Oluwatosin Olaoluwa Department of Chemistry, Federal University of Technology, Akure, Nigeria

Thankgod Onuoha Department of Biological Sciences, Novena University, Ogume, Delta State, Nigeria

Marjorie Jazmin Salazar Orellana Centro de Investigación, Innovación y Transferencia de Tecnología, Universidad Católica de Cuenca, Cuenca, Ecuador
Laboratorio de Ecología Microbiana y Principios Activos, Universidad Católica de Cuenca, Cuenca, Ecuador
Geociencias, Ambiente y Recursos Naturales - GARN - Grupo de Investigación, Universidad Católica de Cuenca, Cuenca, Ecuador

Cyril Chinedu Otali Department of Biological Sciences, Novena University, Ogume, Delta State, Nigeria

Joshua Akinropo Oyetade School of Materials, Energy, water, and Environmental Science, Nelson Mandela African Institution of Science and Technology, Arusha, Tanzania

Olumoye Ezekiel Oyetunji Department of Biosciences and Biotechnology, University of Medical Sciences, Ondo, Nigeria

Gopi Krishna Pitchika Department of Zoology, Vikrama Simhapuri University College, Kavali, Andhra Pradesh, India

Olaniyi JohnPaul Popoola Department of Geological Sciences, Achievers University, Owo, Ondo State, Nigeria

Saloni Prasad University College of Engineering Science and Technology Hyderabad, Jawaharlal Nehru Technological University Kukatpally, Hyderabad, Telangana, India

M. Ramakrishna Department of Botany, Vikrama Simhapuri University College, Kavali, Andhra Pradesh, India

Bellamkonda Ramesh Molecular Biology, University of Nebraska Medical Center, Omaha, NE, USA

Kondakindi Venkateswar Reddy University College of Engineering Science and Technology Hyderabad, Jawaharlal Nehru Technological University Kukatpally, Hyderabad, Telangana, India

Yahaya Yunusa Riko Department of Microbiology, Faculty of Natural and Applied Sciences, Umaru Musa 'Yar'adua University, Katsina, Nigeria

Juan Carlos González Rojas Centro de Investigación, Innovación y Transferencia de Tecnología, Universidad Católica de Cuenca, Cuenca, Ecuador
Laboratorio de Entomología Agrícola Fitosanitario, Universidad Católica de Cuenca, Jardín Botánico de Cuenca, Cuenca, Ecuador
Geociencias, Ambiente y Recursos Naturales - GARN - Grupo de Investigación, Universidad Católica de Cuenca, Cuenca, Ecuador

Saheed Kayode Saheed Department of I.T and Computing, American University of Nigeria, Yola, Nigeria

D. Sai Vaishnavi University College of Engineering Science and Technology Hyderabad, Jawaharlal Nehru Technological University, Kukatpally, Hyderabad, Telangana, India

Job Oloruntoba Samuel Department of Microbiology, Federal University of Technology, Minna, Niger State, Nigeria

Samson Senbore Department of Geology, University of Pretoria, Pretoria, South Africa

P. Sudhakar Reddy Department of Zoology, Vikrama Simhapuri University College, Kavali, Andhra Pradesh, India

B. Swapna Department of Botany, Vikrama Simhapuri University College, Kavali, Andhra Pradesh, India

Fanen Terdoo Department of Geography, Federal University Dutsin-Ma, Katsina State, Nigeria

Umar Aliyu Umar Department of Biochemistry, Faculty of Life Sciences, Ahmadu Bello University Zaria, Kaduna State, Nigeria

Priyanka Verma Kamla Raja Girls Government Post Graduate College, Gwalior, Madhya Pradesh, India

Jhonny Javier Zambrano Zambrano Departamento de Ciencias Biológicas, Facultad de Ciencias de la Salud, Universidad Técnica de Manabí, Portoviejo, Ecuador

About the Editors

Sesan Abiodun Aransiola is a Ph.D. holder in Environmental Microbiology in the Department of Microbiology, Federal University of Technology, Minna, Nigeria. He obtained his bachelor's degree (B.Tech) and master's degree (M.Tech) from the same department in 2009 and 2014, respectively. He has demonstrated his research expertise in the production of vermicasts from vermicomposting (Vermitechnology) of organic wastes to assist plants in the remediation of polluted soil with heavy metals. Currently, he is a Lecturer at the Department of Microbiology, University of Abuja, Nigeria. His area of interest is Environmental Microbiology with research area in phytoremediation, vermicomposting, biosorption and bioremediation of soil-contaminated environment. He is an award-winning researcher and has over 70 publications including book chapters, research and review articles of good international repute with high impact factor. He has co-edited scientific books of global interest which include *Microbial Biotechnology for Bioenergy* (Elsevier); *Ecological Interplays in Microbial Enzymology* and *Prospects for Soil Regeneration and Its Impact on Environmental Protection* (Springer); *Microbial Biofilms: Applications and Control*, *Marine Greens: Environmental, Agricultural, Industrial and Biomedical Applications*, *Phytoremediation in Food Safety: Risks and Prospects*, *Emerging Contaminants in Food and Food Products* (CRC Press) with many more in the process. He has worked to the rank of an assistant chief scientific officer at Bioresources Development Centre, National Biotechnology Research and Development Agency, Nigeria, where he was involved in the uses of vinasse (a by-product of ethanol) as bio-fertilizer to reclaim polluted soil for agricultural purposes. Abiodun is a member of Nigerian Society for Microbiology and American Society for Microbiology, among others.

Habiba Iliyasu Atta is an Associate Professor in the Department of Microbiology in Ahmadu Bello University, Zaria, Nigeria. Her career as an academician spans a period of 18 years. Her areas of research include Environmental Microbiology, Bioremediation, Microbial Ecology, Biofuel production, WASH and Biocontrol of pathogens. She is currently holding a sabbatical appointment in Baze University, Abuja, Nigeria. She was a Fulbright Visiting Scholar in the year 2021, at Rutgers University, USA, where she conducted postdoctoral research on studying aromatic hydrocarbon degrading bacteria using whole genome sequencing and other molecular tools. The research was targeted at harnessing beneficial soil bacteria in the removal of pollutants from the environment as a sustainable approach in remediating polluted ecosystems. She was a beneficiary of the Petroleum Technology Development Fund postgraduate scholarship, for her doctorate research on using molecular tools in identifying soil bacteria with ability to utilize polycyclic aromatic hydrocarbons. The scholarship also funded her doctorate research at Cardiff University, Wales, UK. Dr. Atta has conducted multidisciplinary research with collaborators in the fields of Soil Science, Chemical Engineering, Environmental Engineering, Polymer Technology, Built Environment and others. She has authored several peer-reviewed journal articles, contributed to book publications, and is an Associate Editor and Book reviewer for the Fulbright Chronicles. Dr. Atta serves as an expert and reviewer with the Intergovernmental Science-Policy Platform on Biodiversity and Ecosystem Services (IPBES). She is passionate about science communication, and is a strong advocate on STEM education and environmental pollution issues. She is currently the Institutional Coordinator of the Organization for

Women in Science for the Developing world (OWSD), Ahmadu Bello University Branch. Dr. Atta is a member of the Nigerian Society for Microbiology, American Society for Microbiology, Society for General Microbiology, Society of Environmental Toxicology and Chemistry (SETAC) and many other professional bodies.

Naga Raju Maddela received his M.Sc. (1996–1998) and Ph.D. (2012) in Microbiology from Sri Krishnadevaraya University, Anantapuramu, India. During his doctoral programme in the area of Environmental Microbiology, he investigated the effects of industrial effluents/insecticides on soil microorganisms and their biological activities. He is working as Faculty in Microbiology since 1998, teaching undergraduate and postgraduate students. He worked on Prometeo Investigator (fellowship received from SENESCYT) at Universidad Estatal Amazónica, Ecuador, during 2013–15, received "Postdoctoral Fellowship" (2016–2018) from Sun Yat-sen University, China. He also received external funding from "China Postdoctoral Science Foundation" in 2017, internal funding from "Universidad Técnica de Manabí" in 2020. He participated in national/international conferences, and presented research data in China, Cuba, Ecuador, India and Singapore. Currently, he is working as a Full Professor at the Facultad de Ciencias de la Salud, Universidad Técnica de Manabí, Portoviejo, Ecuador. He has been actively publishing scientific articles, books (authored and edited) and chapters since 2007. As of now, he published 100 articles, 25 books and 50 book chapters.

Part I
Soil as a Habitat to Microbes

Soil Microbes and Soil Contamination

Musa Ojeba Innocent, Adamu Mustapha, Mustapha Abdulsalam, Miracle Uwa Livinus, Job Oloruntoba Samuel, Shehu-Alimi Elelu, Salami Olaitan Lateefat, and Auwal Sagir Muhammad

Abstract Soil contamination, caused by human activities, has a negative effect on the health of soil and the functioning of ecosystems. This paper analyzes the crucial functions that soil microorganisms perform in the process of pollutant degradation and the subsequent impacts on soil microbiomes. Soil microbes efficiently break down organic pollutants, such as petroleum hydrocarbons and pesticides, via enzymatic pathways. They also convert inorganic contaminants, such heavy metals, through processes like biosorption and metal precipitation. Advanced bioremediation methods, such as bioaugmentation and biostimulation, utilize the ability of microorganisms to accelerate the degradation of contaminants. Contamination has a negative effect on soil microbiomes as it decreases the variety of microorganisms present, changes the makeup of the community, and interferes with metabolic processes. These alterations negatively affect the process of nutrient cycling, the fertility of the soil, the interactions between plants and microorganisms, and the biogeochemical processes. Contaminants have a negative impact on the structure and stability of soil by hindering the processes of microbial-mediated aggregation and affecting soil fauna, which are crucial for nutrient cycling and the breakdown of organic waste. The review emphasizes the significance of microbial variety and

M. O. Innocent (✉) · M. Abdulsalam
Department of Microbiology, Skyline University Nigeria, Kano, Nigeria

A. Mustapha · J. O. Samuel
Department of Microbiology, Federal University of Technology, Minna, Niger State, Nigeria

M. U. Livinus
Department of Biochemistry, School of Sciences and Information Technology, Skyline University Nigeria, Kano, Nigeria

S.-A. Elelu
Department of Chemistry, Howard University, Washington, DC, USA

S. O. Lateefat
Department of Chemistry and Molecular Biology, University of Gothenburg, Gothenburg, Sweden

A. S. Muhammad
School of Informatics, Xiamen University, Xiamen, China

S. A. Aransiola et al. (eds.), *Soil Microbiome in Green Technology Sustainability*,
https://doi.org/10.1007/978-3-031-71844-1_1

activity in preserving soil health and resilience, promoting the adoption of sustainable land management approaches to limit the impacts of contamination. Future research should prioritize the investigation of microbial-mediated degradation mechanisms, the refinement of bioremediation technologies, and the promotion of activities that improve soil health in contaminated environments. Through comprehending and utilizing the inherent capabilities of soil microorganisms, one can attain efficient remediation and ensure the long-term sustainability of soil ecosystems.

Keyword Contamination · Bioaugumentation · Biostimulation · Biogeochemical

1 Introduction

1.1 Overview on Soil Microbes and Soil Contamination

Numerous tiny organisms found in large numbers in the soil are referred to as soil microorganisms. Among the creatures are bacteria, fungi, archaea, and protists, all of which are essential to maintaining the health of the soil and the functioning of the ecosystem (Rani, 2022). Large quantities and ubiquitous, bacteria are essential to the recycling of nutrients, the breakdown of organic materials, and the suppression of dangerous infections. Through their vast mycelial networks, fungus enhance soil structure, make it easier for plants to absorb nutrients, and help break down organic materials. Though less well understood, archaea are known to live in extreme environments and to be heavily involved in the nitrogen cycle and methane production. Amoebae and flagellates are examples of protozoa that participate in both the control of other microbe populations and the recycling of nutrients (Parray et al., 2023). These bacteria have complex interactions with one another that are often advantageous to one another, enabling important activities like the nitrogen and carbon cycles. Knowing the range and functions of soil microorganisms is essential to using sustainable soil management methods since they are essential to soil fertility, plant health, and resistance to environmental stresses.

According to Thepbandit and Athinuwat (2024), soil contamination is the presence of hazardous substances in soil that are higher than normal levels and damage ecosystems, human health, and agricultural productivity. Many contaminants that are introduced by industrial activities, agricultural practices, urban growth, and improper garbage disposal can pollute soil. Typical pollutants are poisons like lead, cadmium, and arsenic that accumulate in soil from mining activities, industrial processes, and the use of particular fertilizers and pesticides (Aransiola et al., 2023). Designed especially to control weeds and pests in agriculture, pesticides and herbicides can persist in soil and harm biodiversity and soil health. Petroleum hydrocarbons contaminated soil via spills, leaks, and improper oil and fuel disposal reduces the soil's ability to support plant growth and microbial activity. Solvents, dioxins and polychlorinated biphenyls (PCBs) are examples of industrial chemicals that seriously jeopardize the quality of groundwater and soil. These chemicals frequently last for lengthy times and cause long-term environmental damage. Because of its several sources and dire effects, soil pollution poses significant hazards to human health and the ecosystem that need for quick response. Human activity including industrial processes, agriculture, mining, waste disposal, and urbanization can all result in soil contamination. Among the many pollutants that can build up in soil settings are heavy metals, herbicides, petroleum hydrocarbons, industrial chemicals, and novel contaminants (Aransiola et al., 2013; Snow, 2020; Abioye et al., 2021).

Both soil fertility and productivity suffer from soil pollution, as does human health. Direct contact, inhaling, or swallowing tainted soil or food grown in it bring these risks. Contaminated soil can also leak pollutants into groundwater, which can contaminate drinking water supplies and exacerbate environmental problems. To reduce the negative effects of soil contamination on ecosystems and human wellbeing, effective remediation strategies and support for sustainable land management techniques must be understood (Aransiola et al., 2021; Neelam & Tabassum, 2023).

1.2 Importance of Soil Microbiomes and Soil Contamination to Environmental Health and Sustainability

Soil microbiomes, the intricate communities of microorganisms living within soil, play a pivotal role in maintaining environmental health and sustainability (Shah et al., 2022). These microscopic organisms, including bacteria, fungi, protozoa, and viruses, engage in complex interactions that regulate nutrient cycling, plant growth, and soil structure. Moreover, they serve as indicators of soil health and are profoundly impacted by soil contamination, which can have far-reaching consequences on environmental well-being. One of the primary functions of soil microbiomes is nutrient cycling. Microbes decompose organic matter, releasing essential nutrients such as nitrogen, phosphorus, and potassium, which are vital for plant growth (Kaur et al., 2021). This process, known as mineralization, ensures the continuous availability of nutrients within the ecosystem. Additionally, certain microbes form symbiotic relationships with plants, aiding in nutrient uptake and enhancing their resilience to environmental stresses. Consequently, healthy soil microbiomes contribute to agricultural productivity, food security, and ecosystem stability. Soil microbiomes play a crucial role in carbon sequestration and climate regulation. Microorganisms store carbon in the soil through the decomposition of organic matter, helping mitigate greenhouse gas emissions and combatting climate change (Banerjee & Van Der Heijden, 2023). Soil acts as a significant carbon sink, with microbial activity influencing the balance between carbon storage and release. Therefore, disruptions to soil microbiomes, such as contamination with pollutants, can disrupt carbon cycling processes, exacerbating climate change impacts. Soil contamination, resulting from various human activities such as industrial processes, agriculture, and improper waste disposal, poses a significant threat to soil microbiomes and environmental health. Contaminants such as heavy metals, pesticides, and petroleum hydrocarbons can alter microbial community composition, inhibit microbial activity, and impair nutrient cycling processes. Moreover, some contaminants exhibit bioaccumulation and biomagnification, posing risks to human health and ecological integrity (Atiatorme et al., 2022). The impacts of soil contamination extend beyond microbial communities to encompass ecosystem services and human well-being. Polluted soils can lead to reduced agricultural productivity, compromised water quality, and loss of biodiversity. Contaminants may leach into groundwater, contaminating drinking water sources and posing risks to human health. Additionally, soil pollution can have economic ramifications, affecting industries reliant on soil resources, such as agriculture, forestry, and construction. Addressing soil contamination requires a multifaceted approach that integrates scientific research, policy interventions, and public awareness initiatives. Remediation techniques such as bioremediation, which harnesses the metabolic capabilities of soil microorganisms to degrade pollutants, offer sustainable solutions for restoring contaminated soils. Furthermore, implementing soil conservation practices, such as organic farming and reduced tillage, can mitigate soil erosion and reduce the influx of pollutants into soil ecosystems. Islam et al. (2020) opined that enhancing soil microbiome resilience is essential for promoting environmental sustainability and mitigating the impacts of soil contamination. Sustainable land management practices that prioritize soil health, biodiversity conservation, and ecosystem resilience are critical for safeguarding soil microbiomes and maintaining ecosystem functions. Additionally, fostering interdisciplinary collaborations between scientists, policymakers, and stakeholders can facilitate the development and implementation of strategies to address soil contamination and promote environmental stewardship. Soil microbiomes play a crucial role in maintaining environmental health and sustainability through their contributions to nutrient cycling, carbon sequestration, and ecosystem resilience. Soil contamination poses significant threats to soil microbiomes and ecosystem integrity, highlighting the importance of addressing soil pollution and promoting sustainable land management practices. By prioritizing soil health and biodiversity conservation, we can safeguard soil microbiomes and ensure the long-term sustainability of terrestrial ecosystems (Wang & Li, 2019).

1.3 Relationship Between Soil Microbiomes and Contamination

Environmental health depends critically on the delicate interplay between soil microbiomes and pollution. Microbial populations can be upset and their diversity and function altered by pollutants like heavy metals and insecticides. The health of the ecosystem is impacted by this disturbance of essential soil processes including nitrogen cycling and organic matter breakdown (Fajardo et al., 2019).

By bioremediation, in which some microorganisms break down pollutants, soil microbiomes, on the other hand, might affect the fate of contaminants (Fajardo et al., 2019).

Frossard et al. (2018) did, however, propose that chronic pollution can result from overuse of microbial repair capabilities. Methods for managing soil and reducing pollution need an understanding of this link. Ecosystem integrity can be preserved and soil resources managed sustainably by promoting healthy soil microbiomes and reducing pollution inputs.

2 Soil Microbiomes: Key Players in Soil Health

The health of soil and the sustainability of ecosystems depend on soil microbiomes, which are diverse communities of microbes (Dubey et al., 2019). These tiny creatures, which include nematodes, viruses, algae, bacteria, fungi, protozoa, and archaea, actinomycetes, and nutrient cycling, interact in complex ways to control plant development and the breakdown of organic materials (Suman et al., 2022).

Maintaining soil fertility, encouraging sustainable agriculture, and slowing down environmental deterioration all depend on an understanding of the function of soil microbiomes (Hartmann & Six, 2023).

Fertilizer cycling is one important job of soil microbiomes. Carbon, nitrogen, phosphorus, and sulfur are among the nutrients that bacteria are vital in cycling between their organic and inorganic forms so that plants may absorb them. Plant development and soil fertility are increased, for instance, by nitrogen-fixing bacteria converting atmospheric nitrogen into forms that plants can use. Similar symbiotic interactions between mycorrhizal fungi and plant roots enable nutrient absorption and increase plant resistance to external challenges. Soil microbiomes help terrestrial ecosystems be productive and sustainable by cycling nutrients (Lehmann et al., 2020).

Organic matter breakdown depends critically on soil microbiomes. As they break down organic stuff, decomposer microorganisms bacteria, fungus, and detritivorous nematodes release carbon dioxide and nutrients into the soil. The production of soil organic matter and recycling of nutrients improve soil structure, water retention, and nutrient availability, and these processes are dependent on one other. Among filamentous bacteria, actinomycetes are especially effective decomposers and generate enzymes that can break down complicated organic molecules (Islam et al., 2020). Soil bacteria help to soil fertility, carbon sequestration, and ecosystem resilience by breaking down organic material.

Islam et al. (2020) concluded from more research that soil microbiomes are essential for plant health and disease prevention. Beneficial microorganisms that improve nutrient absorption, water absorption, and disease resistance include rhizobacteria and mycorrhizal fungi. Plant-microbe relationships are a class of advantageous interactions that help to promote plant development and inhibit diseases, therefore lowering the need for chemical inputs in agriculture. In addition, certain bacteria generate enzymes and antibacterial substances that stop plant pathogens from growing, therefore helping soil ecosystems naturally manage illness.

But perturbations to soil microbiomes include changes in land use, soil pollution, and climate variability. Heavy metal, pesticide, and petroleum hydrocarbon contamination of soil can upset the structure and operation of the microbial community, therefore affecting the breakdown of organic matter and nutrient cycling. The need of reducing soil pollution and encouraging sustainable land management techniques is underscored by the potential cascade impact this disturbance might have on soil health, plant productivity, and ecosystem functioning (Timmis & Ramos, 2021).

2.1 Types of Soil Microbiomes

Soil microbiomes encompass a diverse array of microorganisms that inhabit the soil environment, playing crucial roles in nutrient cycling, organic matter decomposition, and plant health. Understanding the types of soil microbiomes is essential for comprehending their functions and contributions to soil health and ecosystem sustainability (Mishra et al., 2023; Musa et al., 2024). The different types of soil microbiomes include.

2.1.1 Bacteria

Bacteria are one of the most abundant and diverse groups of microorganisms in soil. They play essential roles in various soil processes, including nitrogen fixation, nutrient cycling, and organic matter decomposition. Nitrogen-fixing bacteria, such as Rhizobium and Azotobacter, convert atmospheric nitrogen into a form that plants can utilize, enhancing soil fertility (Galperin, 2018; Innocent et al., 2024). Additionally, bacteria contribute to soil aggregation, which improves soil structure and water retention.

2.1.2 Fungi

Fungi are another significant component of soil microbiomes, comprising both beneficial mycorrhizal fungi and decomposer fungi. Mycorrhizal fungi form symbiotic relationships with plant roots, enhancing nutrient uptake and improving plant resilience to environmental stresses (Chung et al., 2019). Decomposer fungi, such as saprophytic fungi, break down organic matter, releasing nutrients for plant uptake and contributing to soil organic matter formation.

2.1.3 Protozoa

Protozoa are single-celled organisms that play important roles in nutrient cycling and microbial biomass regulation. They primarily consume bacteria and other microorganisms, contributing to the regulation of microbial populations and the release of nutrients through microbial grazing. Protozoa also facilitate nutrient cycling by mineralizing organic matter and releasing nutrients in plant-available forms (Estrada et al., 2016).

2.1.4 Algae

Algae are photosynthetic microorganisms that contribute to soil fertility and organic matter decomposition. Although they are more commonly associated with aquatic environments, certain species of algae can thrive in soil, especially in moist and nutrient-rich conditions. Algae contribute to soil organic matter formation through photosynthesis and can enhance soil fertility by fixing atmospheric carbon and nitrogen (Abinandan et al., 2019).

2.1.5 Archaea

Archaea are prokaryotic microorganisms that inhabit diverse environments, including soil. While less studied than bacteria, archaea play essential roles in nitrogen cycling, methane metabolism, and organic matter decomposition in soil ecosystems. Methanogenic archaea, for example, produce methane through anaerobic decomposition processes, influencing soil greenhouse gas emissions (Baker et al., 2020).

2.1.6 Actinomycetes

Actinomycetes are a group of filamentous bacteria with characteristics of both bacteria and fungi. They are renowned for their ability to produce bioactive compounds, such as antibiotics and enzymes, with significant implications for soil health and ecosystem functioning (Bhatti et al., 2017) Actinomycetes contribute to organic matter decomposition, nutrient cycling, and disease suppression in soil environments.

2.1.7 Nematodes

Soil nematodes are microscopic roundworms that inhabit soil ecosystems and play diverse ecological roles. They are categorized into various functional groups based on their feeding habits, including bacterial feeders, fungal feeders, and plant parasites. Nematodes influence soil nutrient cycling, microbial community dynamics, and plant health, making them essential components of soil food webs (Sorribas et al., 2020).

2.2 *Functions of Soil Microbiomes*

The microscopic organisms that inhabit soil ecosystems, known as soil microbiomes, perform a multitude of crucial functions that are essential to the health of the soil, the long-term sustainability of ecosystems, and the general chemical cycle throughout the biosphere of the Earth. The complex interactions between the numerous microbes—bacteria, fungi, archaea, protozoa, algae, and viruses—control the cycle of nutrients, the breakdown of organic material, plant growth, and soil structure (Ossowicki et al., 2021). Understanding the functions of soil microorganisms is necessary to use sustainable soil management methods, progress agriculture, and protect the environment.

2.2.1 Nutrient Cycling

Soil microbes play a central role in nutrient cycling, facilitating the transformation of organic and inorganic nutrients essential for plant growth and ecosystem functioning. Decomposer microorganisms, such as bacteria and fungi, break down complex organic compounds present in dead plant and animal material, releasing nutrients like carbon, nitrogen, phosphorus, and sulfur into the soil. These nutrients are then taken up by plants or recycled within the soil ecosystem through microbial processes like mineralization, nitrification, and denitrification (Bahram et al., 2018).

2.2.2 Organic Matter Decomposition

Decomposer microbes are key players in the decomposition of organic matter, including plant litter, root exudates, and animal remains. Bacteria, fungi, and detritivorous nematodes secrete enzymes that break down complex organic compounds into simpler forms, releasing nutrients and carbon dioxide into the soil. This decomposition process contributes to soil fertility, organic matter turnover, and soil structure improvement Golchin et al. (2018).

2.2.3 Plant Nutrition and Health

Soil microbes form symbiotic relationships with plants, enhancing nutrient uptake, water absorption, and disease resistance. Mycorrhizal fungi, for example, form associations with plant roots, extending their nutrient-absorbing capacity and improving plant health (El-Ramady et al., 2022). Additionally, rhizobacteria fix atmospheric nitrogen and produce growth-promoting compounds like auxins and cytokinins, contributing to plant growth promotion and disease suppression.

2.2.4 Soil Structure and Aggregation

Soil microbes, particularly fungi and bacteria, play a crucial role in soil aggregation and structure formation (Yudina & Kuzyakov, 2023). Fungi produce substances like glomalin and polysaccharides that bind soil particles together, forming aggregates. These aggregates improve soil structure, porosity, and water infiltration, reducing erosion and enhancing soil fertility and resilience to environmental stresses (Lehmann et al., 2017).

2.2.5 Carbon Sequestration

Soil microbes contribute to carbon sequestration by incorporating organic carbon into soil organic matter. Decomposer microorganisms break down plant and animal residues, releasing carbon dioxide into the atmosphere. However, a portion of this carbon is stabilized in the soil as humus and microbial biomass, contributing to long-term carbon storage and mitigating climate change Tiefenbacher et al. (2021).

2.2.6 Disease Suppression

Some soil microbes have the ability to suppress soil-borne plant diseases through various mechanisms. Certain bacteria and fungi produce antimicrobial compounds and enzymes that inhibit the growth of plant pathogens, contributing to natural disease control in soil ecosystems (Miljaković et al., 2020). Additionally, beneficial microbes can induce systemic resistance in plants, enhancing their resilience to pathogen attacks.

2.2.7 Contaminant Degradation

Soil microbes play a vital role in the degradation and detoxification of pollutants and contaminants present in soil environments. Microbial degradation processes, including bioremediation, involve the transformation or mineralization of pollutants such as pesticides, heavy metals, and petroleum hydrocarbons into less harmful or inert forms. This microbial activity helps restore contaminated soils and ecosystems, mitigating environmental pollution (Tufail et al., 2021).

2.2.8 Nitrogen Fixation

Nitrogen-fixing bacteria and archaea convert atmospheric nitrogen gas into ammonia and other plant-available forms of nitrogen through biological nitrogen fixation. These microbes form symbiotic associations with plants or exist as free-living organisms in the soil (Tufail et al., 2021). Nitrogen fixation enhances soil fertility and reduces the need for synthetic nitrogen fertilizers in agriculture, contributing to sustainable soil management practices.

3 Types of Soil Contaminants

Human health, ecosystem integrity, and environmental quality are all at danger from compounds found in soil settings at quantities greater than normally occurring levels, known as soil contaminants (Bayabil et al., 2022). Agricultural methods, industrial operations, garbage disposal, and urbanization are among the many sources of these pollutants. Protecting soil resources, putting remediation plans into action, and evaluating environmental risks all depend on knowing the kinds of soil contaminants (Kowalska et al., 2018).

3.1 Heavy Metals

Heavy metals are metallic elements with high atomic weights that can accumulate in soil ecosystems, posing significant risks to environmental health (Ali et al., 2021). Common heavy metals found in soil include lead (Pb), cadmium (Cd), mercury (Hg), arsenic (As), chromium (Cr), and nickel (Ni). These metals can originate from industrial emissions, mining activities, agricultural inputs such as fertilizers and pesticides, and improper waste disposal. Heavy metals are persistent in the environment and can bioaccumulate in soil organisms, leading to adverse effects on human health and ecosystem functioning (Masindi & Muedi, 2018; Aransiola et al., 2024).

3.2 Pesticides

Pesticides are chemical substances used to control pests, including insects, weeds, and fungi, in agricultural and urban environments. These chemicals can persist in soil ecosystems for extended periods, posing risks to soil health, biodiversity, and water quality. Common types of pesticides found in soil include organochlorines, organophosphates, carbamates, and pyrethroids. Pesticides can leach into groundwater, contaminate surface water bodies, and harm non-target organisms, including beneficial soil microbes and pollinators (Meftaul et al., 2020). Kuppusamy et al. (2020) in their research find that petroleum hydrocarbons are persistent pollutants that can accumulate in soil, posing risks to human health and ecosystem functioning. They can inhibit soil microbial activity, reduce soil fertility, and contaminate groundwater resources, leading to long-term environmental impacts.

3.3 Petroleum Hydrocarbons

Petroleum hydrocarbons are organic compounds derived from crude oil and petroleum products, including gasoline, diesel, and lubricating oils. These hydrocarbons can contaminate soil environments through accidental spills, leaks, and improper disposal practices (Varjani, 2017).

3.4 Persistent Organic Pollutants (POPs)

POPs are organic compounds that are resistant to environmental degradation and can persist in soil ecosystems for extended periods. These compounds include polychlorinated biphenyls (PCBs), dioxins, and certain pesticides like dichlorodiphenyltrichloroethane (DDT) Jones (2021). POPs can enter soil environments through atmospheric deposition, industrial emissions, and improper waste disposal. They bioaccumulate in soil organisms, posing risks to human health and wildlife, including reproductive and developmental effects, immune system suppression, and cancer.

3.5 Industrial Chemicals

Industrial chemicals encompass a wide range of synthetic compounds used in manufacturing processes, including solvents, plasticizers, flame retardants, and pharmaceuticals. These chemicals can enter soil environments through industrial discharges, waste disposal, and land application of biosolids. Industrial chemicals can persist in soil ecosystems, posing risks to soil health, biodiversity, and human health. They can leach into groundwater, contaminate surface water bodies, and accumulate in food chains, leading to adverse effects on aquatic and terrestrial organisms (Bobbitt et al., 2017).

3.6 *Radionuclides*

Radionuclides are radioactive isotopes of elements that emit ionizing radiation, posing risks to human health and environmental quality. Radionuclides can enter soil environments through nuclear accidents, mining activities, and atmospheric fallout from nuclear weapons testing. Common radionuclides found in soil include uranium (U), radium (Ra), thorium (Th), and cesium-137 (Cs-137). These radionuclides can accumulate in soil, leading to radiation exposure for soil organisms and human populations (Adeola et al., 2023).

3.7 *Acidic or Alkaline Substances*

Acidic or alkaline substances can alter soil pH levels, affecting soil chemistry, nutrient availability, and microbial activity. Acidic substances, such as sulfuric acid and acid mine drainage, can lower soil pH, leading to soil acidification and aluminum toxicity. Alkaline substances, such as lime and cement kiln dust, can raise soil pH, affecting nutrient solubility and plant growth. Improper application of acidic or alkaline substances can degrade soil quality and impair ecosystem functioning Neina (2019).

4 Sources of Soil Contamination

The causes of soil pollution are many and include industrial activity, farming methods, waste management, mining operations, urbanization, and natural processes (Lü et al., 2018). The integrity of ecosystems, human health, and environmental quality are all at danger when these sources release a variety of pollutants into soil settings. Evaluating environmental hazards, carrying out remediation plans, and safeguarding soil resources for next generations all depend on knowing the causes of soil contamination (Li et al., 2019).

4.1 *Industrial Activities*

Significant contributors to soil pollution are the industrial, mining, energy, and chemical processing industries. By use of emissions, spills, leaks, and waste disposal procedures, industrial activities can discharge a range of contaminants into soil habitats. Typical pollutants linked to industrial activity include industrial chemicals, petroleum hydrocarbons, persistent organic pollutants (POPs), and heavy metals (Weldeslassie et al., 2018). Lead, mercury, and cadmium, for instance, can be released into soil and water bodies by mining operations through runoff, acid mine drainage, and tailings. Similarly, by handling, storing, and disposing of industrial chemicals and hazardous waste improperly, manufacturing facilities can leak them into soil and groundwater.

4.2 Agricultural Practices

Due to the fact that it uses fertilizers, pesticides, herbicides, and animal manure, agriculture is a major source of soil contamination. Along with trace elements like cadmium and arsenic, which can build up in soil over time and endanger both the health of the soil and the purity of the water, chemical fertilizers also include nitrogen, phosphorus, and potassium. Organochlorines, organophosphates and glyphosate are among the pesticides and herbicides that can seep into soil and harm surface water bodies, groundwater and non-target species. If improperly managed, livestock waste—such as manure and slurry—contains hormones, antibiotics, pathogens, and minerals that can contaminate soil and water.

4.3 Waste Disposal

Disposing of waste improperly can contaminate soil with a variety of pollutants through landfilling, open dumping, and burning. Because organic compounds, toxic substances, and heavy metals seep from solid waste, landfills are a major source of soil contamination. Pollutants from industrial and municipal trash dumped openly into soil and groundwater can endanger both human health and the ecosystem. Waste materials can be burned to produce airborne pollutants like heavy metals, furans, and dioxins that can settle on soil surfaces and pollute soil ecosystems (Cachada et al., 2018).

4.4 Mining Operations

A variety of pollutants may be released into soil habitats by mining operations, including the extraction, processing, and disposal of waste. Mining operations produce waste rock, acid mine drainage, and mine tailings, all of which can include sulfuric acid, heavy metals, and other hazardous materials (Long et al., 2021). Leaching into soil and water bodies, these pollutants can contaminate terrestrial and aquatic ecosystems and endanger animal and human health. Because pollutants and the history of mining operations linger, abandoned mines and mine sites continue to provide difficulties for environmental restoration and soil remediation (Ikhumetse et al., 2019).

4.5 Urbanization

By way of land disturbance, building operations, road runoff, and sewage discharges, urbanization and development can contaminate soil. Elevated concentrations of heavy metals, petroleum hydrocarbons, polycyclic aromatic hydrocarbons (PAHs), and other pollutants linked to industrial activity, urban runoff, and vehicle emissions can be found in urban soils (Gollin et al., 2016). Legacy pollutants from former land uses, such industrial sites, gas stations, and brownfield regions, may also contaminate urban soils (Gollin et al., 2016).

4.6 Natural Processes

Because they release naturally occurring pollutants into soil ecosystems, natural processes such as weathering, erosion, and volcanic activity can contaminate soil. Arsenic, selenium, and fluoride are examples of trace elements that can be released into soil and groundwater by the weathering of rocks and minerals (Hack, 2020). Volcanic eruptions, according to Zhang and Wang (2020), can throw volcanic ash and gases containing sulfur dioxide, hydrogen sulfide, and other harmful materials onto soil surfaces, therefore impairing soil fertility and ecosystem function.

4.7 Interactions Between Soil Microbiomes and Contaminants

In soil ecosystems, the fate, transport, and toxicity of pollutants are mostly determined by interactions between soil bacteria and contaminants. Different metabolic capacities of soil microbes—bacteria, fungus, archaea, protozoa, and algae—can affect the biotransformation, breakdown, immobilization, or mobilization of pollutants (Gonzalez et al., 2018). Appraising environmental hazards, creating remediation plans, and safeguarding soil resources all depend on an understanding of these relationships.

4.7.1 Biodegradation

Many pollutants, including pesticides, industrial chemicals, petroleum hydrocarbons, and persistent organic pollutants (POPs), can be biodegraded by soil bacteria. Enzymes found in some bacteria, such species of Bacillus, Rhodococcus, and Pseudomonas, can break down complicated organic molecules into simpler, less harmful ones. By means of oxidation, hydroxylation, and ring breakage, for instance, aerobic bacteria are able to metabolize hydrocarbons and produce carbon dioxide, water, and microbial biomass. Comparably, it is well known that fungi like white rot fungus may break down resistant substances like lignin and polycyclic aromatic hydrocarbons (PAHs) (Ahmed & Fakhruddin, 2018).

4.7.2 Bioremediation

Bioremediation is a procedure that utilizes the metabolic ability of soil microbes to clean up polluted soils and groundwater, with the help of microorganisms (Pande et al., 2020). Bioremediation techniques encompass in situ and ex situ methods, including biostimulation, bioaugmentation, and phytoremediation. Biostimulation refers to the process of augmenting microbial activity and promoting the breakdown of contaminants in soil settings through the provision of nutrients, oxygen, and electron acceptors. Bioaugmentation is the process of introducing certain groups of microorganisms or genetically modified microbes into polluted areas in order to improve the breakdown of pollutants. Phytoremediation is the process of employing plants to help absorb, move, and break down contaminants in soil environments by interacting with bacteria in the rhizosphere (Pande et al., 2020).

4.7.3 Sorption and Sequestration

Soil microbes can interact with contaminants through sorption and sequestration processes, whereby contaminants are adsorbed, bound, or immobilized onto microbial cell surfaces, extracellular polymeric substances (EPS), or soil organic matter. Microbial cells can serve as sorbents or sequestering agents for metals, metalloids, and organic compounds, reducing their bioavailability and toxicity to soil organisms and ecosystems. Additionally, microbial-mediated processes such as biomineralization and biosorption can precipitate metals into insoluble forms or bind organic contaminants onto microbial surfaces, facilitating their removal from soil environments (Saravanan et al., 2020).

4.7.4 Transformation and Mineralization

Soil bacteria have the ability to convert pollutants into various chemical forms by undergoing metabolic activities such oxidation, reduction, methylation, and demethylation. These changes can result in the production of metabolites, intermediates, or end products that may be more or less harmful, depending on the metabolic pathways of microorganisms and the surrounding environmental conditions. For instance, some bacteria have the ability to convert mercury into methylmercury, which is an extremely poisonous and bioaccumulative variant of mercury. Methylmercury can increase in concentration as it moves up the food chain, posing threats to both human health and animals. In contrast, several bacteria have the ability to convert methylmercury into less harmful forms through a process called demethylation, thereby decreasing its negative effects on the environment.

4.7.5 Mutualistic Interactions

The symbiotic relationships between soil microorganisms and plants might impact the destiny and movement of pollutants in soil settings. Mycorrhizal fungi establish symbiotic relationships with plant roots, hence augmenting nutrient assimilation, water absorption, and plant development (Saravanan et al., 2020). The presence of mutualistic interactions can impact the availability and movement of pollutants in soil by modifying soil pH, redox conditions, and microbial activity. In addition, specific rhizosphere microorganisms have the ability to enhance plant growth and phytoremediation through the production of growth-promoting chemicals, stimulation of root development, and improvement of plant resistance to pollutants.

4.7.6 Competition and Predation

Soil microorganisms engage in competition for resources such as nutrients, energy sources, and space, which in turn affects the processes of contaminant biotransformation and degradation. Microbial populations engaging in competitive interactions can impact the makeup and functioning of soil microbial communities, leading to changes in the rates and methods of pollutant degradation. In addition, specific microorganisms have the ability to feed on other microorganisms, such as bacteria that degrade contaminants, protozoa, and nematodes. This has an effect on the amount, variety, and functioning of microorganisms in soil habitats.

4.8 Microbial Degradation of Contaminants

The process of microbial degradation involves the breakdown of toxins by microorganisms, including bacteria, fungi, and algae, into simpler and less hazardous compounds. This natural occurrence exhibits considerable potential for environmental remediation, providing a sustainable and economically efficient method for the purification of contaminated locations (Pande et al., 2022).

The metabolic flexibility of bacteria is crucial to microbial degradation. They have enzyme machinery that can catalyze a variety of metabolic reactions, enabling them to use different contaminants as sources of energy and carbon. Hydrocarbon-degrading bacteria, such as Pseudomonas and Alcanivorax, employ enzymes called oxygenases to chemically transform hydrocarbons into carbon dioxide and water. White rot fungi, such as fungi of the genus Phanerochaete, utilize ligninolytic enzymes to break down complex organic pollutants such as polycyclic aromatic hydrocarbons (PAHs) and pesticides (Pande et al., 2022).

The efficacy of microbial degradation is contingent upon multiple factors, such as the nature of the pollutant, the composition of the microbial population, environmental parameters (e.g., temperature, pH, oxygen concentration), and the accessibility of vital nutrients. Engineered microbial consortia, consisting of numerous species with complementing metabolic capabilities, are sometimes used to improve degradation efficiency.

Krasucka et al. (2021) defined bioremediation as a method that utilizes microbial decomposition to address environmental contamination, and it includes a range of different techniques. In situ bioremediation refers to the treatment of pollutants directly at the location of contamination, without the need for excavation. On the other hand, ex situ bioremediation entails the removal of contaminated materials from the site and their treatment at a different location. Both methodologies have been effectively implemented in various environments, including instances such as oil spills, industrial sites, agricultural fields, and wastewater treatment plants.

An exemplary instance of microbial degradation in progress is the remediation of oil spills. After a spill, indigenous bacteria that can break down hydrocarbons multiply and metabolize the oil components, helping to break it down and eventually remove it from the environment. In addition, bioremediation strategies, such as the introduction of specialized microbial strains (bioaugmentation) and the provision of nutrients or oxygen (biostimulation), are used to speed up the process and improve effectiveness (Krasucka et al., 2021; Musa et al., 2024).

Although microbial degradation holds potential, it also encounters obstacles. Concerns include slow degradation rates, limited efficacy against specific pollutants, and potential unforeseen ecological consequences. Furthermore, the ability to scale up and maintain the effectiveness of bioremediation systems over a long period of time justifies the need for additional research and development.

4.9 *Factors Affecting Microbial Degradation*

Microbial degradation is influenced by a wide range of factors, including environmental, chemical, and biological characteristics. These factors affect the efficiency and effectiveness of biodegradation processes. Gaining a comprehensive understanding of these characteristics is essential for enhancing bioremediation procedures and effectively utilizing microorganisms to reduce environmental pollution (Pande et al., 2022; Musa et al., 2024).

4.9.1 Type of Contaminant

The nature and complexity of the contaminant significantly influence microbial degradation. Microorganisms exhibit varying capabilities to degrade different classes of pollutants, such as hydrocarbons, pesticides, chlorinated compounds, heavy metals, and pharmaceuticals. For instance, hydrocarbon-degrading bacteria like

Pseudomonas are well-adapted to metabolize aliphatic and aromatic hydrocarbons, while certain fungi possess ligninolytic enzymes capable of degrading complex organic pollutants like polycyclic aromatic hydrocarbons (PAHs). The chemical structure, stability, solubility, and toxicity of contaminants also affect their susceptibility to microbial degradation (Varjani & Upasani, 2017).

4.9.2 Microbial Community Composition

The structure, variety, and quantity of microbial communities in the polluted environment are essential factors in influencing the effectiveness of degradation. Indigenous microorganisms that have adjusted to the particular pollutants found in the environment frequently have a crucial function in the biodegradation procedures. In addition, the use of specialized microbial consortia or genetically modified microorganisms can improve the speed at which degradation occurs by offering additional metabolic capacities or overcoming restrictions associated with the existing populations of bacteria.

4.9.3 Environmental Conditions

Microbial degradation processes are significantly influenced by environmental parameters such as temperature, pH, moisture content, oxygen availability, and nutrition levels. Optimal environmental conditions facilitate microbial activity and enzymatic function, hence increasing degradation rates. For instance, mesophilic bacteria generally display their highest level of metabolic activity within a moderate temperature range of 20–40 °C, but thermophilic microbes flourish in temperatures over 50 °C. Enzyme activity and microbial development can be influenced by pH levels. Most microbes tend to thrive in circumstances that are slightly acidic to neutral, often ranging from pH 6 to 8. Sufficient moisture is crucial for promoting microbial growth and sustaining metabolic activity, while the presence of oxygen determines whether breakdown pathways are aerobic or anaerobic. The presence of nutrients, including carbon, nitrogen, phosphorus, and trace metals, is crucial for the growth of microorganisms and the production of enzymes that facilitate breakdown processes (Pande et al., 2022).

4.9.4 Availability of Electron Acceptors

The presence of electron acceptors, such as oxygen, nitrate, sulfate, and carbon dioxide, affects the redox potential of the environment and dictates the main microbial degradation routes. Aerobic degradation processes are more common in situations with abundant oxygen, where bacteria use oxygen as the final electron acceptor in aerobic respiration. Microorganisms in oxygen-deprived environments utilize different electron acceptors, such as nitrate, sulfate, or carbon dioxide, through various processes such denitrification, sulfate reduction, or methanogenesis. The presence of electron acceptors has a direct influence on the effectiveness and orientation of microbial degradation processes.

4.9.5 Bioavailability and Accessibility of Contaminants

The extent to which pollutants may be absorbed and reached by microbial populations is a critical factor in determining the speed at which they can be broken down. Complicated or insoluble pollutants can be trapped within soil aggregates, attached to mineral surfaces, or enclosed within organic substances, which restricts microbial access and enzymatic breakdown. The bioavailability of contaminants can be increased through physical and chemical processes such as desorption, solubilization, and emulsification. These processes release pollutants that are bound or trapped, allowing them to enter the water phase and making it easier for microbial degradation to occur.

4.9.6 Interactions with Co-contaminants

Co-contaminants in the environment can affect microbial degradation processes through mechanisms such as resource competition, cross-feeding interactions, and metabolic inhibition. Co-contaminants can engage in competition with target pollutants for microbial attachment sites, nutrients, or electron acceptors, resulting in a decrease in breakdown rates. Alternatively, the cooperative interactions between several groups of microorganisms might increase the efficiency of degradation by enabling the simultaneous breakdown of various pollutants or by supplying intermediate compounds that can be used as fuel for degradation processes (Livinus et al., 2024).

4.9.7 Environmental Stressors

Environmental stressors, such as elevated salinity, harsh temperatures, pollution with heavy metals, and exposure to xenobiotic chemicals, can have an impact on microbial degradation processes by limiting the development of microorganisms, modifying their metabolic activity, or triggering cellular stress responses. Certain microorganisms possess stress tolerance mechanisms, such as the synthesis of osmolytes, heat shock proteins, or metal chelators, to counteract the negative impacts of environmental stresses and sustain metabolic stability.

4.9.8 Presence of Inhibitors and Toxicity

Inhibitory chemicals and hazardous metabolites that arise during microbial degradation can impede degradation processes and suppress microbial activity. Inhibitors might arise from the pollutant itself, intermediary degradation products, or co-contaminants found in the environment. Toxicity has the capacity to disturb the cell membranes of microorganisms, hinder the functioning of enzymes, or alter metabolic pathways. This can result in reduced rates of degradation and lower survivability of the microorganisms.

4.9.9 Genetic and Physiological Characteristics of Microorganisms

The genetic and physiological attributes of microbes, such as their metabolic variety, range of enzymes, adaptation mechanisms, and ability to withstand stress, determine their capacity to break down particular pollutants and flourish under various environmental circumstances. Microorganisms that have adaptable metabolic capacities, a wide range of substrate specificity, and efficient pathways for utilizing nutrients are frequently more suitable for applications involving biodegradation.

4.10 Role of Soil Microbes in Contaminant Remediation

4.10.1 Biodegradation of Organic Contaminants

Soil microbes are adept at degrading a wide range of organic contaminants, including petroleum hydrocarbons, polycyclic aromatic hydrocarbons (PAHs), pesticides, and industrial chemicals. These microorganisms possess enzymatic machinery capable of catalyzing biochemical reactions that break down complex organic molecules into simpler, less toxic compounds (Ite & Ibok, 2019). For example, hydrocarbon-degrading bacteria like Pseudomonas and Alcanivorax utilize oxygenases and other enzymes to oxidize hydrocarbons, converting them into carbon dioxide and water. Fungi such as white rot fungi produce ligninolytic enzymes that can degrade recalcitrant compounds like PAHs. Through these metabolic processes, soil microbes play a crucial role in reducing the bioavailability and toxicity of organic contaminants in soil environments (Timmis & Ramos, 2021).

4.10.2 Transformation of Inorganic Contaminants

In addition to organic pollutants, soil microbes are involved in the transformation and immobilization of inorganic contaminants such as heavy metals and metalloids.

Certain bacteria and fungi possess metal-reducing and metal-binding capabilities, allowing them to precipitate metals as insoluble compounds or sequester them within microbial biomass. For instance, sulfate-reducing bacteria can biologically reduce sulfate to sulfide, which then reacts with metal ions to form insoluble metal sulfides. Similarly, metal-tolerant fungi like Aspergillus and Penicillium can accumulate metals within their mycelia through processes like biosorption and intracellular sequestration. These microbial-mediated transformations play a crucial role in reducing the mobility and bioavailability of toxic metals in contaminated soils.

4.10.3 Enhanced Bioremediation Techniques

Soil microbes are harnessed in various enhanced bioremediation techniques to accelerate the degradation of contaminants and enhance remediation efficiency. Bioaugmentation involves the introduction of exogenous microbial strains or consortia with specialized metabolic capabilities into contaminated soils to supplement indigenous microbial populations and enhance degradation rates. For example, genetically engineered bacteria or fungi expressing specific catabolic pathways can be introduced to degrade target contaminants more effectively. Biostimulation involves the modification of environmental conditions to promote the growth and activity of indigenous microbial communities capable of degrading contaminants. Strategies such as the addition of organic amendments, nutrients, electron acceptors, or oxygen can stimulate microbial activity and enhance degradation processes. These enhanced bioremediation techniques leverage the metabolic potential of soil microbes to accelerate contaminant remediation and restore soil quality.

4.10.4 Rhizosphere Interactions

The rhizosphere, the zone of soil surrounding plant roots, represents a hotspot of microbial activity and organic matter turnover. Soil microbes in the rhizosphere interact closely with plant roots through symbiotic associations, nutrient exchange, and the release of root exudates. These interactions influence microbial community structure, metabolic activity, and contaminant degradation potential in the rhizosphere. Plants can exude a variety of compounds, including organic acids, sugars, and amino acids, which serve as carbon and energy sources for soil microbes. Certain root exudates can also stimulate the expression of microbial genes involved in contaminant degradation pathways. In return, soil microbes can enhance plant growth and health through nutrient cycling, phytohormone production, and the suppression of soil-borne pathogens. Harnessing these rhizosphere interactions can enhance the effectiveness of phytoremediation, a remediation strategy that utilizes plants and associated soil microbes to uptake, transform, or degrade contaminants (Timmis & Ramos, 2021; Musa et al., 2024).

4.10.5 Long-Term Soil Health and Resilience

Beyond their immediate role in contaminant remediation, soil microbes contribute to the long-term health and resilience of soil ecosystems. Microbial communities are involved in critical ecosystem processes such as nutrient cycling, soil organic matter decomposition, and soil structure formation. By degrading contaminants and reducing their impact on soil microbial communities, soil microbes help maintain ecosystem functioning and resilience. Furthermore, mi crobial diversity and activity are closely linked to soil health indicators such as soil structure stability, nutrient availability, and plant productivity (Timmis & Ramos, 2021). Promoting soil microbial diversity and activity through sustainable land management practices can enhance soil resilience to environmental stressors and improve long-term soil quality.

5 Impacts of Contamination on Soil Microbiomes

The impacts of contamination on soil microbiomes are multifaceted, affecting microbial diversity, community composition, metabolic activity, and ecosystem functioning. Soil microbiomes play crucial roles in nutrient cycling, organic matter decomposition, plant-microbe interactions, and soil structure formation. Therefore, disturbances to soil microbiomes due to contamination can have far-reaching consequences for soil health, ecosystem resilience, and human well-being (Fajardo et al., 2019).

5.1 Loss of Microbial Diversity

Contamination can lead to a loss of microbial diversity in soil ecosystems. Pollutants such as heavy metals, pesticides, petroleum hydrocarbons, and industrial chemicals can exert selective pressures on microbial communities, favoring the proliferation of pollutant-tolerant or -resistant microorganisms while suppressing sensitive taxa. This shift in microbial community composition can reduce overall microbial diversity and alter the functional redundancy of soil microbiomes. Reduced microbial diversity may compromise ecosystem resilience and stability, making soil ecosystems more vulnerable to environmental stressors and disturbances (Timmis & Ramos, 2021; Abdulsalam et al., 2024).

5.2 Shifts in Microbial Community Composition

Contamination can induce shifts in microbial community composition, altering the relative abundance and distribution of microbial taxa in soil ecosystems. Certain contaminants may selectively enrich or deplete specific microbial groups, leading to changes in community structure and diversity. For example, heavy metal contamination can select for metal-tolerant microorganisms capable of metal resistance mechanisms such as efflux pumps, metal-binding proteins, and enzymatic detoxification pathways. Similarly, pesticide contamination can influence the abundance of pesticide-degrading microorganisms and non-target microbial taxa in soil environments. These shifts in microbial community composition can have cascading effects on ecosystem functioning, nutrient cycling, and biogeochemical processes (Livinus et al., 2024).

5.3 Disruption of Microbial Metabolic Pathways

Contaminants can disrupt microbial metabolic pathways and enzymatic activities, impairing critical ecosystem processes such as nutrient cycling, organic matter decomposition, and contaminant degradation. For example, heavy metals can inhibit enzyme activities involved in nutrient acquisition, respiration, and carbon metabolism, leading to decreased microbial biomass and metabolic activity. Similarly, pesticides and industrial chemicals may interfere with enzymatic reactions involved in cellular respiration, photosynthesis, and metabolic pathways, affecting microbial growth and activity. These disruptions to microbial metabolic pathways can compromise soil fertility, organic matter turnover, and ecosystem productivity (Fajardo et al., 2019).

5.4 Impaired Nutrient Cycling and Soil Fertility

Contamination can impair nutrient cycling and soil fertility by disrupting microbial-mediated processes such as nitrogen fixation, nitrification, denitrification, and mineralization. Microorganisms play key roles in cycling nutrients between organic and inorganic forms, facilitating nutrient uptake by plants and maintaining soil fertility. However, contamination-induced changes in microbial diversity, community composition, and metabolic activity can disrupt nutrient cycling processes, leading to imbalances in nutrient availability and soil fertility. For example, heavy metal contamination can inhibit nitrogen fixation by nitrogen-fixing bacteria and archaea, leading to nitrogen deficiencies in soil ecosystems. Similarly, pesticide contamination can suppress soil microbial communities involved in nitrogen cycling, reducing nitrogen mineralization rates and nutrient availability for plants (Musa et al., 2024).

5.5 Altered Plant-Microbe Interactions

Contamination can alter plant-microbe interactions and symbiotic relationships, affecting plant growth, health, and productivity. Soil microbes play crucial roles in nutrient acquisition, disease suppression, and stress tolerance in plants through symbiotic associations such as mycorrhizal symbiosis, nitrogen fixation, and rhizosphere interactions. However, contamination-induced changes in soil microbiomes can disrupt these beneficial interactions, leading to impaired plant growth and increased susceptibility to biotic and abiotic stressors. For example, heavy metal contamination can inhibit mycorrhizal colonization of plant roots, reducing nutrient uptake and plant growth. Similarly, pesticide contamination can disrupt beneficial rhizosphere interactions, altering microbial community dynamics and plant-microbe signaling pathways (Innocent et al., 2024).

5.6 Shifts in Soil Biogeochemical Processes

Contamination can induce shifts in soil biogeochemical processes, altering the rates and pathways of carbon, nitrogen, and phosphorus cycling in soil ecosystems. Microorganisms drive key biogeochemical processes such as carbon mineralization, nitrogen fixation, and phosphorus solubilization, which regulate soil fertility, greenhouse gas emissions, and ecosystem productivity. However, contamination-induced disruptions to microbial communities and metabolic activities can alter the dynamics of biogeochemical cycles, leading to imbalances in nutrient availability and greenhouse gas fluxes. For example, petroleum hydrocarbon contamination can stimulate microbial respiration and carbon mineralization rates, leading to increased carbon dioxide emissions from soil ecosystems. Similarly, pesticide contamination can inhibit microbial nitrogen fixation and nitrification processes, leading to nitrogen losses and greenhouse gas emissions.

5.7 Impaired Soil Structure and Stability

Contamination can impair soil structure and stability by affecting microbial-mediated processes such as aggregation, soil organic matter turnover, and root-microbe interactions (Abdulsalam et al., 2024). Soil aggregates play critical roles in soil structure formation, water infiltration, nutrient retention, and plant root growth. However, contamination-induced changes in microbial communities and enzymatic activities can disrupt soil aggregation processes, leading to decreased soil porosity, increased compaction, and reduced water infiltration rates. For example, heavy metal contamination can inhibit microbial production of extracellular polymeric substances (EPS), which serve as binding agents in soil aggregates, leading to soil erosion and loss of soil structure. Similarly, pesticide contamination can suppress soil microbial communities involved in organic matter decomposition, reducing the formation of stable soil aggregates.

5.8 Ecotoxicological Effects on Soil Fauna

Contamination can have ecotoxicological effects on soil fauna, including earthworms, nematodes, microarthropods, and soil-dwelling insects, which play crucial roles in soil nutrient cycling, organic matter decomposition, and soil ecosystem functioning. Soil organisms may be directly exposed to contaminants through ingestion, dermal contact, or inhalation, leading to physiological stress, biochemical alterations, and reproductive impairments. Additionally, contamination-induced changes in soil microbiomes and food web dynamics can indirectly impact soil fauna through alterations in food availability, habitat quality, and trophic interactions. For example, heavy metal contamination can accumulate in earthworm tissues, impairing nutrient uptake and reproductive capacity, while pesticide contamination can disrupt soil nematode communities, affecting nutrient cycling and plant health (Abdulsalam et al., 2024).

6 Mitigating Soil Contamination and Promoting Soil Microbial Health

Sustainable land management practices are essential for mitigating soil contamination and promoting soil microbial health. Soil contamination arises from various sources, including agricultural activities, industrial processes, urban development, and improper waste disposal. Contaminants such as heavy metals, pesticides, petroleum hydrocarbons, and industrial chemicals can adversely impact soil microbial communities, disrupting essential ecosystem functions and compromising soil fertility. Sustainable land management practices aim to minimize soil contamination risks, restore soil health, and promote the resilience of soil ecosystems (Alori et al., 2017).

6.1 Contaminant Source Control

Contaminant source control is a fundamental principle of sustainable land management. By preventing the release of pollutants into soil ecosystems, land managers can minimize soil contamination risks and protect soil microbial communities. Contaminant source control strategies may include implementing best management practices (BMPs) for agriculture, industry, and urban development, as well as regulatory measures to limit the use and discharge of hazardous substances. For example, farmers can adopt integrated pest management (IPM) practices to reduce reliance on chemical pesticides and herbicides, thereby minimizing the accumulation of pesticide residues in soil. Similarly, industries can implement pollution prevention measures, such as waste minimization, recycling, and treatment, to reduce the release of toxic substances into soil and water resources. By prioritizing contaminant source control, land managers can effectively reduce soil contamination risks and protect soil microbial health (Livinus et al., 2024).

6.2 Phytoremediation

Phytoremediation is a sustainable remediation technique that utilizes plants to remove, degrade, or immobilize contaminants from soil environments (Aransiola et al., 2019). Certain plant species, known as hyperaccumulators, have the ability to uptake and accumulate high concentrations of metals or organic pollutants in their tissues. By planting hyperaccumulator species in contaminated areas, land managers can effectively reduce soil contaminant levels and promote soil microbial health (Innocent et al., 2024). Phytoremediation processes may involve phytoextraction, where plants uptake contaminants from soil and translocate them to aboveground tissues, phytodegradation, where plants metabolize contaminants within their tissues, or phytostabilisation, where plants immobilize contaminants in soil through root uptake and binding processes. Phytoremediation offers a cost-effective, environmentally friendly approach to soil remediation, with minimal disturbance to soil ecosystems (Abdulsalam et al., 2024).

6.3 Biochar Amendment

Biochar is a carbon-rich material produced from the pyrolysis of biomass, such as agricultural residues, forestry waste, or organic byproducts. Biochar has been shown to improve soil quality, enhance nutrient retention, and promote microbial activity in soil ecosystems. When applied as a soil amendment, biochar can also adsorb contaminants, such as heavy metals and organic pollutants, reducing their bioavailability and mobility in soil. By incorporating biochar into contaminated soils, land managers can effectively mitigate the impacts of soil contamination while simultaneously promoting soil microbial health. Biochar amendment can enhance soil structure, increase water retention capacity, and provide habitat for beneficial soil microorganisms. Additionally, biochar can serve as a long-term carbon sink, helping to mitigate climate change and promote soil carbon sequestration (Livinus et al., 2024).

6.4 Cover Cropping

Cover cropping involves planting non-cash crops, such as legumes, grasses, or cover crops, during fallow periods or between cash crop rotations. Cover crops help protect soil from erosion, suppress weed growth, and improve soil structure and fertility. Additionally, certain cover crop species can release exudates that promote beneficial microbial activity in soil ecosystems. By incorporating diverse cover crop mixtures into agricultural rotations, land managers can effectively promote soil microbial diversity and function. Cover cropping can enhance nutrient cycling, increase soil organic matter content, and stimulate microbial activity in soil ecosystems. Additionally, cover cropping can help reduce the need for synthetic fertilizers and pesticides, thereby minimizing soil contamination risks and promoting sustainable soil management practices (Musa et al., 2024).

6.5 Crop Rotation and Diversification

Crop rotation and diversification are essential components of sustainable agriculture that promote soil health and reduce the risk of soil contamination. Rotating crops with different nutrient requirements, root structures, and pest susceptibilities can help break pest and disease cycles, improve soil structure, and enhance nutrient cycling. Additionally, diversifying crop rotations with legumes, forages, or green manure crops can increase soil organic matter inputs, promote nitrogen fixation, and stimulate microbial activity in soil ecosystems. By implementing diverse and resilient crop rotations, land managers can effectively reduce reliance on chemical inputs, mitigate soil contamination risks, and support soil microbial health (Innocent et al., 2024).

6.6 Organic Farming Practices

Organic farming practices prioritize soil health and biodiversity conservation by minimizing the use of synthetic fertilizers, pesticides, and genetically modified organisms (GMOs). Instead, organic farmers rely on natural inputs, crop rotations, and biological pest control methods to maintain soil fertility and suppress pests and diseases. Organic farming practices promote soil microbial diversity, enhance nutrient cycling, and improve soil structure and water retention capacity. By adopting organic farming practices, land managers can effectively reduce soil contamination risks, promote soil microbial health, and support sustainable agriculture (Innocent et al., 2024).

6.7 Soil Conservation Techniques

Soil conservation techniques, such as contour plowing, terracing, and conservation tillage, help minimize soil erosion, conserve soil moisture, and protect soil from degradation. By reducing soil erosion rates, these techniques help prevent the transport of contaminants off-site and mitigate the spread of soil pollution. Additionally, soil conservation practices promote soil microbial health by maintaining soil structure, organic matter content, and microbial habitat availability. By implementing soil conservation techniques, land managers can protect soil quality, preserve soil microbial communities, and enhance ecosystem resilience to contamination (Tack & Egene, 2019; Musa et al., 2024).

7 Conclusion

Soil microbiomes are essential for regulating the effects of soil contamination on the health and functioning of ecosystems. In this chapter, we have examined the complex connections between soil microbiomes and pollutants, emphasizing the varied reactions of microbial communities to various forms and concentrations of pollution. Soil pollutants, such as heavy metals, pesticides, petroleum hydrocarbons, and industrial chemicals, can have significant impacts on microbial diversity, community composition, metabolic activity, and ecosystem processes. Although soil contamination presents issues, there is an increasing acknowledgment of the resilience and adaptive capabilities of soil microbiomes in mitigating the effects of pollution (Singh et al., 2020). Microorganisms have exceptional metabolic flexibility and enzymatic ability, enabling them to break down, neutralize, and isolate pollutants via several biochemical routes. In addition, soil microbial communities possess functional redundancy and compensation mechanisms, which allow them to sustain vital ecological activities despite disturbances. Interdisciplinary research and collaborative efforts are necessary to enhance our comprehension of soil microbiomes and their reactions to pollution. The integration of molecular biology, microbial ecology, biogeochemistry, and environmental science will improve our capacity to forecast and control the effects of soil contamination on microbial populations and ecosystem services. Through the utilization of soil microbiomes, we can create novel approaches for soil remediation, ecosystem restoration, and sustainable land management. Enhancing soil health and resilience is crucial for tackling the intricate issues of soil contamination and guaranteeing the long-term viability of our food, water, and environmental assets. Implementing sustainable land management practices, such as controlling contaminant sources, using phytoremediation, incorporating biochar, practicing cover cropping and crop rotation, adopting organic farming methods, and prioritizing soil conservation, shows great potential in reducing soil contamination risks and enhancing soil microbial health (Singh et al., 2020).

References

Abdulsalam, M., Elehu, S.-A., Innocent, M. O., Livinus, M. U., Latifat, S. O., & Mohammed, A. S. (2024). Women empowerment in environmental conservation. Chapter (17). In *Prospect for soil regeneration and its impact on environmental protection* (pp. 351–370). Springer. https://doi.org/10.1007/978-3-031-53270-2_7. ISBN: 978-3-031-53270-2.

Abinandan, S., Subashchandrabose, S. R., Venkateswarlu, K., & Megharaj, M. (2019). Soil microalgae and cyanobacteria: The biotechnological potential in the maintenance of soil fertility and health. *Critical Reviews in Biotechnology, 39*(8), 981–998.

Abioye, O. P., Ijah, U. J. J., Aransiola, S. A., Auta, S. H., & Ojeba, M. I. (2021). Bioremediation of toxic pesticides in soil using microbial products. In R. Prasad, S. C. Nayak, R. N. Kharwar, & N. K. Dubey (Eds.), *Mycoremediation and environmental sustainability. Fungal biology.* Springer. https://doi.org/10.1007/978-3-030-54422-5_1

Adeola, A. O., Iwuozor, K. O., Akpomie, K. G., Adegoke, K. A., Oyedotun, K. O., Ighalo, J. O., Amaku, J. F., Olisah, C., & Conradie, J. (2023). Advances in the management of radioactive wastes and radionuclide contamination in environmental compartments: A review. *Environmental Geochemistry and Health, 45*(6), 2663–2689.

Ahmed, F., & Fakhruddin, A. N. M. (2018). A review on environmental contamination of petroleum hydrocarbons and its biodegradation. *International Journal of Environmental Sciences & Natural Resources, 11*(3), 1–7.

Ali, M. M., Hossain, D., Al-Imran, A., Khan, M. S., Begum, M., & Osman, M. H. (2021). Environmental pollution with heavy metals: A public health concern. In *Heavy metals-their environmental impacts and mitigation* (pp. 771–783). IntechOpen.

Alori, E. T., Dare, M. O., & Babalola, O. O. (2017). Microbial inoculants for soil quality and plant health. In *Sustainable agriculture reviews* (pp. 281–307). Springer.

Aransiola, S. A., Ijah, U. J. J., & Abioye, O. P. (2013). Phytoremediation of lead polluted soil by *Glycine max L. Applied and Environmental Soil Science, 2013*, Article ID 631619. https://doi.org/10.1155/2013/631619. https://www.hindawi.com/journals/aess/2013/631619/abs/

Aransiola, S. A., Ijah, U. J. J., Abioye, O. P., & Bala, J. D. (2019). Microbial-aided phytoremediation of heavy metals contaminated soil: A review. *European Journal of Biological Research, 9*(2), 104–125. http://www.journals.tmkarpinski.com/index.php/ejbr/article/view/157

Aransiola, S. A., Ijah, U. J. J., Abioye, O. P., & Victor-Ekwebelem, M. O. (2021). ANAMMOX in wastewater treatment. In N. R. Maddela, L. C. García Cruzatty, & S. Chakraborty (Eds.), *Advances in the domain of environmental biotechnology. Environmental and microbial biotechnology*. Springer. https://doi.org/10.1007/978-981-15-8999-7_15

Aransiola, S. A., Ijah, U. J. J., Abioye, O. P., & Bala, J. D. (2023). Vermicompost-assisted phytoremediation of toxic trace element-contaminated soil in Madaka, Nigeria using *Melissa officinalis* L and *Sida acuta. International Journal of Environmental Science and Technology.* https://doi.org/10.1007/s13762-022-04105-y

Aransiola, S. A., Ijah, U. J. J., Abioye, O. P., Bala, J. D., Rivadeneira-Mendoza, R. F., Luque, R., Rodríguez-Díaz, J. M., & Maddela, N. R. (2024). Micro and vermicompost assisted remediation of heavy metal contaminated soils using phytoextractors. *Case Studies in Chemical and Environmental Engineering., 9*(2024), 100755. https://doi.org/10.1016/j.cscee.2024.100755

Atiatorme, E., Bramhachari, P. V., Kariali, E., & Sudhakar, P. (2022). A paradigm shift in the role of the microbiomes in environmental health and agriculture sustainability. In *Understanding the microbiome interactions in agriculture and the environment* (pp. 83–101). Springer Nature Singapore.

Bahram, M., Hildebrand, F., Forslund, S. K., Anderson, J. L., Soudzilovskaia, N. A., Bodegom, P. M., Bengtsson-Palme, J., Anslan, S., Coelho, L. P., Harend, H., & Huerta-Cepas, J. (2018). Structure and function of the global topsoil microbiome. *Nature, 560*(7717), 233–237.

Baker, B. J., De Anda, V., Seitz, K. W., Dombrowski, N., Santoro, A. E., & Lloyd, K. G. (2020). Diversity, ecology and evolution of archaea. *Nature Microbiology, 5*(7), 887–900.

Banerjee, S., & Van Der Heijden, M. G. (2023). Soil microbiomes and one health. *Nature Reviews Microbiology, 21*(1), 6–20.

Bayabil, H. K., Teshome, F. T., & Li, Y. C. (2022). Emerging contaminants in soil and water. *Frontiers in Environmental Science, 10*, 873499.

Bhatti, A. A., Haq, S., & Bhat, R. A. (2017). Actinomycetes benefaction role in soil and plant health. *Microbial Pathogenesis, 111*, 458–467.

Bobbitt, N. S., Mendonca, M. L., Howarth, A. J., Islamoglu, T., Hupp, J. T., Farha, O. K., & Snurr, R. Q. (2017). Metal–organic frameworks for the removal of toxic industrial chemicals and chemical warfare agents. *Chemical Society Reviews, 46*(11), 3357–3385.

Cachada, A., Rocha-Santos, T., & Duarte, A. C. (2018). Soil and pollution: An introduction to the main issues. In *Soil pollution* (pp. 1–28). Academic Press.

Chung, D., Kim, H., & Choi, H. S. (2019). Fungi in salterns. *Journal of Microbiology, 57*, 717–724.

Dubey, A., Malla, M. A., Khan, F., Chowdhary, K., Yadav, S., Kumar, A., Sharma, S., Khare, P. K., & Khan, M. L. (2019). Soil microbiome: A key player for conservation of soil health under changing climate. *Biodiversity and Conservation, 28*, 2405–2429.

El-Ramady, H., Hajdú, P., Törős, G., Badgar, K., Llanaj, X., Kiss, A., Abdalla, N., Omara, A. E. D., Elsakhawy, T., Elbasiouny, H., & Elbehiry, F. (2022). Plant nutrition for human health: A pictorial review on plant bioactive compounds for sustainable agriculture. *Sustainability, 14*(14), 8329.

Estrada, R. M., Santos, M. R., Aguilar, M. E. V., & Aguilar, R. A. (2016). Protozoarios (Protozoa). *La biodiversidad en la Ciudad de México, 2*, 204–207.

Fajardo, C., Costa, G., Nande, M., Botías, P., García-Cantalejo, J., & Martín, M. (2019). Pb, Cd, and Zn soil contamination: Monitoring functional and structural impacts on the microbiome. *Applied Soil Ecology, 135*, 56–64.

Frossard, A., Donhauser, J., Mestrot, A., Gygax, S., Bååth, E., & Frey, B. (2018). Long-and short-term effects of mercury pollution on the soil microbiome. *Soil Biology and Biochemistry, 120*, 191–199.

Galperin, M. Y. (2018). What bacteria want. *Environmental Microbiology, 20*(12), 4221–4229.

Golchin, A., Baldock, J. A., & Oades, J. M. (2018). A model linking organic matter decomposition, chemistry, and aggregate dynamics. In *Soil processes and the carbon cycle* (pp. 245–266). CRC Press.

Gollin, D., Jedwab, R., & Vollrath, D. (2016). Urbanization with and without industrialization. *Journal of Economic Growth, 21*, 35–70.

Gonzalez, E., Pitre, F. E., Pagé, A. P., Marleau, J., Guidi Nissim, W., St-Arnaud, M., Labrecque, M., Joly, S., Yergeau, E., & Brereton, N. J. (2018). Trees, fungi and bacteria: Tripartite metatranscriptomics of a root microbiome responding to soil contamination. *Microbiome, 6*, 1–30.

Hack, H. R. G. (2020). Weathering, erosion, and susceptibility to weathering. In *Soft rock mechanics and engineering* (pp. 291–333). Springer.

Hartmann, M., & Six, J. (2023). Soil structure and microbiome functions in agroecosystems. *Nature Reviews Earth & Environment, 4*(1), 4–18.

Ikhumetse, A. A., Abioye, O. P., & Aransiola, S. A. (2019). Biosorption potential of bacteria on Lead and chromium in groundwater obtained from mining community. *Acta Scientific Microbiology, 2*(6), 123–137. https://actascientific.com/ASMI/pdf/ASMI-02-0252.pdf

Innocent, M. O., Oloruntoba, S. J., Mustapha, A., Mustapha, A., Vivian, N., Maude, A. M., Adedayo, O. A., & Tunde, T. A.'. G. (2024). Soil erosion, mineral depletion and regeneration. Chapter (7). In *Prospect for soil regeneration and its impact on environmental protection* (pp. 159–172). Springer. https://doi.org/10.1007/978-3-031-53270-2_7. ISBN: 978-3-031-53270-2.

Islam, W., Noman, A., Naveed, H., Huang, Z., & Chen, H. Y. (2020). Role of environmental factors in shaping the soil microbiome. *Environmental Science and Pollution Research, 27*, 41225–41247.

Ite, A. E., & Ibok, U. J. (2019). Role of plants and microbes in bioremediation of petroleum hydrocarbons contaminated soils. *International Journal of Environmental Bioremediation & Biodegradation, 7*, 1–19.

Jones, K. C. (2021). Persistent organic pollutants (POPs) and related chemicals in the global environment: Some personal reflections. *Environmental Science & Technology, 55*(14), 9400–9412.

Kaur, T., Devi, R., Kour, D., Yadav, A., Yadav, A. N., Dikilitas, M., Abdel-Azeem, A. M., Ahluwalia, A. S., & Saxena, A. K. (2021). Plant growth promoting soil microbiomes and their potential implications for agricultural and environmental sustainability. *Biologia, 76*(9), 2687–2709.

Kowalska, J. B., Mazurek, R., Gąsiorek, M., & Zaleski, T. (2018). Pollution indices as useful tools for the comprehensive evaluation of the degree of soil contamination–A review. *Environmental Geochemistry and Health, 40*, 2395–2420.

Krasucka, P., Pan, B., Ok, Y. S., Mohan, D., Sarkar, B., & Oleszczuk, P. (2021). Engineered biochar–A sustainable solution for the removal of antibiotics from water. *Chemical Engineering Journal, 405*, 126926.

Kuppusamy, S., Maddela, N. R., Megharaj, M., Venkateswarlu, K., Kuppusamy, S., Maddela, N. R., Megharaj, M., & Venkateswarlu, K. (2020). Ecological impacts of total petroleum hydrocarbons. In *Total petroleum hydrocarbons: Environmental fate, toxicity, and remediation* (pp. 95–138). Springer.

Lehmann, A., Leifheit, E. F., & Rillig, M. C. (2017). Mycorrhizas and soil aggregation. In *Mycorrhizal mediation of soil* (pp. 241–262). Elsevier.

Lehmann, J., Bossio, D. A., Kögel-Knabner, I., & Rillig, M. C. (2020). The concept and future prospects of soil health. *Nature Reviews Earth & Environment, 1*(10), 544–553.

Li, C., Zhou, K., Qin, W., Tian, C., Qi, M., Yan, X., & Han, W. (2019). A review on heavy metals contamination in soil: Effects, sources, and remediation techniques. *Soil and Sediment Contamination: An International Journal, 28*(4), 380–394.

Livinus, M. U., Bala, S. Z., Abdulsalam, M., Innocent, M. O., Hassan, M., & Kini, P. (2024). Natural occurrences of soil dilapidation. Chapter (9). In *Prospect for soil regeneration and its impact on environmental protection* (pp. 205–223). Springer. https://doi.org/10.1007/978-3-031-53270-2_7. ISBN: 978-3-031-53270-2.

Long, Z., Huang, Y., Zhang, W., Shi, Z., Yu, D., Chen, Y., Liu, C., & Wang, R. (2021). Effect of different industrial activities on soil heavy metal pollution, ecological risk, and health risk. *Environmental Monitoring and Assessment, 193*, 1–12.

Lü, H., Mo, C. H., Zhao, H. M., Xiang, L., Katsoyiannis, A., Li, Y. W., Cai, Q. Y., & Wong, M. H. (2018). Soil contamination and sources of phthalates and its health risk in China: A review. *Environmental Research, 164*, 417–429.

Masindi, V., & Muedi, K. L. (2018). Environmental contamination by heavy metals. *Heavy metals, 10*(4), 115–133.

Meftaul, I. M., Venkateswarlu, K., Dharmarajan, R., Annamalai, P., & Megharaj, M. (2020). Pesticides in the urban environment: A potential threat that knocks at the door. *Science of the Total Environment, 711*, 134612.

Miljaković, D., Marinković, J., & Balešević-Tubić, S. (2020). The significance of Bacillus spp. in disease suppression and growth promotion of field and vegetable crops. *Microorganisms, 8*(7), 1037.

Mishra, A., Singh, L., & Singh, D. (2023). Unboxing the black box—One step forward to understand the soil microbiome: A systematic review. *Microbial Ecology, 85*(2), 669–683.

Musa, O. I., Samuel, J. O., Ilyasu, U. S., Akande, S. A., Tiamiyyu, A. T., Ijah, U. J. J., Abioye, O. P., Maude, A. M., & Mustapha, A. (2024). Distribution of biomass sources for bioenergy production: Challenges and benefits. Chapter (4). In *Microbial biotechnology for bioenergy* (pp. 65–79). Elsevier. https://doi.org/10.1016/B978-0-443-14112-6.00004-3

Neelam, A., & Tabassum, S. (2023). Optical sensing technologies to elucidate the interplay between plant and microbes. *Micromachines, 14*(1), 195.

Neina, D. (2019). The role of soil pH in plant nutrition and soil remediation. *Applied and Environmental Soil Science, 2019*, 1–9.
Ossowicki, A., Raaijmakers, J. M., & Garbeva, P. (2021). Disentangling soil microbiome functions by perturbation. *Environmental Microbiology Reports, 13*(5), 582–590.
Pande, V., Pandey, S. C., Sati, D., Pande, V., & Samant, M. (2020). Bioremediation: An emerging effective approach towards environment restoration. *Environmental Sustainability, 3*, 91–103.
Pande, V., Pandey, S. C., Sati, D., Bhatt, P., & Samant, M. (2022). Microbial interventions in bioremediation of heavy metal contaminants in agroecosystem. *Frontiers in Microbiology, 13*, 824084.
Parray, J. A., Shameem, N., & Egamberdieva, D. (Eds.). (2023). *Microbiome drivers of ecosystem function*. Elsevier.
Rani, M. (2022). Pedosphere: A hot spot of the largest and most complex diversity of microorganisms among terrestrial ecosystems. In *Structure and functions of pedosphere* (pp. 83–101). Springer Nature Singapore.
Saravanan, A., Jeevanantham, S., Narayanan, V. A., Kumar, P. S., Yaashikaa, P. R., & Muthu, C. M. (2020). Rhizoremediation–A promising tool for the removal of soil contaminants: A review. *Journal of Environmental Chemical Engineering, 8*(2), 103543.
Shah, A. M., Khan, I. M., Shah, T. I., Bangroo, S. A., Kirmani, N. A., Nazir, S., Malik, A. R., Aezum, A. M., Mir, Y. H., Hilal, A., & Biswas, A. (2022). Soil microbiome: A treasure trove for soil health sustainability under changing climate. *Land, 11*(11), 1887.
Singh, C., Tiwari, S., & Singh, J. S. (2020). Biochar: A sustainable tool in soil pollutant bioremediation. In *Bioremediation of industrial waste for environmental safety: Volume II: Biological agents and methods for industrial waste management* (pp. 475–494). Springer.
Snow, T. (2020). Ecosystem metabolisms and functions: An eco-literacy framework and model for designers. *FormAkademisk, 13*(4), 4–4.
Sorribas, F. J., Djian-Caporalino, C., & Mateille, T. (2020). Nematodes. In *Integrated pest and disease management in greenhouse crops* (pp. 147–174). Springer.
Suman, J., Rakshit, A., Ogireddy, S. D., Singh, S., Gupta, C., & Chandrakala, J. (2022). Microbiome as a key player in sustainable agriculture and human health. *Frontiers in Soil Science, 2*, 821589.
Tack, F. M., & Egene, C. E. (2019). Potential of biochar for managing metal contaminated areas, in synergy with phytomanagement or other management options. In *Biochar from biomass and waste* (pp. 91–111). Elsevier.
Thepbandit, W., & Athinuwat, D. (2024). Rhizosphere microorganisms supply availability of soil nutrients and induce plant defense. *Microorganisms, 12*(3), 558.
Tiefenbacher, A., Sandén, T., Haslmayr, H. P., Miloczki, J., Wenzel, W., & Spiegel, H. (2021). Optimizing carbon sequestration in croplands: A synthesis. *Agronomy, 11*(5), 882.
Timmis, K., & Ramos, J. L. (2021). The soil crisis: The need to treat as a global health problem and the pivotal role of microbes in prophylaxis and therapy. *Microbial Biotechnology, 14*(3), 769–797.
Tufail, A., Price, W. E., Mohseni, M., Pramanik, B. K., & Hai, F. I. (2021). A critical review of advanced oxidation processes for emerging trace organic contaminant degradation: Mechanisms, factors, degradation products, and effluent toxicity. *Journal of Water Process Engineering, 40*, 101778.
Varjani, S. J. (2017). Microbial degradation of petroleum hydrocarbons. *Bioresource Technology, 223*, 277–286.
Varjani, S. J., & Upasani, V. N. (2017). A new look on factors affecting microbial degradation of petroleum hydrocarbon pollutants. *International Biodeterioration & Biodegradation, 120*, 71–83.
Wang, L., & Li, X. (2019). Steering soil microbiome to enhance soil system resilience. *Critical Reviews in Microbiology, 45*(5–6), 743–753.

Weldeslassie, T., Naz, H., Singh, B., & Oves, M. (2018). Chemical contaminants for soil, air and aquatic ecosystem. In *Modern age environmental problems and their remediation* (pp. 1–22). Springer.

Yudina, A., & Kuzyakov, Y. (2023). Dual nature of soil structure: The unity of aggregates and pores. *Geoderma, 434*, 116478.

Zhang, Q., & Wang, C. (2020). Natural and human factors affect the distribution of soil heavy metal pollution: A review. *Water, Air, & Soil Pollution, 231*, 1–13.

Soil Microbial Products: Specific Roles in Green Vegetations

Ebenezer Adewuyi Ademola, Olumoye Ezekiel Oyetunji, Adeyemi Ojutalayo Adeeyo, and Joshua Akinropo Oyetade

Abstract Soil microorganisms inhabit a unique ecological niche, characterized by intimate associations with other diverse living organisms, driving ecosystem processes and plant fertility. The microorganisms synthesize a range of metabolites ranging from enzymes to phytohormones, antimicrobial substances, organic acids, siderophores, volatile organic acids, extracellular polysaccharide substances, among others. The microbial products contribute to green vegetation in terms of plant growth and productivity directly or indirectly by improving soil nutrients and its uptake by plants, controlling certain physiological and metabolic processes in the plants, inhibiting plant pathogens, and increasing the plant response to certain biotic and abiotic stresses. This has led to some innovations in the use of soil microbes in biofertilizers, inoculant technology, in environmentally friendly green vegetation, and sustainable agriculture.

Keywords Ecosystem · Plant · Microorganism · Technology · Agriculture

E. A. Ademola (✉)
Department of Biological Sciences and Biotechnology, Caleb University, Imota, Lagos, Nigeria
e-mail: ebenezer.ademola@calebuniversity.edu.ng

O. E. Oyetunji
Department of Biosciences and Biotechnology, University of Medical Sciences, Ondo, Nigeria

A. O. Adeeyo
Institute of Nanotechnology and Water Sustainability, College of Science, Engineering and Technology (CSET), University of South Africa, The Science Campus, Roodeport, Johannesburg, South Africa

J. A. Oyetade
School of Materials, Energy, water, and Environmental Science, Nelson Mandela African Institution of Science and Technology, Arusha, Tanzania

S. A. Aransiola et al. (eds.), *Soil Microbiome in Green Technology Sustainability*,
https://doi.org/10.1007/978-3-031-71844-1_2

1 Introduction

Soil is the uppermost layer of the earth's crust, formed by physical, chemical, and biological processes generally referred to as weathering. It is a complex environment and is composed of organic matters, mineral particles, water, living organisms, and air, organized into horizons (Costa et al., 2018; Bekele & Getaneh, 2022). Soil living organisms are a major part of the soil ecosystem and include plants, animals, and microorganisms. The interactions among these organisms and other components of The soil environment make it a unique ecosystem. Of all the groups of soil living organisms, soil microorganisms are the dominant type, and these include archaea, bacteria, actinomycetes, fungi, algae, protists, and viruses (Kumar et al., 2022; Costa et al., 2018). Bacteria and fungi being the most dominant of all soil microbiota, vary with soil depth, soil components, and other soil properties (Costa et al., 2018; Dell'Abate and Biro, 2015). According to Kumar et al. (2022) and Glick and Gamalero (2021), an estimate of 10^6 to 10^9 bacterial cells can be found in a gram of soil sample, and ten-fold of these can be found in the soil zone that surround plant roots (Babaniyi et al., 2023).

Soil microbes play significant roles in global cycling and the availability of nutrients, water, organic matter and energy. They are also involved in soil formation and aggregation, maintenance of soil health and quality, enhancement of soil productivity, stimulation of plant growth, and inhibition of plant pathogens (Idowu et al., 2023; Bekele & Getaneh, 2022; Nielsen et al., 2002; Costa et al., 2018). Soil microorganisms have been reported to be of significant relevance in crop productivity and sustainable agriculture (Hayat et al., 2010). Beyond these, they also promote plant health and productivity, and maintain animal health. They achieve these through various biological processes such as mineralization, decomposition, degradation, solubilization, immobilization/chelation, suppression of pathogens, nitrogen fixation, nitrification, denitrification, and so on (Bekele & Getaneh, 2022). However, these processes are driven and regulated by certain microbial metabolic products or metabolites ranging from enzymes to antimicrobials, organic acids, volatile organic compounds (VOC), and phytostimulators, among others. This chapter addressed the various classes of soil microbial products with specific examples, their roles and mechanisms of actions in green vegetation and sustainable agriculture. The current innovations for soil microbial products and their applications in green technology were reported.

2 Classes of Soil Microbial Products

Soil microbial products are various metabolites produced by soil microorganisms, that contribute to green vegetation in terms of plant growth and health, either directly or indirectly. The interaction of soil with the root of plant species (rhizosphere) heavily enriches the plant root with microorganisms, and results in several activities needed for the biosynthesis of different metabolites from microbes such as fungi and

Table 1 Classification of soil microbial products

Class of microbial product	Examples	Mechanism
Volatile Organic compounds (VOCs)	Ketones, Alcohols, pyrazines, Benzonoids, Terpenes, Geosmin, Phenols etc.	Indirect
Phytohormones/ Phytostimulators	Indole Acetic Acid (Auxin), Cytokinin, Gibberellins, Ethylene Abscisic Acid (ABA),	Direct
Bioactive metabolites (Antimicrobial compounds/biocontrol agents)	Hydrogen cyanide, Phenazine, 2,4-Diacetylphloroglucinol (DAPG), Cyclic Lipopeptides, Antibiotics, antifungal	Direct, Indirect
Enzymes	AAC Deaminase, Chitinase, β-Glucanase, Phosphatase and/or Phosphohydrolase, Phytase, Dehydrogenase, Proteases, Cellulases, Lipases, Esterases, Amylases, Xylanases, Nitrogenase, Peroxidases	Direct, Indirect
Organic Acids	Glycolic acid, Oxalic acid, Isovaleric acid, Isobutyric acid, Ketogluconic acid, Citric, Propionic, Acetic acid, Lactic acid, Oxalic acid, Malonic acid etc	Direct and Indirect
Siderophores	hydroxamate, catecholate and carboxylate	Direct, Indirect
Nitrogen and Biotransformed Nitrogen containing compounds	Nitrogen, Nitrite, Nitrate	Direct
Extracellular polymeric substances (Exopolysaccharides)	Dextran, Alternan, Reuteran, Mutan, Fructans, Cellulose, and Polygalactans.; alginate, Kefiran, Gellan, Xanthan	Indirect

bacteria (Egamberdieva et al., 2017). This is because the exudates from the plant root are richly packed with nutrients, consequent to the heavy colonization of the root noodles with microorganisms (Hashem et al., 2016; Schlaeppi & Bulgarelli, 2015). Similarly, Vorholt (2012) added that the rhizosphere region of the plant is enriched with nutrients that aid the microbial synthesis of useful metabolites that facilitate the growth and developmental process of plants while providing succinct resistance to abiotic and biotic stressors (Grover et al., 2013).

Soil microbial products can be classified based on their roles or functions in green vegetation. However, there are overlaps, as most of the microbial products have multiple roles. They can also be grouped based on the class of chemical compounds or chemical composition, and the mechanism(s) of their roles in plant growth and green vegetation, whether directly or indirectly. A summary of the classification is given in Table 1.

3 Roles of Soil Microbial Products in Green Vegetation

3.1 Volatile Organic Compounds

Volatile organic compounds (VOCs) are a broad class of low-molecular-weight secondary metabolites that evaporate when produced. VOCs are categorized into eight main basic classes, comprising hydrocarbons (alkanes, alkenes, and alkynes), alcohols, aldehydes, ketones, benzenoids, pyrazines, sulfides, and terpenes (Schulz & Dickschat, 2007). It has been discovered that microbial volatile organic compounds (MVOCs) considerably improve plant development (Poveda, 2021) and stress tolerance in a range of crops (Srikamwang et al., 2023; Basu et al., 2022). According to Basu et al. (2022), it has been demonstrated that in certain crops, nutrient absorption and root development are stimulated to promote plant growth directly. It has been demonstrated that under stress, exposure to certain MVOCs increases biomass, yield, and the formation of secondary metabolites (Fadiji et al., 2022). Moreover, MVOCs are naturally insect repellent and have the potential to reduce crop losses due to insect pests (Srikamwang et al., 2023). They are eco-friendly and can be used as a sustainable, cost-effective strategy for applications in agricultural practises as plant growth enhancers and sources of natural products to improve plant productivity and disease resistance (Poveda, 2021).

According to Lemfack et al. (2014), over 350 distinct bacterial species produce about 846 different possible volatile organic compounds. *Pseudomonas* species, *Bacillus* species, *Stenotrophomonas* species, *Serratia* species, *Arthrobacter* species, *Trichoderma species, and Cladosporium sphaerospermum* are among VOCs producing soil microorganisms (Bailly & Weisskopf, 2012; Raza et al., 2016; Ryu et al., 2003; Li et al., 2019; Lee et al., 2019). Certain bacteria release a variety of volatile substances that enable them to remotely control and impede the growth of plants. The VOC hexenal isoprenoids, which include mono- and sesqui-terpenes, promote plant growth by reducing weed growth through the prevention of root development and seed germination in weeds (Nishida et al., 2005). As a result of this, competition with the plant is reduced and nutrients, water, and light is delivered effectively to plants (Issar et al., 2021). Park et al. (2015) showed that *Pseudomonas fluorescens* SS101 generated 13-tetradecadien-1-ol, 2-butanone, and 2-methyl-n-1-tridecene, which promoted *Nicotiana tabacum* growth. Plants may sense and react to microbial volatile organic compounds (VOCs) in their surroundings, which can control their growth and lead to the development of systemic resistance (Issar et al., 2021). Also, the volatile organic compounds, isoprenoids cooperate with cytokinins and carotenoids to control senescence in plants (Dani et al., 2016).

In addition to increasing plant growth, microbial VOCs can also produce resistance to abiotic stress (Liu & Zhang, 2015). Plants frequently experience dehydration as a result of salinity, dryness, and osmotic conditions, and osmoprotectants, which regulate osmotic tension and physiological changes, naturally counteract this stress (Zulfiqar et al., 2019). According to a study by Cho et al. (2008), *Pseudomonas chlororaphis* O6's 2R,3R-butanediol enhanced *A. thaliana*'s resistance to drought by raising the proportion of closed stomata. Exposure of *A. thaliana* seedlings to *B. subtilis* GB03's VOC emissions resulted in faster development and accumulation of less Na^+ by *A. thaliana* seedlings under osmotic and ionic stress caused by elevated sodium (Na^+) concentrations (Zhang et al., 2008). Also, Zhou et al. (2017) found that *B. amyloliquefaciens* strain SAY09's volatile organic compounds (VOCs) helped *A. thaliana* seedlings reduce their sensitivity to cadmium by boosting auxin production. The findings showed that plants' ability to absorb iron was enhanced and that cadmium was deposited in the cell walls of roots.

Similarly, MVOCs have been reported to be involved in gene regulation in plants. *B. subtilis* GB03's VOCs affected the gene linked to cell expansion in *A. thaliana* leaves, during which genes related to cell-wall rigidity (pectinases), cell-wall losing (UDP-glucose-4-epimerase), and flavonoid formation (phenylalanine lyase) were down-regulated (Zhang et al., 2007). Another study conducted by Kim et al. (2017) showed that *B. subtilis* JS's volatile organic compounds (VOCs) alter the transcription of genes on tobacco seedlings. These genes are linked to cellular processes (87%), metabolisms (92%), and stimulus response (31%). Furthermore, genes related to cellular elements (such as organelles and cells), molecular activities (such as binding and catalysis), and biological processes (such as response to stimuli, metabolisms, cellular processes, and biological regulation) were regulated by volatile organic compounds (VOCs) secreted by *B. amyloliquefaciens* FZB42 on *A. thaliana* (Hao et al., 2016). Research has also confirmed that VOCs released by *S. plymutica* and *Stenotrophomonas maltophilia* regulate genes linked to molecular functions, cellular components, biological processes (such as transporter activity, transcription factor, and nucleotide binding), and response to biotic and abiotic stimuli (Wenke et al., 2012). Another body of research indicated that *Trichoderma's* VOC emissions alter genes linked to tension, immune reaction, and chemical, hormone, and/or chemical stimulus reactions (Lee et al., 2019).

3.2 *Phytohormones*

Phytohormones are compounds identified as small organic molecules that have a premium role in the growth and development of green vegetation. They are vital organic compounds that coordinate plant activities and their corresponding responses to biotic and abiotic stressors. The abiotic stressors are salt, drought, cold and heat, while the biotic stressors come from pathogenic organisms and herbivores (Fahad et al., 2015; Zhao et al., 2021). At low concentrations, phytohormones dictate plant growth by influencing development, stress resistance, cell division, germination, roots, flowering, fruiting, sex determination, dormancy, and organ shedding. Examples of these organic compounds are auxins, cytokinins (Cks), gibberellins (GAs), abscisic acid (ABA), and ethylene (Eth), brassinosteroids (BRs), jasmonic acid (JA), salicylic acid (SA), strigolactones (SLs) and peptides. Studies have presented the interconnective roles of phytohormones in the plant life cycle and their appreciable action in facilitating plant response. Research has also identified other vital phytohormones, polyamines (PAs) and karrikins (KARs), known with specialized roles as plant regulators. Phytohormones roles in plant growth and development, as well as in green vegetation, span through different parts of the plant either directly or indirectly (Fig. 1).

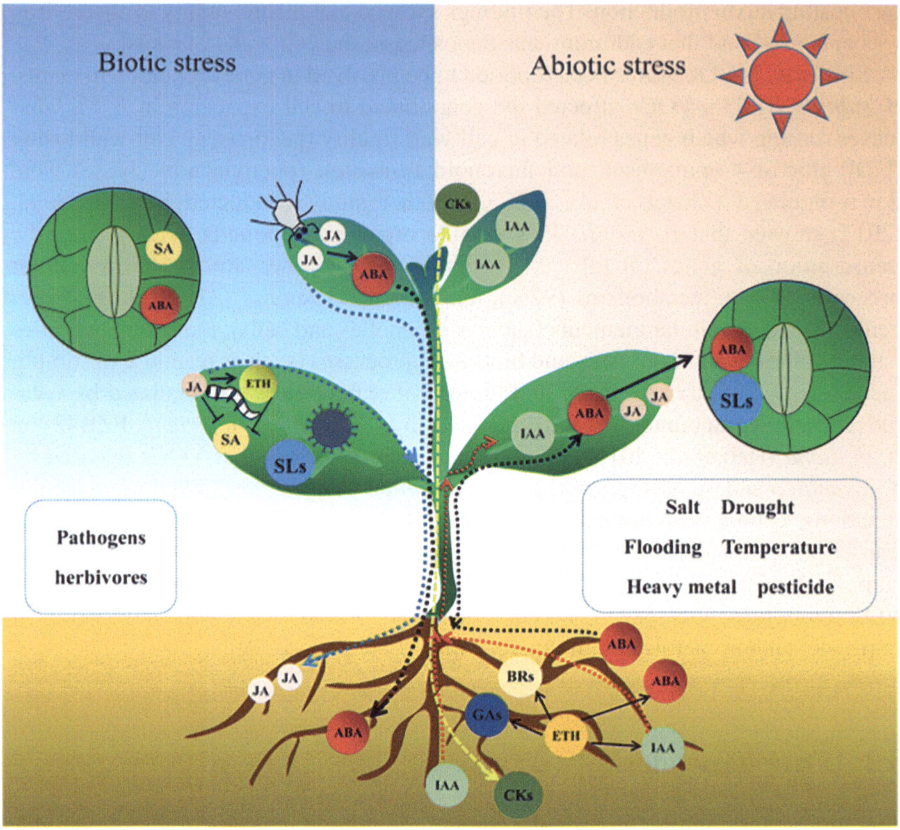

Fig. 1 Site of activities of microbially produced phytohormones in plants

Diverse microbial species, especially plant growth-promoting microorganisms (bacteria and fungi) within the rhizosphere region, are known to be phytohormone producers (Table 2). Some of these microorganisms are listed in Table 3.

Auxins are often described as the most investigated and naturally occurring phytohormone whose chemical name is indole-3-acetic acid or indoleacetic acid (IAA) (Glick, 2012; Orozco-Mosqueda et al., 2023; Patten & Glick, 1996). Studies have proposed four pathways for their synthesis. These are via the indole-3-acetamide pathway, indole-3-pyruvic acid pathway, tryptamine pathway, and indole-3-acetaldoxime pathway. The plant growth-promoting bacteria (PGPB) (Table 3) are major organisms known to produce auxins, which largely influenced by plant-bacterial interaction (Glick, 2012; Keswani et al., 2022; Morales-Cedeño et al., 2021). They play a premium role in plant growth, root nodulation and mediate the response of plants to light and stressful conditions (Luziatelli et al., 2020).

The phytohormones, cytokinins (CK) are structurally related to adenine. However, the common form is Zeatin, which was isolated from corn in the 1950s (Kieber & Schaller, 2010). Studies have revealed the biosynthesis of high levels of CK from

Table 2 Classes of phytohormones and their specific roles

Phytohormone	Specific roles
Auxins	Participate in leaf development Regulation of plant growth Major internal signaling Major shoot signaling Stimulation of large number of gene transcription Pigment formation Biosynthesis of various metabolites
Cytokinins Gibberellins	Control plant-cell division and proliferation Chloroplast biogenesis Apical dominance Vascular and shoot differentiation Enhances nutrient mobilization facilitate production of anthocyanin production Regulate the process of aging in plant's leaf (leaf senescence) Account for leaf complexity Photomorphogenic development
Abscisic acid Ethylene	Facilitate stress resistance of plant under adverse conditions Serves as plant growth inhibitor while promoting maturity and aging Regulation of seed germination Acts as endogenous messenger Long-distance water stress signal
Brassinosteroids	Plant adaptability and reproductive growth. Seed germination Vascular formation Cell growth Production of flowers and fruit
Jasmonic acid (JA) Salicylic acid (SA),	Enhance stress resistance of plant Seed germination and transpiration Glycolysis Flowering Fruit yield Ion uptake and transport, photosynthetic rate and stomatal conductance
Strigolactones	Enhances plant responses to abiotic stress

46 strains of bacteria of the genus *Methylobacterium, with* trans-Zeatin (tZ) being the most active form of it. Cytokinins play vital roles in the facilitation of plant cell division, xylem and chloroplast differentiation, fruit and flow development, and nutrition signaling, among others (Morales-Cedeño et al., 2021).

There are over 126 known gibberellins which are structurally described as tetracyclic di-terpenoid compounds. However, four of these compounds are reported to be biologically active (GA1, GA3, GA4, and GA7) (Morales-Cedeño et al., 2021), while the others that are non-active contribute to biosynthesis or degradation of the active gibberellins (Yamaguchi, 2008). They play key specific roles in influencing vital processes in plants, such as improvement of the photosynthetic efficiency of plants, leaf area index, light interception, efficiency of nutrient usage, and

Table 3 Prominent phytohormones and microorganisms involved in their synthesis

	Microbes responsible for biosynthesis	Phytohormones
Bacteria	*Mycobacterium* species, *Pseudomonas putida* *Azospirillum brasilense,* Plant Growth-Promoting Bacteria*: Acetobacter, Acinetobacter, Azospirillum, Arthrobacter, Azotobacter, Bacillus, Bradyrhizobium, Burkholderia, Herbaspirillum, Klebsiella, Mesorhizobium, Paenibacillus, Pantoea, Pseudomonas, Rhizobium, Rhodococcus, Serratia, Strenotophomonas, Streptomyces, and Rouxiella* *Halomonas desiderata, Bacillus megaterium, Bacillus cereus, Bacillus subtilis, Escherichia coli, Pseudomonas savastanoi and A. tumefaciens* *A. brasilense, B. japonicum strains* *Pseudomonas, Xanthomonas and Erwinia*	Indole-3-Acetic Acid or Indoleacetic Acid (IAA) Indole-3-Acetic Acid or Indoleacetic Acid (IAA), Gibberellins (GB) Cytokinins (CKs) Abscisic acid (ABA) Ethylene (ETh)
Fungi	*Trichoderma* species, *Fusarium strains, Streptomyce, Rhizopus suinu* *Fusarium monilifomra, Gibberella fujikuroi, Aspergillus, Cladosporium, Neurospora, Penicillium, Phoma, Colletotrichum gloeosporioides*	Indole-3-Acetic Acid or Indoleacetic Acid (IAA) Gibberellins (GB), Cytokinins (CKs)

Source: Egamberdieva et al., 2017; Nazli et al., 2020; Spaepen, 2014

modulation of diverse plant development processes (Fahad et al., 2015; Roychoudhury & Aftab, 2021).

Abscisic acid (ABA) is a vital isoprenoid phytohormone belonging to the terpenoid metabolite family having a characteristic of a C_{15}-carbon chain molecule. It significantly influences the physiological processes and responses of plants (Table 2). However, under stressed situations, their physiological responses are depicted in the form of morphological, cytological, and anatomical alterations. This is a result of either synergistic or antagonistic interaction with multiple phytohormones (Singh & Roychoudhury, 2023). Abscisic acid coupled with salicylic acid, brassinosteroids, and gibberellic acid enhances plant defense against hemibiotrophic pathogens like *Magnaporthe oryzae* and *Pseudomonas syringae* (Yang et al., 2019).

Ethylene is a ubiquitous gaseous hormone whose biosynthetic pathway is methionine via S-adenosyl-L-methionine (AdoMet) and the cyclic non-protein amino acid ACC. It is characterized by a wide spectrum of biological activities. It is found in higher amounts in ripening fruit but exists in lower levels at the early stage of plant growth and development, and it serves as a notable aid for plant response to various biotic and abiotic stresses (Althiab-Almasaud et al., 2021; Vanderstraeten et al., 2019). Studies showed the involvement of ethylene in plant defense responses against pathogens, and it also works coordinatively with jasmonate, thereby enhancing the resistance of plants to necrotrophic pathogens (Pieterse et al., 2012).

Brassinosteroids (BRs) are a known phytosteroidial hormone that can regulate plant growth and development by producing an array of physiological changes (Table 2) (Nolan et al., 2020). They are categorized as either C_{27}, C_{28} or C_{29} BRs based on the number of carbons in a given structure (Bajguz, 2009; Vardhini, 2013).

Although, over 60 compounds have been identified, the most active ones studied are brassinolide (BL), 28-homobrassinolide (28-HomoBL), and 24-epibrassinolide (24-EpiBL). A study has shown the nexus of this compound with GAs, especially in the cooperative role played in plant physiological processes and biosynthesis regulation of GAs (Castro-Camba et al., 2022). They participate in plant responses to light by enhancing the transcriptional activity of phytochrome interacting factors (Vardhini & Anjum, 2015).

The phenolic phytohormone salicylic acid (SA) is descriptively known to regulate plant growth, development, and interaction with other species. The major role of the hormone is in the regulation of biotic stress responses and several abiotic stresses like drought and many others (Ahmad et al., 2021; Aires et al., 2022). Koo et al. (2020) added that this compound has the potential to stimulate flowering, ion absorption, nutrient transfer, the movement of plant stomata, and protein synthesis. Also, SA acts as a binder to amino acids like arginine and proline, which is consequent to the stability of plants against environmental stressors (Morales-Cedeño et al., 2021).

Jasmonic acid (JA) is often studied with its derivatives, such as methyl ester (MeJA) and isoleucine conjugate (JA-Ile), which are commonly referred to as jasmonates (JAs) (Ruan et al., 2019). This phytohormone is known for its key role as a mediator of plant responses to abiotic stressors via JA signaling pathways. The process often involves transcription regulation of the stress-responsive genes and antioxidants subjected to stressed conditions (Kim et al., 2021). Apart from its active participation in plant growth and development, JA can initiate a defensive response to counteract the impact of biotic stress such as insect wounding, pathogen attack on plants, and abiotic stressors like drought, low temperature, and salinity (He et al., 2017; Wang et al., 2021).

3.3 Antimicrobial Substances and Bioactive Metabolites

Agricultural studies have validated the invasive impacts of insects and pests as the major cause of low yields in plant productivity. However, research has also shown the tremendous potential of some soil microorganisms to produce bioactive, antimicrobial metabolites (Dimkić et al., 2022; Khatoon et al., 2020). The microbial strains play a vital antagonistic role against plant pathogens and harmful insects by reducing the rate of plant contagion (Basu et al., 2021; Prasad et al., 2019). Generally, they act as biocontrol for plant growth and development while improving the health status of the soil. The major strains of microorganisms are from the genera *Bacillus*, *Lysinibacillus, Paenibacillus, Brevibacillus, Microbacterium, Cupriavidus, Pseudomonas, Photorhadbus, Azospirillum, Rhizobium*, *Serratia,* and *Streptomyces,* which produce useful bioactive antibiotics such as phenazine-1-carboxylic acid, 2,4-diacetyl phloroglucinol, oomycin, pyrrolnitrin, pyoluteorin, polyketide synthase, lipopeptides, fengycin, pyrrolnitrin, bacillomycins, kanosamine, zwittermicin-A, hydrogen cyanide, non-ribosomal peptide, phenolics, phenazine, and pantocin (Table 4) (Chin-A-Woeng

Table 4 Examples of biocontrol agents synthesized by soil microorganisms

Microorganisms	Synthesized Bioactive	Targeted pathogenic organism	References
Pseudomonas spp.	2,4-diacetylphloro glucinol (2,4-DAPG)	*Aphanomyces cochlioides* *Gaeumannomyces graminis var. tritici*	de Werra et al. (2009); Islam and Fukushi (2010)
	Pyoluteorin	*Xanthomonas oryzae pv. Oryzae* *Rhizoctonia solani*	Brodhagen et al. (2004); Velusamy et al. (2006)
	Pyrrolnitrin	*Gaeumannomyces graminis var tritici* *Sclerotinia sclerotiorum*	Ajouz et al. (2010)
	Aerugine	*Colletotrichum orbiculare* *Phytophthora capsici* *Pythium ultimum*	Lee et al. (2003); Lee et al. (2017)
	Phenazines Pyoluteorin, 2,4-DAPG Phenazine-Carboxamide Pyrrolnitrin Phenazines	*Pythium myriotylum* *P. spendens* *Sclerotinia sclerotiorum* *Pythium ultimum* *Fusariumoxysporumf. sp.* *Radicis-lycopersici* *Pyrrolnitrin* *Phenazines* *Sclerotinia sclerotiorum*	Athukorala et al. (2010); Perneel et al. (2007) Flury et al. (2017); Kang et al. (2018, 2019); Lee et al. (2011); Park et al. (2011); Ruffner et al. (2013)
Bacillus spp.	Surfactins Bacillomycin D Fengycins Azalomycin F Iturins Biosurfactant	*Sclerotinia sclerotiorum* *Alternaria alteranta* *Fusarium graminearum* *Botrytis cinerea* *Fusarium solani* *Rhizoctonia solani* *Phytophthora capsic* *Botrytis cinerea* *Fusarium solani* *Rhizoctonia solani* *Phytophthora capsica* *Fusarium solani* *Rhizoctonia bataticola* *Rhizoctonia solani*	Athukorala et al. (2009); Kim et al. (2010); Zhao et al. (2010)
Paenibacillus spp	Chitinase Methyl-2,3dihy droxybenzoate (phenolic compound)	*Botrytis cinerea* *Cladosporium Sphaerospermum* *Botrytis cinerea* *Rhizoctonia solani*	Kim et al. (2017); Lee et al. (2017); Neung et al. (2014); Nguyen et al. (2013)

(continued)

Table 4 (continued)

Microorganisms	Synthesized Bioactive	Targeted pathogenic organism	References
	Protocatechuic acid 1-Octen-3-ol benzothiazole	*Fusarium oxysporum f. sp. Lycopersici*	
Streptomyces spp	Flavensomycin	*Saccharomyces sp* *Penicillium sp.*	Kaur et al. (2016)

et al., 2003; Kenawy et al., 2019; Khatoon et al., 2020; Morales-Cedeño et al., 2021; Saeed et al., 2021). These bioactive metabolites act through various mechanistic pathways such as direct antibiosis, plant growth promotion and the induction of systemic resistance. It is needful to add that the antagonistic behaviour of these biocontrol agents is based on the secretion of biogenic metabolites with inhibitory characteristics to pathogenic bacterial and fungal diseases with deleterious effects on plants (Narayanasamy, 2013). Antifungal activity has been reported against *Magnaporthe oryzae, Fusarium graminearum, F. moniliforme, F. oxysporum* and *Rhizoctonia solani* (Khaskheli et al., 2020; Cao et al., 2020).

These microbes surround the plant's rhizosphere while producing antagonistic metabolites to form a formidable protective shield for the plant against the invasions of pests and targeted pathogenic microbes, eliciting induced systemic resistance (ISR) through their antibiotic metabolites (Lee et al., 2022). The process ensured the generation of patchy biofilms as protective barriers for the root surface. The aqueous diffusion in soil pore fluids spreads these antagonistic metabolites further into the rhizosphere while dispersing these antagonistic biocontrol-active volatiles to the bulk soil and air space. These mechanisms help the biocontrol bacterium survive while enhancing the movement of infected nematodes and insect larvae, and transporting inocula away from the root.

Létoffé et al. (2022) and Anand et al. (2023) have reported the biocontrol role of *Pseudomonas* species secreted hydrogen cyanide against *Phytophthora infestans.* Also, Chin-A-Woeng et al. (2003) revealed the impacts of bio-secreted phenolics and phenazine groups as metabolites responsible for oxidative stress leading to the death of targeted pathogenic organisms. These metabolites could also act as an iron chelator in the presence of hydrolytic enzymes (chitinases, proteases, and lipases) to orchestrate the damage of cell membrane structure and the polymeric cell wall of plant invasive organisms such as the egg coats of nematodes and fungi (Singh & Arya, 2019).

3.4 Enzymes

Microbial enzymes have multifaceted roles in supporting green vegetation, encompassing nutrient cycling, disease resistance and biocontrol, and the breakdown of organic matter (Ramakrishna et al., 2019; Mehmood et al., 2023). These roles showcase the diverse ways in which microbial enzymes contribute to the overall health and resilience of green vegetation, ranging from stress tolerance to environmental remediation. Understanding microbial enzymes and their roles is crucial for sustainable agriculture and ecosystem management, emphasizing the intricate relationship between microorganisms and green vegetation.

The likes of nitrogen-fixing and cellulolytic microbes are a few microbial groups with positive impact on plant growth and soil fertilization (Mehmood et al., 2023; Bautista-Cruz et al., 2024). For example, cellulase and indole acetic acid producing microbes like *Serratia nematodiphila* enhance soil fertility by breaking down plant residues and solubilizing tricalcium phosphate, releasing nutrients for plant uptake (Tang et al., 2020). Similarly, other phosphatase-producing bacteria may also contribute to phosphorus availability for plants, positively impacting their growth. Certain microbes produce chitinase and glucanase enzymes, which play a role in the degradation of fungal cell walls, acting as biocontrol or plant defence agents against plant pathogens (Prasetyawan et al., 2018). These enzymes are crucial components of induced systemic resistance, where plants respond to microbial colonization by upregulating defense mechanisms to protect against potential threats. The intricate interaction between plants and beneficial microbes highlights the significance of microbial enzymes in enhancing plant resilience. Cellulases and ligninases produced by fungi, such as *Trichoderma* species facilitate the breakdown of complex plant polymers, contributing to the formation of humus and the recycling of carbon in soil. Microbial enzymes also play a role in the decomposition of plant residues and the cycling of carbon in ecosystems.

Microorganisms, such as mycorrhizal fungi, produce enzymes like phosphatases and nitrogenase, that aid in the breakdown of organic matter, releasing essential nutrients like phosphorus and nitrogen, which are vital for plant nutrition (Nall, 2010). The mycorrhizal symbiosis enhances nutrient uptake by plants, contributing to improved growth and overall plant health. Furthermore, microbial enzymes contribute to the synthesis of plant growth-promoting substances (Khare & Yadav, 2017). For instance, certain bacteria produce enzymes like 1-aminocyclopropane-1-carboxylate (ACC) deaminase, which decreases ethylene levels in plants, mitigating the inhibitory effects of ethylene on root elongation and promoting overall plant growth. The involvement of specific microbial enzymes in enhancing plant growth exemplifies the intricate web of interactions between microorganisms and vegetation, emphasizing the significance of microbial enzymatic activities in promoting sustainable plant development. Some cell wall degrading extracellularly produced lytic enzymes such as cellulases, chitinases, glucanases, and proteases have been reported to play significant roles as biocontrol agents (Mabood et al., 2014; Jeyanthi & Kanimozhi, 2018). Similarly, termiticidal activity of chitinase, lipase, and protease has been reported by Rakshiya et al. (2016). Some of the roles and products of microbial enzymes in green vegetation are reported in Table 5.

Table 5 Selected enzymes and associated roles in green vegetation

Enzyme	Microbe(s)	Role in green vegetation	Mechanism of impact	References
Cellulase	Cellulolytic bacteria, Fungi	Breakdown of cellulose in plant cell walls, releasing sugars for plant use.	Hydrolysis of cellulose into glucose monomers	Halimah et al. (2019)
Phosphatase	*Bacillus spp.*, *Pseudomonas spp.* Mycorrhizal Fungi	Releases bound phosphate, enhancing nutrient uptake	Hydrolysis of organic phosphates into inorganic forms	Tazisong et al. (2015)
Nitrate Reductase	*Pseudomonas*, *Bacillus*	Converts nitrate to nitrite, facilitating nitrogen assimilation	Reduction of nitrate to nitrite in the nitrogen cycle	Campbell (1999)
Polygalacturonase	*Rhizopus*, *Aspergillus*	Breaks down pectin in plant cell walls, aiding in decomposition	Cleaves polygalacturonic acid in the plant cell wall	Haile and Ayele (2022) Tucker (2004)
Lignin Peroxidase Dehalogenase	*Phanerochaete chrysosporium* Actinomycetes	Facilitates lignin degradation, promoting organic matter decomposition. Improving Phytoremediation	Oxidative cleavage of lignin bonds	Khindaria et al. (1995) Kumar and Chandra (2020)
Catalase, Superoxide Dismutase (SOD)	Various Microbes	Alleviating Environmental Stress	Catalase and SOD detoxify reactive oxygen species, mitigating oxidative stress in plants.	Rajput et al. (2021)
Nitrogenase	*Rhizobium* spp., *Frankia* spp.	Nitrogen Fixation	Conversion of atmospheric nitrogen to ammonia, promoting nitrogen availability.	Bennett et al. (2023) Boyd and Peters (2013)
Glutathione-S-transferase	*Pseudomonas* spp., *Bacillus* spp.	Herbicide Detoxification	Detoxification of herbicides, reducing their impact on plant health.	Marrs (1996)
Ligninases	White rot fungi, *Actinomycetes*	Lignin Modification	Modification of lignin, enhancing decomposition of plant residues.	Dashtban et al. (2010)

(continued)

Table 5 (continued)

Enzyme	Microbe(s)	Role in green vegetation	Mechanism of impact	References
Urease	Ureolytic bacteria	Enhanced Nutrient Uptake	Hydrolysis of urea to release nitrogen, supporting plant nutrition.	Joseph et al. (2022) Sigurdarson et al. (2018)
Proteases, Chitinases	Plant-associated bacteria and fungi	Indirect Plant Defense	Degrades proteins and chitin in potential pathogens, enhancing plant defense.	Jashni et al. (2015) Kumar et al. (2018)
Nitrous oxide reductase	Denitrifying bacteria	Nitrogen Denitrification	Converts nitrate to nitrogen gas, affecting nitrogen availability for plants.	Zhang et al. (2019)
Ammonia lyase	Ammonia-producing bacteria	Ammonia Production	Converts amino acids into ammonia, influencing nitrogen availability in the soil.	Parmeggiani et al. (2018)
Pectinases	Pectinolytic bacteria and fungi	Pectin Degradation	Breaks down pectin in plant cell walls, influencing cell structure and nutrient release.	Haile and Ayele (2022)

3.5 *Organic Acids*

Microbially produced organic acids are low molecular weight, slightly soluble, weak acids produced and released by different species of soil bacteria and fungi (Silva et al., 2023; Adeleke et al., 2017). They are clearly distinguished from high molecular weight, insoluble, naturally occurring organic acids, humic and fulvic acids (Adeleke et al., 2017). These include gluconic acid, 2-keto gluconic acid, citric acid, oxalic acid, succinic acid, acetic acid, lactic acid, oxalic acid, maleic acid, tartaric acid, and fumaric acid (Silva et al., 2023; Adeleke et al., 2017).

Organic acids contribute significantly to green vegetation and agricultural productivity and sustainability directly and indirectly by improving soil nutrient availability and uptake through mineralization and solubilization, biocontrol of plant pathogens, bio-stimulation of other beneficial soil microbes (microbial chemotaxis), and detoxification of metals (Adeleke et al., 2017; Jones et al., 2003; Macias-Benitez et al., 2020). As a result of the chelating properties of organic acids, they form complexes with positively charged soil mineral ions (e.g. Ca^{2+}, Al^{3+}, Fe^{3+}), resulting in the release of unavailable nutrients for plant uptake, growth, and development (Morgan et al., 2005; Posso et al., 2017). Organic acid secretion is characteristics of phosphate-solubilizing microorganisms (PSM), potassium-solubilizing microorganisms (KSM), and zinc-solubilizing microorganisms (ZSM), such as *Pseudomonas*,

Bacillus, *Serratia*, Citrobacter, *Aspergillus*, Arbuscular mycorrhizae, *Alternaria*, *Trichoderma*, *Streptomyces*, *Micrococcus*, among others (Patel et al., 2008; Tandon et al., 2020; Sindhu et al., 2016; Ijaz et al., 2021; Kumar et al., 2024).

Phosphate solubilization and mineralization in soil are carried out by PSM, which converts the organic form of phosphate into inorganic form through the activity of organic acids. Phosphate is an essential nutrient for plant growth and development, and it exists in insoluble tricalcium phosphate, dicalcium phosphate, and mineral forms, but the soluble, inorganic orthophosphate is very low in the soil. Organic acid secretion by soil microbes results in acidification of the rhizosphere, low pH, and consequent solubilization of organic reservoirs of phosphorus (Ku et al., 2018; Andrino et al., 2021; Kumar et al., 2024). Organic acids chelate the cations attached to phosphorus from their hydroxyl and carboxyl groups. In the same vein, organic acids secretion by KSM converts the insoluble form of potassium, the third most important plant nutrient to the soluble form. ZSM also produce organic acids that result in decrease of pH. The organic acids act on zinc complexes in soil, thus causing sequestering of zinc cations (Kumar et al., 2024). Overall, production of organic acids by soil microbes promotes nutrient solubilization, mobilization, and availability for plant growth, green vegetation, and crop yield. This prevents the use of phosphorus- and potassium-containing inorganic fertilizer and the associated environmental contamination.

Beyond the nutrient mobilization and availability, organic acids by soil microbes have also demonstrated biostimulant potentials, antimicrobial properties and act as biocontrol agents, inhibiting plant pathogens (Macias-Benitez et al., 2020; Oburger et al., 2009).

3.6 Siderophores

Siderophores are low molecular weight, small peptides, secondary microbial metabolites that have affinity for inorganic iron and function as metal chelators (Silva et al., 2023; Gómez-Godínez et al., 2023). There are four major classes of siderophores based on their structural features, iron-coordinating functional groups, and types of ligands: catecholates, carboxylates, hydroxamates, and pyoverdines (Jeyanthi & Kanimozhi, 2018; Bekele & Getaneh, 2022).

Iron is a major element required for plant metabolism, deficiency of which can affect plant respiration and photosynthesis; that are of significance in the formation of green vegetation. Siderophores are secreted by microbes to circumvent iron limitation and provide plants with iron in rhizospere (Vejan et al., 2016). They promote the bioavailability of iron by influencing their low solubility by attaching to the insoluble ferric (Fe^{3+}) state and facilitating their dissolution into the soluble ferrous (Fe^{2+}) state or complexes, which can be taken up directly by active transport into the cell (Kumar et al., 2022). Siderophores are secreted by members of the bacterial genera *Pseudomonas*, *Bacillus*, *Azotobacter*, *Rhizobium*, *Streptomyces*, *Azadirachta*, *Serratia*, *Aeromonas*, *Enterobacter*, and *Bulkholderia* (Sultana et al.,

Table 6 Selected examples of siderophore-producing microorganisms

Siderophore producing organisms	Roles	References
Escherichia coli Trichoderma asperellum, Aspergillus niger, Trichoderma harzianum, Streptomyces species, Penicillium citrinum,	Nutrient mobilization and enhancement of plant growth	Rungin et al. (2012); Yadav et al. (2011); Gangwar and Kaur (2009
Pseudomonas fluorescens, Bacillus subtilis	Biological control of phyto-pathogens	Yu et al. (2011); Pal et al. (2001)
Pseudomonas azotoformans Pseudomonas aeruginosa Agrobacterium radiobacter	Removal/bioremediation of heavy metals	Nair et al. (2006); Braud et al. (2009); Wang et al. (2011); Ines et al. (2012)

2021; Kumar et al., 2022; Gómez-Godínez et al., 2023). Among the fungal species, *Aspergillus niger*, *Trichoderma harzianum*, and *Penicillium citrinum* have been reported (Yadav et al., 2011; Kumar et al., 2024). The binding of siderophores to metals increases their solubility and thus their available concentration in the soil (Murugappan et al., 2012; Gómez-Godínez et al., 2023). Selected examples of siderophore-producing microbes and their roles are presented in Table 6.

Siderophores are also involved in the vanadium and molybdenum (nitrogenase cofactors) uptake by *Azotobacter vinelandii* (McRose et al., 2017; Kumar et al., 2024). Some siderophores have affinity for heavy metals of environmental significance such as lead, cadmium, cobalt, arsenic, copper, and zinc, as well as uranium (Gómez-Godínez et al., 2023), thus contributing to their removal from soil.

Furthermore, siderophores producing soil microorganisms play significant roles in plant growth promotion, and have also been reported to be involved in biocontrol of plant pathogens. Siderophore produced by *Pseudomonas* species has been reported to control pathogens and diseases in some plants, such as Vigna radiata (green gram) (Sindhu & Sahu, 2011), and chilli and paddy (Sasirekha & Srividya, 2016). Yu et al. (2011) also reported that siderophore-producing *Bacillus subtilis* controls *Fusarium oxysporum* wilt disease of pepper. Co-inoculation of *Pseudomonas koreensis* and *Bacillus subtilis* has also been reported to suppress *Cephalosporium maydis* damping-off diseases with an increase in the activities of catalase, polyphenol oxidase, and peroxidase, as well as total chlorophyl and carotenoids, which are essential for photosynthesis (Ghazy & El-Nahrawy, 2021).

3.7 Extracellular Polymeric Substances/Exopolysaccharides (EPS)

Extracellular Polymeric Substances (EPS) are high molecular weight compounds that are composed of polysaccharides, proteins, lipids, organic and inorganic compounds, DNA, and metal ions (Paul et al., 2024; Bhagat et al., 2021). They are produced by a wide range of microorganisms like bacteria, yeasts, molds, cyanobacteria, and microalgae. The type of microorganism, growth condition, cultivation time, and substrate availability influence the composition and quantity of EPS production, as well as the structure and functions of the EPS (Bhagat et al., 2021). EPS are majorly classified into two types: homopolysaccharides and heteropolysaccharides based on their composition and mechanism of synthesis. Homopolysaccharides are polymers with one type of monosaccharide such as dextran, alternan, reuteran, mutan, fructans, cellulose, and polygalactans.; while heteropolysaccharides are polymers that contain two or more monosaccharide units such as alginate, kefiran, gellan, hyaluronic acid, and xanthan (Netrusov et al., 2023).

Extracellular Polymeric Substances production by soil microbes is significant and presents enormous prospects for sustainable soil management, agricultural practices, crop productivity, and green vegetation. EPS adhere and bind to soil particles and soil surfaces, acting as conditioner, facilitating soil aggregation, reducing soil wetting and swelling, as well as run-off. They also improve soil aeration, porosity, permeability, and root penetration (Bhagat et al., 2021). EPS serve as chelators and mediate in nutrient interactions, thereby enhancing soil fertility, nutrient availability, and uptake by plant roots. These improve plant chlorophyll content, growth, and biomass development. Diverse EPS-producing soil microbes also play specific roles in maintaining the metabolic and physiological activities of plants during abiotic stress conditions like drought, salinity, metals, and temperature. High salinity reduces plant nutrient uptake; however, EPS bind with sodium ions in the soil and prevent them from reaching the stem, which in turn decreases sodium absorption and increases nutrient uptake by roots. Apart from this, EPS impart defense against heavy metals such as arsenic, chromium, cadmium, cobalt, and lead, among others, from natural processes and anthropogenic activities (Bhagat et al., 2021). Heavy metals are removed by EPS through biosorption due to electrostatic interaction between EPS anionic functional group and the cationic heavy metals, resulting in the formation of complexes (Sayyed et al., 2015; Mishra et al., 2017). This limits the mobility, bioavailability, and absorption of the heavy metals to plants, thereby preventing the oxidative stress and associated negative impact on plants such as reduced chlorophyll contents, blockage of gas exchange factors, and production of reactive oxygen species (Mishra et al., 2017; Bhagat et al., 2021). EPS have also been reported to result in bioremediation and biomineralization of metal ions (Bhagat et al., 2021; Cao et al., 2023). Under drought conditions, the hygroscopic nature of soil microbial polymers induces plant survival by facilitating soil aggregation, water permeability, root adherence to soil, nutrient uptake, and increased antioxidant levels (Bhagat et al., 2021; Khan et al., 2019). Similarly, EPS prevent plants against high temperature-induced stress by increasing the production of heat shock proteins, antioxidants, chlorophyll content, and abscisic acid content in the leaf (Abd El-Daim et al., 2018; Bhagat et al., 2021).

Beyond these, microbe-microbe interaction and microbe-plant interaction have been reported to be facilitated by EPS by increasing biomass quality, rhizosphere bacterial community, enzyme activities, and microbiome concentration (Cao et al., 2013). Production of certain lipopeptides by soil microbes have also been linked to protection against plant pathogens (Andrić et al., 2021). Some microbially produced carbohydrates, lipids, proteins, aminoacids, etc. have been reported to induce osmoregulatory activities in plants (Van Oosten et al., 2017; Khoshru et al., 2020; Lopes et al., 2021). They prevent membrane plasmolysis, regulate enzyme mechanisms, maintain homeostasis, and increase the production of heat shock proteins which improve plant tolerance, survival, and growth under extreme conditions caused by abiotic stress (Khoshru et al., 2020; Lopes et al., 2021).

3.8 Nitrogen and Biotransformed Nitrogen Contain Compounds e.g. Nitrogen, Ammonia, Nitrite, Nitrate

Nitrogen is a major plant nutrient, and it is freely available in the atmosphere, constituting 78%, but this form of nitrogen cannot be utilized by plants. It can only be utilized by plants in the form of nitrate and ammonia (and other organic nitrogen compounds) (Idowu et al., 2023). Biotransformation of volatile organic and inorganic nitrogen by microbes into usable forms that support plant growth and development has significantly contributed to sustainable agriculture and green vegetation. Certain free-living and symbiotic soil microbes produce ammonia for plant use by converting atmospheric nitrogen to the soluble, useable nitrogen compound in a process called nitrogen fixation (Grzyb et al., 2021). The amount of nitrogen fixed varies with the strains of bacteria, plant species, and environmental factors. The microbes include members of genera Rhizobium, Bradyrhizobium, Azolla, Azospirillium, Acetobacter, Frankia, Blue Green Algae, Azotobacter, *Burkholderia*, *Ralstonia,* etc. (Hayat et al., 2010; Grzyb et al., 2021). *Nitrobacter* and *Nitrosomonas* species also cooperatively produce nitrate for plant use by transformation of ammonia to nitrate in a two-step process called nitrification (Sorokin et al., 2012).

In a similar vein, organic nitrogen embedded in dead plant and animal cells is released by microorganisms during decomposition in the form of ammonia or oxidized to nitrates. Mineral nitrogen in soil is also biotransformed by microbes. This nitrogen biotransformation involves mineralization and immobilization. During nitrogen mineralization, nitrogen containing organic complexes are decomposed and converted into more mobile inorganic forms for plant use. However, during nitrogen immobilization, nitrogen-containing inorganic compounds are assimilated (Grzyb et al., 2021; Amoo & Babalola, 2017). As a result of these properties, these organisms are used as major components of biofertilizer formulation, and *Rhizobium* inoculant technology for organic agriculture, as they promote green vegetation through nutrient availability, thereby preventing environmental degradation and other associated consequences of inorganic nitrogen-containing fertilizer.

4 Recent Advances and Future Innovations on the Roles of Microbial Products in Green Vegetation

Recent advancements in microbial products for sustainable agriculture and green vegetation have revolutionized agricultural practices. These innovations offer diverse solutions to enhance soil health, nutrient availability, plant growth and crop yield, while promoting eco-friendly farming methods. Examples include Rhizobium inoculants that enhance nitrogen availability, mycorrhizal fungi products improving nutrient uptake, and biofertilizers with phosphatase enzymes promoting phosphorus availability. Plant growth-promoting bacteria (PGPB) inoculants, enzyme-based soil conditioners, biological pest control products, and enzyme-enhanced compost accelerators demonstrate the versatility and effectiveness of microbial products in sustainable agriculture (Table 7). These innovations underscore the potential for continued research and development in this field to advance eco-friendly agricultural solutions and contribute to a more sustainable future. Soil microbes and their associated products play diverse and critical roles in green vegetation, from nutrient cycling and plant defense to carbon decomposition and growth promotion. Mycorrhizal fungi and bacteria such as *Trichoderma* species have assisted in the understanding of some of the intricate mechanisms through which microbial metabolites are involved. Improved understanding of microbial-plant interactions, i.e., through the omics techniques is essential for harnessing the potential of microorganisms and their metabolites in sustainable agriculture and

Table 7 Selected innovation of soil microbial products used in green vegetation

Microbial enzyme Product	Innovation	Reference
Rhizobium Inoculants	Enhance nitrogen availability in soil by forming symbiotic relationships with leguminous plants	Gopalakrishnan et al. (2015)
Mycorrhizal Fungi Products	Improve nutrient uptake, water absorption, and overall plant health by forming associations with plant roots	Wahab et al. (2023)
Biofertilizers with Phosphatase Enzymes	Enhance phosphorus availability by releasing phosphates from insoluble forms, promoting plant growth and reducing reliance on chemical fertilizers	Alori et al. (2017)
Plant Growth-Promoting Bacteria (PGPB) Inoculants	Enhance plant growth through the production of phytohormones, nutrient solubilization, and disease suppression	Lopes et al. (2021)
Enzyme-Based Soil Conditioners	Break down organic matter, improving soil structure, water retention, and nutrient availability	Li et al. (2021)
Biological Pest Control Products	Degrade insect proteins, providing an eco-friendly alternative to chemical pesticides	Ayilara et al. (2023)
Enzyme-Enhanced Compost Accelerators	Speed up the composting process by breaking down complex organic matter, producing nutrient-rich compost	Rastogi et al. (2020)

environmental management. Overall, these microbial products and the innovations showcase the diverse applications in sustainable agriculture, promoting environmentally friendly practices, reducing chemical inputs, and enhancing overall crop productivity. Continued research and development in this field will contribute to the advancement of green vegetation, sustainable and eco-friendly agricultural solutions.

5 Conclusion

This chapter highlights the importance of wide arrays of soil microorganisms and their metabolites in maintaining soil health and supporting plant productivity through different mechanisms towards achieving green technology for sustainable agriculture. Therefore, efforts should be intensified to support various innovations that will harness the use of microorganisms and/or their products as an integral part of agricultural activities and green vegetation for improved food production, environmental sustenance, and human health in achieving the sustainable development goal.

References

Abd El-Daim, I. A., Bejai, S., Fridborg, I., & Meijer, J. (2018). Identifying potential molecular factors involved in *Bacillus amyloliquefaciens* 5113 mediated abiotic stress tolerance in wheat. *Plant Biology, 20*, 271–279. https://doi.org/10.1111/plb.12680

Adeleke, R., Nwangburuka, C., & Oboirien, B. (2017). Origins, roles and fate of organic acids in soils: A review. *South African Journal of Botany, 108*, 393–406. https://doi.org/10.1016/j.sajb.2016.09.002

Ahmad, A., Aslam, Z., Naz, M., Hussain, S., Javed, T., Aslam, S., Raza, A., Ali, H. M., Siddiqui, M. H., Salem, M. Z. M., Hano, C., Shabbir, R., Ahmar, S., Saeed, T., & Jamal, M. A. (2021). Exogenous salicylic acid-induced drought stress tolerance in wheat (Triticum aestivum L.) grown under hydroponic culture. *PloS One, 16*(12), e0260556. https://doi.org/10.1371/journal.pone.0260556. (Retraction published PLoS One. 2023 Apr 19;18(4):e0284669. https://doi.org/10.1371/journal.pone.0284669)

Aires, E. S., Ferraz, A. K. L., Carvalho, B. L., Teixeira, F. P., Putti, F. F., de Souza, E. P., Rodrigues, J. D., & Ono, E. O. (2022). Foliar application of salicylic acid to mitigate water stress in tomato. *Plants (Basel, Switzerland), 11*(13), 1775. https://doi.org/10.3390/plants11131775

Ajouz, S., Walker, A. S., Fabre, F., Leroux, P., Nicot, P., & Bardin, M. (2010). Variability of Botrytis cinerea sensitivity to pyrrolnitrin, an antibiotic produced by biological control agents. *BioControl, 56*(3), 353–363. https://doi.org/10.1007/s10526-010-9333-7

Alori, E. T., Glick, B. R., & Babalola, O. O. (2017). Microbial phosphorus Solubilization and its potential for use in sustainable agriculture. *Frontiers in Microbiology, 8*. https://doi.org/10.3389/fmicb.2017.00971

Althiab-Almasaud, R., Sallanon, H., Chang, C., & Chervin, C. (2021). 1-Aminocyclopropane-1-carboxylic acid stimulates tomato pollen tube growth independently of ethylene receptors. *Physiologia Plantarum, 173*(4), 2291–2297. https://doi.org/10.1111/ppl.13579

Amoo, A. E., & Babalola, O. O. (2017). Ammonia-oxidizing microorganisms: Key players in the promotion of plant growth. *Journal of Soil Science and Plant Nutrition, 17*(4), 935–947. https://doi.org/10.4067/s0718-95162017000400008

Anand, A., Falquet, L., Abou-Mansour, E., L'Haridon, F., Keel, C., & Weisskopf, L. (2023). Biological hydrogen cyanide emission globally impacts the physiology of both HCN-emitting and HCN-perceiving *Pseudomonas*. *MBio, 14*(5), e0085723. https://doi.org/10.1128/mbio.00857-23

Andrić, S., Meyer, T., Rigolet, A., Prigent-Combaret, C., Höfte, M., Balleux, G., Steels, S., Hoff, G., De Mot, R., McCann, A., De Pauw, E., Argüelles Arias, A., & Ongena, M. (2021). Lipopeptide interplay mediates molecular interactions between soil bacilli and pseudomonads. *Microbiology Spectrum, 9*, e02038–e02021. https://doi.org/10.1128/spectrum.02038-21

Andrino, A., Guggenberger, G., Kernchen, S., Mikutta, R., Sauheitl, L., & Boy, J. (2021). Production of organic acids by Arbuscular Mycorrhizal fungi and their contribution in the mobilization of phosphorus bound to iron oxides. *Frontiers in Plant Science, 12*. https://doi.org/10.3389/fpls.2021.661842

Athukorala, S. N., Fernando, W. G., & Rashid, K. Y. (2009). Identification of antifungal antibiotics of Bacillus species isolated from different microhabitats using polymerase chain reaction and MALDI-TOF mass spectrometry. *Canadian Journal of Microbiology, 55*(9), 1021–1032. https://doi.org/10.1139/w09-067

Athukorala, S. N., Fernando, W. D., Rashid, K. Y., & De Kievit, T. (2010). The role of volatile and non-volatile antibiotics produced byPseudomonas chlororaphisstrain PA23 in its root colonization and control ofSclerotinia sclerotiorum. *Biocontrol Science and Technology, 20*(8), 875–890. https://doi.org/10.1080/09583157.2010.484484

Ayilara, M. S., Adeleke, B. S., Akinola, S. A., Fayose, C. A., Adeyemi, U. T., Gbadegesin, L. A., Omole, R. K., Johnson, R. M., Uthman, Q. O., & Babalola, O. O. (2023). Biopesticides as a promising alternative to synthetic pesticides: A case for microbial pesticides, phytopesticides, and nanobiopesticides. *Frontiers in Microbiology, 14*. https://doi.org/10.3389/fmicb.2023.1040901

Babaniyi, B. R., Ogundele, O. D., Bisi-omotosho, A., Babaniyi, E. E., & Aransiola, S. A. (2023). Remediation approaches in environmental sustainability. In N. R. Maddela, L. K. W. Eller, & R. Prasad (Eds.), *Microbiology for cleaner production and environmental sustainability*. CRC Press. ISBN: 9781032496061. https://www.routledge.com/Microbiology-for-Cleaner-Production-and-Environmental-Sustainability/Maddela-Eller-Prasad/p/book/9781032496061

Bailly, A., & Weisskopf, L. (2012). The modulating effect of bacterial volatiles on plant growth: Current knowledge and future challenges. *Plant Signaling & Behavior, 7*(1), 79–85. https://doi.org/10.4161/psb.7.1.18418

Bajguz, A. (2009). Brassinosteroid enhanced the level of abscisic acid in Chlorella vulgaris subjected to short-term heat stress. *Journal of Plant Physiology, 166*(8), 882–886. https://doi.org/10.1016/j.jplph.2008.10.004

Basu, A., Prasad, P., Das, S. N., Kalam, S., Sayyed, R. Z., Reddy, M. S., & Enshasy, H. E. (2021). Plant Growth Promoting Rhizobacteria (PGPR) as green bioinoculants: Recent developments, constraints, and prospects. *Sustainability, 13*(3), 1140. https://doi.org/10.3390/su13031140

Basu, S., Priyadarshini, P., Prasad, R., & Kumar, G. (2022). *Effects of microbial signaling in plant growth and development beneficial microorganisms in agriculture* (pp. 329–348). Springer Nature Singapore. https://doi.org/10.1007/978-981-19-0733-3_14

Bautista-Cruz, A., Aquino-Bolaños, T., Hernández-Canseco, J., & Quiñones-Aguilar, E. E. (2024). Cellulolytic aerobic bacteria isolated from agricultural and Forest soils: An overview. *Biology, 13*(2), 102. https://doi.org/10.3390/biology13020102

Bekele, M., & Getaneh, S. (2022). Function of microorganisms on soil health maintenance: A review article. *International Journal of Advanced Research in Biological Sciences, 9*(4), 82–93. https://doi.org/10.22192/ijarbs.2022.09.04.009

Bennett, E. M., Murray, J. W., & Isalan, M. (2023). Engineering Nitrogenases for synthetic nitrogen fixation: From pathway engineering to directed evolution. *Biodesign Research, 5*, 0005. https://doi.org/10.34133/bdr.0005

Bhagat, N., Raghav, M., Dubey, S., & Bedi, N. (2021). Bacterial exopolysaccharides: Insight into their role in plant abiotic stress tolerance. *Journal of Microbiology and Biotechnology, 31*(8), 1045–1059. https://doi.org/10.4014/jmb.2105.05009

Boyd, E. S., & Peters, J. W. (2013). New insights into the evolutionary history of biological nitrogen fixation. *Frontiers in Microbiology, 4*, 201. https://doi.org/10.3389/fmicb.2013.00201

Braud, A., Hannauer, M., Mislin, G. L., & Schalk, I. J. (2009). The Pseudomonas aeruginosa pyochelin-iron uptake pathway and its metal specificity. *Journal of Bacteriology, 191*(11), 3517–3525. https://doi.org/10.1128/JB.00010-09

Brodhagen, M., Henkels, M. D., & Loper, J. E. (2004). Positive autoregulation and signaling properties of pyoluteorin, an antibiotic produced by the biological control organism Pseudomonas fluorescens Pf-5. *Applied and Environmental Microbiology, 70*(3), 1758–1766. https://doi.org/10.1128/AEM.70.3.1758-1766.2004

Campbell, W. H. (1999). Nitrate reductase structure, function and regulation: Bridging the gap between biochemistry and physiology. *Annual Review of Plant Physiology and Plant Molecular Biology, 50*, 277–303. https://doi.org/10.1146/annurev.arplant.50.1.277

Cao, C., Peng, F., & Cui, B. K. (2013). Chemical characterization and structure of exopolysaccharides from submerged culture of new medicinal mushroom from China, Phellinus mori (higher basidiomycetes). *International Journal of Medicinal Mushrooms, 15*(1), 57–69. https://doi.org/10.1615/intjmedmushr.v15.i1.70

Cao, P., Li, C., Wang, H., Yu, Z., Xu, X., Wang, X., Zhao, J., & Xiang, W. (2020). Community structures and antifungal activity of root-associated Endophytic Actinobacteria in healthy and diseased cucumber plants and Streptomyces sp. HAAG3-15 as a promising biocontrol agent. *Microorganisms, 8*(2), 236. https://doi.org/10.3390/microorganisms8020236

Cao, R., Zhang, Y., Ju, Y., Wang, W., Zhao, Y., Liu, N., Zhang, G., Wang, X., Xie, X., Dai, C., Liu, Y., Yin, H., Shi, K., He, C., Wang, W., Zhao, L., Jeon, C. O., & Hao, L. (2023). Exopolysaccharide-producing bacteria enhanced Pb immobilization and influenced the microbiome composition in rhizosphere soil of pakchoi (Brassica chinensis L.). *Frontiers in Microbiology, 14*. https://doi.org/10.3389/fmicb.2023.1117312

Castro-Camba, R., Sánchez, C., Vidal, N., & Vielba, J. (2022). Interactions of gibberellins with Phytohormones and their role in stress responses. *Horticulturae, 8*(3), 241. https://doi.org/10.3390/horticulturae8030241

Chin-A-Woeng, T. F. C., Bloemberg, G. V., & Lugtenberg, B. J. J. (2003). Phenazines and their role in biocontrol by Pseudomonas bacteria. *The New Phytologist, 157*(3), 503–523. https://doi.org/10.1046/j.1469-8137.2003.00686.x

Cho, S. M., Kang, B. R., Han, S. H., Anderson, A. J., Park, J. Y., Lee, Y. H., Cho, B. H., Yang, K. Y., Ryu, C. M., & Kim, Y. C. (2008). 2R,3R-butanediol, a bacterial volatile produced by Pseudomonas chlororaphis O6, is involved in induction of systemic tolerance to drought in Arabidopsis thaliana. *Molecular Plant-Microbe Interactions: MPMI, 21*(8), 1067–1075. https://doi.org/10.1094/MPMI-21-8-1067

Costa, J., Oliveira, R. S., Tiago, I., Ma, Y., Galhano, C., Freitas, H., & Castro, P. (2018). Soil microorganisms. In A. M. Sánchez-Moreiras & M. J. Reigosa (Eds.), *Advances in plant ecophysiology techniques*. Springer International Publishing AG, part of Springer Nature. https://doi.org/10.1007/978-3-319-93233-0_27

Dani, K. G., Fineschi, S., Michelozzi, M., & Loreto, F. (2016). Do cytokinins, volatile isoprenoids and carotenoids synergically delay leaf senescence? *Plant, Cell & Environment, 39*(5), 1103–1111. https://doi.org/10.1111/pce.12705

Dashtban, M., Schraft, H., Syed, T. A., & Qin, W. (2010). Fungal biodegradation and enzymatic modification of lignin. *International Journal of Biochemistry and Molecular Biology, 1*(1), 36–50.

Dell'Abate, M. T., & Biro, B. (2015). Soil microbes and soil organic matter dynamics. EIP-AGRI, 1–14. https://ec.europa.eu/eip/agriculture/sites/default/files/fg5_mp_om_microbial_comm_2015_en.pdf

de Werra, P., Péchy-Tarr, M., Keel, C., & Maurhofer, M. (2009). Role of gluconic acid production in the regulation of biocontrol traits of Pseudomonas fluorescens CHA0. *Applied and Environmental Microbiology, 75*(12), 4162–4174. https://doi.org/10.1128/AEM.00295-09

Dimkić, I., Janakiev, T., Petrović, M., Degrassi, G., & Fira, D. (2022). Plant-associated Bacillus and Pseudomonas antimicrobial activities in plant disease suppression via biological control mechanisms – A review. *Physiological and Molecular Plant Pathology, 117*, 101754. https://doi.org/10.1016/j.pmpp.2021.101754

Egamberdieva, D., Wirth, S. J., Alqarawi, A. A., Abd Allah, E. F., & Hashem, A. (2017). Phytohormones and beneficial microbes: Essential components for plants to balance stress and fitness. *Frontiers in Microbiology, 8*, 2104. https://doi.org/10.3389/fmicb.2017.02104

Fadiji, A. E., Santoyo, G., Yadav, A. N., & Babalola, O. O. (2022). Efforts towards overcoming drought stress in crops: Revisiting the mechanisms employed by plant growth-promoting bacteria. *Frontiers in Microbiology, 13*. https://doi.org/10.3389/fmicb.2022.962427

Fahad, S., Hussain, S., Bano, A., Saud, S., Hassan, S., Shan, D., Khan, F. A., Khan, F., Chen, Y., Wu, C., Tabassum, M. A., Chun, M. X., Afzal, M., Jan, A., Jan, M. T., & Huang, J. (2015). Potential role of phytohormones and plant growth-promoting rhizobacteria in abiotic stresses: Consequences for changing environment. *Environmental Science and Pollution Research International, 22*(7), 4907–4921. https://doi.org/10.1007/s11356-014-3754-2

Flury, P., Vesga, P., Péchy-Tarr, M., Aellen, N., Dennert, F., Hofer, N., Kupferschmied, K. P., Kupferschmied, P., Metla, Z., Ma, Z., Siegfried, S., de Weert, S., Bloemberg, G., Höfte, M., Keel, C. J., & Maurhofer, M. (2017). Antimicrobial and insecticidal: Cyclic Lipopeptides and hydrogen cyanide produced by plant-beneficial *Pseudomonas* strains CHA0, CMR12a, and PCL1391 contribute to insect killing. *Frontiers in Microbiology, 8*, 100. https://doi.org/10.3389/fmicb.2017.00100

Gangwar, M., & Kaur, G. (2009). Isolation and characterization of endophytic bacteria from endorhizosphere of sugarcane and ryegrass. *International Journal of Microbiology, 7*(1), 139–144. https://doi.org/10.5580/181d

Ghazy, N., & El-Nahrawy, S. (2021). Siderophore production by Bacillus subtilis MF497446 and Pseudomonas koreensis MG209738 and their efficacy in controlling Cephalosporium maydis in maize plant. *Archives of Microbiology, 203*(3), 1195–1209. https://doi.org/10.1007/s00203-020-02113-5

Glick, B. R. (2012). Plant growth-promoting bacteria: Mechanisms and applications. *Scientifica, 2012*, 1–15. https://doi.org/10.6064/2012/963401

Glick, B. R., & Gamalero, E. (2021). Recent developments in the study of plant microbiomes. *Microorganisms, 9*(7), 1533. https://doi.org/10.3390/microorganisms9071533

Gómez-Godínez, L. J., Aguirre-Noyola, J. L., Martínez-Romero, E., Arteaga-Garibay, R. I., Ireta-Moreno, J., & Ruvalcaba-Gómez, J. M. (2023). A look at plant-growth-promoting bacteria. *Plants, 12*(8), 1668. https://doi.org/10.3390/plants12081668

Gopalakrishnan, S., Sathya, A., Vijayabharathi, R., Varshney, R. K., Gowda, C. L., & Krishnamurthy, L. (2015). Plant growth promoting rhizobia: Challenges and opportunities. *3. Biotech, 5*(4), 355–377. https://doi.org/10.1007/s13205-014-0241-x

Grover, A., Mittal, D., Negi, M., & Lavania, D. (2013). Generating high temperature tolerant transgenic plants: Achievements and challenges. *Plant Science: An International Journal of Experimental Plant Biology, 205–206*, 38–47. https://doi.org/10.1016/j.plantsci.2013.01.005

Grzyb, A., Wolna-Maruwka, A., & Niewiadomska, A. (2021). The significance of microbial transformation of nitrogen compounds in the light of integrated crop management. *Agronomy, 11*(7), 1415. https://doi.org/10.3390/agronomy11071415

Haile, S., & Ayele, A. (2022). Pectinase from microorganisms and its industrial applications. *The Scientific World Journal, 2022*, 1881305. https://doi.org/10.1155/2022/1881305

Halimah, N., Baktir, A., & Purkan, P. (2019). Exploration of cellulolytic microorganism as a biocatalyst candidate for liquid fertilizer production. *IOP Conference Series: Earth and Environmental Science, 217*, 012021. https://doi.org/10.1088/1755-1315/217/1/012021

Hao, H. T., Zhao, X., Shang, Q. H., Wang, Y., Guo, Z. H., Zhang, Y. B., Xie, Z. K., & Wang, R. Y. (2016). Comparative digital gene expression analysis of the Arabidopsis response to volatiles emitted by bacillus amyloliquefaciens. *PLoS One, 11*(8), e0158621. https://doi.org/10.1371/journal.pone.0158621

Hashem, A., Abd-Allah, E. F., Alqarawi, A. A., Al-Huqail, A. A., Wirth, S., & Egamberdieva, D. (2016). The interaction between Arbuscular Mycorrhizal fungi and Endophytic bacteria enhances plant growth of Acacia gerrardii under salt stress. *Frontiers in Microbiology, 7*. https://doi.org/10.3389/fmicb.2016.01089

Hayat, R., Ali, S., Amara, U., Khalid, R., & Ahmed, I. (2010). Soil beneficial bacteria and their role in plant growth promotion: A review. *Annals of Microbiology, 60*, 579–598. https://doi.org/10.1007/s13213-010-0117-1

He, Y., Zhang, H., Sun, Z., Li, J., Hong, G., Zhu, Q., Zhou, X., MacFarlane, S., Yan, F., & Chen, J. (2017). Jasmonic acid-mediated defense suppresses brassinosteroid-mediated susceptibility to Rice black streaked dwarf virus infection in rice. *The New Phytologist, 214*(1), 388–399. https://doi.org/10.1111/nph.14376

Idowu, J., Pietrasiak, N., & Hoellrich, M. (2023). *Soil biological processes [Guide A-153]*. New Mexico State University Cooperative Extension Service. https://pubs.nmsu.edu/_a/A153.pdf

Ijaz, A., Mumtaz, M. Z., Wang, X., Ahmad, M., Saqib, M., Maqbool, H., Zaheer, A., Wang, W., & Mustafa, A. (2021). Insights into manganese solubilizing bacillus spp. for improving plant growth and manganese uptake in Maize. *Frontiers in Plant Science, 12*. https://doi.org/10.3389/fpls.2021.719504

Ines, M., Amel, K., Yousra, T., Neila, S., Imen, D., Marie, M. J., & Abdennasseur, H. (2012). Effect of dose-response of zinc and manganese on Siderophores production. *American Journal of Environmental Sciences, 8*(2), 143–151. https://doi.org/10.3844/ajessp.2012.143.151

Islam, M. T., & Fukushi, Y. (2010). Growth inhibition and excessive branching in Aphanomyces cochlioides induced by 2,4-diacetylphloroglucinol is linked to disruption of filamentous actin cytoskeleton in the hyphae. *World Journal of Microbiology & Biotechnology, 26*(7), 1163–1170. https://doi.org/10.1007/s11274-009-0284-z

Issar, S., Yadav, N., & Gaur, R. K. (2021). Role of volatile organic compounds in plant growth, communication and defense. *Agrica/Agrica: An International Journal for Plant Science and Related Industries, 10*(2), 111–119. https://doi.org/10.5958/2394-448x.2021.00017.1

Jashni, M. K., Mehrabi, R., Collemare, J., Mesarich, C. H., & de Wit, P. J. (2015). The battle in the apoplast: Further insights into the roles of proteases and their inhibitors in plant-pathogen interactions. *Frontiers in Plant Science, 6*, 584. https://doi.org/10.3389/fpls.2015.00584

Jeyanthi, V., & Kanimozhi, S. (2018). Plant growth promoting Rhizobacteria (PGPR) – Prospective and mechanisms: A review. *Journal of Pure and Applied Microbiology, 12*(2), 733–749. https://doi.org/10.22207/jpam.12.2.34

Jones, D., Dennis, P., Owen, A., & Van Hees, P. (2003). Organic acid behavior in soils—Misconceptions and knowledge gaps. *Plant and Soil, 248*(1/2), 31–41. https://doi.org/10.1023/a:1022304332313

Joseph, P. S., Musa, D. A., Egwim, E. C., & Uthman, A. (2022). *Function of urease in plants with reference to legumes: A review*. IntechOpen eBooks. https://doi.org/10.5772/intechopen.102646

Kang, B. R., Anderson, A. J., & Kim, Y. C. (2018). Hydrogen cyanide produced by *pseudomonas chlororaphis* O6 exhibits Nematicidal activity against *Meloidogyne hapla*. *The Plant Pathology Journal, 34*(1), 35–43. https://doi.org/10.5423/PPJ.OA.06.2017.0115

Kang, B. R., Anderson, A. J., & Kim, Y. C. (2019). Hydrogen cyanide produced by Pseudomonas chlororaphis O6 is a key aphicidal metabolite. *Canadian Journal of Microbiology, 65*(3), 185–190. https://doi.org/10.1139/cjm-2018-0372

Kaur, T., Jasrotia, S., Ohri, P., & Manhas, R. K. (2016). Evaluation of in vitro and in vivo nematicidal potential of a multifunctional streptomycete, Streptomyces hydrogenans strain DH16 against Meloidogyne incognita. *Microbiological Research, 192*, 247–252. https://doi.org/10.1016/j.micres.2016.07.009

Kenawy, A., Dailin, D. J., Abo-Zaid, G. A., Malek, R. A., Ambehabati, K. K., Zakaria, K. H. N., Sayyed, R. Z., & Enshasy, H. A. E. (2019). Biosynthesis of antibiotics by PGPR and their roles in biocontrol of plant diseases. In *Microorganisms for sustainability* (pp. 1–35). https://doi.org/10.1007/978-981-13-6986-5_1

Keswani, C., Singh, S. P., García-Estrada, C., Mezaache-Aichour, S., Glare, T. R., Borriss, R., Rajput, V. D., Minkina, T. M., Ortiz, A., & Sansinenea, E. (2022). Biosynthesis and beneficial effects of microbial gibberellins on crops for sustainable agriculture. *Journal of Applied Microbiology, 132*(3), 1597–1615. https://doi.org/10.1111/jam.15348

Khan, N., Bano, A., Rahman, M. A., Guo, J., Kang, Z., & Babar, M. A. (2019). Comparative physiological and metabolic analysis reveals a complex mechanism involved in drought tolerance in chickpea (Cicer arietinum L.) induced by PGPR and PGRs. *Scientific Reports, 9*(1). https://doi.org/10.1038/s41598-019-38702-8

Khare, E., & Yadav, A. (2017). The role of microbial enzyme systems in plant growth promotion. *Climate Change and Environmental Sustainability, 5*(2), 122. https://doi.org/10.5958/2320-642x.2017.00013.8

Khaskheli, M. A., Wu, L., Chen, G., Chen, L., Hussain, S., Song, D., Liu, S., & Feng, G. (2020). Isolation and characterization of root-associated bacterial endophytes and their biocontrol potential against major fungal Phytopathogens of Rice (*Oryza sativa* L.). *Pathogens (Basel, Switzerland), 9*(3), 172. https://doi.org/10.3390/pathogens9030172

Khatoon, Z., Huang, S., Rafique, M., Fakhar, A., Kamran, M. A., & Santoyo, G. (2020). Unlocking the potential of plant growth-promoting rhizobacteria on soil health and the sustainability of agricultural systems. *Journal of Environmental Management, 273*, 111118. https://doi.org/10.1016/j.jenvman.2020.111118

Khindaria, A., Grover, T. A., & Aust, S. D. (1995). Reductive dehalogenation of aliphatic halocarbons by lignin peroxidase of Phanerochaete chrysosporium. *Environmental Science & Technology, 29*(3), 719–725. https://doi.org/10.1021/es00003a020

Khoshru, B., Mitra, D., Khoshmanzar, E., Myo, E. M., Uniyal, N., Mahakur, B., Mohapatra, P. K. D., Panneerselvam, P., Boutaj, H., Alizadeh, M., Cely, M. V. T., Senapati, A., & Rani, A. (2020). Current scenario and future prospects of plant growth-promoting rhizobacteria: An economic valuable resource for the agriculture revival under stressful conditions. *Journal of Plant Nutrition, 43*(20), 3062–3092. https://doi.org/10.1080/01904167.2020.1799004

Kieber, J. J., & Schaller, G. E. (2010). The perception of cytokinin: a story 50 years in the making. *Plant Physiology, 154*(2), 487–492. https://doi.org/10.1104/pp.110.161596

Kim, P. I., Ryu, J., Kim, Y. H., & Chi, Y. T. (2010). Production of Biosurfactant Lipopeptides Iturin A, Fengycin and Surfactin A from Bacillus subtilis CMB32 for control of Colletotrichum gloeosporioides. *Journal of Microbiology and Biotechnology, 20*(1), 138–145. https://doi.org/10.4014/jmb.0905.05007

Kim, Y. C., Kang, B. R., Kim, Y. H., & Park, S. K. (2017). An effective and practical strategy for biocontrol of plant diseases using on-site mass cultivation of chitin-degrading bacteria. *Singmulbyeong Yeon-gu, 23*(1), 19–34. https://doi.org/10.5423/rpd.2017.23.1.19

Kim, H., Seomun, S., Yoon, Y., & Jang, G. (2021). Jasmonic acid in plant abiotic stress tolerance and interaction with Abscisic acid. *Agronomy, 11*(9), 1886. https://doi.org/10.3390/agronomy11091886

Koo, Y. M., Heo, A. Y., & Choi, H. W. (2020). Salicylic acid as a safe plant protector and growth regulator. *The Plant Pathology Journal, 36*(1), 1–10. https://doi.org/10.5423/PPJ.RW.12.2019.0295

Ku, Y., Xu, G., Tian, X., Xie, H., Yang, X., Cao, C., & Chen, Y. (2018). Root colonization and growth promotion of soybean, wheat and Chinese cabbage by Bacillus cereus YL6. *PloS one, 13*(11), e0200181. https://doi.org/10.1371/journal.pone.0200181

Kumar, A., & Chandra, R. (2020). Ligninolytic enzymes and its mechanisms for degradation of lignocellulosic waste in environment. *Heliyon, 6*(2), e03170. https://doi.org/10.1016/j.heliyon.2020.e03170

Kumar, M., Brar, A., Yadav, M., Chawade, A., Vivekanand, V., & Pareek, N. (2018). Chitinases—Potential candidates for enhanced plant resistance towards fungal pathogens. *Agriculture, 8*(7), 88. https://doi.org/10.3390/agriculture8070088

Kumar, S., Diksha, S., & S. S., & Kumar, R. (2022). Biofertilizers: An ecofriendly technology for nutrient recycling and environmental sustainability. *Current Research in Microbial Sciences, 3*, 100094. https://doi.org/10.1016/j.crmicr.2021.100094

Kumar, C. S., D'Silva, S., Praveena, R., Kaprakkaden, A., Krishnan, L. A., Rajkumar, M. B., Srinivasan, V., & Dinesh, R. (2024). Zinc solubilization and organic acid production by the entomopathogenic fungus, Metarhizium pingshaense sheds light on its key ecological role in the environment. *Science of the Total Environment, 923*, 171348. https://doi.org/10.1016/j.scitotenv.2024.171348

Lee, J. Y., Moon, S. S., & Hwang, B. K. (2003). Isolation and antifungal and antioomycete activities of aerugine produced by Pseudomonas fluorescens strain MM-B16. *Applied and Environmental Microbiology, 69*(4), 2023–2031. https://doi.org/10.1128/AEM.69.4.2023-2031.2003

Lee, J. H., Ma, K. C., Ko, S. J., Kang, B. R., Kim, I. S., & Kim, Y. C. (2011). Nematicidal activity of a nonpathogenic biocontrol bacterium, Pseudomonas chlororaphis O6. *Current Microbiology, 62*(3), 746–751. https://doi.org/10.1007/s00284-010-9779-y

Lee, Y. S., Nguyen, X. H., Cho, J. Y., Moon, J. H., & Kim, K. Y. (2017). Isolation and antifungal activity of methyl 2,3-dihydroxybenzoate from Paenibacillus elgii HOA73. *Microbial Pathogenesis, 106*, 139–145. https://doi.org/10.1016/j.micpath.2016.01.007

Lee, S., Behringer, G., Hung, R., & Bennett, J. (2019). Effects of fungal volatile organic compounds on Arabidopsis thaliana growth and gene expression. *Fungal Ecology, 37*, 1–9. https://doi.org/10.1016/j.funeco.2018.08.004

Lee, J. H., Anderson, A. J., & Kim, Y. C. (2022). Root-associated bacteria are biocontrol agents for multiple plant pests. *Microorganisms, 10*(5), 1053. https://doi.org/10.3390/microorganisms10051053

Lemfack, M. C., Nickel, J., Dunkel, M., Preissner, R., & Piechulla, B. (2014). mVOC: A database of microbial volatiles. *Nucleic Acids Research, 42*(Database issue), D744–D748. https://doi.org/10.1093/nar/gkt1250

Létoffé, S., Wu, Y., Darch, S. E., Beloin, C., Whiteley, M., Touqui, L., & Ghigo, J. M. (2022). *Pseudomonas aeruginosa* production of hydrogen cyanide leads to airborne control of *Staphylococcus aureus* growth in biofilm and *in vivo* lung environments. *mBio, 13*(5), e0215422. https://doi.org/10.1128/mbio.02154-22

Li, Z. T., Janisiewicz, W. J., Liu, Z., Callahan, A. M., Evans, B. E., Jurick, W. M., 2nd, & Dardick, C. (2019). Exposure *in vitro* to an environmentally isolated strain TC09 of *Cladosporium sphaerospermum* triggers plant growth promotion, early flowering, and fruit yield increase. *Frontiers in Plant Science, 9*, 1959. https://doi.org/10.3389/fpls.2018.01959

Li, Y., Rao, P., Wang, J., Song, S., Wang, R., Liu, S., & Che, L. (2021). Study on preparation and application of a multifunctional microspheric soil conditioner based on Arabic gum, gelatin, chitosan and β-cyclodextrin. *International Journal of Biological Macromolecules, 183*, 1851–1860. https://doi.org/10.1016/j.ijbiomac.2021.05.205

Liu, X. M., & Zhang, H. (2015). The effects of bacterial volatile emissions on plant abiotic stress tolerance. *Frontiers in Plant Science, 6*, 774. https://doi.org/10.3389/fpls.2015.00774

Lopes, M. J. D. S., Dias-Filho, M. B., & Gurgel, E. S. C. (2021). Successful plant growth-promoting microbes: Inoculation methods and abiotic factors. *Frontiers in Sustainable Food Systems, 5*. https://doi.org/10.3389/fsufs.2021.606454

Luziatelli, F., Gatti, L., Ficca, A. G., Medori, G., Silvestri, C., Melini, F., Muleo, R., & Ruzzi, M. (2020). Metabolites secreted by a plant-growth-promoting *Pantoea agglomerans* strain improved rooting of *Pyrus communis* L. cv Dar Gazi cuttings. *Frontiers in Microbiology, 11*, 539359. https://doi.org/10.3389/fmicb.2020.539359

Mabood, F., Zhou, X., & Smith, D. L. (2014). Microbial signaling and plant growth promotion. *Canadian Journal of Plant Science/Canadian Journal of Plant Science, 94*(6), 1051–1063. https://doi.org/10.4141/cjps2013-148

Macias-Benitez, S., Garcia-Martinez, A. M., Jimenez, P. C., Gonzalez, J. M., Moral, M. T., & Rubio, J. P. (2020). Rhizospheric organic acids as biostimulants: Monitoring feedbacks on soil microorganisms and biochemical properties. *Frontiers in Plant Science, 11*, 633. https://doi.org/10.3389/fpls.2020.00633

Marrs, K. A. (1996). The functions and regulation of glutathione s-transferases in plants. *Annual Review of Plant Physiology and Plant Molecular Biology, 47*(1), 127–158. https://doi.org/10.1146/annurev.arplant.47.1.127

McRose, D. L., Baars, O., Morel, F. M. M., & Kraepiel, A. M. L. (2017). Siderophore production in Azotobacter vinelandii in response to Fe-Mo- and V-limitation. *Environmental Microbiology, 19*(9), 3595–3605. https://doi.org/10.1111/1462-2920.13857

Mehmood, N., Saeed, M., Zafarullah, S., Hyder, S., Rizvi, Z. F., Gondal, A. S., Jamil, N., Iqbal, R., Ali, B., Ercisli, S., & Kupe, M. (2023). Multifaceted impacts of plant-beneficial *Pseudomonas* spp. in managing various plant diseases and crop yield improvement. *ACS Omega, 8*(25), 22296–22315. https://doi.org/10.1021/acsomega.3c00870

Mishra, J., Singh, R., & Arora, N. K. (2017). Alleviation of heavy metal stress in plants and remediation of soil by rhizosphere microorganisms. *Frontiers in Microbiology, 8*, 1706. https://doi.org/10.3389/fmicb.2017.01706

Morales-Cedeño, L. R., Orozco-Mosqueda, M. D. C., Loeza-Lara, P. D., Parra-Cota, F. I., de Los Santos-Villalobos, S., & Santoyo, G. (2021). Plant growth-promoting bacterial endophytes as biocontrol agents of pre- and post-harvest diseases: Fundamentals, methods of application and future perspectives. *Microbiological Research, 242*, 126612. https://doi.org/10.1016/j.micres.2020.126612

Morgan, J. A., Bending, G. D., & White, P. J. (2005). Biological costs and benefits to plant-microbe interactions in the rhizosphere. *Journal of Experimental Botany, 56*(417), 1729–1739. https://doi.org/10.1093/jxb/eri205

Murugappan, R., Karthikeyan, M., Aravinth, A., & Alamelu, M. (2012). Siderophore-mediated iron uptake promotes yeast-bacterial symbiosis. *Applied Biochemistry and Biotechnology, 168*(8), 2170–2183. https://doi.org/10.1007/s12010-012-9926-y

Nair, A., Juwarkar, A. A., & Singh, S. K. (2006). Production and characterization of Siderophores and its application in arsenic removal from contaminated soil. *Water, Air and Soil Pollution/Water, Air & Soil Pollution, 180*(1–4), 199–212. https://doi.org/10.1007/s11270-006-9263-2

Nall V. (2010). *Role of mycorrhizae in rhizosphere processes and phosphorus dynamics* (Thesis). Lincoln University, Christchurch, pp. 204. https://hdl.handle.net/10182/1803

Narayanasamy, P. (2013). Mechanisms of action of bacterial biological control agents. In *Biological management of diseases of crops. Progress in biological control* (Vol. 15). Springer. https://doi.org/10.1007/978-94-007-6380-7_5

Nazli, F., Mustafa, A., Ahmad, M., Hussain, A., Jamil, M., Wang, X., Shakeel, Q., Imtiaz, M., & El-Esawi, M. A. (2020). A review on practical application and potentials of Phytohormone-producing plant growth-promoting Rhizobacteria for inducing heavy metal tolerance in crops. *Sustainability, 12*(21), 9056. https://doi.org/10.3390/su12219056

Netrusov, A. I., Liyaskina, E. V., Kurgaeva, I. V., Liyaskina, A. U., Yang, G., & Revin, V. V. (2023). Exopolysaccharides producing bacteria: A review. *Microorganisms, 11*(6), 1541. https://doi.org/10.3390/microorganisms11061541

Neung, S., Nguyen, X. H., Naing, K. W., Lee, Y. S., & Kim, K. Y. (2014). Insecticidal potential of Paenibacillus elgii HOA73 and its combination with organic sulfur pesticide on diamondback moth, Plutella xylostella. *Han'gug Eung'yong Saengmyeong Hwa Haghoeji/Journal of the Korean Society for Applied Biological Chemistry, 57*(2), 181–186. https://doi.org/10.1007/s13765-013-4273-4

Nguyen, X. H., Naing, K. W., Lee, Y. S., Jung, W. J., Anees, M., & Kim, K. Y. (2013). Antagonistic potential of Paenibacillus elgii HOA73 against the root-knot nematode, Meloidogyne incognita. *Nematology, 15*(8), 991–1000. https://doi.org/10.1163/15685411-00002737

Nielsen, M. N., Winding, A., Binnerup, S., & Hansen, B. M. (2002). *Microorganisms as indicators of soil health*. https://www.dmu.dk/1_viden/2_Publikationer/3_fagrapporter/rapporter/FR388.pdf

Nishida, N., Tamotsu, S., Nagata, N., Saito, C., & Sakai, A. (2005). Allelopathic effects of volatile monoterpenoids produced by Salvia leucophylla: Inhibition of cell proliferation and DNA synthesis in the root apical meristem of Brassica campestris seedlings. *Journal of Chemical Ecology, 31*(5), 1187–1203. https://doi.org/10.1007/s10886-005-4256-y

Nolan, T. M., Vukašinović, N., Liu, D., Russinova, E., & Yin, Y. (2020). Brassinosteroids: Multidimensional regulators of plant growth, development, and stress responses. *The Plant Cell, 32*(2), 295–318. https://doi.org/10.1105/tpc.19.00335

Oburger, E., Kirk, G. J., Wenzel, W. W., Puschenreiter, M., & Jones, D. L. (2009). Interactive effects of organic acids in the rhizosphere. *Soil Biology and Biochemistry/Soil Biology & Biochemistry, 41*(3), 449–457. https://doi.org/10.1016/j.soilbio.2008.10.034

Orozco-Mosqueda, M. D. C., Santoyo, G., & Glick, B. R. (2023). Recent advances in the bacterial Phytohormone modulation of plant growth. *Plants (Basel, Switzerland), 12*(3), 606. https://doi.org/10.3390/plants12030606

Pal, K. K., Tilak, K. V., Saxena, A. K., Dey, R., & Singh, C. S. (2001). Suppression of maize root diseases caused by Macrophomina phaseolina, Fusarium moniliforme and Fusarium graminearum by plant growth promoting rhizobacteria. *Microbiological Research, 156*(3), 209–223. https://doi.org/10.1078/0944-5013-00103

Park, J. Y., Oh, S. A., Anderson, A. J., Neiswender, J., Kim, J. C., & Kim, Y. C. (2011). Production of the antifungal compounds phenazine and pyrrolnitrin from Pseudomonas chlororaphis O6 is differentially regulated by glucose. *Letters in Applied Microbiology, 52*(5), 532–537. https://doi.org/10.1111/j.1472-765X.2011.03036.x

Park, Y. S., Dutta, S., Ann, M., Raaijmakers, J. M., & Park, K. (2015). Promotion of plant growth by *Pseudomonas fluorescens* strain SS101 via novel volatile organic compounds. *Biochemical and Biophysical Research Communications, 461*(2), 361–365. https://doi.org/10.1016/j.bbrc.2015.04.039

Parmeggiani, F., Weise, N. J., Ahmed, S. T., & Turner, N. J. (2018). Synthetic and therapeutic applications of ammonia-lyases and Aminomutases. *Chemical Reviews, 118*(1), 73–118. https://doi.org/10.1021/acs.chemrev.6b00824

Patel, D. K., Archana, G., & Kumar, G. N. (2008). Variation in the nature of organic acid secretion and mineral phosphate solubilization by Citrobacter sp. DHRSS in the presence of different sugars. *Current Microbiology, 56*(2), 168–174. https://doi.org/10.1007/s00284-007-9053-0

Patten, C. L., & Glick, B. R. (1996). Bacterial biosynthesis of indole-3-acetic acid. *Canadian Journal of Microbiology, 42*(3), 207–220. https://doi.org/10.1139/m96-032

Paul, S., Parvez, S. S., Goswami, A., & Banik, A. (2024). Exopolysaccharides from agriculturally important microorganisms: Conferring soil nutrient status and plant health. *International Journal of Biological Macromolecules, 262*, 129954. https://doi.org/10.1016/j.ijbiomac.2024.129954

Perneel, M., Heyrman, J., Adiobo, A., De Maeyer, K., Raaijmakers, J. M., De Vos, P., & Höfte, M. (2007). Characterization of CMR5c and CMR12a, novel fluorescent Pseudomonas strains from the cocoyam rhizosphere with biocontrol activity. *Journal of Applied Microbiology, 103*(4), 1007–1020. https://doi.org/10.1111/j.1365-2672.2007.03345.x

Pieterse, C. M., Van der Does, D., Zamioudis, C., Leon-Reyes, A., & Van Wees, S. C. (2012). Hormonal modulation of plant immunity. *Annual Review of Cell and Developmental Biology, 28*, 489–521. https://doi.org/10.1146/annurev-cellbio-092910-154055

Posso, E. J. S., De Prager, M. S., Rojas, C., & a. C. (2017). Organic acids production by rhizosphere microorganisms isolated from a Typic Melanudands and its effects on the inorganic phosphates solubilization. *Acta Agronómica/Acta AgronóMica, 66*(2). https://doi.org/10.15446/acag.v66n2.56148

Poveda, J. (2021). Beneficial effects of microbial volatile organic compounds (MVOCs) in plants. *Applied Soil Ecology, 168*, 104118. https://doi.org/10.1016/j.apsoil.2021.104118

Prasad, M., Srinivasan, R., Chaudhary, M., Choudhary, M., & Jat, L. K. (2019). Plant growth promoting Rhizobacteria (PGPR) for sustainable agriculture. In *PGPR amelioration in sustainable agriculture* (pp. 129–157). Elsevier. https://doi.org/10.1016/b978-0-12-815879-1.00007-0

Prasetyawan, S., Sulistyowati, L., & Aulanni'am, N. (2018). Glucanase and Chitinase from some isolates of Endophytic fungus Trichoderma spp. *IOP Conference Series Materials Science and Engineering, 299*, 012026. https://doi.org/10.1088/1757-899x/299/1/012026

Rajput, V. D., Harish, Singh, R. K., Verma, K. K., Sharma, L., Quiroz-Figueroa, F. R., Meena, M., Gour, V. S., Minkina, T., Sushkova, S., & Mandzhieva, S. (2021). Recent developments in enzymatic antioxidant Defence mechanism in plants with special reference to abiotic stress. *Biology, 10*(4), 267. https://doi.org/10.3390/biology10040267

Rakshiya, Y., Verma, M., & Sindhu, S. (2016). Efficacy of antagonistic soil bacteria in management of subterranean termites (Isoptera). *Research in Environment and Life Sciences, 9*, 949–955.

Ramakrishna, W., Yadav, R., & Li, K. (2019). Plant growth promoting bacteria in agriculture: Two sides of a coin. *Applied Soil Ecology, 138*, 10–18. https://doi.org/10.1016/j.apsoil.2019.02.019

Rastogi, M., Nandal, M., & Khosla, B. (2020). Microbes as vital additives for solid waste composting. *Heliyon, 6*(2), e03343. https://doi.org/10.1016/j.heliyon.2020.e03343

Raza, W., Yousaf, S., & Rajer, F. U. (2016). Plant growth promoting activity of volatile organic compounds produced by biocontrol strains. *Science Letter, 4*, 40–43.

Roychoudhury, A., & Aftab, T. (2021). Phytohormones, plant growth regulators and signaling molecules: Cross-talk and stress responses. *Plant Cell Reports, 40*(8), 1301–1303. https://doi.org/10.1007/s00299-021-02755-9

Ruan, J., Zhou, Y., Zhou, M., Yan, J., Khurshid, M., Weng, W., Cheng, J., & Zhang, K. (2019). Jasmonic acid signaling pathway in plants. *International Journal of Molecular Sciences, 20*(10), 2479. https://doi.org/10.3390/ijms20102479

Ruffner, B., Péchy-Tarr, M., Ryffel, F., Hoegger, P., Obrist, C., Rindlisbacher, A., Keel, C., & Maurhofer, M. (2013). Oral insecticidal activity of plant-associated pseudomonads. *Environmental Microbiology, 15*(3), 751–763. https://doi.org/10.1111/j.1462-2920.2012.02884.x

Rungin, S., Indananda, C., Suttiviriya, P., Kruasuwan, W., Jaemsaeng, R., & Thamchaipenet, A. (2012). Plant growth enhancing effects by a siderophore-producing endophytic streptomycete isolated from a Thai jasmine rice plant (Oryza sativa L. cv. KDML105). *Antonie Van Leeuwenhoek, 102*(3), 463–472. https://doi.org/10.1007/s10482-012-9778-z

Ryu, C. M., Farag, M. A., Hu, C. H., Reddy, M. S., Wei, H. X., Paré, P. W., & Kloepper, J. W. (2003). Bacterial volatiles promote growth in Arabidopsis. *Proceedings of the National Academy of Sciences of the United States of America, 100*(8), 4927–4932. https://doi.org/10.1073/pnas.0730845100

Saeed, Q., Xiukang, W., Haider, F. U., Kučerik, J., Mumtaz, M. Z., Holatko, J., Naseem, M., Kintl, A., Ejaz, M., Naveed, M., Brtnicky, M., & Mustafa, A. (2021). Rhizosphere bacteria in plant growth promotion, biocontrol, and bioremediation of contaminated sites: A comprehensive review of effects and mechanisms. *International Journal of Molecular Sciences, 22*(19), 10529. https://doi.org/10.3390/ijms221910529

Sasirekha, B., & Srividya, S. (2016). Siderophore production by Pseudomonas aeruginosa FP6, a biocontrol strain for Rhizoctonia solani and Colletotrichum gloeosporioides causing diseases in chilli. *Agriculture and Natural Resources, 50*(4), 250–256. Retrieved from https://li01.tci-thaijo.org/index.php/anres/article/view/244038

Sayyed, R. Z., Patel, P. R., & Shaikh, S. S. (2015). Plant growth promotion and root colonization by EPS producing Enterobacter sp. RZS5 under heavy metal contaminated soil. *Indian Journal of Experimental Biology, 53*(2), 116–123.

Schlaeppi, K., & Bulgarelli, D. (2015). The plant microbiome at work. *Molecular Plant-Microbe Interactions: MPMI, 28*(3), 212–217. https://doi.org/10.1094/MPMI-10-14-0334-FI

Schulz, S., & Dickschat, J. S. (2007). Bacterial volatiles: The smell of small organisms. *Natural Product Reports, 24*(4), 814–842. https://doi.org/10.1039/b507392h

Sigurdarson, J. J., Svane, S., & Karring, H. (2018). The molecular processes of urea hydrolysis in relation to ammonia emissions from agriculture. *Reviews in Environmental Science and Bio-technology/Reviews in Environmental Science and Bio/Technology, 17*(2), 241–258. https://doi.org/10.1007/s11157-018-9466-1

Silva, L. I. D., Pereira, M. C., Carvalho, A. M. X. D., Buttrós, V. H., Pasqual, M., & Dória, J. (2023). Phosphorus-solubilizing microorganisms: A key to sustainable agriculture. *Agriculture, 13*, 462. https://doi.org/10.3390/agriculture13020462

Sindhu, S., & Sahu, G. (2011). Disease control and plant growth promotion of green gram by Siderophore producing Pseudomonas sp. *Journal of Microbiology, 6*(10), 735–749. https://doi.org/10.3923/jm.2011.735.749

Sindhu, S. S., Parmar, P., Phour, M., & Sehrawat, A. (2016). Potassium-Solubilizing Microorganisms (KSMs) and its effect on plant growth improvement. In *Springer eBooks* (pp. 171–185). https://doi.org/10.1007/978-81-322-2776-2_13

Singh, G., & Arya, S. K. (2019). Antifungal and insecticidal potential of chitinases: A credible choice for the eco-friendly farming. *Biocatalysis and Agricultural Biotechnology, 20*, 101289. https://doi.org/10.1016/j.bcab.2019.101289

Singh, A., & Roychoudhury, A. (2023). Abscisic acid in plants under abiotic stress: Crosstalk with major phytohormones. *Plant Cell Reports, 42*(6), 961–974. https://doi.org/10.1007/s00299-023-03013-w

Sorokin, D. Y., Lücker, S., Vejmelkova, D., Kostrikina, N. A., Kleerebezem, R., Rijpstra, W. I., Damsté, J. S., Le Paslier, D., Muyzer, G., Wagner, M., van Loosdrecht, M. C., & Daims, H. (2012). Nitrification expanded: Discovery, physiology and genomics of a nitrite-oxidizing bacterium from the phylum Chloroflexi. *The ISME Journal, 6*(12), 2245–2256. https://doi.org/10.1038/ismej.2012.70

Spaepen, S. (2014). Plant hormones produced by microbes. In *Springer eBooks* (pp. 247–256). https://doi.org/10.1007/978-3-319-08575-3_26

Srikamwang, C., Onsa, N. E., Sunanta, P., Sangta, J., Chanway, C. P., Thanakkasaranee, S., & Sommano, S. R. (2023). Role of microbial volatile organic compounds in promoting plant growth and disease resistance in horticultural production. *Plant Signaling & Behavior/Plant Signalling & Behavior, 18*(1). https://doi.org/10.1080/15592324.2023.2227440

Sultana, S., Alam, S., & Karim, M. M. (2021). Screening of siderophore-producing salt-tolerant rhizobacteria suitable for supporting plant growth in saline soils with iron limitation. *Journal of Agriculture and Food Research, 4*, 100150. https://doi.org/10.1016/j.jafr.2021.100150

Tandon, A., Fatima, T., Anshu, N., Shukla, D., Tripathi, P., Srivastava, S., & Singh, P. C. (2020). Phosphate solubilization by Trichoderma koningiopsis (NBRI-PR5) under abiotic stress conditions. *Journal of King Saud University. Science/Mağallaẗ Ğāmi'aẗ Al-malik Sa'ūd. al-'Ulūm, 32*(1), 791–798. https://doi.org/10.1016/j.jksus.2019.02.001

Tang, A., Haruna, A. O., Majid, N. M. A., & Jalloh, M. B. (2020). Potential PGPR properties of cellulolytic, nitrogen-fixing, phosphate-solubilizing bacteria in rehabilitated tropical forest soil. *Microorganisms, 8*(3), 442. https://doi.org/10.3390/microorganisms8030442

Tazisong, I. A., Senwo, Z. N., & He, Z. (2015). Phosphatase hydrolysis of organic phosphorus compounds. *Advances in Enzyme Research, 03*(02), 39–51. https://doi.org/10.4236/aer.2015.32005

Tucker, G. (2004). Improving fruit and vegetable texture by genetic transformation. In *Texture in food* (pp. 321–341). https://doi.org/10.1533/9781855538362.3.321

Van Oosten, M. J., Pepe, O., De Pascale, S., Silletti, S., & Maggio, A. (2017). The role of biostimulants and bioeffectors as alleviators of abiotic stress in crop plants. *Chemical and Biological Technologies in Agriculture, 4*(1). https://doi.org/10.1186/s40538-017-0089-5

Vanderstraeten, L., Depaepe, T., Bertrand, S., & Van Der Straeten, D. (2019). The ethylene precursor ACC affects early vegetative development independently of ethylene signaling. *Frontiers in Plant Science, 10*, 1591. https://doi.org/10.3389/fpls.2019.01591

Vardhini, B. V. (2013). Comparative study of Sorghum vulgare Pers. Grown in two experimental sites by brassinolide application at vegetative, flowering and grain filling stage. *Proceedings of Andhra Pradesh Akademi of Sciences, 15*, 75–79.

Vardhini, B. V., & Anjum, N. A. (2015). Brassinosteroids make plant life easier under abiotic stresses mainly by modulating major components of antioxidant defense system. *Frontiers in Environmental Science, 2*. https://doi.org/10.3389/fenvs.2014.00067

Vejan, P., Abdullah, R., Khadiran, T., Ismail, S., & Nasrulhaq Boyce, A. (2016). Role of plant growth promoting Rhizobacteria in agricultural sustainability-A review. *Molecules (Basel, Switzerland), 21*(5), 573. https://doi.org/10.3390/molecules21050573

Velusamy, P., Immanuel, J. E., Gnanamanickam, S. S., & Thomashow, L. (2006). Biological control of rice bacterial blight by plant-associated bacteria producing 2,4-diacetylphloroglucinol. *Canadian Journal of Microbiology, 52*(1), 56–65. https://doi.org/10.1139/w05-106

Vorholt, J. A. (2012). Microbial life in the phyllosphere. *Nature Reviews. Microbiology, 10*(12), 828–840. https://doi.org/10.1038/nrmicro2910

Wahab, A., Muhammad, M., Munir, A., Abdi, G., Zaman, W., Ayaz, A., Khizar, C., & Reddy, S. P. P. (2023). Role of Arbuscular Mycorrhizal fungi in regulating growth, enhancing productivity, and potentially influencing ecosystems under abiotic and biotic stresses. *Plants (Basel, Switzerland), 12*(17), 3102. https://doi.org/10.3390/plants12173102

Wang, Q., Xiong, D., Zhao, P., Yu, X., Tu, B., & Wang, G. (2011). Effect of applying an arsenic-resistant and plant growth-promoting rhizobacterium to enhance soil arsenic phytoremediation by Populus deltoides LH05-17. *Journal of Applied Microbiology, 111*(5), 1065–1074. https://doi.org/10.1111/j.1365-2672.2011.05142.x

Wang, Y., Mostafa, S., Zeng, W., & Jin, B. (2021). Function and mechanism of Jasmonic acid in plant responses to abiotic and biotic stresses. *International Journal of Molecular Sciences, 22*(16), 8568. https://doi.org/10.3390/ijms22168568

Wenke, K., Wanke, D., Kilian, J., Berendzen, K., Harter, K., & Piechulla, B. (2012). Volatiles of two growth-inhibiting rhizobacteria commonly engage AtWRKY18 function. *The Plant Journal: For Cell and Molecular Biology, 70*(3), 445–459. https://doi.org/10.1111/j.1365-313X.2011.04891.x

Yadav, S., Kaushik, R., Saxena, A. K., & Arora, D. K. (2011). Diversity and phylogeny of plant growth-promoting bacilli from moderately acidic soil. *Journal of Basic Microbiology, 51*, 98–106. https://doi.org/10.1002/jobm.201000098

Yamaguchi, S. (2008). Gibberellin metabolism and its regulation. *Annual Review of Plant Biology, 59*, 225–251. https://doi.org/10.1146/annurev.arplant.59.032607.092804

Yang, J., Duan, G., Li, C., Liu, L., Han, G., Zhang, Y., & Wang, C. (2019). The crosstalks between jasmonic acid and other plant hormone signaling highlight the involvement of jasmonic acid as a core component in plant response to biotic and abiotic stresses. *Frontiers in Plant Science, 10*, 1349. https://doi.org/10.3389/fpls.2019.01349

Yu, X., Ai, C., Xin, L., & Zhou, G. (2011). The siderophore-producing bacterium, Bacillus subtilis CAS15, has a biocontrol effect on Fusarium wilt and promotes the growth of pepper. *European Journal of Soil Biology, 47*(2), 138–145. https://doi.org/10.1016/j.ejsobi.2010.11.001

Zhang, H., Kim, M. S., Krishnamachari, V., Payton, P., Sun, Y., Grimson, M., Farag, M. A., Ryu, C. M., Allen, R., Melo, I. S., & Paré, P. W. (2007). Rhizobacterial volatile emissions regulate auxin homeostasis and cell expansion in Arabidopsis. *Planta, 226*(4), 839–851. https://doi.org/10.1007/s00425-007-0530-2

Zhang, H., Kim, M. S., Sun, Y., Dowd, S. E., Shi, H., & Paré, P. W. (2008). Soil bacteria confer plant salt tolerance by tissue-specific regulation of the sodium transporter HKT1. *Molecular Plant-Microbe Interactions: MPMI, 21*(6), 737–744. https://doi.org/10.1094/MPMI-21-6-0737

Zhang, L., Wüst, A., Prasser, B., Müller, C., & Einsle, O. (2019). Functional assembly of nitrous oxide reductase provides insights into copper site maturation. *Proceedings of the National*

Academy of Sciences of the United States of America, 116(26), 12822–12827. https://doi.org/10.1073/pnas.1903819116

Zhao, Z., Wang, Q., Wang, K., Brian, K., Liu, C., & Gu, Y. (2010). Study of the antifungal activity of Bacillus vallismortis ZZ185 in vitro and identification of its antifungal components. *Bioresource Technology, 101*(1), 292–297. https://doi.org/10.1016/j.biortech.2009.07.071

Zhao, B., Liu, Q., Wang, B., & Yuan, F. (2021). Roles of Phytohormones and their signaling pathways in leaf development and stress responses. *Journal of Agricultural and Food Chemistry, 69*(12), 3566–3584. https://doi.org/10.1021/acs.jafc.0c07908

Zhou, C., Zhu, L., Ma, Z., & Wang, J. (2017). Bacillus amyloliquefaciens SAY09 increases cadmium resistance in plants by activation of auxin-mediated signaling pathways. *Genes, 8*(7), 173. https://doi.org/10.3390/genes8070173

Zulfiqar, F., Akram, N. A., & Ashraf, M. (2019). Osmoprotection in plants under abiotic stresses: New insights into a classical phenomenon. *Planta, 251*(1), 3. https://doi.org/10.1007/s00425-019-03293-1

Interaction Between Soil Microbiomes and Plants

Srinivasan Kameswaran, Bellamkonda Ramesh, Gopi Krishna Pitchika, Manjunatha Bangeppagari, B. Swapna, M. Ramakrishna, and P. Sudhakar Reddy

Abstract Over the years, the reckless use of chemical pesticides and fertilizers has significantly harmed the ecosystem. However, with the projected increase in food consumption in the coming years, more productive and efficient agriculture is necessary. Recent research has shown that plant growth-promoting microorganisms (PGPMs) can be a sustainable and environmentally friendly alternative to chemical products. The effectiveness of PGPM applications can vary due to the complex relationships formed with various living and non-living elements, such as the environment, soil, and microorganism interactions. This chapter highlights certain applications suitable for an "agroecological transition" and looks at the key variables affecting the effectiveness of PGPM applications.

Keywords Organic farming · PGPR · Microorganisms · Biocontrol · Sustainability · Plant-microbe interaction

1 Introduction

To ensure food security, the world's population is growing faster than ever. This requires increased agricultural output to achieve higher crop yields and overall production (FAO, 2017). Agricultural practices are among the human endeavors most significantly responsible for the rise in chemical-based contaminants, as the excessive use of synthetic chemical pesticides and fertilizers exacerbates

S. Kameswaran (✉) · B. Swapna · M. Ramakrishna
Department of Botany, Vikrama Simhapuri University College, Kavali, Andhra Pradesh, India

B. Ramesh
Molecular Biology, University of Nebraska Medical Center, Omaha, NE, USA

G. K. Pitchika · P. S. Reddy
Department of Zoology, Vikrama Simhapuri University College, Kavali, Andhra Pradesh, India

M. Bangeppagari
Department of Cell Biology and Molecular Genetics, Sri Devaraj Urs Academy of Higher Education and Research, Kolar, Karnataka, India

S. A. Aransiola et al. (eds.), *Soil Microbiome in Green Technology Sustainability*,
https://doi.org/10.1007/978-3-031-71844-1_3

environmental degradation and presents health risks to humans. Nitrous oxide (N_2O), a primary source of greenhouse gases that contribute to global warming, is created when excessive amounts of nitrogen fertilizers are used. This is just one of the chemical pollutants that arise from agricultural activities. The management of agricultural land was responsible for up to 74% of the US's overall N_2O emissions in 2013 (EPA, 2016). For sustainable agriculture, it is essential for crops to have the ability to withstand disease, salinity, drought, and heavy metal stress, as well as increased nutritional content. One practical way to achieve these goals is by using plant growth-promoting microorganisms (PGPMs), such as bacteria and fungi. These microorganisms can enhance a plant's nutrient-absorption capacity, water-use efficiency, and resistance to plant diseases (Armada et al., 2014; Kumar & Verma, 2018). The rhizosphere of plants or soil contains plant growth-promoting rhizobacteria (PGPR), which could be employed as biostimulants, bio-fertilizers, and stimulants of "resistance against" a range of biotic and abiotic challenges (Kumar & Verma, 2018). However, the complex relationships formed between the crop, the microorganisms, and the soil environment—particularly in the rhizosphere—limit the effectiveness of PGPM application and use (Fig. 1). The interactions between microorganisms and plants can be influenced by various variables, such as genotype, existing soil microorganisms, and environmental factors like soil pH and temperature. This paper aims to provide an overview of these

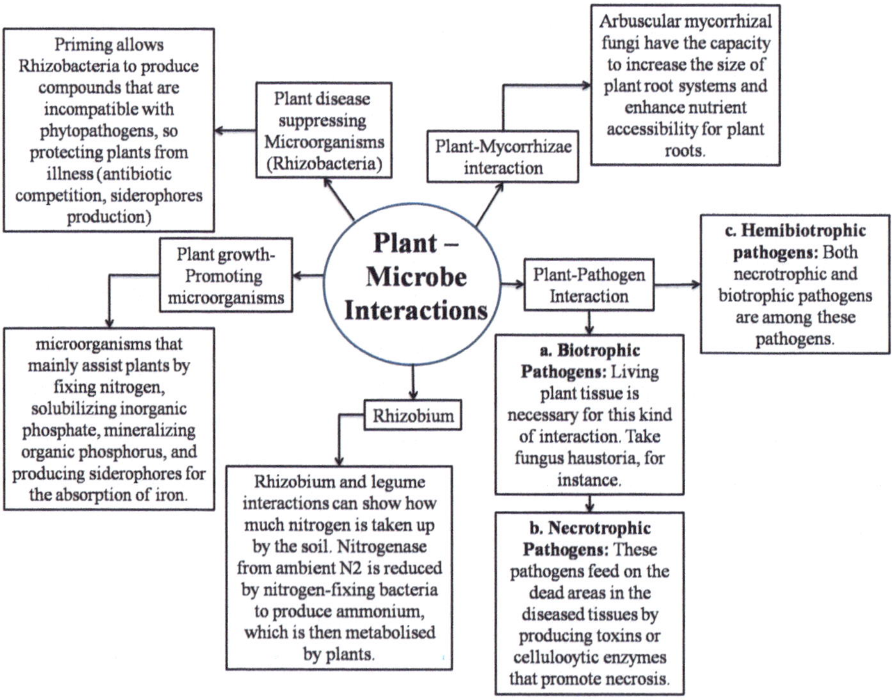

Fig. 1 Types of microbes and plants interactions

variables and the necessary considerations to maximize the positive benefits of PGPM application.

2 Plant Growth Promoting Fungi (PGPF)

2.1 What Are They?

Non-pathogenic fungi, known as PGPF, make up a diverse collection that promotes plant growth by acting on the rhizosphere. They are primarily saprotrophic, meaning they feed on decaying organic matter. However, they can also be necrotrophic, consuming dead and mostly decomposed organic matter, or biotrophic, feeding on living organic matter (Hossain et al., 2017). The soil-dwelling fungus can be further classified into free-living PGPF found in the rhizosphere, free-living epiphytes residing on the surface of the roots and exchanging metabolites with the plant, and endophytes residing inside the roots in the inter- and intracellular spaces. Taxonomically, this fungus belongs to the Ascomycetes, Basidiomycetes, and Oomycetes phyla. *Trichoderma* is the most isolated genus from various soils, followed by *Aspergillus, Fusarium, Gliocladium, Penicillium*, and *Phoma*. Additionally, PGPF is recognised for sporeless mycelial isolates classified as "sterile red fungus" ("SRF"), "sterile dark fungus" ("SDF"), and "sterile black fungus" ("SBF"). These isolates are frequently challenging to identify due to their undefined taxonomic state. Ectomycorrhiza and endomycorrhiza are two types of fungi commonly found in plant growth-promoting fungi (PGPF) and are not always reliant on their host plant for mutualism to exist. Distinguishing them from arbuscular mycorrhizal fungi (AMF), which are often found in the phylum Glomeromycota, is important. Unlike ectomycorrhiza and endomycorrhiza, AMFs create an unavoidable mutualism with their host plant throughout their life cycle, resulting in the production of asexual spores (Vorholt, 2012).

2.2 Enhancing Plant Growth

The ability of fungi living in the rhizosphere, such as mycorrhiza or saprophytic mycoflora, to improve plant growth and development has been extensively researched. Plants exposed to PGPF show positive effects on various factors, including yield, photosynthetic efficiency, germination, sprout and root growth, and flowering (Cornejo et al., 2017). One of the key factors contributing to their successful use in the field is the duration of the PGPF biofunctional activities in plants. Early in a plant's growth cycle, they often have beneficial effects, and in some cases, they have also shown promise in medium and late ontogenetic phases, leading to increased yield (Murali et al., 2021). The reasons for the consistent positive effects of PGPF on plants are not fully understood. However, one theory suggests that fungi colonize the root system and maintain it by diffusing their materials throughout the rhizosphere via plant exudates (Hossain et al., 2017).

2.3 "The Order and Method of Root Colonization"

The key strategy for establishing a strong relationship and initiating the process of enhancing plant growth is the colonization of the root system (Compant et al., 2021). The fungus's ability to survive and thrive over time despite the presence of native microflora enables this process (Grace & Saldajeno, 2012). This capability of fungi is referred to as rhizospheric competence, and it is essential for effective PGPF (Jahagirdar et al., 2019). Re-isolation from the rhizome has been highlighted in multiple studies as a proxy for the fungus's colonizing ability and, thus, its "rhizospheric competence". It appears that some *Penicillium* strains isolated from *Arabidopsis thaliana* after three weeks have been shown to maintain a high presence rate (>90%), as stated by Hossain (Hossain et al., 2007). The same has been demonstrated for *Aspergillus* in cucumbers. Research suggests that, despite the evidence of high reisolation rates, large rates of root colonization, such as those observed in the case of *Phoma* spp. on cucumber plants, are necessary to achieve rhizospheric competence 10 weeks after soil inoculation (Elsharkawy, 2022). Non-sporulating sterile fungi, in particular, are PGPF that can still promote plant growth even when they are unable to colonize. The evidence suggests that PGPF root colonization may not be necessary for growth enhancement (Cornejo et al., 2017). Different sections of the root have different amounts of PGPF, and the colonization of the root network is irregular (Wilkes et al., 2021; Crossay et al., 2019). It appears that the top and center of the roots have higher colonization compared to the basal and middle portions, which could be due to a radical growth rate greater than the PGPF colonization rate (Siddiqui & Pichtel, 2008). In contrast, arbuscular mycorrhizal fungi (AMF) colonize the root system more uniformly than PGPF by growing shrubs and branched structures that, when combined with adhesive (glomalin) soil particles, increase the surface area of radical contact with the soil and create a true network within the soil (Singh, 2012; Begum et al., 2019). Following the recognition of the plant by the fungus, the fungus is able to create the appressorium and penetrate the root tissue, leading to colonization. This establishes a close bond and the obligatory mutualism that characterizes this group of fungi (Al-Ani, 2018).

2.4 Interactions Between Microorganisms and the Host

The preference of the host plant for PGPF-AMF or other PGPMs (including bacteria) is a crucial factor in the application of PGPF. It has been observed that certain plants and fungus communities interact in a preferential manner. Multiple studies have shown that specific plant species benefit from the presence of one or more well-defined fungal species, but the effects on other species may differ. The different inherent properties of previously tested fungi may influence how saprophytic mycelium and AMF interact in the rhizosphere. Species within the same genus, and even within the same species, can have opposing impacts on the host plant (Aransiola et al., 2023). These often antagonistic groupings of microorganisms work together to inhibit some plant illnesses and generally support plant growth. By occupying distinct or complementary intervention niches and providing protection for crops in a variety of growth phases and environments, "ad hoc" mixtures of such fungi could improve the overall efficiency of plants (Grace & Saldajeno, 2012; Wilkes et al., 2021). The interaction between growth-promoting bacteria and fungi in the rhizosphere is fascinating. Fungal hyphae have the ability to create biofilms from various secretions, as shown in several studies, to support and nurture bacterial growth (Singh, 2012; Spatafora et al., 2016; Guo et al., 2018a, b). Mycorrhizal radical exudates also function similarly, providing substrates for bacterial growth and converting into nutrients for the plant (Guo et al., 2018a, b). One well-known relationship is endosymbiosis, particularly between bacteria and basidiomycetes. During the senescence stage of these symbioses, several growth-promoting enzymes are produced between bacterial and fungal hyphae, demonstrating intriguing benefits (Meena et al., 2018).

2.5 Mechanisms that Foster Growth

There are various ways in which PGPF can influence plant growth and development, both directly and indirectly. The fungus produces readily available nutrients that aid in the plant's growth, which are known as direct mechanisms. Additionally, it creates chemicals such as antioxidants, enzymes, and volatile organic compounds (VOC). Indirect methods include suppressing infections and reducing stress (salt, water, heat, and metals) that plants experience. By utilizing one or more of these pathways, PGPF can significantly impact plant growth and development (Hossain & Sultana, 2020).

2.5.1 "Solubilization of Phosphorus"

Phosphorus is an essential part of many critical macromolecules in living cells. It is necessary for a broad range of processes that are required for living things to survive and grow. Although phosphorus is present in significant quantities in agricultural soils, it is mostly in insoluble forms. This occurs because phosphorus becomes insoluble and unavailable to plants when it forms complexes with iron, aluminum, or calcium, depending on the type of soil. To address this issue, PGPF can be quite helpful in rendering insoluble phosphorus in the soil soluble. They generate the enzymes phytase, phosphatase, and organic acids, which solubilize phosphates and liberate soluble phosphorus. The genera *Aspergillus, Trichoderma, Fusarium*, and *Penicillium* are known to produce significant amounts of phytase and phosphatase (Radhakrishnan et al., 2015; Chuang et al., 2007). In acidic soil conditions, these fungi have a higher solubility of phosphorus compared to bacteria. Research in this area has shown that these plant growth-promoting fungi (PGPFs) use different methods to solubilize phosphorus (Radhakrishnan et al., 2015; Chuang et al., 2007; Janardan et al., 2011; Li et al., 2016; Gujar et al., 2013; Rai et al., 2021). According to an Ethiopian study, the fungi "*Aspergillus niger*" and several "*Penicillium*" sp. had the most significant amounts of "phosphorus solubilization". "*Brassica integrifolia*", "*Vicia faba*", "*Phaseolus vulgaris*", "*Saccharum officinarum*", and "*Lycopersicon esculentum*" the root system isolates were examined in the study (Wakelin et al., 2007).

2.5.2 Soil Organic Matter Mineralization

The microbial mineralization of organic materials is a crucial process for plant growth in soil. Certain beneficial microorganisms, known as PGPFs (Plant Growth-Promoting Fungi), like *Trichoderma*, accelerate soil mineralization, which in turn promotes plant growth (Yedidia et al., 2001; Yadav et al., 2009; Singh et al., 2010; Contreras-Cornejo et al., 2009; Ruess & Ferris, 2004; Behie et al., 2012). Fungi are particularly efficient at breaking down polyaromatic compounds such as lignin, humic acids, and phenolic acids, with the highest substrate absorption efficiency among all types of microbes (Waqas et al., 2012). Furthermore, through nitrification and ammonization, PGPF directly allows the degradation of organic nitrogenous compounds (Hossain et al., 2017). Studies have revealed that among bacteria, fungi are the most effective decomposers. In experiments, *Metarhizium robertsii* has been found to translocate nitrogen from dead insects to common bean host plants once it establishes itself as a root endophyte (Behie et al., 2012).

2.5.3 Phytohormone Synthesis

The beneficial connections between plants and PGPF are supported by phytohormones, which play a role in various plant-microbe interactions. The two most common hormone classes found in PGPF are auxins (IAAs) and gibberellins (GAs). IAAs limit root growth, leading to increased formation of lateral and adventitious roots, thus influencing various aspects of plant growth, particularly root morphology. Two species of *Trichoderma*, *T. harzianum* and *T. virens*, were observed to promote the growth and multiplication of lateral roots in *Arabidopsis* when their interaction with the plant was examined. Tests using markers sensitive to auxin produced by *Arabidopsis* were employed to reach this conclusion (Contreras-Cornejo et al., 2009). Development and growth of plant processes, including "stem elongation", "germination", "flower formation", and "flowering duration", are well-established to be influenced by GAs. Several studies have shown that certain

Penicillium species produce this hormone class (Waqas et al., 2012; Foo et al., 2013). Cytokinins, particularly zeatin, promote cell division and affect various plant processes such as cell expansion, flowering, and prevention of senescence. PGPF uses this class of hormones to enhance plant growth. Zeatin has been identified and recorded in *T. harzianum*, *Phoma* sp., and *Piriformospora indica*. Their ability to produce these kinds of botanical hormones has been shown in several studies, frequently in conjunction with one another (Hamayun et al., 2009; Khan et al., 2008; You et al., 2012; Sofo et al., 2012; Harman et al., 2004a, b; Jogaiah et al., 2013; Hossain et al., 2013; Vujanovic et al., 2012; Hossain et al., 2014; Waweru et al., 2014; Akhter et al., 2015; Verma et al., 2019).

2.5.4 Microbial Deaminase ACC

Trichoderma harzianum (Glick, 2014) has been found to produce 1-aminocyclopropane-1-carboxylic acid (ACC deaminase), an essential enzyme. This enzyme helps plants to produce less ethylene by converting the ethylene precursor, 1-aminocyclopropane-1-carboxylic acid, into ammonia (NH3) and butyric acid (α-ketobutyrate). Excessive ethylene production can lead to senescence. The inducible enzyme is encoded by the AcdS genes found in bacteria and fungi (Li et al., 2015).

3 "Plant Growth-Promoting Rhizobacteria" ("PGPR")

3.1 What Are They?

The term "plant growth-promoting rhizobacteria" (PGPR) was first used by Kloepper in the 1970s (Kloepper & Schroth, 1981). It covers a wide range of soil bacterial families that are crucial to soil-plant systems, such as *"Arthrobacter"*, *"Bacillus"*, *"Bradyrhizobium"*, *"Burkholderia"*, *"Azobacter"*, *"Klebsiella"*, *"Mesorhizobium"*, *"Enterobacter"*, *"Flavobacterium"*, *"Pseudomonas"*, *"Rhodococcus"*, *"Serratia"*, *"Streptomyces"*, *"Variovorax"*, etc. In these systems, the rhizosphere is where these bacteria interact intensely, influencing plant growth and yield. Plant growth regulators, phytohormones, and other biologically active substances are released by PGPR. This affects endogenous phytohormone levels, improves nutrient availability and uptake, lessens the negative effects of pathogenic microorganisms on plants, and utilizes various other mechanisms of action (Bishnoi, 2015). (1) Symbiotic rhizobacteria, which enter the cell from the inside (intracellular PGPR, like "nodule bacteria"), and (2) "free-living rhizobacteria", those are located beyond the "plant cells" ("extracellular PGPR", like "*Azotobacter*") are the two primary groups of rhizobacteria (Kenneth, 2017). Plants can benefit from PGPR inoculation through various mechanisms, including nitrogen fixation, phosphate solubilization, siderophore generation, phytohormone production, and enhanced tolerance to both biotic and abiotic stressors (Nazir et al., 2018; Bukhat et al., 2020).

3.2 The Action Mechanisms

The production of different chemicals alters the overall microbial community in the rhizosphere niche, as explained by Kloepper J. W. (Kloepper & Schroth, 1981). This is the process through which PGPR-mediated plant growth promotion operates. Plant Growth-Promoting Rhizobacteria (PGPR) either directly by "biologically fixing nitrogen", "phosphate solubilizing" it, and sequestering iron through "siderophores", or indirectly by regulating the amounts of plant hormones like "auxins", "gibberellins" ("GAs"), "cytokinins" ("CK"), and "nitric oxide" ("NO"). Furthermore, "jasmonic acid" ("JA") and "cadaverine" ("Cad") "biosynthesis", "rhizosphere competition", "induced systemic resistance"("ISR"), and an enzyme called "1-aminocyclopropane-1-carboxylate ("ACC") deaminase" are some of the indirect ways that PGPR can stimulate plant growth (Nazir et al., 2018).

3.2.1 Factors Associated with Nitrogen Fixation

One of the main nutrients that limit the growth rate in natural systems is nitrogen (Nazir et al., 2018; Bukhat et al., 2020). Nitrogen makes up around 2% of the dry matter of the plant and is essential for the food chain (Schausberger et al., 2017). Despite being abundant in the atmosphere, nitrogen in the form of N_2 gas is not readily available to plants. Instead, plants must absorb fixed nitrogen from the soil through their roots in the form of ammonium and nitrates, which are biologically available (Schausberger et al., 2017). By using PGPR-mediated biological fixation of N_2, plants can convert nitrogen into ammonia through a sophisticated enzyme system called nitrogenase, allowing them to utilize the molecular N_2 found in the atmosphere (Santi et al., 2013). Microorganisms that naturally fix nitrogen can thrive in a wide range of temperatures and are predominantly found in the natural world. There are two main types of nitrogen-fixing microorganisms: (i) "symbiotic nitrogen-fixing forms", includes "*Cyanobacteria*", "*Azoarcus*", "*Azotobacter*", "*Acetobacter*", "*Azospirillum*", "*Burkholderia*", "*Diazotrophicus*", "*Enterobacter*", "*Pseudomonas*", and "*Gluconacetobacter*" ("*Anabaena*", "*Nostoc*") (Basile & Lepek, 2021). "Symbiotic N_2-fixing bacteria" include the family "Rhizobiaceae" ("*Rhizobium*", "*Bradyrhizobium*", "*Sinorhizobium*", "*Rhizobium*", and "*Frankia*"), which form associaation with legume plants (e.g., "*Rhizobia*") (Glick, 2012) and non-leguminous plants (e.g., *Frankia*) (Santi et al., 2013). Diazotrophs are all known to be prokaryotes, and they are all nitrogen-fixing organisms. Nitrogen fixation is a broadly and paraphyletically distributed ability in both the bacterial and archaeal realms. In these species, the "nitrogenase" enzyme mechanism is the only source of nitrogen fixation capacity. It fixes one "N_2

molecule" by hydrolyzing "16 ATP molecules". This process is among the most metabolically costly biological processes known to science (Raymond et al., 2004). PGPR initiates the colonization process through molecular communication with the plant host (Bukhat et al., 2020). Soil bacteria emit chemical cues, typically flavonoids, during the initial stages of colonization. By attaching to a specific bacterial receptor (*NodD*), these chemicals function as "transcriptional activators" for additional "nodulation genes" ("*nodA*", "*nodB*", "*nodC*", and "*nodFE*") (Raymond et al., 2004).

The Process of Nitrogen Fixation in Legumes
Nodules in the roots are created when bacteria come into touch with root hairs and start the nodulation process. The bacteria live in organs known as "symbiosomes" inside those structures, where they differentiate into nitrogen-fixing bacteria (Basile & Lepek, 2021; Provorov et al., 2018). In a symbiotic relationship, the bacteria receive carbon molecules from the plant, while the plant benefits from the nitrogen supply. Although this mechanism is generally non-specific, it has been observed that specific rhizobia species can induce the development of nodules in particular legumes. Additionally, multiple rhizobia species can nodulate various legumes, such as *Glycine max* and *Phaseolus vulgaris* (Provorov et al., 2018; Ji et al., 2017). Nodulation of various legume species has been observed with certain strains of rhizobia (Pueppke & Broughton, 1999). *Rhizobia symbionts* coexist in "nodules with non-nodulating rhizobia" and "non-rhizobial species", according to research conducted in vitro and, more recently, using metagenomics ("Busby et al., 2016"; "Lu et al., 2017"). This observation can impact the symbiotic process and nitrogen fixation (Basile & Lepek, 2021). While legumes benefit from this process, they also use energy for it, which is subsequently regulated (Sachs et al., 2018). Plants have developed multiple mechanisms to facilitate the colonization of compatible bacteria while inhibiting the colonization of incompatible prokaryotes (Basile & Lepek, 2021). Additionally, the plant may tolerate nodules colonized by rhizobia with limited ability to fix nitrogen (Sanjuan et al., 2021; Kiers et al., 2003). The bacteria must overcome the plant's defenses against pathogens and environmental challenges. Consequently, rhizobia have developed defense mechanisms and processes that promote nodulation, enabling them to thrive in specific unfavorable soil conditions (Westhoek et al., 2021; Feng et al., 2021). Several factors can impact nodulation, the interaction between bacteria and plants, and ultimately, nitrogen fixation in plants. Examples of these factors include nodule cysteine-rich (NCR) peptides, plant defense responses, bacterial survival in the rhizosphere, autoregulation of nodule number, and penalties for inadequate nitrogen fixation.

Rhizosphere Bacterial Survival
Soil bacterial communities lose some of their biodiversity near the roots. In the rhizosphere—the soil that is impacted by plant roots—root exudates alter the makeup of the microbial population (Vives-Peris et al., 2020; Eng et al., 2020). The ability of a bacterial population to employ various substances released by the roots as substrates and to chemotaxis towards them is a factor in the population's persistence in the "rhizosphere" (Feng et al., 2021; Chaparro et al., 2013; Bais et al.,

2006; Badri et al., 2009). To reach the root and affect nodulation, bacterial species must possess the ability to move in the soil, which means they must have flagella (Velázquez et al., 2019). Microbe–microbe interactions, which can be cooperative or competitive and contribute to the overall community structure in the rhizosphere, also have an impact on the makeup of microbial communities (Hassani et al., 2018; Han et al., 2020). The competitive effect in these types of interactions can be influenced by the growth rate, the capacity to seize limited nutrients, including iron, and the existence of harmful substances (Eng et al., 2020; Hassani et al., 2018).

Particular Appreciation for Starting the Nodulation Process
Radical exudates contain both flavonoids and isoflavonoids (Mazurek et al., 2015; Shah & Smith, 2020). A particular flavonoid released from a suitable legume triggers the rhizobia's *NodD* transcription factor (Gerardo et al., 2020; Broughton et al., 2000; Chen et al., 2005; Peck et al., 2006). Several *NodD* copies and corresponding transcription factors are activated by distinct flavonoids that are perceived differently by various varieties of rhizobia (Chen et al., 2015). Given their potential for varied roles, the various isoforms may participate in different stages of symbiotic infection or may be transcribed under various environmental circumstances. These latter scenarios frequently involve enhancing nodulation competitiveness or expanding the host population through sensing distinct plant signals (Chen et al., 2015; Acosta-Jurado et al., 2019; Kelly et al., 2018; Del Cerro et al., 2015, 2017, 2020; Demont et al., 1994). Particularly, on the root of a suitable legume, "lipochitin oligosaccharides" with unique alternatives create a unique molecule that is detected by particular "receptors" ("LysM receptor-like kinase") (Basile & Lepek, 2021). According to Gerardo et al. (2020), the "*NodD* gene" stimulates the amplification of "bacterial genes" that regulate the manufacture of "Nod factors".

Upon identification of the distinct Nod components, a series of signals initiate that facilitate the morphological, physiological, and molecular alterations characteristic of the nodulation process. Once within the nodule cell, bacteria that have undergone nitrogen-fixing differentiation are encircled by a membrane derived from plants known as the peribacterial membrane (Brewin, 1998; Geurts & Bisseling, 2002; Gage, 2004). There have been two different kinds of nodules identified: determinate and indeterminate. The spherical, determinate nodules are shaped by cell multiplication in the exterior cortex and have a reversible differentiation process for their microorganisms. They lack a permanent meristem. The elongated, persistent apical meristem of the indeterminate nodules, on the other hand, is derived from inner cortical cells and has distinct sections that correspond to different developmental phases (Patriarca et al., 2002; Crespi & Gálvez, 2000; Roy et al., 2020).

Bacterial differentiation in indeterminate nodules is irreversible, in contrast to the previous ones. The infection of plant tissues through a crack-entry process is another, less common nodulation mechanism (Oldroyd & Downie, 2008; Markmann et al., 2012). Numerous transcriptional factors, including micro RNA, enzymes, and phytohormones, are involved in the molecular plant signaling cascade that Nod factors cause. This reprogramming of transcription and translation occurs (Libault et al., 2010; Breakspear et al., 2014; "Liu et al., 2018"; "Huang et al., 2019";

"Traubenik et al., 2020"). Exopolysaccharide" (EPS) is "another component" of bacteria that plays a role in the particular recognition process (Kawaharada et al., 2015). The structures of the extracellular polymeric substances (EPSs) of several bacterial species and strains of the same species were shown to differ (Skorupska et al., 2006; Chai et al., 2018).

Response of Plant Defenses

Plants identify molecules known as generic elicitors known as disease/microbial-associated molecular structures (PAMPs/MAMPs) to initiate an initial defense response in response to pathogen attacks (Wanke et al., 2021). Different polysaccharides, bacterial surface proteins, flagella, thermo-unstable elongation factor (EF-Tu), and other structures can contain compounds that the plant recognizes as PAMPs (Newman et al., 2013; Chisholm et al., 2006; Cerekovic & Poltronieri, 2017). Rhizobia displaying MAMPs causes a plant's defense mechanism to activate, just like infections do. On the other hand, rhizobia flagella are not identified as MAMPs, in contrast to infections (Lopez-Gomez et al., 2012). Plants identify PAMPs utilizing certain receptors (kinases that resemble the LysM receptor), which subsequently initiate the initial defense mechanism known as PAMP-triggered immunity (PTI). The type three secretion systems (T3SS) of bacteria allow them to translocate effector proteins into plant cells, which decreases PTI (Azmina et al., 2020). However, the plant is also able to create proteins that can identify these effectors, or molecules that have been altered by them, and cause an immune response known as effector-triggered immunity (ETI) (Azmina et al., 2020). It has been observed that bacteria can release more effectors that inhibit HR through T3SS (Chisholm et al., 2006; O'malley et al., 2021). Several rhizobes exhibit a T3SS in addition to inducing the expression of the nod genes (Ratet et al., 2017; Bajunaid et al., 2020). In addition to upregulating NodD function, flavonoids that activate TtsI also trigger the transcription of T3SS-secreted components and effectors that inhibit plant defenses (Bajunaid et al., 2020; Dong & Song, 2020).

The type IV and type VI secretion systems (T4SS and T6SS, respectively) of bacteria have also been reported to be involved in the symbiotic process, in addition to T3SS (Nelson & Sadowsky, 2015; Lin et al., 2018; Salinero-Lanzarote et al., 2019).

Nodule Cysteine-Rich (CR) Peptides

Activities that characterize the compatibility between hosts and Rhizobes also take place in the latter phases of the nodulation process, that is, when bacteria differentiate into Bacteroides. When legumes exhibit indeterminate nodules, a variety of translated peptides are seen. These peptides have several cysteine residues, which is why they are called nodular cysteine-rich (NCR) peptides (Mergaert et al., 2003; Alunni et al., 2007; Lima et al., 2020). A significant number of these "peptides" have demonstrated antibacterial action in vitro (Van de Velde et al., 2010), which can lead to bacterial cell death, DNA duplication, and cell elongation via altering the membrane's permeability. The innate immune mechanism of plants is supported by these NCR peptides, which are descended from defensins (Sathoff & Samac, 2019). But

NCR peptide activity in the nodule can also cause irreversible terminal bacterial differentiation, or nitrogen fixation, by making the membrane more permeable, making the bacteria polyploid, making the bacteria bigger, and changing the shape of the bacteria (Kondorosi et al., 2013). Therefore, a combination of factors including the degree of "NCR antibacterial activity", its reactivity towards the "rhizobial strain", the relationship among NCR in the nodules, and the processes that bacteria possess to tolerate or evacuate these "NCR peptides" determines the nitrogen-fixing phenotype of the NCR-formed nodule (Feng et al., 2021).

Nodule Number Autoregulation
To prevent energy loss, Fabaceae species have a systemic mechanism in place to control the amount of nodules they have (Caetano-Anollés & Gresshoff, 1991; Sulieman & Tran, 2013). Long-distance communications between shoots and roots mediate the negative feedback mechanism known as autoregulation of nodulation (AON). The system works with short peptides (12–13 amino acids), which are the signals produced at the roots when the nodulation process begins (Mortier et al., 2012).

Particular kinase receptors in the bud are involved in the negative regulation of *CLAVATA/ESR*-related (CLE) peptides upon nodulation (Liu et al., 2018). CLE peptides are thought to travel through the xylem "from roots to the shoot", where they bind to a specific "receptor" ("Okamoto et al., 2013"). The production of a sprout-derived inhibitor (SDI) is triggered by CLE peptide perception in sprouts, and this leads to further nodule formation (Sasaki et al., 2014). Both the AON mechanism and the signal transduction cascade generated by the Nod factor are responsible for the formation of nodules (Mortier et al., 2012). Moreover, Nod factor signaling has been reported to trigger the production of additional factors that impede the AON process (Lei et al., 2019). This implies that a complex equilibrium involving many signaling pathways mediated by rhizobia governs the regulation of nodule number (Lei et al., 2019). Particular CLE peptides are also involved in the process that prevents the formation of nodules in soils that contain high nitrate levels (Reid et al., 2011).

Improper Nitrogen Fixation Sanctioned
Rhizobia that fix nitrogen has been found to cause nodules to grow regularly, but nodules that are ineffectual or inefficient at fixing nitrogen typically stay tiny (Basile & Lepek, 2021; Schumpp & Deakin, 2010; Zhou et al., 2020). Sanction is a term used to describe the capacity that legumes have to identify and suppress the growth of less favorable genotypes of rhizobia as well as to limit the presence of ineffective and less effective rhizobia (Kiers et al., 2003; Oono et al., 2011). In both definite and indeterminate nodules, this phenomenon has been recorded. Through the induction of cellular senescence, the plant typically sanctioned the useless microorganisms found in the nodule (Regus et al., 2017). This is essentially a nodule aging process, often known as a nodule organ rupture (Thomas & Reddy, 2013). Distinct bacterial genotypes may be present in distinct sections of nodules colonized by both ineffective and effective rhizobia. Only cells with inadequate rhizobia within them exhibit

senescence (Basile & Lepek, 2021). We still don't know the exact process by which high senescence is caused by low nitrogen fixation. Nodules inhabited by less productive rhizomes are seen in nature despite the plant's capacity to eradicate uncooperative strains; this is most likely due to genetic variations in the host's degree of control over these rhizomes (Wendlandt et al., 2019).

3.2.2 Solubilization of Phosphate

Nitrogen is the primary limiting nutrient, with phosphorus (P) coming in second. Although this nutrient is incredibly abundant in the soil (found in fertilizer, for example) and natural phosphorus (found in plants), it is unfortunately not readily available to plants due to its rapid process of fixation and association with other soil-based components (Rodríguez & Fraga, 1999). As part of the soil phosphorus cycle, PGPR is crucial in supplying plants with soil-immobilized phosphorus (Gouda et al., 2018). By immediately solubilizing and mineralizing inorganic phosphorus and facilitating its movement through a biogeochemical cycle, these microorganisms—also known as phosphate-solubilizing bacteria, or PSB—can improve root uptake efficiency (Tian et al., 2021). The most important PSBs are said to include *Azotobacter, Bacillus, Beijerinckia, Burkholderia, Enterobacter, Erwinia, Flavobacterium, Microbacterium, Pseudomonas, Rhizobium,* and *Serratia*, among other genera to which these bacteria belong (Verma et al., 2018). They can solubilize inorganic P sources and promote the development and yield of plants. In certain instances, they may also have a close relationship with a multitude of agricultural products, including potatoes, tomatoes, wheat, and radishes (Zaidi et al., 2009). It has been shown that the action of low molecular weight organic acids produced by the different soil bacterial strains causes the solubilization of inorganic phosphorus (Das & Pradhan, 2016). On the other side, the mineralization of organic phosphorus happens when bacteria synthesize different phosphatases, which catalyze the hydrolysis of phosphoric esters (Glick, 2012). Similar to *P. polymyxa*, the PGPR *Paenibacillus mucilaginosus* may fix nitrogen and is capable of breaking down insoluble soil minerals by releasing nutrient ions including potassium and phosphorus (Goswami et al., 2015; Lugtenberg, 2016).

3.2.3 Production of Siderophores

Although one of the most abundant minerals in the world, iron is not easily absorbed by plants (Boopathi & Rao, 1999). Plants frequently lack this element, which occurs naturally as the very insoluble form Fe^{3+} in soils polluted by heavy metals (Wuana et al., 2011; Islam et al., 2021). In the process of chelating iron (Fe^{3+}), siderophores—substances with a low molecular weight—may be produced by PGPR (Ahmed & Holmström, 2014). When the environment lacks iron, microbic siderophores enable iron to be reduced to the soluble form Fe^{2+}, which is subsequently released into the plant cell (Verma et al., 2019; Bishnoi, 2015; Pecoraro et al., 2021). Additionally, PGPR promotes plant growth by generating extracellular siderophores that are highly effective in controlling a variety of plant diseases by depriving pathogens of their source of iron (Eng et al., 2020). Siderophores can build stable structures with iron as well as other metals that are considered environmentally hazardous, including Al, Cd, Cu, Pb, and Zn (Das Ghosh et al., 2014). According to several studies (Verma et al., 2019; Kuffner et al., 2008; Carrillo-Castañeda et al., 2002), "siderophore-producing bacteria" from the genera "*Bradyrhizobium*", "*Pseudomonas*", "*Rhizobium*", "*Serratia*", and "*Streptomyces*" were isolated from the rhizosphere. After the siderophore-producing bacteria *Pseudomonas, Rhizobium,* and *Azospirillum* were inoculated into plants under iron-limiting conditions, the beneficial effects on the plants were recorded (Carrillo-Castañeda et al., 2002; De Serrano, 2017).

3.2.4 Production of Phytohormones

Growth regulators can be synthesized and released by the majority of microorganisms that have been isolated from the rhizosphere of different crops (Patten & Glick, 1996). Specifically, PGPR can change the plant's pre-existing pool of phytohormones and enhance the creation of new ones (Nazir et al., 2018; Grover et al., 2021). Indole-3-acetic acid, or IAA, is the phytohormone that bacteria secrete most of all of them. It typically affects the cell division, extension, and differentiation levels of plants; promotes germination and speeds up the development of xylem and roots; regulates growth and vegetative processes, conveys "responses to light", "gravity", and flowering; and "affects pigmentation", "photosynthesis", "biosynthesis of various metabolites", and resistance to various stressors ("Nazir et al., 2018"). Most of the time, changes brought about specifically by the pool of phytohormones produced by soil bacteria can have an impact on the pool of endogenous phytohormones that the plants already possess. The interactions between rhizobacteria and plants are therefore greatly influenced by IAA (Ahemad & Kibret, 2014). Stressors in the environment that alter these chemicals' manufacture in various bacteria have a discernible effect on this production. Oxidative and matrix stresses, acid pH, and carbon limitation are examples of these stressors (Nazir et al., 2018; Molina et al., 2018). Additionally, certain aerobic sporogenous may synthesize other phytohormones including "auxins" and "gibberellins", and they are linked to the control of "ethylene" synthesis utilizing the production of biocatalyst such as "ACC-deaminase" (Kudoyarova et al., 2019). Within this particular context, it has been demonstrated that the *Bacillus* strains *B. pumilus* and *B. licheniformis* exhibit bio-stimulating activity utilizing gibberellin synthesis (Verma et al., 2018). By generating plant hormones including IAA, cytokinins, gibberellins, and ethylene as well as volatile chemicals, the PGPR *Paenibacillus polymyxa*, formerly known as *Bacillus polymyxa*, can encourage plant development (Kudoyarova et al., 2019).

4 Stress Management for Abiotic and Biotic Conditions

4.1 Abiotic Stresses

Numerous significant crops have lower yields as a result of abiotic stressors because they have various effects on plants at different phases of growth. Plant growth, development, and reproduction are primarily regulated by light, temperature, water, carbon, and mineral nutrients (Lugtenberg, 2016). Plant growth and development are restricted by environments that exceed or fall short of the ideal concentrations of one or more of these variables (Lata et al., 2018). Plants respond to severe levels by changing morphologically and physiologically to cope with these sudden shifts (Nievola et al., 2017). Drought, salt, extremes of temperature (both high and low), and acidity are the most hazardous abiotic stressors for plants (Shukla et al., 2012; He et al., 2018). In addition to impeding photosynthesis, causing a loss of cell membrane integrity, and increasing the generation of ROS (reactive oxygen species), drought circumstances have a deleterious effect on when seeds germinate (Ali, 2020). In general, the growth proper development, and productivity of plants are suppressed by drought and excessive soil salinity levels, which also decrease leaf area, water potential, and stomatal opening (Khan et al., 2019). Different plant species have different effects from different temperatures. Plants die as a result of denaturation and aggregation of cellular proteins caused by high temperatures, whereas the activity of enzymes is affected by changes in protein conformation, interaction, and other membrane properties caused by low temperatures (Xu et al., 2016). A plethora of hitherto underappreciated elements may be very real allies for crops in their fight against abiotic stressors, according to recent research. Thousands of distinct microbial constituents can be found in agricultural fields (Khan et al., 2017; Roesch et al., 2007), and it has been demonstrated that many of these soil bacteria preferentially connect with plant roots (Lundberg et al., 2012; Thompson et al., 2017). These bacteria connected to roots live inside and around plant roots, as was previously explained. But as one gets closer to the root system, the diversity of microorganisms drops; there are more microbes in the rhizosphere, which is the closest part of the root, and fewer in the endosphere, which is inside the root (Peiffer et al., 2013). The microbiome's composition varies not only within or along the root tissue, but also during the many growth and development stages, as multiple studies indicate (Edwards et al., 2015; Chaparro et al., 2014). Notably, there are notable variations among the members of the previously mentioned community. On the other hand, abiotic stresses also affect the diversity and composition of microorganisms in the soil (Xu et al., 2018; Bouskill et al., 2016; Hartmann et al., 2017). Consequently, the availability of microorganisms that can be obtained from plants, like plant-genetically modified microbes (PGPMs), is dependent upon the amount of microorganisms present in the pool. This results in a feedback loop that is circular and affects both the surrounding microbiome and the plant (Martiny et al., 2017). Plants can trigger a broad variety of molecular, physiological, and biochemical responses in response to particular abiotic stresses (Caddell et al., 2019; Zhalnina et al., 2018), and theories have been put forth regarding how these may draw in and favor particular microbial populations in the root zone (Giovannetti et al., 1996; Hu et al., 2018). It has been suggested by recent research that crops may have a multitude of hitherto unrecognized allies that might help them fight against abiotic challenges (Timmusk et al., 2014; Ilyas et al., 2020; Mayak et al., 2004; Niu et al., 2018; Bohm et al., 2018). Thousands of distinct microbial constituents have been found in agricultural areas (Khan et al., 2017; Roesch et al., 2007), and it has been demonstrated that many of these soil microbes connect preferentially with plant roots (Sagar et al., 2020; Kudoyarova et al., 2019; Thompson et al., 2017). The microbiome's composition varies not only within or along the root tissue, but also during the many growth and development stages, as multiple studies indicate (Lundberg et al., 2012; Edwards et al., 2015). Notably, there are notable variations among the members of the previously known community. The microbial diversity and composition in the soil are also affected by abiotic stresses (Xu et al., 2018; Bouskill et al., 2016; Hartmann et al., 2017). As a result, the availability of microorganisms that plants can recruit, like PGPMs, is dependent on the overall population of microorganisms in the soil. This creates a positive feedback loop between the plant and the surrounding microbiome (Martiny et al., 2017). In addition to stimulating the synthesis of phyto-constituents in the plant, the microbes use their metabolism to create a variety of combinations of carbon-based gaseous chemicals known as volatile organic compounds (VOCs) (Hossain et al., 2017; Guler et al., 2016; Naznin et al., 2014; Whipps, 2001). One of the most extensively studied unique characteristics of PGPF is its impact on how plants respond to abiotic stressors such as drought, salt, heavy metal toxicity, high temperatures, and oxidative stress (Ahmad et al., 2015). Because of their high ability to recycle oxidized "ascorbates" and "glutathiones", as well as scavenge "reactive oxygen species" (ROS), plants treated with PGPF have a greater tolerance to abiotic stress, in other words, this trait applies to all phytochemicals that a plant synthesizes under stress (Rasool et al., 2012; Mahato et al., 2018). Furthermore, they raise the concentrations of proline and glycine-betaine, which boost the plant's ability to withstand stress. In general, they also trigger the production of antioxidants, both non-enzymatic (ascorbic acid, reduced glutathione, and oxidized glutathione) and "enzymatic" ("peroxidase", "catalase", "superoxide dismutase", "ascorbate peroxidase", "glutathione reductase", "monodehydroascorbate reductase", "dehydroascorbate reductase", "glutathione S-transferase", and "gaucol peroxidase"), similar to what is also frequently observed for PGPR (Timmusk et al., 2014; Ilyas et al., 2020; Mayak et al., 2004; Sagar et al., 2020; Lundberg et al., 2012; Niu et al., 2018; Bohm et al., 2018), some research have shown that different PGPF species boost crops' tolerance to abiotic stressors (Murali et al., 2021; Khan & Lee, 2013; Guler et al., 2016; Özbay et al., 2018; Gajera & Vakharia, 2012).

4.2 Biotic Stresses

PGPMs can inhibit or lessen plant pathogen inoculum and development, and consequently, their invasion, via various methods (Hossain et al., 2017; Bishnoi, 2015). PGPF-specific mycoparasitism, antibiosis systemic resistance induction (ISR), and competition for nutrients and space are the key mechanisms of action (Ahmad et al., 2015). Producing lytic enzymes such chitinases, proteases, and β-1,3 glucanases is a significant mode of action of PGPF (Vujanovic et al., 2012; Naznin et al., 2014; Gajera & Vakharia, 2012). Several PGPF, including *T. harzianum* (Murali et al., 2021; Ghasemi et al., 2020; Saldajeno & Hyakumachi, 2011), have demonstrated antagonistic effect against pathogen growth in rhizosphere-isolated fungi (Harman et al., 2004a, b; Jogaiah et al., 2013; Hossain et al., 2013, 2014; Waweru et al., 2014; Hermosa et al., 2012; Sood et al., 2020). The growth of *Pythium ultimum* is inhibited by the synthesis of antibiotics such gliovirin by *Gliocladium virens* (Waweru et al., 2014). By competing with pathogenic *Fusarium* strains for resources, reducing chlamydospore germination, and competing to colonize infection sites (roots), non-pathogenic strains of *Fusarium oxysporum* isolated from rhizospheric soil suppressed their growth, according to Kaur et al. (2010). The antifungal activity of *Aspergillus fischeri* against *Botrytis cinerea* was also quite strong (Tóth et al., 2020). *Trichoderma* strains make for roughly 90% of the fungal biocontrol agents used to combat pathogenic bacteria (Sood et al., 2020; Zin & Badaluddin, 2020; Harman et al., 2004a, b). Plants may form symbiotic associations with *Trichoderma* species. They invade soil quickly and invasively, promoting plant development and controlling plant diseases such as bacteria, nematodes, and, most importantly, phytopathogenic fungi in soil contaminated by pathogens (Macías-Rodríguez et al., 2020). *Trichoderma* has been documented to employ competition and antibiosis as direct antagonistic strategies against mycoparasitism in fungi (Ghorbanpour et al., 2018). It has been demonstrated that introducing diverse *Trichoderma* species into the rhizosphere enhances the plant's defense mechanism against a range of phytopathogens, including bacteria, fungi, and viruses. "Induced systemic resistance" ("ISR"), "hypersensitivity response" ("HR"), and "acquired systemic resistance" ("SAR") are some of the methods by which this is accomplished (Contreras-Cornejo et al., 2011). *Trichoderma* spp. have been widely researched and used economically as biocontrol agents, soil improvers, and biofertilizers due to their proven effectiveness against plant diseases (Contreras-Cornejo et al., 2011). There are three possible modes of action for PGPR biocontrol activity: antibiosis, induction of plant defense responses (ISR), and competition with pathogens for nutrients and habitats (Wang et al., 2021). Many other genera, including *Streptomyces, Pantoea, Pseudomonas, Bacillus,* and *Serratia*, have been shown to produce antimicrobial metabolites, primarily with broad-spectrum activity, as biocontrol bacteria (Sumi et al., 2015; Lee & Kim, 2015). Among these are several antimicrobial chemicals that have been described, including lipopeptides (LPs), phenazine, pyoluteorin, pyrrolnitrin, tensin, tropolone, 2,4-diacetylphloroglucinol (DAPG), iturin, fengycin, oomycin-A, amphisin, and phenazine. Permeabilization and rupture of the cell membrane are the mechanisms by which these LPs, which are mostly active against microorganisms, have been reported to be potent antibacterial agents. Particular research into the use of antibiotics in biocontrol has focused on "*Bacillus*" and "*Pseudomonas*" because they may produce morphologically diverse phytochemicals with a wide range of antibacterial action (Sumi et al., 2015; Lee & Kim, 2015; Maheshwari, 2013).

5 PGPMs and Organic Matter in the Soil

In addition to producing environmental benefits like carbon sequestration (C) for mitigating climate change, soil organic matter (SOM) is linked to soil fertility and is crucial to the cycling of nutrients (Minasny et al., 2017). Higher concentrations of SOM are found in the top 20 to 30 cm of the main soil profiles (Condron et al., 2010; Yarwood, 2018). Since agricultural activities have been intensive for millennia, the amount of organic matter in the soil has decreased, especially in agricultural soils as opposed to pastures (Sanderman et al., 2017). It was specifically designed to reduce erosion and SOM loss and boost the water retention capacity that conservative agriculture returned to, involving little tillage and little soil disturbances (Giller et al., 2015). Utilizing soil residue from crops and additional "fresh organic matter" ("FOM") sources, which include compost, is advised in this conservative agriculture to restore sufficient levels of soil organic matter (SOM). Growing focus is being placed on the microbial degradation of the FOM and the integration of some of the microbial residues into the SOM (Coonan et al., 2020). Microbial neuromas, comprising extracellular polymeric components, metabolites, nucleic acids, cell walls and membranes, and extracellular enzymes, are what characterize microbial detritus (Coonan et al., 2020). A portion of the carbon that remains in living tissue is assimilated by microorganisms known as microbial biomass and is breathed as carbon dioxide (CO_2) when they metabolize C-based substrates such as plant residues, roots, and root exudates. Microbial waste (biomass) is recycled or stored in the SOM after death and turnover (Coonan et al., 2020). This suggests that the breakdown of plant-based products, or more broadly, products of organic origin, and microbial detritus, with microbial activities regulating the relative input of each, are the two sources of SOM (Cotrufo et al., 2013). With organic molecules now mineralized and vital nutrients made available for plants, it is clear to see how the decomposition process and the soil organic matter cycle are tightly related. Furthermore, it makes sense to believe that the diversity, quantity, and/or makeup of soil microbes affect how well the ecosystem functions (Coonan et al., 2020). Diverse variety levels must be taken into account as they may impact or affect microbial decomposition rates in different ways. Microbial abundance and species diversity, respectively, are defined as the entire amount of individuals in an organism and the entirety of its biomass. Other terms that describe the number and frequency of species include species richness, species uniformity, species identity or composition community, and diversity of species or ecosystems. In contrast to variations in the makeup of the microbial community, alterations in microbial abundance are more significant for soil carbon mineralization (Dong et al., 2021).

6 The Properties of Soil Affecting Microbial Populations

The variable field results following PGPM treatment pose a major challenge to PGPM research. These results can be attributed to abiotic factors such as soil pH, light intensity, nutrients, and heavy metal concentration, as well as inoculation techniques and/or abiotic factors. By nitrogen-fixing atmospheric N and solubilizing soil minerals like phosphorus and potassium, PGPMs are generally utilized as biofertilizers, improving the availability of nutrients. Abiotic circumstances have the power to drastically alter the rhizosphere, which has an impact on PGPM inoculants' ability to survive and flourish, as well as their ability to increase microbial diversity and enhance plant health and growth (De Souza et al., 2015; Venturi & Keel, 2016; Hernández-Montiel et al., 2017; Mahmood et al., 2019; Etesami, 2020). Climate can affect soil composition, structure, and humidity, as well as alter C and N transformations, metabolic activities, and microbial survival (Lopes et al., 2021). Temperature and water availability can also affect these qualities. The number of nutrients available and the toxicity of minerals, such as iron and aluminum, can be greatly influenced by the pH changes caused by the fertilizers applied and the organic material that is added or already there. The DNA of bacteria can be altered or destroyed when their surface is exposed to sunlight (Mahmood et al., 2019; Etesami, 2020; Lopes et al., 2021; Varma et al., 2019). When applying PGPM inoculants to the soil, all of these variables may have favorable effects, depending on the degree of each. PGPMs can increase plant tolerance to severely limiting stresses for agricultural production, such as water availability, water scarcity (drought), or water excess (flood) (Enebe & Babalola, 2018; Ipek et al., 2019; Danish et al., 2019). "*Azotobacter chroococcum*" and "*Azospirillum brasilense*" in "*Mentha pulegium*" L. (Asghari et al., 2020), "*Pseudomonas*" sp. and "*Azotobacter*" sp. in "*Cymbopogon citrates*" (Mirzaei et al., 2020), and "*Zea mays*" (Danish et al., 2019) are just a few instances of how PGPMs might increase drought stress tolerance (Fleming et al., 2019; Asghari et al., 2020).

According to reports in Glycine max (Kim et al., 2017) and "*Zea mays*" (Danish et al., 2019), "*Klebsiella variicola*" and "*Azospirillum*" sp. can increase "flood stress tolerance" through "adaptations", like "adventitious root" development brought on by endogenous hormonal modifications. Plant-microbe interactions, however, can be impacted by water stress. Drought raises the soil's warmth, which can prevent helpful bacteria from multiplying. Flooding limits the amount of oxygen that can be found in the soil, which inhibits the growth of microorganisms that cannot respire anaerobically (Enebe & Babalola, 2018; Ipek et al., 2019; Hartman & Tringe, 2019). Moreover, the plant is impacted by both drought and flooding because they interfere with the generation of "root exudates" (Enebe & Babalola, 2018; Danish et al., 2019; Fleming et al., 2019; Hartman & Tringe, 2019).

The solubility of various metal ions, nutritional availability, and physical characteristics of the soil are all dependent on the pH of the soil (Dutta & Bora, 2019; Msimbira & Smith, 2020). High pH levels in alkaline soils impact nutritional availability, leading to osmotic stress, nutrient shortage, and elevated reactive oxygen species (ROS) generation (Dutta & Bora, 2019; Salwan et al., 2019). Acidic soils with poor pH and elevated aluminum ion levels are hazardous and cause "phosphoric acid complexes" to form, which inhibit plants from getting phosphorus (Dutta & Bora, 2019; Zerrouk et al., 2019). The ideal pH range for plant growth and the rise in the microbial production of root exudates is 5.5–6.5. It can be generally stated that fungi prefer acidic pH circumstances, whereas bacteria prefer near-neutral or slightly alkaline pH conditions (Msimbira & Smith, 2020). Microorganisms that are helpful to plants, like *Bacillus cereus* in *Solanum lycopersicum* (Mukhtar et al., 2020) and *Pseudomonas putida* in *Triticum* sp. (Ali, 2020), can be effectively injected to improve plant growth and mitigate harmful effects produced by severe temperatures. On the other hand, *Burkholderia* sp. boosts plant yields in *Vitis vinifera* and builds tolerance to low temperatures by altering the metabolism of carbohydrates (Fernandez et al., 2012). As recurrent stresses in agriculture, extreme temperatures alter the composition of root exudates, interfering with plant–PGPM interactions (Ali, 2020; Ipek et al., 2019; Meena et al., 2015).

By altering the quantity and chemical makeup of root exudates, light can disrupt the relationship between plants and microorganisms (Venturi & Keel, 2016; Lopes et al., 2018). In *Ophiopogon japonicus* and *Lolium perenne*, for example, the inoculation of beneficial bacteria can enhance plant growth in low-light settings by enhancing their tolerance to shade (Fu et al., 2020). But in low light, the "microorganisms themselves"—including "*Glomus*" sp., "*Paraglomus*" sp., "*Rhizophagus*" sp., and "*Rhizobium*" sp. in "*Phaseolus lunatus*"—may put the plant under more stress (Ballhorn et al., 2016).

7 From PGPMs Inoculants to Microbiome

Numerous bacterial taxa are present in the plant-associated microbiome, and many of them strengthen plant development, and tolerance to biotic and abiotic stimuli, inhibit plant diseases, break down xenobiotic chemicals, and have a favorable impact on yields (Berg et al., 2014). The significance of the plant-associated microbiome is illustrated by a comprehensive study (Carrión et al., 2019). They used metagenomic research and structure inference to show the trend of *Rhizoctonia solani* infections in the roots of sugar beetroot plants, which are rich in Flavobacteriaceae and Chitinophagaceae in the root endosphere. Along with additional unknown biosynthetic gene clusters that account for the synthesis of polyketide synthases (PKS) and non-ribosomal peptide synthetases (NRPS), the study also uncovered many genes encoding chitinases. According to a different study by Ghadamgahi et al. (2022), *Pseudomonas aeruginosa* FG106 isolated from tomato plant rhizospheres exhibits antagonistic activity against the following pathogens: *Phytophthora colocasiae,*

P. infestans, Rhizoctonia solani, Botrytis cinerea, Alternaria alternata, and *Clavibacter michiganensis* subsp. *michiganensis*. It has also been demonstrated that *P. aeruginosa* (strain FG106) produces lipases and proteases, induces high phosphate solubilization, generates siderophores, ammonia, hydrogen cyanide (HCN), and indole acetic acid (IAA), and forms biofilms that aid in biocontrol and plant growth. Genes linked to growth promotion and biocontrol were also discovered in this strain using genome mining techniques (Ghadamgahi et al., 2022).

When looking for a source of beneficial bacteria, extreme settings are of special interest. For instance, According to Inostroza et al. (2017), stimulation of growth was provided to plants of wheat in slightly dehydrated soils which had low phosphorus levels by "microbial consortia" obtained from the "rhizosphere" of "*Atriplex*" sp. raised in soils of an unaffected arid "Chilean ecosystem" lacking any prior "anthropogenic disturbance". This demonstrates how microbial inoculants suitable for agroecosystems deficient in nutrients and/or vulnerable to unfavorable weather occurrences can be obtained from arid habitats (Inostroza et al., 2017).

Isolating beneficial microorganisms from the halophytic environment is also an intriguing prospect (Salwan et al., 2019). In a comprehensive study conducted by Christakis et al. (2021), 115 endophytic isolates were identified as originating from "three crops wild-related" ("CWR") "halophytes": "*Acacia maritima*", "*Matthiola tricuspidata*", and "*Crithmum maritimum*". This selection produced several strains that exhibited antagonistic activity against *Aspergillus fumigatus*, a human pathogen, and the plant diseases *Verticillium dahliae, Ralstonia solanacearum,* and *Clavibacter michiganensis*. Moreover, growth under elevated salinity and plant growth enhancement was confirmed by several strains (Christakis et al., 2021).

"Commercially available PGPMs" are primarily employed in farming systems because according to studies on the "native microbiome", they function well and complement certain natural processes related to the nutrient cycle, plant defense against pathogens and parasites, and other related biological processes (Adeleke et al., 2019). Positive reports indicate that PGPM applications in the agro-environment primarily benefit the ecosystem as a whole, either fully or partially. Relationships can take many forms, some of which have been discussed in this chapter's earlier paragraphs. In mutualistic relationships, the microorganisms use a variety of techniques to control plant growth and provide biological defense against potential pathogens and predators, while the plant provides nutrients and/or shelter for the microbes (Adeleke et al., 2019). It is vital to consider whether to use a microbial consortium or a single strain due to the complexity of the interactions that occur both between microbes and plants. In a single study (Bradáčová et al., 2019), the effectiveness of microbial consortium products vs specific single-strain commercial microbial inoculants with demonstrated plant growth-promoting ability was evaluated in large-scale tomato growing systems subjected to a range of environmental difficulties. The production circumstances were real-life. Different single-strain bacterial and fungal inoculants as well as microbial consortia have demonstrated remarkably comparable efficiency in a protected greenhouse production environment. The enhanced effectiveness of microbial consortia in more adverse environments was one particularly fascinating finding. This was in an Israeli tomato production system that had "sandy soil" in the Negev desert, "low fertility", and "mineral fertilization" above a "pH of 7.9". Indeed, under situations of limited phosphate (P), the microbial consortia treatments demonstrated improved P acquisition, increased stimulation of the vegetative shoots' biomass production, and a larger yield of the final fruits. Consortia inoculation has been linked to specific alterations in the rhizosphere bacterial community's structure, particularly about *Sphingobacteria* and *Flavobacteria*, which are recognized to be drought stress defenders, according to the same study. The inoculation of microbial consortia reversed the reduction in bacterial population diversity on the root surface (rhizoplane) that occurred when there was a phosphate shortage, indicating an improvement in the P status of the plants. This seems to confirm the theory that, particularly under challenging environmental conditions, the utilization of microbial consortia can improve the repeatability and efficiency of PGPM-assisted agricultural production strategies (Bradáčová et al., 2019). Baldi et al. (2021) treated two different kinds of bio-fertilizers ("AMF" and "*Trichoderma*" spp.) in an Italian apricot orchard while assessing the outcomes using an "unfertilized control". This work provides more evidence about the modification of soil microbiome effects of PGPM application. The experiment's objective was to ascertain whether the use of biofertilizers could affect various models of the organic material's breakdown processes, hence potentially stimulating the native microbiota in different ways. It was feasible to see a notable shift in the chemical breakdown process over time and faster organic matter degradation when *Trichoderma* spp. was present using near-infrared (NIR) spectroscopy and thermogravimetry-differential thermal analysis (TG-DTA) were used. Additionally, several groups of *Trichoderma* spp. were identified through the use of metabarcoding in next-generation sequencing and were noted to have altered the final organic matter's microbial and fungal makeup, leading to the degradation of litterbags (Baldi et al., 2021).

8 Conclusions and Discussion

The agriculture industry will see an increase in demand for food as the world's population continues to rise. Over a billion people were spared starvation and malnutrition when the Green Revolution succeeded in raising agricultural productivity worldwide a few years ago. However, these actions have additionally resulted in the development and improper use of "chemically" synthesized "pesticides" and fertilizers, having harmed the environment and reduced agrobiodiversity and soil ecosystems. PGPMs have been demonstrated in numerous studies to have favorable effects on agricultural yield while maintaining ecosystem integrity. These effects include biofertilization, biocontrol, and bioremediation. PGPMs have the potential to be effective substitutes for conventional "agrochemical industry products" in sustainable agriculture in this specific setting. They can also encourage the transition from traditional farming to "integrated", "organic", and environmentally "sustainable management of agro-ecosystems". Unfortunately, because of the environment, soil, and crops, the tremendous complexity of ecosystems and how they interact causes uneven outcomes when PGPM is applied. It is evident from reading through the research included in this review that to precisely characterize the treatments based on soil characteristics, environmental circumstances, crops, and cultivated genotypes, preliminary investigations must be conducted before applying PGPMs. The utilization of PGPMs is to be better understood and more knowledge on plant-soil-microorganism interactions has been made possible by the recent and rapid cements in the omics sciences. Enhancing crop stability, productivity, and the general sustainability of the agroecosystem is made possible by this significant step.

References

Acosta-Jurado, S., Rodríguez-Navarro, D. N., Kawaharada, Y., Rodríguez-Carvajal, M. A., Gil-Serrano, A., Soria-Díaz, M. E., Pérez-Montaño, F., Fernández-Perea, J., Niu, Y., Alias-Villegas, C., & Jiménez-Guerrero, I. (2019). Sinorhizobium fredii HH103 nolR and nodD2 mutants gain capacity for infection thread invasion of Lotus japonicus Gifu and Lotus burttii. *Environmental Microbiology, 21*(5), 1718–1739.

Adeleke, R. A., Nunthkumar, B., Roopnarain, A., & Obi, L. (2019). Applications of plant–microbe interactions in agro-ecosystems. In *Microbiome in plant health and disease: Challenges and opportunities* (pp. 1–34). Springer.

Ahemad, M., & Kibret, M. (2014). Mechanisms and applications of plant growth promoting rhizobacteria: Current perspective. *Journal of King saud University-science, 26*(1), 1–20.

Ahmad, P., Hashem, A., Abd-Allah, E. F., Alqarawi, A. A., John, R., Egamberdieva, D., & Gucel, S. (2015). Role of Trichoderma harzianum in mitigating NaCl stress in Indian mustard (Brassica juncea L) through antioxidative defense system. *Frontiers in Plant Science, 6*, 868.

Ahmed, E., & Holmström, S. J. (2014). Siderophores in environmental research: Roles and applications. *Microbial Biotechnology, 7*(3), 196–208.

Akhter, W., Bhuiyan, M. K. A., Sultana, F., & Hossain, M. M. (2015). Integrated effect of microbial antagonist, organic amendment and fungicide in controlling seedling mortality (Rhizoctonia solani) and improving yield in pea (Pisum sativum L.). *Comptes Rendus Biologies, 338*(1), 21–28.

Al-Ani, L. K. T. (2018). Trichoderma: Beneficial role in sustainable agriculture by plant disease management. *Plant microbiome: stress response*, 105–126.

Ali, S. (2020). Plant growth promoting and stress mitigating abilities of soil born microorganisms. *Recent Patents on Food, Nutrition & Agriculture, 11*, 96–104.

Alunni, B., Kevei, Z., Redondo-Nieto, M., Kondorosi, A., Mergaert, P., & Kondorosi, E. (2007). Genomic organization and evolutionary insights on GRP and NCR genes, two large nodule-specific gene families in Medicago truncatula. *Molecular Plant-Microbe Interactions, 20*(9), 1138–1148.

Aransiola, S. A., Ijah, U. J. J., Abioye, O. P., & Bala, J. D. (2023). Vermicompost-assisted phytoremediation of toxic trace element-contaminated soil in Madaka, Nigeria using *Melissa officinalis* L and *Sida acuta. International Journal of Environmental Science and Technology.* https://doi.org/10.1007/s13762-022-04105-y

Armada, E., Portela, G., Roldán, A., & Azcón, R. (2014). Combined use of beneficial soil microorganism and agrowaste residue to cope with plant water limitation under semiarid conditions. *Geoderma, 232*, 640–648.

Asghari, B., Khademian, R., & Sedaghati, B. (2020). Plant growth promoting rhizobacteria (PGPR) confer drought resistance and stimulate biosynthesis of secondary metabolites in pennyroyal (*Mentha pulegium* L.) under water shortage condition. *Scientia Horticulturae, 263*, 109132.

Azmina, N., Malik, A., Kumar, S., & Nadarajah, K. (2020). Molecular sciences elicitor and receptor molecules: Orchestrators of plant defense and immunity. *International Journal of Molecular Sciences, 21*, 963.

Badri, D. V., Weir, T. L., van der Lelie, D., & Vivanco, J. M. (2009). Rhizosphere chemical dialogues: Plant-microbe interactions. *Current Opinion in Biotechnology, 20*, 642–650.

Bais, H. P., Weir, T. L., Perry, L. G., Gilroy, S., & Vivanco, J. M. (2006). The role of root exudates in rhizosphere interactions with plants and other organisms. *Annual Review of Plant Biology, 57*, 233–266.

Bajunaid, W., Haidar-Ahmad, N., Kottarampatel, A. H., Ourida Manigat, F., Silué, N., Tchagang, C. F., Tomaro, K., & Campbell-Valois, F. X. (2020). The T3SS of Shigella: Expression, structure, function, and role in vacuole escape. *Microorganisms, 8*(12), 1933.

Baldi, E., Gioacchini, P., Montecchio, D., Mocali, S., Antonielli, L., Masoero, G., & Toselli, M. (2021). Effect of biofertilizers application on soil biodiversity and litter degradation in a commercial apricot orchard. *Agronomy, 11*(6), 1116.

Ballhorn, D. J., Schädler, M., Elias, J. D., Millar, J. A., & Kautz, S. (2016). Friend or foe—Light availability determines the relationship between mycorrhizal fungi, rhizobia and lima bean (Phaseolus lunatus L.). *PLoS One, 11*(5), e0154116.

Basile, L. A., & Lepek, V. C. (2021). Legume–rhizobium dance: An agricultural tool that could be improved? *Microbial Biotechnology, 14*(5), 1897–1917.

Begum, N., Qin, C., Ahanger, M. A., Raza, S., Khan, M. I., Ashraf, M., Ahmed, N., & Zhang, L. (2019). Role of arbuscular mycorrhizal fungi in plant growth regulation: Implications in abiotic stress tolerance. *Frontiers in Plant Science, 10*, 1068.

Behie, S. W., Zelisko, P. M., & Bidochka, M. J. (2012). Endophytic insect-parasitic fungi translocate nitrogen directly from insects to plants. *Science, 336*(6088), 1576–1577.

Berg, G., Grube, M., Schloter, M., & Smalla, K. (2014). Unraveling the plant microbiome: Looking back and future perspectives. *Frontiers in Microbiology, 5*, 71409.

Bishnoi, U. (2015). PGPR interaction: An ecofriendly approach promoting the sustainable agriculture system. *Advances in Botanical Research, 75*, 81–113.

Bohm, K., Gerards, S., Hundscheid, M., Melenhorst, J., de Boer, W., & Garbeva, P. (2018). Calling from distance: Attraction of soil bacteria by plant root volatiles. *The ISME Journal, 12*, 1252–1262.

Boopathi, E., & Rao, K. S. (1999). A siderophore from Pseudomonas putida type A1: Structural and biological characterization. *Biochimica et biophysica Acta (BBA)-Protein Structure and Molecular Enzymology, 1435*(1–2), 30–40.

Bouskill, N. J., Wood, T. E., Baran, R., Ye, Z., Bowen, B. P., Lim, H., Zhou, J., Nostrand, J. D. V., Nico, P., Northen, T. R., & Silver, W. L. (2016). Belowground response to drought in a tropical forest soil. I. Changes in microbial functional potential and metabolism. *Frontiers in microbiology, 7*, 525.

Bradáčová, K., Florea, A. S., Bar-Tal, A., Minz, D., Yermiyahu, U., Shawahna, R., Kraut-Cohen, J., Zolti, A., Erel, R., Dietel, K., & Weinmann, M. (2019). Microbial consortia versus single-strain inoculants: An advantage in PGPM-assisted tomato production? *Agronomy, 9*(2), 105.

Breakspear, A., Liu, C., Roy, S., Stacey, N., Rogers, C., Trick, M., Morieri, G., Mysore, K. S., Wen, J., Oldroyd, G. E. D., & Downie, J. A. (2014). The root hair "infectome" of Medicago truncatula uncovers changes in cell cycle genes and reveals a requirement for auxin signaling in rhizobial infection. *The Plant Cell, 26*, 4680–4701.

Brewin, N. J. (1998). Tissue and cell invasion by rhizobium: The structure and development of infection threads and symbiosomes. In H. P. Spaink, A. Kondorosi, & P. J. J. Hooykaas (Eds.), *The Rhizobiaceae: Molecular biology of model plant-associated bacteria* (pp. 417–429). Springer. ISBN 978-94-011-5060-6.

Broughton, W. J., Jabbouri, S., & Perret, X. (2000). Keys to symbiotic harmony. *Journal of Bacteriology, 182*, 5641–5652.

Bukhat, S., Imran, A., Javaid, S., Shahid, M., Majeed, A., & Naqqash, T. (2020). Communication of plants with microbial world: Exploring the regulatory networks for PGPR mediated defense signaling. *Microbiological Research, 238*, 126486.

Busby, R. R., Rodriguez, G., Gebhart, D. L., & Yannarell, A. C. (2016). Native Lespedeza species harbor greater non-rhizobial bacterial diversity in root nodules compared to the coexisting invader, L. cuneata. *Plant and Soil, 401*, 427–436.

Caddell, D. F., Deng, S., & Coleman-Derr, D. (2019). Role of the plant root microbiome in abiotic stress tolerance. In *Seed endophytes* (pp. 273–311). Springer.

Caetano-Anollés, G., & Gresshoff, P. M. (1991). Plant genetic control of nodulation. *Annual Review of Microbiology, 45*, 345–382.

Carrillo-Castañeda, G., Muños, J. J., Peralta-Videa, J., Gomez, E., Tiemannb, K., Duarte-Gardea, M., & Gardea-Torresdey, J. (2002). Alfalfa growth promotion by bacteria grown under iron limiting conditions. *Advances in Environmental Research, 6*, 391–399.

Carrión, V. J., Perez-Jaramillo, J., Cordovez, V., Tracanna, V., De Hollander, M., Ruiz-Buck, D., Mendes, L. W., van Ijcken, W. F., Gomez-Exposito, R., Elsayed, S. S., & Mohanraju, P. (2019). Pathogen-induced activation of disease-suppressive functions in the endophytic root microbiome. *Science, 366*, 606–612.

Cerekovic, N., & Poltronieri, P. (2017). Plant signaling pathways activating defence response and interfering mechanisms by pathogen effectors, protein decoys and bodyguards. *AIMS Molecular Science, 4*, 370–388.

Chai, Y., Cai, P., Gillor, O., Kuramae, E. E., Costa, A. O. Y., & Raaijmakers, J. M. (2018). Microbial extracellular polymeric substances: Ecological function and impact on soil aggregation. *Frontiers in Microbiology, 23*, 1636.

Chaparro, J. M., Badri, D. V., Bakker, M. G., Sugiyama, A., Manter, D. K., & Vivanco, J. M. (2013). Root exudation of phytochemicals in arabidopsis follows specific patterns that are developmentally programmed and correlate with soil microbial functions. *PLoS One, 8*, e55731.

Chaparro, J. M., Badri, D. V., & Vivanco, J. M. (2014). Rhizosphere microbiome assemblage is affected by plant development. *The ISME Journal, 8*, 790–803.

Chen, X. C., Feng, J., Hou, B. H., Li, F. Q., Li, Q., & Hong, G. F. (2005). Modulating DNA bending affects NodD-mediated transcriptional control in Rhizobium leguminosarum. *Nucleic Acids Research, 33*(8), 2540–2548.

Chen, D. S., Liu, C. W., Roy, S., Cousins, D., Stacey, N., & Murray, J. D. (2015). Identification of a core set of rhizobial infection genes using data from single cell-types. *Frontiers in Plant Science, 6*, 575.

Chisholm, S. T., Coaker, G., Day, B., & Staskawicz, B. J. (2006). Host-microbe interactions: Shaping the evolution of the plant immune response. *Cell, 124*, 803–814.

Christakis, C. A., Daskalogiannis, G., Chatzaki, A., Markakis, E. A., Mermigka, G., Sagia, A., Rizzo, G. F., Catara, V., Lagkouvardos, I., Studholme, D. J., & Sarris, P. F. (2021). Endophytic bacterial isolates from halophytes demonstrate phytopathogen biocontrol and plant growth promotion under high salinity. *Frontiers in Microbiology, 12*, 681567. [CrossRef].

Chuang, C. C., Kuo, Y. L., Chao, C. C., & Chao, W. L. (2007). Solubilization of inorganic phosphates and plant growth promotion by *Aspergillus niger*. *Biology and Fertility of Soils, 43*, 575–584.

Compant, S., Cambon, M. C., Vacher, C., Mitter, B., Samad, A., & Sessitsch, A. (2021). The plant endosphere world—Bacterial life within plants. *Environmental Microbiology, 23*, 1812–1829.

Condron, L., Stark, C., O'callaghan, M., Clinton, P., & Huang, Z. (2010). The role of microbial communities in the formation and decomposition of soil organic matter. In *Soil microbiology and sustainable crop production* (pp. 81–118). Springer Science and Business Media LLC.

Contreras-Cornejo, H. A., Macías-Rodríguez, L., Cortés-Penagos, C., & Ló Pez-Bucio, J. (2009). *Trichoderma virens*, a plant beneficial fungus, enhances biomass production and promotes lateral root growth through an auxin-dependent mechanism in *Arabidopsis*. *Plant Physiology, 149*, 1579–1592.

Contreras-Cornejo, H. A., Macías-Rodríguez, L., Beltrán-Peña, E., Herrera-Estrella, A., & López-Bucio, J. (2011). *Trichoderma*-induced plant immunity likely involves both hormonal- and camalexin-dependent mechanisms in Arabidopsis thaliana and confers resistance against necrotrophic fungi Botrytis cinerea. *Plant Signaling & Behavior, 6*, 1554–1563.

Coonan, E. C., Kirkby, C. A., Kirkegaard, J. A., Amidy, M. R., Strong, C. L., & Richardson, A. E. (2020). Microorganisms and nutrient stoichiometry as mediators of soil organic matter dynamics. *Nutrient Cycling in Agroecosystems, 117*, 273–298.

Cornejo, P., Seguel, A., Aguilera, P., Meier, S., Larsen, J., & Borie, F. (2017). Arbuscular mycorrhizal fungi improve tolerance of agricultural plants to cope abiotic stress conditions. In D. Singh, H. Singh, & R. Prabha (Eds.), *Plant-microbe interactions in agro-ecological perspectives: Volume 2: Microbial interactions and agro-ecological impacts* (pp. 55–80). Singapore. ISBN 978-981-10-6593-4.

Cotrufo, M. F., Wallenstein, M. D., Boot, C. M., Denef, K., & Paul, E. (2013). *The Microbial Efficiency-Matrix Stabilization (MEMS) framework integrates plant litter decomposition with soil organic matter stabilization: Do labile plant inputs form stable soil organic matter?* (p. 19). Global Chang.

Crespi, M., & Gálvez, S. (2000). Molecular mechanisms in root nodule development. *Journal of Plant Growth Regulation, 19*, 155–166.

Crossay, T., Majorel, C., Redecker, D., Gensous, S., Medevielle, V., Durrieu, G., Cavaloc, Y., & Amir, H. (2019). Is a mixture of arbuscular mycorrhizal fungi better for plant growth than single-species inoculants? *Mycorrhiza, 29*, 325–339.

Danish, S., Zafar-Ul-Hye, M., Hussain, S., Riaz, M., & Qayyum, M. F. (2019). Mitigation of drought stress in maize through inoculation with drought tolerant ACC deaminase containing PGPR under axenic conditions. *Pakistan Journal of Botany, 52*, 52.

Das Ghosh, U., Saha, C., Maiti, M., Lahiri, S., Ghosh, S., Seal, A., & Ghosh, M. M. (2014). Root associated iron oxidizing bacteria increase phosphate nutrition and influence root to shoot partitioning of iron in tolerant plant *Typha angustifolia*. *Plant and Soil, 381*, 279–295.

Das, I., & Pradhan, M. (2016). Potassium-solubilizing microorganisms and their role in enhancing soil fertility and health. In V. Meena, B. Maurya, J. Verma, & R. Meena (Eds.), *Potassium solubilizing microorganisms for sustainable agriculture* (pp. 281–291). Springer.

De Serrano, L. O. (2017). Biotechnology of siderophores in high-impact scientific fields. *Biomolecular Concepts, 8*, 169–178.

De Souza, R., Ambrosini, A., & Passaglia, L. M. P. (2015). Plant growth-promoting bacteria as inoculants in agricultural soils. *Genetics and Molecular Biology, 38*, 401–419.

Del Cerro, P., Rolla-Santos, A. A. P., Gomes, D. F., Marks, B. B., Pérez-Montaño, F., Rodríguez-Carvajal, M. Á., Nakatani, A. S., Gil-Serrano, A., Megías, M., Ollero, F. J., & Hungria, M. (2015). Regulatory nodD1 and nodD2 genes of Rhizobium tropici strain CIAT 899 and their roles in the early stages of molecular signaling and host-legume nodulation. *BMC Genomics, 16*, 251.

Del Cerro, P., Pérez-Montaño, F., Gil-Serrano, A., López-Baena, J. F., Megías, M., Hungria, M., & Ollero, F. J. (2017). The *Rhizobium tropici* CIAT 899 NodD2 protein regulates the production of Nod factors under salt stress in a flavonoid-independent manner. *Scientific Reports, 7*, 46712.

Del Cerro, P., Ayala-García, P., Buzón, P., Castells-Graells, R., Javier López-Baena, F., Ollero, F. J., & Pérez-Montaño, F. (2020). OnfD, an AraC-type transcriptional regulator encoded by Rhizobium tropici CIAT 899 and involved in Nod factor synthesis and symbiosis. *Applied and Environmental Microbiology, 86*, e01297–e01320.

Demont, N., Ardourel, M., Maillet, F., Promé, D., Ferro, M., Promé, J. C., & Dénarié, J. (1994). The Rhizobium meliloti regulatory nodD3 and syrm genes control the synthesis of a particular class of nodulation factors N-acylated by (omega-1)-hydroxylated fatty acids. *The EMBO Journal, 13*, 2139–2149.

Dong, W., & Song, Y. (2020). The significance of flavonoids in the process of biological nitrogen fixation. *International Journal of Molecular Sciences, 21*, 5926.

Dong, X., Liu, C., Ma, D., Wu, Y., Man, H., Wu, X., Li, M., & Zang, S. (2021). Organic carbon mineralization and bacterial community of active layer soils response to short-term warming in the Great Hing'an Mountains of Northeast China. *Frontiers in Microbiology, 12*, 802213.

Dutta, J., & Bora, U. (2019). Role of PGPR for alleviating aluminum toxicity in acidic soil. In *Microbial rejuvenation of polluted environment* (pp. 309–326). Springer Science and Business Media LLC.

Edwards, J., Johnson, C., Santos-Medellin, C., Lurie, E., Podishetty, N. K., Bhatnagar, S., Eisen, J. A., & Sundaresan, V. (2015). Structure, variation, and assembly of the root-associated microbiomes of rice. *Proceedings of the National Academy of Sciences of the United States of America, 112*, E911–E920.

Elsharkawy, M. M. (2022). Plant growth-promoting Phoma spp. In M. Rai, B. Zimowska, & G. J. Kovics (Eds.), *Phoma: Diversity, taxonomy, bioactivities, and nanotechnology* (pp. 301–309). Springer International Publishing. ISBN 978-3-030-81218-8.

Enebe, M. C., & Babalola, O. O. (2018). The influence of plant growth-promoting rhizobacteria in plant tolerance to abiotic stress: A survival strategy. *Applied Microbiology and Biotechnology, 102*, 7821–7835.

Eng, T., Herbert, R. A., Martinez, U., Wang, B., Chen, J. C., Brown, J. B., Deutschbauer, A. M., Bissell, M. J., Mortimer, J. C., & Mukhopadhyay, A. (2020). Iron supplementation eliminates antagonistic interactions between root-associated bacteria. *Frontiers in Microbiology, 11*, 1742.

EPA. U.S. (2016). Greenhouse gas inventory report: 1990–2014. In *U.S. Enviromental Protection Agency*.

Etesami, H. (2020). Plant-microbe interactions in plants and stress tolerance. In *Plant life under changing environment* (pp. 355–396). Elsevier BV.

FAO. (2017). The future of food and agriculture—Trends and challenges. *Annual Report, 296*, 1–180.

Feng, H., Fu, R., Hou, X., Lv, Y., Zhang, N., Liu, Y., Xu, Z., Miao, Y., Krell, T., Shen, Q., & Zhang, R. (2021). Chemotaxis of beneficial rhizobacteria to root exudates: The first step towards root–microbe rhizosphere interactions. *International Journal of Molecular Sciences, 22*, 6655.

Fernandez, O., Theocharis, A., Bordiec, S., Feil, R., Jacquens, L., Clément, C., Fontaine, F., & Barka, E. A. (2012). Burkholderia phytofirmans PsJN acclimates grapevine to cold by modulating carbohydrate metabolism. *Molecular Plant-Microbe Interactions, 25*, 496–504.

Fleming, T. R., Fleming, C. C., Levy, C., Repiso, C., Hennequart, F., Nolasco, J. B., & Liu, F. (2019). Biostimulants enhance growth and drought tolerance in *Arabidopsis thaliana* and exhibit chemical priming action. *The Annals of Applied Biology, 174*, 153–165.

Foo, E., Ross, J. J., Jones, W. T., & Reid, J. B. (2013). Plant hormones in arbuscular mycorrhizal symbioses: An emerging role for gibberellins. *Annals of Botany, 111*, 769–779.

Fu, J., Luo, Y., Sun, P., Gao, J., Zhao, D., Yang, P., & Hu, T. (2020). Effects of shade stress on turfgrasses morphophysiology and rhizosphere soil bacterial communities. *BMC Plant Biology, 20*, 92.

Gage, D. J. (2004). Infection and invasion of roots by symbiotic, nitrogen-fixing rhizobia during nodulation of temperate legumes. *Microbiology and Molecular Biology Reviews, 68*, 280–300.

Gajera, H. P., & Vakharia, D. N. (2012). Production of lytic enzymes by *Trichoderma* isolates during in vitro antagonism with Aspergillus Niger, the causal agent of collar rot of peanut. *Brazilian Journal of Microbiology, 43*, 43–52.

Gerardo, N. M., Dicenzo, G. C., Ardley, J., Gifford, M. L., Walker, L., & Lagunas, B. (2020). Determinants of host range specificity in legume-rhizobia symbiosis. *Frontiers in Microbiology, 11*, 585749.

Geurts, R., & Bisseling, T. (2002). Rhizobium Nod factor perception and signalling. *Plant Cell, 14*, S239–S249.

Ghadamgahi, F., Tarighi, S., Taheri, P., Saripella, G. V., Anzalone, A., Kalyandurg, P. B., Catara, V., Ortiz, R., & Vetukuri, R. R. (2022). Plant growth-promoting activity of *Pseudomonas aeruginosa* FG106 and its ability to act as a biocontrol agent against potato, tomato and taro pathogens. *Biology, 11*, 140.

Ghasemi, S., Safaie, N., Shahbazi, S., Shams-Bakhsh, M., & Askari, H. (2020). The role of cell wall degrading enzymes in antagonistic traits of *Trichoderma virens* against *Rhizoctonia solani*. *Iranian Journal of Biotechnology, 18*, e2333.

Ghorbanpour, M., Omidvari, M., Abbaszadeh-Dahaji, P., Omidvar, R., & Kariman, K. (2018). Mechanisms underlying the protective effects of beneficial fungi against plant diseases. *Biological Control, 117*, 147–157.

Giller, K. E., Andersson, J., Corbeels, M., Kirkegaard, J., Mortensen, A. D., Erenstein, O., & Vanlauwe, B. (2015). Beyond conservation agriculture. *Frontiers in Plant Science, 6*, 870.

Giovannetti, M., Sbrana, C., Citernesi, A. S., & Avio, L. (1996). Analysis of factors involved in fungal recognition responses to host-derived signals by arbuscular mycorrhizal fungi. *The New Phytologist, 133*, 65–71.

Glick, B. R. (2012). Plant growth-promoting bacteria: Mechanisms and applications. *Scientifica, 2012*, 963401.

Glick, B. R. (2014). Bacteria with ACC deaminase can promote plant growth and help to feed the world. *Microbiological Research, 169*, 30–39.

Goswami, D., Parmar, S., Vaghela, H., Dhandhukia, P., & Thakker, J. (2015). Describing Paenibacillus mucilaginosus strain N3 as an efficient plant growth promoting rhizobacteria (PGPR). *Cogent Food & Agriculture, 1*, 1.

Gouda, S., Kerry, R. G., Das, G., Paramithiotis, S., Shin, H. S., & Patra, J. K. (2018). Revitalization of plant growth promoting rhizobacteria for sustainable development in agriculture. *Microbiological Research, 206*, 131–140.

Grace, M., & Saldajeno, B. (2012). *Investigation on the interactions between Arbuscular Mycorrhizal Fungus (AMF) and plant GrowthPromoting Fungi (PGPF) in their root colonization and disease suppression*. (Ph.D. Thesis). The United Graduate School of Agricultural Science, Gifu University.

Grover, M., Bodhankar, S., Sharma, A., Sharma, P., Singh, J., & Nain, L. (2021). PGPR mediated alterations in root traits: Way toward sustainable crop production. *Front. Sustain. Food Syst., 4*, 287.

Gujar, P. D., Bhavsar, K. P., & Khire, J. M. (2013). Effect of phytase from *Aspergillus Niger* on plant growth and mineral assimilation in wheat (*Triticum aestivum* linn.) and its potential for use as a soil amendment. *Journal of the Science of Food and Agriculture, 93*, 2242–2247.

Guler, N. S., Pehlivan, N., Karaoglu, S. A., Guzel, S., & Bozdeveci, A. (2016). *Trichoderma atroviride* ID20G inoculation ameliorates drought stress-induced damages by improving antioxidant defence in maize seedlings. *Acta Physiologiae Plantarum, 38*, 1–9.

Guo, Y., Matsuoka, Y., Miura, T., Nishizawa, T., Ohta, H., & Narisawa, K. (2018a). Complete genome sequence of *Agrobacterium pusense* vsbac-y9, a bacterial symbiont of the dark septate endophytic fungus *Veronaeopsis simplex* y34 with potential for improving fungal colonization in roots. *Journal of Biotechnology, 284*, 31–36.

Guo, Y., Matsuoka, Y., Nishizawa, T., Ohta, H., & Narisawa, K. (2018b). Effects of rhizobium species living with the dark septate endophytic fungus *Veronaeopsis simplex* on organic substrate utilization by the host. *Microbes and Environments, 33*, 102–106.

Hamayun, M., Khan, S. A., Khan, M. A., Khan, A. L., Kang, S. M., Kim, S. K., Joo, G. J., & Lee, I. J. (2009). Gibberellin production by pure cultures of a new strain of Aspergillus fumigatus. *World Journal of Microbiology and Biotechnology, 25*, 1785–1792.

Han, Q., Ma, Q., Chen, Y., Tian, B., Xu, L., Bai, Y., Chen, W., & Li, X. (2020). Variation in rhizosphere microbial communities and its association with the symbiotic efficiency of rhizobia in soybean. *The ISME Journal, 14*, 1915–1928.

Harman, G. E., Howell, C. R., Viterbo, A., Chet, I., & Lorito, M. (2004a). *Trichoderma* species—Opportunistic, avirulent plant symbionts. *Nature Reviews. Genetics, 2*, 43–56.

Harman, G. E., Petzoldt, R., Comis, A., & Chen, J. (2004b). Interactions between *Trichoderma harzianum* strain T22 and maize inbred line Mo17 and effects of these interactions on diseases caused by *Pythium ultimum* and *Colletotrichum graminicola*. *Phytopathology, 94*, 147–153.

Hartman, K., & Tringe, S. G. (2019). Interactions between plants and soil shaping the root microbiome under abiotic stress. *The Biochemical Journal, 476*, 2705–2724.

Hartmann, M., Brunner, I., Hagedorn, F., Bardgett, R. D., Stierli, B., Herzog, C., Chen, X., Zingg, A., Graf-Pannatier, E., Rigling, A., & Frey, B. (2017). A decade of irrigation transforms the soil microbiome of a semi-arid pine forest. *Molecular Ecology, 26*, 1190–1206.

Hassani, M. A., Durán, P., & Hacquard, S. (2018). Microbial interactions within the plant holobiont. *Microbiome, 6*, 58.

He, M., He, C. Q., & Ding, N. Z. (2018). Abiotic stresses: General defenses of land plants and chances for engineering multistress tolerance. *Frontiers in Plant Science, 9*, 1771.

Hermosa, R., Viterbo, A., Chet, I., & Monte, E. (2012). Plant-beneficial effects of *Trichoderma* and of its genes. *Microbiol., 158*, 17–25.

Hernández-Montiel, L. G., Contreras, C. J. C., Amador, B. M., Hernández, L. V., Aguilar, E. E. Q., & Contreras, R. G. C. (2017). Efficiency of two inoculation methods of *Pseudomonas putida* on growth and yield of tomato plants. *Journal of Soil Science and Plant Nutrition, 17*, 1003–1012.

Hossain, M. M., & Sultana, F. (2020). Application and mechanisms of plant growth promoting fungi (PGPF) for phytostimulation. In *Organic Agriculture*. IntechOpen.

Hossain, M. M., Sultana, F., Kubota, M., Koyama, H., & Hyakumachi, M. (2007). The plant growth-promoting fungus *Penicillium simplicissimum* gp17-2 induces resistance in *Arabidopsis thaliana* by activation of multiple defense signals. *Plant & Cell Physiology, 48*, 1724–1736.

Hossain, M. M., Hossain, N., Sultana, F., Islam, S. M. N., Islam, M. S., & Bhuiyan, M. K. A. (2013). Integrated management of fusarium wilt of chickpea (*Cicer arietinum* L.) caused by *Fusarium oxysporum* f. sp. ciceris with microbial antagonist, botanical extract. *African Journal of Biotechnology, 12*, 4699–4706.

Hossain, M. M., Sultana, F., Miyazawa, M., & Hyakumachi, M. (2014). The plant growth-promoting fungus *Penicillium* spp. gp 15-1 enhances growth and confers protection against damping-off and anthracnose in the cucumber. *Journal of Oleo Science, 63*, 391–400.

Hossain, M., Sultana, F., & Islam, S. (2017). Plant Growth-Promoting Fungi (PGPF): Phytostimulation and induced systemic resistance. In D. Singh, H. Singh, & R. Prabha (Eds.), *Plant-microbe interactions in agro-ecological perspectives: Volume 2: Microbial interactions and agro-ecological impacts* (pp. 135–191). Singapore. ISBN 978-981-10-6593-4.

Hu, L., Robert, C. A., Cadot, S., Zhang, X. I., Ye, M., Li, B., Manzo, D., Chervet, N., Steinger, T., Van Der Heijden, M. G., & Schlaeppi, K. (2018). Root exudate metabolites drive plant-soil feedbacks on growth and defense by shaping the rhizosphere microbiota. *Nature Communications, 9*, 1–13.

Huang, C. Y., Wang, H., Hu, P., Hamby, R., & Jin, H. (2019). Small RNAs—Big players in plant-microbe interactions. *Cell Host & Microbe, 26*, 173–182.

Ilyas, N., Mazhar, R., Yasmin, H., Khan, W., Iqbal, S., El Enshasy, H., & Dailin, D. J. (2020). Rhizobacteria isolated from saline soil induce systemic tolerance in wheat (*Triticum aestivum* L.) against salinity stress. *Agronomy, 10*, 989.

Inostroza, N. G., Barra, P. J., Wick, L. Y., Mora, M. L., & Jorquera, M. A. (2017). Effect of rhizobacterial consortia from undisturbed arid- and agro-ecosystems on wheat growth under different conditions. *Letters in Applied Microbiology, 64*, 158–163.

Ipek, M., Arıkan, Ş., Pırlak, L., & Eşitken, A. (2019). Sustainability of crop production by PGPR under abiotic stress conditions. In *Plant growth promoting Rhizobacteria for agricultural sustainability* (pp. 293–314). Springer Science and Business Media LLC.

Islam, M. S., Kormoker, T., Idris, A. M., Proshad, R., Kabir, M. H., & Ustaoglu, F. (2021). Plant-microbe-metal interactions for heavy metal bioremediation: A review. *Crop & Pasture Science, 73*, 181–201.

Jahagirdar, S., Kambrekar, D. N., Navi, S. S., & Kunta, M. (2019). Plant growth-promoting fungi: Diversity and classification. In S. Jogaiah & M. Abdelrahman (Eds.), *Bioactive molecules in plant defense: Signaling in growth and stress* (pp. 25–34). Springer International Publishing. ISBN 978-3-030-27165-7.

Janardan, Y., Verma, J. P., & Tiwari, K. N. (2011). Plant growth promoting activities of fungi and their effect on chickpea plant growth. *Asian Journal of Biological Sciences, 4*, 291–299.

Ji, Z. J., Yan, H., Cui, Q. G., Wang, E. T., Chen, W. F., & Chen, W. X. (2017). Competition between rhizobia under different environmental conditions affects the nodulation of a legume. *Systematic and Applied Microbiology, 40*, 114–119.

Jogaiah, S., Abdelrahman, M., Phan Tran, L.-S., & Shin-ichi, I. (2013). Characterization of rhizosphere fungi that mediate resistance in tomato against bacterial wilt disease. *Journal of Experimental Botany, 64*, 3829–3842.

Kaur, R., Kaur, J., & Singh, R. S. (2010). Non-pathogenic *Fusarium* as a biological control agent. *Plant Pathology Journal, 9*, 79–91.

Kawaharada, Y., Kelly, S., Nielsen, M. W., Hjuler, C. T., Gysel, K., Muszyński, A., Carlson, R. W., Thygesen, M. B., Sandal, N., Asmussen, M. H., & Vinther, M. (2015). Receptor-mediated exopolysaccharide perception controls bacterial infection. *Nature, 523*, 308–312.

Kelly, S., Sullivan, J. T., Kawaharada, Y., Radutoiu, S., Ronson, C. W., & Stougaard, J. (2018). Regulation of Nod factor biosynthesis by alternative NodD proteins at distinct stages of symbiosis provides additional compatibility scrutiny. *Environmental Microbiology, 20*, 97–110.

Kenneth, C. O. (2017). Plant Growth Promoting Rhizobacteria (PGPR): A bioprotectant bioinoculant for sustainable agrobiology. A review. *International Journal of Advanced Research in Biological Sciences, 4*, 123–142.

Khan, A. L., & Lee, I. J. (2013). Endophytic *Penicillium funiculosum* LHL06 secretes gibberellin that reprograms *Glycine max* L. growth during copper stress. *BMC Plant Biology, 13*, 1–14.

Khan, S. A., Hamayun, M., Yoon, H., Kim, H. Y., Suh, S. J., Hwang, S. K., Kim, J. M., Lee, I. J., Choo, Y. S., Yoon, U. H., & Kong, W. S. (2008). Plant growth promotion and *Penicillium citrinum*. *BMC Microbiology, 8*, 1–10.

Khan, N., Ali, S., Shahid, M. A., & Kharabian-Masouleh, A. (2017). Advances in detection of stress tolerance in plants through metabolomics approaches. *Plant Omics, 10*, 153–163.

Khan, N., Bano, A., Rahman, M. A., Rathinasabapathi, B., & Babar, A. (2019). UPLC-HRMS-based untargeted metabolic profiling reveals changes in chickpea (*Cicer arietinum*) metabolome following long-term drought stress. *Plant, Cell & Environment, 42*, 115–132.

Kiers, E. T., Rousseau, R. A., West, S. A., & Denlson, R. F. (2003). Host sanctions and the legume-rhizobium mutualism. *Nature, 425*, 78–81.

Kim, A. Y., Shahzad, R., Kang, S. M., Seo, C. W., Park, Y. G., Park, H. J., & Lee, I. J. (2017). IAA-producing Klebsiella variicola AY13 reprograms soybean growth during flooding stress. *Journal of Crop Science and Biotechnology, 20*, 235–242.

Kloepper, J. W., & Schroth, M. N. (1981). Plant growth-promoting rhizobacteria and plant growth under gnotobiotic conditions. *Phytopathology, 71*, 642–644.

Kondorosi, E., Mergaert, P., & Kereszt, A. (2013). A paradigm for endosymbiotic life: Cell differentiation of rhizobium bacteria provoked by host plant factors. *Annual Review of Microbiology, 67*, 611–628.

Kudoyarova, G., Arkhipova, T., Korshunova, T., Bakaeva, M., Loginov, O., & Dodd, I. C. (2019). Phytohormone mediation of interactions between plants and non-symbiotic growth promoting bacteria under edaphic stresses. *Frontiers in Plant Science, 10*, 1368.

Kuffner, M., Puschenreiter, M., Wieshammer, G., Gorfer, M., & Sessitsch, A. (2008). Rhizosphere bacteria affect growth and metal uptake of heavy metal accumulating willows. *Plant and Soil, 304*, 35–44.

Kumar, A., & Verma, J. P. (2018). Does plant—Microbe interaction confer stress tolerance in plants: A review? *Microbiological Research, 207*, 41–52.

Lata, R., Chowdhury, S., Gond, S. K., & White, J. (2018). Induction of abiotic stress tolerance in plants by endophytic microbes. *Letters in Applied Microbiology, 66*, 268–276.

Lee, D. W., & Kim, B. S. (2015). Antimicrobial cyclic peptides for plant disease control. *Plant Pathology Journal, 31*, 1–11.

Lei, Y., Su, S., He, L., Hu, X., & Luo, D. (2019). A member of the alog gene family has a novel role in regulating nodulation in *Lotus japonicus*. *Journal of Integrative Plant Biology, 61*, 463–477.

Li, Z., Chang, S., Ye, S., Chen, M., Lin, L., Li, Y., Li, S., & An, Q. (2015). Differentiation of 1-Aminocyclopropane-1-carboxylate (ACC) deaminase from its homologs is the key for identifying bacteria containing ACC deaminase. *FEMS Microbiology Ecology, 91*, 112.

Li, Z., Bai, T., Dai, L., Wang, F., Tao, J., Meng, S., Hu, Y., Wang, S., & Hu, S. (2016). A study of organic acid production in contrasts between two phosphate solubilizing fungi: *Penicillium oxalicum* and *Aspergillus niger*. *Scientific Reports, 6*, 1–8.

Libault, M., Farmer, A., Brechenmacher, L., Drnevich, J., Langley, R. J., Bilgin, D. D., Radwan, O., Neece, D. J., Clough, S. J., May, G. D., & Stacey, G. (2010). Complete transcriptome of the soybean root hair cell, a single-cell model, and its alteration in response to *Bradyrhizobium japonicum* infection. *Plant Physiology, 152*, 541–552.

Lima, R. M., Kylarová, S., Mergaert, P., & Kondorosi, É. (2020). Unexplored arsenals of legume peptides with potential for their applications in medicine and agriculture. *Frontiers in Microbiology, 11*, 13071.

Lin, H.-H., Huang, H.-M., Yu, M., Lai, E.-M., Chien, H.-L., & Liu, C.-T. (2018). Functional exploration of the bacterial type VI secretion system in mutualism: *Azorhizobium caulinodans* ORS571–Sesbania rostrata as a research model. *Molecular Plant-Microbe Interactions, 31*, 856–867.

Liu, H., Zhang, C., Yang, J., Yu, N., Wang, E., & Liu, C.-M. (2018). Hormone modulation of legume-rhizobial symbiosis. *Journal of Integrative Plant Biology, 60*, 632–648.

Lopes, M. J. S., Dias-Filho, M. B., Castro, T. H. R., & Silva, G. B. (2018). Light and plant growth-promoting rhizobacteria effects on *Brachiaria brizantha* growth and phenotypic plasticity to shade. *Grass and Forage Science, 73*, 493–499.

Lopes, M. J. D. S., Dias-Filho, M. B., & Gurgel, E. S. C. (2021). Successful plant growth-promoting microbes: Inoculation methods and abiotic factors. *Frontiers in Sustainable Food Systems, 5*, 5.

Lopez-Gomez, M., Sandal, N., Stougaard, J., & Boller, T. (2012). Interplay of flg22-induced defence responses and nodulation in *Lotus japonicus*. *Journal of Experimental Botany, 63*, 393–401.

Lu, J., Yang, F., Wang, S., Ma, H., Liang, J., & Chen, Y. (2017). Co-existence of rhizobia and diverse non-rhizobial bacteria in the rhizosphere and nodules of *Dalbergia odorifera* seedlings inoculated *with Bradyrhizobium elkanii, Rhizobium multihospitium*-like and *Burkholderia pyrrocinia*-like strains. *Frontiers in Microbiology, 8*, 2255.

Lugtenberg, B. (2016). *Principles of plant-microbe interactions*. Springer.

Lundberg, D. S., Lebeis, S. L., Paredes, S. H., Yourstone, S., Gehring, J., Malfatti, S., Tremblay, J., Engelbrektson, A., Kunin, V., Rio, T. G. D., & Edgar, R. C. (2012). Defining the core *Arabidopsis thaliana* root microbiome. *Nature, 488*, 86–90.

Macías-Rodríguez, L., Contreras-Cornejo, H. A., Adame-Garnica, S. G., Del-Val, E., & Larsen, J. (2020). The interactions of *Trichoderma* at multiple trophic levels: Inter-kingdom communication. *Microbiological Research, 240*, 126552.

Mahato, S., Bhuju, S., & Shrestha, J. (2018). Effect of *Trichoderma viride* as biofertilizer on growth and yield of wheat. *Malaysian Journal of Sustainable Agriculture, 2*, 1–5.

Maheshwari, D. K. (2013). *Bacteria in agrobiology: Disease management*. Springer.

Mahmood, I., Rizvi, R., Sumbul, A., & Ansari, R. A. (2019). Potential role of plant growth promoting Rhizobacteria in alleviation of biotic stress. In *Plant health under biotic stress* (pp. 177–188). Springer Science and Business Media LLC.

Markmann, K., Radutoiu, S., & Stougaard, J. (2012). Infection of *Lotus japonicus* roots by *Mesorhizobium loti*. In P. Silvia & F. Baluška (Eds.), *Signaling and communication in plant symbiosis* (pp. 31–50). Springer. ISBN 978-3-642-20966-6.

Martiny, J., Martiny, A. C., Weihe, C., Lu, Y., Berlemont, R., Brodie, E., Goulden, M., Treseder, K., & Allison, S. D. (2017). Microbial legacies alter decomposition in response to simulated global change. *The ISME Journal, 11*, 490–499.

Mayak, S., Tirosh, T., & Glick, B. R. (2004). Plant growth-promoting bacteria confer resistance in tomato plants to salt stress. *Plant Physiology and Biochemistry, 42*, 565–572.

Mazurek, P., Bojarski, J., Marcinkowski, K., Furmanowa, M., Gutkowska, B., Kaliszan, R., Pachecka, J., Pawlaczyk, J., Pluta, J., Wieniawski, W., et al. (2015). Drug research. *Acta Poloniae Pharmaceutica, 72*, 3–11.

Meena, R. K., Singh, R. K., Singh, N. P., Meena, S. K., & Meena, V. S. (2015). Isolation of low temperature surviving plant growth–promoting rhizobacteria (PGPR) from pea (*Pisum sativum* L.) and documentation of their plant growth promoting traits. *Biocatalysis and Agricultural Biotechnology, 4*, 806–811.

Meena, R. S., Vijayakumar, V., Gulab, S. Y., & Mitran, T. (2018). Response and interaction of *Bradyrhizobium japonicum* and arbuscular mycorrhizal fungi in the soybean rhizosphere. *Plant Growth Regulation, 84*, 207–223.

Mergaert, P., Nikovics, K., Kelemen, Z., Maunoury, N., Vaubert, D., Kondorosi, A., & Kondorosi, E. (2003). A novel family in *Medicago truncatula* consisting of more than 300 nodule-specific genes coding for small, secreted polypeptides with conserved cysteine motifs. *Plant Physiology, 132*, 161–173.

Minasny, B., Malone, B. P., McBratney, A. B., Angers, D. A., Arrouays, D., Chambers, A., Chaplot, V., Chen, Z. S., Cheng, K., Das, B. S., & Field, D. J. (2017). Soil Carbon 4 per Mille. *Geoderma, 292*, 59–86.

Mirzaei, M., Moghadam, A. L., Hakimi, L., & Danaee, E. (2020). Plant Growth Promoting Rhizobacteria (PGPR) improve plant growth, antioxidant capacity, and essential oil properties of lemongrass (*Cymbopogon citratus*) under water stress. *Iranian Journal of Plant Physiology, 10*, 3155–3166.

Molina, R., Rivera, D., Mora, V., Lopez, G., Rosas, S., Spaepen, S., Vanderleyden, J., & Cassán, F. (2018). Regulation of IAA biosynthesis in *Azospirillum brasilense* under environmental stress conditions. *Current Microbiology, 75*, 1408–1418.

Mortier, V., Holsters, M., & Goormachtig, S. (2012). Never too many? How legumes control nodule numbers. *Plant, Cell & Environment, 35*, 245–258.

Msimbira, L. A., & Smith, D. L. (2020). The roles of plant growth promoting microbes in enhancing plant tolerance to acidity and alkalinity stresses. *Frontiers in Sustainable Food Systems, 4*, 4.

Mukhtar, T., Rehman, S. U., Smith, D., Sultan, T., Seleiman, M. F., Alsadon, A. A., Amna, Ali, S., Chaudhary, H. J., Solieman, T. H., & Ibrahim, A. A. (2020). Mitigation of heat stress in *Solanum lycopersicum* L. by ACC-deaminase and exopolysaccharide producing *Bacillus cereus*: Effects on biochemical profiling. *Sustainability, 12*, 2159.

Murali, M., Naziya, B., Ansari, M. A., Alomary, M. N., AlYahya, S., Almatroudi, A., Thriveni, M. C., Gowtham, H. G., Singh, S. B., Aiyaz, M., & Kalegowda, N. (2021). Bioprospecting of rhizosphere-resident fungi: Their role and importance in sustainable agriculture. *Journal of Fungi, 7*, 314.

Nazir, N., Kamili, A. N., & Shah, D. (2018). Mechanism of plant growth promoting Rhizobacteria (PGPR) in enhancing plant growth-A review. *International Journal of Engineering Research and Management Technology, 8*, 709–721.

Naznin, H. A., Kiyohara, D., Kimura, M., Miyazawa, M., Shimizu, M., & Hyakumachi, M. (2014). Systemic resistance induced by volatile organic compounds emitted by plant growth-promoting fungi in *Arabidopsis thaliana*. *PLoS One, 9*, e86882.

Nelson, M. S., & Sadowsky, M. J. (2015). Secretion systems and signal exchange between nitrogen-fixing rhizobia and legumes. *Frontiers in Plant Science, 6*, 491.

Newman, M.-A., Sundelin, T., Nielsen, J. T., Erbs, G., & Tör, M. (2013). MAMP (microbe-associated molecular pattern) triggered immunity in plants. *Frontiers in Plant Science, 4*, 139.

Nievola, C. C., Carvalho, C. P., Carvalho, V., & Rodrigues, E. (2017). Rapid responses of plants to temperature changes. *Temperature, 4*, 371–405.

Niu, X., Song, L., Xiao, Y., & Ge, W. (2018). Drought-tolerant plant growth-promoting Rhizobacteria associated with foxtail millet in a semi-arid agroecosystem and their potential in alleviating drought stress. *Frontiers in Microbiology, 8*, 2580.

O'malley, M. R., Anderson, J. C., & Sessa, G. (2021). Microorganisms regulation of the *pseudomonas syringae* type III secretion system by host environment signals. *Microorganisms, 9*, 1227.

Okamoto, S., Shinohara, H., Mori, T., Matsubayashi, Y., & Kawaguchi, M. (2013). Root-derived CLE glycopeptides control nodulation by direct binding to HAR1 receptor kinase. *Nature Communications, 4*, 1–7.

Oldroyd, G. E. D., & Downie, J. A. (2008). Coordinating nodule morphogenesis with rhizobial infection in legumes. *Annual Review of Plant Biology, 59*, 519–546.

Oono, R., Anderson, C. G., & Denison, R. F. (2011). Failure to fix nitrogen by non-reproductive symbiotic rhizobia triggers host sanctions that reduce fitness of their reproductive clonemates. *Proceedings of the Royal Society B: Biological Sciences, 278*, 2698–2703.

Özbay, N., Ergun, M., & Demirkiran, A. R. (2018). Ticari Mikrobiyal Gübre Sim Derma® (*Trichoderma harzianum*, Kuen 1585) Uygulamasının Ispanakta Çimlenme, Geli¸sme ve Verim Üzerine Etkisi. *Türk Tarım ve Doğa Bilimleri Dergisi, 5*, 482–491.

Patriarca, E. J., Tatè, R., & Iaccarino, M. (2002). Key role of bacterial NH_4^+ metabolism in rhizobium-plant symbiosis. *Microbiology and Molecular Biology Reviews, 66*, 203–222.

Patten, C. L., & Glick, B. R. (1996). Bacterial biosynthesis of indole-3-acetic acid. *Canadian Journal of Microbiology, 42*, 207–220.

Peck, M. C., Fisher, R. F., & Long, S. R. (2006). Diverse flavonoids stimulate nodd1 binding to nod gene promoters in *Sinorhizobium meliloti*. *Journal of Bacteriology, 188*, 5417–5427.

Pecoraro, L., Wang, X., Shah, D., Song, X., Kumar, V., Shakoor, A., Tripathi, K., Ramteke, P. W., & Rani, R. (2021). Biosynthesis pathways, transport mechanisms and biotechnological applications of fungal Siderophores. *Journal of Fungi, 8*, 21.

Peiffer, J. A., Spor, A., Koren, O., Jin, Z., Tringe, S. G., Dangl, J. L., Buckler, E. S., & Ley, R. E. (2013). Diversity and heritability of the maize rhizosphere microbiome under field conditions. *Proceedings of the National Academy of Sciences of the United States of America, 110*, 6548–6553.

Provorov, N., Wienkoop, S., Yurievna Shtark, O., Mercedes Lucas, M., Coba de la Peña, T., Fedorova, E., & Pueyo, J. J. (2018). The symbiosome: Legume and rhizobia co-evolution toward a nitrogen-fixing organelle? *Frontiers in Plant Science, 1*, 2229.

Pueppke, S. G., & Broughton, W. J. (1999). *Rhizobium* sp. strain NGR234 and *R. fredii* USDA257 share exceptionally broad, nested host ranges. *Molecular Plant-Microbe Interactions, 12*, 293–318.

Radhakrishnan, R., Khan, A. L., Kang, S. M., & Lee, I. J. (2015). A comparative study of phosphate solubilization and the host plant growth promotion ability of *Fusarium verticillioides* rk01 and *Humicola* sp. knu01 under salt stress. *Annales de Microbiologie, 65*, 585–593.

Rai, A. K., Dinkar, A., Basak, N., Dixit, A. K., Das, S. K., Dev, I., Sundha, P., Chandra, P., & Kumar, S. (2021). Phosphorus nutrition of oats genotypes in acidic soils: Exploiting responsive plant-microbe partnership. *Applied Soil Ecology, 167*, 104094.

Rasool, S., Ahmad, A., Siddiqi, T. O., & Ahmad, P. (2012). Changes in growth, lipid peroxidation and some key antioxidant enzymes in chickpea genotypes under salt stress. *Acta Physiologiae Plantarum, 35*, 1039–1050.

Ratet, P., Javier López-Baena, F., Kvitko, B. H., Teaumroong, N., Giraud, E., Songwattana, P., Noisangiam, R., Teamtisong, K., Prakamhang, J., Teulet, A., et al. (2017). Type 3 secretion system (T3SS) of Bradyrhizobium sp. DOA9 and its roles in legume symbiosis and rice endophytic association. *Frontiers in Microbiology, 8*, 1810.

Raymond, J., Siefert, J. L., Staples, C. R., & Blankenship, R. E. (2004). The natural history of nitrogen fixation. *Molecular Biology and Evolution, 21*, 541–554.

Regus, J. U., Quides, K. W., O'Neill, M. R., Suzuki, R., Savory, E. A., Chang, J. H., & Sachs, J. L. (2017). Cell autonomous sanctions in legumes target ineffective rhizobia in nodules with mixed infections. *American Journal of Botany, 104*, 1299–1312.

Reid, D. E., Ferguson, B. J., Hayashi, S., Lin, Y. H., & Gresshoff, P. M. (2011). Molecular mechanisms controlling legume autoregulation of nodulation. *Annals of Botany, 108*, 789–795.

Rodríguez, H., & Fraga, R. (1999). Phosphate solubilizing bacteria and their role in plant growth promotion. *Biotechnology Advances, 17*, 319–339.

Roesch, L. F. W., Fulthorpe, R. R., Riva, A., Casella, G., Hadwin, A. K. M., Kent, A. D., Daroub, S. H., Camargo, F. A. O., Farmerie, W. G., & Triplett, E. W. (2007). Pyrosequencing enumerates and contrasts soil microbial diversity. *The ISME Journal, 1*, 283–290.

Roy, P., Achom, M., Wilkinson, H., Lagunas, B., & Gifford, M. L. (2020). Symbiotic outcome modified by the diversification from 7 to over 700 nodule-specific cysteine-rich peptides. *Genes, 11*, 348.

Ruess, L., & Ferris, H. (2004). Decomposition pathways and successional changes. *Nematology Monographs and Perspectives, 2*, 547–556.

Sachs, J. L., Quides, K. W., & Wendlandt, C. E. (2018). Legumes versus rhizobia: A model for ongoing conflict in symbiosis. *The New Phytologist, 219*, 1199–1206.

Sagar, A., Sayyed, R. Z., Ramteke, P. W., Sharma, S., Marraiki, N., Elgorban, A., & Syed, A. (2020). ACC deaminase and antioxidant enzymes producing halophilic *Enterobacter* sp. PR14 promotes the growth of rice and millets under salinity stress. *Physiology and Molecular Biology of Plants, 26*, 1847–1854.

Saldajeno, M., & Hyakumachi, M. (2011). The plant growth-promoting fungus *Fusarium equiseti* and the arbuscular mycorrhizal fungus *Glomus mosseae* stimulate plant growth and reduce severity of anthracnose and damping-off diseases in cucumber (*Cucumis sativus*) seedlings. *The Annals of Applied Biology, 159*, 28–40.

Salinero-Lanzarote, A., Pacheco-Moreno, A., Domingo-Serrano, L., Durán, D., Ormeño-Orrillo, E., Martínez-Romero, E., Albareda, M., Palacios, J. M., & Rey, L. (2019). The type VI secretion system of *rhizobium etli* mim1 has a positive effect in symbiosis. *FEMS Microbiology Ecology, 95*, fiz054.

Salwan, R., Sharma, A., & Sharma, V. (2019). Microbes mediated plant stress tolerance in saline agricultural ecosystem. *Plant and Soil, 442*, 1–22.

Sanderman, J., Hengl, T., & Fiske, G. J. (2017). Soil carbon debt of 12,000 years of human land use. *Proceedings of the National Academy of Sciences of the United States of America, 114*, 9575–9580.

Sanjuan, J., Janczarek, M., Sánchez-Cañizares, C., Mendoza-Suárez, M., Andersen, S. U., & Poole, P. S. (2021). Competition, nodule occupancy, and persistence of inoculant strains: Key factors in the rhizobium-legume symbioses. *Frontiers in Plant Science, 1684*.

Santi, C., Bogusz, D., & Franche, C. (2013). Biological nitrogen fixation in non-legume plants. *Annals of Botany, 111*, 743–767.

Sasaki, T., Suzaki, T., Soyano, T., Kojima, M., Sakakibara, H., & Kawaguchi, M. (2014). Shoot-derived cytokinins systemically regulate root nodulation. *Nature Communications, 5*, 1–7.

Sathoff, A. E., & Samac, D. A. (2019). Antibacterial activity of plant defensins. *Molecular Plant-Microbe Interactions, 32*, 507–514.

Schausberger, P., Souza, V., Khaitov, B., Leitch, I. J., Guignard, M. S., Leitch, A. R., Acquisti, C., Eizaguirre, C., Elser, J. J., Hessen, D. O., et al. (2017). Impacts of nitrogen and phosphorus: From genomes to natural ecosystems and agriculture. *Frontiers in Ecology and Evolution, 5*, 70.

Schumpp, O., & Deakin, W. J. (2010). How inefficient rhizobia prolong their existence within nodules. *Trends in Plant Science, 15*, 189–195.

Shah, A., & Smith, D. L. (2020). Flavonoids in agriculture: Chemistry and roles in, biotic and abiotic stress responses, and microbial associations. *Agronomy, 10*, 1209.

Shukla, P. S., Agarwal, P. K., & Jha, B. (2012). Improved salinity tolerance of *Arachis hypogaea* (L.) by the interaction of halotolerant plant-growth-promoting Rhizobacteria. *Journal of Plant Growth Regulation, 31*, 195–206.

Siddiqui, Z. A., & Pichtel, J. (2008). Mycorrhizae: An overview. In Z. A. Siddiqui, M. S. Akhtar, & K. Futai (Eds.), *Mycorrhizae: Sustainable agriculture and forestry* (pp. 1–35). Springer. ISBN 978-1-4020-8770-7.

Singh, P. K. (2012). Role of glomalin related soil protein produced by arbuscular mycorrhizal fungi: A review. *Agricultural Science Research Journals, 2*, 119–125.

Singh, V., Singh, P. N., Yadav, R. L., Awasthi, S. K., Joshi, B. B., Singh, R. K., Lal, R. J., & Duttamajumder, S. K. (2010). Increasing the efficacy of *Trichoderma harzianum* for nutrient uptake and control of red rot in sugarcane. *Journal of Horticultural Sciences, 2*, 66–71.

Skorupska, A., Janczarek, M., Marczak, M., Mazur, A., & Król, J. (2006). Microbial cell factories rhizobial exopolysaccharides: Genetic control and symbiotic functions. *Microbial Cell Factories, 5*, 7.

Sofo, A., Tataranni, G., Xiloyannis, C., Dichio, B., & Scopa, A. (2012). Direct effects of *Trichoderma harzianum* strain t-22 on micropropagated shoots of gisela6® (*Prunus cerasus* × *Prunus canescens*) rootstock. *Environmental and Experimental Botany, 76*, 33–38.

Sood, M., Kapoor, D., Kumar, V., Sheteiwy, M. S., Ramakrishnan, M., Landi, M., Araniti, F., & Sharma, A. (2020). *Trichoderma*: The "secrets" of a multitalented biocontrol agent. *Plants, 9*, 762.

Spatafora, J. W., Chang, Y., Benny, G. L., Lazarus, K., Smith, M. E., Berbee, M. L., Bonito, G., Corradi, N., Grigoriev, I., Gryganskyi, A., & James, T. Y. (2016). A phylum-level phylogenetic classification of zygomycete fungi based on genome-scale data. *Mycologia, 108*, 1028–1046.

Sulieman, S., & Tran, L. S. P. (2013). Asparagine: An amide of particular distinction in the regulation of symbiotic nitrogen fixation of legumes. *Critical Reviews in Biotechnology, 33*, 309–327.

Sumi, C. D., Yang, B. W., Yeo, I. C., & Hahm, Y. T. (2015). Antimicrobial peptides of the genus *bacillus*: A new era for antibiotics. *Canadian Journal of Microbiology, 61*, 93–103.

Thomas, P., & Reddy, K. M. (2013). Microscopic elucidation of abundant endophytic bacteria colonizing the cell wall-plasma membrane peri-space in the shoot-tip tissue of banana. *AoB Plants, 5*, plt011.

Thompson, L. R., Sanders, J. G., McDonald, D., Amir, A., Ladau, J., Locey, K. J., Prill, R. J., Tripathi, A., Gibbons, S. M., Ackermann, G., & Navas-Molina, J. A. (2017). A communal catalogue reveals Earth's multiscale microbial diversity. *Nature, 551*, 457–463.

Tian, J., Ge, F., Zhang, D., Deng, S., & Liu, X. (2021). Roles of phosphate solubilizing microorganisms from managing soil phosphorus deficiency to mediating biogeochemical P cycle. *Biology, 10*, 158.

Timmusk, S., El-Daim, I. A., Copolovici, L., Tanilas, T., Kännaste, A., Behers, L., Nevo, E., Seisenbaeva, G., Stenström, E., & Ülo, N. (2014). Drought-tolerance of wheat improved by rhizosphere bacteria from harsh environments: Enhanced biomass production and reduced emissions of stress volatiles. *PLoS One, 9*, e96086.

Tóth, L., Váradi, G., Éva, B., Borics, A., Ficze, H., Nagy, I., Tóth, G. K., Rákhely, G., Marx, F., & Galgóczy, L. (2020). Biofungicidal potential of Neosartorya (*Aspergillus*) Fischeri antifungal protein NFAP and novel synthetic γ-Core peptides. *Frontiers in Microbiology, 11*, 820.

Traubenik, S., Reynoso, M. A., Hobecker, K., Lancia, M., Hummel, M., Rosen, B., Town, C., Bailey-Serres, J., Blanco, F., & Zanetti, M. E. (2020). Reprogramming of root cells during nitrogen-fixing symbiosis involves dynamic polysome association of coding and noncoding RNAs. *Plant Cell, 32*, 352–373.

Van de Velde, W., Zehirov, G., Szatmari, A., Debreczeny, M., Ishihara, H., Kevei, Z., Farkas, A., Mikulass, K., Nagy, A., Tiricz, H., & Satiat-Jeunemaitre, B. (2010). Plant peptides govern terminal differentiation of bacteria in symbiosis. *Science, 327*, 1122–1126.

Varma, A., Tripathi, S., & Prasad, R. (2019). *Plant biotic interactions: State of the art*. Berlin/Heidelberg, Germany.

Velázquez, E., Martinez-Romero, E., Ormeño-Orrillo, E., Lodeiro, A. R., Covelli, J. M., Alvarez, F., Iturralde, E. T., Pérez-Giménez, J., & Arrese-Igor, C. (2019). Soybean-nodulating strains with low intrinsic competitiveness for nodulation, good symbiotic performance, and stress-tolerance isolated from soybean-cropped soils in Argentina. *Frontiers in Microbiology, 10*, 1061.

Venturi, V., & Keel, C. (2016). Signaling in the rhizosphere. *Trends in Plant Science, 21*, 187–198.

Verma, M., Mishra, J., & Arora, N. K. (2018). Plant growth-promoting Rhizobacteria: Diversity and applications. In *Environmental biotechnology: For sustainable future* (pp. 129–173). Springer.

Verma, P. P., Shelake, R. M., Das, S., Sharma, P., & Kim, J. Y. (2019). Plant Growth-Promoting Rhizobacteria (PGPR) and Fungi (PGPF): Potential biological control agents of diseases and pests. In D. Singh, V. Gupta, & R. Prabha (Eds.), *Microbial interventions in agriculture and environment: Volume 1: Research trends, priorities and prospects* (pp. 281–311). Singapore. ISBN 978-981-13-8391-5.

Vives-Peris, V., De Ollas, C., Gómez-Cadenas, A., & Pérez-Clemente, R. M. (2020). Root exudates: From plant to rhizosphere and beyond. *Plant Cell Reports, 39*, 3–17.

Vorholt, J. A. (2012). Microbial life in the phyllosphere. *Nature Reviews. Microbiology, 10*, 828–840.

Vujanovic, V., Yit, K., & Goh, K. (2012). qPCR quantification of Sphaerodes mycoparasitica biotrophic mycoparasite interaction with *Fusarium graminearum*: In vitro and in planta assays. *Archives of Microbiology, 194*, 707–717.

Wakelin, S. A., Gupta, V. V. S. R., Harvey, P. R., & Ryder, M. H. (2007). The effect of *Penicillium* fungi on plant growth and phosphorus mobilization in neutral to alkaline soils from southern Australia. *Canadian Journal of Microbiology, 53*, 106–115.

Wang, H., Liu, R., You, M. P., Barbetti, M. J., & Chen, Y. (2021). Pathogen biocontrol using Plant Growth-Promoting Bacteria (PGPR): Role of bacterial diversity. *Microorganisms, 9*, 1988.

Wanke, A., Malisic, M., Wawra, S., & Zuccaro, A. (2021). Unraveling the sugar code: The role of microbial extracellular glycans in plant-microbe interactions. *Journal of Experimental Botany, 72*, 15–35.

Waqas, M., Khan, A. L., Kamran, M., Hamayun, M., Kang, S. M., Kim, Y. H., & Lee, I. J. (2012). Endophytic fungi produce gibberellins and indoleacetic acid and promotes host-plant growth during stress. *Molecules, 17*, 10754–10773.

Waweru, B., Turoop, L., Kahangi, E., Coyne, D., & Dubois, T. (2014). Non-pathogenic *Fusarium oxysporum* endophytes provide field control of nematodes, improving yield of banana (*Musa* sp.). *Biological Control, 74*, 82–88.

Wendlandt, C. E., Regus, J. U., Gano-Cohen, K. A., Hollowell, A. C., Quides, K. W., Lyu, J. Y., Adinata, E. S., & Sachs, J. L. (2019). Host investment into symbiosis varies among genotypes of the legume Acmispon strigosus, but host sanctions are uniform. *The New Phytologist, 221*, 446–458.

Westhoek, A., Clark, L. J., Culbert, M., Dalchau, N., Griffiths, M., Jorrin, B., Karunakaran, R., Ledermann, R., Tkacz, A., Webb, I., & James, E. K. (2021). Conditional sanctioning in a legume–*Rhizobium mutualism. Proceedings of the National Academy of Sciences of the United States of America, 118*, e2025760118.

Whipps, J. M. (2001). Microbial interactions and biocontrol in the rhizosphere. *Journal of Experimental Botany, 52*, 487–511.

Wilkes, T. I., Warner, D. J., Edmonds-Brown, V., Davies, K. G., & Denholm, I. (2021). The tripartite rhizobacteria and fungal-host plant relationship in winter wheat: Impact of multi-species inoculation, tillage regime and naturally occurring rhizobacteria species. *Plants, 10*, 1357.

Wuana, R. A., Okieimen, F. E., Montuelle, B., & Steinman, A. D. (2011). Heavy metals in contaminated soils: A review of sources, chemistry, risks and best available strategies for remediation. *International Scholarly Research Notices (Netw. ISRN Ecol.), 2011*, 20.

Xu, Z., Jiang, Y., Jia, B., & Zhou, G. (2016). Elevated-CO_2 response of stomata and its dependence on environmental factors. *Frontiers in Plant Science, 7*, 7.

Xu, L., Naylor, D., Dong, Z., Simmons, T., Pierroz, G., Hixson, K. K., Kim, Y. M., Zink, E. M., Engbrecht, K. M., Wang, Y. I., & Gao, C. (2018). Drought delays development of the sorghum root microbiome and enriches for monoderm bacteria. *Proceedings of the National Academy of Sciences of the United States of America, 115*, E4284–E4293.

Yadav, R. L., Shukla, S. K., Suman, A., & Singh, P. N. (2009). *Trichoderma* inoculation and trash management effects on soil microbial biomass, soil respiration, nutrient uptake and yield of ratoon sugarcane under subtropical conditions. *Biology and Fertility of Soils, 45*, 461–468.

Yarwood, S. A. (2018). The role of wetland microorganisms in plant-litter decomposition and soil organic matter formation: A critical review. *FEMS Microbiology Ecology, 94*. https://doi.org/10.1093/femsec/fiy175

Yedidia, I., Srivastva, A. K., Kapulnik, Y., & Chet, I. (2001). Effect of *Trichoderma harzianum* on microelement concentrations and increased growth of cucumber plants. *Plant and Soil, 235*, 235–242.

You, Y. H., Yoon, H., Kang, S. M., Shin, J. H., Choo, Y. S., Lee, I. J., Lee, J. M., & Kim, J. G. (2012). Fungal diversity and plant growth promotion of endophytic fungi from six halophytes in suncheon bay. *Journal of Microbiology and Biotechnology, 22*, 1549–1556.

Zaidi, A., Khan, M., Ahemad, M., & Oves, M. (2009). Plant growth promotion by phosphate solubilizing bacteria. *Acta Microbiologica et Immunologica Hungarica, 56*, 263–284.

Zerrouk, I. Z., Rahmoune, B., Khelifi, L., Mounir, K., Baluska, F., & Ludwig-Müller, J. (2019). Algerian Sahara PGPR confers maize root tolerance to salt and aluminum toxicity via ACC deaminase and IAA. *Acta Physiologiae Plantarum, 41*, 91.

Zhalnina, K., Louie, K. B., Hao, Z., Mansoori, N., Da Rocha, U. N., Shi, S., Cho, H., Karaoz, U., Loqué, D., Bowen, B. P., & Firestone, M. K. (2018). Dynamic root exudate chemistry and microbial substrate preferences drive patterns in rhizosphere microbial community assembly. *Nature Microbiology, 3*, 470–480.

Zhou, G., Mengoni, A., Eugenia Zanetti, M., Liu, J., Yu, X., Qin, Q., Dinkins, R. D., & Zhu, H. (2020). The impacts of domestication and breeding on nitrogen fixation symbiosis in legumes. *Frontiers in Genetics, 11*, 00973.

Zin, N. A., & Badaluddin, N. A. (2020). Biological functions of *Trichoderma* spp. for agriculture applications. *Annals of Agricultural Science, 65*, 168–178.

Role of Microbes in Soil Food Webs and Vegetation Development

Miracle Uwa Livinus, Sunday Zeal Bala, Mustapha Abdulsalam, Musa Ojeba Innocent, Madinat Hassan, Shehu-Alimi Elelu, and Priscilla Kini

Abstract Soil ecosystems are complex networks of interspecific relationships linking a vast array of organisms; microbes are central to the process of nutrient cycling, decomposition of organic matter, and plant-microbe associations. Microbiome as a line of scientific study has been advancing in the context of the last several decades and is currently recognized by both the scientific community and the public. Contrary to expectations, despite the increased interest in the topic across disciplinary boundaries, there is still no consensus definition of the word "microbiome." The distribution, functions, and connections of soil microorganisms throughout the soil food chain, as well as their influence on plant growth, yield, and ecosystem stability, are all covered in this work. The chapter also examines how microbial populations influence and govern nutrient dynamics, soil structure, and plant-microbe interactions that govern plant succession and dissemination. Recent developments in ecological modelling and molecular biology demonstrate the abundance and diversity of soil microorganisms. A thorough understanding of these microbial processes is necessary to support healthy ecosystems and land rehabilitation techniques.

M. U. Livinus (✉)
Department of Biochemistry, School of Sciences and Information Technology, Skyline University Nigeria, Kano, Nigeria

S. Z. Bala
Department of Education, University of Maryland Eastern Shore, Princess Anne, MD, USA

M. Abdulsalam · M. O. Innocent
Department of Microbiology, School of Sciences and Information Technology, Skyline University Nigeria, Kano, Nigeria

M. Hassan
Foresight Institute of Research and Translation (FIRAT), Kigali, Rwanda

S.-A. Elelu
Department of Chemistry, Howard University, Washington, DC, USA

P. Kini
Department of Toxicology, University of Maryland Eastern Shore, Princess Anne, MD, USA

S. A. Aransiola et al. (eds.), *Soil Microbiome in Green Technology Sustainability*,
https://doi.org/10.1007/978-3-031-71844-1_4

Keywords Diversity · Ecosystem · Interaction · Microbe · Microbiome · Soil food web · Vegetation

1 Introduction

Multiple species interact and flourish in the intricate and dynamic ecosystem of soil, the thin layer of the Earth's crust that supports life. Microorganisms are the most significant and common of these organisms (Wagg et al., 2019). The great majority of microorganisms found in the soil, known as the microbial soil biota, range from bacteria and fungi to archaea and protists. These microbial populations are essential to the operation of the terrestrial biomes, participating in a multitude of soil system functions (Jiao et al., 2018). The primary actors in the soil food webs, microbes control the flow of energy, the turnover of organic matter, and the dynamics of nutrients. Moreover, microorganisms and plant roots engage in intricate symbiotic interactions that influence plant physiology, resilience to biotic and abiotic stress, and plant performance (Pugnaire et al., 2019).

A complex of feeding relationships between different soil-dwelling creatures, such as bacteria, fungi, microfauna, mesofauna, and macrofauna, is referred to as the "soil food web." According to Scheu et al. (2005) and Sofo et al. (2020), microbes play a major part in this complexity by cycling nutrients through their metabolism and breaking down organic materials. The breakdown of complex organic materials into simpler molecules by bacteria and fungi, for example, is well-recognised for releasing nutrients that are essential for both plant development and ecosystem productivity. Additionally, microbes build food pyramids and nutritional networks that support life forms in the soil habitat through their interactions with other species such as nematodes, protozoa, and microarthropods (Potapov et al., 2022).

Soil microbiology is an intriguing topic of research, with one of the most fascinating features being the symbiotic relationships between microbes and plant roots. Thus, several physiological processes in plants, including nutrition absorption, hormone regulation, and defence against pathogens, are impacted by rhizosphere interactions, which occur when microorganisms attach themselves to plant roots (Chamkhi et al., 2022). As an example, most terrestrial plants grow in mutualism with mycorrhizal fungi, which thicken the roots and increase the uptake efficiency of nitrogen and phosphorus. The plants receive carbon molecules from the fungi in return, which helps the mycorrhizal networks develop and function. In addition to enabling plants to grow in a variety of environments and increase their resistance to biotic and abiotic stresses, these mutualistic connections are essential for the control of plant nutrients and diseases (Saeed et al., 2021; Thepbandit & Athinuwat, 2024).

Therefore, the functional role of microbes is not just seen in individual plants but also has tremendous impacts on entire ecosystems and landscapes. Microbes are very important in the development and succession of vegetation whereby they affect the vegetation heterogeneity or plant species and their arrangements in concern to space and time (Poorter et al., 2024). Through the regulation of the nutrient demands and supply systems, the structure formation within the soil, and the relationship between the plant and microbes, microbes control the competitiveness stakes from one plant species to another, initiating a succession with the pioneer plants and culminating into the development of a multistrata ecosystem. In addition, microbial communities change temporally in any developing ecosystem by responses to variations in environmental conditions and interactions with disturbance regimes and land uses. These microbial functions that shape ecosystem properties are fundamental to anticipating how ecosystems may or may not adapt to environmental change and how better to protect or rehabilitate ecosystems (Philippot et al., 2021).

Scientific progress in the last decade in molecular and microbial biology, and ecological network modelling has elaborately changed the paradigm of the soil microbiome and its functions (Garg et al., 2023; Mishra et al., 2023). Due to advances in sequencing technologies, the investigation of the structure and function of soil microbes has become much easier and more efficient, and this has led to the discovery of diverse and potential functions of unnoticed soil microbes (Lahlali et al., 2021). Xiao et al. (2021) studied the effect of Tamarix shrubs on the structure and function of bacterial communities in the Tarim River Basin soil using high throughput sequencing and reported that there were differences in properties of the soils and compositions of bacterial communities in different parts of the shrubs and the key soil factors affected the bacterial structure and functional properties. By employing the metagenomic and metatranscriptomic techniques, studies established the existence as well as the functional potential of different microbes in the soil that contributed to the ecosystem functions. In addition, the introduction of microbial processes into the ecological models has provided a better possibility of predicting the changes in the characteristics of the soil and the dynamics of the plant cover in response to changes in the environment (Aguiar-Pulido et al., 2016; Garg et al., 2023).

The relationships between soil microbes, the soil food web, and vegetation dynamics are still mostly unexplored, though. Understanding the relationships between microbial activities are challenging due to the diversity and complexity of soil ecosystems (Lladó et al., 2017; Chu et al., 2020). Furthermore, additional difficulties in comprehending and managing soil ecosystems are brought about by the effects of global changes, such as changes in land use, climate change, or the introduction of invasive species (Pugnaire et al., 2019). Innovative research approaches, interprofessional collaboration, and management frameworks that comprehend the intricate relationships among soil microbes, plant life, and ecosystem processes are necessary to address such concerns (Diwan et al., 2022).

This chapter will concentrate on the various roles that bacteria play in soil food webs and how they significantly impact the growth of vegetation and the ecosystem's overall biota. The different kinds of soil microorganisms, their functions and connections within the food chain, and their impacts on plant development, yield, and environmental stability will all be covered. Furthermore, the impact of microbial communities on soil structure, nutrient cycling, and plant-microbe interactions—which in turn control plant dispersion patterns and ecosystem development—will be taken into account. Through the review of existing literature and the assessment of knowledge gaps, the presented research is expected to offer a better understanding of the relationships and dynamics of soil ecosystems and support future advances in managing soil and conserving ecosystems.

2 Diversity and Functions of Soil Microbes

Comprising a variety of microorganisms from various taxonomic and functional categories, the soil microbiome is tailored to certain soil types and environments. The soil microbiome is a heterogeneous community of organisms that range in taxonomy and functional distribution over a range of soil habitats and ecological niches, encompassing bacteria, fungi, archaea, and protists (Berg et al., 2020). Although fungi are the second most varied category of microorganisms and may adapt to a wide range of tasks, bacteria generally exhibit the highest number of microorganisms. The processes in soil involve archaea as well as other microbial species. These bacteria carry out essential tasks that are critical to the health, structure, nutrient cycling, pathogen suppression, and ecosystem activities. They are impacted by other microbes as well as the immediate surrounding environment (Gupta et al., 2017; Odelade & Babalola, 2019).

2.1 Bacterial Diversity and Functions

In soil habitats, bacteria are among the most diverse and widely distributed microbial populations; in fact, there can be billions of bacteria per gram of soil. They are very functionally diverse and are involved in almost all of the biochemical reactions that can affect the fertility of the soil and the health of plants. Like *Bacillus subtilis*, *Pseudomonas fluorescens*, *Streptomyces* spp., *Clostridium* spp., *Actinomyces* spp., *Azotobacter* spp., *Cytophaga* spp., these bacteria which are referred to as decomposers are involved in the decomposition of organic matter since they break down complex organic compounds and turn them into simpler forms through processes like mineralization and humification. This makes nutrients such as nitrogen, phosphorus, and sulfur available to plants and other organisms (Gupta et al., 2017; Odelade & Babalola, 2019).

Nitrogen fixation, nitrification, and denitrification are three further biogeochemical transformations and nutrient cycling processes that require certain specific kinds of bacteria. For instance, the nitrogen-fixing bacteria Rhizobium and Bradyrhizobium interact with leguminous plants and transform atmospheric nitrogen gas into ammonia, which plants can use for growth and development (Pajares and Bohannan). Nitrosomonas and Nitrobacter are two kinds of nitrifying bacteria that can convert ammonium into nitrate form of nitrogen, which is readily absorbed by plants. On the other hand, denitrifying bacteria reduce nitrate into nitrogen gas thus playing the role of cycling nitrogen back to the atmosphere (Huey et al., 2020; Ling et al., 2022). In addition to their involvement in the cycling of nutrients, bacteria also play a role in mutualistic endophytic and mycorrhizal relationships with plant roots, as well as harmful interactions (Koza et al., 2022). For instance, mycorrhizal bacteria are a type of bacteria that has a mutualistic relationship with the roots of some plants and helps in the uptake of nutrients and water. However, there are specific pathogenic bacteria that are capable of affecting plant production through contracting diseases, which affects the yields (Jain et al., 2019).

2.2 *Fungal Diversity and Functions*

Another group of soil microorganisms is fungi which is a vast group of species that serves different functions in the ecosystem. Fungi are well adapted for the breakdown of complex organic matter especially the structural components of plants such as lignin and cellulose. Saprophytic fungi including Aspergillus and Penicillium release enzymes that degrade these complex polymers to liberate carbon and other nutrients into the soil. This process of decomposition is the main driver of nutrient cycling and the recycling of organic matter within soil ecosystems (Clocchiatti et al., 2020). Aside from decomposition, fungi can also be associated with plant roots where a mutualistic relationship termed mycorrhizae is formed between the fungi and plants. Mycorrhizal fungi are a type of plant symbionts which increase the area of the roots and enhance the plants' access to both nutrients and moisture within the earth. In return, plants supply the fungi with carbon compounds necessary for their growth and activity. These associations are more common in nutrient-poor environments because mycorrhizae help to improve nutrient uptake and improve plant growth and health (Gabaldón, 2019). Some categories of fungi are also implicated in plant diseases affecting many agricultural and ornamental plants. For instance, Fusarium and Phytophthora species are known to be responsible for root rot and damping-off of seedlings, which results in massive crop losses (Jain et al., 2019).

2.3 *Archaeal and Protist Diversity and Functions*

Compared to bacteria and fungi, archaea and protists are less explored; however, both play a significant role in the functionality of the soil ecosystem. One of the most important groups of the population is ammonia-oxidizing archaea (AOA) which perform the first step of nitrification, the process of ammonia conversion into nitrite in the soil context. This is an important process in the nitrogen cycling and availability of nitrogen in the plant system. Furthermore, some archaea are Fextremophiles that are adapted to live in extreme conditions like hot springs and acidic soils and perform specific ecological functions (Zhalnina et al., 2012; Wang et al., 2017). Amoebae, flagellates and ciliates are a class of protists that feed on bacteria and fungi and help in controlling soil microbial and nutrient dynamics. Some protists are also involved in mutualistic interactions with plants where they assist in nutrient acquisition and plant development (Pajares & Bohannan, 2016; Paśmionka et al., 2021).

3 Soil Food Webs

Soil food webs therefore show the intricate ecosystem of the soil that includes microbes, microfauna and mesofauna, macrofauna etc. as shown in Fig. 1. These food webs are organized into trophic levels, the base of which is primary producers (plants), then detritivores, predators and decomposers. Soil microbes are central players in soil food chains as the initial consumer of organic matter and the major energy source to the upper levels of the food chain. They are involved in the cycling of nutrients and energy flows that maintain and determine the stability and production of the ecosystem (Wu et al., 2024). Soil food webs depict complex and dynamic systems hosting an enormous number of organisms, which are intertwined in numerous and complicated trophic chains and communities. At the centre of these ecosystems is the soil food web, where organisms interact with each other through a series of feeding relationships that range from producers to decomposers. Soil food webs are involved in nutrient cycling, decomposition of organic matter, energy flow and structural properties of terrestrial ecosystems (Du et al., 2018).

3.1 *Structure of Soil Food Webs*

Soil food webs are organized in tiers and include several trophic levels that distinguish functional populations. Therefore, at the lowest level of the ecosystem is the population of the primary consumers which comprise the plants and photosynthesizing bacteria and algae (Fig. 2). These organisms harness light energy to produce chemical energy through photosynthesis and serve as the base of the

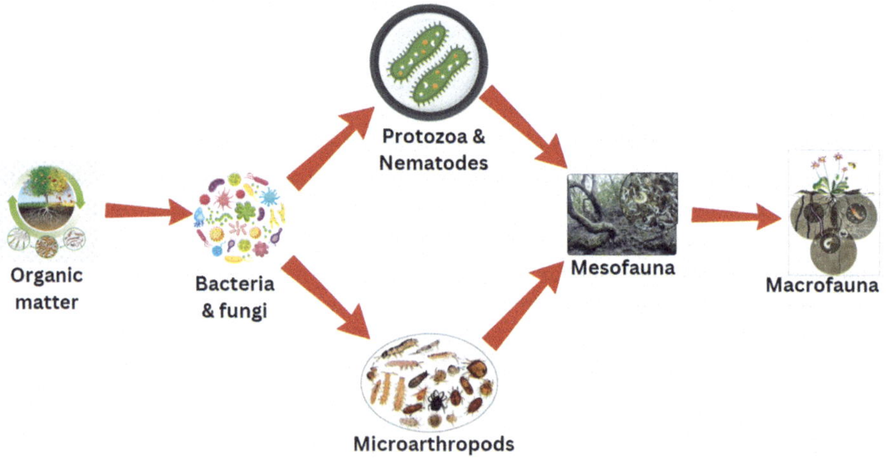

Fig. 1 Trophic interactions within the soil food web: from organic matter decomposition to macrofauna predation

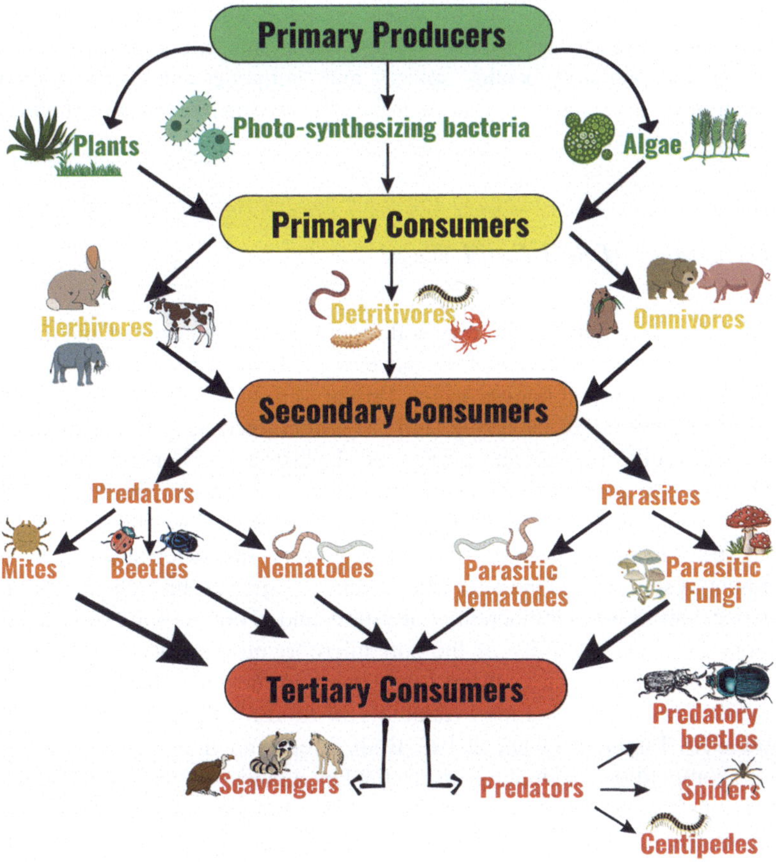

Fig. 2 Structure of soil food webs

decomposer trophic network by releasing energy and nutrients to the upper tiers (Briones, 2014; Potapov, 2022). Above the primary producers are the primary consumers, that is the animals that feed directly on the plants or organic matter such as herbivores, detritivores and omnivores amongst others. Herbivores feed on plant tissue while detritivores feed on dead organic matter like leaves, plant parts and dead animals. Omnivores are broad-spectrum feeders that feed on both plant and animal matter and therefore, perform nutrient cycling and decomposition of organic matter functions (Briones, 2014).

The second tier of the food chain is the secondary consumers which includes the predators and parasites who feed on the primary consumers. Some of the common carnivores include; predators like nematodes, beetles and mites that feed on herbivores and detritivores to ensure there is a check on their growth and to balance the ecosystem. Host organisms are infected and utilized by parasites such as parasitic nematodes and fungi to feed thereby altering their behaviour, physiology and population densities (Potapov et al., 2022). The third trophic level of the soil food web is the tertiary consumers which include predators and scavengers that feed on secondary consumers and decomposers. These organisms, which may include larger predators such as predatory beetles, spiders, and centipedes can effectively manage the lower trophic levels and recycle materials through the ecosystem (Palevsky & Ruess, 2022).

3.2 Dynamics of Soil Food Webs

Community structure and dynamics of soil food webs are determined and influenced by organism interactions and assimilated energy flows, which result in complex feedback mechanisms and nonlinearities. Primary producers, in this connection, are involved in the process of capturing sunlight and converting it to chemical energy through photosynthesis (Moënne-Loccoz et al., 2014). This energy is then transferred through the food web as organisms feed on other organisms, and in the process lose some of this energy through respiration, excretion, and death (Franklin et al., 2016). The dynamics of the soil food webs are determined such as the environment, resources and the interactions between species through several factors. For instance, variations in temperature, moisture, and pH of the soil result in changes in the abundance and activity of the soil microorganisms, thereby changing the organization and performance of the food chains in the soil. Plant growth rates or species shifts may also affect the availability of food for the consumers of a given soil food web (Pugnaire et al., 2019). Both direct and indirect connections like predation, competition and mutualism are the key elements determining the nutrient dynamics of the soil food web. Some of the functions of predators include maintaining the numbers of prey species to avoid overproduction of the species that cause imbalance in the ecosystem (Malmstrom, 2010). Fundamentally, species within a soil food web compete for resources including; food and space can affect the distribution and abundance of species within a given area leading to niche differentiation. Symbiotic relationships, for instance between plants and mycorrhizal fungi, play a crucial role in nutrient recycling and plant growth thus improving the ecosystem stability and production (Moënne-Loccoz et al., 2014).

3.3 Functions of Soil Food Webs

Soil food webs thus have several roles that are vital for ecosystem functions and stability. Nutrient cycling is one of the most important roles of the soil food webs, as organic matter is broken down and nutrients made available to the organisms within the soil and used by plants and other organisms. When complex macromolecules are broken down into simpler forms by bacteria, fungi, and detritivores acting as decomposers, nutrients including sulphur, phosphorus, and nitrogen are released back into the soil (Jiao et al., 2018; Wu et al., 2024). Other important functions of soil food webs include the breakdown of organic matter, which is the process by which decayed plant and animal remains are reabsorbed into the soil. Bacteria, fungi, earthworms, and arthropods are a few of the decomposers; they use mechanical, chemical, and biological processes to break down organic materials. In addition to their general involvement in maintaining the health of terrestrial ecosystems, these species have important characteristics that enhance soil fertility, structure, and production (Sofo et al., 2020). Additionally, they contribute to the preservation of species richness and plant health by way of rhizosphere and root activities. For example, in addition to helping plants withstand illnesses and other pressures, mycorrhizal fungi attach themselves to plant roots and aid in the absorption of nutrients and water (Clocchiatti et al., 2020). Similarly, hormones and chemicals that promote plant growth and development are secreted by bacterial and fungal species from the soil, which increases ecosystem productivity and sustainability (Jain et al., 2019; Odelade & Babalola, 2019).

3.4 Human Impacts on Soil Food Webs

Agricultural practices, land-use intensification, and habitat conversion to urban or agricultural land all can affect soil food webs and ecosystem processes. Agricultural practices that use synthetic fertilizers and pesticides are particularly damaging to soil food webs through changes in soil pH, nutrient cycling, and microbial habitats (Pugnaire et al., 2019). Likewise, urbanization and habitat fragmentation may split up the soil habitats and affect the complexity of the interacting species with negative effects on a decrease in the soil species' variety and their services (Haddad et al., 2015). Climate change will also have several effects on the organization and functioning of soil food webs because variation in temperature, quantity and concentration of atmospheric CO2 affects the numbers and geographical distribution of soil organisms (Franklin et al., 2016). Certain disturbances, such as changes in soil moisture regimes, impact the quantity and quality of the soil microorganisms that are involved in the decomposition of the organic matter and hence the rates of nutrient cycling and ecosystem processes. In addition, the effect of rising temperatures may influence the competitive ability of different organisms within a soil food web and hence cause changes in species composition and position within a particular food chain (Lladó et al., 2017; Potapov, 2022).

4 Biotic Interactions

Biotic interactions relate to the activities between plants and other organisms of different trophic levels including plants-plant, plants-animal and plant-microbe activities. These biotic interferences could be either beneficial or detrimental to the plants and hence impact the plant species' performance and density (Moënne-

Loccoz et al., 2014). This means that the processes of biotic interactions may have significant impacts on succession and leave behind various impressions of the environment.

4.1 Plant–Plant

Plant–plant interactions affect a target plant either positively (facilitation) or negatively (competition, inhibition, allelopathy) (Hischier et al., 2023; Kong et al., 2023). These include the enhancement of climates through sheltering other plants from unfavourable conditions, improvement of resource status through adding litter input, hydraulic uplift and nitrogen fixation, and protection of plants from herbivores (Brooker, 2006). This is especially important in the initial stages of primary succession, as well as the early stages of secondary succession which enhances succession and results in uniform vegetation cover. This is a situation where plants compete for resources, which are scarce in the environment. In early successional states, competition is relatively low for the first reason that vegetation cover density is low, but it increases with an increase in plant cover density. The initial struggle for factors of production is mainly light and water while nutrients become the second after light becomes limited. Clonal grasses and ferns which are early successional plants are effective competitors because they establish themselves immediately and form compact stands making it difficult for other plant species to penetrate. With successional time, competition for light increases and taller organisms dominate, leading to an orderly replacement sequence from short-lived to long-lived species (Hischier et al., 2023).

Prevention involves situations where one plant has denied space to other plants through competition. A pioneer species often undergo frequent turnover while long-lived species become more dominant over time, explaining the dynamics of successional changes in community structure (Lambers et al., 2008). Therefore, early-successional communities are comprised of short-lived species and have high rates of turnover and invasibility compared to later successional communities comprised of long-lived species that have low rates of succession and invisibility (Kneitel, 2010). Allelopathy is a phenomenon wherein plants release secondary compounds that suppress the growth of other plants. For instance, Solidago, Aster, and grass species exert negative interactions upon Prunus trees hindering succession from grass savannas to other successional stages. Altogether, these interactions determine the nature and succession processes of plant populations (Kong et al., 2023).

4.2 Plant-Animal

Plant-animal interactions may be mutualistic by enhancing plant reproduction and regeneration through pollination and dispersal services, or antagonistic by reducing plant fitness through seed predation or herbivory. In tropical rainforests, seed predation by insects, birds, and small rodents is high during early succession because of shelter offered by dense tangle of lianas (Paine & Beck, 2007). This predation can have a positive impact on species succession with predation rates that vary between 50% to 100% among tree species (Haavik & Stephen, 2023).

The process where insects or animals feed on plant materials is known as herbivory and can be classified as grazing where the animal feeds on grass and herbs while browsing feeds on twigs and leaves. It has a significant role in successional dynamics, especially for herbaceous vegetation with a higher rate than woody species as the former is more nutritious (Pacini et al., 2008). Early-successional species are more preferred by herbivores because they are usually more nutritious, palatable and less resistant to herbivores than the later-successional species. In such environments, herbivores can quickly reduce the biomass of plants to the stage of early succession and recycle nutrients in the late-successional species, which nourishes early-successional species. On the other hand, herbivores in infertile systems may act as catalysts of succession to eliminate palatable species and enhance the growth of unpalatable species with poor-quality leaves and litter (Kool et al., 2011).

The density of herbivores plays an important role in plant-animal dynamics and outcomes of successional processes. High livestock density or landscape integrity and conservation status influencing wild herbivore numbers can modify these interactions (Ciccazzo et al., 2016). Some pests, which include bark beetles and moths can pose serious disturbances through the reduction of most canopy trees within a short span. These are mostly associated with the boreal and northern temperate biomes, and the stands contain few associated tree species. For instance, the mountain pine beetle targets *Pinus contorta* in the USA while the European spruce bark beetle targets *Picea abies* in Europe (Haavik & Stephen, 2023; Thomas, 2023).

4.3 Plant-Microbe

Plants and microbes engage in intricate symbiotic relationships that profoundly influence plant growth, health, and resilience (Lyu et al., 2021). Mycorrhizal fungi form symbioses with the roots of most terrestrial plants, enhancing nutrient uptake, water retention, and stress tolerance. The interaction between rhizosphere bacteria and plant roots affects hormone signalling, defence mechanisms, and nutrient availability (Dastogeer et al., 2020; Huey et al., 2020). According to Scheu et al. (2005), microorganisms associated with plants can produce chemicals that promote plant development, improve soil aggregation, and provide protection against pathogens. Plant-microbe interactions, in which microorganisms and plants participate in intricate symbiotic relationships that have a significant impact on plant growth, health, and ecosystem functioning, are an intriguing and complex feature of terrestrial ecosystems. The interactions between plants and microorganisms influence the dynamics of plant communities and the well-being of terrestrial ecosystems, ranging from mutualistic partnerships that improve nutrient uptake to pathogenic interactions which trigger diseases (Chauhan et al., 2022). Plants and microbes interact in a wide range of ways. Some of these interactions are covered here, including symbiotic relationships, microbial-mediated food cycling, and the function of microorganisms in plant defence systems.

4.3.1 Symbiotic Associations

There are many symbiotic relationships in nature between bacteria and plants, with mutualistic interactions being important for the uptake of nutrients, resilience to stress, and ecology. The mutualistic connection between plants and mycorrhizal fungi is among the most researched instances of symbiotic relationships (Chauhan et al., 2022). Most terrestrial plants' roots have symbiotic relationships with mycorrhizal fungi, which together create structures called mycorrhizae. These relationships increase the efficiency of nutrient absorption and increase the effective root surface area, which improves nutrient uptake, especially of nitrogen and phosphorus (Huey et al., 2020).

Ectomycorrhizae, which envelop the plant's roots, and endomycorrhizae, which pierce the plant's root cells, are the two primary forms of mycorrhizal relationships (Huey et al., 2020). In temperate forests, ectomycorrhizal fungi are typically linked with trees, whereas endomycorrhizal fungi are present in a variety of plant species, such as grasses, legumes, and cereals. Mycorrhizal associations, in their various forms, improve nutrient uptake, water absorption, and tolerance to environmental challenges, hence augmenting plant growth and productivity (Dastogeer et al., 2020). Nitrogen-fixing bacteria are a significant class of symbiotic associations. They establish symbiotic interactions with the roots of specific plant species, especially legumes. These bacteria, referred to as rhizobia, settle inside the root nodules of leguminous plants and use nitrogen fixation to change atmospheric nitrogen into ammonia. The carbon molecules and ideal growing and activity environment that the plants give the bacteria in exchange. According to Pajares and Bohannan (2016), Lyu et al. (2021), and Paśmionka et al. (2021), nitrogen-fixing bacteria are essential for providing nitrogen to plants, boosting their development and production, and lowering the demand for synthetic fertilisers.

4.3.2 Microbial-Mediated Nutrient Cycling

In terrestrial ecosystems, microorganisms are important for nutrient cycling and biogeochemical transformations. They affect the availability and cycling of vital nutrients like phosphorus, nitrogen, and carbon. Through processes like mineralization, humification, and ammonification, bacteria and fungi are key players in the decomposition of organic matter, converting complex organic compounds into simpler forms. In the process, they discharge nutrients into the soil, where plants and other living things can absorb them, including phosphorus, sulphur, and nitrogen

(Wu et al., 2024). Apart from breaking down, specific types of microorganisms participate in the cycling of nitrogen, a process that transforms nitrogen into various chemical states like ammonium, nitrate, and nitrogen gas. As was previously mentioned, bacteria that fix nitrogen transform atmospheric nitrogen into ammonia, whereas bacteria that nitrate nitrogen convert ammonium into nitrite and nitrate (Jiao et al., 2018). On the other hand, nitrate is transformed into nitrogen gas by denitrifying bacteria, which completes the nitrogen cycle and releases nitrogen back into the atmosphere. Maintaining soil fertility, promoting plant growth, and preserving ecosystem functioning all depend on these microbially driven activities (Paśmionka et al., 2021). Similarly, microorganisms are crucial to the cycling of phosphorus, which releases phosphorus from minerals and organic materials and makes it available to plants and other living things. Enzymes produced by phosphorus-solubilizing bacteria and fungi degrade organic phosphorus compounds and release phosphorus from insoluble forms like apatite and rock phosphate. According to Tian et al. (2021), these bacteria increase the availability of phosphorus in the soil, which in turn stimulates plant growth and productivity in ecosystems where phosphorus is scarce.

4.3.3 Microbes in Plant Defense Mechanisms

Microbes are helpful, but they also have a significant impact on plant defence mechanisms, shielding plants from environmental stressors, diseases, and herbivores. Plant illnesses can be brought on by soil-borne pathogens including nematodes, bacteria, and fungi, which can lower crop yields and cause crop losses. To fend off pathogen invasion, plants have developed sophisticated defensive mechanisms, such as chemical defences, inducible immune responses, and physical barriers (Koza et al., 2022).

Induced systemic resistance (ISR), a phenomenon where plants respond to helpful bacteria by activating defence mechanisms that protect against pathogens, is one of the most well-known examples of plant-microbe interactions in defence mechanisms (Choudhary et al., 2007). Beneficial microorganisms, such as certain bacterial and fungal strains, colonise the rhizosphere, causing the plant to mount a defence against pathogens by inducing systemic reactions (Fig. 3). According to Nishad et al. (2020), this priming effect includes the synthesis of antimicrobial chemicals, the reinforcing of physical barriers, and the activation of defense-related genes, all of which lead to increased resistance to infections.

Plants have symbiotic relationships with endophytic fungi and bacteria in addition to ISR, which shields them from pathogens and herbivores. Endophytic microorganisms create secondary compounds that prevent pathogen growth and discourage herbivores, all while colonising the internal tissues of plants without causing illness. By enhancing plant resilience and health, these endophytic relationships support sustainable agricultural methods and lessen the need for chemical pesticides (White et al., 2019; Wilson et al., 2023).

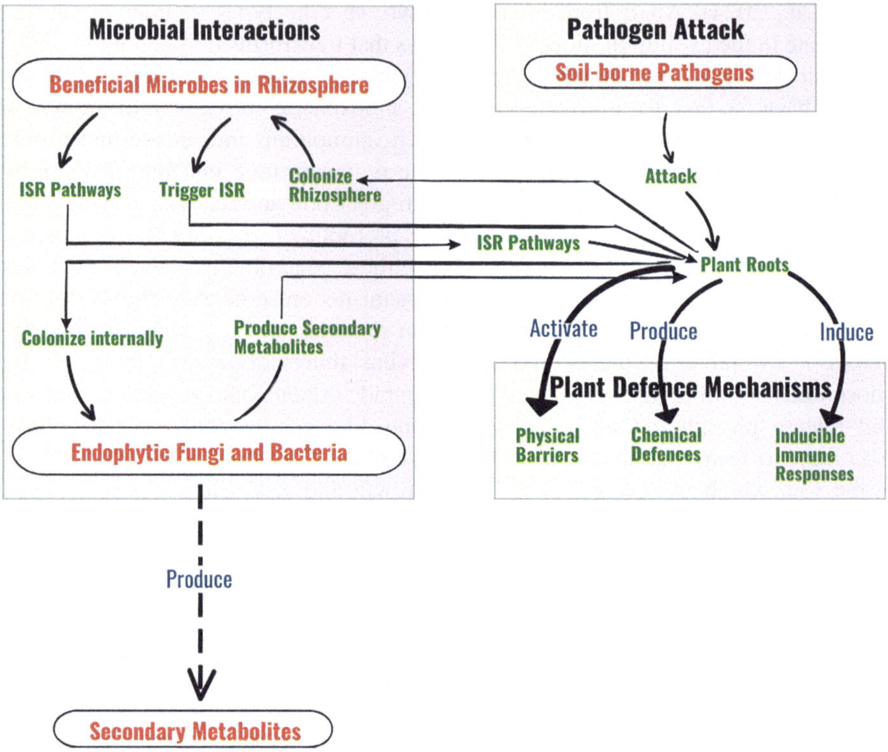

Fig. 3 Plant defense mechanisms against soil-borne pathogens

5 Impacts on Vegetation Development

The interactions between soil microbes, soil food webs, and plants exert significant influences on vegetation dynamics, composition, and succession. Microbial-mediated processes, such as nutrient cycling, organic matter decomposition, and nitrogen fixation, contribute to soil fertility and support plant establishment and growth (Fig. 4) (Wang et al., 2022). Moreover, microbial symbioses enhance plant resilience to environmental stresses, facilitating vegetation adaptation and succession in disturbed or degraded ecosystems (Wang et al., 2024). Vegetation development refers to the process by which plant communities evolve and change over time in response to various environmental factors, including climate, soil conditions, disturbance regimes, and biotic interactions. The development of vegetation communities is influenced by a myriad of factors, including the availability of resources, competition for space and light, and the presence of mutualistic or antagonistic interactions with other organisms (Sritharan et al., 2021). The impacts of various factors on vegetation development, including natural processes such as succession, as well as anthropogenic activities such as land use change and climate change were explored in this chapter.

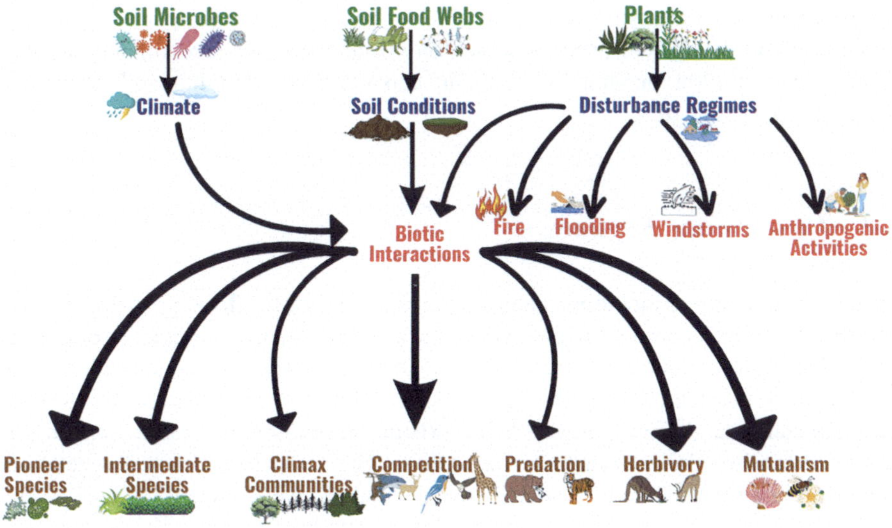

Fig. 4 Interactions and relationships in vegetation development and ecosystem dynamics

5.1 Succession

Succession is the process by which plant communities undergo predictable and orderly changes in species composition and structure over time in response to environmental disturbances or changes in environmental conditions. Succession can be initiated by a variety of factors, including disturbances such as fire, flooding, or volcanic eruptions, as well as gradual changes in climate or soil conditions (Poorter et al., 2024). Succession typically proceeds through a series of stages, beginning with pioneer species that are well adapted to colonize disturbed or newly formed habitats, followed by intermediate species that thrive in more stable conditions, and culminating in climax communities that represent the stable endpoint of succession under prevailing environmental conditions (Kneitel, 2010; Poorter et al., 2024).

One of the key drivers of vegetation development during succession is the availability of resources, particularly light, water, and nutrients. Pioneer species are often adapted to colonize open or disturbed habitats with abundant light and space, while later successional species may be better adapted to compete for limited resources in more mature or stable environments (Koza et al., 2022). Changes in soil conditions, such as the accumulation of organic matter or the development of soil horizons, can also influence vegetation development by altering nutrient availability, soil moisture, and pH (Sritharan et al., 2021).

Succession can have profound impacts on ecosystem structure and functioning, including changes in species diversity, productivity, and nutrient cycling (Ciccazzo et al., 2016). As plant communities transition through different stages of succession, they may exhibit changes in species composition, with early successional species being replaced by later successional species that are better adapted to prevailing environmental conditions (Poorter et al., 2024). These changes in species composition can affect ecosystem processes such as nutrient cycling, decomposition, and carbon sequestration, leading to shifts in ecosystem functioning and resilience (Ciccazzo et al., 2016).

5.2 Disturbance Regimes

Disturbance regimes, including natural events such as fire, flooding, windstorms, and insect outbreaks, as well as anthropogenic activities such as logging, agriculture, and urbanization, play important roles in shaping vegetation development and ecosystem dynamics (Burton et al., 2020). Disturbances can create opportunities for colonization and growth by removing existing vegetation or creating gaps in the canopy that allow light to reach the forest floor. They can also influence vegetation composition and structure by selecting species that are well-adapted to survive and reproduce under specific disturbance conditions (Philippot et al., 2021). The frequency, intensity, and severity of disturbances can vary widely depending on factors such as climate, topography, and land use history. In fire-prone ecosystems, for example, frequent wildfires may select for species that are adapted to survive and regenerate following fire, such as fire-adapted conifers or species with serotinous cones that require heat to release their seeds (McLauchlan et al., 2020). In floodplains, periodic flooding events may create dynamic and diverse habitats that support a wide range of plant species adapted to fluctuating water levels and nutrient availability (Burton et al., 2020). Anthropogenic activities can also have significant impacts on disturbance regimes and vegetation development, particularly through activities such as deforestation, land conversion, and habitat fragmentation. Deforestation, for example, can remove large areas of forest cover, disrupting natural disturbance regimes and altering species composition and structure. Similarly, urbanization can fragment natural habitats and create barriers to dispersal, leading to declines in species diversity and ecosystem resilience (Haddad et al., 2015; Burton et al., 2020).

5.3 Climate Change

Climate change is increasingly recognized as a major driver of vegetation development, with rising temperatures, altered precipitation patterns, and changing CO_2 levels influencing the distribution, composition, and productivity of plant communities around the world (Weiskopf et al., 2020). Climate changes can affect vegetation development through direct effects on plant physiology, such as changes in photosynthesis, water use efficiency, and growth rates, as well as indirect effects on habitat suitability, disturbance regimes, and species interactions (Pugnaire et al., 2019).

One of the most significant impacts of climate change on vegetation development is the shift in species distributions and phenology, with many plant species moving poleward or to higher elevations in response to warming temperatures. Alpine and Arctic ecosystems, for example, are experiencing rapid changes in vegetation composition and structure as cold-adapted species are replaced by more thermophilic species moving upslope in search of suitable habitat (Vanneste et al., 2017). Similarly, changes in precipitation patterns can influence the distribution of water-sensitive species and alter the productivity of ecosystems such as grasslands and savannas (Franklin et al., 2016).

Climate change can also interact with other drivers of vegetation development, such as disturbance regimes and invasive species, to amplify or mitigate their impacts on plant communities. For example, changes in fire regimes associated with climate change may increase the frequency and severity of wildfires in some regions, leading to shifts in vegetation composition and structure. Similarly, invasive species may thrive under changing climate conditions, outcompeting native species and altering ecosystem dynamics (Franklin et al., 2016; Weiskopf et al., 2020).

5.4 Biotic Interactions

Biotic interactions, including competition, predation, herbivory, and mutualism, play important roles in shaping vegetation development by influencing species distributions, population dynamics, and community structure (Wisz et al., 2013). Competition for resources such as light, water, and nutrients can influence the distribution and abundance of plant species, leading to changes in species composition and community structure. Similarly, predation and herbivory can affect plant growth and reproduction by reducing herbivore damage or altering plant defenses (Moënne-Loccoz et al., 2014). Mutualistic interactions, such as those between plants and pollinators, seed dispersers, or mycorrhizal fungi, can also influence vegetation development by enhancing plant reproduction, dispersal, or nutrient uptake. Pollinators such as bees, butterflies, and birds play essential roles in pollinating flowering plants and promoting seed production, while seed dispersers such as birds, mammals, and ants help to disperse seeds and colonize new habitats. Mycorrhizal fungi form symbiotic associations with plant roots, enhancing nutrient uptake and promoting plant growth and health (Huey et al., 2020).

6 Advancements in Molecular Biology and Ecological Modeling in Understanding the Role of Microbes in Soil Food Webs and Vegetation Development

The intricate dynamics of soil food webs and vegetation development are profoundly influenced by the diverse microbial communities residing in the soil. These microorganisms, which include bacteria, fungi, archaea, and protists, perform essential functions that drive nutrient cycling, organic matter decomposition, and energy flow within ecosystems (Wu et al., 2024). Recent advancements in molecular biology and ecological modelling have significantly enhanced our understanding of these microbial roles, providing detailed insights into the complex interactions within soil ecosystems and their impacts on vegetation (Braga et al., 2016; Lahlali et al., 2021; Lemke & DeSalle, 2023).

6.1 High-Throughput Sequencing Techniques

One of the most transformative advancements in studying soil microbiomes has been the development of high-throughput sequencing technologies. Techniques such as 16S rRNA gene sequencing, metagenomics, and metatranscriptomics have allowed scientists to explore the composition and functional potential of soil microbial communities with unprecedented depth and accuracy.

The 16S rRNA Gene Sequencing technique targets the 16S ribosomal RNA gene, a highly conserved genetic marker in bacteria and archaea, to identify and classify microorganisms present in soil samples (Dweh et al., 2023). By amplifying and sequencing this gene, researchers can generate a comprehensive profile of microbial diversity, revealing the presence and relative abundance of different taxa. This method has unveiled the immense diversity of soil microbes, many of which were previously unknown or unculturable, highlighting their potential roles in soil processes (Nam et al., 2023). Unlike 16S rRNA gene sequencing, metagenomics analyzes the entire genetic content of soil samples, providing a broader perspective on microbial community functions. This approach allows researchers to identify genes involved in critical ecological processes such as nitrogen fixation, phosphorus solubilization, and organic matter degradation. By examining these genetic blueprints, scientists can infer the functional capabilities of soil microbes and their contributions to nutrient cycling and soil health (Nam et al., 2023).

While metagenomics provides a snapshot of potential functions, metatranscriptomics reveals the active metabolic pathways within microbial communities by analyzing RNA transcripts. This technique captures the dynamic responses of soil microbes to environmental conditions, shedding light on how microbial activity varies with factors such as soil moisture, temperature, and nutrient availability. Metatranscriptomic studies have been instrumental in understanding how microbes adapt to changing environments and their roles in real-time ecological processes (Mukherjee & Reddy, 2020; Garg et al., 2023).

6.2 Functional Insights and Ecological Roles

The application of these molecular techniques has led to significant discoveries about the functional roles of soil microbes. For example, high-throughput sequencing has identified key microbial players in the decomposition of complex organic compounds, such as lignin and cellulose, which are crucial for nutrient recycling and soil organic matter formation. Fungi, particularly mycorrhizal species, have been found to form symbiotic relationships with plant roots, extending their hyphal networks to enhance nutrient absorption, especially phosphorus and nitrogen. In return, plants supply carbon compounds to the fungi, fostering a mutualistic relationship that supports plant growth and soil health (Lahlali et al., 2021). Bacteria, such as those in the genera Bacillus and Pseudomonas, have been shown to produce enzymes that break down organic matter, releasing essential nutrients that plants can readily absorb. These bacteria also engage in biocontrol activities, producing antibiotics and secondary metabolites that protect plants from pathogens (Aransiola et al., 2022). Additionally, nitrogen-fixing bacteria, such as Rhizobium and Azotobacter, convert atmospheric nitrogen into forms usable by plants, playing a vital role in maintaining soil fertility and supporting plant growth (Aasfar et al., 2021; Timofeeva et al., 2023). Advancements in molecular biology have also highlighted the importance of microbial interactions within soil food webs. These interactions form complex networks of predator-prey relationships, nutrient exchanges, and cooperative behaviours that sustain soil ecosystems. For example, protozoa and nematodes feed on bacteria and fungi, regulating microbial populations and influencing nutrient availability through their excretory processes. This trophic regulation ensures a balanced microbial community structure, promoting soil stability and resilience (Garg et al., 2023).

6.3 Ecological Modeling and Predictive Insights

Complementing molecular techniques, ecological modelling has become a powerful tool in understanding and predicting the roles of microbes in soil food webs and vegetation development. Models that integrate microbial processes with environmental variables provide valuable insights into how soil ecosystems respond to various factors, including climate change, land use practices, and disturbances.

Dynamic Simulation Models These models simulate the interactions between soil microbes, plants, and environmental conditions over time (Chandel et al., 2023). By incorporating data from molecular studies, such as metagenomic and metatranscriptomic profiles, dynamic models can predict changes in microbial community structure and function in response to environmental perturbations. For instance, models can forecast how shifts in temperature and precipitation patterns due to climate change will impact microbial-mediated nutrient cycling and plant growth (Rubio & Barreto-Hernández, 2022).

Nutrient Cycling Models These models focus on the pathways and rates of nutrient transformations within soil ecosystems. By integrating microbial activity data, nutrient cycling models can predict how changes in microbial community composition affect the availability of essential nutrients, such as nitrogen and phosphorus, for plant uptake. This information is crucial for developing sustainable agricultural practices and managing soil health (Wang & Goll, 2021; Chandel et al., 2023).

Plant-Microbe Interaction Models These models explore the symbiotic relationships between plants and soil microbes, such as mycorrhizal associations and nitrogen-fixing symbioses. By simulating the exchange of nutrients and signalling compounds between plants and microbes, these models provide insights into how these interactions influence plant productivity, resilience, and adaptation to environmental stresses (Afkhami et al., 2020; Shi et al., 2024).

6.4 Applications and Future Direction

The integration of molecular biology and ecological modelling has profound implications for ecosystem management, agriculture, and conservation, offering insights that can enhance soil health, optimize nutrient management, and promote sustainable land use practices. In sustainable agriculture, understanding soil microbial functions facilitates the development of microbial-based solutions to boost crop productivity and soil fertility, such as inoculating crops with beneficial microbes like mycorrhizal fungi and nitrogen-fixing bacteria, thereby improving nutrient uptake and reducing reliance on chemical fertilizers (Tahat et al., 2019; Sprunger & Martin, 2022). This approach supports sustainable agriculture while mitigating the environmental impact of intensive farming practices. In ecosystem restoration, molecular and modelling insights guide efforts by identifying key microbial communities that support ecosystem recovery, such as reintroducing specific microbial consortia to degraded soils to accelerate nutrient cycle restoration and enhance ecosystem resilience and stability. For climate change mitigation, predictive models incorporating microbial dynamics assess impacts on soil ecosystems and vegetation, forecasting shifts in microbial-mediated processes and informing adaptive management strategies to protect soil health and plant productivity (Tiedje et al., 2022). Despite significant progress, challenges remain in fully understanding microbes' intricate roles in soil food webs and vegetation development due to soil environment complexity and microbial functional diversity. Future research should focus on interdisciplinary collaboration among microbiologists, ecologists, and soil scientists to integrate molecular data with ecological models, establish long-term monitoring programs to track microbial community changes and functions over time and continue developing advanced molecular techniques like single-cell genomics and proteomics to further unravel microbial interactions and their ecological roles.

7 Conclusions

A soil microbiome is a complex assemblage of microorganisms that is capable of sustaining terrestrial ecosystems, plant growth and productivity. These microbes control vital activities including nutrient cycling, decomposition of organic matter as well as energy flow in the soil food web affecting plant growth and the ecosystem.

Increased recognition of the variety and versatility of soil microbes, as well as molecular tools and ecological modelling, have indicated the importance of these microorganisms in ecosystems and the potential for predicting the effects of environmental change. Nevertheless, there are still some limitations in the actual comprehension of the multiple interactions occurring in a soil ecosystem, and how these may be affected by climate change, changes in land use, and invasions of alien species. To manage these challenges, it is crucial to involve cross-disciplinary research and to seek new approaches to management. By enhancing our existing knowledge base of how the numerous kinds of microbes within soil function and interact, scientists can design and recommend proper management practices for the land whilst promoting resistance and stability in the natural world.

References

Aasfar, A., Bargaz, A., Yaakoubi, K., Hilali, A., Bennis, I., Zeroual, Y., & Kadmiri, I. M. (2021). Nitrogen fixing Azotobacter species as potential soil biological enhancers for crop nutrition and yield stability. *Frontiers in Microbiology, 12*. https://doi.org/10.3389/fmicb.2021.628379

Afkhami, M. E., Almeida, B. K., Hernandez, D. J., Kiesewetter, K. N., & Revillini, D. P. (2020). Tripartite mutualisms as models for understanding plant–microbial interactions. *Current Opinion in Plant Biology, 56*, 28–36. https://doi.org/10.1016/j.pbi.2020.02.003

Aguiar-Pulido, V., Huang, W., Suarez-Ulloa, V., Cickovski, T., Mathee, K., & Narasimhan, G. (2016). Metagenomics, metatranscriptomics, and metabolomics approaches for microbiome analysis. *Evolutionary Bioinformatics Online, 12*(Suppl 1), 5–16. https://doi.org/10.4137/EBO.S36436

Aransiola, S. A., Victor-Ekwebelem, M. O., & Maddela, N. R. (2022). Microbial enzymes for sustainable development – Future guidelines. In N. R. Maddela, S. A. Aransiola, & R. Prasad (Eds.), *Ecological interplays in microbial enzymology* (1st ed.). Springer Nature Singapore Pte Ltd. ISSN: 2662-1681. https://www.springer.com/series/16324

Berg, G., Rybakova, D., Fischer, D., Cernava, T., Vergès, C. C., Charles, T., Chen, X., Cocolin, L., Eversole, K., Corral, G. H., Kazou, M., Kinkel, L., Lange, L., Lima, N., Loy, A., Macklin, J. A., Maguin, E., Mauchline, T., McClure, R., et al. (2020). Microbiome definition re-visited: Old concepts and new challenges. *Microbiome, 8*. https://doi.org/10.1186/s40168-020-00875-0

Braga, R. M., Dourado, M. N., & Araújo, W. L. (2016). Microbial interactions: Ecology in a molecular perspective. *Brazilian Journal of Microbiology, 47*, 86–98. https://doi.org/10.1016/j.bjm.2016.10.005

Briones, M. J. (2014). Soil fauna and soil functions: A jigsaw puzzle. *Frontiers in Environmental Science, 2*, 85955. https://doi.org/10.3389/fenvs.2014.00007

Brooker, R. W. (2006). Plant–plant interactions and environmental change. *New Phytologist, 171*(2), 271–284. https://doi.org/10.1111/j.1469-8137.2006.01752.x

Burton, P. J., Jentsch, A., & Walker, L. R. (2020). The ecology of disturbance interactions. *Bioscience, 70*(10), 854–870. https://doi.org/10.1093/biosci/biaa088

Chamkhi, I., Omari, N. E., Balahbib, A., Menyiy, N. E., Benali, T., & Ghoulam, C. (2022). Is the rhizosphere a source of applicable multi-beneficial microorganisms for plant enhancement? *Saudi Journal of Biological Sciences, 29*(2), 1246–1259. https://doi.org/10.1016/j.sjbs.2021.09.032

Chandel, A. K., Jiang, L., & Luo, Y. (2023). Microbial models for simulating soil carbon dynamics: A review. *Journal of Geophysical Research: Biogeosciences, 128*(8), e2023JG007436. https://doi.org/10.1029/2023JG007436

Chauhan, P., Sharma, N., Tapwal, A., Kumar, A., Verma, G. S., Meena, M., Seth, C. S., & Swapnil, P. (2022). Soil microbiome: Diversity, benefits and interactions with plants. *Sustainability, 15*(19), 14643. https://doi.org/10.3390/su151914643

Choudhary, D. K., Prakash, A., & Johri, B. N. (2007). Induced systemic resistance (ISR) in plants: Mechanism of action. *Indian Journal of Microbiology, 47*(4), 289–297. https://doi.org/10.1007/s12088-007-0054-2

Chu, H., Gao, F., Ma, Y., Fan, K., & Delgado-Baquerizo, M. (2020). Soil microbial biogeography in a changing world: Recent advances and future perspectives. *MSystems, 5*(2). https://doi.org/10.1128/mSystems.00803-19

Ciccazzo, S., Esposito, A., & Borruso, L. (2016). Microbial communities and primary succession in high altitude mountain environments. *Annales de Microbiologie, 66*, 43–60. https://doi.org/10.1007/s13213-015-1130-1

Clocchiatti, A., Hannula, S. E., Van den Berg, M., Korthals, G., & De Boer, W. (2020). The hidden potential of saprotrophic fungi in arable soil: Patterns of short-term stimulation by organic amendments. *Applied Soil Ecology, 147*, 103434. https://doi.org/10.1016/j.apsoil.2019.103434

Dastogeer, K. M., Tumpa, F. H., Sultana, A., Akter, M. A., & Chakraborty, A. (2020). Plant microbiome–an account of the factors that shape community composition and diversity. *Current Plant Biology, 23*, 100161. https://doi.org/10.1016/j.cpb.2020.100161

Diwan, D., Rashid, M. M., & Vaishnav, A. (2022). Current understanding of plant-microbe interaction through the lenses of multi-omics approaches and their benefits in sustainable agriculture. *Microbiological Research, 265*, 127180. https://doi.org/10.1016/j.micres.2022.127180

Du, X. F., Li, Y. B., Liu, F., Su, X. L., & Li, Q. (2018). Structure and ecological functions of soil micro-food web. *The Journal of Applied Ecology, 29*(2), 403–411. https://doi.org/10.13287/j.1001-9332.201802.033

Dweh, T. J., Pattnaik, S., & Sahoo, J. P. (2023). Assessing the impact of meta-genomic tools on current cutting-edge genome engineering and technology. *International Journal of Biochemistry and Molecular Biology, 14*(4), 62–75. https://www.ncbi.nlm.nih.gov/pmc/articles/PMC10509535/

Franklin, J. M. J., Syphard, A. D., & Regan, H. M. (2016). Global change and terrestrial plant community dynamics. *Proceedings of the National Academy of Sciences, 113*(14), 3725–3734. https://doi.org/10.1073/pnas.1519911113

Gabaldón, T. (2019). Fungal evolution: Major ecological adaptations and evolutionary transitions. *Biological Reviews of the Cambridge Philosophical Society, 94*(4), 1443–1476. https://doi.org/10.1111/brv.12510

Garg, D., Patel, N., Rawat, A., & Rosado, A. S. (2023). Cutting-edge tools in the field of soil microbiology. *Current Research in Microbial Sciences, 6*, 100226. https://doi.org/10.1016/j.crmicr.2024.100226

Gupta, A., Gupta, R., & Singh, R. L. (2017). Microbes and environment. In *Principles and applications of environmental biotechnology for a sustainable future* (pp. 43–84). Springer. https://doi.org/10.1007/978-981-10-1866-4_3

Haavik, L. J., & Stephen, F. M. (2023). Insect ecology. In D. J. Allison, T. D. Paine, B. Slippers, & M. J. Wingfield (Eds.), *Forest entomology and pathology*. Springer. https://doi.org/10.1007/978-3-031-11553-0_4

Haddad, N. M., Brudvig, L. A., Clobert, J., Davies, K. F., Gonzalez, A., Holt, R. D., Lovejoy, T. E., Sexton, J. O., Austin, M. P., Collins, C. D., Cook, W. M., Damschen, E. I., Ewers, R. M., Foster, B. L., Jenkins, C. N., King, A. J., Laurance, W. F., Levey, D. J., Margules, C. R., et al. (2015). Habitat fragmentation and its lasting impact on Earth's ecosystems. *Science Advances, 1*(2), e1500052. https://doi.org/10.1126/sciadv.1500052

Hischier, C. M., Hille Ris Lambers, J., & Iseli, E. (2023). Positive and negative plant−plant interactions influence seedling establishment at both high and low elevations. *Alpine Botany, 134*, 15–27. https://doi.org/10.1007/s00035-023-00302-8

Huey, C. J., Gopinath, S. C., Uda, M. N., Zulhaimi, H. I., Jaafar, M. N., Kasim, F. H., & Wan Yaakub, A. R. (2020). Mycorrhiza: A natural resource that assists plant growth under varied soil conditions. *3 Biotech, 10*(5), 204. https://doi.org/10.1007/s13205-020-02188-3

Jain, A., Sarsaiya, S., Wu, Q., Lu, Y., & Shi, J. (2019). A review of plant leaf fungal diseases and its environment speciation. *Bioengineered, 10*(1), 409–424. https://doi.org/10.1080/21655979.2019.1649520

Jiao, S., Chen, W., & Wang, J. (2018). Soil microbiomes with distinct assemblies through vertical soil profiles drive the cycling of multiple nutrients in reforested ecosystems. *Microbiome, 6*, 146. https://doi.org/10.1186/s40168-018-0526-0

Kneitel, J. (2010). Successional changes in communities. *Nature Education Knowledge, 3*(10), 41.

Kong, C., Li, Z., Li, F., Xia, X., & Wang, P. (2023). Chemically mediated plant–plant interactions: Allelopathy and allelobiosis. *Plants, 13*(5), 626. https://doi.org/10.3390/plants13050626

Kool, A., Meijer, S. S., Kohi, E., Heitkönig, I. M., Grant, R. C., Peel, M. J., Slotow, R., & Prins, H. H. (2011). Large herbivores may alter the vegetation structure of semi-arid savannas through soil nutrient mediation. *Oecologia, 165*(4), 1095–1107. https://doi.org/10.1007/s00442-010-1899-3

Koza, N. A., Adedayo, A. A., Babalola, O. O., & Kappo, A. P. (2022). Microorganisms in plant growth and development: Roles in abiotic stress tolerance and secondary metabolites secretion. *Microorganisms, 10*(8). https://doi.org/10.3390/microorganisms10081528

Lahlali, R., Ibrahim, S. S., Belabess, Z., Kadir Roni, M. Z., Radouane, N., Vicente, S. L., Menéndez, E., Mokrini, F., Barka, E. A., & Peng, G. (2021). High-throughput molecular technologies for unraveling the mystery of soil microbial community: Challenges and future prospects. *Heliyon, 7*(10), e08142. https://doi.org/10.1016/j.heliyon.2021.e08142

Lambers, H., Chapin, F. S., & Pons, T. L. (2008). Interactions among plants. In *Plant physiological ecology*. Springer. https://doi.org/10.1007/978-0-387-78341-3_16

Lemke, M., & DeSalle, R. (2023). The next generation of microbial ecology and its importance in environmental sustainability. *Microbial Ecology, 85*(3), 781–795. https://doi.org/10.1007/s00248-023-02185-y

Ling, N., Wang, T., & Kuzyakov, Y. (2022). Rhizosphere bacteriome structure and functions. *Nature Communications, 13*, 836. https://doi.org/10.1038/s41467-022-28448-9

Lladó, S., López-Mondéjar, R., & Baldrian, P. (2017). Forest soil bacteria: Diversity, involvement in ecosystem processes, and response to global change. *Microbiology and Molecular Biology Reviews, 81*(2). https://doi.org/10.1128/MMBR.00063-16

Lyu, D., Msimbira, L. A., Nazari, M., Antar, M., Pagé, A., Shah, A., Monjezi, N., Zajonc, J., Tanney, C. A., Backer, R., & Smith, D. L. (2021). The coevolution of plants and microbes underpins sustainable agriculture. *Microorganisms, 9*(5). https://doi.org/10.3390/microorganisms9051036

Malmstrom, C. (2010). Ecologists study the interactions of organisms and their environment. *Nature Education Knowledge, 3*(10), 88.

McLauchlan, K. K., Higuera, P. E., Miesel, J., Rogers, B. M., Schweitzer, J., Shuman, J. K., Tepley, A. J., Varner, J. M., Veblen, T. T., Adalsteinsson, S. A., Balch, J. K., Baker, P., Batllori, E., Bigio, E., Brando, P., Cattau, M., Chipman, M. L., Coen, J., Crandall, R., et al. (2020). Fire as a fundamental ecological process: Research advances and frontiers. *Journal of Ecology, 108*(5), 2047–2069. https://doi.org/10.1111/1365-2745.13403

Mishra, A., Singh, L., & Singh, D. (2023). Unboxing the black box—One step forward to understand the soil microbiome: A systematic review. *Microbial Ecology, 85*(2), 669–683. https://doi.org/10.1007/s00248-022-01962-5

Moënne-Loccoz, Y., Mavingui, P., Combes, C., Normand, P., & Steinberg, C. (2014). Microorganisms and biotic interactions. In *Environmental microbiology: Fundamentals and applications* (pp. 395–444). Springer. https://doi.org/10.1007/978-94-017-9118-2_11

Mukherjee, A., & Reddy, M. S. (2020). Metatranscriptomics: An approach for retrieving novel eukaryotic genes from polluted and related environments. *3 Biotech, 10*(2), 71. https://doi.org/10.1007/s13205-020-2057-1

Nam, N. N., Khoa Do, H. D., Loan Trinh, K. T., & Lee, N. Y. (2023). Metagenomics: An effective approach for exploring microbial diversity and functions. *Food, 12*(11). https://doi.org/10.3390/foods12112140

Nishad, R., Ahmed, T., Rahman, V. J., & Kareem, A. (2020). Modulation of plant defense system in response to microbial interactions. *Frontiers in Microbiology, 11*, 1298. https://doi.org/10.3389/fmicb.2020.01298

Odelade, K. A., & Babalola, O. O. (2019). Bacteria, fungi and archaea domains in Rhizospheric soil and their effects in enhancing agricultural productivity. *International Journal of Environmental Research and Public Health, 16*(20). https://doi.org/10.3390/ijerph16203873

Pacini, E., Viegi, L., & Franchi, G. G. (2008). Types, evolution and significance of plant-animal interactions. *Rendiconti Lincei. Scienze Fisiche e Naturali, 19*, 75–101. https://doi.org/10.1007/s12210-008-0005-9

Paine, C. E., & Beck, H. (2007). Seed predation by Neotropical rainforest mammals increases diversity in seedling recruitment. *Ecology, 88*(12), 3076–3087. https://doi.org/10.1890/06-1835.1

Pajares, S., & Bohannan, B. J. (2016). Ecology of nitrogen fixing, nitrifying, and denitrifying microorganisms in tropical forest soils. *Frontiers in Microbiology, 7*. https://doi.org/10.3389/fmicb.2016.01045

Palevsky, E., & Ruess, L. (2022). Soil nematodes as a means of conservation of soil predatory mites for biocontrol. *Agronomy, 13*(1), 32. https://doi.org/10.3390/agronomy13010032

Paśmionka, I. B., Bulski, K., & Boligłowa, E. (2021). The participation of microbiota in the transformation of nitrogen compounds in the soil—A review. *Agronomy, 11*(5), 977. https://doi.org/10.3390/agronomy11050977

Philippot, L., Griffiths, B. S., & Langenheder, S. (2021). Microbial community resilience across ecosystems and multiple disturbances. *Microbiology and Molecular Biology Reviews, 85*(2). https://doi.org/10.1128/MMBR.00026-20

Poorter, L., Amissah, L., Bongers, F., Hordijk, I., Kok, J., Laurance, S. G., Martínez-Ramos, M., Matsuo, T., Meave, J. A., Muñoz, R., Peña-Claros, M., Herault, B., Jakovac, C. C., Lebrija-Trejos, E., Norden, N., & Lohbeck, M. (2024). A comprehensive framework for vegetation succession. *Ecosphere, 15*(4), e4794. https://doi.org/10.1002/ecs2.4794

Potapov, A. M. (2022). Multifunctionality of belowground food webs: Resource, size and spatial energy channels. *Biological Reviews, 97*(4), 1691–1711. https://doi.org/10.1111/brv.12857

Potapov, A. M., Beaulieu, F., Birkhofer, K., Bluhm, S. L., Degtyarev, M. I., Devetter, M., Goncharov, A. A., Gongalsky, K. B., Klarner, B., Korobushkin, D. I., Liebke, D. F., Maraun, M., Mc Donnell, R. J., Pollierer, M. M., Schaefer, I., Shrubovych, J., Semenyuk, I. I., Sendra, A., Tuma, J., et al. (2022). Feeding habits and multifunctional classification of soil-associated consumers from protists to vertebrates. *Biological Reviews, 97*(3), 1057–1117. https://doi.org/10.1111/brv.12832

Pugnaire, F. I., Morillo, J. A., Peñuelas, J., Reich, P. B., Bardgett, R. D., Gaxiola, A., & Wardle, D. A. (2019). Climate change effects on plant-soil feedbacks and consequences for biodiversity and functioning of terrestrial ecosystems. Science. *Advances, 5*(11). https://doi.org/10.1126/sciadv.aaz1834

Rubio, F. D., & Barreto-Hernández, E. (2022). Simulating microbial functional diversity dynamics in agricultural soils: An individual based modeling approach. *Advances in Bioscience and Biotechnology, 13*, 159–174. https://doi.org/10.4236/abb.2022.133008

Saeed, Q., Xiukang, W., Haider, F. U., Kučerik, J., Mumtaz, M. Z., Holatko, J., Naseem, M., Kintl, A., Ejaz, M., Naveed, M., Brtnicky, M., & Mustafa, A. (2021). Rhizosphere bacteria in plant growth promotion, biocontrol, and bioremediation of contaminated sites: A comprehensive review of effects and mechanisms. *International Journal of Molecular Sciences, 22*(19). https://doi.org/10.3390/ijms221910529

Scheu, S., Ruess, L., & Bonkowski, M. (2005). Interactions between microorganisms and soil micro- and mesofauna. In F. Buscot & A. Varma (Eds.), *Microorganisms in soils: Roles in genesis and functions* (pp. 253–278). Springer.

Shi, X., Zhao, Y., Xu, M., Ma, L., Adams, J. M., & Shi, Y. (2024). Insights into plant–microbe interactions in the rhizosphere to promote sustainable agriculture in the new crops era. *New Crops, 1*, 100004. https://doi.org/10.1016/j.ncrops.2023.11.002

Sofo, A., Mininni, A. N., & Ricciuti, P. (2020). Soil macrofauna: A key factor for increasing soil fertility and promoting sustainable soil use in fruit orchard Agrosystems. *Agronomy, 10*(4), 456. https://doi.org/10.3390/agronomy10040456

Sprunger, C. D., & Martin, T. K. (2022). An integrated approach to assessing soil biological health. *Advances in Agronomy, 182*, 131–168. https://doi.org/10.1016/bs.agron.2023.06.003

Sritharan, M. S., Scheele, B. C., Blanchard, W., & Lindenmayer, D. B. (2021). Spatial associations between plants and vegetation community characteristics provide insights into the processes influencing plant rarity. *PLoS One, 16*(12), e0260215. https://doi.org/10.1371/journal.pone.0260215

Tahat, M. M., Alananbeh, K. M., & Othman, Y. A. (2019). Soil health and sustainable agriculture. *Sustainability, 12*(12), 4859. https://doi.org/10.3390/su12124859

Thepbandit, W., & Athinuwat, D. (2024). Rhizosphere microorganisms supply availability of soil nutrients and induce plant defense. *Microorganisms, 12*(3), 558. https://doi.org/10.3390/microorganisms12030558

Thomas, S. C. (2023). Insects and Forest succession. In D. J. Allison, T. D. Paine, B. Slippers, & M. J. Wingfield (Eds.), *Forest entomology and pathology*. Springer. https://doi.org/10.1007/978-3-031-11553-0_8

Tian, J., Ge, F., Zhang, D., Deng, S., & Liu, X. (2021). Roles of phosphate solubilizing microorganisms from managing soil phosphorus deficiency to mediating biogeochemical P cycle. *Biology, 10*(2). https://doi.org/10.3390/biology10020158

Tiedje, J. M., Bruns, M. A., Casadevall, A., Criddle, C. S., Eloe-Fadrosh, E., Karl, D. M., Nguyen, N. K., & Zhou, J. (2022). Microbes and climate change: A research prospectus for the future. *mBio, 13*, e0080022. https://doi.org/10.1128/mbio.00800-22

Timofeeva, A. M., Galyamova, M. R., & Sedykh, S. E. (2023). Plant growth-promoting soil bacteria: Nitrogen fixation, phosphate solubilization, siderophore production, and other biological activities. *Plants, 12*(24). https://doi.org/10.3390/plants12244074

Vanneste, T., Michelsen, O., & Graae, B. J. (2017). Impact of climate change on alpine vegetation of mountain summits in Norway. *Ecological Research, 32*, 579–593. https://doi.org/10.1007/s11284-017-1472-1

Wagg, C., Schlaeppi, K., Banerjee, S., & Kuramae, E. E. (2019). Fungal-bacterial diversity and microbiome complexity predict ecosystem functioning. *Nature Communications, 10*, 4841. https://doi.org/10.1038/s41467-019-12798-y

Wang, P., & Goll, D. S. (2021). Modelling of land nutrient cycles: Recent progress and future development. *Faculty Reviews, 10*. https://doi.org/10.12703/r/10-53

Wang, J., Kan, J., Zhang, X., Xia, Z., Zhang, X., Qian, G., Miao, Y., Leng, X., & Sun, J. (2017). Archaea dominate the ammonia-oxidizing community in deep-sea Sediments of the eastern Indian Ocean—From the equator to the Bay of Bengal. *Frontiers in Microbiology, 8*, 415. https://doi.org/10.3389/fmicb.2017.00415

Wang, G., Ren, Y., Bai, X., Su, Y., & Han, J. (2022). Contributions of beneficial microorganisms in soil remediation and quality improvement of medicinal plants. *Plants, 11*(23). https://doi.org/10.3390/plants11233200

Wang, X., Chi, Y., & Song, S. (2024). Important soil microbiota's effects on plants and soils: A comprehensive 30-year systematic literature review. *Frontiers in Microbiology, 15*. https://doi.org/10.3389/fmicb.2024.1347745

Weiskopf, S. R., Rubenstein, M. A., Crozier, L. G., Gaichas, S., Griffis, R., Halofsky, J. E., Hyde, K. J., Morelli, T. L., Morisette, J. T., Muñoz, R. C., Pershing, A. J., Peterson, D. L., Poudel, R., Staudinger, M. D., Sutton-Grier, A. E., Thompson, L., Vose, J., Weltzin, J. F., & Whyte, K. P. (2020). Climate change effects on biodiversity, ecosystems, ecosystem services, and natural resource management in the United States. *Science of the Total Environment, 733*, 137782. https://doi.org/10.1016/j.scitotenv.2020.137782

White, J. F., Kingsley, K. L., Zhang, Q., Verma, R., Obi, N., Dvinskikh, S., Elmore, M. T., Verma, S. K., Gond, S. K., & Kowalski, K. P. (2019). Review: Endophytic microbes and their potential applications in crop management. *Pest Management Science, 75*(10), 2558–2565. https://doi.org/10.1002/ps.5527

Wilson, S. K., Pretorius, T., & Naidoo, S. (2023). Mechanisms of systemic resistance to pathogen infection in plants and their potential application in forestry. *BMC Plant Biology, 23*, 404. https://doi.org/10.1186/s12870-023-04391-9

Wisz, M. S., Pottier, J., Kissling, W. D., Pellissier, L., Lenoir, J., Damgaard, C. F., Dormann, C. F., Forchhammer, M. C., Grytnes, A., Guisan, A., Heikkinen, R. K., Høye, T. T., Kühn, I., Luoto, M., Maiorano, L., Nilsson, C., Normand, S., Öckinger, E., Schmidt, N. M., et al. (2013). The role of biotic interactions in shaping distributions and realised assemblages of species: Implications for species distribution modelling. *Biological Reviews of the Cambridge Philosophical Society, 88*(1), 15–30. https://doi.org/10.1111/j.1469-185X.2012.00235.x

Wu, H., Cui, H., Fu, C., Li, R., Qi, F., Liu, Z., Yang, G., Xiao, K., & Qiao, M. (2024). Unveiling the crucial role of soil microorganisms in carbon cycling: A review. *Science of the Total Environment, 909*, 168627. https://doi.org/10.1016/j.scitotenv.2023.168627

Xiao, F., Li, Y., Li, G., He, Y., Lv, X., Zhuang, L., & Pu, X. (2021). High throughput sequencing-based analysis of the soil bacterial community structure and functions of Tamarix shrubs in the lower reaches of the Tarim River. *PeerJ, 9*, e12105. https://doi.org/10.7717/peerj.12105

Zhalnina, K., Dörr de Quadros, P., Camargo, F. A. O., & Triplett, E. W. (2012). Drivers of archaeal ammonia-oxidizing communities in soil. *Frontiers in Microbiology, 3*, 210. https://doi.org/10.3389/fmicb.2012.00210

Soil Microbial Community Dynamics and Plant Proliferation

Anjaneyulu Musini, P Gnana Deepu, Kathuroju Harikrishna, Patnam Prashanth Kumar, Surisetty Navyatha, and Jogipeta Harihara

Abstract In Agricultural sector, with the change in environment and climates has affected the diverse ecosystem and microbial life present in the soil. Soil microbial communities have directly or indirectly impacted the proliferation of plant systems. These communities contribute uniquely to ecological balance of soil, due to presence of variety of organisms such as bacteria, archaea, protozoa, fungi, etc. These communities can directly impact plant growth and development by producing growth-promoting hormones, nutrient acquisition, and protection against pathogens. Various factors like Temperature, pH, organic matter, and nutrient availability effects the action and mechanism of these microbial communities. Understanding their dynamics and interactions with plants is crucial for sustainable agriculture, nutrient cycling and ecosystem stability. In this chapter, we explore the role of soil microorganisms in plant proliferation, interactions between plants and soil microbial community, management strategies for soil microbial community, etc.

Keywords Soil microbial communities · Plant proliferation · Bacteria · Fungi · Archaea

1 Introduction

Within Earth's diverse ecosystems, soil teems with a multitude of microbial life. These microorganisms engage in complex interactions that significantly impact plant health and productivity. Soil microbial communities are intricate and captivating components of soil ecosystems. They play essential roles in decomposition of organic matter, nutrient cycling, and overall soil health. These communities consist of diverse organisms such as bacteria, archaea, protozoa and fungi each contributing uniquely to the ecological balance of the soil (Yue et al., 2024). Understanding the dynamics of soil microbial communities is crucial for unraveling the intricate

A. Musini (✉) · P G. Deepu · K. Harikrishna · P. P. Kumar · S. Navyatha · J. Harihara
Center for Biotechnology, University College of Engineering Science and Technology, JNTUH, Hyderabad, Telangana, India
e-mail: manjaneyulu@jntuh.ac.in

S. A. Aransiola et al. (eds.), *Soil Microbiome in Green Technology Sustainability*,
https://doi.org/10.1007/978-3-031-71844-1_5

workings of ecosystems. They are known to demonstrate dynamic responses to changes in environmental conditions, such as temperature, pH, soil moisture, availability of nutrients and organic carbon and the presence of plants (Chu et al., 2024). Moreover, soil microbes play a crucial role as intermediaries in plant-soil interactions, coordinating a wide range of biochemical processes that influence plant growth and proliferation (Berendsen et al., 2012).

These microorganisms engage with plants within the rhizosphere, a confined soil region influenced by root secretions and the presence of associated soil microorganisms (Chu et al., 2024). Plants impact the soil microbial community by releasing organic litter and root exudates, which serve as nourishment for microbes. This dynamic interaction establishes a feedback loop where plant health and growth are influenced by soil microbes, and, in turn, the microbes are affected by the presence of plants (Ke & Miki, 2015). Soil microbial communities have a multifaceted impact on plant proliferation, as microorganisms play vital roles in supporting plant growth and development. For instance, mycorrhizal fungi establish symbiotic relationships with plant roots, enhancing the plants' capacity to absorb water and nutrients, especially phosphorus, from the soil (Ames & Bethlenfalvay, 1987). Microorganisms establish interactions through various mechanisms. Plant residues, such as roots and leaves, act as a primary carbon source, sustaining microbial communities in the soil. These microorganisms play a crucial role in releasing nutrients through metabolic processes, thereby facilitating soil development and supporting plant growth (Wang et al., 2024). Additionally, through mycorrhizal symbiosis, stress signaling and hormone secretion, soil microorganisms directly influence plant growth and development (Kaur et al., 2021). The complex relationships between soil microbial communities and plant growth have been the subject of extensive research in the field of plant ecology, with numerous studies exploring the intricate mechanisms by which these interactions shape the success and resilience of terrestrial ecosystems (Hu et al., 2018).

Recent advances in molecular ecology have unveiled intricate microbial mechanisms that impact plant growth and development. These mechanisms range from the production of growth-promoting hormones to the synthesis of bioactive compounds. Additionally, emerging evidence indicates that soil microbes enhance plant resilience to environmental stresses like drought and salinity, contributing to adaptation in our rapidly changing climate (Bardgett & Van Der Putten, 2014; Vandenkoornhuyse et al., 2015). Moreover, they contribute to plant proliferation by solubilizing nutrients like potassium, zinc, and phosphorus, chelating iron and fixing nitrogen, while also indirectly regulating pathogen growth and mitigating abiotic stress. Soil microbial communities are dynamic and responsive, influenced by environmental factors such as moisture content, soil type, and plant species composition. Understanding their temporal and spatial dynamics is crucial for unraveling their intricate relationship with plant proliferation. Investigating seasonal shifts in microbial diversity and spatial variations along root gradients represents a frontier in contemporary ecosystem science. Moreover, anthropogenic activities, such as agriculture intensification and land-use change profoundly impact soil microbial communities. These alterations cascade to affect plant biodiversity and ecosystem function. Similarly, urbanization and deforestation can diminish microbial diversity, fragment habitats, thereby compromising ecosystem resilience to environmental changes (Vandenkoornhuyse et al., 2015). Given the existing challenges, tapping into the potential of soil microbial communities offers significant prospects for ecosystem rehabilitation and sustainable agriculture. Researchers are exploring innovative approaches by utilizing native microbial consortia to boost crop productivity, address climate change impacts and minimize fertilizer usage. From biofertilizers enriched with beneficial microbes to soil amendments based on microbiomes, the burgeoning field of microbial ecology is driving a shift toward more resilient and regenerative agricultural practices (Singh et al., 2010).

This chapter aims to delve into the current understanding of the dynamics of soil microbial communities and their influence on plant proliferation, highlighting the key factors that contribute to this intricate relationship and the potential implications for agricultural and ecological management strategies.

2 Soil Microbial Community Composition

2.1 Description of the Types of Microbes Present in Soil

Numerous microorganisms are abundant in soil, and each one is essential to preserve the fertility and health of the soil. The most prevalent microorganisms in soil are, bacteria they are distinguished by their remarkable diversity and range of functions. They play a crucial role in the cycling of nutrients, such as phosphorus, sulfur, and nitrogen. Numerous functional groups can be distinguished among bacteria, including decomposers, nitrogen-fixing bacteria (e.g. Rhizobium and Azotobacter), nitrifiers (e.g. Nitrosomonas and Nitrobacter), and denitrifiers (e.g. Pseudomonas and Clostridium) (Fierer, 2017).

A further essential element of the soil microbiome is fungi. Yeasts, molds, and mushrooms are among them. Aspergillus and Penicillium are examples of saprophytic fungi that degrade complex organic components, while mycorrhizal fungi, such as Glomeromycotan, create symbiotic relationships with plants and contribute to the structure of the soil. In addition to causing illnesses in plants, pathogenic fungi regulate the dynamics of plant populations (Peay et al., 2016). Although archaea are less common than bacteria, they are important because of their functions in harsh environments and certain metabolic activities like ammonia oxidation and methanogenesis (Methanogens, e.g., Methanobacterium) (Ammonia-oxidizing archaea e.g., Nitrosopumilus) (Offre et al., 2013). Protozoa are single-celled eukaryotes that feed on bacteria and other microorganisms. As a result, they are essential for controlling microbial populations in processes including nutrition cycling, phagocytosis by amoebae (e.g. Acanthamoeba), ciliates (e.g. Paramecium), and flagellates (e.g. Giardia) (Geisen et al., 2018).

Soil viruses, sometimes known as bacteriophages (e.g., T4 phage), affect the dynamics of microbial communities and genetic exchange through horizontal gene

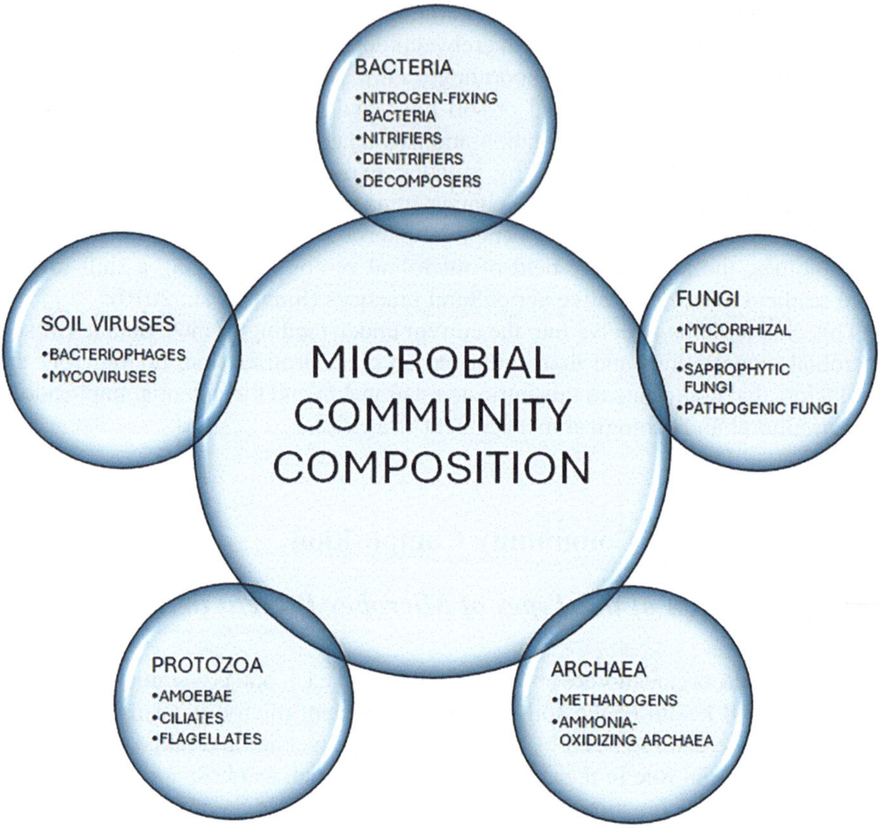

Fig. 1 Soil microbial community composition

transfer. They can infect soil bacteria, archaea, and even fungi (e.g., Mycoviruses). (Williamson et al., 2017). Figure 1 provides microbial community composition in soil and few examples for each type.

2.2 *Factors Influencing Soil Microbial Community Composition*

Many different aspects of the soil interact in intricate ways to determine the composition of the soil microbial community. The soil's pH, organic matter concentration, texture, moisture content, temperature, and nutrient availability are important variables. Figure 2 provides image of Factors Effecting Microbial Community Composition and examples. One of the most important elements influencing the makeup of the microbial community is the pH of the soil. It affects the

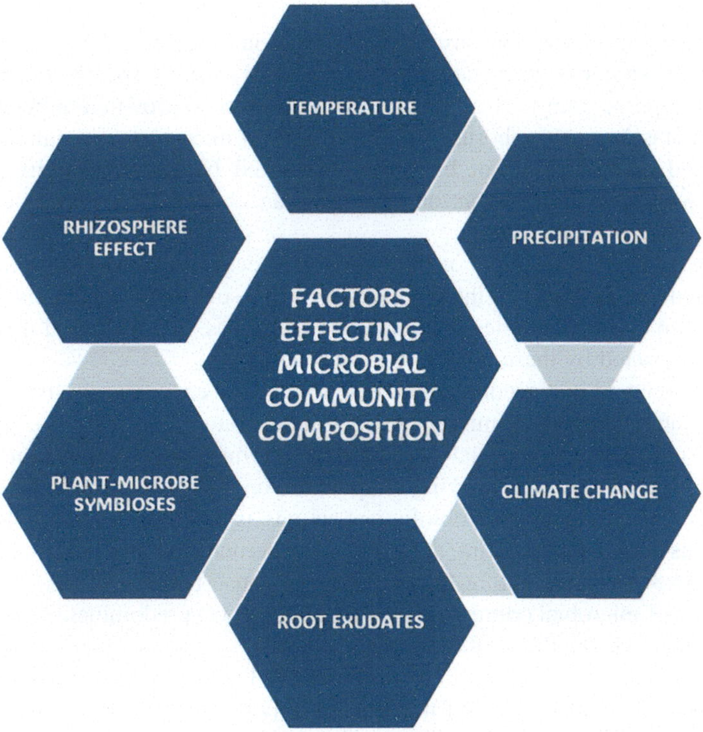

Fig. 2 Factors effecting microbial community composition

availability of nutrients, enzyme activity, and microbial development. Different microbial taxa have different pH preferences. For example, although fungus can grow in more acidic soils, bacteria typically prefer neutral to slightly alkaline environments (Rousk et al., 2010).

Organic matter in the soil supplies vital carbon and energy sources for the growth and metabolism of microorganisms. A higher microbial biomass and diversity are generally supported by a high amount of organic materials. Microbes' breakdown of organic matter adds to the cycling of nutrients (Nielsen et al., 2017).

The ratios of sand, silt, and clay in the soil dictate its texture, which in turn impacts the physical structure of the soil and, ultimately, microbial habitats, water retention, and aeration. Because finer textured soils, like clay, retain water and nutrients better, they frequently have higher microbial biomass (Bünemann et al., 2004).

The physical environment and nutrition and oxygen availability for microorganisms are influenced by soil moisture. Microbial activity can be inhibited by both surplus and deficit moisture content. Generally speaking, optimal moisture levels encourage more microbial diversity and activity (Barnard et al., 2013).

Enzymatic activity, community makeup, and microbial metabolic rates are all impacted by temperature. The structure and function of microbial communities can be altered by seasonal temperature fluctuations, as various species have distinct temperature optimal ranges (Karhu et al., 2014). Microbial growth requires nutrients including potassium, phosphorus, and nitrogen. Soil microbial communities' functionality and composition can be greatly impacted by the availability of these nutrients. Microbial populations and activity can change as a result of nutrient imbalances (Fierer et al., 2009).

Climate Through its effects on temperature, precipitation, and seasonality, climate has a major influence on the composition of the soil microbial community. The general health of soil ecosystems, community dynamics, and microbial metabolism can all be impacted by these climate variables.

An important climatic component that influences enzyme activity, microbial metabolic rates, and community makeup is temperature. Temperature variations can cause changes in microbial populations and functions because various soil microbial communities have varying temperature optima.

Impact on Metabolism: Microbiological activity is normally increased by greater temperatures until a certain point, at which point thermal stress may cause activity to decrease. Seasonal temperature fluctuations can cause temporal alterations in the composition of microbial communities, with distinct taxa predominating at different periods of the year (Karhu et al., 2014).

Rainfall affects the amount of moisture in the soil, which in turn affects microbial activity, nutrient availability, and the physical environment in which bacteria live. Soil microbial populations can be significantly impacted by both drought and abundant rainfall. Access to Moisture Microbial operations require a certain amount of moisture, yet soggy circumstances can result in anaerobic habitats, which promote distinct microbial populations than those found in soils with adequate air circulation. Drought Stress Although some bacteria may be evolved to withstand protracted dry times, drought conditions can lower microbial activity and biomass (Barnard et al., 2013).

Dynamic shifts in soil microbial communities can result from differences in temperature and precipitation caused by seasonal changes in the climate. The functional activities, community structure, and microbial diversity can all be impacted by these cyclical cycles. Seasonal Dynamics Variations in the availability of resources and shifting environmental conditions can lead to distinct microbial species being prevalent at different periods of the year. Adaptability and Sturdiness Certain soil bacteria are resilient and quick to react to favorable conditions because they have evolved to seasonal variations (Lauber et al., 2013).

Soil microbial populations are anticipated to be significantly impacted by climate change, which is defined by increased frequency of extreme weather events, modified precipitation patterns, and global warming. Warming Effects: Prolonged global warming may cause changes in the composition of the microbial community, favoring fast-growing and thermophilic microbial species. Changes in precipitation patterns can have an impact on soil moisture regimes, which in turn can have an impact on microbial activity and community structure. Abrupt Occurrences

Droughts and floods occurring more frequently can cause disruptions to occur more frequently, which can test the resilience of microbial communities (Castro et al., 2010).

Plant species affect the composition of the soil microbial community through a number of methods, including as interactions between microorganisms and plants, root exudates, and litter quality. The rhizosphere habitats created by various plant species influence the composition and activities of soil microbial populations. Root exudates are the various organic chemicals that plants emit through their roots. These substances function as substrates and signaling molecules for soil microorganisms. They include sugars, amino acids, organic acids, and secondary metabolites. Selectivity: varying plant species selectively enrich particular microbial communities in the rhizosphere by exuding varying types and volumes of root exudates. Microbial Recruitment: Root exudates have the power to draw in advantageous microorganisms, including rhizobacteria (PGPR), mycorrhizal fungus, and bacteria that fix nitrogen in plants (Bais et al., 2006).

The decomposition process and nutrient cycling of plant litter, or dead plant waste, are influenced by several factors that have a substantial impact on the soil microbial population. The decomposer microbial population is impacted by the variations in the chemical composition of litter resulting from diverse plant species. These variations include variations in lignin, cellulose, and nutrient content. Decomposition Rate: Compared to low-quality litter (high lignin, low nutrient content), high-quality litter (low lignin, high nutrient content) decomposes more quickly and supports a different microbial population (Strickland & Rousk, 2010).

The rhizosphere, a confined area of soil affected by soil microorganisms and root secretions, is a hotspot for microbial variety and activity. Plant species exhibit different effects from the rhizosphere. Diversity of Microbes In comparison to bulk soil, the rhizosphere often has a greater diversity of microorganisms, including microbial assemblages particular to individual plant species. Functional Groups: Microbes engaged in pathogen suppression, nutrition acquisition, and stress tolerance are examples of functional groups that plants can affect differently (Philippot et al., 2013).

An ecosystem's plant variety can affect the composition of the community and the diversity of soil microbes. A wider variety of microbial species are supported by varied habitats created by diverse plant groups. Various Root Habitats: A range of root habitats and exudates are provided by different plant species, which in turn support a variety of microbial communities. Consistency and Adaptability Increased plant diversity can improve the soil microbial communities' stability and resistance to shocks (Zak et al., 2003).

2.3 Methods for Studying Soil Microbial Communities

Various techniques that can capture the richness, structure, and function of these intricate ecosystems are needed to study soil microbial communities. Primarily separated into two categories: methods of community analysis that depend on culture and methods of community analysis that don't (Liu et al., 2006).

The variety, composition, and function of soil microbes are extensively studied using DNA-based methods. Microbial taxa can be identified and measured using these techniques. 16S rRNA Gene The 16S ribosomal RNA gene, which is highly conserved in bacteria and archaea, is sequenced using this method, enabling the phylogenetic analysis and identification of microbial communities (Caporaso et al., 2012). Metagenomics involves the direct sequencing of DNA from soil samples to analyze the collective genome of the microbial community. This approach provides insights into the functional potential and diversity of soil microbes (Howe et al., 2014).

RNA-based approaches provide insights into the active portion of the microbial community by targeting messenger RNA (mRNA) and ribosomal RNA (rRNA). Meta transcriptomics involves sequencing the total RNA from soil samples to study gene expression and functional activity within microbial communities (Carvalhais et al., 2012).

Protein-based approaches, such as meta proteomics, analyze the proteins present in soil samples, providing insights into the functional activities and metabolic pathways of soil microbes. Meta proteomics involves the extraction and identification of proteins from soil samples, allowing for the study of microbial functions and interactions at the protein level (Holmfeldt et al., 2014).

Culture-based Methods Traditional culture-based methods involve isolating and culturing microbes from soil samples to study their physiological properties and interactions. Isolation and Cultivation this method involves growing soil microbes on selective media under controlled laboratory conditions to study their morphology, physiology, and biochemical characteristics (Stewart, 2012).

Biochemical methods, such as phospholipid fatty acid (PLFA) analysis, provide insights into the microbial community structure based on the lipid profiles of microbial cell membranes. PLFA analysis involves extracting and analyzing phospholipid fatty acids from soil samples to determine the biomass and composition of microbial communities (Frostegård et al., 2011). Stable Isotope Probing (SIP) is used to study microbial activity by tracking the incorporation of isotopically labeled substrates (e.g., ^13C or ^15N) into microbial biomass, DNA, or RNA (Dumont & Murrell, 2005).

Microscopy and imaging techniques, such as fluorescence in situ hybridization (FISH) and confocal laser scanning microscopy (CLSM), allow for the visualization and spatial analysis of microbial communities in soil. FISH uses fluorescently labeled probes that bind to specific nucleic acid sequences within microbial cells, enabling the identification and localization of microbes within soil samples (Wagner et al., 2006).

3 Role of Soil Microbes in Plant Proliferation

Plants have become one of the major food sources to satisfy the hunger of the day-by-day increase in world population. To tackle the huge demand for food for the growing population, the yield of the crops should keep up with the market. In Agriculture, soil, soil is identified as one of the key factors for improving the yield. It houses various microorganisms, chemicals, compounds, organisms, etc. The presence of a variety of microorganisms provides various benefits for plant proliferation. Some of these microorganisms play a vital role in recycling various essential micronutrients like nitrogen, carbon, phosphorous, etc. through biogeochemical cycles to improve the soil quality. Few microorganisms produce Exopolysaccharides which help in the aggregation of soil particles thereby enhancing the soil structure, Water retention capacity, and Soil aeration. And few plant growth-promoting microorganisms (PGPM) help in disease suppression and Stress tolerance, etc., and allelopathic interactions can help in plant proliferation by suppressing weeds. Apart from these, there are various other ways in which microorganisms in soil help in proliferating plant growth. Some of these are discussed below.

3.1 Nutrient Cycling and Availability

Nutrient Cycling is a process through which organic compounds and fragments of insoluble forms of nutrients decompose into nutrient elements such as Carbon, Nitrogen, Phosphorus, etc., from living/ dead organisms to the environment. These nutrients are essential for plant growth and development.

3.1.1 Nitrogen Recycling

Nitrogen is one of the important elements present inside the living organisms. It also exists as amino acids, Proteins, and Nucleic acids (Deoxyribonucleic acid (DNA) and Ribonucleic acid (RNA)), and plays a key role in plant proliferation. Though Nitrogen is present in abundant quantities in the environment, most of it is unable to be used by plants as it is present in the unusable form ($N \equiv N$). This unusable form of nitrogen can be converted to a usable form with the help of Nitrogen-fixing bacteria through the enzyme Nitrogenase. Some of the Species of microorganisms like Rhizobium, Mesorhizobium, Bradyrhizobium, Azorhizobium, Allorhizobium, and Sinorhizobium can form a symbiotic association with the roots of the plants to form root nodules (Dakora, 1995, 2003; Lhuissier, 2001; Matiru & Dakora, 2004) where nitrogen fixation takes place. The fixed nitrogen gets used by plants for the synthesis of their organic molecules and when they die this fixed nitrogen can be recycled to the environment by Denitrifying bacteria such as Pseudomonas and Thiobacillus, etc.

3.1.2 Phosphorus Recycling

Phosphorus is one of the essential micronutrients for plant proliferation. The phosphorus reserves are found in rocks. Soil contains phosphate (Both organic and inorganic phosphates) that dissolves when rocks weather. According to some research, bacteria may release phosphate from organic phosphate sources, but the amount of phosphate in soil is often quite low (Goldstein, 1994; McGrath et al., 1995, 1998; Ohtake et al., 1996; Rodriguez & Fraga, 1999) The first strains identified as phosphate-solubilizing bacteria (PSB) were four strains of *Arthrobacter ureafaciens, Phyllobacterium myrsinacearum, Rhodococcus erythropolis*, and Delftia species. These strains are absorbed by the plant through its roots and provide phosphate ions necessary for the plant's growth (Chen et al., 2006).

3.2 Soil Structure Modification

3.2.1 Physical Modifications

Aggregates are created when some bacteria and fungi mix soil particles through the secretion of polysaccharides and glycoproteins. Through increased aeration, water infiltration, and root penetration, these aggregates strengthen the structure of the soil. Plant growth is directly impacted by changes in soil structure, which is mostly influenced by soil bacteria. Their actions improve the soil's capacity to support plant development by improving its biological, chemical, and physical characteristics. The hyphae, which are filamentous structures found in fungi, create networks within soil particles that aid in the creation and consolidation of soil aggregates.

3.2.2 Chemical Modifications

The effects on the pH of the soil are often seen during the actions of microorganisms. Some microorganisms decompose the organic materials and break them down into organic acids, which lowers the pH of the soil. On the other hand, some reactions in these microorganisms lead to a rise in the pH and create a more organized environment for plant development. Numerous enzymes that are produced by soil-dwelling microbes aid in the digestion of both organic and inorganic materials to provide nutrients that are beneficial to plants.

3.3 Plant Growth Promotion and Disease Suppression

By cycling nutrients, solubilizing phosphate and zinc, moving potassium, moving trace elements (micronutrients), and producing plant hormones like auxins, gibberellins, cytokinins, and compounds that are similar to these hormones, soil microorganisms help plants grow faster (Akhtar et al., 2010; Akhtar & Siddiqui, 2010).

Soils that inhibit disease can offer efficient defense against illnesses resulting from soil-borne pathogens. They provide plants with microbiome-mediated defense against several root diseases brought on by soil microorganisms. A few fungal and bacterial species, such as pseudomonas, are important in controlling the spread of soil-borne pathogens and shielding plants from the illnesses they can spread. The primary strategies of disease suppression in pseudomonas are competition and antibiosis via lytic enzyme production.

3.4 Stress Tolerance Enhancement in Plants

Stress, whether biotic or abiotic, can significantly impact the delicate balance of interactions between plants and soil. It frequently causes the soil to lose its microbial diversity, get less fertile, and become more competitive for the few available nutrient supplies (Afkhami, 2023). However, as they coexist with a wide variety of helpful microorganisms, such as bacteria that promote plant development and mycorrhizal fungi, plants are remarkably resilient and can adjust to unfavorable environments.

According to Vimal et al. (2017), these plant-associated microbial communities are essential for enabling plants to withstand and flourish under challenging conditions. It has been demonstrated that beneficial rhizobacteria and mycorrhizae, which are important elements of the soil microbiota, improve plant fitness and preserve soil health under stressful environments.

3.4.1 Drought Stress Tolerance

Drought-stressed microorganisms can improve plant growth and development by adapting to low water availability. Inoculating plants with plant growth-promoting rhizobium (PGPR) increases abscisic acid levels and modulates drought tolerance via producing drought-related proteins. (Jiang et al., 2013).

3.4.2 Salinity Tolerance

Plant growth-promoting bacteria can produce biofilms under salinity stress, which can assist in reducing the negative consequences of salinity stress (Kasim et al., 2016).

3.5 Allelopathic Interactions

Allelopathy refers to the interactions between plants and microorganisms (Weir et al., 2004; Cipollini et al., 2012). This phenomenon has substantial implications for forest ecosystem management (Mallik, 2003; Blanco, 2007; Mosina & Merzlaya, 2013). According to recent studies, substances such as terpenoids, alkaloids, steroids, phenolic compounds, and other metabolites have allelopathic actions. Plant roots create a variety of chemicals that offer nutrients to microorganisms, promoting relationships between the roots and the surrounding soil (Boronin, 1998). Microbes can metabolize phenolic chemicals, lowering soil phytotoxicity (Jilani et al., 2008). Furthermore, several bacteria produce antibiotics that assist in reducing the presence of phytopathogens (Glick et al., 2007).

4 Plant Proliferation and Soil Microbial Communities

Plant proliferation describes an increase in the population size of plants in a specific region. This phenomenon can occur naturally or as a result of human intervention, such as land management techniques or climate change. It is influenced by the makeup and activity of soil microbial populations that include bacteria, fungi, archaea, and viruses. These microbes interact with plants via a variety of processes, like nutrition cycling, disease control, and symbiotic associations, etc. Additionally, they can directly impact plant growth and development by producing growth-promoting hormones, nutrient acquisition, and protection against pathogens.

4.1 Influence of Soil Microbial Communities on Plant Growth and Proliferation

Soil microbial communities have a significant influence on plant development and proliferation, and their association aids in comprehending the complex dynamics of soil-plant interactions. Depending on the microbial species and their roles, these interactions can have both positive and negative consequences on plant health. Some microbial species contribute to the breakdown of organic matter and the release of critical nutrients such as nitrogen, phosphorus, and potassium, making them available for plant absorption. Few microbial species can fix atmospheric nitrogen and turn it into a form that plants can absorb. Few soil microbial species have mutualistic interactions with plants, supplying them with critical nutrients while also protecting them from diseases. Mycorrhizal fungi create symbiotic associations with plant roots, allowing them to absorb water and nutrients, notably phosphorus, in return for carbon from the plant (Smith & Read, 2008). This mutualistic relationship contributes greatly to plant growth and development.

Soil microbial communities play an important role in maintaining soil health and productivity in agroecosystems by cycling major elements. They can indirectly benefit plant growth by increasing soil fertility, producing growth-promoting substances, suppressing plant pathogens, and improving overall ecosystem function.

They interact by chemical signals (e.g., flavonoids, strigolactones), which can cause beneficial symbioses such as nodulation in legumes and activate plant defense mechanisms, making plants more resistant to diseases and pests via ISR pathways. They also create biofilms on plant roots, which improve nutrient exchange and defend against diseases. Genes encoding advantageous features, such as antibiotic synthesis or nutrient solubilization, can be exchanged across soil microorganisms, increasing their combined potential to promote plant development.

4.2 Mechanisms Through Which Soil Microbes Promote Plant Health and Productivity

Soil microbes significantly enhance plant health and productivity through various intricate mechanisms. These include mycorrhizal associations, rhizosphere interactions, microbial-mediated signaling pathways, induced systemic resistance (ISR), quorum sensing, and microbial communication. Each of these mechanisms contributes uniquely to plant growth and resilience.

4.2.1 Mycorrhizal Associations

It plays a vital role in enhancing plant health and productivity by facilitating nutrient uptake, particularly phosphorus, from the soil. Mycorrhizal associations are mutualistic relationships between plant roots and certain fungi, known as mycorrhizal fungi. The fungi form a network of hyphae that extend into the soil, greatly increasing the surface area for nutrient absorption.

Examples:

- Arbuscular Mycorrhizal Fungi (AMF): AMF, such as Glomus spp., invade the root cortex and create arbuscules that facilitate nutrient exchange. This connection is critical for phosphorus, zinc, and other micronutrient absorption, especially in nutrient-poor soils (Smith & Read, 2008). AMF in crops such as maize and wheat boosts phosphorus and micronutrient intake, hence enhancing growth and resistance. AMF's large hyphal network enhances the surface area available for nutrient absorption and allows access to nutrient pools outside the roots' depletion zone. In exchange, plants provide the fungus with carbohydrates obtained from photosynthesis.
- Ectomycorrhizal Fungi: These fungi wrap around roots and penetrate between the cells. In forest ecosystems, ectomycorrhizal interactions with trees such as pines and oaks improve nutrient absorption and protect against soil-borne diseases

(Courty et al., 2010). These fungi help plant roots absorb nutrients like nitrogen and phosphate and create antimicrobial chemicals that protect them from dangerous microorganisms.

4.2.2 Rhizosphere Interactions

The rhizosphere, the soil area impacted by root secretions, is a hub of microbial activity where roots, soil microbes, and nutrients interact. Rhizosphere interactions are complex networks of interaction and communication between plants and soil microbes. Root exudates, which contain sugars, amino acids, and organic acids, provide nutrition for soil bacteria, encouraging microbial development and activity in the rhizosphere. This, in turn, improves nutrient availability and helps to reduce dangerous microorganisms, resulting in a healthy root environment (Dakora & Phillips, 2002). Microbes in the rhizosphere can enhance plant health and productivity through several mechanisms. They can solubilize nutrients, fix nitrogen, and produce growth-promoting chemicals including phytohormones and enzymes.

Examples:

- Nutrient Cycling: Rhizosphere bacteria such as Pseudomonas and Bacillus spp. contributes to nutrient solubilization and cycling. For example, phosphate-solubilizing bacteria (PSB) transforms insoluble phosphates into plant-accessible forms (Richardson et al., 2009).
- Biocontrol Agents: Some rhizosphere bacteria generate antibiotics and siderophores that suppress the growth of soil-borne pathogens, therefore protecting plants from illness.
- Plant growth-promoting rhizobacteria (PGPR) can solubilize minerals, create phytohormones, and reduce soil-borne pathogens, all of which help to improve plant health and production.

4.2.3 Microbial-Mediated Signaling Pathways

Soil microorganisms can influence plant development and stress responses by releasing signaling chemicals such as auxins, cytokinins, and gibberellins. These microbial-mediated signaling pathways affect plant growth, root architecture, and stress tolerance, boosting plant health and production. These microbial-mediated signaling pathways are also known as chemical signals between plants and bacteria. These chemical signals can trigger a variety of plant reactions, such as changes in gene expression or the synthesis of defensive chemicals, which promotes symbiosis and plant development. Key signaling molecules include flavonoids, strigolactones, and nod factors.

Examples:

- Rhizobium-Legume Symbiosis: Legume plants produce flavonoids, which stimulate the development of nod factors in Rhizobium bacteria. These nod factors cause nodule development on roots, where nitrogen fixation occurs.
- Mycorrhizal Signaling: Strigolactones generated by plant roots attract AMF, resulting in symbiosis and improved nutrient absorption (Bais et al., 2004).

4.2.4 Induced Systemic Resistance (ISR)

Soil microorganisms can elicit systemic resistance in plants, preparing them to defend against diseases and pests. This process involves the activation of plant defensive responses by microbial elicitors, which leads to increased resistance to diseases and pests. ISR includes the synthesis of signaling chemicals such as jasmonic acid and ethylene, which activate plant defense genes and pathways. This priming effect provides a quicker and more robust response to subsequent pathogen assaults, lowering illness frequency and severity (Van Wees et al., 2008).

Examples:

- PGPR-Induced ISR: Plant growth-promoting rhizobacteria (PGPR), including Pseudomonas fluorescens, can cause ISR in plants. This causes increased resistance to diseases such as Pythium and Fusarium (Van Wees et al., 2008).
- Trichoderma-Induced ISR: Beneficial fungus Trichoderma spp. induce ISR in plants, increasing resistance to many diseases (Harman et al., 2004).

4.2.5 Quorum Sensing and Microbial Communication

Soil bacteria also contribute to plant health and productivity via quorum sensing and microbial communication pathways. Quorum sensing is a process in which bacteria interact by chemical signals known as autoinducers. This communication regulates gene expression in response to population density, coordinating collective actions such biofilm formation and pathogenicity (Venturi, 2006). This communication enables microbial populations to impact plant health and productivity by regulating nutrient availability, disease suppression, and stress tolerance in the rhizosphere. These systems work together to provide a comprehensive strategy to improving plant health and production.

Examples:

- Pseudomonas Quorum Sensing: Pseudomonas aeruginosa employs quorum sensing to control the synthesis of secondary metabolites such as antibiotics and siderophores, giving it a competitive advantage and inhibiting pathogens (Venturi, 2006).

- Rhizobium Quorum Sensing: Rhizobium spp. employ quorum sensing to control nodulation and nitrogen fixation, hence improving their symbiotic interaction with legumes.
- Bacteria in the rhizosphere can communicate using quorum sensing to coordinate enzyme synthesis, which degrades organic debris and releases nutrients for plant absorption.

4.3 Soil Microbial Community Structure and Continuous Cropping

Continuous cropping, the practice of growing the same crop year after year on the same land, has significant impacts on soil microbial community structure. This practice can lead to shifts in the diversity and composition of soil microbes, with potential consequences for soil health and agricultural sustainability (Kodaparthi et al., 2024). Below is an exploration of these impacts, methods used to study soil microbial communities in the context of continuous cropping, and the implications for agricultural management.

4.3.1 Impact of Continuous Cropping on Soil Microbial Communities

The impact of continuous cropping on soil microbial communities is not limited to diversity; it can also influence the overall composition and structure of these communities, with potential implications for agricultural sustainability and soil health. For example, long-term monoculture of crops such as coffee has been shown to alter the soil's chemical properties and lead to a reduction in bacterial and fungal diversity (Dai et al., 2018; Kodaparthi et al., 2024). This loss of microbial diversity can have detrimental effects on soil functions, including disease suppression and plant growth promotion. Similarly, the cultivation of oilseed rape in short rotations has been associated with a decline in the diversity of the rhizosphere microbial community, potentially due to the limited concentration and diversity of organic materials available to support a more diverse soil biome (Hilton et al., 2013). The lack of crop rotation limits the variety of root exudates, which are critical for sustaining a diverse microbial community. This homogeneity can lead to an imbalance in the soil ecosystem, reducing resilience against diseases and pests.

Understanding the mechanisms behind these changes in soil microbial communities under continuous cropping regimes is crucial for developing sustainable agricultural management strategies. By studying the methods used to assess soil microbial diversity and community structure in the context of continuous cropping, researchers can provide valuable insights into the implications for agricultural sustainability and soil health. These findings can inform the development of crop rotation strategies, the use of cover crops, and other practices that promote a diverse and resilient soil microbial community, ultimately enhancing the long-term productivity and environmental sustainability of agricultural systems. For instance, the studies on maize cultivation have demonstrated that sustainable conservation agriculture practices, which involve reduced tillage and crop residue retention, can increase the diversity of soil microbial species and functions compared to more intensive conventional practices. These shifts in microbial community composition can have significant implications for the ecosystem services provided by the soil, such as nutrient cycling, organic matter decomposition, and plant-microbe interactions.

4.3.2 Significance of Plant Growth-Promoting Rhizobacteria (PGPR)

The introduction of PGPR can enhance soil microbial diversity by promoting beneficial microbial populations and suppressing pathogens. PGPR can help mitigate the negative impacts of continuous cropping by enhancing soil health, promoting plant growth, and suppressing pathogens, and can reduce the need for chemical fertilizers and pesticides, leading to more sustainable agricultural practices (Patten & Glick, 2002).

Mechanisms of Action

- Nutrient Solubilization: PGPR such as Pseudomonas and Bacillus species can solubilize phosphate and produce siderophores that chelate iron, making these nutrients more available to plants (Richardson et al., 2009).
- Phytohormone Production: PGPR produce phytohormones like indole-3-acetic acid (IAA), which promote root growth and improve plant nutrient uptake (Patten & Glick, 2002).
- Induced Systemic Resistance (ISR): PGPR can trigger ISR in plants, enhancing their resistance to pathogens through the activation of defense pathways (Van Wees et al., 2008).

Example: In continuous wheat cropping systems, the application of PGPR has been shown to increase microbial diversity and improve soil health, leading to better plant growth and yield.

5 Management Strategies for Soil Microbial Community Dynamics and Proliferation

Soil microbial communities play a crucial role in maintaining soil health, nutrient cycling, and overall ecosystem functioning. Effective management strategies for promoting microbial community dynamics and proliferation are essential for sustainable agriculture, ecosystem restoration, and environmental conservation. Here are various approaches and techniques for managing soil microbial communities.

5.1 Organic Matter Management

Organic matter serves as a primary source of energy and nutrients for soil microorganisms. Incorporating organic amendments such as compost, manure, or crop residues into the soil can significantly enhance microbial diversity and abundance (Bünemann et al., 2006). These organic materials provide a wide range of carbon and nitrogen sources, promoting the growth and activity of various microbial groups, including bacteria, fungi, and protozoa (Kallenbach & Grandy, 2011).

5.2 Conservation Tillage Practices

Conventional tillage practices can disrupt soil structure, aerate the soil excessively, and expose organic matter to rapid decomposition, negatively impacting microbial communities (Lupwayi et al., 2004). Conservation tillage practices, such as no-till or reduced tillage, minimize soil disturbance and preserve soil aggregates, providing a stable habitat for microorganisms to thrive (Doran et al., 1998). These practices also help maintain soil moisture and temperature, further supporting microbial growth and activity (Govaerts et al., 2007).

5.3 Crop Rotation and Cover Cropping

Diversifying crop rotations and incorporating cover crops can enhance microbial diversity and activity in the soil (McDaniel et al., 2014). Different plant species release unique root exudates and residues, creating a diverse array of carbon and nutrient sources for microorganisms (Berendsen et al., 2012). Cover crops, particularly legumes, can also contribute to nitrogen fixation and improve soil fertility, benefiting microbial communities (Blanco-Canqui et al., 2013).

5.4 Biochar Application

Biochar, a carbon-rich material produced from the pyrolysis of biomass, has gained attention for its potential to improve soil properties and promote microbial growth (Lehmann et al., 2011). Biochar can increase soil porosity, water-holding capacity, and nutrient availability, creating a favorable environment for microorganisms (Kolton et al., 2011). Additionally, biochar can serve as a habitat for microbes, protecting them from predation and environmental stresses (Ameloot et al., 2013).

5.5 *Inoculation with Beneficial Microorganisms*

Introducing beneficial microorganisms, such as plant growth-promoting rhizobacteria (PGPR) or arbuscular mycorrhizal fungi (AMF), can enhance microbial diversity and functionality in the soil (Adesemoye et al., 2009). These beneficial microbes can improve nutrient uptake, promote plant growth, and increase resistance to biotic and abiotic stresses (Smith & Read, 2010). Inoculation can be achieved through seed coating, soil amendment, or direct application to the rhizosphere.

5.6 *Minimizing Synthetic Pesticide and Fertilizer Use*

Excessive use of synthetic pesticides and fertilizers can have detrimental effects on soil microbial communities (Imfeld & Vuilleumier, 2012). These chemicals can disrupt the delicate balance of microbial populations, reducing diversity and potentially favoring certain groups over others (Muñoz-Leoz et al., 2011). Adopting integrated pest management strategies and using organic or biofertilizers can help maintain a healthy and diverse microbial community (Franke-Whittle, 2009).

5.7 *Monitoring and Assessment*

Effective management of soil microbial communities requires regular monitoring and assessment. Various techniques, such as soil microbial biomass analysis, community profiling using molecular methods (e.g., metagenomics, meta transcriptomics), and functional assays, can provide valuable insights into the composition, diversity, and activity of microbial communities (Bastida et al., 2008). This information can guide adaptive management strategies tailored to specific soil conditions and ecosystem requirements.

It is important to note that soil microbial communities are complex and dynamic, and their responses to management practices can vary depending on factors such as soil type, climate, and ecosystem characteristics. Therefore, a combination of multiple strategies, tailored to local conditions, may be necessary to achieve optimal results.

Furthermore, ongoing research and collaboration among scientists, farmers, and land managers are crucial for developing and refining soil microbial management strategies. Interdisciplinary approaches that integrate knowledge from microbiology, soil science, agronomy, and ecosystem ecology will be essential for advancing our understanding and promoting sustainable soil management practices.

6 Conclusion

In conclusion, the dynamic interactions between soil microbial populations and plants are vital for plant growth. Microbial communities such as fungi, bacteria, and archaea substantially impact plant growth, development, and proliferation. These microbes improve nutrient cycling, create plant growth-promoting chemicals, and inhibit illnesses, allowing for healthier and more vigorous plant development. Conversely, imbalances or disturbances in the soil microbiome can stunt plant development and render them more vulnerable to abiotic and biotic challenges. Researchers advocate applying sustainable agriculture methods such as biofertilizer usage, cover cropping, and other health-promoting strategies to maximize soil microbial populations and increase plant proliferation. Maintaining a healthy and diverse soil microbiome can be achieved through the use of biofertilizers, which introduce beneficial microbes directly into the soil.

Understanding and regulating soil microbial populations has far-reaching consequences for agricultural sustainability and ecosystem health. Plant productivity may be enhanced by cultivating a varied and active soil microbiome, resulting in better yields and more robust crops. This, in turn, increases ecosystem resilience because good soils can better endure environmental pressures like drought and disease. Furthermore, sustainable soil management approaches help to ensure environmental sustainability by lowering the need for chemical inputs, increasing soil carbon sequestration, and fostering biodiversity. This chapter's findings emphasize the crucial function of soil microbial communities in plant proliferation and the need to implement management techniques that promote a healthy soil microbiome. These tactics not only increase plant production, but they also improve the overall health and sustainability of agricultural ecosystems, making them critical considerations for future agricultural practices and land management regulations.

References

Adesemoye, A. O., Torbert, H. A., & Kloepper, J. W. (2009). Plant growth-promoting Rhizobacteria allow reduced application rates of chemical fertilizers. *Microbial Ecology, 58*(4), 921–929. https://doi.org/10.1007/s00248-009-9531-y

Afkhami, M. E. (2023). Past microbial stress benefits tree resilience. *Science, 380*(6647), 798–799. https://doi.org/10.1126/science.adi1594

Akhtar, M. S., & Siddiqui, Z. A. (2010). Role of plant growth promoting Rhizobacteria in biocontrol of plant diseases and sustainable agriculture. In *Microbiology monographs* (pp. 157–195). https://doi.org/10.1007/978-3-642-13612-2_7

Akhtar, M. S., Shakeel, U., & Siddiqui, Z. A. (2010). Biocontrol of Fusarium wilt by bacillus Pumilus, pseudomonas Alcaligenes and rhizobium Sp. on lentil. *Turkish Journal of Biology*. https://doi.org/10.3906/biy-0809-12

Ameloot, A. N., Graber, E. R., Verheijen, F. G. A., & De Neve, S. (2013). Interactions between biochar stability and soil organisms: Review and research needs. *European Journal of Soil Science, 64*(4), 379–390. https://doi.org/10.1111/ejss.12064

Ames, R. N., & Bethlenfalvay, G. J. (1987). Mycorrhizal fungi and the integration of plant and soil nutrient dynamics. *Journal of Plant Nutrition, 10*(9–16), 1313–1321. https://doi.org/10.1080/01904168709363663

Bais, H. P., Park, S. W., Weir, T. L., Callaway, R. M., & Vivanco, J. M. (2004). How plants communicate using the underground information superhighway. *Trends in Plant Science, 9*(1), 26–32. https://doi.org/10.1016/j.tplants.2003.11.008

Bais, H. P., Weir, T. L., Perry, L. G., Gilroy, S., & Vivanco, J. M. (2006). The role of root exudates in rhizosphere interactions with plants and other organisms. *Annual Review of Plant Biology, 57*(1), 233–266. https://doi.org/10.1146/annurev.arplant.57.032905.105159

Bardgett, R. D., & Van Der Putten, W. H. (2014). Belowground biodiversity and ecosystem functioning. *Nature, 515*(7528), 505–511. https://doi.org/10.1038/nature13855

Barnard, R. L., Osborne, C. A., & Firestone, M. K. (2013). Responses of soil bacterial and fungal communities to extreme desiccation and rewetting. *The ISME Journal, 7*(11), 2229–2241. https://doi.org/10.1038/ismej.2013.104

Bastida, F., Zsolnay, A., Hernández, T., & García, C. (2008). Past, present and future of soil quality indices: A biological perspective. *Geoderma, 147*(3–4), 159–171. https://doi.org/10.1016/j.geoderma.2008.08.007

Berendsen, R. L., Pieterse, C. M. J., & Bakker, P. A. H. M. (2012). The rhizosphere microbiome and plant health. *Trends in Plant Science, 17*(8), 478–486. https://doi.org/10.1016/j.tplants.2012.04.001

Blanco, J. A. (2007). The representation of allelopathy in ecosystem-level forest models. *Ecological Modelling, 209*(2–4), 65–77. https://doi.org/10.1016/j.ecolmodel.2007.06.014

Blanco-Canqui, H., Holman, J. D., Schlegel, A. J., Tatarko, J., & Shaver, T. M. (2013). Replacing fallow with cover crops in a semiarid soil: Effects on soil properties. *Soil Science Society of America Journal, 77*(3), 1026–1034. https://doi.org/10.2136/sssaj2013.01.0006

Boronin, A. M. (1998). Rhizosphere bacteria of the genus Pseudomonas enabling plant growth and development. *Sorovsky Educational Magazine, 10*, 25–31.

Bünemann, E., Bossio, D., Smithson, P., Frossard, E., & Oberson, A. (2004). Microbial community composition and substrate use in a highly weathered soil as affected by crop rotation and P fertilization. *Soil Biology and Biochemistry/Soil Biology & Biochemistry, 36*(6), 889–901. https://doi.org/10.1016/j.soilbio.2004.02.002

Bünemann, E. K., Schwenke, G. D., & Van Zwieten, L. (2006). Impact of agricultural inputs on soil organisms—A review. *Soil Research, 44*(4), 379. https://doi.org/10.1071/sr05125

Caporaso, J. G., Lauber, C. L., Walters, W. A., Berg-Lyons, D., Huntley, J., Fierer, N., Owens, S. M., Betley, J., Fraser, L., Bauer, M., Gormley, N., Gilbert, J. A., Smith, G., & Knight, R. (2012). Ultra-high-throughput microbial community analysis on the Illumina HiSeq and MiSeq platforms. *The ISME Journal, 6*(8), 1621–1624. https://doi.org/10.1038/ismej.2012.8

Carvalhais, L. C., Dennis, P. G., Tyson, G. W., & Schenk, P. M. (2012). Application of metatranscriptomics to soil environments. *Journal of Microbiological Methods, 91*(2), 246–251. https://doi.org/10.1016/j.mimet.2012.08.011

Castro, H. F., Classen, A. T., Austin, E. E., Norby, R. J., & Schadt, C. W. (2010). Soil microbial community responses to multiple experimental climate change drivers. *Applied and Environmental Microbiology, 76*(4), 999–1007. https://doi.org/10.1128/aem.02874-09

Chen, Y., Rekha, P., Arun, A., Shen, F., Lai, W. A., & Young, C. (2006). Phosphate solubilizing bacteria from subtropical soil and their tricalcium phosphate solubilizing abilities. *Applied Soil Ecology, 34*(1), 33–41. https://doi.org/10.1016/j.apsoil.2005.12.002

Chu, Q., Feng, W., Tian, Y., Zhang, L., Kang, F., Zhao, Y., Yuan, W., et al. (2024). Soil microbial community contributes more to plant-soil feedback and plant-plant interactions than root traits

under warming and drought. *Plant and Soil*, 1–15. https://doi.org/10.1007/s11104-024-06606-w
Cipollini, D., Rigsby, C. M., & Barto, E. K. (2012). Microbes as targets and mediators of Allelopathy in plants. *Journal of Chemical Ecology, 38*(6), 714–727. https://doi.org/10.1007/s10886-012-0133-7
Courty, P. E., Buée, M., Diedhiou, A. G., Frey-Klett, P., Tacon, F. L., Rineau, F., Turpault, M. P., Uroz, S., & Garbaye, J. (2010). The role of ectomycorrhizal communities in forest ecosystem processes: New perspectives and emerging concepts. *Soil Biology and Biochemistry/Soil Biology & Biochemistry, 42*(5), 679–698. https://doi.org/10.1016/j.soilbio.2009.12.006
Dai, Z., Su, W., Chen, H., Barberán, A., Zhao, H., Yu, M., Yu, L., Brookes, P. C., Schadt, C. W., Chang, S. X., & Xu, J. (2018). Long-term nitrogen fertilization decreases bacterial diversity and favors the growth of Actinobacteria and Proteobacteria in agro-ecosystems across the globe. *Global Change Biology, 24*(8), 3452–3461. https://doi.org/10.1111/gcb.14163
Dakora, F. (1995). Plant flavonoids: Biological molecules for useful exploitation. *Functional Plant Biology, 22*(1), 87. https://doi.org/10.1071/pp9950087
Dakora, F. D. (2003). Defining new roles for plant and rhizobial molecules in sole and mixed plant cultures involving symbiotic legumes. *New Phytologist, 158*(1), 39–49. https://doi.org/10.1046/j.1469-8137.2003.00725.x
Dakora, F. D., & Phillips, D. A. (2002). Root exudates as mediators of mineral acquisition in low-nutrient environments. *Plant and Soil, 245*(1), 35–47. https://doi.org/10.1023/a:1020809400075
Doran, J., Elliott, E., & Paustian, K. (1998). Soil microbial activity, nitrogen cycling, and long-term changes in organic carbon pools as related to fallow tillage management. *Soil & Tillage Research, 49*(1–2), 3–18. https://doi.org/10.1016/s0167-1987(98)00150-0
Dumont, M. G., & Murrell, J. C. (2005). Stable isotope probing — Linking microbial identity to function. *Nature Reviews. Microbiology, 3*(6), 499–504. https://doi.org/10.1038/nrmicro1162
Fierer, N. (2017). Embracing the unknown: Disentangling the complexities of the soil microbiome. *Nature Reviews. Microbiology, 15*(10), 579–590. https://doi.org/10.1038/nrmicro.2017.87
Fierer, N., Strickland, M. S., Liptzin, D., Bradford, M. A., & Cleveland, C. C. (2009). Global patterns in belowground communities. *Ecology Letters, 12*(11), 1238–1249. https://doi.org/10.1111/j.1461-0248.2009.01360.x
Franke-Whittle, I. H. (2009). Application of COMPOCHIP microarray to investigate the Franke-Whittle, I. H., Knapp, B. A., Fuchs, J., Kaufmann, R., & Insam, H. (2008). Application of COMPOCHIP microarray to investigate the bacterial communities of different composts. *Microbial Ecology, 57*(3), 510–521. https://doi.org/10.1007/s00248-008-9435-2
Frostegård, S., Tunlid, A., & Bååth, E. (2011). Use and misuse of PLFA measurements in soils. *Soil Biology and Biochemistry/Soil Biology & Biochemistry, 43*(8), 1621–1625. https://doi.org/10.1016/j.soilbio.2010.11.021
Geisen, S., Mitchell, E. A. D., Adl, S., Bonkowski, M., Dunthorn, M., Ekelund, F., Fernández, L. D., Jousset, A., Krashevska, V., Singer, D., Spiegel, F. W., Walochnik, J., & Lara, E. (2018). Soil protists: a fertile frontier in soil biology research. *FEMS Microbiology Reviews, 42*(3), 293–323. https://doi.org/10.1093/femsre/fuy006
Glick, B. R., Cheng, Z., Czarny, J., & Duan, J. (2007). Promotion of plant growth by ACC deaminase-producing soil bacteria. *European Journal of Plant Pathology, 119*(3), 329–339. https://doi.org/10.1007/s10658-007-9162-4
Goldstein, A. H. (1994). Involvement of the quinoprotein glucose dehydrogenase in the solubilization of exogenous phosphates by gram-negative bacteria. In A. Torriani-Gorini, E. Yagil, & S. Silver (Eds.), *Phosphate in microorganisms: Cellular and molecular biology* (pp. 197–203). ASM Press.
Govaerts, B., Mezzalama, M., Unno, Y., Sayre, K. D., Luna-Guido, M., Vanherck, K., Dendooven, L., & Deckers, J. (2007). Influence of tillage, residue management, and crop rotation on soil microbial biomass and catabolic diversity. *Applied Soil Ecology, 37*(1–2), 18–30. https://doi.org/10.1016/j.apsoil.2007.03.006

Harman, G. E., Howell, C. R., Viterbo, A., Chet, I., & Lorito, M. (2004). Trichoderma species — Opportunistic, avirulent plant symbionts. *Nature Reviews. Microbiology, 2*(1), 43–56. https://doi.org/10.1038/nrmicro797

Hilton, T., Ferracane, J., Mancl, L., Baltuck, C., Barnes, C., Beaudry, D., Shao, J., Lubisich, E., Gilbert, A., & Lowder, L. (2013). Comparison of CaOH with MTA for direct pulp capping. *Journal of Dental Research, 92*(7_suppl), S16–S22. https://doi.org/10.1177/0022034513484336

Holmfeldt, K., Howard-Varona, C., Solonenko, N., & Sullivan, M. B. (2014). Contrasting genomic patterns and infection strategies of two co-existingBacteroidetespodovirus genera. *Environmental Microbiology, 16*(8), 2501–2513. https://doi.org/10.1111/1462-2920.12391

Howe, A. C., Jansson, J. K., Malfatti, S. A., Tringe, S. G., Tiedje, J. M., & Brown, C. T. (2014). Tackling soil diversity with the assembly of large, complex metagenomes. *Proceedings of the National Academy of Sciences of the United States of America, 111*(13), 4904–4909. https://doi.org/10.1073/pnas.1402564111

Hu, L., Robert, C. A. M., Selma Cadot, X. I., Zhang, M. Y., Li, B., Manzo, D., et al. (2018). Root exudate metabolites drive plant-soil feedbacks on growth and defense by shaping the rhizosphere microbiota. *Nature Communications, 9*(1), 2738. https://doi.org/10.1038/s41467-018-05122-7

Imfeld, G., & Vuilleumier, S. (2012). Measuring the effects of pesticides on bacterial communities in soil: A critical review. *European Journal of Soil Biology, 49*, 22–30. https://doi.org/10.1016/j.ejsobi.2011.11.010

Jiang, S., Zhang, D., Wang, L., Pan, J., Liu, Y., Kong, X., Zhou, Y., & Li, D. (2013). A maize calcium-dependent protein kinase gene, ZmCPK4, positively regulated abscisic acid signaling and enhanced drought stress tolerance in transgenic Arabidopsis. *Plant Physiology and Biochemistry, 71*, 112–120. https://doi.org/10.1016/j.plaphy.2013.07.004

Jilani, G., Mahmood, S., Chaudhry, A. N., Hassan, I., & Akram, M. (2008). Allelochemicals: Sources, toxicity and microbial transformation in soil —A review. *Annals of Microbiology, 58*(3), 351–357. https://doi.org/10.1007/bf03175528

Kallenbach, C., & Grandy, A. S. (2011). Controls over soil microbial biomass responses to carbon amendments in agricultural systems: A meta-analysis. *Agriculture, Ecosystems & Environment, 144*(1), 241–252. https://doi.org/10.1016/j.agee.2011.08.020

Karhu, K., Auffret, M. D., Dungait, J. A. J., Hopkins, D. W., Prosser, J. I., Singh, B. K., Subke, J. A., Wookey, P. A., Ågren, G. I., Sebastià, M. T., Gouriveau, F., Bergkvist, G., Meir, P., Nottingham, A. T., Salinas, N., & Hartley, I. P. (2014). Temperature sensitivity of soil respiration rates enhanced by microbial community response. *Nature, 513*(7516), 81–84. https://doi.org/10.1038/nature13604

Kasim, W. A., Gaafar, R. M., Abou-Ali, R. M., Omar, M. N., & Hewait, H. M. (2016). Effect of biofilm forming plant growth promoting rhizobacteria on salinity tolerance in barley. *Annals of Agricultural Science, 61*(2), 217–227. https://doi.org/10.1016/j.aoas.2016.07.003

Kaur, T., Devi, R., Kour, D., Yadav, A., Yadav, A. N., Dikilitas, M., Abdel-Azeem, A. M., Ahluwalia, A. S., & Saxena, A. K. (2021). Plant growth promoting soil microbiomes and their potential implications for agricultural and environmental sustainability. *Biologia, 76*(9), 2687–2709. https://doi.org/10.1007/s11756-021-00806-w

Ke, P.-J., & Miki, T. (2015). Incorporating the soil environment and microbial community into plant competition theory. *Frontiers in Microbiology, 6*, 133134. https://doi.org/10.3389/fmicb.2015.01066

Kodaparthi, A., Ranjit, P., Deepu, P. G., Kaushik, D. M., Valli, L. A., Ashrutha, P., Harihara, J., & Chepuri, K. (2024). Impact of regenerative agriculture on soil erosion. In *Earth and environmental sciences library* (pp. 269–295). Springer. https://doi.org/10.1007/978-3-031-53270-2_12

Kolton, M., Harel, Y. M., Pasternak, Z., Graber, E. R., Elad, Y., & Cytryn, E. (2011). Impact of biochar application to soil on the root-associated bacterial community structure of fully developed greenhouse pepper plants. *Applied and Environmental Microbiology, 77*(14), 4924–4930. https://doi.org/10.1128/aem.00148-11

Lauber, C. L., Ramirez, K. S., Aanderud, Z., Lennon, J., & Fierer, N. (2013). Temporal variability in soil microbial communities across land-use types. *The ISME Journal, 7*(8), 1641–1650. https://doi.org/10.1038/ismej.2013.50

Lehmann, J., Rillig, M. C., Thies, J., Masiello, C. A., Hockaday, W. C., & Crowley, D. (2011). Biochar effects on soil biota—A review. *Soil Biology and Biochemistry/Soil Biology & Biochemistry, 43*(9), 1812–1836. https://doi.org/10.1016/j.soilbio.2011.04.022

Lhuissier, F. (2001). Time course of cell biological events evoked in legume root hairs by rhizobium nod factors: State of the art. *Annals of Botany, 87*(3), 289–302. https://doi.org/10.1006/anbo.2000.1333

Liu, B. R., Jia, G. M., Chen, J., & Wang, G. (2006). A review of methods for studying microbial diversity in soils. *Pedosphere, 16*(1), 18–24. https://doi.org/10.1016/s1002-0160(06)60021-0

Lupwayi, N. Z., Harker, K. N., Clayton, G. W., Turkington, T. K., Rice, W. A., & O'Donovan, J. T. (2004). Soil microbial biomass and diversity after herbicide application. *Canadian Journal of Plant Science/Canadian Journal of Plant Science, 84*(2), 677–685. https://doi.org/10.4141/p03-121

Mallik, A. U. (2003). Conifer regeneration problems in boreal and temperate forests with ericaceous understory: Role of disturbance, seedbed limitation, and Keytsone species change. *Critical Reviews in Plant Sciences, 22*(3–4), 341–366. https://doi.org/10.1080/713610860

Matiru, V. N., & Dakora, F. D. (2004). Potential use of rhizobial bacteria as promoters of plant growth for increased yield in landraces of African cereal crops. *African Journal of Biotechnology, 3*, 1–7. https://doi.org/10.5897/AJB2004.000-2002

McDaniel, M. D., Tiemann, L. K., & Grandy, A. S. (2014). Does agricultural crop diversity enhance soil microbial biomass and organic matter dynamics? A meta-analysis. *Ecological Applications, 24*(3), 560–570. https://doi.org/10.1890/13-0616.1

McGrath, J. W., Wisdom, G. B., McMullan, G., Larkin, M. J., & Quinn, J. P. (1995). The purification and properties of Phosphonoacetate hydrolase, a novel carbon-phosphorus bond-cleavage enzyme from Pseudomonas Fluorescens 23F. *European Journal of Biochemistry, 234*(1), 225–230. https://doi.org/10.1111/j.1432-1033.1995.225_c.x

McGrath, J. W., Hammerschmidt, F., & Quinn, J. P. (1998). Biodegradation of Phosphonomycin by Rhizobium huakuii PMY1. *Applied and Environmental Microbiology, 64*(1), 356–358. https://doi.org/10.1128/aem.64.1.356-358.1998

Mosina, L. V., & Merzlaya, G. E. (2013). Ecological assessment of the influence of organic and mineral fertilizers on the microflora of soddy-podzolic soil and the productivity of agrocenoses in extreme weather conditions. *Izv. Timiryazevsk. S-kh. Akad., 5*, 5–18.

Muñoz-Leoz, B., Ruiz-Romera, E., Antigüedad, I., & Garbisu, C. (2011). Tebuconazole application decreases soil microbial biomass and activity. *Soil Biology and Biochemistry/Soil Biology & Biochemistry, 43*(10), 2176–2183. https://doi.org/10.1016/j.soilbio.2011.07.001

Nielsen, K. E., Irizar, A., Nielsen, L. P., Kristiansen, S. M., Damgaard, C., Holmstrup, M., Petersen, A. R., & Strandberg, M. (2017). In situ measurements reveal extremely low pH in soil. *Soil Biology and Biochemistry/Soil Biology & Biochemistry, 115*, 63–65. https://doi.org/10.1016/j.soilbio.2017.08.003

Offre, P., Spang, A., & Schleper, C. (2013). Archaea in biogeochemical cycles. *Annual Review of Microbiology, 67*(1), 437–457. https://doi.org/10.1146/annurev-micro-092412-155614

Ohtake, H., Wu, H., Imazu, K., Anbe, Y., Kato, J., & Kuroda, A. (1996). Bacterial phosphonate degradation, phosphite oxidation and polyphosphate accumulation. *Resources, Conservation and Recycling, 18*(1–4), 125–134. https://doi.org/10.1016/s0921-3449(96)01173-1

Patten, C. L., & Glick, B. R. (2002). Role of pseudomonas putida Indoleacetic acid in development of the host plant root system. *Applied and Environmental Microbiology, 68*(8), 3795–3801. https://doi.org/10.1128/aem.68.8.3795-3801.2002

Peay, K. G., Kennedy, P. G., & Talbot, J. M. (2016). Dimensions of biodiversity in the earth mycobiome. *Nature Reviews. Microbiology, 14*(7), 434–447. https://doi.org/10.1038/nrmicro.2016.59

Philippot, L., Raaijmakers, J. M., Lemanceau, P., & Van Der Putten, W. H. (2013). Going back to the roots: The microbial ecology of the rhizosphere. *Nature Reviews. Microbiology, 11*(11), 789–799. https://doi.org/10.1038/nrmicro3109

Richardson, A. E., Barea, J. M., McNeill, A. M., & Prigent-Combaret, C. (2009). Acquisition of phosphorus and nitrogen in the rhizosphere and plant growth promotion by microorganisms. *Plant and Soil, 321*(1–2), 305–339. https://doi.org/10.1007/s11104-009-9895-2

Rodriguez, H., & Fraga, R. (1999). Phosphate solubilizing bacteria and their role in plant growth promotion. *Journal of Biotechnology Advances, 17*, 319–339. https://doi.org/10.1016/S0734-9750(99)00014-2

Rousk, J., Brookes, P. C., & Bååth, E. (2010). The microbial PLFA composition as affected by pH in an arable soil. *Soil Biology and Biochemistry/Soil Biology & Biochemistry, 42*(3), 516–520. https://doi.org/10.1016/j.soilbio.2009.11.026

Singh, B. K., Bardgett, R. D., Smith, P., & Reay, D. S. (2010). Microorganisms and climate change: Terrestrial feedbacks and mitigation options. *Nature Reviews Microbiology, 8*(11), 779–790. https://doi.org/10.1038/nrmicro2439

Smith, S. E., & Read, D. J. (2008). *Mycorrhizal symbiosis*. Academic Press. https://doi.org/10.1016/b978-0-12-370526-6.x5001-6

Smith, S. E., & Read, D. J. (2010). *Mycorrhizal symbiosis*. Academic.

Stewart, E. J. (2012). Growing unculturable bacteria. *Journal of Bacteriology, 194*(16), 4151–4160. https://doi.org/10.1128/jb.00345-12

Strickland, M. S., & Rousk, J. (2010). Considering fungal: Bacterial dominance in soils—Methods, controls, and ecosystem implications. *Soil Biology and Biochemistry/Soil Biology & Biochemistry, 42*(9), 1385–1395. https://doi.org/10.1016/j.soilbio.2010.05.007

Van Wees, S. C., Van Der Ent, S., & Pieterse, C. M. (2008). Plant immune responses triggered by beneficial microbes. *Current Opinion in Plant Biology, 11*(4), 443–448. https://doi.org/10.1016/j.pbi.2008.05.005

Vandenkoornhuyse, P., Quaiser, A., Duhamel, M., Le Van, A., & Dufresne, A. (2015). The importance of the microbiome of the plant holobiont. *New Phytologist, 206*(4), 1196–1206. https://doi.org/10.1111/nph.13312

Venturi, V. (2006). Regulation of quorum sensing inPseudomonas. *FEMS Microbiology Reviews, 30*(2), 274–291. https://doi.org/10.1111/j.1574-6976.2005.00012.x

Vimal, S. R., Singh, J. S., Arora, N. K., & Singh, S. (2017). Soil-plant-microbe interactions in stressed agriculture management: A review. *Pedosphere, 27*(2), 177–192. https://doi.org/10.1016/s1002-0160(17)60309-6

Wagner, M., Nielsen, P. H., Loy, A., Nielsen, J. L., & Daims, H. (2006). Linking microbial community structure with function: Fluorescence in situ hybridization-micro autoradiography and isotope arrays. *Current Opinion in Biotechnology, 17*(1), 83–91. https://doi.org/10.1016/j.copbio.2005.12.006

Wang, X., Chi, Y., & Song, S. (2024). Important soil microbiota's effects on plants and soils: A comprehensive 30-year systematic literature review. *Frontiers in Microbiology, 15*, 1347745. https://doi.org/10.3389/fmicb.2024.1347745

Weir, T. L., Park, S. W., & Vivanco, J. M. (2004). Biochemical and physiological mechanisms mediated by allelochemicals. *Current Opinion in Plant Biology, 7*(4), 472–479. https://doi.org/10.1016/j.pbi.2004.05.007

Williamson, K. E., Fuhrmann, J. J., Wommack, K. E., & Radosevich, M. (2017). Viruses in soil ecosystems: An unknown quantity within an unexplored territory. *Annual Review of Virology, 4*(1), 201–219. https://doi.org/10.1146/annurev-virology-101416-041639

Yue, Y., Hao, H., Wang, Q., Xiao, T., Zhang, Y., Chen, Q., Chen, H., & Zhang, J. (2024). Dynamics of the soil microbial community associated with Morchella cultivation: Diversity, assembly mechanism and yield prediction. *Frontiers in Microbiology, 15*, 1345231. https://doi.org/10.3389/fmicb.2024.1345231

Zak, D. R., Holmes, W. E., White, D. C., Peacock, A. D., & Tilman, D. (2003). Plant diversity, soil microbial communities, and ecosystem function: Are there any links? *Ecology, 84*(8), 2042–2050. https://doi.org/10.1890/02-04331

Challenges of Soil Microbiomes and Food Safety

Srinivasan Kameswaran, Bellamkonda Ramesh, Gopi Krishna Pitchika, and Manjunatha Bangeppagari

Abstract The serious agricultural issues we currently face pose a threat to the future of global food security. The potential for soil microorganisms to offer economically and environmentally sound remedies may facilitate the introduction of novel techniques to enhance agricultural operations and, in turn, increase crop yields. In addition to examining novel techniques and technologies, the chapter describes the current state of microbial studies. We also share our knowledge of basic studies on the microbiome to help improve our understanding of the movement of microbial communities and how they interact with various plant hosts and the environment. We want to link these strategies to possible uses and consider how to include microbes ecosystems in contemporary farming practices to achieve a more tailored and sustainable utilization of these essential resources.

Keywords Plant interactions · Microbial community · Bacterial function · Microbiome · Omics · Farming systems

1 Introduction

Perhaps the more different environment on the Planet is the soil, which is home to bigger animals like worms, ants, and moles as well as millions of microbial species and fungi representing hundreds of genera (Bardgett & van der Putten, 2014). Bacterial diversification is the main cause of the expanding Tree of Life, which is

S. Kameswaran (✉)
Department of Botany, Vikrama Simhapuri University College, Kavali, Andhra Pradesh, India

B. Ramesh
Molecular Biology, University of Nebraska Medical Center, Omaha, NE, USA

G. K. Pitchika
Department of Zoology, Vikrama Simhapuri University College, Kavali, Andhra Pradesh, India

M. Bangeppagari
Department of Cell Biology and Molecular Genetics, Sri Devaraj Urs Academy of Higher Education and Research, Kolar, Karnataka, India

S. A. Aransiola et al. (eds.), *Soil Microbiome in Green Technology Sustainability*,
https://doi.org/10.1007/978-3-031-71844-1_6

being produced by an increasing number of microbial culture libraries and new advancements in high throughput sequencing techniques (Hug et al., 2016). This enormous diversity results from bacteria's ability to transfer genes between phylogenetic groups that are very different from one another (McDonald & Currie, 2017). In addition, microbial communities can proliferate more quickly than higher organisms like humans, possess substantial rates of mutation (from "10–4" in Escherichia coli), as well as are made up of many individuals per species (Robbins et al., 2016; Kibota & Lynch, 1996; Boe et al., 2000; Denamur & Matic, 2006). Even within the same species, individual microorganisms may have different genetic makeups and, as a result, functional characteristics. These characteristics increase the diversity of bacteria and microbial communities.

Apart from providing defense against biotic and abiotic stressors, soil microbes are crucial for the cycling of nutrients such as nitrogen and phosphorus (Bender et al., 2016; Lladó et al., 2017). Although crop yields have increased as a result of intensive farming, the biological and physical characteristics of the soil have suffered as a result (Pimentel et al., 1995; Bouwman et al., 2009). Although plowing disrupts microbe populations, fertilizer application can replenish decreased fertility in the soil in intensively farmed agricultural areas (Johnson et al., 1997). The way that crops are currently produced, is especially crucial because over half of the world's agricultural land is degrading and disruption of the nitrogen and phosphorus cycles presents serious issues. Given the likelihood of global warming, it is quite likely that this condition will get worse (Rockström et al., 2009; Yuan et al., 2018). The UN has therefore suggested restoring sustainable land management approaches to lessen land degradation (Sanz et al., 2017). Utilizing locally adapted species, intercropping, and crop diversity are some of the strategies used to maintain soil fertility, sequester carbon, maintain nutrient cycling, and manage soil erosion (Sanz et al., 2017). It is noteworthy that these techniques also enhance the general management of soil diseases (Weller et al., 2002; Bonilla et al., 2015). It can also support ecological processes including regulating nutrient cycles to preserve the diversity, organization, and makeup of the microbial population (Fig. 1).

Because of research on the microbiome, perceptions of the complexity and organization of microbial communities have changed during the past ten years. Nevertheless, little is still known about the composition of these complex communities and their interactions with the biotic and abiotic worlds. New perspectives on the changes occurring inside microorganisms as well as how they communicate with

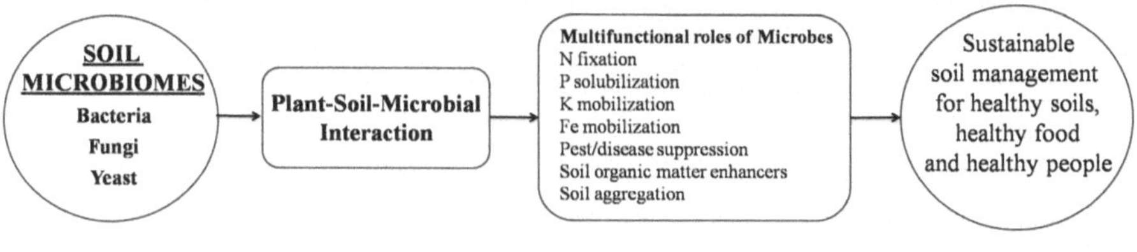

Fig. 1 The role of soil microorganisms in enhancing plant health, food crop productivity, and nutritional quality and safety throughout time

the hosts have been made possible by the development of new instruments and the growing demand for different experimental strategies (Goodrich et al., 2017). The intricacy of microbial connections and the significance of investigating them separately for every person has led to the development of novel therapeutics through the research of the microbiome of humans. One might pursue a customized application of microbiome, considering the unique microbiome profile of every host. By treating each case individually and better tailoring the microbiome to the needs of the host, we would be able to maximize the potential of the host and improve treatment efficiency (Human Microbiome Project Consortium, 2012).

The single agricultural crop variety's uniformity of genes in host genomes would make it easier to apply "personalized" microbial methods in farming. Analogous studies have been conducted not only on plant and soil microbial communities but also on human microbiomes (Bulgarelli et al., 2012; Lundberg et al., 2012). Being a role of an impending "green revolution", these efforts aim to lead the way in advances in organic agriculture operations (Jez et al., 2016). But to use microbiomes to their full potential, we must develop novel techniques for analysis to understand the range of beneficial properties that microbial populations have (Bashiardes et al., 2018). A varied and harmonious microbiome along the connection between soil and crops is crucial for agricultural productivity. Microbiomes can be used to improve soil fertility, health, and nutrient availability—three critical aspects that impact agricultural output (Syed Ab Rahman et al., 2018) (Fig. 1). Here, the primary concern is translating lab-created knowledge into real-world applications. Not only must we figure out the composition of the plant/soil microbiome, but it is also imperative, as Sánchez-Cañizares et al. (2017) note, that we link the dynamics of microbial communities to the functioning of the microbiome (Schlaeppi & Bulgarelli, 2015). This chapter outlines the challenges and latest efforts aimed at expanding our understanding of the various facets of microbiomes (including dynamics and structure) and their effects on plants. We also present emerging approaches to maximize crop productivity through the use of the soil microbiome, which includes beneficial bacteria.

2 Research on Microbiomes Is Expanding in the "OMICS" Era

The decline in sequencing costs, together with advances in sequencing technology and bigger computers, have allowed for a plethora of microbiome-related studies concerning soil and rhizosphere (Prosser, 2015; Fierer, 2017). Three primary forms of sequencing are commonly used by scientists: (1) "Metataxonomic primer-based amplicon sequencing", generally known as the intergenic spacer (ITS) region (Eukaryotes) or 16S rRNA (Bacteria and Archaea), focuses on amplifying specific regions of ubiquitous genetic markers. (2) Singleton or whole genome sequencing that includes metagenomics and metatranscriptomics information inside a sample

(3) The study of separated and fragmented proteins is known as metaproteomics, and it frequently combines mass spectrometry (LC-MS-MS) with nuclear magnetic resonance (NMR) or mass spectrometry (MS) for the identification of metabolites (metabolomics). The technical limits of amplicon sequencing and metabolomics have been extensively reviewed in several studies. These limitations include sample mistakes, biases in primers or processing, computer capacity, and suitable analytical techniques. (Hirsch et al., 2010; Carvalhais et al., 2012; Pinto & Raskin, 2012; Scholz et al., 2012; Temperton & Giovannoni, 2012; Tkacz & Poole, 2015) These reviews will not be reproduced here. It is heartening that significant progress has been achieved in addressing these restrictions. Though several "perfect" metaomics approaches have been developed, there are still many conceptual limits, such as the requirement for careful experimental design to extract useful information from metaomics datasets. For example, a significant obstacle to projecting results from "omics datasets" is the misannotation of "genes in the datasets" (Lidbury et al., 2014; Fierer, 2017). Moreover, a great deal of genetic information that could be able to perform new functions is usually overlooked in favor of genes with known functions, particularly in the context of metagenomics. This is particularly true for genes that encode essential extracellular proteins for the environment (Christie-Oleza et al., 2015).

3 Innovative Methods for Illustrating the Characteristics of Complex Microbial Communities

The ability to sequence many samples simultaneously in a single sequencing experiment is one of the many benefits of amplicon sequencing, which improves spatio-temporal resolution and makes assessing multivariate variables easier. According to several studies (Bulgarelli et al., 2012; Lundberg et al., 2012; Peiffer et al., 2013; Tkacz et al., 2015), amplified sequencing has therefore shown to be quite beneficial in detecting general patterns in the variety of microorganisms within the plant microbiome. These days, to contribute to the resolution of particular issues about the recruitment and operation of bacteria in the rhizosphere by plants, amplicon sequencing is frequently used in conjunction with more sophisticated techniques like microscopy (Rybakova et al., 2017), crop mutant line experiments (Senga et al., 2017), "multi-generation plant trait selection studies (Haney et al., 2015; Panke-Buisse et al., 2015), and "exometabolic profiling of plant exudates" (Badri et al., 2013; Zhalnina et al., 2018) (Fig. 1). The use of "16S rRNA gene data", according to Dawson et al. (2017), additionally indicates that organisms belonging to the "rare" "biosphere are actively" sought after "in the rhizosphere", indicating that their rarity may not hinder them from carrying out significant activities. Additionally, Bulgarelli et al. (2015) proposed that to access and characterize a greater variety of microorganisms along to obtain an enhanced illustration understanding of the variety and organization of microbes, it is necessary to use markers other than "16S rRNA," like "18S rRNA" or "internal transcribed spacers" ("ITSs"). Simply "sequencing" each of the "genetic material" in the specimen at once, "metagenomics" or "metatranscriptomics" can somewhat alleviate this issue (Chaparro et al., 2014; Bulgarelli et al., 2015).

4 Difficulties and Novel Approaches in Determining the Role of Microorganisms

Every microbial group is generally thought of as a functional group, although it is exceedingly difficult to assign a role to individual microbiome individuals or groups. It is not enough to just identify the architecture of the microbiome in different environments. However, even within the same genus, animals may lead completely different lives due to horizontal gene transfer or environmental factors (Hacquard et al., 2016; Hiruma et al., 2016). Examples of these scenarios include parasite to mutualist (Qiu et al., 2009). The bacterial morphologies of favorable features, including phosphorus mobilization, may undergo notable alterations due to this diversity (Lidbury et al., 2016). According to Schleppi and Bulgarelli et al. (2015), large-scale, effective techniques are needed for improved functional characterization of each species as well as more sensitive methodologies to characterize the microbiome beyond the genus level. With the rate at which technology is advancing, combining modeling and computational techniques, for instance, holds much promise. "Metaphenomics," for instance, is a field that is developing from "metagenomics." It combines the output of the "metagenome," or "expressed functions encoded in microbial genomes," with the external environment, taking into consideration all factors that may affect the dynamics of interactions between communities and their surroundings (Jansson & Hofmockel, 2018). The field of "metagenomics" has significant potential as it can provide insights into the unique gene sets (such as metabolic pathways) and broad functional abilities (such as the generation of secondary metabolites and the utilization of carbohydrates) of the microbiome (Bulgarelli et al., 2015; Wang et al., 2015). New analytical methods may help us comprehend the various taxa that have important functional traits, according to Prosser (2015). Since basic commercial kits have been available, it is now simple to extract DNA from a given sample. Yet the functional "potential" of a microbial community is all that DNA can offer. Furthermore, a sizable amount of genetic material, or microbial biomass, is latent or inactive, particularly in soils (Fierer, 2017 offers a comprehensive summary). This number is anticipated to drop sharply in the "rhizosphere" and root, though, as a consequence of plant-mediated mechanisms that selectively select microorganisms in these niches, where more bacteria are active in metabolism (Bulgarelli et al., 2013). RNA isolated from rhizosphere samples has been used in several studies to try and identify metabolically active bacteria and the "genetic pathways" that activate when reacting to plant and microbial "stimuli" (Turner et al., 2013b; Chaparro et al., 2014; Yergeau et al.,

2014). To better understand the connections between the microbes and crop host, 13C-labeled CO_2 has thus been added and used in conjunction with "metatranscriptomics" in several studies to identify bacteria responding to plant exudates (Haichar et al., 2016; Lueders et al., 2016).

When compared to "proteins", particularly "exoproteins", that remain less volatile in the surroundings, RNA has a short change period and decreases its complexity and robustness of sampling efficiency (Prosser, 2015; Armengaud et al., 2012). Because of this, the sample is more error-prone and methodologically simpler. As a means of obtaining an overview of protein expression and, by extension, ("metabolic") reactions within a specific specimen, metaproteomics provides another fascinating direction for omics research (Heyer et al., 2015). Extracellular hydrolytic enzymes and transporter systems are two examples of the more ecologically significant proteins that are enriched by exoproteomics, also referred to as "exometaproteomics," which is involved in the uptake of nutrients and the interactions between microbes and their hosts (Armengaud et al., 2012; Lidbury et al., 2016). Nevertheless, "metaproteomics" is not without its limitations. To be more precise, a significant amount of beginning material—up to 100 grams of soil may occasionally be needed—and computer capacity is needed to correctly match peptides found with the appropriate proteins. The metagenome's enormous database is primarily to blame for these problems (Timmins-Schiffman et al., 2017; Muth et al., 2013, 2016). This may be the cause of the paucity of meta(exo)proteomics research on rhizospheres in comparison to studies of other less complex microbial niches, like enhanced chitin-degrading sandy soil samples (Johnson-Rollings et al., 2014) and anaerobic digestion systems and seawater (Sowell et al., 2009; Williams et al., 2012; Heyer et al., 2015). This technique can also be used to successfully identify the major extracellular enzymes responsible for phosphate mobilization in the rhizosphere (Lidbury et al., 2016; Lidbury et al., unpublished data).

5 Utilizing, Isolating, and Determining Microbial Strains in Artificial Environments: An Attempt

Numerous recent studies have returned to a stringent isolation process that is followed by gene sequencing and phenotypic characterization because the four main phyla of rhizobacteria—"Actinobacteria", "Proteobacteria", "Bacteroidetes", and "Firmicutes"—can be grown (Bai et al., 2015; Mauchline et al., 2015; Levy et al., 2018). Identification of important species and recruitment patterns in the rhizosphere is commonly achieved by combining them with the restoration of artificial communities (Bai et al., 2015; Niu et al., 2017). Plant development-influencing "microbe-microbe" associations can be found by reconstructing microbial populations (Hartman et al., 2017). Compared to producing "metagenomes" obtained "in situ", the isolation strategy has the advantage of "sequencing" and constructing specific genomes considerably more easily, and frequently at a "greater resolution of data". Moreover, Levy et al. (2018) state that the isolation strategy is a sophisticated method of confirming isolates' interactions with host plants and/or their functional validity within a community. It is simple to use isolates exhibiting either distinct or increased activity for additional research aimed at pinpointing the exact chemical processes and related "rate kinetics of" important biocatalysts. Reports by Mauchline et al. (2006), Lidbury et al. (2016), Martino et al. (2018), and Fernández et al. (2013) have reported significant insights gained from examining the transcriptome or proteome responses of specific microbes or fungi strains to the crop microbes or related nutritional stressors. Genes implicated in potentially important pathways related to plant growth promotion (PGP) and the attraction of advantageous microorganisms have been found in several of these investigations. The identification of novel features linked to PGP activities mediated by microorganisms can also benefit greatly from this strategy (Bruto et al., 2014). Finkel et al. (2017) suggest that the application of "genomic screening" methods may facilitate a thorough and quick testing procedure while encouraging the identification of novel PGP features and genes. This is because in vitro screening methods for PGP features may not always display phenotypes linked to plants. By combining these techniques with complementary molecular methods, like "bio-reporter expression systems" and mutagenic plants (Wetmore et al., 2015; Cheng et al., 2016; Pini et al., 2017), we can better understand the function of these "PGP traits/genes" and make "predictions about the mechanisms" underlying "interactions within plant microbiomes".

6 Tributary Set-Up Microbiome Investigation

6.1 The Importance of Comprehending Interactions

The composition of a plant's root microbiome can be influenced by infections or other harmful microorganisms to the detriment of beneficial microorganisms (Berendsen et al., 2012; Bulgarelli et al., 2015). To take advantage of "the host's" metabolism, microorganisms can also alter it (Jones & Dangl, 2006; Engelstädter & Hurst, 2009; Hacquard et al., 2015; Levy et al., 2018). Furthermore, the microbial and plant communities are impacted by the abiotic environment (e.g., edaphic variables), which increases the complexity of ecosystems (Bakker et al., 2014). The significance of "microbial hubs," or highly interconnected microbial taxa that have a significant impact on communities and are believed to be essential to comprehending microbiome dynamics, as well as the impact of individual microbes on the composition of microbial communities, has been demonstrated by network analyses (Agler et al., 2016). Niu et al. (2017), For example, synthetic communities using seven relevant isolates of bacteria obtained from all three most prevalent taxa from "maize roots". Using this distilled community, their goal was to identify processes governing this system's behavior. Notably, the elimination of a single strain resulted in the community's total collapse, underscoring the significance of individual microbiome members (Niu et al., 2017). It also implies that minor adjustments, or even slight ones, can have a big impact on the architecture of the microbiome. Stable synthetic community validation may thus depend on our ability to understand the underlying inter-microbial processes that shape community architecture. New attempts are being undertaken to analyze these complicated interactions and create reliable systems to (i) overcome the complexity of the soil's environment and (ii) increase our basic knowledge of bacterial and plant-microbe relationships.

7 Progress Towards Resolving Complexity in Soil Ecosystem

Since the discovery of the "Arabidopsis thaliana core root microbiome", has offered a greater understanding of crop microbiota structures, a plethora of research has together highlighted the importance of microbiomes in ecosystem functioning (Bulgarelli et al., 2012; Agler et al., 2016). Additionally, the establishment of well-defined gnotobiotic systems in conjunction with the isolation and characterization of microbial species (Lebeis et al., 2015; Finkel et al., 2017) enables the specific functional characterization of each member of the plant microbiome (Bai et al., 2015). The importance of "gnotobiotic systems" in particular for studying the microbiome has been acknowledged since they make it possible to differentiate between the impact of individual bacteria or microbial combinations and the environment (by imposing specific conditions) on plant phenotypes. It provides the opportunity to link genotypes to phenotypes by illuminating specific mechanisms in the multicomponent interactions between plants, microbes, and environments. However the "artificial" or "de novo" synthesis of microbiomes, combined with the systems' reduced complexity, may hinder natural systems' ability to recapitulate and hence operate fully (Vorholt et al., 2017).

A novel use of microbes was introduced by Hartman et al. (2017) in an attempt to comprehend intricate interactions within microbiomes. Following the extraction of microorganisms from *Trifolium pratense* roots, they chose a single strain to grow in "sterile microcosms" from each of the four OTUs (the majority common microbe groupings). They observed that while *Flavobacterium* had a detrimental influence on *Trifolium* growth, the presence of any of the additional three bacteria species among "Pseudomonas", "*Janthinobacterium*", and "*Microbacterium*" lessened this effect. Remarkably, not one of the three bacteria had an impact on the Flavobacterium's abundance inside the synthetic community. Thus, according to Hartman et al. (2017), *Pseudomonas, Janthinobacterium*, and *Microbacterium* "buffered" the detrimental activity of *Flavobacterium*. A reductionist method like this can identify new "keystone" actors that control the microbiome's activity and how it interacts with plants.

Given their importance for the growth of microbiome research, all these analytical methodologies have benefited substantially from the developments in computer analysis and machine learning (Knight et al., 2018). By connecting bacteria to specific plants or inhabitants through the use of pre-existing datasets, bioinformatics tools such as COREMIC facilitate the creation and validation of theoretical models (Rodrigues et al., 2018). Regarding additional omics-based analyses, attention has been drawn to the necessity of reproducibility and the creation of "golden standards" to enhance experiment consistency and comparability (Knight et al., 2018). In addition, network analyses have provided microbiome researchers with advanced instruments to examine and interpret the intricacy of bacterial populations (Adair & Douglas, 2017; Wang et al., 2017b). Network analyses have certain limitations in terms of finding synergistic, additive, and antagonistic effects, even though they frequently lay the groundwork for exposing the nature of microbial interactions and the function of microbial taxa (Poudel et al., 2016). Therefore, the important roles of some low-abundance microorganisms may go unnoticed or unappreciated (Shade et al., 2014; Shi et al., 2016). Sensitive methods that reveal correlative relationships that can be confirmed in biological tests are therefore necessary to achieve the ultimate goal of these studies, which is to identify interactions among microbial communities on plants as a consequence of inter-microbial communication.

8 Developments in the Basic Science of the Interactions Between Plants and Microbes

Beyond an understanding of the intricacy of microbial communities and interactions, fundamental principles of the regulation (communicative) of interactions must be identified to guide experimental design. To research microbial interactions, cytology-based systems have recently been developed. Using a microplate reader, Hennessy et al. (2017) developed a system to measure the movements and interactions of living microorganisms. With live imaging of fluorescing metabolites and microbial growth, this approach enables the identification and tracking of specific gene expression in defined microbial communities and hence constitutes a promising high-throughput screen. The rate at which *Pseudomonas fluorescens* produced fluorescent compounds in response to the presence of *Fusarium graminearum* was assessed in the study as a measure of the interaction between the two (Hennessy et al., 2017). An alternative method for imaging plant root–microbe interactions in vivo is a microfluidics-based system that was recently published by Massalha et al. (2017). They were able to record the root zone—ideally colonized by a bacterium with a fluorophore tag—by using a clear chamber. They were able to examine microbe–bacterium interactions in real-time by introducing a second microbe into the system. These investigations, despite their simple setup, provide important insights into underlying principles that influence interactions between microbes and roots. Accordingly, the use of transparent soil provides a novel way to investigate and see microorganisms on plant roots in a setting that simulates various soil textures (Downie et al., 2012). It enables the investigation of the mechanisms driving the dispersion of microorganisms in large quantities of the substrate through the root (Downie et al., 2014) and the behavior of microbes and communities as influenced by major root pests, such as nematodes (O'Callaghan et al., 2018).

Apart from the microbiomes of bacteria and fungi, various additional species such as viruses, archaea, nematodes, and insects also have a direct impact on soil and plant processes. Though their exact function is unknown, viruses are crucial to soil biochemical processes and serve as gene reservoirs for horizontal gene transfer (Pratama & van Elsas, 2018). In a similar vein, Archaea and nematodes have a major role in the variety of the microbiome, as well as in the interaction of other microorganisms with crop-soil mechanism and environment functioning (Adam et al., 2017; Castillo et al., 2017; Elhady et al., 2017). Benítez et al. (2017) have provided a fascinating perspective on the interactions between plants, microbes, and insects in this regard. They stated that by regulating the release of plant volatiles, soil microbes can influence aboveground interactions between plants and insects (Pineda et al., 2015; Beck & Vannette, 2017; Benítez et al., 2017). This research suggests that to comprehend which biotic and edaphic elements impact the structure and, consequently, function of soil and plant microbiome systems, we need to conduct more thorough, comprehensive investigations on multitrophic interactions.

It will take knowledge from several domains to fully utilize microorganisms and microbial communities. We need to create new ecological systems with increasing complexity in addition to utilizing methods based on plant biology and microbiology to better comprehend complicated interactions between plants and microbes as well as multitrophic interactions. Most importantly, for this knowledge to be effectively applied to agriculture, it is necessary to comprehend how various cultivation methods impact the microbial ecosystem and the way it relates to crop health and, ultimately, agricultural productivity. To bridge the gap between the lab and the farm, it is also essential to assess the function of microbial communities in a highly diverse and complicated environment, such as a field.

9 Mitigation of the Lab-Field Gap

9.1 Restrictions on the Trials Conducted Under Controlled Conditions

Plant-microbe interactions can be advantageous or disadvantageous for the microbial symbiont or host, depending on how related biotic and abiotic factors balance out, according to mounting data. Although research on pairwise interactions conducted in controlled environments has expanded our understanding of the gene and metabolite expression profiles involved in plant-microbe interactions, it has not provided us with much insight into the role of microbes in a natural ecosystem (de Boer, 2017). This holds for many other relationships as well, including those between plants and their mycorrhizal or rhizobia, even though de Boer (2017) made this statement regarding fungal-bacterial interactions. In some circumstances, such as low light, low temperature, or low phosphorus (P) availability, it has been demonstrated that certain species of arbuscular mycorrhizae fungi (AMF), which are known to promote plant growth, instead hinder plant growth (Smith & Smith, 1996; Johnson et al., 1997). Furthermore, it has been observed that the soil microbiota can inhibit AMF activity (Svenningsen et al., 2018). This underscores the necessity of doing field studies to gain a comprehensive understanding of microbial behavior.

The complexity of interactions between microbes in a natural environment is often not captured in lab investigations, which is the primary cause of the lab-farm gap. Currently, however, it is well acknowledged that interactions between plants and microbes as well as microbial communities rely heavily on the ecosystem as a whole (Bulgarelli et al., 2013; de Boer, 2017; Lewis et al., 2018). For example, plant microbial communities have been demonstrated to be influenced by host genotypes (Bulgarelli et al., 2013; Horton et al., 2014). Furthermore, a "genome-wide association study" ("GWAS") research showed that the communities of both bacteria and fungus are shaped by the same host biological processes (such as signal transduction or the defensive response).

However, other genes seem to be involved in the relationship (Horton et al., 2014), as well as microbe communities undergo additional modification during plant growth based on the requirements of the host (Chaparro et al., 2014). The microbial population is probably managed by the transport of nutrients through "the root to the rhizosphere", also known as "rhizodeposition" (Chaparro et al., 2014; Baran et al., 2015). This process has been demonstrated to draw in beneficial microorganisms and deter pathogen attacks. A beautiful illustration of rhizodeposits choosing advantageous microorganisms is the legume-rhizobia symbiosis. The host produces flavonoid chemicals in response to low nitrogen, which starts a molecular conversation with rhizobia that fixes nitrogen. This conversation leads to root nodulation and nitrogen fixation (Oldroyd, 2013). Furthermore, according to Jones et al. (2009), rhizodeposition serves as a chemical signal for the initiation of interactions between roots and microbes. As a result, various hosts with varying gene sets will react differently to the same inoculant. Additionally, the microbial ecology and soil nutrients will affect the root exudates released by the same host. The cyclic feedback between the plant, soil, and microbial components of this ecosystem is exemplified by each of these instances. This probably clarifies why field microbial inoculants don't last very long (Finkel et al., 2017). Furthermore, since the formation of the root microbiome is dependent on the soil environment (Edwards et al., 2015; Zarraonaindia et al., 2015), it is not a given that the inoculant would colonize the plant host even if it does survive within the soil community.

Any advantages that bacteria mediate and that are seen in controlled environments must eventually be functional in the field. This suggests that they have remained in the field for an extended period and have colonized plants successfully under a variety of biotic, abiotic, and climatic conditions. Since it will be extremely rare to find a single inoculant that works in all of these conditions, microbial precision agriculture—which is modeled after the idea of human personalized medicine—must be developed (Hamburg & Collins, 2010). The significance of the root microbiome in regulating plant fitness is demonstrated by the remarkable resemblances between the human gut and plant root microbiomes (Berendsen et al., 2012). The synergistic effects of a community have been demonstrated, for example, by the ability of complex microbial inoculums to better stimulate growth and improve plant disease resistance than individual inoculums (reviewed in Finkel et al., 2017). Still, to address their possible application as soil supplements, these discoveries must take the soil context into account. Functional research where the ecosystem is tested by various variables (such as temperature, light, and humidity) would also be necessary to have a greater knowledge of a microbe's role within a community and a host. To support more field trials, those studies would need to verify their community interactions and their favorable or unfavorable effects on plants (Fig. 1).

9.2 *Taking Care of Field-Based Microbiome Research*

Many scales are involved in the regulation of the huge microbial variety found in soil. For instance, alterations in soil texture, biotic interactions, or exudates from plant roots can have a significant impact at the smallest (microbial) scales. Variations in both chemical and physical characteristics of "soil", including "pH", "nutrient distribution", or "water retention", possess an effect on the biological diversity of soil on medium (field) scales. According to Fierer and Jackson (2006), the diversity of microbes is significantly influenced by soil pH, which also links the structure of microbe populations to the availability and recycling of nutrients in the soil (Li et al., 2017). Lastly, at pilot scales, climate and "geo-localization" may be more important factors in regulating soil biodiversity (Bardgett & van der Putten, 2014).

Living evidence that microbial communities not only support plant growth but also offer defense against plant diseases is the presence of disease-suppressive soils (Chaparro et al., 2012; Van Der Heijden et al., 2008). (Alabouvette, 1986; Andrivon, 1994; Shiomi et al., 1999). General suppression of disease can result from competing with native soil fauna, while specialized suppression can be caused by the presence of particular subsets of bacteria. Animal immune techniques are analogous to these soil protection strategies (Raaijmakers & Mazzola, 2016). General suppression is as ubiquitous in all soils as it is in complex ecosystems. Conversely, specific suppression is eliminated by soil pasteurization and can be transplanted to different soils. Regrowth of vulnerable hosts and pathogens in the soil can restore its suppressive properties; the growth of non-host plants can cause the soil to lose this property (Wiseman et al., 1996; Berendsen et al., 2012; Raaijmakers & Mazzola, 2016).

One important technique for plants to thrive in adverse environments may be their ability to draw in helpful root microorganisms. According to several research (Mueller & Sachs, 2015; Panke-Buisse et al., 2015; Herrera Paredes et al., 2018; Orozco-Mosqueda et al., 2018), it may be possible to engineer microbiomes to control plant features that can be utilized to boost and sustain plant productivity.

According to Berendsen et al. (2018), plants that are infected with pathogens can draw in beneficial bacteria from the rhizosphere and endosphere, which can alter the host's immune response. By manipulating the microbial structure of the banana rhizosphere, bioorganic fertilizers have been developed that reduce the occurrence of Panama disease (Xue et al., 2015). As a sustainable way to boost yields and raise resistance to both biotic and abiotic stress, all of these studies have shown that engineering plant microbiomes is feasible.

The examination of seed microbiomes is a recent breakthrough discovery that is extremely significant to microbiome research. Plants that provide desirable features that have driven and continue to drive plant development are thought to have co-selected and evolved with it (Puente et al., 2009; Johnston-Monje & Raizada, 2011; Turner et al., 2013a; Bouffaud et al., 2014; Delaux et al., 2014; Hardoim et al., 2015). A small range of microbial species appear to be present in seed microbiomes, and this small number is likely because these organisms must survive all stages of seed development, even the most severe ones like desiccation (Truyens et al., 2015). These investigations align with recent experiments that demonstrate the potential to pass on plant microbiomes to the following "generation" (Mitter et al., 2017). Crop microorganism breeding as well as engineering are currently at the leading edge of organic farming as a result of all of those discoveries (Wei & Jousset, 2017).

Finally, we must recognize that agricultural practices impact microbiomes and microbial diversity in field trials, which are not vulnerable to them in glasshouse research. This understanding is crucial for closing the lab-field divide. Regarding Sanz et al.'s (2017) findings, the UN says that to lessen soil degradation, plowing, and sustainable agriculture should be resumed, along with sustainable land management practices like the rotation of crops, "intercropping", along with the cultivation of "native plant species". Uncertainty exists on the best way to restore these methods to developed-world agricultural systems without compromising or surpassing crop yields. Additionally, these activities play a significant role in the structure and function of the microbiome community (Oberson et al., 1993; Mäder et al., 2002; Dossa et al., 2012; Debenport et al., 2015; Reganold & Wachter, 2016; Hartman et al., 2017, 2018; Wang et al., 2017a). As such, they may be pertinent factors to take into account when planning field studies on the microbiome or assessing microbial inoculants used in commercial applications.

9.3 *Utilizing Local Plant Varieties, Rotating Crops, and Intercropping*

Around the 1940s, monocropping and the use of synthetic fertilizers greatly enhanced crop yields, which led to the decline in the importance of some traditional agricultural methods. To feed the world's expanding population, the Green Revolution included several significant agricultural modifications. Several agricultural approaches were rendered impractical in industrialized nations along with the emergence of input-intensive agriculture systems for distinct income crops. The problem of creating sustainable agricultural solutions, however, is feeding the world's expanding population with the least amount of negative ecological and economic effects in the face of land degradation, a lack of fertilizer, and global warming.

The stability and composition of soil and root microbial communities, and subsequently the activities associated with the microbiome, are greatly influenced by land management (Hartman et al., 2017, 2018). Because of an excess of harmful bacteria relative to beneficial microbes, crops cultivated in monoculture or short rotations frequently experience a loss in production (Bennett et al., 2012; Hilton et al., 2013; Santhanam et al., 2015). Thus, using local microbial resources to boost plant productivity and fitness in sustainable agriculture is possible. The introduction of illnesses linked to repeated cropping can be considerably decreased in a farming environment by inoculating using "native root-associated bacterial isolates" (Santhanam et al., 2015). As was previously indicated, continued cropping of the same species appears to be directly tied to this trait as well, since soil disease suppression is lost when a different host plant is cultivated. These two contradicting outcomes for agricultural productivity indicate both the significant influence that crop hosts possess on soil microflora and the potential impact of the latter on the species of plants that can efficiently colonize an ecological system under natural settings.

Apart from the general edaphic parameters, the release of unique root exudates is influenced by plant species (Rovira, 1969), plant ecotypes (Micallef et al., 2009), and even the microenvironment (Pinton et al., 2007) in which a root system is located. Thus, the type of plant planted determines the form of the microbial communities in the soil. The advent of contemporary intensive agricultural methods forced the abandonment of intercropping, an antiquated agricultural technique. Intercropping, on the other hand—growing various plant species adjacent to one another—remains a popular technique in developing nations. According to studies conducted in the Sahel region of Africa, intercropping boosts crop productivity, soil organic carbon levels, and the variety of bacterial and fungal communities (Debenport et al., 2015). Furthermore, integrating with native species improves soil fertility and preserves nitrogen (Dossa et al., 2012), highlighting the importance of using native plants that have already adapted to make use of the natural resources of an ecosystem. One substitute for sustainable agricultural output has been proposed: intercropping. Nevertheless, several obstacles would need to be overcome before they could spread throughout industrialized nations, like the creation of planting systems tailored to this agricultural technique.

Further investigation is necessary in the field of microbiome research to fully understand the significance of cropping techniques on the soil microbiota and to determine whether cropping strategies and microbiome engineering may eventually support sustainable agriculture.

9.4 Tillage and Soil Farming

Nowadays, land tilling is done more often because it reduces the growth of weeds and makes the seedbed more compatible with field machines. There has been research on no-tillage systems as a conservation agriculture technique ever since plant growth regulators were introduced in the 1940s (Bagavathiannan & Davis, 2018). Nevertheless, many farmers and consumers object to no-tilling systems since they involve the use of cover crops and greater dosages of herbicides. This approach, in turn, reduces the disturbance of soil particles, raises the amount of organic carbon in the soil, improves soil aggregation, and facilitates "water infiltration" (Álvaro-Fuentes et al., 2008; Hobbs et al., 2008; Li et al., 2017; Wang et al., 2017a). The soil pH (which is somewhat higher in organic systems) and nutrient transport from the soil matrix to the soil solution are affected by long-term no-tillage and organic input management techniques (Mäder et al., 2002). Microbiome research has shown that "organic farming" and "no-tilling" practices are linked to improvements in "soil microbial diversity", "biomass", and "microbial community stability" (Oberson et al., 1993; Mäder et al., 2002; Reganold & Wachter, 2016; Wang et al., 2017a). The increase in organic matter (which serves as a food supply for the microbial population), the decrease in physical disturbances, and the improvement in soil aggregate stability are probably the causes of these beneficial impacts on the soil microbiota (Wang et al., 2017a; Siegrist et al., 1998).

10 Conclusion

The creation of microbial communities with tailored (beneficial) functions may prove to be a formidable strategy to improve crop health, fight plant diseases, and use less fertilizer—all of which will contribute to more sustainable agricultural output. To this end, ongoing investigations that provide a greater knowledge of the organization of the soil microbial ecosystem over time—as well as its flexibility and reactivity to environmental changes—are required. It's also essential to have a basic understanding of how microbes and plants interact to form the crop's microbiota. Further thorough research on community dynamics employing these individual microorganisms and their soil microbial communities would help advance the field, as individual microbes are essential for controlling the stability and structure of microbial communities. This information may contribute to a comprehensive understanding of the effects these keystone organisms have on global nutrient cycles, disease resistance, and crop yields, as well as to the development of techniques for microbiome engineering.

References

Adair, K. L., & Douglas, A. E. (2017). Making a microbiome: The many determinants of host-associated microbial community composition. *Current Opinion in Microbiology, 35*, 23–29.

Adam, P. S., Borrel, G., Brochier-Armanet, C., & Gribaldo, S. (2017). The growing tree of Archaea: New perspectives on their diversity, evolution and ecology. *The ISME Journal, 11*, 2407–2425.

Agler, M. T., Ruhe, J., Kroll, S., Morhenn, C., Kim, S. T., Weigel, D., et al. (2016). Microbial hub Taxa link host and abiotic factors to plant microbiome variation. *PLoS Biology, 14*, e1002352.

Alabouvette, C. (1986). Fusarium-wilt suppressive soils from the Châteaurenard region: Review of a 10-year study. *Agronomie, 6*, 273–284.

Álvaro-Fuentes, J., Lopez, M. V., Cantero-Martinez, C., & Arrue, J. L. (2008). Tillage effects on soil organic carbon fractions in Mediterranean dryland agroecosystems. *Soil Science Society of America Journal, 72*, 541–547.

Andrivon, D. (1994). Dynamics of the survival and infectivity to potato tubers of sporangia of Phytophthora infestans in three different soils. *Soil Biology and Biochemistry, 26*, 945–952.

Armengaud, J., Christie-Oleza, J. A., Clair, G., Malard, V., & Duport, C. (2012). Exoproteomics: Exploring the world around biological systems. *Expert Review of Proteomics, 9*, 561–575.

Badri, D. V., Chaparro, J. M., Zhang, R., Shen, Q., & Vivanco, J. M. (2013). Application of natural blends of phytochemicals derived from the root exudates of arabidopsis to the soil reveal that phenolic-related compounds predominantly modulate the soil microbiome. *The Journal of Biological Chemistry, 288*, 4502–4512.

Bagavathiannan, M. V., & Davis, A. S. (2018). An ecological perspective on managing weeds during the great selection for herbicide resistance. *Pest Management Science, 74*(10), 2277–2286.

Bai, Y., Müller, D. B., Srinivas, G., Garrido-Oter, R., Potthoff, E., Rott, M., et al. (2015). Functional overlap of the Arabidopsis leaf and root microbiota. *Nature, 528*, 364–369.

Bakker, M. G., Schlatter, D. C., Otto-Hanson, L., & Kinkel, L. L. (2014). Diffuse symbioses: Roles of plant-plant, plant-microbe and microbe-microbe interactions in structuring the soil microbiome. *Molecular Ecology, 23*, 1571–1583.

Baran, R., Brodie, E. L., Mayberry-Lewis, J., Hummel, E., Da Rocha, U. N., Chakraborty, R., et al. (2015). Exometabolite niche partitioning among sympatric soil bacteria. *Nature Communications, 6*, 8289.

Bardgett, R. D., & van der Putten, W. H. (2014). Belowground biodiversity and ecosystem functioning. *Nature, 515*, 505–511.

Bashiardes, S., Godneva, A., Elinav, E., & Segal, E. (2018). Towards utilization of the human genome and microbiome for personalized nutrition. *Current Opinion in Biotechnology, 51*, 57–63.

Beck, J. J., & Vannette, R. L. (2017). Harnessing insect-microbe chemical communications to control insect pests of agricultural systems. *Journal of Agricultural and Food Chemistry, 65*, 23–28.

Bender, S. F., Wagg, C., & van der Heijden, M. G. A. (2016). An underground revolution: Biodiversity and soil ecological engineering for agricultural sustainability. *Trends in Ecology & Evolution, 31*, 440–452.

Benítez, E., Paredes, D., Rodríguez, E., Aldana, D., González, M., Nogales, R., et al. (2017). Bottom-up effects on herbivore-induced plant defences: A case study based on compositional patterns of rhizosphere microbial communities. *Scientific Reports, 7*, 6251.

Bennett, A. J., Bending, G. D., Chandler, D., Hilton, S., & Mills, P. (2012). Meeting the demand for crop production: The challenge of yield decline in crops grown in short rotations. *Biological Reviews, 87*, 52–71.

Berendsen, R. L., Pieterse, C. M. J., & Bakker, P. A. (2012). The rhizosphere microbiome and plant health. *Trends in Plant Science, 17*, 478–486.

Berendsen, R. L., Vismans, G., Yu, K., Song, Y., De Jonge, R., Burgman, W. P., et al. (2018). Disease-induced assemblage of a plant-beneficial bacterial consortium. *The ISME Journal, 12*, 1496–1507.

Boe, L., Danielsen, M., Knudsen, S., Petersen, J. B., Maymann, J., & Jensen, P. R. (2000). The frequency of mutators in populations of *Escherichia coli*. *Mutation Research, 448*, 47–55.

Bonilla, N., Vida, C., Martínez-Alonso, M., Landa, B. B., Gaju, N., Cazorla, F. M., et al. (2015). Organic amendments to avocado crops induce suppressiveness and influence the composition and activity of soil microbial communities. *Applied and Environmental Microbiology, 81*, 3405–3418.

Bouffaud, M. L., Poirier, M. A., Muller, D., & Moënne-Loccoz, Y. (2014). Root microbiome relates to plant host evolution in maize and other Poaceae. *Environmental Microbiology, 16*, 2804–2814.

Bouwman, A. F., Beusen, A. H. W., & Billen, G. (2009). Human alteration of the global nitrogen and phosphorus soil balances for the period 1970–2050. *Global Biogeochemical Cycles, 23*, 1–16.

Bruto, M., Prigent-Combaret, C., Muller, D., & Moënne-Loccoz, Y. (2014). Analysis of genes contributing to plant-beneficial functions in plant growth-promoting rhizobacteria and related Proteobacteria. *Scientific Reports, 4*, 6261.

Bulgarelli, D., Rott, M., Schlaeppi, K., Loren, V., van Themaat, E., Ahmadinejad, N., Assenza, F., et al. (2012). Revealing structure and assembly cues for Arabidopsis root-inhabiting bacterial microbiota. *Nature, 488*, 91–95.

Bulgarelli, D., Schlaeppi, K., Spaepen, S., van Themaat, E. V. L., & SchulzeLefert, P. (2013). Structure and functions of the bacterial microbiota of plants. *Annual Review of Plant Biology, 64*, 807–838.

Bulgarelli, D., Garrido-Oter, R., Münch, P. C., Weiman, A., Dröge, J., Pan, Y., et al. (2015). Structure and function of the bacterial root microbiota in wild and domesticated barley. *Cell Host & Microbe, 17*, 392–403.

Carvalhais, L. C., Dennis, P. G., Tyson, G. W., & Schenk, P. M. (2012). Application of metatranscriptomics to soil environments. *Journal of Microbiological Methods, 91*, 246–251.

Castillo, J. D., Vivanco, J. M., & Manter, D. K. (2017). Bacterial microbiome and nematode occurrence in different potato agricultural soils. *Microbial Ecology, 74*, 888–900.

Chaparro, J. M., Sheflin, A. M., Manter, D. K., & Vivanco, J. M. (2012). Manipulating the soil microbiome to increase soil health and plant fertility. *Biology and Fertility of Soils, 48*, 489–499.

Chaparro, J. M., Badri, D. V., & Vivanco, J. M. (2014). Rhizosphere microbiome assemblage is affected by plant development. *The ISME Journal, 8*, 790–803.

Cheng, G., Karunakaran, R., East, A. K., & Poole, P. S. (2016). Multiplicity of sulfate and molybdate transporters and their role in nitrogen fixation in rhizobium leguminosarum bv. viciae Rlv3841. *Molecular Plant-Microbe Interactions, 29*, 143–152.

Christie-Oleza, J. A., Armengaud, J., Guerin, P., & Scanlan, D. J. (2015). Functional distinctness in the exoproteomes of marine Synechococcus. *Environmental Microbiology, 17*, 3781–3794.

Dawson, W., Hör, J., Egert, M., van Kleunen, M., & Peste, M. (2017). A small number of low-abundance bacteria dominate plant species-specific responses during rhizosphere colonization. *Frontiers in Microbiology, 8*, 975.

de Boer, W. (2017). Upscaling of fungal–bacterial interactions: From the lab to the field. *Current Opinion in Microbiology, 37*, 35–41.

Debenport, S. J., Assigbetse, K., Bayala, R., Chapuis-Lardy, L., Dick, R. P., & McSpadden Gardener, B. B. (2015). Association of shifting populations in the root zone microbiome of millet with enhanced crop productivity in the Sahel Region (Africa). *Applied and Environmental Microbiology, 81*, 2841–2851.

Delaux, P. M., Varala, K., Edger, P. P., Coruzzi, G. M., Pires, J. C., & Ané, J. M. (2014). Comparative phylogenomics uncovers the impact of symbiotic associations on host genome evolution. *PLoS Genetics, 10*, e1004487.

Denamur, E., & Matic, I. (2006). Evolution of mutation rates in bacteria. *Molecular Microbiology, 60*, 820–827.

Dossa, E. L., Diedhiou, I., Khouma, M., Sene, M., Lufafa, A., Kizito, F., et al. (2012). Crop productivity and nutrient dynamics in a shrub (*Guiera senegalensis*)- based farming system of the Sahel. *Agronomy Journal, 104*, 1255–1264.

Downie, H., Holden, N., Otten, W., Spiers, A. J., Valentine, T. A., & Dupuy, L. X. (2012). Transparent soil for imaging the rhizosphere. *PLoS One, 7*, e44276.

Downie, H. F., Valentine, T. A., Otten, W., Spiers, A. J., & Dupuy, L. X. (2014). Transparent soil microcosms allow 3D spatial quantification of soil microbiological processes in vivo. *Plant Signaling & Behavior, 9*, e970421.

Edwards, J., Johnson, C., Santos-Medellín, C., Lurie, E., Podishetty, N. K., Bhatnagar, S., et al. (2015). Structure, variation, and assembly of the rootassociated microbiomes of rice. *Proceedings. National Academy of Sciences. United States of America, 112*, E911–E920.

Elhady, A., Giné, A., Topalovic, O., Jacquiod, S., Sørensen, S. J., Sorribas, F. J., et al. (2017). Microbiomes associated with infective stages of root-knot and lesion nematodes in soil. *PLoS One, 12*, e0177145.

Engelstädter, J., & Hurst, G. D. D. (2009). The ecology and evolution of microbes that manipulate host reproduction. *Annual Review of Ecology, Evolution, and Systematics, 40*, 127–149.

Fernández, M., Conde, S., Duque, E., & Ramos, J. L. (2013). In vivo gene expression of Pseudomonas putida KT2440 in the rhizosphere of different plants. *Microbial Biotechnology, 6*, 307–313.

Fierer, N. (2017). Embracing the unknown: Disentangling the complexities of the soil microbiome. *Nature Reviews. Microbiology, 15*, 579–590.

Fierer, N., & Jackson, R. B. (2006). The diversity and biogeography of soil bacterial communities. *Proceedings. National Academy of Sciences. United States of America, 103*, 626–631.

Finkel, O. M., Castrillo, G., Herrera Paredes, S., Salas González, I., & Dangl, J. L. (2017). Understanding and exploiting plant beneficial microbes. *Current Opinion in Plant Biology, 38*, 155–163.

Goodrich, J. K., Davenport, E. R., Clark, A. G., & Ley, R. E. (2017). The relationship between the human genome and microbiome comes into view. *Annual Review of Genetics, 51*, 413–433.

Hacquard, S., Garrido-Oter, R., González, A., Spaepen, S., Ackermann, G., Lebeis, S., et al. (2015). Microbiota and host nutrition across plant and animal kingdoms. *Cell Host & Microbe, 17*, 603–616.

Hacquard, S., Kracher, B., Hiruma, K., Münch, P. C., Garrido-Oter, R., Thon, M. R., et al. (2016). Survival trade-offs in plant roots during colonization by closely related beneficial and pathogenic fungi. *Nature Communications, 7*, 11362.

Haichar, F. E. Z., Heulin, T., Guyonnet, J. P., & Achouak, W. (2016). Stable isotope probing of carbon flow in the plant holobiont. *Current Opinion in Biotechnology, 41*, 9–13.

Hamburg, M. A., & Collins, F. S. (2010). The path to personalized medicineperspective. *The New England Journal of Medicine, 363*, 301–304.

Haney, C. H., Samuel, B. S., Bush, J., & Ausubel, F. M. (2015). Associations with rhizosphere bacteria can confer an adaptive advantage to plants. *Nature Plants, 1*, 15051.

Hardoim, P. R., van Overbeek, L. S., Berg, G., Pirttilä, A. M., Compant, S., Campisano, A., et al. (2015). The hidden world within plants: Ecological and evolutionary considerations for defining functioning of microbial endophytes. *Microbiology and Molecular Biology Reviews, 79*, 293–320.

Hartman, K., van der Heijden, M. G. A., Roussely-Provent, V., Walser, J. C., & Schlaeppi, K. (2017). Deciphering composition and function of the root microbiome of a legume plant. *Microbiome, 5*, 2.

Hartman, K., van der Heijden, M. G. A., Wittwer, R. A., Banerjee, S., Walser, J. C., & Schlaeppi, K. (2018). Cropping practices manipulate abundance patterns of root and soil microbiome members paving the way to smart farming. *Microbiome, 6*, 14.

Hennessy, R. C., Stougaard, P., & Olsson, S. (2017). A microplate readerbased system for visualizing transcriptional activity during in vivo microbial interactions in space and time. *Scientific Reports, 7*, 281.

Herrera Paredes, S., Gao, T., Law, T. F., Finkel, O. M., Mucyn, T., Teixeira, P. J. P. L., et al. (2018). Design of synthetic bacterial communities for predictable plant phenotypes. *PLoS Biology, 16*, e2003962.

Heyer, R., Kohrs, F., Reichl, U., & Benndorf, D. (2015). Metaproteomics of complex microbial communities in biogas plants. *Microbial Biotechnology, 8*, 749–763.

Hilton, S., Bennett, A. J., Keane, G., Bending, G. D., Chandler, D., Stobart, R., et al. (2013). Impact of shortened crop rotation of oilseed rape on soil and rhizosphere microbial diversity in relation to yield decline. *PLoS One, 8*, e59859.

Hirsch, P. R., Mauchline, T. H., & Clark, I. M. (2010). Culture-independent molecular techniques for soil microbial ecology. *Soil Biology and Biochemistry, 42*, 878–887.

Hiruma, K., Gerlach, N., Sacristán, S., Nakano, R. T., Hacquard, S., Kracher, B., et al. (2016). Root endophyte colletotrichum tofieldiae confers plant fitness benefits that are phosphate status dependent. *Cell, 165*, 464–474.

Hobbs, P. R., Sayre, K., & Gupta, R. (2008). The role of conservation agriculture in sustainable agriculture. *Philosophical Transactions of the Royal Society of London. Series B, Biological Sciences, 363*, 543–555.

Horton, M. W., Bodenhausen, N., Beilsmith, K., Meng, D., Muegge, B. D., Subramanian, S., et al. (2014). Genome-wide association study of Arabidopsis thaliana leaf microbial community. *Nature Communications, 5*, 5320.

Hug, L. A., Baker, B. J., Anantharaman, K., Brown, C. T., Probst, A. J., Castelle, C. J., et al. (2016). A new view of the tree of life. *Nature Microbiology, 1*, 16048.

Human Microbiome Project Consortium. (2012). Structure, function and diversity of the healthy human microbiome. *Nature, 486*, 207–214.

Jansson, J. K., & Hofmockel, K. S. (2018). The soil microbiome — From metagenomics to metaphenomics. *Current Opinion in Microbiology, 43*, 162–168.

Jez, J. M., Lee, S. G., & Sherp, A. M. (2016). The next green movement: Plant biology for the environment and sustainability. *Science, 353*, 1241–1244.

Johnson, N. C., Graham, J. H., & Smith, F. A. (1997). Functioning of mycorrhizal associations along the mutualism-parasitism continuum. *The New Phytologist, 135*, 575–585.

Johnson-Rollings, A. S., Wright, H., Masciandaro, G., Macci, C., Doni, S., CalvoBado, L. A., et al. (2014). Exploring the functional soil-microbe interface and exoenzymes through soil metaexoproteomics. *The ISME Journal, 8*, 2148–2150.

Johnston-Monje, D., & Raizada, M. N. (2011). Conservation and diversity of seed associated endophytes in Zea across boundaries of evolution, ethnography and ecology. *PLoS One, 6*, e20396.

Jones, J. D. G., & Dangl, J. L. (2006). The plant immune system. *Nature, 444*, 323–329.

Jones, D. L., Nguyen, C., & Finlay, R. D. (2009). Carbon flow in the rhizosphere: Carbon trading at the soil-root interface. *Plant and Soil, 321*, 5–33.

Kibota, T. T., & Lynch, M. (1996). Estimate of the genomic mutation rate deleterious to overall fitness in *E. coli*. *Nature, 381*, 694–696.

Knight, R., Vrbanac, A., Taylor, B. C., Aksenov, A., Callewaert, C., Debelius, J., et al. (2018). Best practices for analysing microbiomes. *Nature Reviews. Microbiology, 16*, 410–422.

Lebeis, S. L., Paredes, S. H., Lundberg, D. S., Breakfield, N., Gehring, J., McDonald, M., et al. (2015). Salicylic acid modulates colonization of the root microbiome by specific bacterial taxa. *Science, 349*, 860–864.

Levy, A., Salas Gonzalez, I., Mittelviefhaus, M., Clingenpeel, S., Herrera Paredes, S., Miao, J., et al. (2018). Genomic features of bacterial adaptation to plants. *Nature Genetics, 50*, 138–150.

Lewis, R., Barth, V., Coffey, T., McFarland, C., Huggins, D., & Sullivan, T. (2018). Altered bacterial communities in long-term no-till soils associated with stratification of soluble aluminum and soil pH. *The Soil, 2*, 7.

Li, Y., Sun, Q., Zhan, J., Yang, Y., & Wang, D. (2017). Soil-covered strategy for ecological restoration alters the bacterial community structure and predictive energy metabolic functions in mine tailings profiles. *Applied Microbiology and Biotechnology, 101*, 2549–2561.

Lidbury, I., Murrell, J. C., & Chen, Y. (2014). Trimethylamine N-oxide metabolism by abundant marine heterotrophic bacteria. *Proceedings. National Academy of Sciences. United States of America, 111*, 2710–2715.

Lidbury, I. D., Murphy, A. R. J., Scanlan, D. J., Bending, G. D., Jones, A. M. E., Moore, J. D., et al. (2016). Comparative genomic, proteomic and exoproteomic analyses of three Pseudomonas strains reveals novel insights into the phosphorus scavenging capabilities of soil bacteria. *Environmental Microbiology, 18*, 3535–3549.

Lladó, S., López-Mondéjar, R., & Baldrian, P. (2017). Forest soil bacteria: Diversity, involvement in ecosystem processes, and response to global change. *Microbiology and Molecular Biology Reviews, 81*, e00063–e00016.

Lueders, T., Dumont, M. G., Bradford, L., & Manefield, M. (2016). RNAstable isotope probing: From carbon flow within key microbiota to target transcriptomes. *Current Opinion in Biotechnology, 41*, 83–89.

Lundberg, D. S., Lebeis, S. L., Paredes, S. H., Yourstone, S., Gehring, J., Malfatti, S., et al. (2012). Defining the core Arabidopsis thaliana root microbiome. *Nature, 488*, 86–90.

Mäder, P., Fließbach, A., Dubois, D., Gunst, L., Fried, P., & Niggli, U. (2002). Soil fertility and biodiversity in organic farming. *Science, 296*, 1694–1697.

Martino, E., Morin, E., Grelet, G. A., Kuo, A., Kohler, A., Daghino, S., et al. (2018). Comparative genomics and transcriptomics depict ericoid mycorrhizal fungi as versatile saprotrophs and plant mutualists. *The New Phytologist, 217*, 1213–1229.

Massalha, H., Korenblum, E., Malitsky, S., Shapiro, O. H., & Aharoni, A. (2017). Live imaging of root–bacteria interactions in a microfluidics setup. *Proceedings. National Academy of Sciences. United States of America, 114*, 4549–4554.

Mauchline, T. H., Fowler, J. E., East, A. K., Sartor, A. L., Zaheer, R., Hosie, A. H. F., et al. (2006). Mapping the Sinorhizobium meliloti 1021 solute-binding proteindependent transportome. *Proceedings. National Academy of Sciences. United States of America, 103*, 17933–17938.

Mauchline, T. H., Chedom-Fotso, D., Chandra, G., Samuels, T., Greenaway, N., Backhaus, A., et al. (2015). An analysis of pseudomonas genomic diversity in take-all infected wheat fields reveals the lasting impact of wheat cultivars on the soil microbiota. *Environmental Microbiology, 17*, 4764–4778.

McDonald, B. R., & Currie, C. R. (2017). Lateral gene transfer dynamics in the ancient bacterial genus Streptomyces. *MBio, 8*, e00644–e00617.

Micallef, S. A., Shiaris, M. P., & Colón-Carmona, A. (2009). Influence of Arabidopsis thaliana accessions on rhizobacterial communities and natural variation in root exudates. *Journal of Experimental Botany, 60*, 1729–1742.

Mitter, B., Pfaffenbichler, N., Flavell, R., Compant, S., Antonielli, L., Petric, A., et al. (2017). A new approach to modify plant microbiomes and traits by introducing beneficial bacteria at flowering into progeny seeds. *Frontiers in Microbiology, 8*, 11.

Mueller, U. G., & Sachs, J. L. (2015). Engineering microbiomes to improve plant and animal health. *Trends in Microbiology, 23*, 606–617.

Muth, T., Benndorf, D., Reichl, U., Rapp, E., & Martens, L. (2013). Searching for a needle in a stack of needles: Challenges in metaproteomics data analysis. *Molecular BioSystems, 9*, 578–585.

Muth, T., Renard, B. Y., & Martens, L. (2016). Metaproteomic data analysis at a glance: Advances in computational microbial community proteomics. *Expert Review of Proteomics, 13*, 757–769.

Niu, B., Paulson, J. N., Zheng, X., & Kolter, R. (2017). Simplified and representative bacterial community of maize roots. *Proceedings. National Academy of Sciences. United States of America, 114*, E2450–E2459.

O'Callaghan, F. E., Braga, R. A., Neilson, R., MacFarlane, S. A., & Dupuy, L. X. (2018). New live screening of plant-nematode interactions in the rhizosphere. *Scientific Reports, 8*, 1440.

Oberson, A., Fardeau, J. C., Besson, J. M., & Sticher, H. (1993). Soil phosphorus dynamics in cropping systems managed according to conventional and biological agricultural methods. *Biology and Fertility of Soils, 16*, 111–117.

Oldroyd, G. E. D. (2013). Speak, friend, and enter: Signalling systems that promote beneficial symbiotic associations in plants. *Nature Reviews. Microbiology, 11*, 252–263.

Orozco-Mosqueda, M. D. C., Rocha-Granados, M. D. C., Glick, B. R., & Santoyo, G. (2018). Microbiome engineering to improve biocontrol and plant growth-promoting mechanisms. *Microbiological Research, 208*, 25–31.

Panke-Buisse, K., Poole, A. C., Goodrich, J. K., Ley, R. E., & Kao-Kniffin, J. (2015). Selection on soil microbiomes reveals reproducible impacts on plant function. *The ISME Journal, 9*, 980–989.

Peiffer, J. A., Spor, A., Koren, O., Jin, Z., Tringe, S. G., Dangl, J. L., et al. (2013). Diversity and heritability of the maize rhizosphere microbiome under field conditions. *Proceedings. National Academy of Sciences. United States of America, 110*, 6548–6553.

Pimentel, D., Harvey, C., Resosudarmo, P., Sinclair, K., Kurz, D., McNair, M., et al. (1995). Environmental and economic costs of soil erosion and conservation benefits. *Science, 267*, 1117–1123.

Pineda, A., Soler, R., Pozo, M. J., Rasmann, S., & Turlings, T. C. J. (2015). Editorial: Above-belowground interactions involving plants, microbes and insects. *Frontiers in Plant Science, 6*, 318.

Pini, F., East, A. K., Appia-Ayme, C., Tomek, J., Karunakaran, R., MendozaSuárez, M., et al. (2017). Bacterial biosensors for in vivo spatiotemporal mapping of root secretion. *Plant Physiology, 174*, 1289–1306.

Pinto, A. J., & Raskin, L. (2012). PCR biases distort bacterial and archaeal community structure in pyrosequencing datasets. *PLoS One, 7*, e43093.

Pinton, R., Varanini, Z., Nannipieri, P., Darrah, P., Simonet, P., Neumann, G., et al. (2007). *The rhizosphere*. CRC Press. https://doi.org/10.1201/9781420005585

Poudel, R., Jumpponen, A., Schlatter, D. C., Paulitz, T. C., Gardener, B. B. M., Kinkel, L. L., et al. (2016). Microbiome networks: A systems framework for identifying candidate microbial assemblages for disease management. *Phytopathology, 106*, 1083–1096.

Pratama, A. A., & van Elsas, J. D. (2018). The "neglected" soil virome – Potential role and impact. *Trends in Microbiology, 26*, 649–662.

Prosser, J. I. (2015). Dispersing misconceptions and identifying opportunities for the use of "omics" in soil microbial ecology. *Nature Reviews. Microbiology, 13*, 439–446.

Puente, M. E., Li, C. Y., & Bashan, Y. (2009). Endophytic bacteria in cacti seeds can improve the development of cactus seedlings. *Environmental and Experimental Botany, 66*, 402–408.

Qiu, X., Kulasekara, B. R., & Lory, S. (2009). Role of horizontal gene transfer in the evolution of Pseudomonas aeruginosa virulence. *Genome Dynamics, 6*, 126–139.

Raaijmakers, J. M., & Mazzola, M. (2016). Soil immune responses. *Science, 352*, 1392–1393.

Reganold, J. P., & Wachter, J. M. (2016). Organic agriculture in the twenty-first century. *Nature Plants, 2*, 15221.

Robbins, R. J., Krishtalka, L., & Wooley, J. C. (2016). Advances in biodiversity: Metagenomics and the unveiling of biological dark matter. *Standards in Genomic Sciences, 11*, 69.

Rockström, J., Steffen, W., Noone, K., Persson, Å., Chapin, F. S., Lambin, E., et al. (2009). Planetary boundaries: Exploring the safe operating space for humanity. *Ecology and Society, 14*, 32.

Rodrigues, R. R., Rodgers, N. C., Wu, X., & Williams, M. A. (2018). COREMIC: A web-tool to search for a niche associated CORE MICrobiome. *PeerJ, 6*, e4395.

Rovira, A. D. (1969). Plant root exudates. *The Botanical Review, 35*, 35–57.

Rybakova, D., Mancinelli, R., Wikström, M., Birch-Jensen, A. S., Postma, J., Ehlers, R. U., et al. (2017). The structure of the Brassica napus seed microbiome is cultivar-dependent and affects the interactions of symbionts and pathogens. *Microbiome, 5*, 104.

Sánchez-Cañizares, C., Jorrín, B., Poole, P. S., & Tkacz, A. (2017). Understanding the holobiont: The interdependence of plants and their microbiome. *Current Opinion in Microbiology, 38*, 188–196.

Santhanam, R., Luu, V. T., Weinhold, A., Goldberg, J., Oh, Y., & Baldwin, I. T. (2015). Native root-associated bacteria rescue a plant from a sudden-wilt disease that emerged during continuous cropping. *Proceedings. National Academy of Sciences. United States of America, 112*, E5013–E5020.

Sanz, M. J., de Vente, J., Chotte, J. L., Bernoux, M., Kust, G., Ruiz, I., (eds)., et al. (2017). Sustainable land management contribution to successful land-based climate change adaptation and mitigation. In *A report of the science-policy interface*. UNCCD.

Schlaeppi, K., & Bulgarelli, D. (2015). The plant microbiome at work. *Molecular Plant-Microbe Interactions, 28*, 212–217.

Scholz, M. B., Lo, C. C., & Chain, P. S. G. (2012). Next generation sequencing and bioinformatic bottlenecks: The current state of metagenomic data analysis. *Current Opinion in Biotechnology, 23*, 9–15.

Senga, R. A., Alegria Terrazas, S., Balbirnie, K., Blank, M., Janiak, A., Szarejko, I., et al. (2017). Root hair mutations displace the barley rhizosphere microbiota. *Frontiers in Plant Science, 8*, 1094.

Shade, A., Jones, S. E., Gregory Caporaso, J., Handelsman, J., Knight, R., Fierer, N., et al. (2014). Conditionally rare taxa disproportionately contribute to temporal changes in microbial diversity. *MBio, 5*, e01371–e01314.

Shi, S., Nuccio, E. E., Shi, Z. J., He, Z., Zhou, J., & Firestone, M. K. (2016). The interconnected rhizosphere: High network complexity dominates rhizosphere assemblages. *Ecology Letters, 19*, 926–936.

Shiomi, Y., Nishiyama, M., Onizuka, T., & Marumoto, T. (1999). Comparison of bacterial community structures in the rhizoplane of tomato plants grown in soils suppressive and conducive towards bacterial wilt. *Applied and Environmental Microbiology, 65*, 3996–4001.

Siegrist, S., Schaub, D., Pfiffner, L., & Mäder, P. (1998). Does organic agriculture reduce soil erodibility? The results of a long-term field study on loess in Switzerland. *Agriculture, Ecosystems and Environment, 69*, 253–264.

Smith, F. A., & Smith, S. E. (1996). Mutualism and parasitism: Diversity in function and structure in the "arbuscular" (VA) mycorrhizal symbiosis. *Advances in Botanical Research, 22*, 1–43.

Sowell, S. M., Wilhelm, L. J., Norbeck, A. D., Lipton, M. S., Nicora, C. D., Barofsky, D. F., et al. (2009). Transport functions dominate the SAR11 metaproteome at low-nutrient extremes in the Sargasso Sea. *The ISME Journal, 3*, 93–105.

Svenningsen, N. B., Watts-Williams, S. J., Joner, E. J., Battini, F., Efthymiou, A., Cruz-Paredes, C., et al. (2018). Suppression of the activity of arbuscular mycorrhizal fungi by the soil microbiota. *The ISME Journal, 12*, 1296–1307.

Syed Ab Rahman, S. F., Singh, E., Pieterse, C. M. J., & Schenk, P. M. (2018). Emerging microbial biocontrol strategies for plant pathogens. *Plant Science, 267*, 102–111.

Temperton, B., & Giovannoni, S. J. (2012). Metagenomics: Microbial diversity through a scratched lens. *Current Opinion in Microbiology, 15*, 605–612.

Timmins-Schiffman, E., May, D. H., Mikan, M., Riffle, M., Frazar, C., Harvey, H. R., et al. (2017). Critical decisions in metaproteomics: Achieving high confidence protein annotations in a sea of unknowns. *The ISME Journal, 11*, 309–314.

Tkacz, A., & Poole, P. (2015). Role of root microbiota in plant productivity. *Journal of Experimental Botany, 66*, 2167–2175.

Tkacz, A., Cheema, J., Chandra, G., Grant, A., & Poole, P. S. (2015). Stability and succession of the rhizosphere microbiota depends upon plant type and soil composition. *The ISME Journal, 9*, 2349–2359.

Truyens, S., Weyens, N., Cuypers, A., & Vangronsveld, J. (2015). Bacterial seed endophytes: Genera, vertical transmission and interaction with plants. *Environmental Microbiology Reports, 7*, 40–50.

Turner, T. R., James, E. K., Poole, P. S., Gilbert, J., Meyer, F., Jansson, J., et al. (2013a). The plant microbiome. *Genome Biology, 14*, 209.

Turner, T. R., Ramakrishnan, K., Walshaw, J., Heavens, D., Alston, M., Swarbreck, D., et al. (2013b). Comparative metatranscriptomics reveals kingdom level changes in the rhizosphere microbiome of plants. *The ISME Journal, 7*, 2248–2258.

Van Der Heijden, M. G. A., Bardgett, R. D., & Van Straalen, N. M. (2008). The unseen majority: Soil microbes as drivers of plant diversity and productivity in terrestrial ecosystems. *Ecology Letters, 11*, 296–310.

Vorholt, J. A., Vogel, C., Carlström, C. I., & Müller, D. B. (2017). Establishing causality: Opportunities of synthetic communities for plant microbiome research. *Cell Host & Microbe, 22*, 142–155.

Wang, Q., Fish, J. A., Gilman, M., Sun, Y., Brown, C. T., Tiedje, J. M., et al. (2015). Xander: Employing a novel method for efficient gene-targeted metagenomic assembly. *Microbiome, 3*, 32.

Wang, Y., Li, C., Tu, C., Hoyt, G. D., DeForest, J. L., & Hu, S. (2017a). Long-term no-tillage and organic input management enhanced the diversity and stability of soil microbial community. *Science of the Total Environment, 609*, 341–347.

Wang, H., Wei, Z., Mei, L., Gu, J., Yin, S., Faust, K., et al. (2017b). Combined use of network inference tools identifies ecologically meaningful bacterial associations in a paddy soil. *Soil Biology and Biochemistry, 105*, 227–235.

Wei, Z., & Jousset, A. (2017). Plant breeding goes microbial. *Trends in Plant Science, 22*, 555–558.

Weller, D. M., Raaijmakers, J. M., Mcspadden Gardener, B. B., & Thomashow, L. S. (2002). Microbial populations responsible for specific soil suppressiveness to plant pathogens. *Annual Review of Phytopathology, 40*, 309–348.

Wetmore, K. M., Price, M. N., Waters, R. J., Lamson, J. S., He, J., Hoover, C. A., et al. (2015). Rapid quantification of mutant fitness in diverse bacteria by sequencing randomly bar-coded transposons. *mBio, 6*, e00306–e00315.

Williams, T. J., Long, E., Evans, F., Demaere, M. Z., Lauro, F. M., Raftery, M. J., et al. (2012). A metaproteomic assessment of winter and summer bacterioplankton from Antarctic Peninsula coastal surface waters. *The ISME Journal, 6*, 1883–1900.

Wiseman, B. M., Neate, S. M., Keller, K. O., & Smith, S. E. (1996). Suppression of Rhizoctonia solani anastomosis group 8 in Australia and its biological nature. *Soil Biology and Biochemistry, 28*, 727–732.

Xue, C., Penton, C. R., Shen, Z., Zhang, R., Huang, Q., Li, R., et al. (2015). Manipulating the banana rhizosphere microbiome for biological control of Panama disease. *Scientific Reports, 5*, 11124.

Yergeau, E., Sanschagrin, S., Maynard, C., St-Arnaud, M., & Greer, C. W. (2014). Microbial expression profiles in the rhizosphere of willows depend on soil contamination. *The ISME Journal, 8*, 344–358.

Yuan, Z., Jiang, S., Sheng, H., Liu, X., Hua, H., Liu, X., et al. (2018). Human perturbation of the global phosphorus cycle: Changes and consequences. *Environmental Science & Technology, 52*, 2438–2450.

Zarraonaindia, I., Owens, S. M., Weisenhorn, P., West, K., Hampton-Marcell, J., Lax, S., et al. (2015). The soil microbiome influences grapevine-associated microbiota. *MBio, 6*, e02527–e02514.

Zhalnina, K., Louie, K. B., Hao, Z., Mansoori, N., da Rocha, U. N., Shi, S., et al. (2018). Dynamic root exudate chemistry and microbial substrate preferences drive patterns in rhizosphere microbial community assembly. *Nature Microbiology, 3*, 470–480.

Effects of Soil Contaminants on Soil Microbiome

Bryan Xavier Daza Daza, Annie Julisa Delgado Mendoza, Jhonny Javier Zambrano Zambrano, Sesan Abiodun Aransiola, and Naga Raju Maddela

Abstract The soil microbiome, which comprises diverse microorganisms such as bacteria, fungi, viruses and archaea; which play a fundamental role in ecosystem functions, from primary production to carbon storage. Likewise, soil microbiomes influence vital processes such as nutrient cycling and water regulation. However, soil health is under threat by different factors, including industrialization, population growth, climate change, and human activities such as erosion and pollution. Heavy metals, hydrocarbons and other contaminants from anthropogenic activities alter microbial communities, harming vital soil functions such as nutrient cycling and the decomposition of organic matter. Additionally, contaminants such as pesticides and polycyclic aromatic hydrocarbons alter the composition of the microbiome, hindering its ability to biodegrade. For centuries, scholars have explored soil microbiomes using '*omics*' technologies to understand their genetic and biochemical makeup. Interaction mechanisms between soil microbiomes and contaminants reveal microbial capabilities to detoxify, sequester or degrade contaminants. Certain bacteria, such as rhizobacteria, that promote plant growth, help in metal chelation, nutrient solubilization and promotion of root growth, mitigating polluting effects. Efforts to restore soil microbiomes are supported by a variety of innovative and effective techniques that seek to comprehensively combat environmental pollution. These strategies range from approaches that use the biological activity of microorganisms to methods that take advantage of the ability of plants to absorb and detoxify soil. Such approaches, developed with the common goal of improving soil quality and mitigating environmental impacts, represent a constantly evolving field of study and exploration of new sustainable solutions for the restoration of terrestrial ecosystems. Understanding the intricate relationship between soil microbiomes and contaminants is vital to designing effective strategies to restore soil health and ensure

B. X. D. Daza · A. J. D. Mendoza · J. J. Z. Zambrano · N. R. Maddela (✉)
Departamento de Ciencias Biológicas, Facultad de Ciencias de la Salud, Universidad Técnica de Manabí, Portoviejo, Ecuador
e-mail: raju.maddela@utm.edu.ec

S. A. Aransiola
Department of Microbiology, Faculty of Science, University of Abuja, Abuja, Nigeria

S. A. Aransiola et al. (eds.), *Soil Microbiome in Green Technology Sustainability*,
https://doi.org/10.1007/978-3-031-71844-1_7

environmental sustainability. Taking advantage of the diverse capacities of the microorganisms present in the soil, the impacts of pollution could be reduced, in order to preserve the vital functions of terrestrial ecosystems. Keeping in view of the importance of soil microbiome in environmental sustainability, following topics were deeply discussed in this chapter: (i) fundamentals of the soil microbiome, (ii) Impact of soil contaminants on microbiome diversity, (iii) mechanisms of interaction of the soil microbiome on contaminants, (iv) rhizobacteria as plant growth promoters (PGPR) in soil pollution mitigation, and (v) tool and strategies for the restoration of the soil microbiome.

Keywords Soil microbiome · Soil contamination · Microbial biodiversity · Rhizobacteria · Soil restoration

1 Introduction

The soil microbiome is the main regulator of vital functions within ecosystems, which influences everything from primary production to carbon storage in the soil (Thakur et al., 2019). This elaborate network of biological and chemical interactions in soil is taken as a pillar of vital importance for human existence (Yang et al., 2022). Being intrinsically linked to hydrological, atmospheric, biospheric and even biogeochemical processes, soil plays a crucial role, as it is responsible for most of the annual production of vegetation and the associated nutrient cycles (Fatichi et al., 2020).

Currently there are various factors that represent serious threats to soil health, among them, the most relevant are: the accelerated growth of industrialization (Sharaff et al., 2020), the increase in population, the imminent demand for food (Kaur et al., 2021), climate change (Dubey et al., 2019). In addition, erosion, pollution, salinization, decrease in diversity, mining, among others have also considerable impact on soil microbiome (Nuñéz et al., 2020). Therefore, omics technology, together with microbial techniques have managed to explore the microbial communities at molecular level, and chemical and biochemical information present in the environment (Sullivan & Gadd, 2019).

The microbiome is one of the most important and essentially complex components of all terrestrial ecosystems, as it houses numerous species (i.e. bacteria, fungi, viruses, protozoa and archaea) that interact with each other (Islam et al., 2020). The main activities of soil microbiome include use of nutrients, productivity and the renewal of organic production (Xun et al., 2021). Likewise, being closely linked to the composition of the soil, soil microbiome is involved in the formation of aggregates, the interconnection of pores, in the regulation of the movement of water, oxygen and nutrients in the system, promoting essential functions in agricultural ecosystems, from soil productivity and crop yields to stress resistance (Hartmann & Six, 2023). Soil microbiome is also potential in suppressing pathogens present in it, as well as the degradation of chemical compounds and improving the resistance of the soil system (Wang & Li, 2019).

Soil microbiomes are highly sensitive to environmental pollution. Soil contaminants adversely affecting soil microbiome; and consequences decreased soil health and functionality of terrestrial ecosystems. Thus, there is a need to understand the effects of impacts and develop effective strategies that allow us to restore the ecosystems and for sustainable environmental management.

2 Fundamentals of the Soil Microbiome

In general, the soil is considered as an unconsolidated natural body, composed of mineral and organic matter, present on the surface of the earth and capable of supporting plant growth. The soil fulfills essential functions in terms of the production of food and fiber through plant growth. Soil microbiome preservation, and the data on soil microbiome are important to predict soil loss (Gisbert & Ibáñez, 2010).

There are numerous functions that the soil fulfills, such as: supporting the growth of plant and animal communities, such as crop production, the soil stores water, nutrients and carbon, as well as regulating the availability of many organic and inorganic contaminants. All of these factors are carried through numerous physical, chemical and biological processes interconnected with each other, which allows ecological preservation and global productivity (Hatfield et al., 2017).

The microbiome is a combination of the words "microbe" and "biome", thus describing the microbial ecosystem and referring to both microorganisms as a community of billions of microorganisms that live in the so-called host (also called microbiota). The microbiome also compasses the entire set of molecules that they produce, including nucleic acids, lipids, proteins, metabolites, polysaccharides, among others, that are produced by organisms that live in the same place and that are influenced by the environmental conditions that surround them. On the other hand, the microbial ecosystems present in the soil play a crucial role in maintaining ecological balance, since they are used as a reserve for microorganisms that can affect the microbiomes of both plants, animals and humans (Tomasulo et al., 2024).

A contaminated soil refers to a portion to which an alteration has been made by the addition of hazardous waste, whether liquid or solid (Marín, n.d. s/n). However, the negative effects that the soil has in the face of contamination are still not fully understood. The reviewed literature accurately indicates that the greater the contamination of the soil, the greater the impact, both in the chemistry of the soil, in the presence of microorganisms, in the health of animals and plants, as well as in human health (Grijalva et al., 2020).

Due to the above, the soil microbiome plays a role in terrestrial systems, the presence of microorganisms is crucial to maintain the health of the soil, so understanding its functioning is essential to seek conservation strategies that allow the preservation of the environment. Scientific productivity (in PubMed data base) in the field of soil microbiome was shown in Fig. 1.

Fig. 1 Scientific articles published in the field of 'Soil Microbiome'. (Data retrieved (19 June 2024) from PubMed using 'soil microbiome' as keyword and CARROT 2.0 program)

3 Types of Soil Contaminants

Soil contaminants have increased in recent decades, this represents a problem for both ecological and human health, one of the main causes of pollution is human activities, which include varieties of organic and inorganic chemicals, which can be due to anthropogenic activities occurring in the soil (Cachada et al., 2017). Snapshot of global soil pollution was presented in Table 1. Any matter could become a pollutant if it is outside its natural environment or exceeds the established concentrations, that is, it is considered polluting when its accumulation has the capacity to affect the environment and living beings. Many pollutants due to insufficient quantities do not pose a threat to life, however, many naturally occurring pollutants are produced in sufficient quantities to affect the environment and human health (Brusseau & Artiola, 2019). Inorganic substances represent the main contaminant that is present in the soil, which is represented by 82.8%, cadmium (Cd) is among the number one among the other heavy metals found in the soil (Li et al., 2019). Data on published scientific articles in the field of 'soil contaminants' retrieved from PubMed data base was shown in Fig. 2.

Table 1 Recent trends in global soil pollution

Global Soil Pollution	
Region	Trends
Eastern Europe, the Caucasus and Central Asia (Eurasia)	The soils of the former Soviet Union have been seriously affected by pollution, the result of intense industrial development, militarization and indiscriminate use of agrochemicals.
Asia and the Pacific	Soil pollution in these places is mainly caused by various human activities, such as industry, agriculture, urban environments, transportation, municipal waste processing, among others.
South Asia	A critical region of arsenic contamination comes from geogenic sources. Japan, for its part, faces problems of soil contamination by radioactive substances, as demonstrated by the case of Fukushima.
Europe	Contaminated soils are affected by vehicle exhaust emissions, inadequate waste disposal and a wide range of household pollutants, including those of organic origin.
Ukraine	Soil contamination is due to the presence of radionuclides, trace elements, petroleum products and by-products of military origin. In addition, abandoned mines from military conflicts cause environmental damage.
Russian Federation and Kazakhstan	Extension of soils contaminated by Polychlorinated Biphenyls (PCB). The inventory of PCBs in industrial waste released tons of PCBs into the soil of the Russian Federation, producing pollution through sediment mobilization.

FAO and UNEP (2022)

4 Classification of Pollutants

Soil contamination by heavy metals has become a global problem. Particularly, heavy metals, we refer to metals or metalloids such as cadmium, mercury, lead, chromium, nickel, zinc and arsenic are of major concern with respect to the soil pollution. This type of contamination represents a threat more than organic pollutants because heavy metals are persistent and irreversible, which leads to a degradation of water, atmosphere and agricultural crops. More importantly, heavy metals considerably affect human health, since there is an accumulation of these contaminants in the food chain (Li et al., 2019). Different classes of pollutants that threaten the soil health were shown in Table 2.

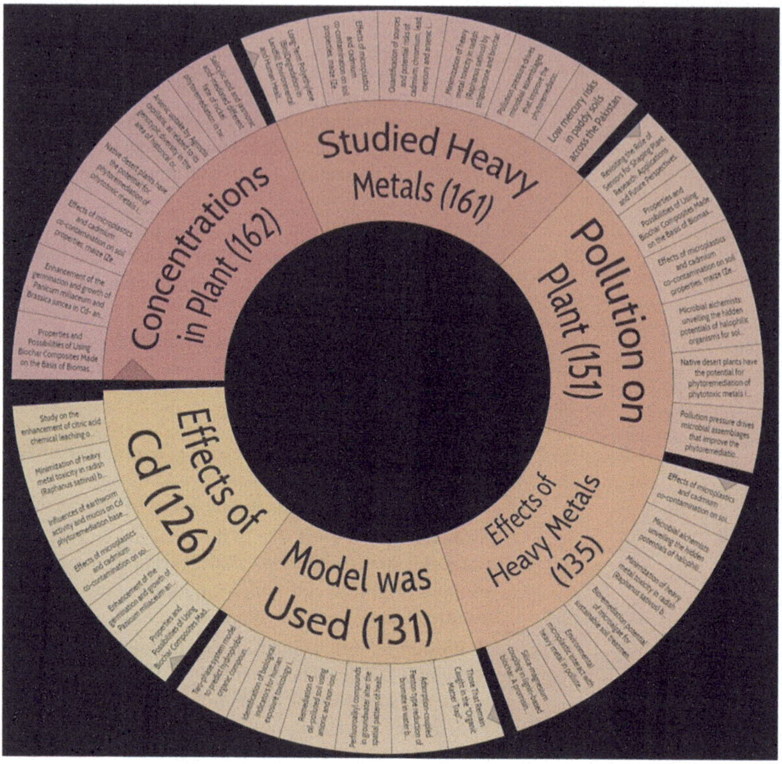

Fig. 2 Scientific articles published in the field of 'Soil Pollutant'. (Data retrieved (19 June 2024) from PubMed using 'soil pollutant' as keyword and CARROT 2.0 program)

Table 2 Principal soil contaminants

Organic contaminants	
Gas stations	Petroleum derivatives: benzene, xylene, polycyclic aromatics, benzene. Aliphatic and aromatic hydrocarbons from petroleum and oil.
Gas factories	Tar, phenols, benzenes, cyanides and polyhydroxyalkanoates (PAHs).
Chemical industries	Pharmaceuticals, food additives, cosmetics: surfactants, dyes, medications. Paints, varnishes, glues and glues Chlorinated solvents (Marín, n.d. s/n): carbon tetrachloride, trichloroethane, trichloroethene, tetrachloroethene.
Agricultural activities	Pesticides: DDT (dichloro-diphenyl-trichloro-ethane), atrazine, 2,4-D (2,4-dichlorophenoxyacetic acid). Polychlorinated biphenyls (PCB): pigments, insulating fluids, plasticizers. Pharmaceutical products. Gaseous compounds: Chlorofluorocarbons (CFCs), hydrochlorofluorocarbons (HCFCs).

(continued)

Table 2 (continued)

Inorganic contaminants	
Metallurgy	Aromatic hydrocarbons, heavy metals and cyanides. Textile industries: Hydrocarbons and heavy metals. Paints, varnishes, glues. Heavy/trace metals: zinc, lead, mercury, arsenic, cadmium, among others. Inorganic salts: sulfate, calcium, nitrate, sodium.
Radioactive contaminants	Elements in solid state: strontium, cobalt, plutonium, uranium. Gaseous elements: radon (Brusseau & Artiola, 2019)

5 Impact of Soil Contaminants on Microbiome Diversity

The presence of contaminants in the soil can have a significant impact on the soil microbiome, since it is capable of altering various factors, such as its diversity, structure and function. Exposure to contaminants such as heavy metals, hydrocarbons, synthetic chemicals, among others, can modify the composition and population of the microbial community present in the soil (Chen et al., 2014). There are numerous biotic and abiotic factors that influence the soil microbiome, in terms of composition, diversity or richness. Among the biotic factors, the stages of development and the genotypes of microbiomes are to be considered important. Among the abiotic factors, rhizospheric deposits, temperature, soil pH and seasonal variation have significant importance in influencing the soil microbiome (Symochko, 2020). Changes in the environment negatively affect the functionality of the soil; the stress produced by various contaminants modifies the composition of the microbiome (Boros et al., 2021). Today, there is a great increase in the number of pollutants that harm the environment, which has caused it to become a major environmental problem on a global scale (Hu et al., 2019).

Anthropogenic activities such as mineral extraction, coke production, incineration and recycling of electronic waste have caused soils to become contaminated with heavy metals and persistent organic substances such as polycyclic aromatic hydrocarbons (Yang et al., 2022). However, when evaluating soils contaminated by diesel from Antarctica, it was revealed that there were heavy metals with high concentrations of cadmium, chromium and lead, as well as contamination with polycyclic aromatic hydrocarbons (Gran et al., 2020). Heavy metals do not undergo degradation due to the action of microbes, however, there are factors such as pH, chelating agents depending on their concentrations, as well as those of inorganic anions and cations that affect their form and also toxicity (Gónzalez & Ghneim, 2021). However, these contaminants can cause alterations in soil enzymes and microbial populations (Boros et al., 2021). Studies show that the presence of heavy metals in the soil, due to human contamination, can generate a decrease in the diversity and activity of the microorganisms present in it; which can negatively affect the soil's ability to perform its basic functions, such as nutrient cycling and the decomposition of organic material (Chen et al., 2014). On the other hand, they can also cause irreversible damage to the microorganisms present in it, which has unfavorable effects on processes that are fundamental, such as microbial respiration, biomass and the decomposition of organic matter and contaminants (Sullivan & Gadd, 2019).

The action of pesticides on the soil affects its physicochemical functions (Kaur et al., 2021). The use of isomers of hexachlorocyclohexane (HCH), which are organic pollutants present in the environment, become important since they have the ability to be absorbed by plants and enter the food chain. There is controversy as to whether this absorption is due to transformation by the microbial action of the soil or enzymatic action of the plant root, despite this, it is known that the action of this contaminant affects the physicochemical properties of the soil not only due to the addition of pesticides with HCH in crops but also that there are concentrations of this in accumulated HCH manure from factories (Liu et al., 2021).

Likewise, there is evidence that polycyclic aromatic hydrocarbons (PAHs) significantly modified the functional structure of the microbiome present in the soil, thus affecting both its capacity to biodegrade contaminants and maintain optimal soil health of terrestrial ecosystem (Sawulski et al., 2014). A study carried out in Argentina examined the impact of pollutants coming from service stations on soil microbial biota. In this research, significant disturbances could be observed in the bacterial structure of the soil due to the introduction of hydrocarbons such as diesel, oil and naphtha (gasoline), separately and in combination. These findings suggest that contaminants will significantly affect the soil and its microbiome, which influences its composition and degradation capacity (Acuña et al., 2020).

Many other studies demonstrate the importance of understanding how soil contaminants affect the terrestrial microbiome and how these changes can lead to significant outcomes for soil health and overall environmental quality; these insights were presented in Table 3.

Table 3 Effects of pollutants on the soil microbiome

Pollutant	Effect on the microbiome	Reference
Pb, Cd, Zn	Great impact on the structure and functionality of the soil microbiome.	Fajardo (2018)
	Decreases the metabolic activity of bacteria.	
	Selective displacement of some bacteria, such as Firmicutes.	
	Zn alone decreases the rate of photosynthesis and stomatal conductance.	Erturk et al. (2015)
	Darkening of the leaves and subsequent death of the plants caused by Zn.	Lee et al. (2019)
Cd, Co, Ni	Modifies the diversity and activity of the soil microbiome.	Boros et al. (2021)
	Produces resistance of certain bacteria to heavy metals.	

(continued)

Table 3 (continued)

Pollutant	Effect on the microbiome	Reference
Cr	Inhibits the germination process of plants.	Nematshahi et al. (2012)
	It has harmful effects on the physiological processes of plants such as photosynthesis, among others.	Dotaniya et al. (2014)
Hg	Reduction in photosynthesis and chlorophyll synthesis.	Fernandez et al. (2018)
mn	It inhibits the development of plants and reduces their ability to synthesize chlorophyll.	Herath et al. (2015)
	It delays the germination of seeds and the correct growth of plants.	Tang et al. (2015)
Cu	Blocking or inhibition of root development.	Jeong et al. (2012)
	Reduces the concentration of chlorophyll.	
	Increase in lipid peroxidation.	
Chemical fertilizers	Reduces microbial diversity.	Rehman et al. (2018)
	Affect the activity of bacteria that promote plant growth.	
Plastics	It produces changes in the composition and structure of the microbial composition.	Guo et al. (2020)
Petrochemicals	Reduction in the diversity and robustness of microbial and plant communities, affecting their functionality in the ecosystem.	Wang et al. (2021)
Heavy metals cadmium (Cd) and silver (Ag)	The microorganisms of the abiotic matrix respond sensitively to these contaminants as they are the first biota affected.	Herruzo et al. (2019)
Glyphosate	Inhibits the proliferation of soil microorganisms in quantities greater than 75 mg/L. It has toxic effects on fungal communities such as *Aspergillus nidulans*.	Gonzáles and Fuente (2022)

6 Mechanisms of Interaction of the Soil Microbiome on Contaminants

Each microbial feeding group has an impact on the processes of the soil ecosystem and its microbiota, which is facilitated by the microbes, which in turn modify both the number of microbial populations and their characteristics. These effects can vary, ranging from increased nutrient circulation to accelerated decomposition and increased plant productivity (Thakur et al., 2019). There are various types of bacteria that interact directly with the plants, such as plant growth-promoting rhizobacteria (PGPR), so called for their action on the metabolism of plant roots, promoting its growth. It also has the ability to grow under abiotic stress thanks to its own metabolism, carrying out the solubilization of phosphates, the production of hormones or with nitrogen fixation. Root associated fungi have a capacity to produce antimicrobial substances that can inhibit the plant pathogens; such microbial functions favor the improvement of the plant or inhibition of plant pathogens (De Andrade et al., 2023).

Within the group of bacteria that help with plant growth are *Gluconacetobacter diazotrophicus* PAL5 and *Azospirillum brasilense* REC3. These bacteria use siderophores of the hydroxamate and catechol type for iron chelating function. It is worth mentioning that plants need iron to carry out physiological processes as part of many important compounds. That is why, in the event that the aforementioned element is out of availability, plants resort to chelation as a way to absorb it (Delaporte et al., 2020).

An analysis was performed with a standard soil test using *Caenorhabditis elegans* nematodes; where it revealed heavy metal toxicity. In one study, it was determined that the soil microbiome adapts to changes produced in them, there were even bacteria resistant to heavy metals (Fajardo, 2018). Heavy metals cause oxidative stress in the microbial environment, thus causing an alteration in the cellular redox state. However, enzymatic mechanisms such as antioxidant enzymes (e.g. oxidase and catalase) have properties of maintaining this cellular redox state and mitigating the damage caused by heavy metals, since they counteract the adverse impacts of reactive oxygen species (Gónzalez & Ghneim, 2021). In a study on heavy metal concentrations (Yang et al., 2022), it was found that the predominant bacteria in these soils were Proteobacteria, Bacteroidetes and Actinobacteria. In Antarctic soils, genera of bacteria such as *Pseudomonas*, *Sphingobium*, *Rhodanobacter*, *Sphingomonas* and *Ferruginibacter* were found in soils contaminated with heavy metals (Gran et al., 2020). Bacterial growth has a mechanism in which it forms a complex with heavy metals, they can also eliminate them and reduce their concentration from the medium in which they are found. Media that preserve their nutrition can even reduce their toxicity, such as *Pseudomonas aeruginosa* with the presence of copper in inhibitory concentrations showed an improvement by increasing nutrients in the medium (Gónzalez & Ghneim, 2021).

In a study carried out on contamination by isomers of hexachlorocyclohexane (HCH) in wheat pots, abundant microbial degraders were detected such as *Sphingomonas* sp. and *Nouosphingobium* sp., which proves that biotransformation occurs in the rhizosphere of the soil in which the action of α—HCH was added (Liu et al., 2020). Another study that was carried out at an industrial site in Ireland investigated the microbiome of soils contaminated by hydrocarbons, in which the results revealed a surprisingly high microbial diversity, among which Alphaproteobacteria and Ascomycota were predominant. The bacterial genera that had the greatest abundance were *Lysobacter*, *Dietzia*, *Pseudomonas* and *Extensimonas*. Although, it is worth mentioning that the genus *Lysobacter* was the most abundant, its participation in the degradation of petrochemicals was not confirmed. However, the genera *Dietzia*, *Pusillimonas* and *Bradyrhizobium* (which were less abundant), showed genes that were related to the degradation of aromatic and aliphatic compounds (Wang & Li, 2019). In the study carried out in Argentina in soils contaminated with hydrocarbons, certain nutrients such as yeast extract, peptone protease, casamino acid, glucose, among others, were added to the control system, causing an increase in the bacterial count, especially in the samples contaminated with diesel. On the other hand, it was found that diesel oil was subjected to highest degradation, while the sample with crude oil demonstrated much less degradation, which was evidenced in the principal component analysis (PCA). The analysis of fatty acids revealed changes in the structure of the bacterial community in response to contaminants and the addition of nutrients, these provide optimal conditions for bacterial growth in the absence of contaminants (Acuña et al., 2020). These findings demonstrate how the soil microbiome represents a natural tool for the remediation of contaminated soils, understanding this relationship is essential for the restoration of contaminated soils and the protection of the environment and health.

7 Rhizobacteria as Plant Growth Promoters (PGPR) in Soil Pollution Mitigation

Rhizosphere bacteria (also called rhizobacteria) are a very large and diverse group that live in the vicinity of plant roots. These establish certain interactions, which can be both beneficial, neutral or, to a lesser extent, even harmful (Rodríguez et al., 2020). The rhizosphere is where the interactions between the roots of the plant, the soil and pathogens exist (Kotoky et al., 2017). Currently, there is an increase in pathogens that cause crop loss. To reduce these losses, the manipulation of plant microbes has been chosen to reduce their diseases. However, the interaction mechanisms among soil microbiome, plants and pathogens are not exactly known. Zhon and collaborators used *Ralstonia solanacearum* as a focal pathogen and tomato plants as hosts; the interactions indicated that the initial microbiome of the soil could predict whether the plant survived or died from the pathogen. Microbiomes of plants such as *Pseudomonas* and *Bacillus* have antimicrobial action, this indicates that a small modification of the soil microbiome can influence the determinants of the interaction between pathogens and the microbiome of the floors (Zhon et al., 2019) (Table 4).

8 Tool and Strategies for the Restoration of the Soil Microbiome

Due to the growing contamination of soil and crops worldwide, studies should focus on efficient tools and strategies that help restore soils. One of these tools is bioremediation with the use of microorganisms capable of cleaning and restoring the environment. This is a technology that promises improved models for the bioremediation of environmental contaminants in the future. Some microbial genera that are potential to be used in the bioremediation, include *Alcaligenes*, *Aspergillus*, *Arthrobacter*, *Trichoderma*, *Pseudomonas*, *Talaromyces*, *Bacillus*, *Stenotrophomonas*, *Mucor*, *Penicillium* and *Burkholderia*, certain archaea such as *Haloferax* and *Natrialba* (Kour et al., 2021). A study highlighted the interaction of bacterial genes associated with

Table 4 Main plant growth promoting rhizobacteria (PGPR) and their functions

Rhizobacteria	Features on soil pollution
IAA-PB	Stimulates the oxidation process of roots to plants; formation of complexes with heavy metals; increase in the amount of plant growth-promoting rhizobacteria (PGPRs) and in the availability of Fe, P, Zn, Cu, among others. Generates stimulation in the growth of plants and their roots. Stimulates the proliferation of roots, the absorption of nutrients and metals is promoted. Promotes the production of enzymes to combat plant oxidation. Induces gene expression in plants to resist heavy metals.
ACCD-PB	Reduces ethylene stress by hydrolysis of ACC to NH_3 and o-ketobutyrate. Stimulates the growth of plants and the root system, increasing root exudate. Promotes the production of enzymes to combat plant oxidation. Induces genetic expression in plants to resist heavy metals.
OA-PB	Formation of complexes with heavy metals. Provides benefits in the acquisition of nutrients such as P, Fe, Ze, etc. Provides tolerance to toxic metals in the formation of metal-oxalate crystals. Promotes the production of enzymes to combat plant oxidation. Induces gene expression in plants to resist heavy metals.
BE-PB	Increases tolerance to heavy metals and helps in the removal of metals from the soil. Preferably binds to toxic metals. Promotes the production of enzymes to combat plant oxidation. Induces gene expression in plants to resist heavy metals.
SPB	Supply of N to plants. Promotes plant growth. Promotes the production of enzymes to combat plant oxidation. Induces genetic expression in plants to resist heavy metals.
PSB	Supply of K to plants. Promotes plant growth. Promotes the production of enzymes to combat plant oxidation. Induces genetic expression in plants to resist heavy metals.
KSB	Supply of P to plants. Promotes plant growth. Reduces the absorption and translocation of heavy metals. Helps in the formation of phosphate complexes with heavy metals. Forms phosphate precipitates of insoluble heavy metals. Produces immobility of heavy metals. Promotes the production of enzymes to combat plant oxidation. Induces gene expression in plants to resist heavy metals.
NFB	Supplies nutrients to plants. Promotes plant growth. Reduces the absorption and translocation of heavy metals. Reduces the generation of free radicals in the root zone of plants. Union of metals with siderophores. Generation/regulation of hormones in plants. Protection of microbial phytohormones (e.g., IAA) from metal-induced oxidative damage. Promotes the production of enzymes to combat plant oxidation. Induces genetic expression in plants to resist heavy metals

Note: *IAA-PB* Indoleacetic Acid Producing Bacteria, *ACCD-PB* 1-Aminocyclopropane-1-Carboxylate Deaminase Producing Bacteria, *OA-PB* Organic Acid Producing Bacteria, *BE-PB* Biosurfactant and Exopolymer Producing Bacteria, *NFB* Nitrogen Fixing Bacteria, *KSB* Potassium Solubilizing Bacteria, *PSB* Phosphate Solubilizing Bacteria, *SPB* Siderophore Producing Bacteria biofilm formation and the direct reduction of pollution. However, many microorganisms have not yet been characterized at molecular level to know their bioremediation potential. For example, fungi such as *Pyronema*, *Tuber* and *Hebeloma*, their genetic function on pollution is not characterized (González et al., 2018).

A strategy to remediate contaminated soils also includes phytoremediation, the use of specific phytoremediator crops such as *Salix* spp., to be able to restore contaminated soils, require a complex microbiome and interaction. Phytoremediation is a technique in which it has been seen those plants such as *Zea mays*, *Helianthus annuus*, *Panicum virgatum*, *Populus tremula*, *Nicotiana tabacum*, and *Oryza sativa* have the ability to absorb and accumulate organic contaminants. On the other hand, it is found that *Salix viminalis*, *Medicago sativa* L. *Taxodium distichum*, *Salix babylonica*, *Phragmites australis*, which in addition to absorbing contaminants, also carry out translocation and volatilization of said contaminants when using pine biochar with bird droppings and bacteria such as *Bradyrhizobium*. Importantly, in this treatment, the bacteria were active for up to one year; it must be taken into consideration that not all biochar are optimal, under conditions they can act as soil contaminants (Xiang et al., 2022). A microbe-assisted phytoremediation strategy, known as "ecopiling", was implemented in a study in Ireland, involving biostimulation of the natural microbial community of contaminated soil to increase its contaminant degradation capacity. To enhance the absorption and decomposition of contaminants by plants, specific microorganisms were introduced, such as *Dietzia*, *Pusillimonas* and *Bradyrhizobium*. Although these bacteria were less abundant, but were effective in degradation of contaminants, since these bacteria showed to have genes related to the degradation of aromatic and aliphatic compounds (Wang et al., 2021).

Soil suppressiveness is another strategy used to improve soil health. This strategy is described as those in which pathogens do not establish or remain; and if they do so they will not cause disease. This is due to mechanisms that reduce infectivity or the number of pathogens such as the edaphic that can be physical, chemical and antagonistic interactions such as antibiosis, parasitism and competition. This technique is based on a general or specific suppressiveness, minimum tillage techniques, crop rotation, organic amendments (e.g. compost) are used (Calvo, 2021). In a bibliographic review, it was observed that organic amendments such as manure, compost, biochar, green manure and stubble in agricultural soils present benefits such as an increase in enzyme activity, and biomass of the soil microbiome. One of the effects of instability in this practice is the increase in temperature which influences the release of CO_2 into the atmosphere, so its use must be monitored to avoid the negative effects (Abarzúa, 2022). The literature review reports essential information on soil restoration strategies and tools; however, it is essential to continue conducting research and developing new sustainable techniques that allow soil restoration and agriculture. Different soil restoration strategies and their functions were presented in Table 5.

Table 5 Soil restoration strategies

Strategies and tools	Function
Bioremediation	It is an effective, profitable and environmentally friendly tool, with low environmental risks. The use of microorganisms presents an opportunity as a biotechnological strategy to decontaminate the environment (Kour et al., 2021).
Phytoremediation	Technique for using plants and associated microbiomes that allow the restoration of contaminated soil, often due to organic contaminants, there are some difficulties in its function due to excess of secondary toxins, lack of microbial populations, among others (Xiang et al., 2022).
Suppressiveness	This strategy refers to soils that are suppressive of phytopathogenic microorganisms, which will remain at low levels without causing disease. We can highlight the natural suppressiveness (physical, chemical and biological) and the induced suppressiveness influenced by cultural practices such as organic amendments such as the use of compost or manure, minimum tillage, crop rotation, among others (Calvo, 2021).
Reforestation and revegetation	Reforestation as a soil restoration strategy has been used for many years. It is used with the objective of recovering the structures and functions of the ecosystem, thereby recovering the soils, avoiding or reversing the degradation that occurs in many ecosystems at the world level (Ventura et al., 2017).
Sustainable agricultural practices	This strategy includes techniques such as the combination of chemical and organic inputs, effective use of soil time and space such as crop associations and rotation strategies, also the cultivation of indigenous or native plants, efficient risk techniques and pest management and diseases (Gonzalez et al., 2024).

References

Abarzúa, B. (2022). *Organic soil amendments in the context of climate change*. Retrieved from https://repositorio.uchile.cl/bitstream/handle/2250/196321/2022_Benjamin_Abarzua_Torres.pdf?sequence=1&isAllowed=y

Acuña, A., et al. (2020). Impact on soil microbial biota due to contaminants from service stations. *Peruvian Journal of Biology, 27*(3), 417–422. https://doi.org/10.15381/rpb.v27i3.17387

Almendros, M., et al. (2021). *Soil and the importance of humus*. Retrieved from DIGITAL.CSIC: http://hdl.handle.net/10261/264023

Argumedo, R., et al. (2009). The fungal genus Trichoderma and its relationship with organic and inorganic pollutants. *Revista Internacional de Contaminación Ambiental, 25*(4), 257–269. Retrieved from https://www.scielo.org.mx/pdf/rica/v25n4/v25n4a6.pdf

Boros, E., et al. (2021). The response of the soil microbiome to contamination with cadmium, cobalt and nickel in soil sown with Brassica napus. *Minerals, 11*(5), 498. https://doi.org/10.3390/min11050498

Brusseau, M., & Artiola, J. (2019). Chemical contaminants. In *Environmental and pollution sciences* (3rd ed., pp. 175–190). Academic Press. https://doi.org/10.1016/B978-0-12-814719-1.00012-4

Cachada, A., et al. (2017). Chapter 1 – Soil and pollution: An introduction to the Main issues. In *Soil pollution. From monitoring to remediation* (1st ed., pp. 1–28). Academic Press. https://doi.org/10.1016/C2016-0-02243-X

Calvo, J. (2021). Suppressive soil and their role in plant disease management. *Environment & Technology, 2*(1), 48–63. https://doi.org/10.56205/ret.2-1.3

Chen, J., et al. (2014). Heavy metal pollution decreases microbial abundance, diversity and activity within particle-size fractions of a paddy soil. *FEMS Microbiology Ecology, 87*(1), 164–181. https://doi.org/10.1111/1574-6941.12212

De Andrade, L., et al. (2023). Plant growth-promoting Rhizobacteria for sustainable agricultural production. *Microorganisms, 11*(4), 1088. https://doi.org/10.3390/microorganisms11041088

Delaporte, P., et al. (2020). The plant growth promoting bacteria Gluconacetobacter diazotrophicus and Azospirillum brasilense contribute to the iron nutrition of strawberry plants through siderophores production. *Plant Growth Regulation, 91*, 185–199. https://doi.org/10.1007/s10725-020-00598-0

Dotaniya, M., Das, H., & Meena, V. (2014). Assessment of chromium efficacy on germination, root elongation, and coleoptile growth of wheat (Triticum aestivum L.) at different growth periods. *Environmental Monitoring and Assessment, 186*, 2957–2963. https://doi.org/10.1007/s10661-013-3593-5

Dubey, A., et al. (2019). Soil microbiome: A key player for conservation of soil health under changing climate. *Springer Link, 28*, 2405–2429. https://doi.org/10.1007/s10531-019-01760-5

Erturk, F., et al. (2015). Analysis of genetic and epigenetic effects of maize seeds in response to heavy metal (Zn) stress. *Environmental Science and Pollution Research, 22*, 10291–10297. https://doi.org/10.1007/s11356-014-3886-4

Fajardo, C. (2018). Pb, Cd, and Zn soil contamination: Monitoring functional and structural impacts on the microbiome. *Applied Soil Ecology*. https://doi.org/10.1016/j.apsoil.2018.10.022

FAO and UNEP. (2022). *Global soil pollution assessment—Summary for policymakers*. FAO. Retrieved from https://doi.org/10.4060/cb4827es

Fatichi, S., et al. (2020). Soil structure is an important omission in earth system models. *Nature Communications, 11*(522). https://doi.org/10.1038/s41467-020-14411-z

Fernandez, R., et al. (2018). Comparison of mercury distribution and mobility in soils affected by anthropogenic pollution around chloralkali plants and ancient mining sites. *Science of the Total Environment, 671*, 1066–1076. Retrieved from https://acortar.link/j0CZiO

Gisbert, J., & Ibáñez, S. (2010). *Soil genesis*. Polytechnic University of Valencia.

Gonzáles, E., & Fuente, M. (2022). Dynamics of glyphosate in soil and its effects on microbiota. *International Journal of Environmental Pollution, 38*, 127–144. https://doi.org/10.20937/rica.54197

Gónzalez, S., & Ghneim, T. (2021). Heavy metals in soils and the remediation potential of bacteria associated with the plant microbiome. *Sec. Toxicology, Pollution and the Environment, 9*. https://doi.org/10.3389/fenvs.2021.604216

González, E., et al. (2018). Trees, fungi and bacteria: Tripartite metatranscriptomics of a root microbiome responding to soil contamination. *Microbiome, 6*(53). https://doi.org/10.1186/s40168-018-0432-5

Gonzalez, F., et al. (2024). Ecointensification of agricultural systems as the potential of soil microorganisms. A meta-analysis. *Revista mexicana de ciencias agrícolas, 14*(8). https://doi.org/10.29312/remexca.v14i8.3322

Gran, A., et al. (2020). Effect of co-contamination by PAHs and heavy metals on bacterial communities of diesel contaminated soils of South Shetland Islands, Antarctica. *Microorganisms, 8*(11), 1749. https://doi.org/10.3390/microorganisms8111749

Grijalva, A., et al. (2020). Water and air pollution by chemical agents. *RECIMUNDO, 4*, 79–93. https://doi.org/10.26820/recimundo/4

Guo, J., et al. (2020). Prospects and applications of plant growth promoting rhizobacteria to mitigate soil metal contamination: A review. *Chemosphere, 246*, 125823. https://doi.org/10.1016/j.chemosphere.2020.125823

Hartmann, M., & Six, J. (2023). Soil structure and microbiome functions in agroecosystems. *Nature Reviews Earth and Environment, 4*, 4–18. https://doi.org/10.1038/s43017-022-00366-w

Hatfield, J. L., Sauer, T. J., & Cruse, R. M. (2017). Chapter one—Soil: The forgotten piece of the water, food, energy nexus. *Advances in Agronomy, 143*, 1–46. https://doi.org/10.1016/bs.agron.2017.02.001

Herath, I., et al. (2015). Immobilization and phytotoxicity reduction of heavy metals in serpentine soil using biochar. *Journal of Soils and Sediments, 15*, 126–138. https://doi.org/10.1007/s11368-014-0967-4

Herruzo, A., Michán, C., & Alhama, J. (2019). Metaproteomic approach to evaluate the effects of cadmium and silver on microbial function in a microcosm experimental design. In F. P. Arturo (Ed.), *Creating doctoral networks: Research and communicate* (pp. 327–330). University of Córdoba. Retrieved from https://dialnet.unirioja.es/servlet/libro?codigo=783250

Hu, D., et al. (2019). Microplastics and nanoplastics: Would they affect global biodiversity change? *Environmental Science and Pollution Research, 26*(19), 19997–20002. https://doi.org/10.1007/s11356-019-05414-5

Islam, W., et al. (2020). Role of environmental factors in shaping the soil microbiome. *Environmental Science and Pollution Research, 27*, 41225–41247. https://doi.org/10.1007/s11356-020-10471-2

Jeong, S., et al. (2012). Application of phosphate-solubilizing bacteria for enhancing bioavailability and phytoextraction of cadmium (Cd) from polluted soil. *Chemosphere, 88*(2), 204–210. https://doi.org/10.1016/j.chemosphere.2012.03.013

Kaur, I., et al. (2021). Plants exerted beneficial influence on soil microbiome in a HCH contaminated soil revealing advantage of microbe-assisted plant-based HCH remediation of a dumpsite. *Chemosphere, 280*. https://doi.org/10.1016/j.chemosphere.2021.130690

Kotoky, R., et al. (2017). The rhizosphere microbiome: Significance in rhizoremediation of polyaromatic hydrocarbon contaminated soil. *Journal of Environmental Management, 217*, 858–870. https://doi.org/10.1016/j.jenvman.2018.04.022

Kour, D., et al. (2021). Beneficial microbiomes for bioremediation of diverse contaminated environments for environmental sustainability: Present status and future challenges. *Environmental Science and Pollution Research, 28*, 24917–24939. https://doi.org/10.1007/s11356-021-13252-7

Lee, D., et al. (2019). Fly ash and zeolite decrease metal uptake but do not improve rice growth in paddy soils contaminated with Cu and Zn. *Environment International, 129*, 551–564. https://doi.org/10.1016/j.envint.2019.04.032

Li, C., et al. (2019). A review on heavy metals contamination in soil: Effects. *Sources, and Remediation Techniques, 28*(4), 380–394. https://doi.org/10.1080/15320383.2019.1592108

Liu, X., et al. (2020). Compound-specific isotope analysis and enantiomer fractionation to characterize the transformation of hexachlorocyclohexane isomers in a soil–wheat pot system. *Environmental Science & Technology, 54*(14), 8690–8698. https://doi.org/10.1021/acs.est.9b07609

Liu, X., et al. (2021). Characterizing the biotransformation of hexachlorocyclohexanes in wheat using compound-specific stable isotope analysis and enantiomer fraction analysis. *Journal of Hazardous Materials, 406*. https://doi.org/10.1016/j.jhazmat.2020.124301

Marín, M. (n.d. s/n). *Chemical analysis of soil and water. Transparencies and problems*. UPV.

Nematshahi, N., Lahouti, M., & Ganjeali, A. (2012). Accumulation of chromium and its effect on growth of (Allium cepa cv. Hybrid). *European Journal of Experimental Biology, 2*(4), 969–974. Retrieved from https://www.primescholars.com/articles/accumulation-of-chromium-and-its-effect-on-growth-ofallium-cepa-cv-hybrid.pdf

Nuñéz, P., et al. (2020). Brief overview of soil: Role, importance, functions, quality and indicators. *APF, 9*(1), 97–114. Retrieved from https://sodiaf.org.do/apf/index.php/apf/article/view/118/102

Rehman, B., Hassan, T., & Bano, A. (2018). Potential of indole-3-acetic acid-producing rhizobacteria to resist Pb toxicity in polluted soil. *Soil & Sediment Contamination, 28*(1), 101–121. https://doi.org/10.1080/15320383.2018.1539947

Rodríguez, A., et al. (2020). Rhizospheric bacteria with potential benefits in agriculture. *Terra Latinoamericana, 38*(2), 333–345. https://doi.org/10.28940/terra.v38i2.470

Sawulski, P., et al. (2014). Effects of polycyclic aromatic hydrocarbons on microbial community structure and PAH ring hydroxylating dioxygenase gene abundance in soil. *Biodegradation, 25*(6), 835–847. https://doi.org/10.1007/s10532-014-9703-4

Sharaff, M., et al. (2020). Mechanistic understanding of the root microbiome interaction for sustainable agriculture in polluted soils. In A. E. Rastegari (Ed.), *New and future developments in microbial biotechnology and bioengineering. Trends of microbial biotechnology for sustainable agriculture and biomedicine systems: Diversity and functional perspectives* (pp. 61–84). Elsevier. https://doi.org/10.1016/B978-0-12-820526-6.00005-1

Sullivan, T., & Gadd, G. (2019). Metal bioavailability and the soil microbiome. *Advances in Agronomy, 155*, 79–120. https://doi.org/10.1016/bs.agron.2019.01.004

Symochko, L. (2020). Soil microbiome; diversity, activity, functional and structural successions. *IJEES, 10*(2), 277–284. https://doi.org/10.31407/ijees10.206

Tang, X., et al. (2015). Effects of inorganic and organic amendments on the uptake of lead and trace elements by Brassica chinensis grown in an acidic red soil. *Chemosphere, 119*, 177–183. https://doi.org/10.1016/j.chemosphere.2014.05.081

Thakur, M., et al. (2019). Trophic regulations of the soil microbiome. *Trends in Microbiology (Regular Ed.), 27*(9), 771–780. https://doi.org/10.1016/j.tim.2019.04.008

Tomasulo, A., et al. (2024). Microbiome one health model for a healthy ecosystem. *Science in One Health, 3*. https://doi.org/10.1016/j.soh.2024.100065

Ventura, R., et al. (2017). Is the reforestation a rehabilitation strategy for pine forest? An experience in Central Mexico. *Fosrest, 38*(1). https://doi.org/10.4067/S0717-92002017000100007

Wang, L., & Li, X. (2019). Steering soil microbiome to enhance soil system resilience. *Critical Reviews in Microbiology*. https://doi.org/10.1080/1040841X.2019.1700906

Wang, M., et al. (2021). Soil microbiome structure and function in Ecopiles used to remediate petroleum-contaminated soil. *Frontiers in Environmental Science, 9*(624070). https://doi.org/10.3389/fenvs.2021.624070

Xiang, L., et al. (2022). Integrating biochar, bacteria, and plants for sustainable remediation of soils contaminated with organic pollutants. *Environmental Science & Technology, 56*(23), 16546–16566. https://doi.org/10.1021/acs.est.2c02976

Xun, W., et al. (2021). Specialized metabolic functions of keystone taxa sustain soil microbiome stability. *Microbiome, 9*(35). https://doi.org/10.1186/s40168-020-00985-9

Yang, Z., et al. (2022). Soil microbiomes divergently respond to heavy metals and polycyclic aromatic hydrocarbons in contaminated industrial sites. *Environmental Science and Ecotechnology*. https://doi.org/10.1016/j.ese.2022.100169

Zhon, W., et al. (2019). Initial soil microbiome composition and functioning predetermine future plant health. *Science Advances, 5*(9). https://doi.org/10.1126/sciadv.aaw0759

Part II
Soil Microbiomes and Agricultural Sustainability

Desertification Impacts on Soil Microbiomes and Plant Growth

S. Abdullahi, A. B. Alhassan, U. Bayero, S. I. Ibrahim, A. K. Abdullahi, A. O. Musa, and H. G. Amir

Abstract Soil serves as the habitat for microorganisms and the medium for plant growth. Soil microbiomes contribute immensely to ecosystem services and maintenance of ecological balance through various beneficial activities such as the breakdown of organic matter, cycling of nutrients, keeping soil fertility and plant health, as well as protecting plants against environmental stresses, diseases, and pathogens. Soil microbial communities can survive different environments including harsh ones like deserts. Extremophiles such as bacteria and archaea are the first groups of soil microbiomes to invade deserts due to their ability to survive extreme conditions. Deserts and the process of desertification have great effects on the soil microbiomes and plant growth which include the decline in diversity, abundance, and functional characteristics of the soil microbial communities and plants or vegetation cover. Understanding the impacts of desertification is important towards predicting and regulating microbial processes. Some microbiomes can adapt and survive desert environments and even have the ability to mitigate the impacts. These have great potential which if properly harnessed can be employed to improve the growth and survival of other organisms, particularly plants in the deserts and other extreme environments.

Keywords Soil microbes · Desert microbes · Desertification · Soil-plant-microbe interaction

S. Abdullahi (✉) · S. I. Ibrahim · A. O. Musa
Department of Botany, Faculty of Life Sciences, Ahmadu Bello University, Zaria, Nigeria
e-mail: sabdullahi@abu.edu.ng

A. B. Alhassan · U. Bayero · A. K. Abdullahi
Department of Biology, Faculty of Life Sciences, Ahmadu Bello University, Zaria, Nigeria

H. G. Amir
School of Biological Sciences, Universiti Sains Malaysia, Pulau Pinang, Malaysia

S. A. Aransiola et al. (eds.), *Soil Microbiome in Green Technology Sustainability*,
https://doi.org/10.1007/978-3-031-71844-1_8

1 Introduction

The diverse community of microorganisms in the soil, known as the soil microbiome is majorly composed of bacteria, fungi, archaea, viruses, and protozoa (Prabhakaran et al., 2022) with bacteria and fungi being regarded as the most common soil microbes (Vandenkoornhuyse et al., 2015). Extremophiles such as bacteria and archaea are the first groups of soil microbiomes to invade deserts due to their ability to survive extreme conditions, after which fungi and other plants like lichens and mosses may resume, and form the biological soil crusts (BSC). BSC stabilizes the soil and regulates nitrogen and carbon cycles for the survival of vascular plants in desert ecosystems (Li et al., 2018; Marasco et al., 2023).

The coexistence and diversity of soil microbes differ with different soil types and the distribution of plant species in an area can be used to identify alteration in the microbiome structure of a soil (Prabhakaran et al., 2022). For example, arbuscular mycorrhizal fungi are found in grasslands extensively, while ectomycorrhizal fungi tend to have mutual interactions with the rhizome of trees in forest soils (Prabhakaran et al., 2022). Tosi et al. (2020) stated that while some plants have developed mechanisms of adapting to different environmental or soil conditions, they still rely on the associated microbiomes for roles such as nutrient acquisition, stress tolerance, protection against diseases and pathogens, and the effects of various forms of land degradation.

The most severe kind of land degradation is referred to as desertification. It is the loss of biological productivity from land and can be brought on by both climate change and human activity (Catania et al., 2022). The degradation of land in dry, semi-arid, and dry sub-humid regions, also referred to as "drylands" is called desertification. Desertification eventually results in barren, unproductive land with limited biological value that cannot be exploited for farming, food production, or other agricultural uses. For sustainable development in northern Nigeria and other African nations, desertification poses a grave threat. According to Catania et al. (2022), desertification has a negative impact on soil microbes, plant growth, human health, food security, economic activity, physical infrastructure, natural resources, and the environment. It also has an impact on national and international security.

Intensified desertification exerts negative influences on soil properties, which may dramatically affect the growth and diversity of soil microorganisms (Cao et al., 2021), and in turn impact the vegetation and ecological restoration of the desertified area (Li et al., 2023). The process modifies the physicochemical characteristics of the soil, including its pH, moisture content, and nutrient availability, and consequently affects the diversity and composition of the microbial population (Li et al., 2023).

1.1 Importance of Soil Microbiomes

Soil microbiomes are mostly found in the region near the plant roots known as the rhizosphere acting in a beneficial, pathogenic, or commensal manner to one another and to the host plants. They play a crucial role in maintaining ecological balance and keeping soil fertility and plant health, some of which are discussed below and presented in Fig. 1. Some of the beneficial role of soil microbiomes towards maintaining ecological balance and keeping soil fertility and plant health are presented in Table 1.

1.2 Soil Microbiomes in Biogeochemical Cycling of Nutrients

Soil microbiomes play a key role in the biogeochemical cycling of nutrients such as nitrogen, carbon, sulphur, and phosphorus (Abdul Rahman et al., 2021). For example, the symbiotic nitrogen-fixing bacteria (*Rhizobium* sp.) that live in the root

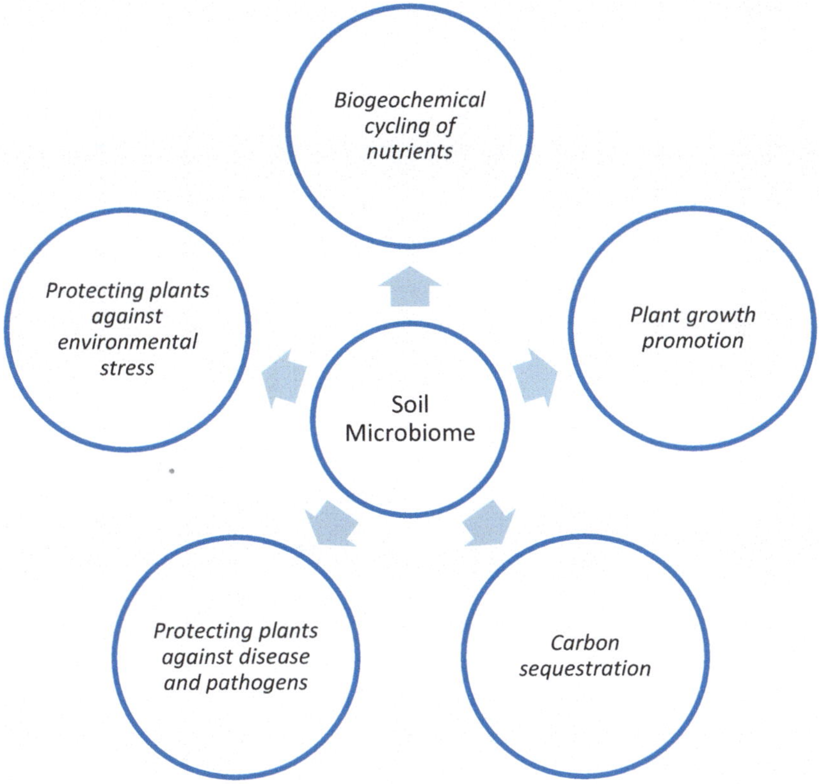

Fig. 1 Beneficial role of soil microbiomes towards maintaining ecological balance and keeping soil fertility and plants health

Table 1 Some beneficial role of soil microbiomes towards maintaining ecological balance and keeping soil fertility and plant health

Microbe(s)	Associated plant(s)	Stress condition(s)	Beneficial role	Reference
Bacillus amyloliquefaciens	*Capsicum annuum* cv. Geumsugangsan	Drought, salinity, and heavy metal	Higher seedling growth rate and improved physiochemical traits High chlorophyll, salicylic acid, sugar, amino acid, and proline contents	Kazerooni et al. (2021)
Bacillus sp. LBF-01	*Capsicum annuum*	*Fusarium oxysporum* infection	Reduction of fungicide use against *Fusarium oxysporum* by 50% Antifungal activity against *Fusarium oxysporum* Enhanced germination potentiality and seedling vigour by 24% and 592.8%, respectively Enhanced plant growth parameters	Chowdhury et al. (2020)
Enterobacter sp. UOM 3; *Pseudomonas hunanensis* IB C7	Barley plants	Presence of petroleum hydrocarbons in the soil	Increase root length, root weight, shoot weight, and leaf surface index Decrease in abscisic acid accumulation in plant Reduced hydrocarbons content in the soil Increased microbiological activity more than when either of the organisms was used individually	Kuzina et al. (2021)
Bacillus thuringiensis, Bacillus pacificus	Tomato plants (*Solanum lycopersicum*)		Improved seed germination (31.93%) and seedling strength (78.59%) Increased stem height, root length, and weight of fresh and dry stems and roots	Kouam et al. (2023)
Arthrobacter arilaitensis, Streptomyces pseudovenezuelae	Maize plants (*Zea mays*)	Drought stress	Significantly higher chlorophyll content index Increased plants heights and root lengths Increased shoot and root dry weights	Chukwuneme et al. (2020)
Pseudomonas spp. B004 and B021, *Bacillus* spp. B005 and B208	Maize and faba bean intercropping	*Fusarium oxysporum*	Reduced gene copies of *F. oxysporum* by 30–84% Increased bacterial richness and Shannon index compared to monocropping Facilitated pathogen suppression and increased resistance of faba bean to *F. oxysporum*	Sun et al. (2022)

nodules of legumes, the free-living nitrogen-fixing bacteria (*Azotobacter* and *Azospirillum*), and several other plant growth-promoting rhizobacteria (PGPR) are involved in the nitrogen cycle. Members of Proteobacteria and arbuscular mycorrhizal fungi (AMF) which are obligatory symbionts are involved in the biogeochemical cycling of C and rely for their carbon on their host plants (Abdul Rahman et al., 2021). Phosphate solubilizing microorganisms (PSMs) and potassium solubilizing microorganisms (KSMs) solubilize insoluble phosphorus and potassium compounds respectively to easily assimilated forms to be used by plants (Tharek et al., 2022). They include well-known genera of bacteria (*Bacillus, Pseudomonas, Burkholderia, Klebsiella,* etc.), fungi (*Aspergillus Penicillium,* etc.), AMF, and actinomycetes.

1.3 Soil Microbiomes in Protecting Plants Against Environmental Stress

Certain soil microbes such as PGPR produce compounds and enzymes that solubilize minerals and other nutrients to help maintain plant growth under various environmental stress conditions like drought, heavy metal stress, salinity stress, and extreme temperatures. Siderophore production by soil microbiomes is one of the mechanisms that provide iron (Fe) for use by plants growing in stressful conditions (Kazerooni et al., 2021). Previous studies showed that iron has a great role in maintaining chloroplast structure and chlorophyll synthesis. Bacteria produce different enzymes involved in degrading organic materials such as pesticides/herbicides, polysaccharides, PAH, and so on. Plant growth-promoting microorganisms such as some soil bacteria through various activities like the production of exopolysaccharides can lead to the formation of hydrophobic films which in turn protect the plants against drought and desiccation (Zhang et al., 2023). Several fungal species integrate themselves with soil aggregates using their hyphae while several bacterial species produce exopolysaccharides and biofilms acting as glue or binding agent, which overall, enhance soil aggregation and stability as well as water retention. Marasco et al. (2023) argued that the microbiota of stressed environments such as drylands that are already adapted to drought stress and oligotrophic conditions have great potential which if properly harnessed can be used for supporting sustainability and survival of plant species under water stress. They further discussed their exploitation and application to the management of conservation-restoration projects of drylands to address soil erosion, revegetation, and general improvement of crop cultivation.

1.4 Soil Microbiomes in Plant Growth Promotion

Soil microbiota play a distinct role in the soil ecosystem, which includes improving plant growth and health, as well as soil fertility which immensely contributes to the yield and quality of agricultural products. Land degradation and climate change have accelerated food insecurity leading to a decline in agricultural productivity putting the world population which is estimated to reach almost 10 billion by 2050 at great risk (Tosi et al., 2020; Marasco et al., 2023). Though plants have at some points developed adaptation mechanisms to sustain and alleviate some environmental stresses, they still rely majorly for many functions on their associated microbiota, as they do not exist as individual entities, but as larger entities or holobiont (Vandenkoornhuyse et al., 2015; Tosi et al., 2020). *Rhizobium* species are the main examples used as inoculants for nitrogen fixation in association with legumes, however, there is increasing interest in harnessing other plant growth-promoting bacteria and fungi for their beneficial roles. Soil microbiomes are reported to improve plant growth and yield through many beneficial functions such as nitrogen fixation; production of growth hormones such as indole acetic acid (IAA); phosphate solubilization; potassium solubilization; production of 1-amino cyclopropane-1-carboxylic acid deaminase (ACC Deaminase); siderophore production and many more (Abdullahi et al., 2020). There is currently increasing interest in the use of soil microbiomes especially bacteria and fungi, which has gone beyond their usage as biofertilizers and biopesticides, to harnessing other beneficial properties of mitigating the deleterious effects of climate change and diseases.

1.5 Soil Microbiomes in Protecting Plants Against Disease and Pathogens

One of the major factors militating against the qualitative and quantitative production of crop plants is disease incidence (Karthika et al., 2020). Several unfavourable consequences affecting the environment and public health when using chemicals for disease and pathogens control, forces legislations to limit the negative impacts of chemical control (Berlin et al., 2018). As a result, many active components of such harmful chemicals are prohibited for agricultural usage, therefore the need for alternative and safer biocontrol methods. The association between many crops and soil microbial communities provides the host plants with protection against several pathogens and diseases. Bacteria and fungi especially are mostly used as biocontrol agents against harmful microbial and other plant disease agents. Use of beneficial microbes against plant pathogens is proving to be cost effective and environmentally friendly than chemical control agents. For example, Chowdhury et al. (2020) in an attempt to assess the biocontrol and disease prevention potential of *Bacillus* sp. LBF-01 in *Capsicum annuum* found the bacterium to exhibit antifungal activity against the fungal pathogen *Fusarium oxysporum* and at the same time reducing the carbendazim fungicide use by 50%.

1.6 Soil Microbiomes in Carbon Sequestration

Sequestration of atmospheric carbon into stable and non-gaseous forms is achieved through both biotic and abiotic processes. Carbon can be made available in the soil by plants through their assimilation of CO_2 and also by autotrophic microorganisms (Marasco et al., 2023). Participation of microbial communities in the biogeochemical and sequestration of carbon is greater in soils with higher diversity, and it involves combined and coordinated activities of majorly bacteria and fungi leading to the formation of carbon polymers and soil aggregates (Li et al., 2018). Activities of single soil microbe or their consortia or the relationship between soil microbes and associated plants can be manipulated and explored for the enhancement of soil carbon storage.

2 Causes of Desertification

There are many different and intricate reasons for desertification; these problems are essentially the outcome of interactions between climatic factors and human activities in the environment, just like many other environmental degradation problems. Some of the causes are discussed below:

2.1 Climatic Variability

A significant cause of many environmental degradation issues is climate variability. Desertification is a naturally occurring phenomenon caused by changes in the climate (Zhang et al., 2023). The amount of greenhouse gases in the atmosphere is rising, which is contributing to global warming and increasing climatic variability. The following are some ways that this change in the climate has shown up:

- A reduction in precipitation in arid regions increases the susceptibility of semi-arid and arid regions to desertification
- Low rainfall and high temperatures could cause a drought and the drying up of water supplies
- Weak vegetation growth resulting in the creation of a desert-like environment

Nigeria is not immune to the effects of climate variability and global warming, with highly localized effects, particularly noticeable in the country's most industrialized towns and Northern Nigeria. These effects have led to the observed degradation of the environment. According to Ayangbenro and Babalola (2021), there use to be increase in temperature and decrease in rainfall which often results in weak vegetation growth, which results in the creation of a desert-like environment.

2.2 Deforestation

According to Catania et al. (2022), deforestation is the process of turning forested regions into empty land. Large-scale forest loss that leaves behind non-forest areas to satisfy diverse human demands is known as deforestation. The main causes of deforestation have been determined to include logging, the growth of crops, urbanization, the collecting of fuel wood, mining and resource extraction, fire hunting, and slash-and-burn techniques. Deforestation of dry lands destroys the trees and vegetation that bind the soil, and because of the prevailing climatic conditions in dry lands, the possibility of regeneration of denuded vegetation is low and hence, the land becomes desertified (Qi et al., 2018).

2.3 Over-Exploitation and Inappropriate Land Use

The practice of inappropriate agricultural practice, overgrazing, over-cultivation, and deforestation are major issues that cause desertification in northern Nigeria. This is due to poverty and total dependency on natural resources for the survival of the poor.

2.4 Poverty-Related Agricultural Practices and Other Land Use Systems

Poor land management practices, overgrazing, a lack of soil and water conservation structures, frequent and continuous irrigation farming without adding better supplements to the soil, and a high frequency of careless bushfires all contribute to land degradation and exacerbate the process of desertification.

2.5 High Population Growth

The number of rural residents in Africa is expected to increase by 325 million (Ayangbenro & Babalola, 2021). Deforestation and overuse of the land result from this. Nigeria's population increased fourfold between 1950 and 2006, from 33 million to 134 million people, prompting farmers to clear marginal land to feed the country's growing population.

2.6 *Insecure and Unclear Land and Other Natural Resources Tenure and Access Rights*

These were found to be some of the primary causes of the end-users of natural resources' reluctance to invest in long-term sustainable land management (SLM). For example, it has been reported that insecurity of land tenure under 10 mailos in parts of the cattle corridor in Uganda and communal land ownership systems discourage farmers from investing in SLM practices.

3 Soil Microbiomes in Desert Soils

Covering almost one-third of the planet's land area, deserts are among the largest and least researched biomes on Earth. They also serve as a major repository for the diversity of microbes on Earth (Fierer et al., 2012). The biogeochemical cycle and ecological stability are significantly influenced by desert microorganisms (Li et al., 2023). Most desert microorganisms live in top soils, where they can coexist alongside adapted vascular plants or as free-living assemblages. But in the driest ecoregions, they assemble into networked groups called biocrusts, which are usually led by phototrophic cyanobacteria and chlorophycean algae (Csotonyi et al., 2010). The establishment of soil heterotrophic communities and biocrusts leads to improved soil stability (for example, through the production of exopolysaccharides, nutrient concentration, and water retention, all of which are beneficial to the growth and development of vascular plants (Maestre et al., 2021).

These soils are home to special microbial communities that have adapted to harsh environmental factors like high heat, little moisture, and a lack of nutrients (Fierer et al., 2012). The vast range of microorganisms found in desert soils, including bacteria, fungi, viruses, and archaea, is abundant despite the severe environment (Leff et al., 2015). Bacteria constitute the dominant microbial group in desert soils, with Actinobacteria and Proteobacteria being the most abundant phyla (Fierer et al., 2012). These microorganisms are important for the breakdown of organic materials, soil stabilization, and nutrient cycling. Specifically, actinobacteria can create secondary metabolites that increase their survival in dry settings and are well-adapted to desiccation (Goodfellow & Williams, 1983).

Fungi in desert soils are mainly represented by Ascomycota and Basidiomycota phyla, with many species forming symbiotic associations with plants, such as mycorrhizae (Kohler et al., 2015). These fungi contribute to nutrient uptake by plants, soil structure improvement, and organic matter decomposition. Archaea and viruses are also present in desert soils, with members of the phyla Thaumarchaeota and Euryarchaeota being the most abundant (Bates et al., 2011). Archaea contribute to nitrogen cycling, particularly through ammonia oxidation, which is crucial for nitrogen availability in nutrient-poor desert soils (Levy-Booth et al., 2014). In desert soils, viruses are common and, through their interactions with bacterial and archaeal hosts, they play important roles in controlling microbial populations and nutrient cycling. Despite their small size, viruses can have profound effects on microbial community structure and function in desert ecosystems (Williamson et al., 2017).

4 Effect of Desertification on Soil Microbiomes

Fertile land becomes desertified as a result of many reasons, including deforestation, overgrazing, and climate change. These variables have a significant impact on soil microbiome (Delgado-Baquerizo et al., 2021). The intensified desertification exerts negative influences on soil properties, which may dramatically affect the growth and diversity of soil microorganisms (Cao et al., 2021), and in turn impact the vegetation and ecological restoration of the desertified area (Li et al., 2023). This process modifies the physicochemical characteristics of the soil, including its pH, moisture content, and nutrient availability. Consequently, these changes affect the diversity and composition of the microbial population (Li et al., 2023).

A decline in microbial diversity and abundance is one of the main consequences of desertification on soil microbiomes. In desertified areas, the species richness and evenness of soil microbial communities are frequently diminished, and a shift towards stress-tolerant taxa is observed. Reductions in the resilience and productivity of ecosystems may result from this loss of microbial diversity (Catania et al., 2022). Desertification also modifies the functional characteristics of soil microbial communities, which impacts the breakdown of soil organic matter and the cycling of nutrients. For instance, microbial biomass and activity may decline as a result of desertification-induced soil degradation, which could slow the pace at which organic matter breaks down and nutrients are mineralized (Cao et al., 2021).

According to Egamberdieva et al. (2022), microbial symbiotic connections with plants, such as mycorrhizal associations, which are critical for nutrient intake and plant growth in arid conditions, can be disrupted by desertification. Therefore, changes in soil microbiomes brought on by desertification may worsen soil degradation and lower ecosystem production. Effective soil conservation and restoration techniques must therefore be implemented in arid and semi-arid environments, and this requires an understanding of how desertification affects soil microbiomes. Enhancing soil fertility, strengthening ecosystem resilience, and lessening the effects of desertification on soil ecosystems can all be achieved by restoring the diversity and function of soil microbes (Leff et al., 2015).

5 Effect of Desertification on Plant Growth

A sizable area of the planet's surface is made up of dry, semi-arid, and hyper arid regions. The severe environmental factors of water scarcity, intense sun radiation, temperature swings, salinity, and nutrient deprivation, which have detrimental effects on plant development and survival, pose enormous challenges to life in these areas (Alsharif et al., 2020).

A wide variety of microorganisms can be found in deserts; they can be free-living, symbiotic, linked to vascular plants, or found in biological soil crusts. Microbiomes support the establishment of fertile islands, nutrient cycling, and climate regulation in drylands. They also serve as the foundation for the ecological succession of vegetation in harsh environments. Therefore, the productivity of dry places could be seriously jeopardized by a potential loss of microbial variety and, as a result, loss of functionality (Leung et al., 2020).

Desertification greatly affects plant growth and ecosystem dynamics in arid and semi-arid regions. Degradation of soil, lack of water, and increasing salinity of soil due to desertification can negatively impact plant groups, resulting in a decrease in vegetation cover and a loss of biodiversity (Schlesinger, 2021).

The reduction in soil fertility and nutrient availability is one of the main consequences of desertification on plant development (Zhang et al., 2023; Li et al., 2023). Plant biomass and species richness are reduced in affected areas as a result of soil degradation processes such as nutrient depletion, compaction, and soil erosion that lower the soil's ability to support plant growth and productivity (Li et al., 2023).

Another significant element preventing plant growth in desertified areas is lack of water. This is because, in desert soils, availability of water is the primary factor influencing the structure and function of the plant community. But generally speaking, these environments only get extremely rare large-scale water inputs (such as precipitation and/or runoff). The availability of soil moisture is decreased due to altered precipitation patterns and elevated evaporation rates, which poses challenges to the survival of numerous plant species. Some effects of desertification on soil microbiomes are presented in Table 2.

6 Adaptation of Soil Microbiomes to Desert Soils

Microbial life faces a great deal of difficulty in desert soils because of their extreme aridity, high temperatures, and restricted nutrient supply. Despite these difficulties, soil microbiomes in desert habitats have demonstrated extraordinary endurance and adaptation (Ebrahimi-Zarandi et al., 2023). Temporary water supply in desert ecosystems can come from a variety of sources, including sporadic precipitation events, dew or fog condensation, and ice or snow melting in polar deserts. These short "metabolic windows" may be essential for desert microorganisms to produce biomass and store reserve chemicals for extended periods of water shortages and the ensuing need for maintenance energy (Noy-Meir, 1973).

Table 2 Effects of desertification and other land degradations on soil microbiomes and plant growth (microbiome-plant interaction)

Microbe	Associated plant	Nature of problem	Effect/role	Reference
Cyanobacteria, Proteobacteria, Acidobacteria, Actinobacteria, Firmicutes, and Nitrospirae		Desertification	Cyanobacteria significantly abundant in non-degraded soils as compared degraded ones, suggesting their activities to be more correlated to soil quality Proteobacteria significantly predominated desertified areas compared to non-desertified ones Proteobacteria, along with other phyla (Acidobacteria, Actinobacteria, Firmicutes, and Nitrospirae), were active in nutrient limiting degraded soils	Qi et al. (2018)
Bacteria (*Chloroflexi, Actinobacteria*, *Conexibacter*) and fungi (*Penicillium, Ilyonectria*)	*Pinus densata*	Desertification	Alpha-diversities of soil bacteria and fungi lower in desertified *P. densata* soils Increased bacterial phyla *Chloroflexi* and *Actinobacteria*, and fungal phylum Ascomycota/Basidiomycota ratio due desertification Relative abundance of pathogenic microbial communities (*Conexibacter*, *Penicillium* and *Ilyonectria*) more in desertified soils	Zhang et al. (2023)
Rhizosphere bacterial communities	*Broussonetia papyrifera* *Ligustrum lucidum*	Mining and smelting	Microbial diversity indices higher in rhizosphere soils, reduced in the bulk soils Enriched bacteria in the rhizosphere are majorly Proteobacteria, Acidobacteria, Actinobacteria, and Bacteroidetes	Guo et al. (2019)
Nitrosomonas, Pirellula, and *Methylobacterium* species		Desertification	Drastic shift in bacterial communities across stages, which decreased towards severe desertification stage Indicator species of *Nitrosomonas, Pirellula*, and *Methylobacterium* could survive wide range of habitats	Fan et al. (2020)
Soil microbes	Meadows	Grassland degradation	Decrease in soil nutrients leading to changes in taxonomic and functional composition of soil microbial and plant communities Relative abundances of cellulose degradation, sugar utilization and chitin degradation genes in degraded meadows significantly higher than those in non-degraded meadows	Luo et al. (2022)

6.1 Absorption of Water from the Atmosphere

Although not well established, it is possible for microorganisms to directly obtain water in their cells from the atmosphere (H_2O (g)), as evidenced by the permeability of the cell membrane to water molecules and the hypertonicity of the cellular cytoplasm. This process is both physically and thermodynamically feasible. The near-surface soil horizon atmosphere, which is found to be in equilibrium with the above-surface atmosphere, appears to be available to the shallow soil microbial communities that can adsorb water, based on the strong correlation observed between the shallow sub-surface relative humidity values and the surface of the desert soil (Glaser et al., 2022). It is quite probable that the biofilm structures that most soil organisms live in can directly absorb atmospheric moisture, regardless of whether microbial cells do so. Most soil microbial communities reside within microbial extracellular polymeric substance (EPS) matrices (Kaur & Dey, 2023). These matrices are composed of high-molecular-weight glycan polymers with diverse makeup, which can account for as much as 90% of a biofilm's biomass. These substances include large amounts of amino (–NH_2) and free hydroxyl (–OH) groups, both of which are highly hydrophilic and increase the ability of EPSs to hold moisture (Wang et al., 2012).

Low-molecular-weight organic solutes, such as polyols, mono- and oligosaccharide sugars, amino acids, and their derivatives, ectoines, and betaines, are accumulated by microorganisms, and at least some of these solutes have been linked to cellular reactions to osmotic stress including desiccation (Da Costa et al., 1998). Trehalose is a hygroscopic disaccharide that can absorb and hold water at atmospheric relative humidity values above 50%. Trehalose accumulates intracellularly during drying in many species. It is potentially possible for dehydrated microbial cells to acquire water through this process in shallow underground microbial communities in both hot and cold desert soils (Bosch et al., 2022). Dew is believed to be a significant source of moisture for arid plants. It has been demonstrated that dew water inputs promote surface microbiological communities and soil surface lichens (Del Prado & Sancho, 2007).

6.2 Source of Water in Polar Deserts

Because water is rarely, if ever, present in an accessible liquid form due to the continuously low temperatures that characterize all but the surface horizons of such habitats, icecaps, and other glacial masses may be considered desert ecosystems. Nonetheless, many glacial masses can contain liquid water on both the upper and bottom surfaces. For instance, the localized melting of numerous glacier masses is linked to biologically active surface microbial communities. Two well-known examples of these groups are cryoconites and snow algae (Hotaling et al., 2021). In ice- and snow-covered algae, primarily Chlamydomonadales and Zygnematales, cells with pigments high in astaxanthin and purpurogallan absorb sunlight, resulting in a heat gain that melts frozen water in the cell surroundings (Hotaling et al., 2021).

6.3 Energy Sources and Conservation

Many microorganisms such as *Arthrobacter* enter a dormant state without undergoing major structural changes. Instead, they use basic tactics that are similar to those of sporulating species, including stopping or slowing growth, shrinking cells, reducing energy-intensive processes such as motility and macromolecule synthesis, altering membrane lipids and components of the cell wall, and increasing macromolecular repair mechanisms (Lebre et al., 2017).

When the soil is wetted, metabolic substrates are made available to heterotrophs. Extracellular soil organic carbon is released when the soil water potential rises (Appel, 1998). The trophic relationships that chemolithoautotrophs have with phototrophs and heterotrophs are also advantageous. Diazotrophic bacteria fix nitrogen to ammonia, a fraction of which is expelled or leaked from cells to other bacterial species. Nitrifying bacteria like Thaumarchaeota, which are usually the predominant archaea in desert soils, are supported by this supply of fixed nitrogen (Magalhães et al., 2014).

6.4 Atmospheric Trace Gas Oxidation (Lithoheterotrophy)

Although light-dependent energy harvesting techniques are crucial physiological processes in desert soil ecosystems (Cary et al., 2010), light penetration constantly restricts these processes. For microorganisms living in and beneath the photic zone of arid soil, atmospheric trace gases might offer a feasible substitute energy source. Trace gases are diffuse high-energy electron donors that are present everywhere, including carbon monoxide (CO), methane (CH_4), and hydrogen (H_2).

7 Mitigating the Effects of Desertification on Soil Microbiomes

7.1 Cover Crops and Vegetation and Soil Amendments

Protecting the soil surface, lowering erosion, and re-vegetating degraded areas with native or drought-resistant plant species can all aid in fostering an environment that is favorable for microbial populations. Through their roots, plants release organic substances that act as substrates for soil microorganisms, increasing their diversity and activity (Veach et al., 2019). Additionally, plant roots help stabilize soil aggregates, reducing erosion and promoting soil microbial habitat (de Vries et al., 2020). Cover crops and native or drought-tolerant plant species can be used to revegetate damaged areas, preserve the soils surface, lessen erosion, and foster the growth of microbial communities. Furthermore, the addition of organic amendments, such as compost, manure, and biochar, can improve the soils structure, increase the amount of organic matter in the soil, and encourage the diversity and activity of microorganisms.

7.2 Microbial Inoculants

In desertified environments, the introduction of beneficial microbial inoculants, such as nitrogen-fixing bacteria and mycorrhizal fungi, can improve soil fertility and plant development. These inoculants can strengthen soil structure, increase nutrient uptake, and encourage symbiotic connections between plants and microbes (Alsharif et al., 2020). For example, nutrients uptake by plants can be enhanced by many plant growth promoting bacteria and arbuscular mycorrhizal fungi, which also promote soil aggregation and contribute to soil fertility and stability (Begum et al., 2019).

7.3 Retention of Crop Residue Mulch

Mulch, such as wood chips or straw, applied to the soil surface can help hold moisture in the soil, lessen temperature swings, and create a home for soil microorganisms. Mulching also keeps the soil microbiomes in a healthy condition by reducing soil erosion and the loss of organic materials (Saha et al., 2020).

7.4 Elimination of Soil Mechanical Disturbances

Conservation tillage techniques, such as reduced or no-till farming, can preserve soil structure, limit soil disturbance, and safeguard soil microbial communities. Minimal soil disturbance contributes to the preservation of microbial populations and soil organic matter, improving soil health and resistance to desertification (Lal, 2015). These practices also reduce erosion and improve water infiltration, contributing to better soil moisture retention and microbial activity.

8 Mitigating the Effects of Desertification on Plant Growth

8.1 *The Use of Plant Growth-Promoting Organisms*

Recently, there has been increased interest in the use of plant growth-promoting (PGP) organisms as a potentially effective way to improve plant growth and development in challenging environmental circumstances (Selim et al., 2019; Ayangbenro & Babalola, 2021). It has been suggested that these rhizosphere-dwelling creatures employ several strategies to reduce plant stress. They produce plant hormones including indole acetic acid, cytokinins, gibberellins, and 1-aminocyclopropane-1-carboxylate deaminase, which give drought tolerance by lowering ethylene levels in plants, and they also encourage the development of root, which improves the absorption of water (Gagné-Bourque et al., 2016).

8.2 *Soil Moisture Management and Water Conservation Techniques*

In desertified environments, improving soil moisture retention is essential for plant growth. Mulching, applying hydrogels, and using soil conditioners are among the methods that can improve the soil ability to retain water. Mulching aids in preserving soil moisture, lowering evaporation, and shielding plants from high temperatures. Because hydrogels are sluggish to absorb and release water, plants can always have access to water (Yang et al., 2014; Demo & Bogale, 2024). Plant water availability can be increased by putting water-saving measures like drip irrigation, check dam construction, and rainwater gathering into practice. By capturing and storing water, these methods lessen runoff and evaporation losses. Water is delivered directly to the plant roots using efficient irrigation techniques, such as drip irrigation, which reduces water waste and increases water use efficiency (Mandal et al., 2020; Gaddikeri et al., 2023). Planting native or drought-resistant plant species that are suited to dry environments can improve vegetation cover and lower the danger of desertification in addition to managing moisture. These species may flourish in soils deficient in nutrients and are frequently more resistant to water stress (Teixeira et al., 2020).

8.3 *Nutrient Management*

Both organic and inorganic fertilizers can be applied to the soil to assist restore nutrients and promote plant development. Compost and manure are examples of organic fertilizers that improve soil structure, raise the amount of organic matter in the soil, and stimulate microbial activity to improve nutrient availability. Certain minerals that are lacking in the soil can be supplied by inorganic fertilizers, fostering plant development and productivity (Ferri et al., 2010).

8.4 Agroforestry Practices

In agroforestry systems, including trees and shrubs with crops and livestock can boost microclimate conditions, improve soil fertility, and encourage plant development (Fahad et al., 2022). Shrubs and trees increase soil organic matter through their roots and leaf litter, give shade, and lessen wind erosion. Additionally, by implementing agroforestry techniques, crop production can be less negatively impacted by drought and more efficiently use of water (Jose, 2009).

8.5 Soil Stabilization

A more conducive environment for plant development can be produced by methods such as contour bounding, terracing, and the use of geotextiles, which stabilize soil and lessen erosion. Techniques for stabilizing soil aid in preserving soil moisture, lowering nutrient loss, and halting topsoil erosion, all of which are critical for the health of plants. Additionally, these methods promote the development and expansion of vegetative cover (Vishnudas et al., 2018).

9 The Use of Beneficial Microbiome in Mitigating the Effect of Desertification

Recent studies on biological methods in mitigating the effect of desertification proved that it was more important to ensure the sustainability of the terrestrial ecosystem. Biological method-mediated soil improvement is defined as a novel approach to green technology sustainability, especially employed to inhibit soil erosion. Plants form intricate relationships within ecosystems, thriving best in their natural habitats. Microbial communities play a pivotal role in bolstering plant resilience against both biotic and abiotic stresses while also augmenting growth and yield through improved nutrient bioavailability in the soil (Peng et al., 2023). Beneficial microbial communities can help mitigate the effects of environmental stress factors such as desertification as forwarded below.

9.1 Phytohormones Production

Phytohormones, organic compounds crucial for plant development, can be modulated by Plant Growth Promoting Microorganisms (PGPM) (Lopez et al., 2021). Microbial inoculants alleviate the effects of stress on plants through the production of various phytohormones such as auxin, cytokinin, gibberellin, ACC deaminase, abscisic acid, jasmonates, brassinosteroids, and strigolactones (Arora et al., 2020). In PGPM selection tests, particular attention is paid to auxin and ACC-deaminase. This is because microbial auxins augment plant auxin levels, promoting growth by improving nutrient and water uptake (Lopez et al., 2021). Additionally, microbial auxins play a beneficial role in regulating cell division, shoot growth, vascular tissue differentiation, adventitious and lateral root formation, as well as elongation and surface area of roots. ACC-deaminase produced by microbiomes is an enzyme that helps reduce ethylene levels, thus mitigating stress in plants. Elevated ethylene levels can lead to leaf chlorosis, necrosis, senescence, decreased fruit yield, impaired root development, leaf expansion, and photosynthesis. Certain bacterial strains such as *Pseudomonas* sp. and *Bacillus* sp. promote plant growth by increasing auxin and ACC-deaminase levels (Khoshru et al., 2020).

9.2 Enhancing the Availability of Plant Nutrients

Numerous agricultural soils approaching desertification suffer from poor quality due to deficiencies in one or more essential plant nutrients, resulting in suboptimal plant growth. To address this issue and enhance crop yields, farmers increasingly rely on inorganic chemical-based fertilizers rich in nitrogen and phosphorus. However, the production of chemical fertilizers is not only expensive but also exhausts non-renewable resources like oil and gas, while also posing adverse effects on both human health and the environment (Shirokov & Tikhnenko, 2021). PGPB plays a crucial role in facilitating nutrient uptake by host plants, including nitrogen, iron, and phosphorus, representing a direct mechanism for promoting plant growth (Peng et al., 2023).

9.3 Siderophores Production

Like phosphorus and potassium, a substantial amount of insoluble iron in soil remains unavailable to plants (Lopez et al., 2021). Under iron-deficient conditions, bacteria employ two effective strategies to bind iron. First, they release protons and organic acids, which lower the soil pH. Second, they secrete iron-chelating complexes (known as siderophores) that specifically bind to ferric ions. Plants can then acquire iron from these siderophores through root chelation degradation or ligand exchange (Sun et al., 2022). Under conditions of iron scarcity, PGPB species such as *G. diazotrophicus* and *A. brasilense* can produce hydroxamate and catechol-type siderophores, respectively. These siderophores play a crucial role in chelating iron and facilitating its absorption by the host plant as observed by Delaporte-Quintana et al. (2020). Additionally, research indicates that certain bacteria secrete growth-inhibitory siderophores, which can alter the interaction dynamics between the rhizosphere microbiome and pathogens.

9.4 Exopolysaccharides Production

Microorganisms can produce exopolysaccharides, which form a protective biofilm on the surface of roots. This biofilm plays a vital role in enhancing water retention within soil particles, thereby maintaining soil moisture levels in the root zone. Consequently, it safeguards root cells against osmotic and ionic stress by regulating the osmotic balance, particularly under fluctuating pH levels, saline conditions, drought, and extreme temperatures (Lopez et al., 2021). PGPM utilizes this mechanism of exopolysaccharide production to alleviate stress, contributing to the stabilization of soil ionic balance by immobilizing Na^+ ions during salinity stress. Additionally, *Bacillus* species produce exopolysaccharides to bolster their antimicrobial activity within the soil (Hashem et al., 2019).

9.5 Antioxidant Enzymes Production

Microbial inoculants act as protective agents, counteracting the detrimental effects of reactive oxygen species (ROS). They achieve this by enhancing the production of antioxidant molecules including Catalase (CAT), superoxide dismutase (SOD), ascorbate peroxidase(AsA), glutathione (GSH), carotenoids, tocopherols, and phenolics (Gouda et al., 2018; Arora et al., 2020). These help secure cellular integrity, protect membranes, and preserve essential biomolecules. Interestingly, environmental stressors can lead to a reduction in CAT activity, accompanied by an increase in hydrogen peroxide (H_2O_2) levels. The deliberate production of catalase by microbiomes serves the primary purpose of enhancing plant tolerance against abiotic stresses. This enzyme efficiently degrades H_2O_2, thereby reducing ROS-induced oxidative stress and ultimately enhancing plant resilience (Lopez et al., 2021).

9.6 Osmoregulants Production

Under stress conditions, microbial inoculants stimulate the production of osmoregulants including carbohydrates, proteins, amino acids, lipids, proline, glycinebetaine, and trehalose (Gouda et al., 2018; Khoshru et al., 2020). This process of osmoregulation helps maintain cellular homeostasis, preventing membrane plasmolysis, enhancing the synthesis of heat shock proteins (HSPs), and regulating essential enzymatic mechanisms (Khoshru et al., 2020). These beneficial microbes contribute to improved plant tolerance by enhancing the accumulation of osmolytes within the plant cell cytoplasm. This, in turn, helps maintain cell turgor and significantly enhance stress tolerance in plants (Khoshru et al., 2020). The osmoregulation mechanism is vital for plant survival and resilience under extreme conditions, effectively reducing cellular damage caused by abiotic stress as discussed by Fernandez et al. (2012).

10 Conclusion and Prospects

Soil microbiomes contribute immensely to ecosystem services and maintenance of ecological balance. They can survive different environments including harsh ones like deserts. Some deleterious environmental processes such as desertification have great effects on the soil microbiomes and plant growth which include the decline in diversity, abundance, and functional characteristics of the soil microbial communities and plants or vegetation cover. Understanding the impacts of desertification is important towards predicting and regulating microbial processes. Some microbiomes can adapt and survive desert environments and even have the ability to mitigate the impacts. These have great potential which if properly harnessed can be employed to improve the growth and survival of other organisms, particularly plants in the deserts and other extreme environments.

References

Abdullahi, S., Haris, H., Zarkasi, K. Z., & Amir, H. G. (2020). Beneficial bacteria associated with Mimosa pudica and potential to sustain plant growth-promoting traits under heavy metals stress. *Bioremediation Journal, 25*(1), 1–21. Taylor and Francis. https://doi.org/10.1080/10889868.2020.1837724

Abdul Rahman, N. S. N., Abdul Hamid, N. W., & Nadarajah, K. (2021). Effects of abiotic stress on soil microbiome. *International Journal of Molecular Sciences, 2021*(22), 9036. https://doi.org/10.3390/ijms22169036

Alsharif, W., Saad, M. M., & Hirt, H. (2020). Desert microbes for boosting sustainable agriculture in extreme environments. *Frontiers in Microbiology, 11*, 496411. https://doi.org/10.3389/fmicb.2020.01666

Appel, T. (1998). Non-biomass soil organic N — The substrate for N mineralization flushes following soil drying–rewetting and for organic N rendered $CaCl_2$-extractable upon soil drying. *Soil Biology and Biochemistry, 30*(10), 1445–1456. https://doi.org/10.1016/S0038-0717(97)00230-7

Arora, N. K., Fatima, T., Mishra, I., & Verma, S. (2020). Microbe-based inoculants: Role in next green revolution. In V. Shukla & N. Kumar (Eds.), *Environmental concerns and sustainable development* (pp. 191–245). Springer. https://doi.org/10.1007/978-981-13-6358-0_9

Ayangbenro, A. S., & Babalola, O. O. (2021). Reclamation of arid and semi-arid soils: The role of plant growth-promoting archaea and bacteria. *Current Plant Biology, 25*, 100173. https://doi.org/10.1016/j.cpb.2020.100173

Bates, S. T., Berg-Lyons, D., Caporaso, J. G., Walters, W. A., Knight, R., & Fierer, N. (2011). Examining the global distribution of dominant archaeal populations in soil. *The ISME Journal, 5*(5), 908–917. https://doi.org/10.1038/ismej.2010.171

Begum, N., Qin, C., Ahanger, M. A., Raza, S., Khan, M. I., Ashraf, M., Ahmed, N., & Zhang, L. (2019). Role of arbuscular mycorrhizal fungi in plant growth regulation: Implications in abiotic stress tolerance. *Frontiers in Plant Science, 10*, 1–15. https://doi.org/10.3389/fpls.2019.01068

Berlin, A., Källström, H. N., Lindgren, A., & Olson, Å. (2018). Scientific evidence for sustainable plant disease protection strategies for the main arable crops in Sweden. A systematic map protocol. *Environmental Evidence, 7*, 31. https://doi.org/10.1186/s13750-018-0141-3

Bosch, J., Marais, E., Maggs-Kölling, G., Ramond, J.-B., Lebre, P. H., Eckardt, F., & Cowan, D. A. (2022). Water inputs across the Namib Desert: Implications for dryland edaphic microbiology. *Frontiers of Biogeography, 14*(2). https://doi.org/10.21425/F5FBG55302

Cao, H., Gao, G., Zhang, Y., Guo, M., Ren, Y., & Ding, G. (2021). Soil bacterial approach to assessing afforestation in the desertfied Northern China. *Journal of Cleaner Production, 292*, 125935. https://doi.org/10.1016/j.jclepro.2021.125935

Cary, S. C., McDonald, I. R., Barrett, J. E., & Cowan, D. A. (2010). On the rocks: The microbiology of Antarctic Dry Valley soils. *Nature Reviews. Microbiology, 8*(2), 129–138. https://doi.org/10.1038/nrmicro2281

Catania, V., Bueno, R. S., Alduina, R., Grilli, E., La Mantia, T., Castaldi, S., & Quatrini, P. (2022). Soil microbial biomass and bacterial diversity in southern European regions vulnerable to desertification. *Ecological Indicators, 145*, 109725. https://doi.org/10.1016/j.ecolind.2022.109725

Chowdhury, S. K., Majumdar, S., & Mandal, V. (2020). Application of *Bacillus* sp. LBF-01 in *Capsicum annuum* plant reduces the fungicide use against *Fusarium oxysporum*. *Biocatalysis and Agricultural Biotechnology, 27*, 101714. https://doi.org/10.1016/j.bcab.2020.101714

Chukwuneme, C. F., Babalola, O. O., Kutu, F. R., & Ojuederie, O. B. (2020). Characterization of actinomycetes isolates for plant growth promoting traits and their effects on drought tolerance in maize. *Journal of Plant Interactions, 15*(1), 93–105. https://doi.org/10.1080/17429145.2020.1752833

Csotonyi, J. T., Swiderski, J., Stackebrandt, E., & Yurkov, V. (2010). A new environment for aerobic anoxygenic phototrophic bacteria: Biological soil crusts. *Environmental Microbiology Reports, 2*(5), 651–656. https://doi.org/10.1111/j.1758-2229.2010.00151.x

Da Costa, M. S., Santos, H., & Galinski, E. A. (1998). An overview of the role and diversity of compatible solutes in bacteria and archaea. In G. Antranikian (Ed.), *Biotechnology of extremophiles* (Vol. 61, pp. 117–153). Springer. https://doi.org/10.1007/BFb0102291

de Vries, F. T., Griffiths, R. I., Knight, C. G., Nicolitch, O., & Williams, A. (2020). Harnessing rhizosphere microbiomes for drought-resilient crop production. *Science, 368*(6488), 270–274. https://doi.org/10.1126/science.aaz5192

Del Prado, R., & Sancho, L. G. (2007). Dew as a key factor for the distribution pattern of the lichen species Teloschistes lacunosus in the Tabernas Desert (Spain). *Flora – Morphology, Distribution, Functional Ecology of Plants, 202*(5), 417–428. https://doi.org/10.1016/j.flora.2006.07.007

Delaporte-Quintana, P., Lovaisa, N. C., Rapisarda, V. A., & Pedraza, R. O. (2020). The plant growth promoting bacteria *Gluconacetobacter diazotrophicus* and *Azospirillum brasilense* contribute to the iron nutrition of strawberry plants through siderophores production. *Plant Growth Regulation, 91*, 185–199. https://doi.org/10.1007/s10725-020-00598-0

Delgado-Baquerizo, M., Eldridge, D. J., Liu, Y.-R., Sokoya, B., Wang, J.-T., Hu, H.-W., He, J.-Z., Bastida, F., Moreno, J. L., Bamigboye, A. R., Blanco-Pastor, J. L., Cano-Díaz, C., Illán, J. G., Makhalanyane, T. P., Siebe, C., Trivedi, P., Zaady, E., Verma, J. P., Wang, L., et al. (2021). Global homogenization of the structure and function in the soil microbiome of urban greenspaces. *Science Advances, 7*(28), eabg5809. https://doi.org/10.1126/sciadv.abg5809

Demo, A. H., & Bogale, G. A. (2024). Enhancing crop yield and conserving soil moisture through mulching practices in dryland agriculture. *Frontiers in Agronomy, 6*, 1–14. https://doi.org/10.3389/fagro.2024.1361697

Ebrahimi-Zarandi, M., Etesami, H., & Glick, B. R. (2023). Fostering plant resilience to drought with Actinobacteria: Unveiling perennial allies in drought stress tolerance. *Plant Stress, 10*, 100242. https://doi.org/10.1016/j.stress.2023.100242

Egamberdieva, D., Alaylar, B., Kistaubayeva, A., Wirth, S., & Bellingrath-Kimura, S. D. (2022). Biochar for improving soil biological properties and mitigating salt stress in plants on salt-affected soils. *Communications in Soil Science and Plant Analysis, 53*(2), 140–152. https://doi.org/10.1080/00103624.2021.1993884

Fahad, S., Chavan, S. B., Chichaghare, A. R., Uthappa, A. R., Kumar, M., Kakade, V., Pradhan, A., Jinger, D., Rawale, G., Yadav, D. K., Kumar, V., Farooq, T. H., Ali, B., Sawant, A. V., Saud, S., Chen, S., & Poczai, P. (2022). Agroforestry systems for soil health improvement and maintenance. *Sustainability, 14*(22), 14677. https://doi.org/10.3390/su142214877

Fan, M., Li, J., Tang, Z., & Shangguan, Z. (2020). Soil bacterial community succession during desertification in a desert steppe ecosystem. *Land Degradation and Development, 31*(13), 1662–1674. https://doi.org/10.1002/ldr.3545

Fernandez, O., Theocharis, A., Bordiec, S., Feil, R., Jacquens, L., & Clément, C. (2012). *Burkholderia phytofirmans* PsJN acclimates grapevine to cold by modulating carbohydrate metabolism. *MPMI, 5*, 496–504. https://doi.org/10.1094/MPMI-09-11-0245

Ferri, D., Montemurro, F., Debiase, G., Fiore, A., Convertini, G., & Mastrangelo, M. (2010). Application of organic amendments to improve land utilization system of fodder crops. *Fresenius Environmental Bulletin, 19*, 1768–1773.

Fierer, N., Leff, J. W., Adams, B. J., Nielsen, U. N., Bates, S. T., Lauber, C. L., Owens, S., Gilbert, J. A., Wall, D. H., & Caporaso, J. G. (2012). Cross-biome metagenomic analyses of soil microbial communities and their functional attributes. *Proceedings of the National Academy of Sciences, 109*(52), 21390–21395. https://doi.org/10.1073/pnas.1215210110

Gaddikeri, V., Jatav, M. S., Pasha, M. M., Gavhane, K., Satpute, A. N., Suna, T., Rajput, J., & Mishra, A. K. (2023). Role of water harvesting and supplemental irrigation in enhancing agriculture productivity of dryland under climate change. In A. Naorem & D. Machiwal (Eds.), *Enhancing resilience of dryland agriculture under changing climate: Interdisciplinary and convergence approaches* (pp. 123–143). Springer Nature. https://doi.org/10.1007/978-981-19-9159-2_8

Gagné-Bourque, F., Bertrand, A., Claessens, A., Aliferis, K. A., & Jabaji, S. (2016). Alleviation of drought stress and metabolic changes in Timothy (*Phleum pratense* L.) colonized with *Bacillus subtilis* B26. *Frontiers in Plant Science, 7*, 584. https://doi.org/10.3389/fpls.2016.00584

Glaser, D. M., Hartnett, H. E., Finn, D. R., Perez-Montaño, S., Cadillo-Quiroz, H., & Desch, S. (2022). Water vapor adsorption provides daily, sustainable water to soils of the Hyperarid Atacama Desert. *Astrobiology, 22*(10), 1222–1238. https://doi.org/10.1089/ast.2021.0171

Goodfellow, M., & Williams, S. T. (1983). Ecology of actinomycetes. *Annual Review of Microbiology, 37*, 189–216. https://doi.org/10.1146/annurev.mi.37.100183.001201

Gouda, S., Kerry, R. G., Das, G., Paramithiotis, S., Shin, H., & Patra, J. K. (2018). Revitalization of plant growth promoting rhizobacteria for sustainable development in agriculture. *Microbiological Research, 206*, 131–140. https://doi.org/10.1016/j.micres.2017.08.016

Guo, D., Fan, Z., Lu, S., Ma, Y., Nie, X., Tong, F., & Peng, X. (2019). Changes in rhizosphere bacterial communities during remediation of heavy metal-accumulating plants around the Xikuangshan mine in southern China. *Scientific Reports, 9*, 1947. https://doi.org/10.1038/s41598-018-38360-2

Hashem, A., Tabassum, B., & Allah, E. F. A. (2019). *Bacillus subtilis*: A plant growth promoting rhizobacterium that also impacts biotic stress. *Saudi Journal of Biological Sciences, 26*, 1291–1297. https://doi.org/10.1016/j.sjbs.2019.05.004

Hotaling, S., Lutz, S., Dial, R. J., Anesio, A. M., Benning, L. G., Fountain, A. G., Kelley, J. L., McCutcheon, J., Skiles, S. M., Takeuchi, N., & Hamilton, T. L. (2021). Biological albedo reduction on ice sheets, glaciers, and snowfields. *Earth-Science Reviews, 220*, 103728. https://doi.org/10.1016/j.earscirev.2021.103728

Jose, S. (2009). Agroforestry for ecosystem services and environmental benefits: An overview. *Agroforestry Systems, 76*(1), 1–10. https://doi.org/10.1007/s10457-009-9229-7

Karthika, S., Midhun, S. J., & Jisha, M. S. (2020). A potential antifungal and growth-promoting bacterium *Bacillus* sp. KTMA4 from tomato rhizosphere. *Microbial Pathogenesis, 142*, 104049. https://doi.org/10.1016/j.micpath.2020.104049

Kaur, N., & Dey, P. (2023). Bacterial exopolysaccharides as emerging bioactive macromolecules: From fundamentals to applications. *Research in Microbiology, 174*(4), 104024. https://doi.org/10.1016/j.resmic.2022.104024

Kazerooni, E. A., Maharachchikumbura, S. S. N., Adhikari, A., Al-Sadi, A. M., Kang, S.-M., Kim, L.-R., & Lee, I.-J. (2021). Rhizospheric *bacillus amyloliquefaciens* protects *Capsicum annuum* cv. Geumsugangsan from multiple abiotic stresses via multifarious plant growth-promoting attributes. *Frontiers in Plant Science, 12*, 669693. https://doi.org/10.3389/fpls.2021.669693

Khoshru, B., Mitra, D., Khoshmanzar, E., Myo, E. M., Uniyal, N., & Mahakur, B. (2020). Current scenario and future prospects of plant growth-promoting rhizobacteria: An economic valuable resource for the agriculture revival under stressful conditions. *Journal of Plant Nutrition, 43*, 3062–3092. https://doi.org/10.1080/01904167.2020.1799004

Kohler, A., Kuo, A., Nagy, L. G., Morin, E., Barry, K. W., Buscot, F., Canbäck, B., Choi, C., Cichocki, N., Clum, A., Colpaert, J., Copeland, A., Costa, M. D., Doré, J., Floudas, D., Gay, G., Girlanda, M., Henrissat, B., Herrmann, S., & Martin, F. (2015). Convergent losses of decay mechanisms and rapid turnover of symbiosis genes in mycorrhizal mutualists. *Nature Genetics, 47*(4), Article 4. https://doi.org/10.1038/ng.3223

Kouam, I. D., Mabah, J., Ntsoli, P. G., Tchamani, L., Yaouba, A., Katte, B., & Bitom, D. (2023). Growth promotion potential of *Bacillus* spp. isolates on two tomato (*Solanum lycopersicum* L.) varieties in the West region of Cameroon. *Open Agriculture, 8*, 20220154.

Kuzina, E., Rafikova, G., Vysotskaya, L., Arkhipova, T., Bakaeva, M., Chetverikova, D., Kudoyarova, G., Korshunova, T., & Chetverikov, S. (2021). Influence of hydrocarbon-oxidizing bacteria on the growth, biochemical characteristics, and hormonal status of barley plants and the content of petroleum hydrocarbons in the soil. *Plants, 10*, 1745. https://doi.org/10.3390/plants10081745

Lal, R. (2015). Restoring soil quality to mitigate soil degradation. *Sustainability, 7*(5), 5875–5895. https://doi.org/10.3390/su7055875

Lebre, P. H., De Maayer, P., & Cowan, D. A. (2017). Xerotolerant bacteria: Surviving through a dry spell. *Nature Reviews. Microbiology, 15*(5), 285–296. https://doi.org/10.1038/nrmicro.2017.16

Leff, J. W., Jones, S. E., Prober, S. M., Barberán, A., Borer, E. T., Firn, J. L., Harpole, W. S., Hobbie, S. E., Hofmockel, K. S., Knops, J. M. H., McCulley, R. L., La Pierre, K., Risch, A. C., Seabloom, E. W., Schütz, M., Steenbock, C., Stevens, C. J., & Fierer, N. (2015). Consistent responses of soil microbial communities to elevated nutrient inputs in grasslands across the globe. *Proceedings of the National Academy of Sciences, 112*(35), 10967–10972. https://doi.org/10.1073/pnas.1508382112

Leung, P. M., Bay, S. K., Meier, D. V., Chiri, E., Cowan, D. A., Gillor, O., & Stegen, J. C. (2020). Energetic basis of microbial growth and persistence in desert ecosystems. *mSystems, 5*, e00495–e00419.

Levy-Booth, D. J., Prescott, C. E., & Grayston, S. J. (2014). Microbial functional genes involved in nitrogen fixation, nitrification and denitrification in forest ecosystems. *Soil Biology and Biochemistry, 75*, 11–25. https://doi.org/10.1016/j.soilbio.2014.03.021

Li, X. R., Jia, R. L., Zhang, Z. S., Zhang, P., & Hui, R. (2018). Hydrological response of biological soil crusts to global warming: A ten-year simulative study. *Global Change Biology, 24*(10), 4960–4971.

Li, S., Lian, W.-H., Han, J.-R., Ali, M., Lin, Z.-L., Liu, Y.-H., Li, L., Zhang, D.-Y., Jiang, X.-Z., Li, W.-J., & Dong, L. (2023). Capturing the microbial dark matter in desert soils using culturomics-based metagenomics and high-resolution analysis. *npj Biofilms and Microbiomes, 9*(1), 1–14. https://doi.org/10.1038/s41522-023-00439-8

Lopez, M. J. S., Dias-Filho, M. B., & Gurgel, E. S. C. (2021). Successful plant growth-promoting microbes: Inoculation methods and abiotic. *Frontiers in Sustainable Food Systems, 5*, 606454. https://doi.org/10.3389/fsufs.2021.606454

Luo, Z., Liu, J., He, L., Du, J., Wang, L., Jia, T., Chai, B., & Wu, T. (2022). Degradation-induced microbiome alterations may aggravate soil nutrient loss in subalpine meadows. *Land Degradation & Development, 33*(15), 2699–2712. https://doi.org/10.1002/ldr.4289

Maestre, F. T., Benito, B. M., Berdugo, M., Concostrina-Zubiri, L., Delgado-Baquerizo, M., Eldridge, D. J., Guirado, E., Gross, N., Kéfi, S., Le Bagousse-Pinguet, Y., Ochoa-Hueso, R., & Soliveres, S. (2021). Biogeography of global drylands. *New Phytologist, 231*(2), 540–558. https://doi.org/10.1111/nph.17395

Magalhães, C. M., Machado, A., Frank-Fahle, B., Lee, C. K., & Cary, S. C. (2014). The ecological dichotomy of ammonia-oxidizing archaea and bacteria in the hyper-arid soils of the Antarctic Dry Valleys. *Frontiers in Microbiology, 5*, 515. https://doi.org/10.3389/fmicb.2014.00515

Mandal, S., Vema, V. K., Kurian, C., & Sudheer, K. P. (2020). Improving the crop productivity in rainfed areas with water harvesting structures and deficit irrigation strategies. *Journal of Hydrology, 586*, 124818. https://doi.org/10.1016/j.jhydrol.2020.124818

Marasco, R., Ramond, J.-P., Goethem, M. W. V., Rossi, F., & Daffonchio, D. (2023). Diamonds in the rough: Dryland microorganisms are ecological engineers to restore degraded land and mitigate desertification. *Microbial Biotechnology, 2023*(16), 1603–1610.

Noy-Meir, I. (1973). Desert ecosystems: Environment and producers. *Annual Review of Ecology and Systematics, 4*(1), 25–51. https://doi.org/10.1146/annurev.es.04.110173.000325

Peng, M., Jiang, Z., Zhou, F., & Wang, Z. (2023). From salty to thriving: Plant growth promoting bacteria as nature's allies in overcoming salinity stress in plants. *Frontiers in Microbiology, 14*, 1169809. https://doi.org/10.3389/fmicb.2023.1169809

Prabhakaran, A., Meenatchi, R., Pal, S., Hassan, S., Bramhachari, P. V., Kiran, G. S., & Selvin, J. (2022). Soil microbiome: Characteristics, impact of climate change and resilience. In P. B. Bramhachari (Ed.), *Understanding the microbiome interactions in agriculture and the environment*. https://doi.org/10.1007/978-981-19-3696-8_15

Qi, D., Wieneke, X., Tao, J., Zhou, X., & Desilva, U. (2018). Soil pH is the primary factor correlating with soil microbiome in karst rocky desertification regions in the Wushan County, Chongqing, China. *Frontiers in Microbiology, 9*, 1027. https://doi.org/10.3389/fmicb.2018.01027

Saha, A., Sekharan, S., & Manna, U. (2020). Superabsorbent hydrogel (SAH) as a soil amendment for drought management: A review. *Soil and Tillage Research, 204*, 104736. https://doi.org/10.1016/j.still.2020.104736

Schlesinger, W. H. (2021). Some thoughts on the biogeochemical cycling of potassium in terrestrial ecosystems. *Biogeochemistry, 154*(2), 427–432. https://doi.org/10.1007/s10533-020-00704-4

Selim, S., Hassan, Y. M., Saleh, A. M., Habeeb, T. H., & AbdElgawad, H. (2019). Actinobacterium isolated from a semi-arid environment improves the drought tolerance in maize (*Zea mays* L.). *Plant Physiology and Biochemistry, 142*, 15–21. https://doi.org/10.1016/j.plaphy.2019.06.029

Shirokov, Y., & Tikhnenko, V. (2021). Analysis of environmental problems of crop production and ways to solve them. In: *E3S Web of Conferences: EDP Sciences*, 01025, 273, https://doi.org/10.1051/e3sconf/202127301025.

Sun, X., Zhang, C., Bei, S., Wang, G., Geisen, S., Bedoussac, L., Christie, P., & Zhang, J. (2022). High bacterial diversity and siderophore producing bacteria collectively suppress *Fusarium oxysporum* in maize/faba bean intercropping. *Frontiers in Microbiology, 13*, 972587. https://doi.org/10.3389/fmicb.2022.972587

Teixeira, H. M., Cardoso, I. M., Bianchi, F. J. J. A., da Cruz Silva, A., Jamme, D., & Peña-Claros, M. (2020). Linking vegetation and soil functions during secondary forest succession in the Atlantic forest. *Forest Ecology and Management, 457*, 117696. https://doi.org/10.1016/j.foreco.2019.117696

Tharek, M., Abdullahi, S, Mia, M. A. B., Najimudin, N., & Ghazali, A. H. (2022). Endophytes as potential biostimulants to enhance plant growth for promoting sustainable agriculture. In *Biostimulants for crop production and sustainable agriculture*. CABI. https://doi.org/10.1079/9781789248098.0026

Tosi, M., Mitter, E. K., Gaiero, J., & Dunfield, K. (2020). It takes three to tango: The importance of microbes, host plant, and soil management to elucidate manipulation strategies for the plant microbiome. *Canadian Journal of Microbiology, 66*, 413–433.

Vandenkoornhuyse, P., Quaiser, A., Duhamel, M., Van, A. L., & Dufresne, A. (2015). The importance of the microbiome of the plant holobiont. *New Phytologist, 206*, 1196–1206. https://doi.org/10.1111/nph.13312

Veach, A. M., Morris, R., Yip, D. Z., Yang, Z. K., Engle, N. L., Cregger, M. A., Tschaplinski, T. J., & Schadt, C. W. (2019). Rhizosphere microbiomes diverge among *Populus trichocarpa* plant-host genotypes and chemotypes, but it depends on soil origin. *Microbiome, 7*(1), 1–15. https://doi.org/10.1186/s40168-019-0668-8

Vishnudas, S., Anil, K. R., Savenije, H. H. G., & Zaag, P. V. D. (2018). Use of coir geotextiles for soil and water conservation: Case studies from India. In *Technological interventions in Management of Irrigated Agriculture* (pp. 161–209). Apple Academic Press.

Wang, L. L., Wang, L. F., Ren, X. M., Ye, X. D., Li, W. W., Yuan, S. J., Sun, M., Sheng, G. P., Yu, H. Q., & Wang, X. K. (2012). pH dependence of structure and surface properties of microbial EPS. *Environmental Science & Technology, 46*(2), 737–744. https://doi.org/10.1021/es203540w

Williamson, K. E., Fuhrmann, J. J., Wommack, K. E., & Radosevich, M. (2017). Viruses in soil ecosystems: An unknown quantity within an unexplored territory. *Annual Review of Virology, 4*, 201–219. https://doi.org/10.1146/annurev-virology-101416-041639

Yang, F., Zhang, G. L., Yang, J. L., Li, D. C., Zhao, Y. G., Liu, F., Yang, R. M., & Yang, F. (2014). Organic matter controls of soil water retention in an alpine grassland and its significance for hydrological processes. *Journal of Hydrology, 519*, 3086–3093. https://doi.org/10.1016/j.jhydrol.2014.10.054

Zhang, Y., Li, X., Zhang, J., Hua, J., Li, J., Liu, D., Bhople, P., Ruan, H., & Yang, N. (2023). Desertification induced changes in soil bacterial and fungal diversity and community structure in a dry-hot valley forest. *Applied Soil Ecology, 189*, 104953. https://doi.org/10.1016/j.apsoil.2023.104953

Soil Microbes and Climate Change Mitigation

Babafemi Raphael Babaniyi, Taiwo Hamidat Olaide, Idowu Blessing Apara, Omosalewa Kikelomo Ajibola, Daramola Oluwatosin Olaoluwa, and Ademola Bisi-Omotosho

Abstract Soil is an invaluable resource for the planet, playing a crucial role in preserving biodiversity and managing ecosystem services in both controlled and natural environments. It harbors a significant portion of microscopic biodiversity, encompassing prokaryotes and microscopic eukaryotes. The soil microbiome plays a vital role in maintaining soil processes, although its contributions are often underestimated in certain systems, such as desert and forest ecosystems. The soil microbiome is heavily influenced by abiotic and biotic factors, including pH, carbon content, soil structure, texture, and vegetation, resulting in significant variations across different ecosystems and populations. The abundance of microorganisms comprising soil microbiomes has a substantial yet frequently overlooked impact on regulating Earth's temperature. As climate change continues, effective mitigation strategies increasingly rely on comprehending the intricate connections between soil microbes and environmental processes. This investigation examines the critical role of soil microbiomes in mitigating climate change. Through various processes, such as organic matter breakdown, nutrient cycling, and the formation of stable soil aggregates, soil microbiomes actively contribute to carbon sequestration. Microbial communities significantly aid in carbon storage by facilitating the conversion of atmospheric carbon dioxide into soil organic carbon, thereby reducing the greenhouse effect. The control of greenhouse gas emissions is largely dependent on the nitrogen cycle, which is managed by soil bacteria. Nitrogen-fixing microorganisms convert atmospheric nitrogen into usable forms for plants, thereby enhancing carbon uptake by promoting vegetation growth. Conversely, denitrifying bacteria mitigate global warming by reducing emissions of nitrous oxide, a potent greenhouse gas.

B. R. Babaniyi
Bioresources Development Centre, National Biotechnology Research and Development Agency, Abuja, Nigeria

T. H. Olaide (✉) · I. B. Apara · O. K. Ajibola · D. O. Olaoluwa
Department of Chemistry, Federal University of Technology, Akure, Nigeria
e-mail: olaidehamidat01@gmail.com

A. Bisi-Omotosho
Department of Sustainability Management and Innovation, London, UK

S. A. Aransiola et al. (eds.), *Soil Microbiome in Green Technology Sustainability*,
https://doi.org/10.1007/978-3-031-71844-1_9

Soil microbiomes exhibit adaptation and tolerance to climate-related stressors, such as temperature fluctuations and extreme weather events. To predict ecosystem resilience and implement adaptive management techniques, understanding microbial responses to various stressors is essential. Furthermore, microbial-mediated processes impact plant-microbe interactions, hydrological cycles, and soil albedo, all of which have cascading effects on climate dynamics. Reforestation and conservation agriculture are two sustainable land management practices that harness the potential of soil microbiomes to mitigate the impacts of climate change.

Keywords Nutrient cycling · Microorganisms · Soil warming · Mulching · Ocean fertilization

1 Introduction to Soil Microbiome

Soil, being the primary reservoir of biological diversity and organic substances on terrestrial ground, supports a wide range of processes crucial for the survival of life on Earth (Delgado-Baquerizo et al., 2016; Nielsen et al., 2015). However, it remains one of the least understood and acknowledged components of the Earth's ecosystem. The biodiversity per unit area in soil surpasses that of any other habitat above ground, making it a hub for biological and molecular interactions (Bach et al., 2020; Thakur & Geisen, 2019). The subsurface networks in soil are intricately connected to above-ground networks through multitrophic communications, nitrogen cycling, and plant-soil feedback. Furthermore, soil plays a critical role in regulating ecosystem functioning and promoting ecosystem services (Bahram et al., 2018; Fierer, 2017). In terms of microscopic biomass, soil often exceeds above-ground biomass, with each hectare of soil containing over a thousand kilos of microscopic biomass carbon (Bar-On et al., 2018; de la Cruz-Amo et al., 2020). Despite the fact that soil microorganisms and plants constitute the majority of the Earth's living biomass, the scientific community and the general public tend to focus more on the macroscopic environment, resulting in limited knowledge about the microscopic biota residing in the heterogeneous soil (Bar-On et al., 2018; Geisen et al., 2019). However, advancements in sequencing tools have gradually improved our understanding of the composition and function of soil microorganisms (Geisen et al., 2019). As awareness of soil microscopic biodiversity grows, it becomes crucial to explore the various soil activities supported by an unknown array of species that play vital roles in soil functions and ecosystem services (Bünemann et al., 2018). Moreover, the soil microbiome also influences climate change processes and the regulation of greenhouse gas emissions (Geisen et al., 2019; Naylor et al., 2020). Comprising a diverse array of microorganisms, the soil microbiome contributes significantly to key ecosystem functions such as nutrient recycling, soil structure protection, and disease suppression (Lladó et al., 2017; Wang et al., 2021). Countless bacteria inhabit the Earth's surface and actively shape their environment. A single gram of soil can harbor billions of microorganisms, including numerous distinct species. Each ecosystem possesses unique soil characteristics that support diverse microbial communities, primarily composed of bacteria, but also including archaea, protists, fungi, viruses, and other microscopic organisms in varying abundances. The composition and health of the surrounding environment, as well as its macroscopic inhabitants, depend on the symbiotic relationship with soil-borne microbial partners. Consequently, any disturbance or imbalance in the soil microbiome can have far-reaching consequences (Omotayo & Babalola, 2021).

The richness and diversity of the soil microbiome are intricately linked to ecosystem functioning, and its composition within the soil environment relies on both the chemical and physical properties of the soil, as well as anthropogenic influences that impact these properties. Edaphic factors, which encompass soil characteristics affecting the resident organisms, are classified as abiotic factors. These edaphic characteristics heavily influence the structure of the soil microbiome and serve as the primary ecological filter (de Gannes et al., 2015). For instance, the structure of the soil, with its abundance of micropores in clay molecules, promotes the development of mesofauna while inhibiting microbe predatory activity. Additionally, different forms of soil biota exhibit varied pH preferences, and it has been observed that soil pH strongly associates with the establishment of a specific soil microbiome assemblage. When acidic soil is limed, notable alterations in the composition and abundance of the soil microbiome, including bacteria, fungi, archaea, nematodes, annelids, and microarthropods, occur as pH and nutrient concentrations rise significantly (Holland et al., 2018). Soil moisture has also been identified as a significant influencer of protistan community composition. Moreover, soil type and structure impact the community makeup of the soil microbiome, with lighter soil structures favoring bacterial development (Furtak & Gałązka, 2019). Crucially, soil nutrient concentrations, such as carbon content and C/N ratio, play a crucial role in governing the diversity and structure of the soil microbiome (Ke et al., 2021). Ecological adaptations of soil microbiota, as evidenced by studies, are driven by changes in life cycle strategies (Li et al., 2022). In dry areas, soil microbiota have been found to adapt to harsh conditions by slowing their growth rate, increasing their reproductive output, and altering their resource utilization efficiency. This highlights the adaptability of soil bacteria to changing environmental conditions, underscoring their vital role in ecosystem functioning across various landscapes. Furthermore, research has revealed that body size serves as a key determinant in shaping soil microbial communities (Wang et al., 2022). The composition of soil microbiota is governed by soil parameters and plant characteristics, emphasizing the significance of soil microbiota in the soil-plant ecosystem. The soil microbiome assumes diverse roles in terrestrial environments, and comprehending these roles can aid in predicting the range of impacts it may exert on both natural and managed ecosystems (Hale et al., 2019; Dodds et al., 2020).

Soil microbiomes are indispensable for preserving the biological and chemical attributes of the soil that support the development of plants and the health of ecosystems. The study of soil microbiomes has gained significant importance in recent times as a means of sustaining soil productivity and environmental sustainability. In a study conducted by Sun et al. (2021), it was observed that microbial communities in the phyllosphere of crops and the endosphere of roots exhibit greater resilience to fertilization compared to soil microbiota. This research highlights the necessity of exploring soil microbiomes in order to gain a better understanding of their impact on plant development and nutrition under diverse soil conditions. Similarly, Massoni et al. (2021) demonstrated that soil bacteria can reach the phyllosphere, underscoring the need for investigating soil microbiomes to comprehend the interactions between plants and microbes. Moreover, soil microbiomes play significant roles in nutrient cycling, disease suppression, and carbon sequestration. Tan et al. (2021) discovered that the fructification of a soil-saprotrophic morel is driven by microbial succession and nutritional accumulation in a semi-synthetic substratum, thereby emphasizing the relevance of studying soil microbiomes to comprehend the dynamics of nutrient cycling. Additionally, Custódio et al. (2021) found that soil microbiota can be manipulated to enhance plant health and productivity. Thus, understanding the role of soil microbiomes in nutrient cycling and disease suppression is crucial for the long-term sustainability of agriculture. The promotion of soil microbial biodiversity through reduced tillage has been proven to benefit soil health and ecosystem function. Frøslev et al. (2021) revealed that reduced tillage methods enhance soil microbial biodiversity, highlighting the significance of studying soil microbiomes to comprehend the impact of agricultural practices on soil health. Consequently, comprehending the importance of soil microbiomes is vital for establishing sustainable farming practices that enhance soil health, ecosystem functioning, and food security (Geisen & Bonkowski, 2018). Human activities, such as the utilization of antibiotics in agriculture, have the potential to adversely affect soil microbiomes. Sun et al. (2021) found that the accumulation of antibiotics in the soil-vegetable system can have negative effects on soil microbiota and resistance genes. Consequently, examining the role of soil microbiomes is crucial for understanding the detrimental effects of human activities on soil health and ecosystem function. The degradation of soil and its biomes is presently being driven by changes in global environmental and climatic conditions. Anthropogenic activities such as changes in land use, pollution, and the introduction of new invasive species into new environments are exacerbating the situation and may have far-reaching consequences for the biological systems of the Earth (de Vries et al., 2012; Liu et al., 2019). Human intervention is one of the primary threats, as humans have altered the local plant diversity through land clearance and the cultivation of specific plant species since the advent of agriculture (de Vries et al., 2012). Pollutants entering the soil as a result of activities such as oil spills and mining are also impacting the soil microbiome and the extensive ecosystem functions provided by the soil (TENG & CHEN, 2019). Furthermore, physical changes to soil, such as compaction, erosion, sealing, habitat fragmentation, and other activities, can have a significant impact on belowground diversity. In recent decades, global climate change has altered precipitation and temperature systems, influencing soil biodiversity both directly and indirectly through their effects on plant variety and primary productivity (Yang et al., 2018). Attention to soil biodiversity and its utilitarian relevance will enable the progress of economic administration methods, which will necessitate a better awareness of potential dangers to soil variety.

1.1 Formation of Microbiome

The soil microbiome is a complex combination of biotic and abiotic elements. Biotic influences include plants, animals, fungi, and bacteria, while abiotic factors consist of temperature, pH, moisture, and nutrient availability (Islam et al., 2020; Rehman et al., 2023). The soil microbiome is shaped by plants through the secretion of root exudates, which provide nutrients to microorganisms and affect soil physicochemical properties (Wei et al., 2019). The composition of the soil microbiome is also impacted by the species and age of plants. Studies suggest that Miscanthus × giganteus has a distinct soil microbiome compared to other plants (Ma et al., 2021). Abiotic factors also play a role in the composition of the soil microbiome. Temperature and moisture levels, for instance, have been found to influence the diversity and abundance of soil microbial communities (Averill et al., 2021). Rehman et al. (2023) found that drought, high temperatures, and salt stress can significantly alter the composition of the soil microbiome. Additionally, the choice of fertilizer can modify nutrient availability and therefore impact the soil microbiome composition (Ma et al., 2021). The formation of the soil microbiome is vital for mitigating climate change. A diverse and healthy soil microbiome can enhance soil quality, facilitate nutrient cycling, and reduce greenhouse gas emissions into the environment (Shi et al., 2019). Research has demonstrated that the soil microbiome has the capacity to capture carbon from the atmosphere and store it in the soil. Mwakilili et al. (2021) found that intercropping maize with Desmodium significantly increased soil carbon sequestration compared to monoculture maize. Furthermore, the use of engineered nanomaterials in soil can enhance carbon sequestration by altering the activity of soil bacteria (Khan et al., 2021).

1.2 Role of Soil Microbiome

Soil bacteria play significant roles in ecosystems, such as enhancing agricultural yield (Dubey et al., 2019), facilitating nutrient cycling, promoting carbon sequestration (Bender et al., 2016), and contributing to environmental restoration (Malla et al., 2018). The soil microbiome is of utmost importance in maintaining ecosystem health and is being increasingly acknowledged in the field of ecology. Establishing specific associations among microbial guilds may optimize the advantages and harness the potential of microbial taxa to provide efficient and prompt solutions to challenges arising from global climate change.

I. **Managing SOM and C sequestration:** Soil microbiome with plant diversity is a huge participant in carbon storage and sequestration; they serve plentiful roles in the pedosphere, critical for the formation and turnover of SOM to fulfil their needs of nutrient and energy for their growth and maintenance (Liang et al., 2017). The prime way that C is stored in soil is in the form of soil organic matter [SOM] with around 80% of C in stable fractions of SOM is constituted of microbial residues and exudates (Wiesmeier et al., 2019). It shows that soil microbial biomass [SMB] holds a binary role in SOM turnover, balancing SOM mineralization and stabilization. A diverse range of organisms underwrite significantly to the living microbial mass of SOM (Murphy, 2015), and the rate of degradation of organic matter is controlled by the microbiome and the abiotic factors like soil pH and temperature and in turn controls the microbial activity(Yadav, 2017). SOC storage is also affected by soil microbiome, influencing the soil structures, e.g., the burrowing activity of earthworms and enchytraeids is changed due to soil macro-porosity. Other than earthworms, millipedes and soil meiofauna such as mites, springtails, and enchytraeids are found to protect the OM from soil aggregation via fecal pellets which hold a huge amount of C in soil ecosystems. Predation by protists leads to a decline in the total bacterial biomass, an increase in the nutrient turnover, and overall stimulation of the soil microbiome (Gao et al., 2019). Prokaryotic components of the microbiome like archaea carry out the two essential metabolisms for nutrient cycling globally, i.e., methanogenesis- and sulfate-dependent anaerobic methane oxidation (Offre et al., 2013). In conclusion, the soil microbiome plays a huge role in the functioning of the ecosystem and performs different activities in partial fulfilment of it as discussed in Table 1 that mentions the representative taxa and the type of functions they perform.

II. **Nutrient cycling:** Soil microbiomes play crucial roles in the cycling of nutrients, which have a direct impact on the growth and development of plants. The microbial population in soil is diverse and complex, consisting of various organisms such as bacteria, fungi, archaea, and viruses. Recent studies have shown that manipulating the microbiota in soil can be an effective method for reducing soil-borne plant diseases within organic farming systems. De Corato (2020) stated that employing microbiome-assisted solutions can enhance plant development and health by increasing the presence of beneficial microorganisms while decreasing the population of harmful ones. Microbial inoculants, biofertilizers, and plant growth-promoting rhizobacteria all contribute to improving soil health and nutrient cycling. The decomposition of plant residues is essential for nitrogen cycling as plant leftovers are a major source of organic matter in soil. Cobo-Díaz et al. (2021) found that maize residues have a significant influence on the soil microbiota and the community of Fusarium spp. The study discovered that the utilization of maize leftovers increased the abundance of beneficial bacteria while reducing the population of harmful fungi. This emphasizes the importance of crop residues in maintaining soil health and facilitating nitrogen cycling. The application of volatiles produced by fungi can potentially impact the soil microbiota. Baroja-Fernández et al. (2021) demonstrated that applying cell-free fungal culture filtrates to pepper plants activated the beneficial soil microbiota, resulting in increased yields. The study concluded that utilizing volatiles produced by fungi can be an effective approach for improving soil health and enhancing nitrogen cycling. There is a symbiotic relationship between plants and soil bacteria, where plants provide nutrients to the microbiota and, in turn, receive benefits from it. Topalović et al.

(2020) discovered that plants and soil microorganisms work together to control plant-parasitic nematodes. The study revealed that specific bacteria in soil can impede the growth and development of nematodes, thereby reducing their population. This underscores the importance of maintaining a healthy soil microbiome in facilitating nutrient cycling and promoting plant growth.

III. **Natural Ecosystem:** The Earth serves as the primary natural reservoir for all the essential elements required by living organisms, encompassing iron, zinc, phosphorus, potassium, sulfur, nitrogen, oxygen, and hydrogen (Gougoulias et al., 2014). These elements constitute the fundamental constituents of minerals present in other organisms, albeit in intricate and stable forms that render them unusable in their natural state. The provision of these constituents is facilitated by beneficial soil bacteria within this enclosed environment (Viles, 2012). By virtue of possessing numerous coding genes for enzymes that contribute to fixation, solubilization, and mineralization, soil bacteria are capable of breaking down intricate substances (Patil et al., 2021). Microbes play a pivotal role in decomposing organic matter within the environment, thereby rendering it accessible for various biological processes. According to Akob and Küsel (2011), soil bacteria are also involved in the cycling of specific elements present in ores and complex configurations, such as carbon, nitrogen, oxygen-hydrogen, phosphorus, potassium, zinc, and iron. Apart from their involvement in biogeochemical cycles, soil microbes facilitate the decomposition of detrimental substances, ultimately contributing to the purification of the Earth's surface.

Soil microorganisms have been recognized as an excellent source of essential nutrients, including Nitrogen, Phosphorous, Potassium, Iron, and Zinc, according to numerous research studies. These microorganisms contribute to the acquisition, solubilization, and retrieval of these minerals from both the soil and the atmosphere, thereby enabling the provision of such nutrients. As stated by Schloter et al. (2017), these bacteria are capable of carrying out various processes, such as acidification, which leads to a decrease in pH, as well as the secretion of extracellular polysaccharides, organic acids, and enzymes that aid in the decomposition of these intricate and insoluble minerals. Furthermore, apart from nutrients, plants also require various plant growth regulators, such as gibberellins, cytokinin, and auxin (Wong et al., 2015). Although these hormones are naturally produced by plants, they are unable to synthesize them during stressful conditions; hence, it is plausible that soil microorganisms may excrete these hormones in order to enhance crop yield.

2 Climate Change and Effect on Soil Microbes

Climate change denotes a statistically noteworthy fluctuation in either the average condition of the climate or its variability, enduring for an extended duration, typically spanning decades or more. This phenomenon may arise from natural internal processes, external influences, or persistent anthropogenic alterations in the composition of the atmosphere or land use (Kumar et al., 2016). Additionally, climate change manifests as a discernible shift in climate patterns primarily induced by emissions of greenhouse gases (GHG). These emissions cause the Earth's atmosphere to trap heat, serving as the principal driver behind global warming. The primary contributors to such emissions encompass both natural systems and human activities, including phenomena such as forest fires, earthquakes, oceans, permafrost, wetlands, mud volcanoes, and volcanoes (Yue & Gao, 2018; Fawzy et al., 2020a, b).

The repercussions of climate change, coupled with alterations in land management practices, can adversely impact soil fertility and soil organic carbon (SOC), thereby influencing the soil microbiome and its overall impact on soil carbon sequestration (Naylor et al., 2020). Forecasts indicate that the climate will continue evolving, leading to more erratic and extreme weather patterns. The response of soil microbiomes to these changes has implications for the atmosphere, both directly through greenhouse gas production and indirectly through effects on ecosystem processes such as plant growth, soil organic matter formation, and nutrient cycling (Jansson & Hofmockel, 2020).

Furthermore, climate change is projected to cause an increase in tropical temperatures by 1.8 °C to 5 °C over the next century, potentially intensifying microbial mineralization of stable soil organic matter (SOM) and resulting in an elevated flux of CO_2 (Wieder et al., 2013). As soil microorganisms play a crucial role in the cycling of soil organic carbon (SOC) and other nutrients, they contribute significantly to climate feedback, involving the production or consumption of greenhouse gases such as CO_2, CH_4, and N_2O. However, predicting whether soil will become a source or sink of greenhouse gases under future climate scenarios remains challenging due to unknown changes in soil carbon and nitrogen pools and variations in microbial responses between soil locations (Friedlingstein et al., 2006; Wang et al., 2017).

A recent report from the Intergovernmental Panel on Climate Change (IPCC) highlights a worsening situation, with 3.3 billion of the world's population highly vulnerable to climate change, and current unsustainable development patterns increasing exposure to climate hazards (Tiedje et al., 2022; IPCC, 2022). The intricate interplay of environmental conditions on microbial dynamics, often influenced by human activities, determines their role as either producers or consumers of gases. Managing certain conditions strategically can favor microbial gas consumption. However, the diverse habitats of microbes engaged in gas production and consumption, characterized by distinct spatial scales and processing times, present challenges in quantifying their contributions and responses to environmental shifts like warming, storms, and drought. Terrestrial, oceanic, and urban ecosystems, each with unique challenges, require robust measurements and models to discern greenhouse gas (GHG) fluxes across various scales and understand their patterns of change in response to human activities (Tiedje et al., 2022).

The profound implications of climate change on both natural and human systems, along with associated risks and vulnerabilities, form a critical foundation for comprehending the current climate emergency. Recent reports from the United Nations Climate Change Secretariat (UNCCS) underscore changes in climate indicators, encompassing temperature, precipitation, sea level rise, ocean acidification, and extreme weather conditions. This includes a spectrum of climate hazards like droughts, floods, hurricanes, severe storms, heatwaves, wildfires, cold spells, and landslides (UNCCS, 2019). Globally, human-induced climate change emerges as a major driver of numerous natural disasters. Climate risks encompass temperature shifts, precipitation variability, altered seasonal patterns, changes in disease distribution, desertification, ocean-related impacts, and degradation of soil and coastal regions, contributing to vulnerability across diverse sectors and countries (UNCCS, 2019).

The impact of climate change on soil microbiome stability and function, and consequently ecosystem sustainability, is a significant concern (Allison & Martiny, 2008; Orwin & Wardle, 2005; Griffiths & Philippot, 2013). Meta-analyses reveal that soil disturbances manifest measurable effects on microbiome stability in around 80% of published studies (Shade et al., 2012). Community stability is typically assessed through metrics such as resistance, resilience (recovery to a stable state), and functional redundancy (Naylor et al., 2020; Griffiths & Philippot, 2013). Climate change poses a considerable disturbance to the soil microbiome, potentially altering soil microbial community and functional profiles. Changes in soil carbon and nitrogen cycling can, in turn, impact climate change, either through positive feedback (e.g., greenhouse gas emissions) or negative feedback (e.g., carbon immobilization into microbial or plant biomass) (Naylor et al., 2020).

The persistent alteration of average weather patterns, i.e., climate change, exerts long-term effects on soil microbiome and biodiversity. Climate-induced shifts in temperature, CO_2 levels, and soil moisture content influence soil biodiversity (Hamidov et al., 2018). The intricate network between the soil microbiome and climate change impacts stability, with different ecosystems exhibiting unique responses to stimuli; for instance, soil warming can lead to a 40–150% increase in soil microbial populations (Mishra et al., 2022). This multifaceted relationship underscores the complex interplay between climate change and soil microbial dynamics, warranting comprehensive investigation and understanding.

2.1 Various Impact of Climate Change on Soil Microbial Community

Decades of dedicated research have delved into unraveling the intricate relationship between microbes and their ecosystems. The emergence of advanced sequencing and metagenomics technologies has unveiled the vast array of microbes inhabiting aquatic, terrestrial, and urban environments. While researchers have made strides in identifying the microbial composition within these environments, a significant knowledge gap exists concerning the specific activities of these microbes in each setting and the anticipated repercussions of climate change on microbial communities and their ecosystem functions. Microbes play a pivotal role in steering elemental cycles, including those of carbon, nitrogen, and phosphorus, across the planet. Their involvement extends to the consumption and production of gases implicated in global warming. Perturbations induced by climate change, such as alterations in temperature, humidity, or elemental flux, can exert profound effects on the structural configuration and metabolic processes of microbial communities (Xue et al., 2016; Zhou et al., 2012; Woodcroft et al., 2018). Comprehending the ramifications of climate change on microbial populations and their carbon dioxide (CO_2), methane (CH_4), and nitrous oxide (N_2O) fluxes across diverse environments is imperative. This understanding is crucial not only for deciphering how microbes will adapt within their immediate ecosystems but also for gaining insights into their broader implications for the human community at large (Xue et al., 2016). This exploration holds promise for shedding light on the dynamic interplay between microbes and their environments in the face of climate change, providing valuable knowledge for both academic and broader readership.

Soil Warming Soil microbial communities exert a discernible influence on climate through soil warming, constituting an inevitable aspect of climate change's impact on the soil microbiome. The effect of soil warming on resident microbial communities is believed to unfold in a sequential manner. This includes an initial enhancement of organic carbon decomposition rates, leading to a short-term increase in microbial biomass. However, microbial respiration tends to decrease over time as labile carbon becomes depleted. The impact of warming on microbial function can occur through direct mechanisms, such as the acceleration of enzymatic rates, or indirect means, such as the stimulation of plant growth, rhizodeposition, and alterations in soil properties. Notably, warming-induced stimulation of phosphorus and sulfur cycling has been observed (Zhou et al., 2012), while drawing definitive conclusions regarding carbon and nitrogen cycling poses greater challenges. Warming has been shown to elevate rates of nitrogen cycling processes, including denitrification, nitrogen fixation, nitrification, and nitrogen mineralization (Zhou et al., 2012; Naylor et al., 2020), though the precise effects depend on the specific gene or process under investigation.

Combinatorial and Indirect Effects The influence of climate change on soil microbes in practical settings is a critical aspect that has been addressed through recent studies. To bridge this knowledge gap, numerous investigations have adopted multifactorial experimental designs, incorporating factors such as elevated CO_2 (eCO_2) coupled with warming (Naylor et al., 2020; Hu et al., 2016), eCO_2 in conjunction with elevated ozone, eCO_2 combined with nitrogen addition, and various other combinations. This approach has revealed divergent outcomes for combined factors compared to individual treatments, underscoring its significance. Notably, in an experiment simulating the impact of warming and/or eCO_2 on field soils within a cotton agroecosystem, the concurrent application of warming and eCO_2 induced alterations in ammonia-oxidizing microbial communities and escalated soil nitrification rates. In contrast, minimal significant effects were observed when considering warming in isolation (Nguyen et al., 2019). Frequently, the simultaneous occurrence of multiple perturbations results in one factor mitigating the effects of the other (Naylor et al., 2020; Hu et al., 2016). This underscores the importance of employing multifactorial designs to comprehensively understand the intricate dynamics of climate change on soil microbial communities.

Effect of Temperature on Soil Microbial Community An elevation in atmospheric temperature corresponds to heightened microbial activity, influencing critical processes such as microbial-mediated nitrification and denitrification in the soil. Changes in soil temperature are expected to modify the dynamics of nitrifiers and denitrifiers populations, potentially leading to shifts in the frequency, intensity, and duration of droughts within the region. The sensitivity of soil microbial community structure and functions to variations in both temperature and water availability is well-documented (Hartel, 2005). Environmental perturbations may induce community shifts and altered metabolic activities in soil microorganisms engaged in nutrient cycling. Additionally, such changes can impact the survival and virulence of soil-mediated pathogenic microorganisms, exemplified by Salmonella typhimurium. The heightened temperatures contribute to accelerated rates of microbial decomposition, resulting in increased CO_2 emissions through soil respiration, thereby creating a positive feedback loop to global warming (Allison & Martiny, 2008). As a consequence of global warming, it is projected that approximately 25% of permafrost could thaw by 2100, potentially releasing around 100 Petagrams (Pg) of carbon available for microbial decomposition (Kumar et al., 2016). The impact of increased temperature extends to the composition of microbial communities, potentially leading to a reduction in the release of soil organic carbon due to the loss of acclimatized microbial groups. However, in certain instances, an elevated temperature may amplify the rate of organic matter decomposition, enhancing microbial activities. Studies have shown that in high-latitude ecosystems, an increase in temperature resulted in a noteworthy up to 50% decrease in bacterial and fungal abundance, as well as soil respiration. This suggests that elevated temperatures do not uniformly lead to enhanced carbon loss to the atmosphere (Kumar et al., 2016; Hartel, 2005). These findings underscore the complex interplay between temperature changes and microbial responses, highlighting the need for a nuanced understanding of these interactions in the context of global environmental shifts.

Effect of Altered Precipitation on Soil Microbial Community A critical factor influencing the structure of terrestrial microbial communities and the rate at which soil organic carbon (SOC) decomposes is soil moisture. The anticipated 20% rise or fall in precipitation, as forecasted by the Intergovernmental Panel on Climate Change (IPCC), stands out as a significant driver. Alterations in the temporal gaps between rainfall events have been observed to impact soil microbial respiration (C-output) in temperate tall grass prairie systems. Interestingly, elongating the intervals between rainfall events reduces microbial respiration but concurrently heightens carbon inputs through leaf photosynthesis, resulting in a net carbon uptake and storage phenomenon (Fay et al., 2008). Extended periods of arid conditions have the potential to impede microbial growth and decomposition processes, leading to a negative feedback impact on carbon fluxes in certain ecosystems. In contrast, within wetlands and peatlands, soil drying might boost oxygen availability, thereby enhancing carbon cycling and creating a positive feedback effect on CO_2 fluxes. Moreover, the cyclic nature of soil wetting and drying also exerts an influence on CO_2 evolution. Upon soil rewetting, the microbial activity, latent in dry soils, surges, contributing to an increased CO_2 evolution (Kumar et al., 2016). These findings underscore the intricate relationship between soil moisture dynamics and microbial processes, offering valuable insights into the potential consequences of projected climate-induced precipitation changes on carbon cycling in diverse ecosystems.

Effect of Increased CO_2 on Soil Microbial Community Approximately 30–40% of anthropogenic carbon dioxide (CO_2) emissions find their way into oceans, rivers, and lakes, contributing to the phenomenon of ocean acidification. The direct influence of heightened CO_2 levels on stimulating above-ground biomass production has been thoroughly examined (Kumar et al., 2016). This augmentation in above-ground net primary production (ANPP) has been demonstrated to enhance carbon supply below-ground, consequently stimulating soil biological activity (Pendall et al., 2004). Elevated concentrations of CO_2 accelerate plant growth rates, enabling increased CO_2 sequestration. However, this growth is accompanied by heightened soil respiration, driven by an augmented availability of nutrients for decomposition, resulting in the release of additional CO_2 into the atmosphere. The amplified levels of atmospheric CO_2 induce quantitative and qualitative shifts in the release of labile sugars, organic acids, and amino acids from plant roots, thereby fostering microbial growth and activity. The rise in atmospheric CO_2, recognized as an outcome of climate change, holds the potential to significantly reshape the soil environment, particularly by altering the distribution of above and below-ground nutrients. For instance, elevated atmospheric CO_2 concentrations may foster increased plant growth, as CO_2 serves as the molecular foundation for photosynthesis. This could lead to heightened litter production rates and modifications in litter chemical composition, subsequently influencing its digestibility. Such alterations are poised to impact the organic matter available for soil microorganisms. Consequently, the adjusted litter production may reshape the overall carbon supply and nitrogen flow dynamics between plants and microorganisms (Kumar et al., 2016). Furthermore, elevated CO_2 levels may induce augmented root growth, thereby exerting a substantial influence on soil structure and significant implications for soil biota.

Effect of Drought Condition on Soil Microbial Community Due to rising temperatures, it is anticipated that mesic ecosystems will experience an increased frequency and duration of drought events (Knapp et al., 2008). This heightened exposure to water stress is expected to impact both microbial and plant communities, disrupting crucial nutrient cycles and plant-microbe feedback. The resulting drought conditions lead to a substantial reduction in soil moisture, creating unfavorable growth conditions and causing a notable 50–80% decline in microbial population size (Sheik et al., 2011). Furthermore, drought stress has been observed to significantly affect both the establishment and activity of the Rhizobium symbiosis in legumes (Kumar et al., 2016). These findings underscore the vulnerability of mesic ecosystems to climate-induced water stress and emphasize the intricate interplay between environmental factors and microbial-plant interactions.

3 Mitigation Strategies of Climate Change on Soil Microbes

Reducing Soil Greenhouse Gas Emissions The quantification of climate mitigation can be done through the rigorous method of measuring the net exchange of major biogenic greenhouse gas emissions (specifically CO_2, CH_4, and N_2O) between the soil and atmosphere in both amended and unamended control field plots mesocosms. However, this approach requires expensive equipment and a high level of user expertise (Novick et al., 2022). Certain carbon amendments, such as biochar, have the ability to slow down the decomposition of soil organic matter (SOM), hinder the generation of methane through anaerobic respiration (methanogenesis), enhance the oxidative consumption of methane (methanotrophy) as an energy source, or facilitate the complete reduction of nitrate to dinitrogen gas (denitrification), which has no global warming potential. These mechanisms collectively contribute to reducing the overall flux of soil greenhouse gases (GHGs) to the atmosphere, providing promising opportunities for climate mitigation (Rubin et al., 2023).

Soil Organic Carbon Sequestration (SOC) SOC sequestration represents the overall change in soil organic carbon stocks following the application of amendments, taking into account the carbon (C) content within the amendment itself. The consideration of C within the original amendment is crucial for determining the genuineness of the SOC increase over time and discerning whether any positive or negative priming effects are at play—priming effects being the activation or suppression of decomposers in response to amendment addition, as elucidated by Kuzyakov and Domanski. In practical terms, SOC sequestration is commonly described as the augmentation of SOC stocks subsequent to an intervention. This is mathematically expressed as (C concentration × soil bulk density at the end of intervention—C concentration × soil bulk density at the beginning of intervention—total C added through amendment application). Alternatively, it is characterized as the disparity between treatment and control plots over an extended period, following the methodology outlined by Rubin et al. (2023). This holistic approach ensures a comprehensive assessment of SOC sequestration dynamics, offering insights into the effectiveness and impact of interventions on soil carbon storage.

Reduced Life Cycle Emissions Life cycle emissions refer to the comprehensive environmental impact of a product from its inception to its disposal, encompassing all stages of production and use. The calculation of life cycle emissions can be subjective, contingent upon how system boundaries are delineated and which "business as usual" scenarios are selected as baseline scenarios. Despite being the least well-established climate mitigation element in the literature, life cycle emissions are crucial for assessing the overall atmospheric benefit of a particular practice. This aspect holds particular significance in the evaluation of climate mitigation potential for various soil amendments, such as biochar and compost. In the case of biochar, the underlying principle is that pyrolysis serves to stabilize carbon that would otherwise be lost to the atmosphere through decomposition or combustion within climate-relevant timescales (less than 100 years). The determination of the offset value involves measuring the carbon content in biochar, with an additional uncertainty adjustment based on the biochar's carbon composition, such as H:C or O:C ratios (CARB, 2019; Rubin et al., 2023). While life cycle emissions present challenges in their quantification, their thorough examination is integral to understanding the climate mitigation potential associated with soil amendments, contributing to the ongoing discourse on sustainable practices in scientific literature.

Mulching/Light Soil Sealing The practice of mulching involves the application of a protective layer on the soil surface, aiming to prevent erosion and improve fertility. Typically implemented at the commencement of the crop growing season, mulch may be re-applied as needed. Its primary functions include the retention of heat and moisture, thereby initially warming the soil. Various materials, such as organic residues (e.g., crop residue, hay, and bark), manures, sewage sludge, compost, as well as rubber or plastic films, can be utilized as effective mulching agents (Kumar et al., 2016). This technique not only safeguards against soil erosion but also contributes to the enhancement of soil quality, making it a versatile and valuable agricultural practice.

Application of Organic Residues (Compost/Manure/Sludge) Utilizing animal manure, sludge, or other carbon-rich organic materials, such as coffee-berry pulp, straw from cereals and legumes, or compost, serves to enhance the organic matter content within the soil. In agricultural contexts, optimizing the decomposition of organic residues over a specified period before their application to the field is generally advisable. This precaution is essential as the introduction of carbon-rich compounds tends to temporarily immobilize nitrogen in the soil. This occurs because microorganisms require both carbon and nitrogen for their growth and developmental processes, as highlighted by Kumar et al. (2016).

The Use of Fertilizers Elevated concentrations of certain inorganic nitrogenous fertilizers supply microbes with readily accessible nitrogen, enhancing their metabolic activity. This heightened microbial activity accelerates the decomposition of low-quality organic inputs and soil organic matter, leading to a reduction in soil carbon levels. Consequently, the continual decline in soil organic matter content ensues, ultimately contributing to the deterioration of soil structure and diminished water holding capacity (Kumar et al., 2016).

3.1 Microbial Communities and Mitigation Options

Managing Microbial Communities to Reduce Carbon Dioxide Emissions Presently, the Earth's soils harbor approximately 2000 petagrams (Pg) of organic carbon, surpassing the carbon content in the atmosphere by twofold and exceeding that in vegetation by threefold (Smith, 2004). The carbon storage capacity of distinct land categories, such as woodlands, pastures, and arable lands, varies, prompting exploration into the prospect of enhancing carbon sequestration in soils through strategic land management. This potential, estimated to reach an additional 1 Pg of carbon annually, has garnered significant scholarly interest (Kumar et al., 2016).

Managing Microbial Communities to Reduce Methane Emissions The regulation of global methane (CH_4) emissions is arguably more directly influenced by microorganisms than carbon dioxide (CO_2) emissions. Microbial methanogenesis, a process predominantly conducted by anaerobic archaea, plays a dominant role in natural CH_4 emissions, accounting for approximately 250 million tonnes per year in wetlands, oceans, rumens, and termite guts. However, human activities, including rice cultivation, landfills, fossil fuel extraction, and livestock farming, surpass these natural sources in contributing to overall CH_4 emissions. In these human-influenced environments, the significant quantities of CH_4 produced are mitigated by methanotrophic bacteria, which function as a crucial buffer. This microbial interaction not only underscores the pivotal role of microorganisms in controlling global CH_4 emissions but also emphasizes the delicate balance maintained by methanotrophic bacteria in environments with heightened anthropogenic impact. This dynamic interplay between microbial communities and human activities adds a nuanced layer to our understanding of the intricate processes shaping global methane emissions.

Soil Carbon Sequestration Soil carbon sequestration involves capturing atmospheric CO_2 by implementing altered land management strategies to enhance soil carbon levels. The quantity of carbon stored in the soil is influenced by the equilibrium between inputs such as residues, litter, roots, and manure, and carbon losses from respiration, primarily affected by soil disturbance. Practices that boost inputs and minimize losses play a pivotal role in driving soil carbon sequestration, leading to improved soil fertility, health, and increased crop yields through the accumulation of organic carbon in soils. Various land management techniques that support soil carbon sequestration include intensifying cropping systems, implementing rotation practices, adopting zero-tillage and conservation tillage methods, managing nutrients effectively, utilizing mulching, crop residues, and manure, integrating biochar, employing organic fertilizers, and optimizing water management. Soil carbon sequestration has a significant impact on perennial cropping systems, with herbaceous crops potentially increasing soil organic carbon levels by 1.14–1.88 tCha−1 year−1 and woody crops contributing 0.63–0.72 tCha−1 to soil organic carbon levels due to the influence of herbaceous and woody perennial cropping systems on soil organic carbon accumulation.

Ocean Fertilization is a promising technique that involves the addition of nutrients, both macro and micro, such as phosphorus, nitrates, and iron, to the upper layer of the ocean to stimulate biological activity and enhance CO_2 uptake. This process is particularly important as it promotes the growth of phytoplankton, microscopic organisms found at the ocean's surface, which play a crucial role in oceanic carbon sequestration. The sequestered CO_2, in the form of organic marine biomass, is then transported to the deep ocean through a process known as "the biological pump." However, it is important to note that this downward flow is balanced to some extent by oceanic carbon respiration. In the natural system, nutrients are available in the ocean as a result of the decomposition of marine life. By utilizing light, CO_2, and nutrients, land-based plants and phytoplankton can grow and contribute to the sequestration of carbon in the ocean. This technique has the potential to significantly reduce atmospheric CO2 levels and mitigate the effects of climate change.

Enhanced Terrestrial Weathering Within the realm of geosystems research, the decomposition of silicate rocks constitutes a phenomenon known as weathering—a chemical process that absorbs atmospheric CO_2 while releasing metallic ions alongside carbonate and/or bicarbonate species into solution. These soluble constituents traverse subterranean aquifers via riverine pathways before ultimately accumulating in marine environments as alkalinity reserves or crystallizing as terrestrial carbonate minerals. To expedite CO_2 sequestration at a significantly faster pace than natural rates, enhanced weathering has emerged as a promising strategy. By mechanically pulverizing silicate rocks, one increases their reactive surface area and enhances their mineral dissolution velocity. Subsequently, these finely ground materials are deployed onto agricultural lands, offering numerous ancillary advantages such as improved soil hydraulic characteristics, augmented nutrient availability, increased soil pH levels, heightened overall soil health, elevated crop yields, and reduced reliance upon traditional agrochemicals. However, the efficacy of this method depends on factors like rock type, application dosage, environmental climate, soil properties, and specific cropping systems.

Aforestation and Reforestation In arboreal ecosystems, carbon dioxide (CO_2) absorption occurs during tree growth, accumulating within living biomass, decayed organic material, and soil deposits. Consequently, afforestation serves as a biogenic form of negative emission technology, contributing substantially to climate change ameliorative endeavors. Afforestation encompasses two primary strategies: aforestation, entailing the establishment of novel woodlands, and reforestation, referring to the restoration of pre-existing forest regions subjected to deforestation or deterioration. The rate at which CO_2 assimilation transpires varies among tree species, typically extending between 20 to 100 years before reaching full maturation; thereafter, sequestration slows considerably. Upon attaining maturity, forest resources may be extracted and employed, thereby offering additional benefits related to ecological sustainability, including enhanced biodiversity, improved food security, and augmented soil, water, and air quality (Harper et al., 2018).

3.2 Application of Different Soil Microbes as Climate Change Mitigation

Numerous microbes have the potential to capture atmospheric CO_2 through diverse mechanisms and pathways, encompassing both photosynthetic and non-photosynthetic processes. These microbial mechanisms play a crucial role in sequestering atmospheric CO_2 into biomass and energy. The advantages associated with CO_2 capture by microbes are manifold, including exceptionally high rates of biofixation, robust capabilities for bioremediating atmospheric CO_2, the potential to produce a wide array of valuable additives, ease of genetic modifications, suitability for industrial bioprocessing, rapid growth and adaptability to continuous culture in bioreactors, and lack of competitive challenges. These attributes make microbial CO_2 capture an attractive and promising avenue for addressing environmental concerns and enhancing sustainable practices.

I. **Bacteria**: are unicellular, microscopic organisms that can be used in mitigation of climate change. Different groups of bacteria such as Archaebacteria, Cyanobacteria, and Eubacteria are autotrophic, which fixes CO_2 for organic carbon production. Clostridium is a class of anaerobic gram-negative bacteria which substantially degraded organic carbon materials, acid synthesis, and the carbon cycle (Migliardini et al., 2014).

II. **Proteobacteria:** Prominent taxonomic groups possess the ability to assimilate atmospheric carbon dioxide through diverse biological pathways, including the Calvin cycle/reductive pentose phosphate cycle, TCA cycle/Krebs cycle, and direct assimilation within the cytoplasm, as elucidated by Ding and Yokota in 2004.

III. **Archaea:** In the realm of microbiology, an extremophile, characterized as a unicellular, prokaryotic microorganism, exhibits remarkable resilience to harsh environmental conditions such as extreme temperatures, pH variations, and the absence of oxygen. Among these extremophiles, Archaea stands out with its diverse groups, including halophiles (capable of withstanding high salt concentrations), thermoacidophiles, and methanogens. Methanogens, a subgroup of Archaea, play a pivotal role in anaerobic environments, producing biofuel in the form of methane. This process involves utilizing carbon dioxide as a raw material and harnessing energy from hydrogen for bioconversion, as outlined by Rittmann et al. (2015). Commercial applications of methanogens include the production of methane through acetoclastic and hydrogenotrophic pathways, showcasing their potential in bioremediation of wastewater and carbon capture. This dual capability allows methanogens to generate methane from sludge while simultaneously recycling carbon dioxide and remediating wastewater. Notable species within this category include Methanobacteriaceae, Methanospirrillaceae, and Methanosarcinaceae spp., as elucidated by Yasin et al. (2015). Further advancements in extremophilic microbiology involve the 3-hydroxy propionate 4-hydroxybutyrate cycle pathway, utilized by specific Archaea such as Cenarchaeum, Archaeoglobus, Metallosphaera, and Sulfolobus sp. This pathway not only demonstrates their ability to thrive in extreme conditions but also serves as a mechanism for fixing atmospheric carbon dioxide. This groundbreaking research, presented by Rehman et al. (2023), unveils the potential applications of extremophiles in addressing environmental challenges through innovative and sustainable processes.

IV. **Cyanobacteria**: Cyanobacteria, commonly known as blue-green bacteria or algae, represent prokaryotic organisms that function as a crucial intermediary between bacteria and green plants. These photoautotrophic gram-negative bacteria exhibit both microscopic and macroscopic forms. Within their cytoplasm, cyanobacteria harbor specialized structures known as carboxysomes, which play a pivotal role in Carbon fixation. Carboxysomes are present in various microorganisms, including both photoautotrophs and chemoautotrophs, such as Bacteria and all cyanobacteria. This adaptation, known as the Carbon Dioxide Concentration Mechanism (CCM), is integral to the Calvin-Benson cycle. The Carbon Dioxide Concentration Mechanism involves the encapsulation of two key enzymes, RuBisCO and carbonic anhydrase, within bacterial microcompartments known as carboxysomes. These carboxysomes are strategically located in the cytosol and contribute to the efficient functioning of the Calvin-Benson cycle (Rehman et al., 2023). This intricate arrangement underscores the significance of cyanobacteria in bridging the gap between bacteria and green plants, offering a unique perspective on the symbiotic relationships within microbial ecosystems.

V. **Algae:** The optimal photosynthetic bio-factory for efficiently incorporating CO_2 into biomass and energy is diverse in size, ranging from macro to micro. Among the macro-sized options, macroalgae stand out for their high lipid content, making them directly suitable for biodiesel production. On a micro scale, various algae types, such as cyanobacteria, diatoms, euglenoids, and algal species with colors spanning green, blue, red, brown, golden, and yellow, demonstrate significant potential. These micro-sized algae exhibit a remarkable capacity for enzymatically fixing CO_2 through the Calvin Benson cycle, facilitated by the enzyme RuBisCO (Rehman et al., 2023). This diversity in size and biochemical processes underscores the multifaceted and promising nature of this photosynthetic bio-factory, presenting an appealing avenue for academic exploration and potential publication.

VI. **Fungi:** Organisms classified as fungi are characterized by being eukaryotic, multicellular entities possessing a rigid cell wall. Functioning as heterotrophs, they play a crucial role as primary decomposers and are pivotal in capturing carbon within terrestrial ecosystems. Saprophytic fungi exhibit the ability to generate enzymes that effectively break down various substances, including cellulose, hemicellulose, pectin, and lignin. These enzymatic activities are integral to processes such as mineralization and the carbon cycle. Notably, mycorrhizal fungi exhibit lower decomposition rates due to the absence of a specific enzyme, as highlighted by Ahmed et al. (2019). Fungi actively contribute to carbon sequestration in the soil by participating in the formation of organic humus, thereby playing a vital role in maintaining the carbon balance. In comparison to bacteria, fungi emerge as significant players in carbon capture and storage (CCS) within terrestrial ecosystems, as emphasized by Rehman et al. (2023). This underscores the importance of understanding the distinct contributions of fungi in terrestrial carbon dynamics, shedding light on their role in environmental sustainability.

VII. **Microorganism and Nitrogen Cycle:** In the Earth's atmosphere, various nitrogen-gaseous molecules, including NH_3, NO, and N_2O, exist alongside the highly stable nitrogen molecule (N_2), which is not readily available to plants and animals without fixation. Nitrogen fixation is the process of converting atmospheric nitrogen into chemical forms that living organisms can utilize. Biological fixation is the primary means by which the biosphere receives N_2. While biological nitrogen fixation is crucial for sustainable agriculture, it cannot entirely replace industrial fixation. Rhizobium bacteria play a vital role in the development of nodules on the roots of legumes such as alfalfa and soybeans. These bacteria exhibit a high degree of specificity towards particular plants, with a type that infects soybeans being unable to infect alfalfa (Rehman et al., 2023). This highlights the importance of understanding the intricate relationships between plants and bacteria in the nitrogen fixation process, which could lead to more efficient and sustainable agricultural practices.

4 Conclusion

Climate change is a multifaceted challenge influenced by natural processes, external forces, and human activities that result in statistically significant variations in the mean state or variability of the climate over an extended period. The primary driver of global warming is the emission of greenhouse gases, primarily from human sources, which leads to diverse impacts on both natural and human systems due to climate change. In order to understand and manage the effects of climate change, it is necessary to conduct comprehensive research on the intricate interplay between environmental conditions and microbial dynamics, particularly in soil ecosystems. The soil microbiome has emerged as a crucial factor in ecosystem stability and function as climate change unfolds. However, the impact of climate change on soil microbiome stability and its role in carbon sequestration, soil fertility, and nutrient cycling present ongoing challenges that need to be investigated and integrated into climate models. Recent reports from the Intergovernmental Panel on Climate Change (IPCC) highlight the urgency of addressing climate-related vulnerabilities, as a significant portion of the global population is highly vulnerable to the impacts of climate change. Mitigation strategies are essential in addressing climate change, with a particular focus on soil management. Rigorous measurement and the application of carbon amendments, such as biochar, show promise in reducing soil greenhouse gas emissions for climate mitigation. Additionally, sustainable soil management practices, including soil organic carbon sequestration, life cycle emissions analysis, and mulching and organic residue application, contribute to addressing climate change. The use of different soil microbes as tools for climate change mitigation demonstrates the diverse potential of microbial communities. Each type of microorganism, such as bacteria, proteobacteria, archaea, cyanobacteria, algae, and fungi, offers unique mechanisms for carbon capture and sequestration, which can address environmental challenges and enhance sustainable practices. Understanding the roles of these microorganisms in nitrogen fixation, methane emission regulation, and soil carbon sequestration is crucial for the development of effective mitigation strategies. In addition to soil management, climate change mitigation also encompasses afforestation and reforestation, enhanced terrestrial weathering, and ocean fertilization. These approaches harness natural processes to capture and store carbon, providing valuable tools in the fight against climate change. In conclusion, a comprehensive and interdisciplinary approach is necessary to understand the complexities of climate change and develop effective mitigation strategies. Ongoing research, informed by the latest scientific reports and advancements, will contribute to a better understanding of the dynamic interactions between environmental conditions, microbial communities, and the impacts of climate change. Ultimately, this knowledge will guide sustainable practices and policies to address the challenges posed by climate change for the benefit of current and future generations.

References

Ahmed, A. A. Q., Odelade, K. A., & Babalola, O. O. (2019). Microbial inoculants for improving carbon sequestration in agroecosystems to mitigate climate change. In *Handbook of climate change resilience* (pp. 1–21). Springer.

Akob, D. M., & Küsel, K. (2011). Where microorganisms meet rocks in the Earth's critical zone. *Biogeosciences, 8*(12), 3531–3543. https://doi.org/10.5194/bg-8-3531-2011

Allison, S. D., & Martiny, J. B. H. (2008). Resistance, resilience, and redundancy in microbial communities. *PNAS, 105*(Supplement 1), 11512–11519.

Averill, C., Werbin, Z. R., Atherton, K. F., Bhatnagar, J. M., & Dietze, M. C. (2021). Soil microbiome predictability increases with spatial and taxonomic scale. *Nature Ecology & Evolution, 5*(6), 747–756. https://doi.org/10.1038/s41559-021-01445-9

Bach, E. M., Ramirez, K. S., Fraser, T. D., & Wall, D. H. (2020). Soil biodiversity integrates solutions for a sustainable future. *Sustainability, 12*(7), 2662. https://doi.org/10.3390/su12072662

Bahram, M., Hildebrand, F., Forslund, S. K., Anderson, J. L., Soudzilovskaia, N. A., Bodegom, P. M., Bengtsson-Palme, J., Anslan, S., Coelho, L. P., Harend, H., Huerta-Cepas, J., Medema, M. H., Maltz, M. R., Mundra, S., Olsson, P. A., Pent, M., Põlme, S., Sunagawa, S., Ryberg, M., & Bork, P. (2018). Structure and function of the global topsoil microbiome. *Nature, 560*(7717), 233–237. https://doi.org/10.1038/s41586-018-0386-6

Baroja-Fernández, E., Almagro, G., Sánchez-López, N. M., Bahaji, A., Gámez-Arcas, S., De Diego, N., Dolezal, K., Muñoz, F. J., Climent Sanz, E., & Pozueta-Romero, J. (2021). Enhanced yield of pepper plants promoted by soil application of volatiles from cell-free fungal culture filtrates is associated with activation of the beneficial soil microbiota. *Frontiers in Plant Science, 12*. https://doi.org/10.3389/fpls.2021.752653

Bar-On, Y. M., Phillips, R., & Milo, R. (2018). The biomass distribution on earth. *Proceedings of the National Academy of Sciences, 115*(25), 6506–6511. https://doi.org/10.1073/pnas.1711842115

Bender, S. F., Wagg, C., & van der Heijden, M. G. (2016). An underground revolution: Biodiversity and soil ecological engineering for agricultural sustainability. *Trends in Ecology & Evolution, 31*(6), 440–452. https://doi.org/10.1016/j.tree.2016.02.016

Bünemann, E. K., Bongiorno, G., Bai, Z., Creamer, R. E., De Deyn, G., de Goede, R., Fleskens, L., Geissen, V., Kuyper, T. W., Mäder, P., Pulleman, M., Sukkel, W., van Groenigen, J. W., &

Brussaard, L. (2018). Soil quality—A critical review. *Soil Biology and Biochemistry, 120*, 105–125. https://doi.org/10.1016/j.soilbio.2018.01.030
California Air Resources Board. (2019). *CA-GREET3.0 model and tier 1 simplified carbon intensity calculators*. https://ww2.arb.ca.gov/resources/documents/lcfs-life-cycle-analysis-models-and-documentation
Cobo-Díaz, J. F., Legrand, F., Le Floch, G., & Picot, A. (2021, June 24). Influence of maize residues in shaping soil microbiota and Fusarium spp. communities. Microbial ecology, consequences of more extreme precipitation regimes for terrestrial ecosystems. *Bioscience, 58*, 811–821.
Custódio, V., Gonin, M., Stabl, G., Bakhoum, N., Oliveira, M. M., Gutjahr, C., & Castrillo, G. (2021). Sculpting the soil microbiota. *The Plant Journal, 109*(3), 508–522. https://doi.org/10.1111/tpj.15568
De Corato, U. (2020). Soil microbiota manipulation and its role in suppressing soil-borne plant pathogens in organic farming systems under the light of microbiome-assisted strategies. *Chemical and Biological Technologies in Agriculture, 7*(1). https://doi.org/10.1186/s40538-020-00183-7
De Gannes, V., Eudoxie, G., Bekele, I., & Hickey, W. J. (2015). Relations of microbiome characteristics to edaphic properties of tropical soils from Trinidad. *Frontiers in Microbiology, 6*. https://doi.org/10.3389/fmicb.2015.01045
De la Cruz-Amo, L., Bañares-de-Dios, G., Cala, V., Granzow-de la Cerda, I., Espinosa, C. I., Ledo, A., Salinas, N., Macía, M. J., & Cayuela, L. (2020). Trade-offs among aboveground, belowground, and soil organic carbon stocks along altitudinal gradients in Andean tropical montane forests. *Frontiers in Plant Science, 11*. https://doi.org/10.3389/fpls.2020.00106
De Vries, F. T., Liiri, M. E., Bjørnlund, L., Bowker, M. A., Christensen, S., Setälä, H. M., & Bardgett, R. D. (2012). Land use alters the resistance and resilience of soil food webs to drought. *Nature Climate Change, 2*(4), 276–280. https://doi.org/10.1038/nclimate1368
Delgado-Baquerizo, M., Maestre, F. T., Reich, P. B., Jeffries, T. C., Gaitan, J. J., Encinar, D., Berdugo, M., Campbell, C. D., & Singh, B. K. (2016). Microbial diversity drives multifunctionality in terrestrial ecosystems. *Nature Communications, 7*(1). https://doi.org/10.1038/ncomms10541
Dodds, W. K., Zeglin, L. H., Ramos, R. J., Platt, T. G., Pandey, A., Michaels, T., Masigol, M., Klompen, A. M. L., Kelly, M. C., Jumpponen, A., Hauser, E., Hansen, P. M., Greer, M. J., Fattahi, N., Delavaux, C. S., Connell, R. K., Billings, S., Bever, J. D., Barua, N., & Agusto, F. B. (2020). Connections and feedback: Aquatic, plant, and soil microbiomes in heterogeneous and changing environments. *Bioscience, 70*(7), 548–562. https://doi.org/10.1093/biosci/biaa046
Dubey, A., Kumar, A., Abd-Allah, E. F., Hashem, A., & Khan, M. L. (2019). Growing more with less: Breeding and developing drought resilient soybean to improve food security. *Ecological Indicators, 105*, 425–437. https://doi.org/10.1016/j.ecolind.2018.03.003
Fawzy, S., Osman, A. I., Doran, J., & Rooney, D. (2020a). Strategies for mitigation of climate change: A review. *Environmental Chemistry Letters, 18*, 2069–2094. https://doi.org/10.1007/s10311-020-01059-w
Fawzy, S., Osman, A. I., Doran, J., & Rooney, D. W. (2020b). Strategies to understand the soil microbiome: A systematic review. *Microbial Ecology, 85*, 669–683. https://doi.org/10.1007/s00248-022-01962-5
Fay, P. A., Kaufman, D. M., Nippert, J. B., Carlisle, J. D., & Harper, C. W. (2008). Changes in grassland ecosystem function due to extreme rainfall events: Implications for responses to climate change. *Global Change Biology, 14*(7), 1600–1608.
Fierer, N. (2017). Embracing the unknown: Disentangling the complexities of the soil microbiome. *Nature Reviews Microbiology, 15*(10), 579–590. https://doi.org/10.1038/nrmicro.2017.87
Friedlingstein, P., Cox, P. M., Betts, R., Bopp, L., Von Bloh, W., Brovkin, V., Cadule, P., Doney, S. C., Eby, M., Fung, I., Bala, G., John, J. G., Jones, C. D., Joos, F., Kato, T., Kawamiya, M., Knorr, W., Lindsay, K., Matthews, H. D., et al. (2006). Climate–carbon cycle feedback analysis:

Results from the C4MIP model intercomparison. *Journal of Climate*. https://doi.org/10.1175/jcli3800.1

Frøslev, T. G., Nielsen, I. B., Santos, S. S., Barnes, C. J., Bruun, H. H., & Ejrnæs, R. (2021). The biodiversity effect of reduced tillage on soil microbiota. *Ambio, 51*(4), 1022–1033. https://doi.org/10.1007/s13280-021-01611-0

Furtak, K., & Gałązka, A. (2019). Edaphic factors and their influence on the microbiological biodiversity of the soil environment. *Postępy Mikrobiologii – Advancements of Microbiology, 58*(4), 375–384. https://doi.org/10.21307/pm-2019.58.4.375

Gao, Z., Karlsson, I., Geisen, S., Kowalchuk, G., & Jousset, A. (2019). Protists: Puppet masters of the rhizosphere microbiome. *Trends in Plant Science, 24*(2), 165–176. https://doi.org/10.1016/j.tplants.2018.10.011

Geisen, S., & Bonkowski, M. (2018). Methodological advances to study the diversity of soil protists and their functioning in soil food webs. *Applied Soil Ecology, 123*, 328–333. https://doi.org/10.1016/j.apsoil.2017.05.021

Geisen, S., Briones, M. J., Gan, H., Behan-Pelletier, V. M., Friman, V. P., de Groot, G. A., Hannula, S., Lindo, Z., Philippot, L., Tiunov, A. V., & Wall, D. H. (2019). A methodological framework to embrace soil biodiversity. *Soil Biology and Biochemistry, 136*, 107536. https://doi.org/10.1016/j.soilbio.2019.107536

Gougoulias, C., Clark, J. M., & Shaw, L. J. (2014). The role of soil microbes in the global carbon cycle: Tracking the below-ground microbial processing of plant-derived carbon for manipulating carbon dynamics in agricultural systems. *Journal of the Science of Food and Agriculture, 94*(12), 2362–2371. https://doi.org/10.1002/jsfa.6577

Griffiths, B. S., & Philippot, L. (2013). Insights into the resistance and resilience of the soil microbial community. *FEMS Microbiology Reviews, 37*(2), 112–129.

Hale, L., Feng, W., Yin, H., Guo, X., Zhou, X., Bracho, R., Pegoraro, E., Penton, C. R., Wu, L., Cole, J., Konstantinidis, K. T., Luo, Y., Tiedje, J. M., Schuur, E. A. G., & Zhou, J. (2019). Tundra microbial community taxa and traits predict decomposition parameters of stable, old soil organic carbon. *The ISME Journal, 13*(12), 2901–2915. https://doi.org/10.1038/s41396-019-0485x

Hamidov, A., Helming, K., Bellocchi, G., Bojar, W., Dalgaard, T., Ghaley, B. B., & Krzeminska, D. (2018). Impacts of climate change adaptation options on soil functions: A review of European casestudies. *Land Degradation and Development, 29*(8), 2378–2389.

Harper, A. B., et al. (2018). Land-use emissions play a critical role in land based mitigation for Paris climate targets. *Nature Communications, 9*, 2938. https://doi.org/10.1038/s41467-018-05340-z

Hartel, P. G. (2005). Soil abiotic environmental factors. In D. M. Sylvia, J. J. Fuhrmann, P. G. Hartel, & D. A. Zuberer (Eds.), *Principles and applications of soil microbiology* (2nd ed., pp. 41–51).

Holland, J., Bennett, A., Newton, A., White, P., McKenzie, B., George, T., Pakeman, R., Bailey, J., Fornara, D., & Hayes, R. (2018). Liming impacts on soils, crops and biodiversity in the UK: A review. *Science of the Total Environment, 610–611*, 316–332. https://doi.org/10.1016/j.scitotenv.2017.08.020

Hu, H., Macdonald, C. A., Trivedi, P., Anderson, I. C., Zheng, Y., Holmes, B. H., Bodrossy, L., Wang, J., He, J., & Singh, B. K. (2016). Effects of climate warming and elevated CO_2 on autotrophic nitrification and nitrifiers in dryland ecosystems. *Soil Biology & Biochemistry., 92*, 1–15. https://doi.org/10.1016/j.soilbio.2015.09.008

IPCC. Climate change. (2022). *Impacts, adaptation, and vulnerability*. Contribution of Working.

Islam, W., Noman, A., Naveed, H., Huang, Z., & Chen, H. Y. H. (2020, August 23). Role of environmental factors in shaping the soil microbiome. *Environmental Science and Pollution Research, 27*(33), 41225–41247. https://doi.org/10.1007/s11356-020-10471-2

Jansson, J. K., & Hofmockel, K. (2020). Soil microbiomes and climate change. *Earth and Biological Sciences Directorate, Pacific Northwest National Laboratory, Richland, Washington, 83*(3), 702–713. https://doi.org/10.1007/s00248-021-01797-6

Ke, J., Wang, B., & Yoshikuni, Y. (2021). Microbiome engineering: Synthetic biology of plant-associated microbiomes in sustainable agriculture. *Trends in Biotechnology, 39*(3), 244–261. https://doi.org/10.1016/j.tibtech.2020.07.008

Khan, S. T., Adil, S. F., Shaik, M. R., Alkhathlan, H. Z., Khan, M., & Khan, M. (2021). Engineered nanomaterials in soil: Their impact on soil microbiome and plant health. *Plants, 11*(1), 109. https://doi.org/10.3390/plants11010109

Knapp, A. K., Beier, C., Briske, D. D., Classen, A. T., Luo, Y., Reichstein, M., Smith, M. D., Smith, S. D., Bell, J. E., Fay, P. A., Heisler, J. L., Leavitt, S. W., Sherry, R. A., Smith, B., & Weng, E. (2008). Consequences of more extreme precipitation regimes for terrestrial ecosystems. *Bioscience, 58*, 811–821. https://doi.org/10.1641/b580908

Kumar, V., Rawat, A. K., & Amule, F. C. (2016). Climate change impact on beneficial soil microbial community: A review. *Nature Environment and Pollution Technology An International Quarterly Scientific Journal, 15*, 619–625.

Li, C., Liao, H., Xu, L., Wang, C., He, N., Wang, J., & Li, X. (2022). The adjustment of life history strategies drives the ecological adaptations of soil microbiota to aridity. *Molecular Ecology, 31*(10), 2920–2934. https://doi.org/10.1111/mec.16445

Liang, C., Schimel, J. P., & Jastrow, J. D. (2017). The importance of anabolism in microbial control over soil carbon storage. *Nature Microbiology, 2*(8). https://doi.org/10.1038/nmicrobiol.2017.105

Liu, T., Hu, F., & Li, H. (2019). Spatial ecology of soil nematodes: Perspectives from global to micro scales. *Soil Biology and Biochemistry, 137*, 107565. https://doi.org/10.1016/j.soilbio.2019.107565

Lladó, S., López-Mondéjar, R., & Baldrian, P. (2017). Forest soil bacteria: Diversity, involvement in ecosystem processes, and response to global change. *Microbiology and Molecular Biology Reviews, 81*(2). https://doi.org/10.1128/mmbr.00063-16

Ma, L., Rocha, F. I., Lee, J., Choi, J., Tejera, M., Sooksa-Nguan, T., Boersma, N., VanLoocke, A., Heaton, E., & Howe, A. (2021). The impact of stand age and fertilization on the soil microbiome of Miscanthus × giganteus. *Phytobiomes Journal, 5*(1), 51–59. https://doi.org/10.1094/pbiomes-01-20-0014-fi

Malla, M. A., Dubey, A., Yadav, S., Kumar, A., Hashem, A., & Abd Allah, E. F. (2018). Understanding and designing the strategies for the microbe-mediated remediation of environmental contaminants using omics approaches. *Frontiers in Microbiology, 9*. https://doi.org/10.3389/fmicb.2018.01132

Massoni, J., Bortfeld-Miller, M., Widmer, A., & Vorholt, J. A. (2021). Capacity of soil bacteria to reach the phyllosphere and convergence of floral communities despite soil microbiota variation. *Proceedings of the National Academy of Sciences, 118*(41). https://doi.org/10.1073/pnas.2100150118

Migliardini, F., De Luca, V., Carginale, V., Rossi, M., Corbo, P., Supuran, C. T., & Capasso, C. (2014). Biomimetic CO2 capture using a highly thermostable bacterial α-carbonic anhydrase immobilized on a polyurethane foam. *Journal of Enzyme Inhibition and Medicinal Chemistry, 29*(1), 146–150.

Mishra, A., Singh, L., & Singh, D. (2022). Unboxing the black box—One step forward to understand the soil microbiome: A systematic review. *Microbial Ecology, 85*, 669–683. https://doi.org/10.1007/s00248-022-01962-5

Murphy, B. W. (2015). Impact of soil organic matter on soil properties—A review with emphasis on Australian soils. *Soil Research, 53*(6), 605. https://doi.org/10.1071/sr14246

Mwakilili, A. D., Mwaikono, K. S., Herrera, S. L., Midega, C. A. O., Magingo, F., Alsanius, B., Dekker, T., & Lyantagaye, S. L. (2021). Long-term maize-Desmodium intercropping shifts structure and composition of soil microbiome with stronger impact on fungal communities. *Plant and Soil, 467*(1–2), 437–450. https://doi.org/10.1007/s11104-021-05082-w

Naylor, D., Sadler, N., Bhattacharjee, A., Graham, E. B., Anderton, C. R., McClure, R., Lipton, M., Hofmockel, K. S., & Jansson, J. K. (2020). Soil microbiomes under climate change and implications for carbon cycling. *Annual Review of Environment and Resources, 45*, 29–59.

Nguyen, L. T., Broughton, K., Osanai, Y., Anderson, I. C., Bange, M. P., Tissue, D. T., & Singh, B. K. (2019). Effects of elevated temperature and elevated CO2 on soil nitrification and ammonia-oxidizing microbial communities in field-grown crop. *Science of the Total Environment, 675*, 81–89. https://doi.org/10.1016/j.scitotenv.2019.04.181

Nielsen, U. N., Wall, D. H., & Six, J. (2015). Soil biodiversity and the environment. *Annual Review of Environment and Resources, 40*(1), 63–90. https://doi.org/10.1146/annurev-environ-102014-021257

Novick, K. A., Metzger, S., Anderegg, W. R. L., Barnes, M. L., Cala, D., Guan, K., Hemes, K. S., Hollinger, D. Y., Kumar, J., Litvak, M. E., Lombardozzi, D., Normile, C. P., Oikawa, P. Y., Runkle, B. R. K., Torn, M. S., & Wiesner, S. (2022). Informing nature-based climate solutions for the United States with the best-available science. *Global Change Biology, 28*, 3778–3794. https://doi.org/10.1111/gcb.16156

Offre, P., Spang, A., & Schleper, C. (2013). Archaea in biogeochemical cycles. *Annual Review of Microbiology, 67*(1), 437–457. https://doi.org/10.1146/annurev-micro-092412-155614

Omotayo, O. P., & Babalola, O. O. (2021). Resident rhizosphere microbiome's ecological dynamics and conservation: Towards achieving the envisioned sustainable development goals, a review. *International Soil and Water Conservation Research, 9*(1), 127–142. https://doi.org/10.1016/j.iswcr.2020.08.002

Orwin, K. H., & Wardle, D. A. (2005). Plant species composition effects on belowground properties and the resistance and resilience of the soil microflora to a drying disturbance. *Plant and Soil, 278*(1–2), 205–221.

Patil, A. G., Kounaina, K., Aishwarya, S., Harshitha, N., Satapathy, P., Hudeda, S., Reddy, K. R., Alrafas, H., Yadav, A. N., Raghu, A. V., & Zameer, F. (2021). Myco-nanotechnology for sustainable agriculture: Challenges and opportunities. *Fungal Biology*. https://doi.org/10.1007/978-3-030-60659-6_20

Pendall, E., Bridgham, S., Hanson, P. J., Hungate, B., Kicklighter, D. W., Johnson, D. W., Law, B. E., Luo, Y. Q., Megonigal, J. P., Olsrud, M., Ryan, M. G., & Wan, S. (2004). Below-ground process responses to elevated CO2 and temperature: A discussion of observations, measurement methods, and models. *New Phytologist, 162*, 311–322.

Rehman, U. M., Muhammad, A., Aziz, U., Sapana, P., & Waleed, I. (2023). Uncovering the hidden world beneath our feet: How soil microbes are key players in mitigating climate change; A. *Review, 19*, 1093–1103.

Rittmann, S., Seifert, A., & Herwig, C. (2015). Essential prerequisites for successful bioprocess development of biological CH4 production from CO_2 and H_2. *Critical Reviews in Biotechnology, 35*(2), 141–151.

Rubin, R., Oldfield, E., Lavallee, J., Griffin, T., Mayers, B., & Sanderman, J. (2023). *Climate mitigation through soil amendments: Quantification, evidence, and uncertainty*. Taylor & Francis Group. https://doi.org/10.1080/17583004.2023.2217785

Schloter, M., Nannipieri, P., Sørensen, S. J., & van Elsas, J. D. (2017). Microbial indicators for soil quality. *Biology and Fertility of Soils, 54*(1), 1–10. https://doi.org/10.1007/s00374-017-1248-3

Shade, A., Peter, H., Allison, S. D., Baho, D. L., Berga, M., et al. (2012). Fundamentals of microbial community resistance and resilience. *Frontiers in Microbiology, 3*, 417.

Sheik, C. S., Beasley, W. H., Elshahed, M. S., Zhou, X., Luo, Y., & Krumholz, L. R. (2011). Effect of warming and drought on grassland microbial communities. *International Society for Microbial Ecology, 5*, 1692–1700.

Shi, W., Li, M., Wei, G., Tian, R., Li, C., Wang, B., Lin, R., Shi, C., Chi, X., Zhou, B., & Gao, Z. (2019). The occurrence of potato common scab correlates with the community composition and function of the geocaulosphere soil microbiome. *Microbiome, 7*(1), 14. https://doi.org/10.1186/s40168-019-0629-2

Smith, P. (2004). Soils as carbon sinks: The global context. *Soil Use Management, 20*, 212–218.

Sun, Y., Guo, Y., Shi, M., Qiu, T., Gao, M., Tian, S., & Wang, X. (2021). Effect of antibiotic type and vegetable species on antibiotic accumulation in soil-vegetable system, soil microbiota, and resistance genes. *Chemosphere, 263*, 128099. https://doi.org/10.1016/j.chemosphere.2020.128099

Tan, H., Liu, T., Yu, Y., Tang, J., Jiang, L., Martin, F. M., & Peng, W. (2021). Morel production related to soil microbial diversity and evenness. *Microbiology Spectrum, 9*(2). https://doi.org/10.1128/spectrum.00229-21

Teng, Y., & Chen, W. (2019). Soil microbiomes—A promising strategy for contaminated soil remediation: A review. *Pedosphere, 29*(3), 283–297. https://doi.org/10.1016/s1002-0160(18)60061-x

Thakur, M. P., & Geisen, S. (2019). Trophic regulations of the soil microbiome. *Trends in Microbiology, 27*(9), 771–780. https://doi.org/10.1016/j.tim.2019.04.008

Tiedje, J. M., Bruns, A. B., Casadevall, A., Criddle, C. S., Eloe-Fadrosh, E., Karl, D. M., Nguyen, N. K., & Zhou, J. (2022). *Microbes and climate change: A research prospectus for the future* (Vol. 13). American Society of Microbiology. https://doi.org/10.1128/mbio.00800-22

Topalović, O., Hussain, M., & Heuer, H. (2020). Plants and associated soil microbiota cooperatively suppress plant-parasitic nematodes. *Frontiers in Microbiology, 11*. https://doi.org/10.3389/fmicb.2020.00313

UNCCS. (2019). *Climate action and support trends, United Nations Climate Change Secretariat*. https://unfccc.int/sites/default/fles/resource/Climate_Action_Support_Trends_2019.pdf

Viles, H. A. (2012). Microbial geomorphology: A neglected link between life and landscape. *Geomorphology, 157–158*, 6–16. https://doi.org/10.1016/j.geomorph.2011.03.021

Wang, K., et al. (2017). Modeling global soil carbon and soil microbial carbon by integrating microbial processes into the ecosystem process model TRIPLEX-GHG. *Journal of Advances in Modeling Earth Systems, 9*, 2368–2384.

Wang, M., Garrido-Sanz, D., Sansegundo-Lobato, P., Redondo-Nieto, M., Conlon, R., Martín, M., Mali, R. S., Liu, X., Dowling, D. N., Rivilla, R., & Germaine, K. J. (2021). Soil microbiome structure and function in Ecopiles used to remediate petroleum-contaminated soil. *Frontiers in Environmental Science, 9*. https://doi.org/10.3389/fenvs.2021.624070

Wang, W., Sun, Z., Mishra, S., Xia, S., Lin, L., & Yang, X. (2022). Body size determines multitrophic soil microbiota community assembly associated with soil and plant attributes in a tropical seasonal rainforest. *Molecular Ecology, 32*(23), 6294–6303. https://doi.org/10.1111/mec.16585

Wei, Z., Gu, Y., Friman, V. P., Kowalchuk, G. A., Xu, Y., Shen, Q., & Jousset, A. (2019). Initial soil microbiome composition and functioning predetermine future plant health. *Science. Advances, 5*(9). https://doi.org/10.1126/sciadv.aaw0759

Wieder, W. R., Bonan, G. B., & Allison, S. D. (2013). Global soil carbon projections are improved by modelling microbial processes. *Nature Climate Change, 3*, 909–912. https://doi.org/10.1038/nclimate1951

Wiesmeier, M., Urbanski, L., Hobley, E., Lang, B., von Lützow, M., Marin-Spiotta, E., van Wesemael, B., Rabot, E., Ließ, M., Garcia-Franco, N., Wollschläger, U., Vogel, H. J., & Kögel-Knabner, I. (2019). Soil organic carbon storage as a key function of soils – A review of drivers and indicators at various scales. *Geoderma, 333*, 149–162. https://doi.org/10.1016/j.geoderma.2018.07.026

Wong, W. C., Tan, S. N., Ge, L., Chen, X., & Yong, J. W. H. (2015). The importance of Phytohormones and microbes in biofertilizers. *Sustainable Development and Biodiversity*. https://doi.org/10.1007/978-3-319-24654-3_6

Woodcroft, B. J., Singleton, C. M., Boyd, J. A., Evans, P. N., Emerson, J. B., Zayed, A. A. F., Hoelzle, R. D., Lamberton, T. O., McCalley, C. K., Hodgkins, S. B., Wilson, R. M., Purvine, S. O., Nicora, C. D., Li, C., Frolking, S., Chanton, J. P., Crill, P. M., Saleska, S. R., Rich, V. I., & Tyson, G. W. (2018). Genome-centric view of carbon processing in thawing permafrost. *Nature, 560*(7716), 49–54. https://doi.org/10.1038/s41586-018-0338-1

Xue, K., Yuan, M. M., Shi, Z. J., Qin, Y., Deng, Y., Cheng, L., Wu, L., He, Z., Van, W., Nostrand, J. D., Bracho, R., Natali, S., Schuur, E. A. G., Luo, C., Konstantinidis, K. T., Wang, Q., Cole, J. R., Tiedje, J. M., Luo, Y., & Zhou, J. Z. (2016). Tundra soil carbon is vulnerable to rapid microbial decomposition under climate warming. *Nature Climate Change, 6*, 595–600. https://doi.org/10.1038/nclimate2940

Yadav, A. N. (2017). Agriculturally important micro biomes: Biodiversity and multifarious PGP attributes for amelioration of diverse abiotic stresses in crops for sustainable agriculture. *Biomedical Journal of Scientific & Technical Research, 1*(4). https://doi.org/10.26717/bjstr.2017.01.000321

Yang, G., Wagg, C., Veresoglou, S. D., Hempel, S., & Rillig, M. C. (2018). How soil biota drive ecosystem stability. *Trends in Plant Science, 23*(12), 1057–1067. https://doi.org/10.1016/j.tplants.2018.09.007

Yasin, N. H. M., Maeda, T., Hu, A., Yu, C. P., & Wood, T. K. (2015). CO2 sequestration by methanogens in activated sludge for methane production. *Applied Energy, 142*, 426–434.

Yue, X. L., & Gao, Q. X. (2018). Contributions of natural systems and human activity to greenhouse gas emissions. *Advances in Climate Change Research, 9*, 243–252. https://doi.org/10.1016/j.accre.2018.12.003

Zhou, J., Xue, K., Xie, J., Deng, Y., Wu, L., Cheng, X., Fei, S., Deng, S., He, Z., Van Nostrand, J. D., & Luo, Y. (2012). Microbial mediation of carbon-cycle feedbacks to climate warming. *Nature Climate Change, 2*, 106–110. https://doi.org/10.1038/nclimate1331

Relationship Between Soil Microbiome, Edaphic Factors, and Green Technology in Emissions Reduction

S. Stella Leh-Togi Zobeashia, Amoge C. Ogu, Tarimoboere Agbalalah, Hadiza Aliyu, and Olabisi Peter Abioye

Abstract The soil microbiome, influenced by various edaphic factors, plays a pivotal role in environmental sustainability, particularly in the context of emissions reduction. Edaphic factors such as soil texture, pH, organic matter content, moisture, temperature, and nutrient availability profoundly affect microbial community composition and function. Green technologies, including biochar application, organic amendments, reduced tillage, cover cropping, and precision agriculture, can optimize these edaphic factors to enhance the soil microbiome's capacity for carbon sequestration and greenhouse gas mitigation. This study reviews the intricate relationships between soil microbiomes, edaphic factors, and green technologies, highlighting how integrated soil management practices can effectively reduce emissions, promote sustainable agriculture and environment which automatically results in the development of strategies to leverage soil microorganisms for environmental benefits, thus, addressing one of the most pressing challenges of the world.

Keywords Soil microbiome · Edaphic factor · Green technology · Microbial community

1 Introduction

The intricate interplay between soil microbiome, edaphic factors, and green technology in emissions reduction has gathered significant attention in the realm of environmental science and sustainability. Soil microbiome, consisting of a diverse array of microorganisms such as bacteria, fungi, and archaea amongst others, plays a

S. S. Leh-Togi Zobeashia (✉) · A. C. Ogu · T. Agbalalah
Naional Biotechnology Research and Development Agency, Abuja, Nigeria

H. Aliyu
Bioresource Development Center National Biotechnology Research and Development Agency, Lugbe, Abuja, Nigeria

O. P. Abioye
Department of Microbiology, Federal University of Technology Minna, Niger State, Nigeria

S. A. Aransiola et al. (eds.), *Soil Microbiome in Green Technology Sustainability*,
https://doi.org/10.1007/978-3-031-71844-1_10

crucial role in nutrient cycling, plant growth promotion, disease suppression, carbon sequestration, soil health, and overall ecosystem functioning. Edaphic factors, encompassing physical, chemical, and biological properties of soil, exert a profound influence on the composition and activity of soil microbiota. The integration of green technology, characterized by eco-friendly practices and innovations, holds promise for mitigating greenhouse gas emissions and promoting sustainable land management practices.

Recent studies have shed light on the dynamic relationships between soil microbiome, edaphic factors, and emissions reduction strategies. Research by Li et al. (2023) demonstrated the significant impact of soil microbial communities on carbon sequestration potential in agroecosystems such as decomposing organic materials and stabilizing soil carbon pools, highlighting the importance of understanding microbial processes for enhancing soil carbon storage. Furthermore, the study by Zhang et al. (2022) elucidated the role of edaphic factors, such as pH and soil moisture, in shaping the microbial diversity and functional potentials in contaminated soils, emphasizing the need for tailored remediation strategies. In the context of emissions reduction, the integration of green technologies presents opportunities to synergistically leverage soil microbiome functions, edaphic properties, and sustainable practices. Studies such as that by Smith et al. (2024) have demonstrated the efficacy of biochar amendments in improving soil fertility, sequestering carbon, and reducing greenhouse gas emissions through interactions with soil microbiota. Similarly, Chen et al. (2023) showcased the benefits of precision agriculture technologies in optimizing nutrient management and minimizing environmental impacts, illustrating the potential of innovative green solutions in emissions reduction strategies.

2 Overview of Soil Microbiome

The soil microbiome encompasses a diverse community of microorganisms residing in soil, crucial for nutrient cycling, plant growth, and ecosystem function. This intricate network includes bacteria, fungi, archaea, protozoa, algae and viruses, interacting in complex ways. Recent years have seen a surge in interest in soil microbiomes, driven by their fundamental importance in agriculture, environmental sustainability, and human health. Studies have shed light on the vast diversity and functional significance of soil microbiomes (Delgado-Baquerizo et al., 2020) High-throughput sequencing techniques like metagenomics and amplicon sequencing have revolutionized our ability to delve into microbial communities in soil at unprecedented depths (Tkacz et al., 2018). These studies reveal extensive taxonomic diversity, with many yet-to-be-identified species. Functional metagenomics has also offered insights into the metabolic potential and ecological roles of soil microorganisms. Soil microbiomes exhibit dynamic responses to environmental factors like pH, moisture, temperature, and organic matter content. For instance, acidic soils host different microbial communities compared to neutral or alkaline soils, affecting

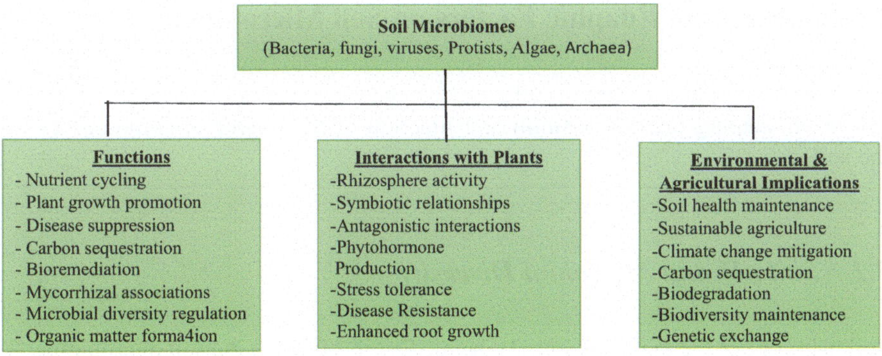

Fig. 1 Overview of soil microbiome

nutrient cycling (Fig. 1). Land-use practices such as agriculture and urbanization significantly alter soil microbial diversity and function, underscoring the need for sustainable land management (Hartmann et al., 2015). Recent research emphasizes soil microbiomes' role in plant health and productivity. Beneficial plant-microbe interactions, such as mycorrhizal associations were fungi forms symbiotic relationships with plant roots, enhancing water and nutrient uptake, antagonistic Relationships where microbes compete with pathogens for resources or produce antimicrobial compounds, protecting plants from diseases thus stress tolerance. Additionally, the interaction also results in nutrient cycling such as Nitrogen fixation and phosphate solubilization, decomposition, bioremediation of environmental pollutant, phytohormone production, disease resistance. Understanding these interactions can lead to innovative agricultural practices, including biofertilization and phytoremediation, to enhance crop yields sustainably. Soil microbiomes are pivotal in carbon and nutrient cycling, affecting global biogeochemical cycles. Microbial decomposition releases carbon dioxide into the atmosphere, contributing to climate change. However, certain microorganisms like the Arhaea produce greenhouse gases like methane. Advancements in omics technologies have unraveled the functional potential of soil microbiomes in unprecedented detail. Metatranscriptomics, metaproteomics, and metabolomics shed light on microbial metabolic activities and ecological functions under varying conditions (Tkacz et al., 2018). Current studies have also highlighted the importance of soil microbiomes in human health and biomedicine. Soil-derived microorganisms produce a vast array of bioactive compounds with pharmaceutical potential, including antibiotics, anticancer agents, and immunomodulators. Moreover, exposure to soil microorganisms has been linked to improved immune function and reduced risk of autoimmune diseases, emphasizing the importance of maintaining healthy soil ecosystems for human well-being.

In essence, soil microbiomes are complex ecosystems with profound implications for agriculture, sustainability, and human health. Advances in omics technologies offer insights into their diversity and potential. However, further research is essential to fully grasp their complexities and leverage their benefits for sustainable agriculture and biotechnological applications.

3 Important of Edaphic Factors on Soil Microbiome

Edaphic factors which relates to soil physical, chemical, and biological properties are pivotal in shaping the composition and function of the microbial communities of the soil.

3.1 Soil pH and Microbial Diversity

Soil pH is one of the most influential edaphic factors affecting the soil microbiome. It determines the solubility of nutrients and the overall soil chemistry, which directly influences microbial activity and diversity. Studies have shown that bacterial diversity tends to decrease as soil pH deviates from neutral, with acidic soils typically harboring fewer bacterial species compared to neutral or slightly alkaline soils (Lauber et al., 2009). Fungi, on the other hand, are generally more tolerant of a wider pH range but also exhibit changes in community composition in response to pH shifts (Rousk et al., 2010a). A study by Zhang et al. (2022) found that soil pH was a major determinant of bacterial community structure, with acidophilic bacteria dominating in low pH soils and neutrophilic bacteria in neutral pH soils. Furthermore, changes in soil pH can affect nutrient availability and metal toxicity, further influencing microbial communities (Rousk et al., 2010a, b).

3.2 Soil Texture and Microbial Habitat

Soil texture, defined by the proportion of sand, silt, and clay, affects the soil's water-holding capacity, aeration, and nutrient availability. These physical properties create distinct microhabitats for microorganisms. For example, clay soils, with their fine particles and high surface area, often support higher microbial biomass due to their ability to retain moisture and nutrients (Liang et al., 2022). Sandy soils, with larger particles and lower nutrient and water retention, typically have lower microbial biomass but can support diverse microbial communities due to better aeration (Tian et al., 2021).

3.3 Organic Matter and Nutrient Availability

The availability of organic matter and nutrients is crucial for the growth and activity of soil microbiomes. Organic matter serves as both a food source and a habitat, influencing microbial abundance and diversity. High organic matter content generally promotes microbial growth by providing essential carbon and energy sources that fosters the formation of soil aggregates, which protect soil microbiomes from predation and environmental stress(Bardgett & van der Putten, 2014). Recent studies by Liu et al. (2023a) shows that soils with higher organic matter content support more diverse and functional microbial communities, enhancing nutrient cycling and soil health. Nutrients such as nitrogen, phosphorus, and sulfur are also critical. For instance, nitrogen availability can affect the balance between bacterial and fungal populations, with nitrogen-rich conditions often favoring bacterial growth (Bardgett et al., 2008). Moreover, soil microbiomes compete with plants for these essential nutrients, and their availability can shape microbial community structure and function. Fertilization practices can alter nutrient levels, affecting microbial diversity and activity. For example, excessive nitrogen fertilization can lead to the dominance of fast-growing, copiotrophic microorganisms at the expense of slow-growing, oligotrophic species (Fierer et al., 2012). Recent research by Geisseler et al. (2022) emphasizes that balanced nutrient management is essential for maintaining diverse and functional soil microbial communities.

3.4 Moisture Content and Microbial Processes

Soil moisture is another crucial edaphic factor affecting the soil microbiome. Water availability influences microbial metabolic activity, community structure, and spatial distribution. Microorganisms require water for nutrient uptake, enzymatic reactions, and cellular processes.

Moisture affects the diffusion of solutes and gases, which in turn impacts microbial respiration and other biochemical processes (Schimel et al., 2007). Drought conditions can lead to reduced microbial activity and shifts in community composition, favoring drought-tolerant species (Barnard et al., 2013). Conversely, waterlogged conditions can create anaerobic environments, promoting the growth of anaerobic microorganisms such as methanogens (Morrissey et al., 2022).

3.5 Temperature and Microbial Metabolism

Temperature directly influences metabolism and growth rates of soil microbiomes. Soil microbial communities can be adapted to a wide range of temperatures, but extreme temperatures often result in reduced diversity and altered community structure. For instance, seasonal temperature fluctuations can lead to shifts in microbial community composition and function. A study by Delgado-Baquerizo et al. (2016) found that temperature was a significant predictor of microbial diversity and enzymatic activity across different biomes. Additionally, temperature affects the decomposition rate of organic matter and nutrient mineralization, indirectly influencing microbial communities.

3.6 Soil Aggregates

Soil aggregates, formed through the binding of soil particles into larger structures, create microenvironments that protect microbes from environmental stress and predation. These aggregates enhance microbial habitat diversity by providing a range of niches with varying oxygen levels, moisture content, and nutrient availability (Six et al., 2004). Aggregates also facilitate the formation of biofilms, where microbial communities can interact more closely, enhancing processes like nutrient cycling and organic matter decomposition (Or et al., 2007).

In conclusion, edaphic factors, including soil texture, pH, organic matter, moisture, temperature, nutrient availability, and structure, plays an important role in shaping the soil microbiome. The interaction of the factors influences the soil microbiome. For example, soil pH can affect nutrient availability, which in turn impacts microbial communities. Similarly, soil moisture and temperature interact to influence microbial metabolic rates and community composition. Sustainable soil management practices that optimize these edaphic factors can enhance soil health, promote beneficial microbial activities, and improve agricultural productivity. Continued research into the interactions between edaphic factors and the soil microbiome is essential for advancing our understanding of soil ecosystems and developing strategies for sustainable land use.

4 Role of Soil Microbiome on Emissions Reduction

The soil microbiome mitigates greenhouse gas emissions, which is essential for combating climate change. The soil microbiome, consisting of a diverse array of dynamic community of microorganisms which interacts with soil organic matter and influences the cycling of key elements such as carbon and nitrogen, thereby affecting the emission of greenhouse gases such as carbon dioxide (CO_2), methane (CH_4), and nitrous oxide (N_2O). Some of the role of soil microbiome on emission reduction include:

4.1 Carbon Sequestration

One of the primary ways the soil microbiome contributes to emissions reduction is through carbon sequestration. Soil microorganisms decompose organic matter, converting it into stable forms of organic carbon that are stored in the soil for extended periods. This process is crucial for removing carbon dioxide (CO_2) from the atmosphere. Certain microbial communities are particularly effective at forming soil aggregates, which protect organic carbon from rapid decomposition and enhance its long-term storage thereby enhancing carbon stabilization in soil aggregates (Liu et al., 2023a, b). Moreover, mycorrhizal fungi, which form symbiotic relationships with plant roots, play a significant role in transporting carbon from plants into the soil, further enhancing carbon transportation and sequestration in the soil (Terrer et al., 2021).

4.2 Nitrous Oxide Reduction

Nitrous oxide, a potent greenhouse gas, produced primarily through microbial processes such as nitrification and denitrification. Recent research emphasizes the role of denitrifying bacteria in reducing N_2O emissions by converting it to N_2 (Butterbach-Bahl et al., 2020). Enhancing soil microbiome activity through soil management practices such as maintaining proper moisture levels and organic carbon inputs, can help reduce N_2O emissions by maintaining optimal soil moisture and pH that can promote the growth of N_2O-reducing bacteria, thereby minimizing N_2O emissions (Cui et al., 2022).

4.3 Methane Oxidation

Methane (CH_4) is another significant greenhouse gas, and soils can act as both sources and sinks for methane, depending on the microbial communities present. Methane emissions from soils are largely influenced by the activity of methanotrophic bacteria, which oxidize methane to carbon dioxide, thus reducing its emission into the atmosphere. Recent advances in understanding the genetics and metabolism of methanotrophs suggest new strategies to enhance their activity in agricultural soils (Leh-Togi Zobeashia et al., 2018; Ho et al., 2022). Practices such as improving soil aeration and incorporating organic amendments can promote the growth of methanotrophs, particularly in methane-prone environments like wetlands and rice paddies (Lee et al., 2021).

4.4 Promoting Soil Health

Maintaining healthy soils with diverse and active microbial communities is essential for reducing greenhouse gas emissions. Sustainable land management practices, including organic farming, reduced tillage, cover cropping, and organic amendments such as application of compost and biochar, have been shown to improve soil structure, increasing organic matter content and soil microbial diversity and function (Wang et al., 2023). These practices not only improve carbon sequestration and reduce emissions of N_2O and CH_4 but also enhance soil fertility and resilience to environmental stresses.

5 Green Technology for Enhancing Soil Microbiome Function

Green technology encompasses a range of innovations and practices designed to minimize environmental impacts and promote sustainability. In agriculture, green technologies aim to enhance soil health, productivity, and the functionality of the soil microbiome. Some green technologies for enhancing the function of the soil microbiome, contributing to more sustainable soil microbiome include biochar, composting, cover cropping, reduced tillage, microbial inoculants, precision agriculture, and phytoremediation amongst others.

5.1 Biochar

Biochar, a stable form of carbon produced from the pyrolysis of organic materials, is increasingly recognized for its potential to enhance soil microbiome function. Recent studies have shown that biochar can improve soil structure, increase water retention, and provide a habitat for beneficial microorganisms (Lehmann & Joseph, 2021). For instance, research by Zhang et al. (2023) demonstrated that biochar amendments increased microbial biomass and activity, particularly of beneficial bacteria involved in nutrient cycling. By enhancing soil properties, biochar creates a conducive environment for microbial growth and diversity, leading to improved soil health and productivity.

5.2 Composting and Organic Amendments

Composting and the use of organic amendments, such as animal manure and plant residues, are traditional but highly effective green technologies for enhancing soil microbiome function. Compost provides a rich source of organic matter and nutrients, which are essential for microbial growth. According to Wang et al. (2022), the application of compost significantly increased soil microbial biomass and enzyme activities, promoting nutrient cycling and soil fertility. Additionally, organic amendments can introduce beneficial microorganisms into the soil, enhancing microbial diversity and resilience (Aransiola et al., 2019; Bernal et al., 2022).

5.3 Cover Cropping and Crop Rotation

Cover cropping and crop rotation are sustainable agricultural practices that can enhance soil microbiome function. Cover crops, such as legumes and grasses, protect the soil from erosion, improve soil structure, and increase organic matter content. They also provide a continuous supply of root exudates, which serve as food for soil microorganisms. A study by Smith et al. (2021) found that cover cropping increased microbial diversity and the abundance of beneficial microbes, such as nitrogen-fixing bacteria and mycorrhizal fungi. Crop rotation, by diversifying the types of plants grown in a field, helps to break pest and disease cycles and promotes a more diverse and stable soil microbiome (Liu et al., 2021).

5.4 Reduced Tillage

Reduced tillage practices, such as no-till or conservation tillage, can significantly enhance soil microbiome function by preserving soil structure and organic matter. Traditional tillage can disrupt soil microbial habitats and reduce microbial diversity. In contrast, reduced tillage maintains soil integrity and fosters a more stable environment for microorganisms. Recent research by Blanco-Canqui et al. (2022) indicates that reduced tillage practices increase microbial biomass and activity, particularly of fungi and other microorganisms that contribute to soil structure and nutrient cycling.

5.5 Microbial Inoculants

Microbial inoculants, also known as biofertilizers, are products containing beneficial microorganisms that can be added to the soil to enhance its microbial function. These inoculants typically include nitrogen-fixing bacteria, phosphorus-solubilizing bacteria, and mycorrhizal fungi. Studies have shown that microbial inoculants can improve plant growth, nutrient uptake, and soil health (Vessey, 2021). For example, a study by Zhang et al. (2023) demonstrated that inoculating soils with mycorrhizal fungi improved plant phosphorus uptake and increased microbial diversity and activity in the rhizosphere.

5.6 Precision Agriculture

Precision agriculture involves using technology to optimize field-level management regarding crop farming. Technologies such as GPS, remote sensing, and soil sensors allow for precise application of inputs like water, fertilizers, and pesticides. By applying inputs only where and when they are needed, precision agriculture reduces the risk of over-application, which can harm the soil microbiome. According to a study by Lobell et al. (2022), precision agriculture practices can enhance soil health by maintaining optimal conditions for microbial activity and reducing the negative impacts of chemical inputs.

5.7 Phytoremediation

Phytoremediation is the use of plants to remove, degrade, or stabilize contaminants in the soil. This green technology not only helps to clean up polluted soils but also enhances soil microbiome function by improving soil health. Plants used in phytoremediation release root exudates that stimulate microbial activity and diversity. Research by Paredes-Páliz et al. (2022) showed that phytoremediation improved soil microbial biomass and enzyme activities, contributing to the recovery of contaminated soils and the restoration of microbial functions (Abioye et al., 2017).

6 Interactions Between Edaphic Factors, Soil Microbiome, and Green Technology

The interactions between soil microbiome, edaphic factors, and green technology are complex and essential in forming ecosystem dynamics and impacting environmental sustainability and emission reduction. These interactions are.

6.1 The Composition of Soil Microbial Community

The nutritional content of the soil is diverse as a result of the rapidly evolving dynamics in the environment and biological components in the humus thereby making the soil home to several microorganisms such as bacteria, Archaea, and fungi. These organisms provide several capabilities, including biodegradation of pollutants, safeguarding soil structure, and disseminating biogenic components that supply nutrients to plants. Edaphic factors like pH, moisture, texture, and nutrient availability are essential environmental channels affecting soil microbiome structure, exhaustive analysis of all microbiome elements integrated with thorough soil characterization is inadequate. Unique types of soil biota have their inclinations for pH and it is noted that soil pH is significantly corresponded with the advancement of a specific set of the soil microbiome. For example, with a rise in pH, the soil bacterial-to-fungal proportion and nematode significantly increase (Van der Wal et al., 2009). The liming of acidic soil often results in shifts in the composition and abundance of soil microbiome including bacteria, fungi, archaea, nematodes, annelids, and microarthropods, as there is a significant increase in pH and nutrient concentrations

((Holland et al., 2018). Understanding the role soil microbiome plays in terrestrial environments would assist in predicting their impact on both the natural and managed ecosystems.

Furthermore, the developing populace requests sustained food production with food safety and environmental sustainability (Lykogianni et al., 2021). Soil microbial communities, nonetheless, vary constantly because of environmental change-prompted stresses and this therefore leads to either acclimatization, lethargy, or death of predominant micro-organisms. The investigations over the past years have untangled the capabilities of microbial mediations in the recovery of soils. The methodologies engulfed in encompass harnessing the native capability of soil microbes to sequester carbon, guaranteeing autotrophic organisms such as algae given changing carbon into lipids, control of adenosine tri-phosphate for the endless capability of carbon sequestration, and regulating different metabolic means in bacteria to orient gene editing of photosynthetic bacteria. Usage of these green technologies executed by particular soil microorganisms enclose photostimulation (Babalola & Glick, 2012), bio fertilization with Nitrogen-fixing microbes, nutrient solubilizing and assembling microbial inoculants (Kalayu, 2019; Park et al., 2010; Smith & Read, 2010; Richardson et al., 2009), rhizome-intervention (Giovannini et al., 2020; Musyoka et al., 2020; Verma & Rawat, 2021) or alteration of stress (Stamenkovic et al., 2018) can tweak the overflow and action of useful microbes, upgrading soil fertility, and the repair of plant growth-promoting microorganisms in the soil.

6.2 *Sustainable Agricultural Practices*

Global hazards to soils remain one of the best concerns and difficulties of the twenty-first century. Most land management practices in metropolitan biological systems compromise soils by reshaping the abiotic climate and increasing erosion, sealing, and compaction of soils that further change natural environmental cycles, for example, water cycling (De Kimpe & Morel, 2000; FAO, 2017). These environmental dangers are generally confounded by altered biotic variables, like low biological activity (Lorenz & Lal, 2009) and soil organic carbon (Bradley et al., 2005), degrading soil state. Soil well-being, or soil quality, in other words, is the ability to work as a living biological system that upholds natural ecological cycles crucial for maintaining healthy ecosystems (Bünemann et al., 2018). Green management practices, like low-influence improvement and green infrastructure (GI), are intended to tackle soil degradation, and support formation, soil well-being, and ecosystem services. Current research evaluating the effects of GI on soil well-being and capabilities essentially centers on bioretention basins, bioswales, or green roofs that utilization designed soil media or stormwater filtration treatment frameworks (Gill et al., 2017; Grover et al., 2013). For instance, several investigations show that GI with designed soil media adjusts soil capability, changing nutrient cycling and expanding greenhouse gas emissions for carbon dioxide, nitrous oxide, and methane (Grover et al., 2013). Other studies contrasting GI utilizing designed soil media with natural soils or unmanaged soils to comprehend influences on soil biological health found higher microbial diversity and activity in GI with engineered soils (Gill et al., 2017). However, few studies have simultaneously evaluated soil physical, chemical, and biological markers in GI that utilize natural, non-designed soils to grasp GI influences on soil well-being and capability, particularly in the dry Southwest (Buzzard et al., 2021). In this way, a comprehensive methodology is important to comprehend what GI frameworks mean for soil well-being through changes in the physical, synthetic, and biological qualities of the soil environment. These practices promise to make good circumstances for beneficial microorganisms, which add to nutrient accessibility, plant growth promotion, disease suppression thus improved soil quality.

6.3 Biogenic Greenhouse Gas (GHG) Mitigation

The greenhouse impact is one of the significant environmental issues problem humans have faced in the twenty-first century. Soil is the biggest carbon repository in the terrestrial environment (Hao et al., 2021); Besides the fact that the soil produces CO_2, the utilization of diesel, gas, and power in farmland practices like cultivating, irrigation, and harvesting additionally causes CO_2 emissions (Deng et al., 2021).

CH_4 is the second most significant GHG after CO_2, with an average lifetime of around 8.75 years in the air and a contribution rate of about 15% of the greenhouse effect. The main emission sources of CH_4 in agriculture are rice and animal husbandry and the anaerobic atmosphere of overflowed rice fields and animal intestines develops positive circumstances for CH_4 production by methanogens. N_2O is another important GHG, representing around 7.9% of the greenhouse effect. Its typical lifetime in the environment is 114 years, and its global warming potential (GWP) is 296 ~ 310 times that of CO_2, which is the fundamental destroyer of stratospheric ozone (Thompson et al., 2019). Nitrogen compost application in horticulture is the principal source of N_2O, and N_2O emissions caused by fertilization account for about 30% of global land emissions. Therefore, decreasing N_2O emissions from farmland soil is critical to ease the greenhouse effect (Cui et al., 2021; Tian et al., 2020).

Currently, scientists have fostered a few measures and green technologies for the decrease in GHG emission in farmland soil, essentially by adding biochar, restoring straw to the field, and applying natural compost or microalgae biofertilizer and soil improvers (such as lime and nitrification inhibitors, etc.) (Zhang et al., 2020; Zuo et al., 2022). Microorganisms assume a vital part in the utilization of these mitigation measures and technologies. An enormous number of studies have shown that temperature is the fundamental element influencing the production and discharges of GHG in soil (Lu et al., 2019; Kellman et al., 2015; Wang et al., 2019). Higher soil temperature can boost the root respiration of crops, speed up the disintegration of

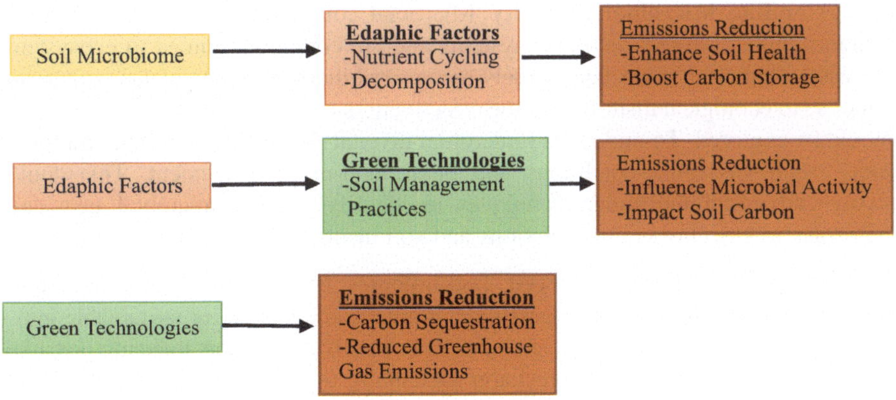

Fig. 2 Interaction of soil microbiome, Edaphic facors and Green

organic matter in the soil, promote the activity of microorganisms in the soil, and thus, step up the diffusion of CO_2 in soil (Carey et al., 2016). Methanogens and methane-oxidizing bacteria mutually decide the emission of CH_4. Within a specific temperature, the metabolic ability of methanogens is positively associated with temperature. At the point when the ambient temperature increases from 20 °C to 35 °C, the emission of CH4 will be twofold. However, current studies have suggested that methanogens have thermal flexibility and microbial action reduces after long-term warming and rises after long-term cooling (Chen et al., 2020). Walker et al. additionally investigated the reaction component of soil microbes to temperature changes through in situ natural warming trials, and the results showed that microbial temperature responsiveness and substrate utilization mutually impacted soil carbon loss by controlling microbial biomass (Walker et al., 2018).

Overall, the interactions between soil microbiome, edaphic factors, and green technology are fundamental to the performance of terrestrial ecosystems and have significant implications for worldwide environmental sustainability. By harnessing these interactions through interdisciplinary research and innovative solutions, can result in the promotion of agricultural resilience, mitigate greenhouse gas emissions, and safeguard ecosystem health for future generations (Fig. 2).

7 Future Directions and Research Needs

The relationship between soil microbiome, edaphic factors, and green technology in emissions reduction is a fascinating area of research with significant implications for sustainable agriculture and environmental management. Future directions and research needs in this field could include:

Understanding Microbial Diversity Microbial communities are responsible for biogeochemical changes in all ecosystems and are complexly linked to ecosystem processes. Although it has been widely recognized that microbiota makeup alters in response to ecological disturbance, the effectiveness of these compositional changes should be explored. To understand microbial diversity's role in the soil microbiome under several edaphic conditions and other land-use management. This could involve advanced molecular techniques to identify specific microbial taxa and their roles in greenhouse gas emissions and soil health.

Assessment of Sustainable Green Technologies Continuous investigation into effective sustainable innovative green technologies is essential for sustainable agriculture and environmental management. This involves utilizing technologies like biosensors, drones, GPS, and precision agriculture to assist farmers in optimizing inputs to curb environmental impacts while increasing productivity and enhancing water usage by directly channelling water to the roots of plants to minimize wastage and improve water retention. Promoting practices such as crop rotation, cover cropping, and boosting soil organic carbon levels to mitigate emissions, enhance soil fertility, and resilience to climate change. Long-term field studies and meta-analyses can provide valuable insights into the practicality and scalability of these approaches.

Incorporation of Omics Approaches To clarify the microbial-mediated processes involved in nutrient cycling and emissions reduction, integrating omics technologies such as metagenomics, meta-transcriptomics, and metabolomics is crucial. This all-inclusive approach can provide deeper insights into microbial community dynamics and their responses to edaphic factors and environmental changes.

Vertical Agriculture Using vertical surfaces to grow crops indoors, using water-based nutrient solutions (hydroponic farming systems) to grow plants without soil to maximize productivity while minimizing land use and water consumption.

Modeling and Prediction Integration of microbial data into predictive models of soil biogeochemical processes and greenhouse gas fluxes. This could enhance our ability to forecast the impacts of climate change and land use practices on soil health and emissions, thereby informing more targeted mitigation strategies.

Knowledge Transfer and Stakeholder Engagement Effective knowledge transfer and stakeholder engagement to bridge the gap between research findings and practical implementation. This involves collaboration between scientists, policymakers, farmers, and industry stakeholders to co-develop and disseminate sustainable soil management practices.

References

Abioye, O. P., Ijah, U. J. J., & Aransiola, S. A. (2017). Phytoremediation of Soil Contaminants by Biodiesel Plant *Jatropha curcas*, Chapter 4. In K. Bauddh et al. (Eds.), *Phytoremediation potential of bioenergy plants*. Springer Nature Singapore Pte Ltd. https://doi.org/10.1007/978-981-10-3084-0_4. https://link.springer.com/book/10.1007%2F978-981-10-3084-0

Aransiola, S. A., Ijah, U. J. J., Abioye, O. P., & Bala, J. D. (2019). Microbial-aided phytoremediation of heavy metals contaminated soil: A review. *European Journal of Biological Research, 9*(2), 104–125. http://www.journals.tmkarpinski.com/index.php/ejbr/article/view/157

Babalola, O. O., & Glick, B. R. (2012). The use of microbial inoculants in African agriculture: Current practice and future prospects. *Journal of Food, Agriculture and Environment, 10*(3&4), 540–549.

Bardgett, R. D., & van der Putten, W. H. (2014). Belowground biodiversity and ecosystem functioning. *Nature, 515*(7528), 505–511.

Bardgett, R. D., Freeman, C., & Ostle, N. J. (2008). Microbial contributions to climate change through carbon cycle feedbacks. *The ISME Journal, 2*(8), 805–814.

Barnard, R. L., Osborne, C. A., & Firestone, M. K. (2013). Responses of soil bacterial and fungal communities to extreme desiccation and rewetting. *The ISME Journal, 7*(11), 2229–2241.

Bernal, M. P., Sommer, S. G., Chadwick, D., Qing, C., Guoxue, L., & Michel, F. C. (2022). Current approaches and future trends in compost quality criteria for agronomic, environmental, and human health benefits. *Advances in Agronomy, 163*, 143–218.

Blanco-Canqui, H., Shaver, T. M., Lindquist, J. L., Shapiro, C. A., Elmore, R. W., Francis, C. A., & Hergert, G. W. (2022). Cover crops and ecosystem services: Insights from studies in temperate soils. *Agronomy Journal, 114*(4), 1–14.

Bradley, R. I., Milne, R., Bell, J., Lilly, A., Jordan, C., & Higgins, A. (2005). A soil carbon and land use database for the United Kingdom. *Soil Use and Management, 21*(4), 363–369.

Bünemann, E. K., Bongiorno, G., Bai, Z., Creamer, R. E., De Deyn, G., de Goede, R., Fleskens, L., Geissen, V., Kuyper, T. W., Mäder, P., Pulleman, M., Sukkel, W., Willem, J., van Groenigen, L., & Brussaard. (2018). Soil quality—A critical review. *Soil Biology and Biochemistry, 120*, 105–125.

Butterbach-Bahl, K., Baggs, E. M., Dannenmann, M., Kiese, R., & Zechmeister-Boltenstern, S. (2020). Nitrous oxide emissions from soils: How well do we understand the processes and their controls? *Philosophical Transactions of the Royal Society B: Biological Sciences, 368*(1621), 20130122.

Buzzard, V., Gil-Loaiza, J., Grachet, N. G., Talkington, H., Youngerman, C., Tfaily, M. M., & Meredith, L. K. (2021). Green infrastructure influences soil health: Biological divergence one year after installation. *Science of the Total Environment, 801*, 149644.

Carey, J. C., Tang, J., Templer, P. H., Kroeger, K. D., Crowther, T. W., Burton, A. J., et al. (2016). Temperature response of soil respiration largely unaltered with experimental warming. *Proceedings of the National Academy of Sciences, 113*(48), 13797–13802.

Chen, H., Zhu, T., Li, B., Fang, C., & Nie, M. (2020). The thermal response of soil microbial methanogenesis decreases in magnitude with changing temperature. *Nature Communications, 11*(1), 5733.

Chen, L., et al. (2023). Precision agriculture technologies for sustainable nutrient management. *Journal of Sustainable Agriculture, 29*(1), 56–69.

Cui, X., Zhou, F., Ciais, P., Davidson, E. A., Tubiello, F. N., Niu, X., et al. (2021). Global mapping of crop-specific emission factors highlights hotspots of nitrous oxide mitigation. *Nature Food, 2*(11), 886–893.

Cui, X., Liang, J., Liu, J., & Tang, H. (2022). Managing soil moisture and organic carbon to mitigate nitrous oxide emissions in agricultural soils: A review. *Agriculture, Ecosystems & Environment, 323*, 107682.

De Kimpe, C. R., & Morel, J. L. (2000). Urban soil management: A growing concern. *Soil Science, 165*(1), 31–40.

Delgado-Baquerizo, M., Maestre, F. T., Reich, P. B., Trivedi, P., Osanai, Y., Liu, Y. R., & Hamonts, K. (2016). Carbon content and climate variability drive global soil bacterial diversity patterns. *Ecological Monographs, 86*(3), 373–390.

Delgado-Baquerizo, M., Oliverio, A. M., Brewer, T. E., Benavent-González, A., Eldridge, D. J., Bardgett, R. D., & Maestre, F. T. (2020). A global atlas of the dominant bacteria found in soil. *Science, 359*(6380), 320–325.

Deng, C., Li, R., Xie, B., Wan, Y., Li, Z., & Liu, C. (2021). Impacts of the integrated pattern of water and land resources use on agricultural greenhouse gas emissions in China during 2006–2017: A water-land-energy-emissions nexus analysis. *Journal of Cleaner Production, 308*, 127221.

FAO, I. (2017). *Global assessment of the impact of plant protection products on soil functions and soil ecosystems*. FAO.

Fierer, N., Strickland, M. S., Liptzin, D., Bradford, M. A., & Cleveland, C. C. (2012). Global patterns in belowground communities. *Ecology Letters, 12*(11), 1238–1249.

Geisseler, D., Lazicki, P. A., & Scow, K. M. (2022). Fertilizer effects on soil microbial communities and soil organic matter. *Soil Biology and Biochemistry, 151*, 108054.

Gill, A. S., Lee, A., & McGuire, K. L. (2017). Phylogenetic and functional diversity of total (DNA) and expressed (RNA) bacterial communities in urban green infrastructure bioswale soils. *Applied and Environmental Microbiology, 83*(16), e00287–e00217.

Giovannini, L., Palla, M., Agnolucci, M., Avio, L., Sbrana, C., Turrini, A., & Giovannetti, M. (2020). Arbuscular mycorrhizal fungi and associated microbiota as plant biostimulants: research strategies for the selection of the best-performing inocula. *Agronomy, 10*(1), 106.

Grover, S. P., Cohan, A., Chan, H. S., Livesley, S. J., Beringer, J., & Daly, E. (2013). Occasional large emissions of nitrous oxide and methane observed in stormwater biofiltration systems. *Science of the Total Environment, 465*, 64–71.

Hao, Z., Zhao, Y., Wang, X., Wu, J., Jiang, S., Xiao, J., et al. (2021). Thresholds in aridity and soil carbon-to-nitrogen ratio govern the accumulation of soil microbial residues. *Communications Earth & Environment, 2*(1), 236.

Hartmann, M., Frey, B., Mayer, J., Mäder, P., & Widmer, F. (2015). Distinct soil microbial diversity under long-term organic and conventional farming. *The ISME Journal, 9*(5), 1177–1194.

Ho, A., Di Lonardo, D. P., & Bodelier, P. L. E. (2022). Revisiting life strategy concepts in environmental microbial ecology. *FEMS Microbiology Ecology, 98*(7), fiac054.

Holland, J. E., Bennett, A. E., Newton, A. C., White, P. J., McKenzie, B. M., George, T. S., et al. (2018). Liming impacts on soils, crops and biodiversity in the UK: A review. *Science of the Total Environment, 610*, 316–332.

Kalayu, G. (2019). Phosphate solubilizing microorganisms: promising approach as biofertilizers. *International Journal of Agronomy, 2019*, 1–7.

Kellman, L., Myette, A., & Beltrami, H. (2015). Depth-dependent mineral soil CO 2 production processes: Sensitivity to harvesting-induced changes in soil climate. *PLoS One, 10*(8), e0134171.

Lauber, C. L., Strickland, M. S., Bradford, M. A., & Fierer, N. (2009). The influence of soil properties on the structure of bacterial and fungal communities across land-use types. *Soil Biology and Biochemistry, 41*(12), 2407–2415.

Lee, S. H., Kim, S. Y., & Kang, H. (2021). Soil methanotrophs: Recent insights on their metabolic potential and microbial ecology. *Soil Biology and Biochemistry, 157*, 108238.

Lehmann, J., & Joseph, S. (Eds.). (2021). *Biochar for environmental management: Science, technology and implementation*. Routledge.

Leh-Togi Zobeashia, S. S., Aransiola, S. A., Ijah, U. J. J., & Abioye, O. P. (2018). Anaerobic digestion and agricultural application of organic wastes. *Advances in Environmental Research, 7*(2), 73–85. http://www.techno-press.org/content/?page=article&journal=aer&volume=7&num=2&ordernum=1

Li, X., et al. (2023). Role of soil microbial communities in carbon sequestration in agroecosystems. *Journal of Environmental Science, 35*(3), 412–425.

Liang, C., Balser, T. C., & Johnson, S. E. (2022). Soil texture and microbial community composition under varied management. *Applied Soil Ecology, 169*, 104234.

Liu, J., Wei, D., Liu, X., Zhao, Y., & Liu, Q. (2021). The effects of crop rotation and intercropping on soil microbial diversity and ecosystem functions. *Applied Soil Ecology, 167*, 104036.

Liu, Y., Guo, Q., Wang, H., & Liu, Y. (2023a). Soil organic matter drives the diversity and function of soil microbial communities. *Soil Biology and Biochemistry, 166*, 108570.

Liu, J., Wang, J., Song, Y., & Zhang, X. (2023b). Microbial community succession and carbon stabilization in soil aggregates under different land uses. *Soil Biology and Biochemistry, 164*, 108469.

Lobell, D. B., Roberts, M. J., Schlenker, W., & Braun, N. (2022). Precision agriculture technologies and climate change adaptation. *Science, 376*(6593), 353–357.

Lorenz, K., & Lal, R. (2009). Biogeochemical C and N cycles in urban soils. *Environment International, 35*(1), 1–8.

Lu, S. B., Xu, Y., Fu, X. P., Xiao, H., Ding, W., & Zhang, Y. J. (2019). Patterns and drivers of soil respiration and vegetation at different altitudes in Southern China. *Applied Ecology & Environmental Research, 17*(2), 3097–3106.

Lykogianni, M., Bempelou, E., Karamaouna, F., & Aliferis, K. A. (2021). Do pesticides promote or hinder sustainability in agriculture? The challenge of sustainable use of pesticides in modern agriculture. *Science of the Total Environment, 795*, 148625.

Morrissey, E. M., Gutknecht, J. L., & McHugh, T. A. (2022). Anaerobic microbial activity in waterlogged soils. *Frontiers in Environmental Science, 10*, 841977.

Musyoka, D. M., Njeru, E. M., Nyamwange, M. M. E., & Maingi, J. M. (2020). Arbuscular mycorrhizal fungi and Bradyrhizobium co-inoculation enhance nitrogen fixation and growth of green grams (Vigna radiata L.) under water stress. *Journal of Plant Nutrition, 43*(7), 1036–1047.

Or, D., Phutane, S., & Dechesne, A. (2007). Extracellular polymeric substances affecting pore-scale hydrologic conditions for bacterial activity in unsaturated soils. *Vadose Zone Journal, 6*(2), 298–305.

Paredes-Páliz, K. I., Bello-Lara, J. E., Pino-Benítez, N., & Pérez-Pérez, C. (2022). Phytoremediation and the soil microbiome: A synergistic relationship to improve soil health. *Science of the Total Environment, 816*, 151634.

Park, J., Bolan, N., Megharaj, M., & Naidu, R. (2010). Isolation of phosphate-solubilizing bacteria and characterization of their effects on lead immobilization (< Special Issue> International symposium: Challenges to soil degradation towards sustaining life and environment, Tokyo Metropolitan University Symposium Series No. 2, 2009). *Pedologist, 53*(3), 67–75.

Richardson, A. E., Barea, J. M., McNeill, A. M., & Prigent-Combaret, C. (2009). Acquisition of phosphorus and nitrogen in the rhizosphere and plant growth promotion by microorganisms. *Plant and Soil, 321*, 305–339.

Rousk, J., Brookes, P. C., & Bååth, E. (2010a). The microbial pH parameter and the link between bacterial growth and soil bacterial community composition. *Soil Biology and Biochemistry, 42*(11), 1979–1986.

Rousk, J., Brookes, P. C., & Bååth, E. (2010b). The microbial PLFA composition as affected by pH in an arable soil. *Soil Biology and Biochemistry, 42*(3), 516–520.

Schimel, J., Balser, T. C., & Wallenstein, M. (2007). Microbial stress-response physiology and its implications for ecosystem function. *Ecology, 88*(6), 1386–1394.

Six, J., Bossuyt, H., Degryze, S., & Denef, K. (2004). A history of research on the link between (micro)aggregates, soil biota, and soil organic matter dynamics. *Soil and Tillage Research, 79*(1), 7–31.

Smith, S. E., & Read, D. J. (2010). *Mycorrhizal symbiosis*. Academic.

Smith, R. G., Atwood, L. W., & Warren, N. D. (2021). The effects of cover crop diversity and management on soil health and multifunctionality. *Agriculture, Ecosystems & Environment, 315*, 107447.

Smith, J., et al. (2024). Biochar amendments for soil fertility improvement and greenhouse gas emissions reduction. *Environmental Technology, 17*(2), 198–210.

Stamenkovic, S., Beskoski, V., Karabegovic, I., Lazic, M., & Nikolic, N. (2018). Microbial fertilizers: a comprehensive review on current findings and future perspectives. *Spanish Journal of Agricultural Research, 16*, e09R01.

Terrer, C., Jackson, R. B., Prentice, I. C., Keenan, T. F., Kaiser, C., Vicca, S., & Franklin, O. (2021). Mycorrhizal association as a primary control of the CO_2 fertilization effect. *Science, 353*(6280), 72–74.

Thompson, R. L., Lassaletta, L., Patra, P. K., Wilson, C., Wells, K. C., Gressent, A., et al. (2019). Acceleration of global N2O emissions seen from two decades of atmospheric inversion. *Nature Climate Change, 9*(12), 993–998.

Tian, H., Xu, R., Canadell, J. G., Thompson, R. L., Winiwarter, W., Suntharalingam, P., et al. (2020). A comprehensive quantification of global nitrous oxide sources and sinks. *Nature, 586*(7828), 248–256.

Tian, H., Shi, W., Song, Z., & Zhang, Z. (2021). Effects of soil texture on the structure and function of microbial communities under aerobic and anaerobic conditions. *Soil Biology and Biochemistry, 156*, 108196.

Tkacz, A., Hortala, M., & Poole, P. S. (2018). Absolute quantitation of microbiota abundance in environmental samples. *Microbiome, 6*(1), 110.

Van der Wal, A., Geerts, R. H. E. M., Korevaar, H., Schouten, A. J., op Akkerhuis, G. J., Rutgers, M., & Mulder, C. (2009). Dissimilar response of plant and soil biota communities to long-term nutrient addition in grasslands. *Biology and Fertility of Soils, 45*, 663–667.

Verma, P., & Rawat, S. (2021). Rhizoremediation of heavy metal-and xenobiotic-contaminated soil: an eco-friendly approach. In *Removal of emerging contaminants through microbial processes* (pp. 95–113). Springer.

Vessey, J. K. (2021). Plant growth-promoting rhizobacteria as biofertilizers. *Plant and Soil, 383*(1–2), 31–45.

Walker, T. W., Kaiser, C., Strasser, F., Herbold, C. W., Leblans, N. I., Woebken, D., et al. (2018). Microbial temperature sensitivity and biomass change explain soil carbon loss with warming. *Nature Climate Change, 8*(10), 885–889.

Wang, Q., Zhao, X., Chen, L., Yang, Q., Chen, S., & Zhang, W. (2019). Global synthesis of temperature sensitivity of soil organic carbon decomposition: Latitudinal patterns and mechanisms. *Functional Ecology, 33*(3), 514–523.

Wang, X., Liu, Q., He, W., & Zhou, B. (2022). Composting: A sustainable approach to enhancing soil microbiome function and health. *Frontiers in Microbiology, 13*, 833978.

Wang, Q., Wu, W., Ding, Q., & Liu, X. (2023). Organic amendments enhance soil microbial diversity and function in agricultural ecosystems. *Agriculture, Ecosystems & Environment, 324*, 107733.

Zhang, B., Zhou, M., Lin, H., Ntacyabukura, T., Wang, Y., & Zhu, B. (2020). Effects of different long-term crop straw management practices on ammonia volatilization from subtropical calcareous agricultural soil. *Atmospheric and Oceanic Science Letters, 13*(3), 232–239.

Zhang, Y., et al. (2022). Influence of edaphic factors on microbial diversity in contaminated soils. *Soil Biology & Biochemistry, 48*(4), 621–635.

Zhang, H., Wu, X., Li, J., Zhao, T., & Zhang, Y. (2023). Biochar amendments enhance soil microbial biomass and activity in agricultural systems: A meta-analysis. *Agriculture, Ecosystems & Environment, 334*, 108003.

Zuo, S., Wu, D., Du, Z., Xu, C., & Wu, W. (2022). Effects of white-rot fungal pretreatment of corn straw return on greenhouse gas emissions from the North China Plain soil. *Science of the Total Environment, 807*, 150837.

The Place of Soil Microbiomes in Carbon-Storing Soil and Green Technology

Ignatius Mzungu, Olalekan Adekola, Fanen Terdoo, and Christian Bemgba Anjembe

Abstract The development of green technology and carbon sequestration are greatly aided by soil microbiomes, which also present viable means of mitigating the effects of climate change. This chapter begins with an introduction to the significance of soil microbiomes in environmental sustainability. It then delves into the diverse and complex nature of soil microbiomes, exploring their composition and functions. The chapter highlights the vital connection between soil microbiomes and carbon sequestration, explaining how these microorganisms contribute to reducing atmospheric carbon dioxide levels. It details the specific mechanisms by which soil microbiomes sequester carbon, including organic matter decomposition, stabilization of soil organic carbon, and symbiotic relationships with plant roots. Techniques for enhancing the capacity of soil microbiomes to sequester carbon are examined, emphasizing practices such as bioaugmentation, organic amendments, crop rotation, and reduced tillage. The integration of soil microbiomes with green technology is explored, showcasing innovations like biochar application, microbial inoculants, and precision agriculture. These techniques enhance soil health and carbon storage while promoting sustainable agricultural practices. Real-world examples of how improving soil microbiomes can result in notable gains in carbon sequestration and general ecosystem health are given through case studies and success stories. Despite the potential, challenges such as the variability of soil microbiomes, the need for tailored strategies, and the long-term sustainability of interventions are discussed. The final section of the chapter highlights the necessity of ongoing studies, interdisciplinary cooperation, and legislative

I. Mzungu (✉)
Department of Microbiology, Federal University Dutsin-Ma, Katsina State, Nigeria
e-mail: imzungu@fudutsinma.edu.ng

O. Adekola
Department of Geography, York St. John's University, York, UK

F. Terdoo
Department of Geography, Federal University Dutsin-Ma, Katsina State, Nigeria

C. B. Anjembe
Department of Soil Science, Joseph Sarwuan Tarka University, Makurdi, Benue State, Nigeria

S. A. Aransiola et al. (eds.), *Soil Microbiome in Green Technology Sustainability*,
https://doi.org/10.1007/978-3-031-71844-1_11

backing in order to fully utilize soil microbiomes' potential for carbon sequestration and green technologies. By leveraging these microorganisms, we can develop effective strategies for climate change mitigation and promote environmental sustainability.

Keywords Soil microbiomes · Carbon sequestration · Green technology · Climate change mitigatio · Bioaugmentation · Sustainable agriculture · Microbial technology

1 Introduction

There has been a steady rise in CO_2 levels with an attendant rise in global temperatures over the past century. This increase, along with weather changes is predicted to continue according to a recent US national climate assessment (Jansson & Hofmockel, 2020), and weather patterns are expected to become extremely erratic. A significant portion of the cycling of various nutrients, including soil organic carbon (SOC), is facilitated by soil microbes. They are essential to climate feedback when it comes to the production or consumption of greenhouse gases like CO_2, CH_4, and N_2O.

An integral part of the soil ecosystem, the soil microbiome can affect a variety of activities, including nutrient turnover, carbon storage, plant growth and stress tolerance, soil formation, and fertility (Naylor et al., 2022). Mitigating climate change requires carbon sequestration, or the storage of carbon by soil microbiomes. Soil microbial biomass is the living mass that makes up the organic matter composition of the soil. Half of the microbial biomass occupy the uppermost 15 cm the top soil. They are important for the decomposition of organic matter and the creation of soil carbon pools, which include fewer than 4 percent of the total organic matter. Estimating soil microbial biomass carbon (MBC) stocks and understanding how these stocks relate to climatic, environmental, and soil biochemical variables can be beneficial for both soil management and adaptation/mitigation of climate change (Das et al., 2023).

Green technology, with its focus on sustainability, intersects significantly with soil health and microbiome function. The carbon (C) and nutrient cycles in Earth's terrestrial ecosystems are primarily influenced by microbes; however, despite mounting evidence that explicitly including microbial processes improves model prediction and lowers uncertainty in terrestrial systems, most Earth system models meant to direct climate change adaptation and mitigation strategies have lacked specific representation of soil microorganisms (Marschmann et al., 2024).

It has been difficult to predict whether the soil would become a source or sink of greenhouse gases under future climate scenarios due to the variety in microbial reactivity between soil locations and the uncertainty around changes in soil carbon and nitrogen stocks (Wang et al., 2018). Thus, it remains challenging to incorporate soil microbial ecology into landscape-scale climate models even though its importance in predicting future climate consequences is widely recognized (Wleder et al., 2013).

In Africa, the importance of soil microbiomes in carbon sequestration and green technology is particularly pronounced due to the continent's unique ecological diversity and socio-economic challenges. African soils are rich in biodiversity, hosting a wide array of microbial species with untapped potential for carbon storage and sustainable development. Leveraging the power of soil microbiomes in Africa holds immense promise for addressing key environmental issues such as soil degradation, deforestation, and food security. By harnessing these natural resources effectively, African nations can not only contribute significantly to global carbon sequestration efforts but also foster innovation and resilience within their economies and ecosystems.

Africa's ecological diversity is a testament to the continent's richness and complexity, encompassing a wide range of ecosystems from dense rainforests to vast savannas, arid deserts, and mountainous regions. This diversity is driven by a combination of factors including climate variations, geological history, and human influence over millennia. With regard to biodiversity, Africa harbors an astounding variety of plant and animal species, many of which are unique to Earth. This diversity extends to microbial life as well, with African soils harboring a multitude of microorganisms crucial for nutrient cycling, soil fertility, and ecosystem functioning.

This ecological richness forms the foundation for Africa's potential for sustainable development. The continent's natural resources, when managed responsibly, can support a wide range of economic activities while preserving environmental integrity. Sustainable agriculture, renewable energy projects, conservation initiatives, and ecotourism are just some examples of sectors where Africa can leverage its biodiversity and ecosystems for long-term growth and prosperity. Moreover, harnessing indigenous knowledge and traditional practices alongside modern technologies can further enhance Africa's capacity for sustainable development, ensuring a harmonious balance between human well-being and environmental conservation.

The chapter aims to delve into the critical role of soil microbiomes in storing carbon and their contributions to advancing green technology, with the objectives to: (i) explore Soil Microbiomes, elucidating their diversity, functions, and mechanisms of action in carbon storage; (ii) highlight Africa's Ecological Diversity by showcasing Africa's vast ecological diversity as a foundation for sustainable development initiatives; (iii) examine Green Technology Innovations by reviewing and discussing innovative green technologies relevant to soil microbiomes, such as biochar production, microbial inoculants, and soil carbon enhancement techniques; (iv) address Challenges and Opportunities by identifying key challenges hindering the harnessing of soil microbiomes in Africa, including technological barriers, funding constraints, and knowledge gaps; and (v) Promote Integration of Indigenous Knowledge. The chapter further emphasizes the importance of integrating Indigenous knowledge and traditional practices into modern soil management strategies, and (vi) Provide Policy Recommendations: Lastly, the chapter aims to offer practical policy recommendations for policymakers, researchers, and stakeholders involved in sustainable development and environmental conservation efforts in Africa.

2 Understanding Soil Microbiomes

Soil microbiomes are made up of a diverse range of microorganisms, such as bacteria, fungi, viruses, archaea, and protozoa. These microorganisms are crucial for maintaining soil structure, supporting plant health, and cycling nutrients. Specific microbial communities—such as bacteria and fungi—are necessary for the soil matrix's ability to sequester carbon. The nonliving component of the soil is also very diverse, containing discrete pore spaces that can hold water, air, or both, as well as materials that can serve as hotspots for the development of microbial life. The soil environment is quite dynamic when you take into account the impact of soil flora and fauna (such as earthworms and insects), temperature variations, and fluctuating redox levels (Jansson & Hofmockel, 2020; Stone et al., 2023).

As one of Earth's most valuable resources, the soil accounts for preservation of biodiversity and management of ecosystem functions in both controlled and natural environments.

Soil microbiomes play important role in controlling soil processes; however, their activities have received less attention in a few systems such as forest and desert ecosystems. The composition of soil microflora greatly depends on several elements that may be physical and biological, including pH, soil structure, carbon content, soil texture, and total plant cover. However, these could greatly vary depending on the ecosystems and the respective inhabitants. Recent advances in molecular microbial ecology such as metagenomics studies provided great insights into the biodiversity of the soil environment, including the non-culturable component of the soil microbiota. Despite the remarkable progress made by molecular biology in clarifying the taxonomy of organisms, the application of this technology in soil ecology has remained a challenge.

First, a multitude of interconnected metabolic pathways that can be carried out by a range of microbial taxonomic groups are responsible for many of the biogeochemical processes occurring in the soil ecosystem rather than being the result of a single metabolic route. (Pepe-Ranney et al., 2016). Second, attempts to connect functional processes to particular soil microbiomes may be thwarted by the abundance of individual genes, such as the 16S and 18S rRNA genes and other functional genes from dormant bacteria and fungi respectively.

Third, there are issues with methods that come with associating certain taxa with specific metabolic processes; these issues include challenges with precise gene annotation (Schnoes et al., 2009), given the quick turnover of proteins, metabolites, and transcripts (Jansson et al., 2012). Fourth, although sequence-based approaches are often employed to count the functional genes or microbial taxa in soil, they typically only provide information about the relative abundances of these elements, not always their absolute abundances.

The many microbial communities in soil can react to varying degrees of environmental change depending on characteristics such as evolution, genetic composition, physiological status with acclimation and adaptation, as well as previous stress exposure. The survival or fitness of specific taxa may be impacted by this reaction

(Zhang et al., 2022; Stone et al., 2023). Additionally, because microorganisms might have variable metabolisms that include using several pathways for energy generation or because they can be quiescent at any one time, their taxonomic diversity does not always correspond to their functional diversity. Crucially, it is extremely difficult to anticipate the fate of carbon with precision due to the absence of knowledge regarding one or more clades or domains of life as well as their simultaneous exposure to many stresses.

3 Carbon Sequestration by Soil Microbiomes and Climate Change

Geological carbon dioxide storage is a rapidly evolving technology that has shown promise to mitigate rising carbon emissions (Hicks et al., 2017). Because soil carbon storage lowers the emission of greenhouse gases, it directly enhances climate change mitigation.

It is important to note that the relevance of carbon-storing soil extends to sustainable agricultural practices, highlighting its multifaceted benefits.

Plant residues were once regarded as the ultimate source of organic carbon in the soil, but more recent researches have established the significant contribution of microbiome to soil organic carbon (SOC) (Kallenbach et al., 2015; Liang et al., 2016). Reported that 50–80% SOC is formed from the necrotic mass of dead microorganisms (Liang et al., 2016). The soil microbiota is clearly the most taxonomically diverse living community in the biosphere, housing at least 25% of Earth's total biodiversity (Sokol et al., 2022).

The changes in the world's soil microbiome brought about by climate change have garnered a lot of attention in recent times (Dubey et al., 2019; Naylor et al., 2020; Hutchins et al., 2019). Some of Earth's most temperature-sensitive biomes, like polar deserts and semi-arid arable lands, have already seen significant changes in their microecology as a result of human-caused climate change (Jansson & Hofmockel, 2020). The atmospheric carbon stocks may be reduced by biotic and/or abiotic activities that sequester carbon into stable, non-gaseous forms. Through the assimilation of atmospheric carbon dioxide, primarily by plants but also by autotrophic soil bacteria, carbon is added to the soil.

Bioavailable carbon forms cycle and create recalcitrant non-bioavailable forms due to biochemical changes aided by microbial metabolism of carbon and further exchange between microbial communities. Soils with greater biodiversity can hold onto carbon more effectively (Lal, 2004). Soil fungi and bacteria collaborate to produce carbon polymers that are necessary for the creation of soil aggregates and occlude soil carbon throughout the carbon sequestration process. One approach to storing carbon in the soil is to harness the unrealized metabolic capacity of the soil microbiota to generate new reactions that promote carbon sequestration. When single or groups of microbes catalyze these processes, the carbon metabolic pathways can be steered toward more stable and resistant end products (Hicks et al., 2017).

Soil microbes may also be modified to improve their ability to absorb and store carbon. It has been demonstrated that microbial residues, which could comprise dead biomass (necromass) or macromolecules from extracellular polymeric compounds, linger as organic carbon in the soil (Kapilan et al., 2018). Jansson et al. (2010) have suggested the addition of pyrolyzed carbon, or biochar, as amendments to the soil to increase the capacity for stable carbon storage. Plant-soil interactions with microbiomes of soil rhizosphere can also be manipulated to promote storage of carbon in the soil (Wallenstein, 2017; Jansson et al., 2018).

4 Mechanisms of Carbon Sequestration by Soil Microbiomes

Soil microorganisms are vital to carbon sequestration, which reduces the consequences of climate change, by storing organic carbon in the soil through a number of activities. Globally, there is significant worry over the rising amounts of carbon dioxide (CO_2) and greenhouse gases (GHGs) in the atmosphere (Grover et al., 2015). Carbon dioxide (CO_2) is a major greenhouse gas (GHG) that makes up 63% of all GHG emissions. However, by sequestering carbon, biological systems may be able to lessen the consequences of increasing atmospheric carbon levels. (Grover et al., 2015). Throughout the whole carbon cycle, microbes serve as gatekeepers, facilitating the breakdown, transformation, and stability of organic matter (Raza et al., 2023).

It has become clear that improving soils, increasing crop productivity, slowing the emission of carbon dioxide into the atmosphere and the rate of climate change, and preserving the health of downstream ecosystems may all be achieved by adding more organic carbon stores to agricultural soils (Basile-Doelsch et al., 2020). The process of taking carbon out of the atmosphere and putting it in a reservoir is called sequestration. It is the means of removing carbon dioxide from the atmosphere and storing it to slow down climate change and global warming (Dhanwantri et al., 2014). The measurement of this is usually in terms of the overall organic carbon storage of a soil (Buragohain, 2019). Carbon pools are another name for these carbon reservoirs or storages. A process or system that has the ability to store or release carbon is called a carbon pool. It may be brought on by humans or be natural (Dhanwantri et al., 2014). Carbon pools in forests are complex mixtures of minerals and both living and dead organic materials.

The two main processes of soil carbon sequestration are (1) physicochemical protection and (2) biochemical change; other mechanisms include land use, various agricultural techniques, crop, water, and soil management. It consists of (i) perennial systems that promote healing and (ii) highly adaptable species. (Buragohain, 2019).

The amount of microbial biomass, microbial community structure, microbial by-products, and soil characteristics like texture, clay mineralogy, pore size distribution, and aggregate dynamics interact to determine the microbial contribution to carbon sequestration (Buragohain, 2019).

4.1 Decomposition Mechanism

Soil microbes decompose organic matter, such as cellulose and lignin biopolymers from dead plant material, roots, and animal residues, through their enzymatic activities. Fungal and bacterial degradation of cellulose is quite common. But lignin is more resilient, and only a certain class of fungi that generate extracellular lignin peroxidases can completely degrade it (Buragohain, 2019). During decomposition, microbes break down these complex organic molecules into simpler compounds, releasing carbon dioxide (CO_2) through microbial respiration. However, a significant portion of the organic carbon become sequestered and transformed into stable forms of soil organic carbon by soil micro biomass. Condensation or polymerization occur after decomposition. During this phase, the simpler molecules created during breakdown are converted into new, more chemically resistant ones. These chemically changed fractions are created while reducing sugars to make melanin-type chemicals or condensation reaction between amino compounds (acids and sugars) and quinones (Buragohain, 2019). Through a variety of direct and indirect processes, soil microbes are essential for the retention and circulation of soil carbon (C) (Plaza et al., 2013).

4.1.1 Direct Mechanisms

Direct mechanisms involve processes that are microbial-mediated which contribute directly to soil carbon content as well as its stabilization. Microorganisms absorb and use plant carbon for growth and emit CO_2 through decomposition processes (Bai & Cotrufo, 2022; Hu et al., 2019). One significant direct route of soil C sequestration mediated by microbes is soil aggregation. As a feature of healthy soil that enhances gas exchange, water infiltration, soil structure, and water holding capacity, soil microaggregates are becoming more widely acknowledged. By aiding in the transfer of C molecules from labile to recalcitrant pools, soil aggregates prolong residence times and prevent SOC from degrading (Hu et al., 2019). Through their secretions and biomass, microorganisms—more especially, fungus and bacteria—bind soil particles together to form micro and macro aggregates. Numerous important taxa were found to be involved in studies, including bacteria (Rhizobium sp.) and fungi (Mucor sp., Fusarium sp., Gibberella sp.) (Hu et al., 2019).

4.1.2 Indirect Mechanisms

Indirect mechanisms involve enhancing growth of plant and resilience through microbial activities, leading to increased biomass input into the soil and subsequent cycling into recalcitrant carbon pools. Mycorrhizal fungi and plant growth-promoting microbes (PGPM) are two significant groups in this regard. Mycorrhizal fungi not only aid in necromass accumulation and soil aggregation but also enhance plant nutrient uptake and availability, thus indirectly influencing soil carbon cycling. Similarly, PGPM supports plant growth and resilience by providing plant-available nutrients, producing phytohormones, and battling phytopathogens as an agent of biological control (Hu et al., 2019). Although, relatively recent, plant growth-promoting microorganisms (PGPMs) is an effective tactic to mitigate the consequences of climate change, enhancing plant growth in stressful situations and pest suppression (Dhanwantri et al., 2014).

5 Techniques for Enhancing Soil Microbiomes for Carbon Sequestration

Recently, the subject of climate change has garnered significant attention from scientists, who are increasingly drawn to soil organic carbon due to its recognition as a significant player in the global carbon cycle and its potential for mitigation (Basile-Doelsch et al., 2020). At the field scale, soil microbial communities are critical to plant nutrition and health as well as to the functioning of ecosystems. They take part in the global element cycling on a bigger scale. Additionally, they contribute to the breakdown of xenobiotics, the movement of organic matter, and the formation of soil aggregates. The ecological state of soils is determined by the makeup and activity of soil microorganisms (Demyanyuk et al., 2020).

Microbes employ several strategies to sequester carbon, including; bacterial and fungal dominance, mycorrhizal association, efficient CO_2 capture by microalgae, and oligotrophs versus copiotrophs (Hu et al., 2019).

5.1 Molecular Techniques to Enhance Carbon Sequestration

One method that shows promise for reducing soil greenhouse gas emissions is microbial engineering. But before being introduced into uncontrolled or natural environments, it would be important to have a solid grasp of microbial populations and effective biosafety regulations (Mukhtar et al., 2023). CRISPR, or clustered regularly interspersed short palindromic repeats, is a gene editing technique that modifies microbial function to improve crop productivity and sequester carbon.

More attention is directed to the CRISPR recently. It can be used to develope specialized microbes, improving the nutritional makeup of the soil and altering gas fluxes within the agricultural environment all at once. For example, changing the biogeochemical process and lowering greenhouse gas emissions from agriculture can be achieved by selective breeding, genetic alteration of symbiotic bacteria in cattle's digestive tracts, and adding a mixture of engineered microorganisms to the soil. (Mukhtar et al., 2023). Fixen et al. (2016) reported that an interesting use of microbial engineering is the modified strain of *Rhodopseudomonas palustris*, which demonstrated the capacity to transform CO_2 into CH_4. Similar to this, *Caulobacter vibrioides* has been altered to promote the subsurface conversion of CO_2 into calcium carbonate ($CaCO_3$) (Krieger, 2012; Mukhtar et al., 2023).

Practices such as organic farming positively impact soil microbiomes and their carbon sequestration potential. The application of biochar and microbial inoculants are effective strategies for enhancing carbon-storing soil.

6 Soil Microbiomes and Green Technology

Integrating soil microbiomes into sustainable land management aligns with green technology goals. Biotechnological applications leverage soil microbiomes to enhance microbial activity, supporting bioenergy production and waste management. Green technologies relevant to soil microbiomes encompass a range of approaches aimed at enhancing soil health, fertility, and carbon sequestration while minimizing environmental impact.

Biochar Production Biochar is a type of charcoal produced from biomass through a process called pyrolysis. It is rich in carbon and has a porous structure that provides a habitat for beneficial soil microbes. Biochar helps improve soil structure, water retention, and nutrient cycling. Additionally, it can sequester carbon in the soil ranging from hundreds to thousands of years, making it a valuable tool for climate change mitigation. When it comes to increasing the organic carbon content in soils impacted by salt and those that are not, biochar is the most effective organic amendment that has been researched. This could have favorable impacts on soil health and crop productivity (Farooqi et al., 2024).

Microbial Inoculants These are formulations containing beneficial microbes like fungi, bacteria, and archaea applied to the soil to enhance its biological activity. Examples include nitrogen-fixing bacteria that improve soil fertility, mycorrhizal fungi that create symbiotic interactions with plant roots, and plant growth-promoting rhizobacteria (PGPR) that increase plant resistance to disease and improve nutrient uptake. Microbial inoculants can improve soil health, nutrient cycling, and crop productivity while reducing the need for synthetic fertilizers and pesticides. As a means of mitigating the effects of drought, microorganisms can enhance soil water retention. For example, in dry situations, they can create extracellular polymeric substances (EPS) to seal soil pores and retain soil water. Plants can operate as carbon sinks for their own carbon by using soil microorganisms to absorb carbon emitted from their roots and store it as cellular biomass or transform it into stable metabolites.

Furthermore, dead microbial biomass, or necromass, can also be a source of stored carbon in soils (Jansson et al., 2018).

Compost and Vermicompost Composting is the microbial decomposition of organic materials, including crop residues, manure, and food waste (Aransiola et al., 2023). However, in vermicomposting, organic material is broken down by earthworms, usually red wigglers (*Eisenia fetida*). The resulting compost is rich in nutrients and beneficial microorganisms, which improve soil structure, fertility, and microbial diversity. Vermicomposting is a variant that uses earthworms to accelerate the composting process and produce nutrient-rich vermicompost. The addition of vermicompost has been shown to significantly increase soil organic carbon content, leading to improvements in crop productivity (Farooqi et al., 2024; Aransiola et al., 2024).

Cover Crops and Crop Rotation By adding organic matter, fixing nitrogen, and lowering erosion, planting cover crops like legumes, grasses, and other non-cash crops can enhance soil health. Crop rotation, which involves growing different crops in succession on the same piece of land, helps preserve soil fertility, break the cycle of pests and diseases, and enhance soil structure.

Precision Agriculture and Soil Sensors Farmers can now more accurately monitor soil moisture, nitrogen levels, and other soil factors through technologies like satellite photography, drones, and soil sensors. This information allows for targeted application of water, fertilizers, and other inputs, optimizing resource use and minimizing environmental impacts.

Agroforestry and Silvopasture Agroforestry systems combine trees with crops or livestock and offer a number of advantages, including better soil health, higher biodiversity, storage of carbon dioxide, and increased climate change resilience. Silvopasture combines trees with grazing pastures, benefiting both livestock and soil health.

Conservation Tillage and No-Till Farming These practices minimize soil disturbance by reducing or eliminating plowing and tillage. They help preserve soil structure, organic matter, and microbial communities, leading to improved water retention, erosion control, and carbon sequestration.

These green technologies work synergistically to promote sustainable agriculture, enhance soil microbiomes, and mitigate climate change as well as enhance adaptation efforts. Integrating approaches such as these into agricultural systems can lead to more resilient and productive food production while safeguarding environmental resources.

7 Case Studies and Success Stories

Case studies demonstrate the practical application of green technology in enhancing soil microbiomes for sustainable outcomes. Here are a few case studies and examples showcasing the successful adoption of green technologies in Africa for soil improvement and carbon sequestration:

7.1 Africa Soil Information Service (AfSIS)—Kenya

This programme draws funds from the Bill and Melinda Gates Foundation, it is a project that has successfully implemented green technologies for soil improvement and carbon sequestration in Kenya and other African countries. By using advanced soil mapping techniques, they identified degraded soils in various regions and implemented soil restoration practices. This includes using agroforestry systems, organic amendments, and biochar to improve soil fertility, health, and carbon storage.

The project not only enhanced agricultural productivity but also contributed to climate change mitigation through carbon sequestration (Hengl et al., 2015).

7.2 Humbo Assisted Natural Regeneration Project—Ethiopia

The BioCarbon Fund of the World Bank is funding this project. Thus far, it has guaranteed an increase in crop yields and regenerated over 2700 hectares of previously deteriorated land. It serves as a template for farmer-led initiatives aimed at natural regeneration in Burkina Faso, Chad, and Niger.

The Humbo Assisted Natural Regeneration Project in Ethiopia is a successful example of using agroforestry and reforestation techniques for soil improvement and carbon sequestration. The project involved restoring degraded lands through tree planting, soil conservation measures, and community engagement. As a result, soil fertility improved, erosion was reduced, and carbon was sequestered in vegetation and soil biomass, contributing to local livelihoods and environmental sustainability.

7.3 Farmer-Managed Natural Regeneration (FMNR)—Niger

FMNR refers to a practice that involves the systematic regeneration of trees and shrubs on farmlands, which has been successfully adopted in Niger and other parts of West Africa. By allowing trees to regenerate and integrating them with crops, farmers improve soil fertility, water retention, and carbon sequestration. FMNR has led to increased crop yields, enhanced resilience to drought, and restoration of degraded landscapes, benefiting both farmers and the environment (Dinesh et al., 2015). The program's efforts to mitigate climate change encompass nearly 5 million hectares of land that have been covered with over 200 million trees and around 4.5 tons of above-ground biomass per hectare.

8 Challenges and Future Directions

Current challenges in maximizing soil microbiome contributions to carbon sequestration require innovative solutions and technological advancements.

Future research should focus on addressing these challenges, with potential collaborations and policy implications for promoting soil health and green technology.

8.1 Limitations of Soil-Based Carbon Storage

Globally, there are hundreds of millions of farmers, the majority of whom work small land parcels. We would need many of them to alter their farming practices both now and for hundreds of years to come in order to fully benefit from soil-based sequestration as a climate solution. Experts disagree on the long-term viability of soil-based sequestration, which is a significant societal and economic concern (Searchinger et al., 2018).

In the end, experts claim that while soil-based carbon sequestration, like other negative emissions technologies, can aid in the fight against climate change, it is unable to remove carbon from the atmosphere at the rate at which it is being added. These carbon storage initiatives must be combined with significant reductions in greenhouse gas emissions in order to halt global warming.

The taxonomic diversity and species richness of soil microorganisms, which can be found in a single gram of soil are estimated to number in the billions.

But most soil microorganisms are difficult to research because they haven't been cultured yet. Furthermore, the makeup of microbial communities and their abundance varied significantly among various soil types and geographical areas. (Thompson et al., 2017). Environmental factors also have an impact on the soil microbiome, or microbial community. Variations in temperature, soil moisture content, and other factors can modify the environment, leading to variations in the kinds and functional activities of soil microorganisms (Jansson & Hofmockel, 2020).

9 Conclusion

Soil microbiomes' significance in carbon storage and green technology underscores the need for sustainable practices. This chapter calls for action to encourage leveraging soil microbiomes for transformative environmental solutions in the areas of carbon storage, green technology, and biodiversity conservation are pivotal to addressing global climate challenges as well as contributing to the resilience and sustainability of natural and agricultural ecosystems in Africa. Consequently, the chapter makes a case for holistic approaches to sustainable development by drawing on the following:

Scientific Knowledge Integrating scientific knowledge, research findings, and technological innovations is crucial for understanding soil microbiomes, carbon dynamics, and sustainable land management practices. This includes advancements in microbial ecology, omics technologies, precision agriculture, and digital solutions for soil health monitoring.

Traditional Wisdom Acknowledging and incorporating traditional ecological knowledge, indigenous practices, and local wisdom is vital for sustainable development. Indigenous communities in Africa have valuable insights into soil management, agroforestry, water conservation, and biodiversity conservation that can inform modern practices and enhance resilience to environmental challenges.

Policy Interventions Implementing supportive policies, regulations, and incentives is essential for promoting sustainable soil management, carbon-neutral initiatives, and green technology adoption. This includes policy frameworks that prioritize soil health, incentivize carbon sequestration, and integrate climate-smart agriculture practices into national development agendas.

Community Engagement Engaging local communities, stakeholders, and civil society organizations is critical for fostering ownership, participation, and collaboration in sustainable development efforts. Community-based approaches, participatory decision-making, and capacity-building initiatives empower individuals and communities to contribute to soil conservation, biodiversity protection, and climate resilience.

Therefore, by recognizing the pivotal role of soil microbiomes in carbon storage and green technology in Africa and globally, sustainable development goals will be attained in no distant time. However, to ensure effective research and practice by academia and practitioners, this chapter underscores a need for embracing holistic approaches that combine scientific knowledge, traditional wisdom, and policy interventions that can foster resilience, promote environmental stewardship, and enhance the well-being of present and future generations.

References

Aransiola, S. A., Ijah, U. J. J., Abioye, O. P., & Bala, J. D. (2023). Vermicompost-assisted phytoremediation of toxic trace element-contaminated soil in Madaka, Nigeria using *Melissa officinalis* L and *Sida acuta. International Journal of Environmental Science and Technology*, 1823–1836. https://doi.org/10.1007/s13762-022-04105-y

Aransiola, S. A., Ijah, U. J. J., Abioye, O. P., Bala, J. D., Rivadeneira-Mendoza, R. F., Luque, R., Rodríguez-Díaz, J. M., & Maddela, N. R. (2024). Micro and vermicompost assisted remediation of heavy metal contaminated soils using phytoextractors. *Case Studies in Chemical and Environmental Engineering, 9*, 100755. https://doi.org/10.1016/j.cscee.2024.100755

Bai, Y., & Cotrufo, M. F. (2022). Grassland soil carbon sequestration: Current understanding, challenges, and solutions. *Science, 377*(6606), 603–608.

Basile-Doelsch, I., Balesdent, J., & Pellerin, S. (2020). Reviews and syntheses: The mechanisms underlying carbon storage in soil. *Biogeosciences, 17*(21), 5223–5242.

Buragohain, P. (2019). Role of microbes on carbon sequestration. *International Journal of Microbiology Research*. ISSN0975-5276.

Das, S., Deb, S., Sahoo, S. S., & Sahoo, U. K. (2023). Soil microbial biomass carbon stock and its relation with climatic and other environmental factors in forest ecosystems: A review. *Acta Ecologica Sinica, 43*(6), 933–945. https://doi.org/10.1016/j.chnaes.2022.12.007

Demyanyuk, O., Symochko, L., & Shatsman, D. (2020). Structure and dynamics of soil microbial communities of natural and transformed ecosystems. *Environmental Research, Engineering and Management, 76*(4), 97–105.

Dhanwantri, K., Sharma, P., Mehta, S., & Prakash, P. (2014). Carbon sequestration, its methods and significance. *Environmental Sustainability: Concepts, Principles, Evidences and Innovations, 151*(2), 151–157.

Dinesh, D., Frid-Nielsen, S., Norman, J., Muntamba, M., Loboguerrero Rodriguez, A. M., & Campbell, B. (2015). *Is climate-smart agriculture effective? A review of cases*. CCAFS Working Paper no.129. CCAFS.

Dubey, A., Malla, M. A., Khan, F., et al. (2019). Soil microbiome: A key player for conservation of soil health under changing climate. *Biodiversity and Conservation, 28*, 2405–2429. https://doi.org/10.1007/s10531-019-01760-5

Farooqi, Z. U. R., Qadir, A. A., Khalid, S., et al. (2024). Greenhouse gas emissions, carbon stocks and wheat productivity following biochar, compost and vermicompost amendments: Comparison of non-saline and salt-affected soils. *Scientific Reports, 14*, 7752. https://doi.org/10.1038/s41598-024-56381-y

Fixen, K. R., Zheng, Y., Harris, D. F., Shaw, S., Yang, Z. Y., Dean, D. R., et al. (2016). Light-driven carbon dioxide reduction to methane by nitrogenase in a photosynthetic bacterium. *Proceedings of the National Academy of Sciences, 113*(36), 10163–10167.

Grover, M., Maheswari, M., Desai, S., Gopinath, K. A., & Venkateswarlu, B. (2015). Elevated CO2: Plant associated microorganisms and carbon sequestration. *Applied Soil Ecology, 95*, 73–85.

Hengl, T., Heuvelink, G. B. M., Kempen, B., Leenaars, J. G. B., Walsh, M. G., Shepherd, K. D., et al. (2015). Mapping soil properties of Africa at 250 m resolution: Random forests significantly improve current predictions. *PLoS One, 10*(6), e0125814. https://doi.org/10.1371/journal.pone.0125814

Hicks, N., Vik, U., Taylor, P., Ladoukakis, E., Park, J., Kolisis, F., & Jakobsen, K. S. (2017). Using prokaryotes for carbon capture storage. *Trends in Biotechnology, 35*(1), 22–32. https://doi.org/10.1016/j.tibtech.2016.06.011

Hu, G., Li, Y., Ye, C., Liu, L., & Chen, X. (2019). Engineering microorganisms for enhanced CO2 sequestration. *Trends in Biotechnology, 37*(5), 532–547.

Hutchins, D. A., Jansson, J. K., Remais, J. V., et al. (2019). Climate change microbiology — Problems and perspectives. *Nature Reviews. Microbiology, 17*, 391–396. https://doi.org/10.1038/s41579-019-0178-5

Jansson, J. K., & Hofmockel, K. S. (2020). Soil microbiomes and climate change. *Nature Reviews. Microbiology, 18*, 35–46. https://doi.org/10.1038/s41579-019-0265-7

Jansson, C., Tuskan, G. A., Wullschleger, S. D., & Kalluri, U. C. (2010). Phytosequestration: Carbon biosequestration by plants and the prospects of genetic engineering. *Bioscience, 60*, 685–696. https://doi.org/10.1525/bio.2010.60.9.6

Jansson, J. K., Neufeld, J. D., Moran, M. A., & Gilbert, J. A. (2012). Omics for understanding microbial functional dynamics. *Environmental Microbiology, 14*(1), 1–3. https://doi.org/10.1111/j.1462-2920.2011.02518.x

Jansson, C., Vogel, J., Hazen, S., Brutnell, T., & Mockler, T. (2018). Climate-smart crops with enhanced photosynthesis. *Journal of Experimental Botany, 69*(16), 3801–3810. https://www.jstor.org/stable/26932831

Kallenbach, C. M., Grandy, S., Frey, S., & Diefendorf, A. F. (2015). M icrobial physiology and necromass regulate agricultural soil carbon accumulation. *Soil Biology and Biochemistry, 91*, 279–290. https://doi.org/10.1016/j.soilbio.2015.09.005

Kapilan, R., Vaziri, M., & Zwiazek, J. J. (2018). Regulation of aquaporins in plants under stress. *Biological Research, 51*, 4. https://doi.org/10.1186/s40659-018-0152-0

Krieger, K. (2012). Genetically engineered bacteria could help fight climate. *Change, 335*, 1026.

Lal, R. (2004). Soil carbon sequestration impacts on global climate change and food security. *Science, 304*, 1623–1627. https://doi.org/10.1126/science.1097396

Liang, J., Qi, X., Souza, L., & Luo, Y. (2016). Processes regulating progressive nitrogen limitation under elevated carbon dioxide: A meta-analysis. *Biogeosciences, 13*, 2689–2699. https://doi.org/10.5194/bg-13-2689-2016

Marschmann, G. L., Tang, J., Zhalnina, K., Karaoz, U., Cho, H., Le, B., et al. (2024). Predictions of rhizosphere microbiome dynamics with a genome-informed and trait-based energy budget model. *Nature Microbiology, 1-13*, 421–433. https://doi.org/10.1038/s41564-023-01582-w

Mukhtar, H., Wunderlich, R. F., Muzaffar, A., Ansari, A., Shipin, O. V., Cao, T. N. D., & Lin, Y. P. (2023). Soil microbiome feedback to climate change and options for mitigation. *Science of the Total Environment, 163412*, 163412.

Naylor, D., Sadler, N., Bhattacharjee, A., et al. (2020). Soil microbiomes under climate change and implications for carbon cycling. *Annual Review of Environment and Resources, 45*(1), 29–59. https://doi.org/10.1146/annurev-environ-012320-082720

Naylor, D., McClure, R., & Jansson, J. (2022). Trends in microbial community composition and function by soil depth. *Microorganisms, 10*(3), 540. https://doi.org/10.3390/microorganisms10030540

Pepe-Ranney, C., Campbell, A. N., Koechli, C. N., Berthrong, S., & Buckley, D. H. (2016). Unearthing the ecology of soil microorganisms using a high-resolution DNA-SIP approach to explore cellulose and xylose metabolism in soil. *Frontiers in Microbiology, 7*, 703. https://doi.org/10.3389/fmicb.2016.00703

Plaza, C., Courtier-Murias, D., Fernández, J. M., Polo, A., & Simpson, A. J. (2013). Physical, chemical, and biochemical mechanisms of soil organic matter stabilization under conservation tillage systems: A central role for microbes and microbial by-products in C sequestration. *Soil Biology and Biochemistry, 57*, 124–134.

Raza, T., Qadir, M. F., Khan, K. S., Eash, N. S., Yousuf, M., Chatterjee, S., et al. (2023). Unrevealing the potential of microbes in decomposition of organic matter and release of carbon in the ecosystem. *Journal of Environmental Management, 344*, 118529.

Schnoes, A. M., Brown, S. D., Dodevski, I., & Babbitt, P. C. (2009). Annotation error in public databases: Misannotation of molecular function in enzyme superfamilies. *PLoS Computational Biology, 5*(12), e1000605. https://doi.org/10.1371/journal.pcbi.1000605

Searchinger, T. D., Wirsenius, S., Beringer, T., et al. (2018). Assessing the efficiency of changes in land use for mitigating climate change. *Nature, 564*, 249–253. https://doi.org/10.1038/s41586-018-0757-z

Sokol, N. W., Slessarev, E., Marschmann, G. L., et al. (2022). Life and death in the soil microbiome: How ecological processes influence biogeochemistry. *Nature Reviews. Microbiology, 20*, 415–430. https://doi.org/10.1038/s41579-022-00695-z

Stone, B. W. G., Dijkstra, P., Finley, B. K., et al. (2023). Life history strategies among soil bacteria-dichotomy for few, continuum for many. *The ISME Journal, 17*(4), 611–619. https://doi.org/10.1038/s41396-022-01354-0

Thompson, L., Sanders, J., McDonald, D., et al. (2017). A communal catalogue reveals Earth's multiscale microbial diversity. *Nature, 551*, 457–463. https://doi.org/10.1038/nature24621

Wallenstein, M. D. (2017). Managing and manipulating the rhizosphere microbiome for plant health: A systems approach. *Rhizosphere, 3*, 230–232. https://doi.org/10.1016/j.rhisph.2017.04.004

Wang, C., Li, B. B., Liang, Q. M., & Wang, J. C. (2018). Has China's coal consumption already peaked? A demand-side analysis based on hybrid prediction models. *Energy, 162*, 272–281.

Wleder, J., Smith, A. B., & Johnson, C. D. (2013). Phylogenetic analysis of tropical rainforest species using molecular data. *Journal of Evolutionary Biology, 26*(5), 987–1003.

Zhang, Q., Qin, W., Feng, J., et al. (2022). Responses of soil microbial carbon use efficiency to warming: Review and prospects. *Soil Ecology Letters, 4*, 307–318. https://doi.org/10.1007/s42832-022-0137-3

Effects of Pesticides and Plastics on Soil Microbiome

Arnab Chakraborty, Priyanka Verma, and Sandip Mondal

Abstract Intensive agriculture relies heavily on external inputs to achieve high productivity and profitability. One commonly used input is plastic mulch, predominantly made from low-density polyethylene (LDPE), which helps reduce evaporation, increase soil warmth, and inhibit weed growth. However, improper cleanup of LDPE mulch leads to soil pollution. Additionally, the traditional use of pesticides in agriculture exacerbates the accumulation of residues in the superficial soil layers. The presence of plastic and chemical residues in agricultural lands can significantly impact on crop sand environments, ultimately affecting geo-micro-biomes. Pesticides are increasingly used on arable land to mitigate pest impacts on crop productivity. However, these unregulated farming practices alter soil characteristics, posing risks to soil micro-fauna, including beneficial bacteria, fungi, algae, and protozoa. These microorganisms play crucial roles in biodegradation, nutrient recycling, nitrogen fixation, plant growth promotion, and disease regulation. The negative effects of pesticides on soil micro-biota can significantly alter soil properties, reducing crop growth and yield. Different pesticides, such as pyrethrins, pyrethroids, carbamates, organochlorines, and organophosphates, affect various microorganisms differently. Understanding the impact of pesticides on microbial populations and metabolic transformation mechanisms is imperative. This chapter reviews the effects of pesticides and micro-plastics on microbial development and activity, highlighting the importance of fostering soil microbial growth for a robust agricultural ecosystem. It explores methods to create a sustainable agro-ecosystem that supports beneficial soil microorganisms while minimizing the use of agrochemicals and pesticides.

A. Chakraborty
School of Pharmaceutical Technology, School of Health and Medical Sciences, Adamas University, Kolkata, India

P. Verma
Kamla Raja Girls Goverment Post Graduate College, Gwalior, Madhya Pradesh, India

S. Mondal (✉)
Amity Institute of Pharmacy, Amity University, Kolkata, West Bengal, India
e-mail: sandip1706@yahoo.com

S. A. Aransiola et al. (eds.), *Soil Microbiome in Green Technology Sustainability*,
https://doi.org/10.1007/978-3-031-71844-1_12

Keywords Agricultural Soil · Low density polyethylene · Micro-plastics · Microbiome · Pesticide

1 Introduction

Conventional plastics, derived from non-renewable resources, are characterized by their extreme resistance to natural breakdown processes, despite their desirable features and affordability. Their half-lives range from 0.035 to 9000 years, indicating prolonged persistence in the environment. Consequently, plastic waste has become a primary source of significant pollution, impacting natural ecosystems. Table 1 reveals that the packaging industry is the largest contributor to global plastic waste each year, producing 142.6 million tons. In stark contrast, marine coatings and colors generate the least plastic waste, with just 541,000 tons annually (Malyan et al., 2019). The building sector falls in between, generating 76.89 million tons of plastic waste per year. Of particular concern are micro-plastics (MPs), which are particles smaller than five millimeters, and nano-plastics (NPs), which are even finer, down to half a micrometer. These particles have emerged as major contributors to pollution due to their widespread distribution across terrestrial, aquatic, and aerial environments (de Souza Machado et al., 2018a; Chamas et al., 2020; Dissanayake et al., 2022; Johansen et al., 2022; Ritter, 2022; Rosenboom et al., 2022; Chowdhury et al., 2023). Globally, MPs have been detected in various aquatic sources, with polyethylene (PE) and polypropylene (PP) being the most frequently found types (Koyuncuoğlu & Erden, 2021; Pelamatti et al., 2021; Aransiola et al., 2023; Musa et al., 2024). Beyond their impact on soil ecosystems, MPs can interact with other contaminants, influencing their fate. Due to their high surface free energy and water repulsion properties, MPs and NPs are effective adsorbents for organic contaminants, potentially serving as vectors for pollutants in various habitats. MPs can also alter the metabolic state and structure of soil microbiota, affecting the degradation and persistence of organic contaminants in soils (Carpenter & Smith Jr, 1972; Horton et al., 2017; Haward, 2018; Warrier et al., 2022). Environmental regulations should address the potential hazards of polymers that could be introduced into soils, providing clear guidelines and measures. Pesticides, a prevalent organic pollutant in modern agriculture, contaminate the environment and pose risks to human health,

Table 1 Annual global plastic waste generation by sectors (Malyan et al. 2019)

Annual global plastic waste generation by sectors of Industries	Million ton
Packaging	142.6
Building	76.89
Transportation	62.17
Textile	43.88
Electronics	17.31
Industrial machinery	2.68
Marine coating and colours	541,000 ton

Table 2 Plastic consumption by different countries (Ritchie & Roser, 2018)

Countries	Pesticides Consumption (Kg/hectare)
China	17
Japan	11
USA	7
South Korea	7
France	5
UK	5
India	0.5

warranting special attention (Stubenrauch & Ekardt, 2020; Conti et al., 2021). According to Table 2, China has the highest pesticide consumption at 17 kg per hectare, which is markedly greater than India's 0.5 kg per hectare. Both the USA and South Korea have a pesticide consumption rate of 7 kg per hectare. Meanwhile, France and the UK each use 5 kg per hectare (Ritchie & Roser, 2018). While the ecological effects of pesticides, plastics, and polymer materials have been documented, the mechanistic pathways leading to the formation of soluble toxins are not thoroughly understood. This review synthesizes recent scientific research on the interactions between plastics and pesticides to enhance understanding of the environmental dispersion and ecotoxicological risks associated with MPs. It examines the etiology, possible chemical reactions, and physical interactions between polymers and pesticides, considering the roles of microbes and small animals. The review is structured into three sections, covering MP entry routes into agricultural soils, their effects on the earth's crust, and their impact on the physiology of living organisms (Martín-Lara et al., 2021; Yusuf et al., 2022).

2 Etiopathogenesis of Soil Microbiota

Microplastics (MPs) significantly impact soil ecosystems in agricultural lands, beaches, urban environments, and landfills. These plastics can enter soils directly from industrial sectors or indirectly through processes like fragmentation and embrittlement. Key contributors to MP contamination include wastewater treatment facilities, air deposition, and municipal solid waste (Dris et al., 2016; Galafassi et al., 2019; Boyle & Örmeci, 2020) (Fig. 1). Recent studies have identified both direct and indirect sources of MP contamination in agricultural soils. These include sludge from toilets, garbage, water supply pipes, plastic mulching, greenhouse films, plastic liners, plastic-encapsulated fertilizers, and polymers from flooded roads. Light exposure and wear and tear, such as plowing in heavily mulched areas, further promote the formation of MPs (Mammo et al., 2020; Wang et al., 2021). The combination of light exposure and wear and tear like as plowing in areas with heavy mulching encourages the growth of MPs (Steinmetz et al., 2016; Lamastra et al., 2018). Wastewater treatment plants (WWTPs) are generally effective at removing MPs, achieving reductions of more than 90%, and in some cases, over

Fig. 1 Key contributors to microplastic pollution

98% (Murphy et al., 2016; Conley et al., 2019). However, the effectiveness of MP removal depends on the nature, size, and shape of the particles, with fibers often being less effectively retained. Despite high removal efficiencies, significant amounts of MPs still enter the environment through the small percentages discharged by WWTPs. Soluble micro-plastics from effluent treatment plants can reveal into the environment via treated water. Another significant indirect source of MP introduction is primary and secondary packaging materials made of plastics. Polyethylene terephthalate (PET), polypropylene (PP), and polyethylene (PE) are commonly used to produce MPs in the form of films and fibers. Biological treatments, such as composting, can cause MPs to fragment further, posing additional environmental risks. Reducing the amount of MPs added to waste before treatment is crucial to mitigate this issue (Corradini et al., 2019; Tan et al., 2022; Yang et al., 2022).Compost is another important transporter of MPs into agro-ecosystems (Weithmann et al., 2018; Vithanage et al., 2021). When composts are used as organic additions to agricultural soil, there is a risk of plastic pollution because, regardless of the source of the materials, plastic to microplastics occurs more quickly in the composting environment. Ten to thirty plastic particles are produced per gram of dried compost from the composting of organic municipal solid wastes, of which five to twenty are produced by MPs (Edo et al., 2022). The scientists carried out an extensive analysis of the many forms and kinds of MPs present in the compost, with a focus on the fiber content. Moreover, they found that 94% of microplastics were composed of the polymers acrylic, polyester, polyvinyl chloride (PVC), polystyrene (PS), PE, and PP, in that order of abundance (Lwanga & Geissen, n.d.; Gui et al., 2021; Scopetani et al., 2022).

3 Effect of Plastic Materials on the Surface Environment

The presence of microplastics (MPs) in soil has been shown to impact crop performance, affecting water use efficiency and root development in various plants (Tourinho et al., 2019; Luo et al., 2020). For example, variations in water use have been observed, and *Lepidium sativum* (garden cress) has exhibited an incapability to produce roots in MP-contaminated soils (de Souza Machado et al., 2018b; Bosker et al., 2019), Studies have documented positive changes in soil pH in some cases, while others have reported negative changes (Palansooriya et al., 2022), or no effects (Qi, Beriot, et al., 2020a), demonstrating the variability in soil response to MP contamination (Feng et al., 2022). The amount of dissolved organic carbon (DOC) that MPs raise can vary depending on how much they degrade (Li et al., 2022a). However, because MPs have a propensity to degrade slowly in the environment due to the relative inertness of polymers, DOC from this source could not be created in considerable quantities (Li et al., 2022a; Schell et al., 2022). Because of their effects on the biomass and activity of soil microorganisms (Chia et al., 2022). High temperature may cause plastic mulches to accelerate the breakdown of soil organic matter, but they have no effect on crop growth and nutrients like phosphate, nitrate, or ammonium levels. Onion leaf nitrogen content and soil nitrogen are both enhanced by the use of polyamide fibers. The plant content and nutrient availability are enhanced by this addition. It's possible that changes in soil structure are the cause of MPs' beneficial effects on Arbuscular mycorrhizal fungi. Mulch films made of biodegradable plastic and LDPE also raise the C: N ratio. Microplastics (MPs) have been observed to impact soil microbial activity; research indicates that while MPs decrease soil microbial diversity, they also significantly increase the quantity and activity of certain microorganisms.

Others, however, have documented detrimental effects of polystyrene (PS), such as elevated basal respiration rate and metabolic quotient, on microbial biomass and enzyme activity. It has been discovered that the activities of urease, dehydrogenase, and alkaline phosphatase are reduced by three different types of MPs: fibrous polypropylene, membrane like polyethylene, and spherical polypropylene. It has been observed that the higher pH of soil can less interact with the MP and activity of geomicrobiome get decreases by inactivation of several enzymes useful for nitrogen fixation in soil mass. These results have not, however, been associated with any changes in the size and soil microbial population or in the diversity of the soil bacterial community. This section addresses the properties and functions of soil directly impacted by the presence of microplastics (MPs). However, these changes can indirectly influence other soil characteristics. For example, soil electrical conductivity has shown varied responses to the addition of MPs, with some studies reporting increases, others decreases, and some finding no alterations. In contrast, soil organic matter typically displays more consistent responses to MP contamination (Qi et al., 2020b).

4 Reactions of Microplastics

4.1 *The Mechanisms and Factors that Control How Pesticides and MPs Interact*

Pesticides and other biological contaminants may harm soil ecosystems through MPs, altering the flow and availability of these toxins. Addressing the final behavior when pesticides and MPs coexist in soils requires an understanding of the interactions between them in aquatic systems.

Recent research has focused on the processes governing the interactions between pesticides and MP in fluid phases (Awet et al., 2018; Yi et al., 2021). Pesticides adhere quickly to MP surfaces, and the pace at which pore diffusion occurs varies based on the properties of the MP and the pesticide. Triazole fungicide sorption to Polystyrene (PS) micro-particles rose quickly in the first six hours and remained essentially constant in the following 24. After just two hours of contact, the adsorption equilibrium of several different organophosphorous pesticides on Polyethylene (PE) was attained. Aging of PE shortened the equilibrium time for several pesticides. While standard MPs reached equilibrium after 48 hours, biodegradable MPs showed faster adsorption rates for fipronil. The time to reach equilibrium for thiacloprid varied with the type of MP, taking 34 hours, 4 hours, or just 1 hour. The Freundlich equation was primarily used to model the adsorption isotherms of different pesticide-MP systems (Fang et al., 2019; Wang et al., 2020b; Mo et al., 2021).

By producing mono or multilayer's, this equation offers a more satisfactory explanation for the adsorption of heterogeneous adsorbents like MP. Some basic physical interactions like hydrogen bonding, pi attractions and inter particular attractions can all have an impact on adsorption mechanisms in pesticides. Depending on the chemical makeup of the pesticide and the micro-particle's (MP) surface chemistry, both processes might coexist. However, hydrogen bonding between the hydroxyl groups in the polymer 'smother moiety may also play a role (Beriot et al., 2020). Due to the aging process, which speeds up oxidation and encourages the formation of oxides and micro-cracks, this mechanism, is mostly suggested in aged MPs. The surface area and surface chemistry of MPs can change as a result of natural weathering in soil conditions or sterilization procedures in WWTPs, which increases the adsorption of pesticides. Atrazine adsorption in immaculate MPs is primarily controlled by partitioning and electrostatic attraction, whereas in elderly MPs, surface diffusion and H-bonding play a major role (Jiang et al., 2020).

According to Lan et al., pesticide adsorption on PE relies on hydrophobic partitioning, but hydrogen bonding must also be considered due to the presence of polar oxygen-containing functionalities in aged PE (Lan et al., 2021). Pentachlorophenol, the most hydrophobic chemical studied, showed a stronger preference for both aged and pristine MPs. Additionally, endosulfan and endrin exhibited increased adsorption on aged polymers, as noted by Concha Graña and colleagues (Seidensticker et al., 2018). No variations were observed in the adsorption of chlorpyrifos and α-endosulfan on aged and pristine PA and a biopolymer, while Bele et al. found that treating PE, PP, and PS with O_3 and O_3/H_2O_2—common in advanced oxidation systems of WWTPs enhanced the polarity of MPs, thereby improving the adsorption of polar insecticides like dichlorvos and chlorpyrifos (Ateia et al., 2020; Chi et al., 2021; Liu et al., 2023).

Many environmental parameters have been investigated in connection to pesticide adsorption, such as pH, ionic strength, and dissolved organic matter (DOM). Jiang et al. report no significant difference in pH among PBS, PE, and PVC in relation to the adsorption of triazole fungicides. However, pH has been extensively studied with mixed results. Notably, an acidic pH in soil sap enhances the adsorption of mycocides on PS (Liu et al., 2022a; Wang et al., 2022). Triadimenol adsorption remained unchanged, while diflubenzuron and difenoconazole adsorption increased and then stabilized, and carbazim and trichlorfon adsorption decreased with increasing pH from 3 to 6.

The ion content of the aqueous phase influences pesticide binding to microplastics (MPs), with low salinity favoring the adsorption of imidacloprid, buprofezin, and difenoconazole to PE. The interaction of α-endosulfan with MPs varies with water's ionic nature. Studies indicate that MPs strongly adsorb hydrophobic DDT but are less likely to collect more polar substances. High ion concentration in river water hinders glyphosate adsorption to PVC due to competition for sorption sites (Jiang et al., 2021; Martín et al., 2021; Fatema & Farenhorst, 2022). The first increase in NaCl concentration causes the adsorption of aged PE and PS on pure PE to rise, but subsequent increases in NaCl concentration cause it to decrease. The highest endosulfan and endrin adsorption was seen in saline water (35% NaCl). Furthermore, tebuconazole adsorption on PA is increased by salt, although it is decreased on PP and PS. Chloride anions can inhibit the adhesion of Cl-containing medicines, such as difenoconazole and diflubenzuron, to the surface of microplastics (MPs).

Mostly research indicates that adsorption increases with rising salt content, likely due to a salting-out affect that decreases pesticide solubility in water and enhances their attraction to MP surfaces. (Liu et al., 2022a).Dissolved oxygen (DO) has been discovered to have varying impacts on the adsorption of insecticides, such as carbendazim and carbofuran, on plastics and polymers (PE and PP). The inclusion of glycine and oxalic acid reduced carbofuran adsorption, possibly due to its reduced binding strength on MPs.

4.2 Interactions Between Microplastics and Pesticides in Agricultural Soils

Understanding the interactions between microplastics (MPs) and pesticides in agricultural soils is crucial, particularly in regions with intensive farming practices. Compared to studies conducted without soil, there are relatively few publications addressing the effects of MPs on pesticide behavior in soil. MPs are primarily found in the upper soil layers due to their limited penetration into deeper layers, although soil movement can redistribute pesticides and microplastics to deeper layers. Soil organisms, such as earthworms, may contribute to the dispersion of pesticides and the movement of MPs throughout the soil profile (Pan et al., 2023). Martín *et al* (Fatema & Farenhorst, 2022) studied the retention of various drugs and the herbicide simazine using colorful microbeads, finding that the type of PE influenced adsorption amounts; simazine results varied, while blue PE adsorbed the most pollutants, such as amoxicillin.

Pesticide adsorption is a key advantage of plastics like PE and PP over traditional polymers, as their rubbery texture at room temperature makes organic pollutants more accessible. Conversely, the glassy nature of polymers like PVC and PS limits pesticide diffusion. The rubbery or glassy texture of these polymers is crucial for adsorbing organo-chlorine insecticides like fipronil. Biodegradable polymers, with their higher specific surface area, porosity, and favorable chemical properties, facilitate stronger pesticide binding and have shown greater adsorption potential for substances like fipronil, difenoconazole, and triadimefon. Additionally, smaller plastic particles typically exhibit higher pesticide adsorption due to more effective absorption sites and greater accessible surface areas (Fang et al., 2019; Wang et al., 2020b). However, an unexpectedly higher DDT adsorption on MPs with bigger diameters was found by Fatema and Farenhorst (2022). Recent observations by multiple authors indicate that increasing the dose of microplastics (MPs) generally has a detrimental effect on pesticide adsorption; however, this effect is dependent on the specific MP/pesticide system (Wang et al., 2020b; Liu et al., 2022b).

In the end, more adsorption will lead to less availability for decomposition. Because of this, the average rate of degradation of around 40 pesticides with a range of chemical properties that were adsorbed on different plastics was a third ratio lower (Beriot et al., 2020). This degradation was linked to the proportion of pesticide interaction with the polymers. Progressive desorption and adsorption on the MP could have contributed to this growth. Interestingly, they showed that PP enhances the stereo selective breakdown of imazapic but not that of the other two herbicides. Similarly, PE shielded deltamethrin from photo-degradation but not from chemical degradation, according to Ramos et al. (Allen et al., 2018; Martín et al., 2021). Given the high concentrations of MPs in soil, it is critical to comprehend their interactions with pesticides. This is particularly true in regions where intensive farming is carried out. Compared to studies done without soil, there are comparatively few publications addressing the effect of MPs on pesticide effects in the soil. As previously noted, MPs are mostly found in the higher soil layers because of less capacity to penetrate in deep layers (Schell et al., 2022).

Moreover, Castan et al. (2021) soil movement may shifts the pesticides and microplastics to deeper layers (Wanner, 2021). Specifically lumbricina may potentially contribute to the dispersion of pesticides as well as movement of MP across the soil profile (Huerta Lwanga et al., 2016; Lwanga et al., 2017). Scientists have recently developed an interest in the interactions that take place in agricultural soils between MPs and pesticides. The 2015 study by Ramos et al. (2015) suggests that plastic residues from PE mulch films could be used in horticultural systems as pesticide collectors. It has been observed that organophosphorous pesticides can penetrate to core of polymer without get solubilized in aqueous medium. Desorption or soil column tests could then be used to release the pesticides that were retained into the surrounding soil. The way these two opposing processes balanced each other out was influenced by the chemical properties of the pesticides and the dimensions of the film segments.In relation to this, it was shown that chlorpyrifos was released into the soil at rates that depended on the size of the particles from contaminated low-density PE (Rodríguez-Seijo et al., 2018).

For instance, Hüffer et al. (2019) found that the adsorption of 4-(2,4-dichlorophenoxy)butyric acid in the soil was reduced in the presence of PE particles. Thus, PE decreased the total amount of pesticide adsorption capability in the soil. It has been identified that the polarity and hydrophilicity -lipophilicity also an important parameter for absorption and adsorption of the pesticides and fertilizers to MP core. The same study did discover, however, that both types of MPs decreased imidacloprid's adsorption, and it was proposed that these differences were caused by the pesticides' Log K_{ow}. The adsorption of very different pesticides on sediment on MP did not differ from the adsorption according to the last observation (Fatema & Farenhorst, 2022). However, the degree to which pesticides adhere to soils may depend on both the properties of MPs and the pesticides themselves. Glass transition temperature (T_g) is an important polymer property in this context. When a polymer's structure is below its T_g, it becomes glassy or inflexible. Organic pollutants are more likely to diffuse. This is the reason of polyurethane pollution (Xu et al., 2020; Wu et al., 2022). Moreover, PE's degree of flexibility is mostly responsible for its malleable qualities in its natural habitat. Its structure may be compared to other common MPs found in soils, such as glassy PVC, to identify a significant factor influencing higher pesticide absorption. PE residues are a frequent pollutant in agricultural fields because PE is very frequently available pollutant from agro industry. Because PE MPs may serve as pesticide transporters, they could therefore have a significant influence on the fate of pesticides in soils (Yang et al., 2018; Kelkar et al., 2019; Wang et al., 2020b).

Eight pesticides' adsorption was shown to be unaffected by any of the PE aging procedures that Wang et al. (2020a) evaluated for sediment, including wet/dry heat, mechanical abrasion, and UV irradiation (Yang et al., 2019; Zhang et al., 2021).The mechanics of reaction have also been the subject of numerous research studies. MPs have been reported to slow down the rate at which pesticides adsorb quickly on soil surfaces, which delays the process of attaining sorption equilibrium (Wu et al., 2022). Like in the preceding section, the strong fit to the Freundlich equation showed that, in the presence of MPs, multilayer adsorption processes influence pesticide interactions with soil through both physical and chemical mechanisms. Numerous interactions between pesticides and MP and soil have been reported in different layers of sediments. Organic matter is the main factor influencing the dynamics of pesticides in soils. Consequently, the metallic component aids in the herbicide's transportation and may influence the simultaneous immobilization of MPs and ionic herbicide inside the soil environment. Metal ions like calcium and iron also complexes with MP to form chelate thus the absorption of herbicides and MP through the microbes occurs as well as by washout. While published research has shown mixed results, co-transport of MPs by soil organisms such as earthworms offers deep penetration. On the one hand, Rodríguez-Seijo et al. (2018) were unable to demonstrate whether chlorpyrifos can be efficiently transported to earthworms by low-density PE.

4.3 Behavior of Pesticides in Soils Contaminated with MPs

Studying the adsorption/desorption and transport behavior of pesticides in soils contaminated with microplastics (MPs) is a critical research area bridging environmental science, soil science, and chemistry (Peña et al., 2023). Adsorption refers to the process where molecules of a substance adhere to the surface of a solid or liquid (Gautam et al., 2024). In the context of pesticides in MP-contaminated soils, pesticides can adhere to the surface of microplastic particles in the soil. The degree of adsorption is influenced by factors like the chemical characteristics of the pesticide, the surface properties of the microplastic, and environmental factors such as pH and temperature (Abioye et al., 2021; Fu et al., 2021). Pesticides that are hydrophobic (water-repellent) tend to have higher adsorption onto MPs due to similarities in their chemical properties (Parashar et al., 2023). Desorption, on the other hand, is the process by which adsorbed molecules are released from the surface of a solid or liquid. Once pesticides are adsorbed onto microplastics in the soil, they can undergo desorption under certain conditions such as changes in pH or the presence of organic matter. Desorption can lead to the release of pesticides back into the soil solution, where they can become available for uptake by plants or leach into ground water (Gonzalez et al., 2019). The transport behavior of pesticides in soils contaminated with MPs refers to how pesticides, once applied to the soil, move through the soil profile. Microplastics can influence the transport behavior of pesticides by altering soil properties such as porosity, hydraulic conductivity, and soil organic matter content (Peña et al., 2023).

Additionally, pesticides that are adsorbed onto microplastics may exhibit different transport behaviors compared to free pesticides. For example, they may move with the microplastic particles as they are transported through the soil, leading to different distribution patterns within the soil profile. Understanding the adsorption/desorption and transport behavior of pesticides in soils contaminated with MPs is crucial for assessing the environmental risks associated with pesticide use and microplastic pollution. It can help inform management practices aimed at minimizing the impact of these contaminants on soil and water quality, as well as human and ecosystem health.

4.4 Spreading of Pesticides in Soils Contaminated with MPs

Xu et al. (Xu et al., 2020)observed no variations in the activities of urease, acid phosphatase, catalase, or thiacloprid washout. Furthermore, thiacloprid degrades rapidly in soils (Rodríguez-Liébana et al., 2017), potentially overlapping with the function of MPs in the degradation process. The effects of glyphosate, a herbicide with varying charges depending on pH, have shown inconsistency. Similar rates of glyphosate breakdown were observed for single or combined treatments in sand with polypropylene (Yang et al., 2018). On the other hand, in the same soil, homopolymer PP improved glyphosate dissipation (Liu et al., 2019; Li et al., 2020)· Moreover, herbicides might also influence MP degradation, potentially accelerating the process. Thus, pesticide residues could act as catalysts for the decomposition of plastics in soil. Liu et al. (Lin et al., 2022) reported the synthesis of carbonyl derivatives, indicative of oxide formation, while Li et al. (2020) did not observe any alterations in MP functional groups. A modest dose of propiconazole would have led to greater enzymatic activity and an increase in MP microorganism attacks in PBAT as opposed to PE. Alkaline soil would facilitate the decomposition of plastic more readily, and the combined effects of insecticides and UV rays would hasten this process.

5 Interactions of Microplastic and Pesticide Complex in Diverse Ecosystems

5.1 Aquatic Ecosystems

The interaction between microplastics and pesticides in aquatic ecosystems can have significant and complex effects on both the environment and the organisms within it. Microplastics can act as carriers for pesticides, facilitating their transport and accumulation in aquatic environments. This can lead to increased exposure of aquatic organisms to pesticides, resulting in higher toxicity levels.Both microplastics and pesticides have the potential to bioaccumulation in aquatic organisms. When present together, they may synergistically enhance the bioaccumulation process, leading to higher concentrations of contaminants in organisms higher up the food chain.

Microplastics can alter the behavior and physiology of aquatic organisms, making them more susceptible to the toxic effects of pesticides. This can disrupt ecological interactions, such as predator-prey relationships and reproductive success, ultimately impacting the structure and function of aquatic ecosystems. Accumulation of microplastics and pesticides in aquatic habitats can lead to physical and chemical alterations, such as changes in sediment composition, water quality, and nutrient cycling. These alterations can further exacerbate the stress on aquatic organisms and their habitats.

The long-term effects of microplastics and pesticides interactions are still not fully understood. However, there is growing concern that chronic exposure to these contaminants may have cumulative and persistent effects on aquatic ecosystems, leading to population declines and ecosystem degradation over time. Overall, the interaction between microplastics and pesticides represents a complex and multifaceted issue that requires interdisciplinary research and holistic management approaches to mitigate its impacts on aquatic ecosystems. Combining MPs with pesticides increases the overall toxicity of each stressor. Several marine species, like Oncorhynchus, also known as rainbow trout, have shown this phenomenon. Changes in protein contents, the composition of amino acids and fatty acids, histopathological and histomorphometrical biomarkers, and other parameters have all been measured after concurrent addition of chlorpyrifos (Felten et al., 2020; Hanachi et al., 2021; Karbalaei et al., 2021). Some fish species shows that the MP and pesticide ross their intestine and goes to systemic circulation by the higher lipophilic nature of the complex. The test was performed on Cyprinuscarpio fish (Chen et al., 2022).

In addition to fish, crustaceans have also been investigated. As a result, it was shown that the number of neonates per surviving female, brood size, and Daphnia magna survival were all adversely affected more by MPs and pyrethroids (Felten et al., 2020; Dolar et al., 2021) another species, *Minucaecuadioriensis*, a crab also get endangered more frequently (Villegas et al., 2021). On the other hand, when high-density PE and the herbicide were used together at lethal quantity, they adversely impacted the Pacific oyster's development and behavior like valve microclosure, valve expansion magnitude and latency) (Ding et al., 2018; Bringer et al., 2021; Hanslik et al., 2022). A study revealed that a mixture of glyphosate and Cosmo-Flux®411F has a negligible toxic effect on the embryos and tadpoles of the species Scinax squalirostris. However, another study found that the combination of polyethylene (PE) with the herbicides glyphosate and glufosinate significantly impacts enzyme activities in Scinax squalirostris tadpoles. Density Functional Theory (DFT) analysis indicates that PE and herbicides spontaneously interact, forming bonds at the herbicide-PE interface. These interactions can increase toxicity when these pollutants are combined, particularly in the case of glyphosate-based herbicides (GBHs). GBHs inhibit the enzyme 5-enolpyruvylshikimate-3-phosphate synthase (EPSPS) in the shikimic acid pathway of plants. This prevents the synthesis of aromatic amino acids, causing plant death by amino acid starvation. PE and herbicides interacted, creating connections between the two substances. Fernández et al. 2022) demonstrated that the effects of exposing the mussel Mytilusgalloprovincialis to algae and MP and chlorpyrifos, were quite similar. As a result, PE might carry the pesticide as effectively as microalgae, if not more so. These results all suggest that the combined application of has a greater effect. However, there have been isolated instances of mild or negative interactions between MPs and the pesticides on occasion.

Li et al. (2022b) found that microplastics (MPs) reduced the organ storage of difenoconazole in zebrafish, thereby lowering the pesticide's lipid peroxidation. Similarly, the reduced bioavailability of glyphosate for the blue-green algae

Microcystis aeruginosa (Zhang et al., 2018), and of chlorpyrifos for zebrafish Danio rerio (Hanslik et al., 2022), has been attributed to significant pesticide sorption on various MPs. In the aforementioned study, glyphosate inadvertently increased the stability of the MP dispersion, which led to increased MP adsorption on the algae and likely resulted in accumulation within the food pyramid. Several studies have explored different aspects of these interactions. For instance, Ziajahromi et al. (Horton et al., 2018; Ziajahromi et al., 2019) examined the effects of polyethylene (PE) and bifenthrin on *Chironomus tepperi* in both synthetic and river waters. Bifenthrin, being highly hydrophobic, showed reduced toxicity in synthetic water as over 92% of it was adsorbed onto the MP. In contrast, there was no reduction in bifenthrin toxicity in river water containing 10 mg/L of dissolved oxygen, likely because bifenthrin interacted with dissolved organic matter (DOM) rather than MP (Zocchi & Sommaruga, 2019).

5.2 Terrestrial Ecosystems

Mishra et al. (Kokalj et al., 2018; Mishra et al., 2022) investigated the effects of the organophosphorous insecticide monocrotophos and the microplastics (MPs) PVC and PP on the earthworm *Eudrilus eugeniae*. The combined adsorption of the pesticide by the two MPs increased oxidative stress in E. eugeniae, resulting in a significant decrease in protein levels, an increase in lipid peroxidation, and heightened enzyme activity. Another study examined *Eisenia fetida*, a different earthworm species, in the presence of soil MPs and the antiviral medication dufulin (Sun et al., 2021). The combination of MPs and the antiviral altered the abundance of three metabolic pathways and 21 metabolites, indicating a decrease in the metabolic profile. Furthermore, MPs and the herbicide 2,4-D were tested on *E. andrei*, with concurrent exposure leading to increased toxicity and oxidative changes, including elevated levels of glutathione S-transferase and catalase and the accumulation of malondialdehyde (Boughattas et al., 2022). The combination of MP and herbicide also increased oxidative stress and caused abnormal gene expression in *E. fetida*, with older MPs having a greater effect than non-aged MPs. A similar pattern of oxidative stress was observed with the coexistence of MPs and atrazine, although the reverse was true for gene expression abnormalities.

Studies on polyethylene's effects on *Lumbricus terrestris* showed that joint treatment negatively impacted gallery volume and dry weight, with higher concentrations (>65%) of smaller MPs (less than 50 μm) found in burrow walls. The burrows contained high levels of glyphosate and its metabolite, suggesting that earthworms might transport pesticides deeper into the soil and increase the mobility of smaller MPs. The size of MPs is important when working with living organisms, although most research is conducted in dirt-free environments. Martín et al. studied the effects of simazine, ibuprofen, and PE microbeads (250–300 μm) on *Caenorhabditis elegans* and Vibrio fischeri, finding negligible impacts as nematodes could not absorb the large MPs. Kiyama et al. suggested that C. elegans can discriminate plastics based on size. MPs may mitigate the harmful effects of toxic substances in plant litter by efficiently holding onto phenolic chemicals.

Rodríguez-Seijo et al. assessed the behavior and enzyme activity of E. fetida in response to a mixture of two MP sizes and chlorpyrifos in soil. They confirmed that the pesticide was released from MPs and inhibited acetylcholinesterase activity, but the earthworms' chlorpyrifos levels remained unchanged as they avoided contaminated plastic. E. fetida might use gustatory and olfactory cues to distinguish between different plastics, as suggested for some helminthes and arthropods.

Lastly, the simultaneous presence of MPs and the organophosphorous pesticide chlorpyrifos was linked to increases in hemoglobin count in the terrestrial arthropod Porcellio scaber (Dolar et al., 2021). The same authors reevaluated the effects of tire particles, PE fibers, and *chlorpyrifos* on *P. scaber* and the springtail *Folsomia candida*, finding that MPs decreased the insecticide's bioavailability depending on the MP type. Tire particles, particularly in high concentrations, reduced chlorpyrifos-caused mortality in P. scaber, but not PE fibers. Differences in arthropod responses could be due to their soil niches; springtails are less exposed to soil toxins than the more sensitive P. scaber.

Although the potential harm of MPs and pesticides to living organisms has not received much attention, recent studies using advanced techniques like Pd-doped nanoplastics, fluorescently tagged PS microbeads, and pyrolysis GC-MS have shown that MPs can be effectively absorbed and translocated by plants. Results on MPs' influence on plants have been mixed. For example, studies on wheat have shown that MPs (PVC and PE) in soil can increase plant biomass by altering soil density and water-holding capacity, which reduces root penetration resistance and increases aeration. However, negative effects on wheat components have also been observed, particularly with biodegradable MPs.

Rice, being more susceptible to MPs than wheat or lettuce, has shown various negative effects, including loss in height and dry weight, reduction in primary root length, elevated oxidative stress, and increased nutrient uptake. Some MPs have minimal to no effect on wheat biomass output, seed germination, or seedling growth, even at high plastic concentrations. Pesticide and plant interactions can also vary. For instance, maize leaf development was initially inhibited but later increased with simultaneous contaminant and MP exposure, while root growth was reduced by 50%. The combination treatment improved the bacterial community and suggested additional effects.

In contrast, individual treatments (pollutants alone or white/blue PE) had more negative impacts on Lactuca sativa than combination treatments, likely due to contaminants adhering to MPs. Together, PS and the herbicide quinclorac avoided rice damage in lettuce by altering carbohydrate and amino acid metabolism and inducing the rice antioxidant enzyme system. The coexistence of MPs and pesticides may negatively impact living organisms, especially plants. The effects of pesticides and plastics on soil microbiomes are complex and multifaceted, with both short-term and long-term implications for soil health and ecosystem functioning (Cheng et al., 2020; Hernández-Gutiérrez et al., 2021; Selonen et al., 2023).

6 Conclusion

Pesticides, although designed to target specific pests, often unintentionally affect non-target soil organisms, disrupting microbial communities, decreasing diversity, and altering microbial functions. Continuous pesticide use can lead to pesticide-resistant microbial populations, negatively impacting beneficial soil microbes and potentially leading to harmful pathogen proliferation. Pesticide residues can persist and accumulate in soil, further disrupting microbial communities and impacting soil nutrient cycling, organic matter decomposition, and overall soil fertility.

Plastics, especially microplastics, can physically alter soil structure, affecting aeration, water retention, and nutrient availability. Microplastics can also serve as substrates for microbial colonization and introduce chemical additives or contaminants. These changes can create persistent microenvironments that alter microbial community composition and function. Some microorganisms can degrade certain types of plastics, influencing their abundance and activity in soil ecosystems. However, the long-term effects of plastic pollution on soil microbiomes are not fully understood and require further research.

Pesticides and plastics can interact synergistically or additively, exacerbating their individual impacts on soil microbiomes. Plastic debris in soil can act as reservoirs for pesticide residues, prolonging their persistence and increasing microbial exposure. Both can indirectly influence soil microbiomes through their effects on higher trophic levels, such as earthworms and arthropods, which can further alter microbial communities. To mitigate these adverse effects, integrated pest management strategies that minimize pesticide use and promote ecological alternatives are necessary. Reducing plastic pollution and improving waste management practices are crucial to prevent plastic accumulation in soil. Continued research is essential to understand better how pesticides and plastics influence soil microbiomes and their interactions with other environmental factors like climate change and land use practices.

In conclusion, the impacts of pesticides and plastics on soil microbiomes present significant challenges for sustainable soil management and agricultural productivity. Addressing these challenges requires a multifaceted approach that incorporates ecological principles, technological innovations, and policy interventions to promote soil health and biodiversity conservation.

References

Abioye, O. P., Ijah, U. J. J., Aransiola, S. A., Auta, S. H., & Ojeba, M. I. (2021). Bioremediation of toxic pesticides in soil using microbial products. In R. Prasad, S. C. Nayak, R. N. Kharwar, & N. K. Dubey (Eds.), *Mycoremediation and environmental sustainability. Fungal biology*. Springer. https://doi.org/10.1007/978-3-030-54422-5_1

Allen, T., Farley, S., Draper, J., Clement, C., & Polidoro, B. (2018). Variations in sorption of organochlorine pesticides and PCBs across six different plastic polymers. *Journal of Environmental and Toxicological Studies, 2*(6). https://doi.org/10.16966/2576-6430.109

Aransiola, S. A., Victor-Ekwebelem, M. O., Ajiboye, A. E., Leh-Togi Zobeashia, S. S., Ijah, U. J. J., & Oyedele, O. J. (2023). Ecological impacts and toxicity of micro and nanoplastics in agroecosystem. In N. R. Maddela, K. V. Reddy, & P. Ranjit (Eds.), *Micro and nanoplastics in soil*. Springer. https://doi.org/10.10007/978-3-031-21195-9_10

Ateia, M., Zheng, T., Calace, S., Tharayil, N., Pilla, S., & Karanfil, T. (2020). Sorption behavior of real microplastics (MPs): insights for organic micropollutants adsorption on a large set of well-characterized MPs. *Science of the Total Environment, 720*, 137634.

Awet, T., Kohl, Y., Meier, F., Straskraba, S., Grün, A.-L., Ruf, T., Jost, C., Drexel, R., Tunc, E., & Emmerling, C. (2018). Effects of polystyrene nanoparticles on the microbiota and functional diversity of enzymes in soil. *Environmental Sciences Europe, 30*, 1–10.

Beriot, N., Zomer, P., Zornoza, R., & Geissen, V. (2020). A laboratory comparison of the interactions between three plastic mulch types and 38 active substances found in pesticides. *PeerJ, 8*, e9876.

Bosker, T., Bouwman, L. J., Brun, N. R., Behrens, P., & Vijver, M. G. (2019). Microplastics accumulate on pores in seed capsule and delay germination and root growth of the terrestrial vascular plant Lepidium sativum. *Chemosphere, 226*, 774–781.

Boughattas, I., Zitouni, N., Hattab, S., Mkhinini, M., Missawi, O., Helaoui, S., Mokni, M., Bousserrhine, N., & Banni, M. (2022). Interactive effects of environmental microplastics and 2, 4-dichlorophenoxyacetic acid (2, 4-D) on the earthworm Eisenia andrei. *Journal of Hazardous Materials, 424*, 127578.

Boyle, K., & Örmeci, B. (2020). Microplastics and nanoplastics in the freshwater and terrestrial environment: A review. *Water, 12*(9), 2633.

Bringer, A., Thomas, H., Dubillot, E., Le Floch, S., Receveur, J., Cachot, J., & Tran, D. (2021). Subchronic exposure to high-density polyethylene microplastics alone or in combination with chlortoluron significantly affected valve activity and daily growth of the Pacific oyster, Crassostrea gigas. *Aquatic Toxicology, 237*, 105880.

Carpenter, E. J., & Smith, K., Jr. (1972). Plastics on the Sargasso Sea surface. *Science, 175*(4027), 1240–1241.

Castan, S., Henkel, C., Hüffer, T., et al. (2021). Microplastics and nanoplastics barely enhance contaminant mobility in agricultural soils. *Communications Earth & Environment, 2*, 193.

Chamas, A., Moon, H., Zheng, J., Qiu, Y., Tabassum, T., Jang, J. H., Abu-Omar, M., Scott, S. L., & Suh, S. (2020). Degradation rates of plastics in the environment. *ACS Sustainable Chemistry & Engineering, 8*(9), 3494–3511.

Chen, J., Rao, C., Yuan, R., Sun, D., Guo, S., Li, L., Yang, S., Qian, D., Lu, R., & Cao, X. (2022). Long-term exposure to polyethylene microplastics and glyphosate interferes with the behavior, intestinal microbial homeostasis, and metabolites of the common carp (Cyprinus carpio L.). *Science of the Total Environment, 814*, 152681.

Cheng, Y., Zhu, L., Song, W., Jiang, C., Li, B., Du, Z., Wang, J., Wang, J., Li, D., & Zhang, K. (2020). Combined effects of mulch film-derived microplastics and atrazine on oxidative stress and gene expression in earthworm (Eisenia fetida). *Science of the Total Environment, 746*, 141280.

Chi, J., Yin, Y., Zhang, W., Putnis, C. V., & Wang, L. (2021). Nanoscale imaging of the simultaneous occlusion of nanoplastics and glyphosate within soil minerals. *Environmental Science: Nano, 8*(10), 2855–2865.

Chia, R. W., Lee, J.-Y., Jang, J., Kim, H., & Kwon, K. D. (2022). Soil health and microplastics: a review of the impacts of microplastic contamination on soil properties. *Journal of Soils and Sediments, 22*(10), 2690–2705.

Chowdhury, S. R., Dey, A., Mondal, S., & Gautam, M. K. (2023). Environmental microplastics and nanoplastics: Effects on cardiovascular system. *Toxicologie Analytique et Clinique., 36*(2), 145–157.

Conley, K., Clum, A., Deepe, J., Lane, H., & Beckingham, B. (2019). Wastewater treatment plants as a source of microplastics to an urban estuary: Removal efficiencies and loading per capita over one year. *Water Research X, 3*, 100030.

Conti, I., Simioni, C., Varano, G., Brenna, C., Costanzi, E., & Neri, L. M. (2021). Legislation to limit the environmental plastic and microplastic pollution and their influence on human exposure. *Environmental Pollution, 288*, 117708.

Corradini, F., Meza, P., Eguiluz, R., Casado, F., Huerta-Lwanga, E., & Geissen, V. (2019). Evidence of microplastic accumulation in agricultural soils from sewage sludge disposal. *Science of the Total Environment, 671*, 411–420.

de Souza Machado, A. A., Kloas, W., Zarfl, C., Hempel, S., & Rillig, M. C. (2018a). Microplastics as an emerging threat to terrestrial ecosystems. *Global Change Biology, 24*(4), 1405–1416.

de Souza Machado, A. A., Lau, C. W., Till, J., Kloas, W., Lehmann, A., Becker, R., & Rillig, M. C. (2018b). Impacts of microplastics on the soil biophysical environment. *Environmental Science & Technology, 52*(17), 9656–9665.

Ding, J., Zhang, S., Razanajatovo, R. M., Zou, H., & Zhu, W. (2018). Accumulation, tissue distribution, and biochemical effects of polystyrene microplastics in the freshwater fish red tilapia (Oreochromis niloticus). *Environmental Pollution, 238*, 1–9.

Dissanayake, P. D., Kim, S., Sarkar, B., Oleszczuk, P., Sang, M. K., Haque, M. N., Ahn, J. H., Bank, M. S., & Ok, Y. S. (2022). Effects of microplastics on the terrestrial environment: a critical review. *Environmental Research, 209*, 112734.

Dolar, A., Selonen, S., van Gestel, C. A., Perc, V., Drobne, D., & Kokalj, A. J. (2021). Microplastics, chlorpyrifos and their mixtures modulate immune processes in the terrestrial crustacean Porcellio scaber. *Science of the Total Environment, 772*, 144900.

Dris, R., Gasperi, J., Saad, M., Mirande, C., & Tassin, B. (2016). Synthetic fibers in atmospheric fallout: A source of microplastics in the environment? *Marine Pollution Bulletin, 104*(1–2), 290–293.

Edo, C., Fernández-Piñas, F., & Rosal, R. (2022). Microplastics identification and quantification in the composted organic fraction of municipal solid waste. *Science of the Total Environment, 813*, 151902.

Fang, S., Yu, W., Li, C., Liu, Y., Qiu, J., & Kong, F. (2019). Adsorption behavior of three triazole fungicides on polystyrene microplastics. *Science of the Total Environment, 691*, 1119–1126.

Fatema, M., & Farenhorst, A. (2022). Sorption of pesticides by microplastics, charcoal, ash, and river sediments. *Journal of Soils and Sediments, 22*(6), 1876–1884.

Felten, V., Toumi, H., Masfaraud, J.-F., Billoir, E., Camara, B. I., & Férard, J.-F. (2020). Microplastics enhance Daphnia magna sensitivity to the pyrethroid insecticide deltamethrin: Effects on life history traits. *Science of the Total Environment, 714*, 136567.

Feng, X., Wang, Q., Sun, Y., Zhang, S., & Wang, F. (2022). Microplastics change soil properties, heavy metal availability and bacterial community in a Pb-Zn-contaminated soil. *Journal of Hazardous Materials, 424*, 127364.

Fernández, B., Campillo, J. A., Chaves-Pozo, E., Bellas, J., León, V. M., & Albentosa, M. (2022). Comparative role of microplastics and microalgae as vectors for chlorpyrifos bioacumulation and related physiological and immune effects in mussels. *Science of the Total Environment, 807*, 150983.

Fu, L., Li, J., Wang, G., Luan, Y., & Dai, W. (2021). Adsorption behavior of organic pollutants on microplastics. *Ecotoxicology and Environmental Safety, 217*, 112207.

Galafassi, S., Nizzetto, L., & Volta, P. (2019). Plastic sources: A survey across scientific and grey literature for their inventory and relative contribution to microplastics pollution in natural environments, with an emphasis on surface water. *Science of the Total Environment, 693*, 133499.

Gautam, M. K., Mondal, T., Nath, R., Mahajon, B., Chincholikar, M., Bose, A., Das, D., Das, R., & Mondal, S. (2024). Harnessing activated hydrochars: A novel approach for pharmaceutical contaminant removal. *C, 10*(1), 8.

Gonzalez, M., Mitton, F. M., Miglioranza, K. S. B., & Peña, A. (2019). Role of a non-ionic surfactant and carboxylic acids on the leaching of aged DDT residues in undisturbed soil columns. *Journal of Soils and Sediments, 19*, 1745–1755.

Gui, J., Sun, Y., Wang, J., Chen, X., Zhang, S., & Wu, D. (2021). Microplastics in composting of rural domestic waste: Abundance, characteristics, and release from the surface of macroplastics. *Environmental Pollution, 274*, 116553.

Hanachi, P., Karbalaei, S., & Yu, S. (2021). Combined polystyrene microplastics and chlorpyrifos decrease levels of nutritional parameters in muscle of rainbow trout (Oncorhynchus mykiss). *Environmental Science and Pollution Research, 28*, 64908–64920.

Hanslik, L., Seiwert, B., Huppertsberg, S., Knepper, T. P., Reemtsma, T., & Braunbeck, T. (2022). Biomarker responses in zebrafish (Danio rerio) following long-term exposure to microplastic-associated chlorpyrifos and benzo (k) fluoranthene. *Aquatic Toxicology, 245*, 106120.

Haward, M. (2018). Plastic pollution of the world's seas and oceans as a contemporary challenge in ocean governance. *Nature Communications, 9*(1), 667.

Hernández-Gutiérrez, E., Rendón-von Osten, J., Escalona-Segura, G., Mendoza-Vega, J., Dzul-Caamal, R., Posthumus, S., Vastenhouw, R., Yang, X., Geissen, V., & Huerta-Lwanga, E. (2021). Morphospecies abundance of above-ground invertebrates in agricultural systems under glyphosate and microplastics in south-eastern Mexico. *Environments, 8*(11), 130.

Horton, A. A., Walton, A., Spurgeon, D. J., Lahive, E., & Svendsen, C. (2017). Microplastics in freshwater and terrestrial environments: Evaluating the current understanding to identify the knowledge gaps and future research priorities. *Science of the Total Environment, 586*, 127–141.

Horton, A. A., Vijver, M. G., Lahive, E., Spurgeon, D. J., Svendsen, C., Heutink, R., van Bodegom, P. M., & Baas, J. (2018). Acute toxicity of organic pesticides to Daphnia magna is unchanged by co-exposure to polystyrene microplastics. *Ecotoxicology and Environmental Safety, 166*, 26–34.

Huerta Lwanga, E., Gertsen, H., Gooren, H., Peters, P., Salánki, T., Van Der Ploeg, M., Besseling, E., Koelmans, A. A., & Geissen, V. (2016). Microplastics in the terrestrial ecosystem: implications for Lumbricus terrestris (Oligochaeta, Lumbricidae). *Environmental Science & Technology, 50*(5), 2685–2691.

Hüffer, T., Metzelder, F., Sigmund, G., Slawek, S., Schmidt, T. C., & Hofmann, T. (2019). Polyethylene microplastics influence the transport of organic contaminants in soil. *Science of the Total Environment, 657*, 242–247.

Jiang, M., Hu, L., Lu, A., Liang, G., Lin, Z., Zhang, T., Xu, L., Li, B., & Gong, W. (2020). Strong sorption of two fungicides onto biodegradable microplastics with emphasis on the negligible role of environmental factors. *Environmental Pollution, 267*, 115496.

Jiang, H., Xiong, Q., Chen, X., Pan, W., & Dai, Y. (2021). Carrier effect of S-metolachlor by microplastics and environmental risk assessment. *Journal of Water Process Engineering, 44*, 102451.

Johansen, M. R., Christensen, T. B., Ramos, T. M., & Syberg, K. (2022). A review of the plastic value chain from a circular economy perspective. *Journal of Environmental Management, 302*, 113975.

Karbalaei, S., Hanachi, P., Rafiee, G., & Seifori, P. (2021). Toxicity of polystyrene microplastics on juvenile Oncorhynchus mykiss (rainbow trout) after individual and combined exposure with chlorpyrifos. *Journal of Hazardous Materials, 403*, 123980.

Kelkar, V. P., Rolsky, C. B., Pant, A., Green, M. D., Tongay, S., & Halden, R. U. (2019). Chemical and physical changes of microplastics during sterilization by chlorination. *Water Research, 163*, 114871.

Kokalj, A. J., Horvat, P., Skalar, T., & Kržan, A. (2018). Plastic bag and facial cleanser derived microplastic do not affect feeding behaviour and energy reserves of terrestrial isopods. *Science of the Total Environment, 615*, 761–766.

Koyuncuoğlu, P., & Erden, G. (2021). Sampling, pre-treatment, and identification methods of microplastics in sewage sludge and their effects in agricultural soils: A review. *Environmental Monitoring and Assessment, 193*, 1–28.

Lamastra, L., Suciu, N. A., & Trevisan, M. (2018). Sewage sludge for sustainable agriculture: Contaminants' contents and potential use as fertilizer. *Chemical and Biological Technologies in Agriculture, 5*(1), 1–6.

Lan, T., Wang, T., Cao, F., Yu, C., Chu, Q., & Wang, F. (2021). A comparative study on the adsorption behavior of pesticides by pristine and aged microplastics from agricultural polyethylene soil films. *Ecotoxicology and Environmental Safety, 209*, 111781.

Li, R., Liu, Y., Sheng, Y., Xiang, Q., Zhou, Y., & Cizdziel, J. V. (2020). Effect of prothioconazole on the degradation of microplastics derived from mulching plastic film: Apparent change and interaction with heavy metals in soil. *Environmental Pollution, 260*, 113988.

Li, H., Liu, L., Xu, Y., & Zhang, J. (2022a). Microplastic effects on soil system parameters: A meta-analysis study. *Environmental Science and Pollution Research*, 1–12.

Li, C., Yuan, S., Zhou, Y., Li, X., Duan, L., Huang, L., Zhou, X., Ma, Y., & Pang, S. (2022b). Microplastics reduce the bioaccumulation and oxidative stress damage of triazole fungicides in fish. *Science of the Total Environment, 806*, 151475.

Lin, Y., Xie, J., Xiang, Q., Liu, Y., Wang, P., Wu, Y., & Zhou, Y. (2022). Effect of propiconazole on plastic film microplastic degradation: Focusing on the change in microplastic morphology and heavy metal distribution. *Science of the Total Environment, 822*, 153609.

Liu, H., Yang, X., Liang, C., Li, Y., Qiao, L., Ai, Z., Xue, S., & Liu, G. (2019). Interactive effects of microplastics and glyphosate on the dynamics of soil dissolved organic matter in a Chinese loess soil. *Catena, 182*, 104177.

Liu, Y., Wu, X., Liu, R., Chen, Y., Fu, J., & Ou, H. (2022a). Modifications of ultraviolet irradiation and chlorination on microplastics: Effect of sterilization pattern. *Science of the Total Environment, 812*, 152541.

Liu, X., Zhou, D.-D., Chen, M., Cao, Y.-W., Zhuang, L.-Y., Lu, Z.-H., & Yang, Z.-H. (2022b). Adsorption behavior of azole fungicides on polystyrene and polyethylene microplastics. *Chemosphere, 308*, 136280.

Liu, W., Pan, T., Liu, H., Jiang, M., & Zhang, T. (2023). Adsorption behavior of imidacloprid pesticide on polar microplastics under environmental conditions: Critical role of photo-aging. *Frontiers of Environmental Science & Engineering, 17*(4), 41.

Luo, Y., Zhang, Y., Xu, Y., Guo, X., & Zhu, L. (2020). Distribution characteristics and mechanism of microplastics mediated by soil physicochemical properties. *Science of the Total Environment, 726*, 138389.

Lwanga, H., & Geissen, G.. Sources of light density microplastic related to two agricultural practices the use of compost and plastic mulch.

Lwanga, E. H., Gertsen, H., Gooren, H., Peters, P., Salánki, T., van der Ploeg, M., Besseling, E., Koelmans, A. A., & Geissen, V. (2017). Incorporation of microplastics from litter into burrows of Lumbricus terrestris. *Environmental Pollution, 220*, 523–531.

Malyan, S. K., Singh, R., Rawat, M., Kumar, M., Pugazhendhi, A., Kumar, A., Kumar, V., & Kumar, S. S. (2019). An overview of carcinogenic pollutants in groundwater of India. *Biocatalysis and Agricultural Biotechnology, 21*, 101288.

Mammo, F., Amoah, I., Gani, K., Pillay, L., Ratha, S., Bux, F., & Kumari, S. (2020). Microplastics in the environment: Interactions with microbes and chemical contaminants. *Science of the Total Environment, 743*, 140518.

Martín, C., Fajardo, C., Costa, G., Sánchez-Fortún, S., San Andrés, M. D., González, F., Nande, M., Mengs, G., & Martín, M. (2021). Bioassays to assess the ecotoxicological impact of polyethylene microplastics and two organic pollutants, simazine and ibuprofen. *Chemosphere, 274*, 129704.

Martín-Lara, M., Godoy, V., Quesada, L., Lozano, E., & Calero, M. (2021). Environmental status of marine plastic pollution in Spain. *Marine Pollution Bulletin, 170*, 112677.

Mishra, C., Samal, S., Samal, R. R., Behera, B. P., Pallavini, P., Dash, P., Brahma, S., Moharana, T., Pradhan, S., & Acharya, P. (2022). Polyvinylchloride and polypropylene as adsorbents of the pesticide monocrotophos enhance oxidative stress in Eudrillus eugeniae (Kinberg). *Chemosphere, 295*, 133837.

Mo, Q., Yang, X., Wang, J., Xu, H., Li, W., Fan, Q., Gao, S., Yang, W., Gao, C., & Liao, D. (2021). Adsorption mechanism of two pesticides on polyethylene and polypropylene microplastics: DFT calculations and particle size effects. *Environmental Pollution, 291*, 118120.

Murphy, F., Ewins, C., Carbonnier, F., & Quinn, B. (2016). Wastewater treatment works (WwTW) as a source of microplastics in the aquatic environment. *Environmental Science & Technology, 50*(11), 5800–5808.

Musa, I. O., Auta, H. S., Ilyasu, U. S., Aransiola, S. A., Makun, H. A., Adabara, N. U., Abioye, O. P., Aziz, A., Jayanthi, B., Maddela, N. R., & Prasad, R. (2024). Micro and nanoplastics in environment—Degradation, detection, ecological impact. *International Journal of Environmental Research, 18*, 1. https://doi.org/10.1007/s41742-023-00551-9

Palansooriya, K. N., Shi, L., Sarkar, B., Parikh, S. J., Sang, M. K., Lee, S. R., & Ok, Y. S. (2022). Effect of LDPE microplastics on chemical properties and microbial communities in soil. *Soil Use and Management, 38*(3), 1481–1492.

Pan, T., Liu, H., Jiang, M., Li, J., Liu, W., Jiao, Q., & Zhang, T. (2023). New insights into the adsorption behavior of thiacloprid at the microfibers/water interface: Role of humic acid. *Chemosphere, 311*, 136938.

Parashar, N., Mahanty, B., & Hait, S. (2023). Microplastics as carriers of per-and polyfluoroalkyl substances (PFAS) in aquatic environment: Interactions and ecotoxicological effects. *Water Emerg. Contam. Nanoplastics, 2*, 15.

Pelamatti, T., Rios-Mendoza, L. M., Hoyos-Padilla, E. M., Galván-Magaña, F., De Camillis, R., Marmolejo-Rodríguez, A. J., & González-Armas, R. (2021). Contamination knows no borders: Toxic organic compounds pollute plastics in the biodiversity hotspot of Revillagigedo Archipelago National Park, Mexico. *Marine Pollution Bulletin, 170*, 112623.

Peña, A., Rodríguez-Liébana, J. A., & Delgado-Moreno, L. (2023). Interactions of microplastics with pesticides in soils and their ecotoxicological implications. *Agronomy, 13*(3), 701.

Qi, Y., Beriot, N., Gort, G., Lwanga, E. H., Gooren, H., Yang, X., & Geissen, V. (2020a). Impact of plastic mulch film debris on soil physicochemical and hydrological properties. *Environmental Pollution, 266*, 115097.

Qi, Y., Ossowicki, A., Yang, X., Lwanga, E. H., Dini-Andreote, F., Geissen, V., & Garbeva, P. (2020b). Effects of plastic mulch film residues on wheat rhizosphere and soil properties. *Journal of Hazardous Materials, 387*, 121711.

Ramos, L., Berenstein, G., Hughes, E. A., Zalts, A., & Montserrat, J. M. (2015). Polyethylene film incorporation into the horticultural soil of small periurban production units in Argentina. *Science of the Total Environment, 523*, 74–81.

Ritchie, H., & Roser, M. (2018). *Plastic pollution*. Published online at OurWorldInData.org.

Ritter, H. (2022). Kunststoffe für die Menschen: Nutzen und Nachteile. *Wiley Online Library, 56*, 147–147.

Rodríguez-Liébana, J. A., ElGouzi, S., & Peña, A. (2017). Laboratory persistence in soil of thiacloprid, pendimethalin and fenarimol incubated with treated wastewater and dissolved organic matter solutions. Contribution of soil biota. *Chemosphere, 181*, 508–517.

Rodríguez-Seijo, A., Santos, B., da Silva, E. F., Cachada, A., & Pereira, R. (2018). Low-density polyethylene microplastics as a source and carriers of agrochemicals to soil and earthworms. *Environmental Chemistry, 16*(1), 8–17.

Rosenboom, J.-G., Langer, R., & Traverso, G. (2022). Bioplastics for a circular economy. *Nature Reviews Materials, 7*(2), 117–137.

Schell, T., Hurley, R., Buenaventura, N. T., Mauri, P. V., Nizzetto, L., Rico, A., & Vighi, M. (2022). Fate of microplastics in agricultural soils amended with sewage sludge: Is surface water runoff a relevant environmental pathway? *Environmental Pollution, 293*, 118520.

Scopetani, C., Chelazzi, D., Cincinelli, A., Martellini, T., Leiniö, V., & Pellinen, J. (2022). Hazardous contaminants in plastics contained in compost and agricultural soil. *Chemosphere, 293*, 133645.

Seidensticker, S., Grathwohl, P., Lamprecht, J., & Zarfl, C. (2018). A combined experimental and modeling study to evaluate pH-dependent sorption of polar and non-polar compounds to polyethylene and polystyrene microplastics. *Environmental Sciences Europe, 30*, 1–12.

Selonen, S., Kokalj, A. J., Benguedouar, H., Petroody, S. S. A., Dolar, A., Drobne, D., & van Gestel, C. A. (2023). Modulation of chlorpyrifos toxicity to soil arthropods by simultaneous exposure to polyester microfibers or tire particle microplastics. *Applied Soil Ecology, 181*, 104657.

Steinmetz, Z., Wollmann, C., Schaefer, M., Buchmann, C., David, J., Tröger, J., Muñoz, K., Frör, O., & Schaumann, G. E. (2016). Plastic mulching in agriculture. Trading short-term agronomic benefits for long-term soil degradation? *Science of the Total Environment, 550*, 690–705.

Stubenrauch, J., & Ekardt, F. (2020). Plastic pollution in soils: Governance approaches to foster soil health and closed nutrient cycles. *Environments, 7*(5), 38.

Sun, W., Meng, Z., Li, R., Zhang, R., Jia, M., Yan, S., Tian, S., Zhou, Z., & Zhu, W. (2021). Joint effects of microplastic and dufulin on bioaccumulation, oxidative stress and metabolic profile of the earthworm (Eisenia fetida). *Chemosphere, 263*, 128171.

Tan, M., Sun, Y., Gui, J., Wang, J., Chen, X., Song, W., & Wu, D. (2022). Distribution characteristics of microplastics in typical organic solid wastes and their biologically treated products. *Science of the Total Environment, 852*, 158440.

Tourinho, P. S., Kočí, V., Loureiro, S., & van Gestel, C. A. (2019). Partitioning of chemical contaminants to microplastics: Sorption mechanisms, environmental distribution and effects on toxicity and bioaccumulation. *Environmental Pollution, 252*, 1246–1256.

Villegas, L., Cabrera, M., & Capparelli, M. V. (2021). Assessment of microplastic and organophosphate pesticides contamination in fiddler crabs from a Ramsar site in the estuary of Guayas River, Ecuador. *Bulletin of Environmental Contamination and Toxicology, 107*(1), 20–28.

Vithanage, M., Ramanayaka, S., Hasinthara, S., & Navaratne, A. (2021). Compost as a carrier for microplastics and plastic-bound toxic metals into agroecosystems. *Current Opinion in Environmental Science & Health, 24*, 100297.

Wang, F., Gao, J., Zhai, W., Liu, D., Zhou, Z., & Wang, P. (2020a). The influence of polyethylene microplastics on pesticide residue and degradation in the aquatic environment. *Journal of Hazardous Materials, 394*, 122517.

Wang, T., Yu, C., Chu, Q., Wang, F., Lan, T., & Wang, J. (2020b). Adsorption behavior and mechanism of five pesticides on microplastics from agricultural polyethylene films. *Chemosphere, 244*, 125491.

Wang, C., Zhao, J., & Xing, B. (2021). Environmental source, fate, and toxicity of microplastics. *Journal of Hazardous Materials, 407*, 124357.

Wang, Y., Liu, C., Wang, F., & Sun, Q. (2022). Behavior and mechanism of atrazine adsorption on pristine and aged microplastics in the aquatic environment: Kinetic and thermodynamic studies. *Chemosphere, 292*, 133425.

Wanner, P. (2021). Plastic in agricultural soils – A global risk for groundwater systems and drinking water supplies? – A review. *Chemosphere, 264*, 128453.

Warrier, A. K., Kulkarni, B., Amrutha, K., Jayaram, D., Valsan, G., & Agarwal, P. (2022). Seasonal variations in the abundance and distribution of microplastic particles in the surface waters of a Southern Indian Lake. *Chemosphere, 300*, 134556.

Weithmann, N., Möller, J. N., Löder, M. G., Piehl, S., Laforsch, C., & Freitag, R. (2018). Organic fertilizer as a vehicle for the entry of microplastic into the environment. *Science Advances, 4*(4), eaap8060.

Wu, C., Pan, S., Shan, Y., Ma, Y., Wang, D., Song, X., Hu, H., Ren, X., Ma, X., & Cui, J. (2022). Microplastics mulch film affects the environmental behavior of adsorption and degradation of pesticide residues in soil. *Environmental Research, 214*, 114133.

Xu, Z., Qian, X., Wang, C., Zhang, C., Tang, T., Zhao, X., & Li, L. (2020). Environmentally relevant concentrations of microplastic exhibits negligible impacts on thiacloprid dissipation and enzyme activity in soil. *Environmental Research, 189*, 109892.

Yang, X., Bento, C. P., Chen, H., Zhang, H., Xue, S., Lwanga, E. H., Zomer, P., Ritsema, C. J., & Geissen, V. (2018). Influence of microplastic addition on glyphosate decay and soil microbial activities in Chinese loess soil. *Environmental Pollution, 242*, 338–347.

Yang, X., Lwanga, E. H., Bemani, A., Gertsen, H., Salanki, T., Guo, X., Fu, H., Xue, S., Ritsema, C., & Geissen, V. (2019). Biogenic transport of glyphosate in the presence of LDPE microplastics: A mesocosm experiment. *Environmental Pollution, 245*, 829–835.

Yang, Z., Lü, F., Hu, T., Xu, X., Zhang, H., Shao, L., Ye, J., & He, P. (2022). Occurrence of macroplastics and microplastics in biogenic waste digestate: Effects of depackaging at source and dewatering process. *Waste Management, 154*, 252–259.

Yi, M., Zhou, S., Zhang, L., & Ding, S. (2021). The effects of three different microplastics on enzyme activities and microbial communities in soil. *Water Environment Research, 93*(1), 24–32.

Yusuf, A., Sodiq, A., Giwa, A., Eke, J., Pikuda, O., Eniola, J. O., Ajiwokewu, B., Sambudi, N. S., & Bilad, M. R. (2022). Updated review on microplastics in water, their occurrence, detection, measurement, environmental pollution, and the need for regulatory standards. *Environmental Pollution, 292*, 118421.

Zhang, Q., Qu, Q., Lu, T., Ke, M., Zhu, Y., Zhang, M., Zhang, Z., Du, B., Pan, X., & Sun, L. (2018). The combined toxicity effect of nanoplastics and glyphosate on Microcystis aeruginosa growth. *Environmental Pollution, 243*, 1106–1112.

Zhang, C., Lei, Y., Qian, J., Qiao, Y., Liu, J., Li, S., Dai, L., Sun, K., Guo, H., & Sui, G. (2021). Sorption of organochlorine pesticides on polyethylene microplastics in soil suspension. *Ecotoxicology and Environmental Safety, 223*, 112591.

Ziajahromi, S., Kumar, A., Neale, P. A., & Leusch, F. D. (2019). Effects of polyethylene microplastics on the acute toxicity of a synthetic pyrethroid to midge larvae (Chironomus tepperi) in synthetic and river water. *Science of the Total Environment, 671*, 971–975.

Zocchi, M., & Sommaruga, R. (2019). Microplastics modify the toxicity of glyphosate on Daphnia magna. *Science of the Total Environment, 697*, 134194.

Soil Microbiomes and Biocontrol of Pathogens

Suresh Babu Bastipati, D. Sai Vaishnavi, Saloni Prasad, P Gnana Deepu, and Kondakindi Venkateswar Reddy

Abstract The soil microbiome plays a crucial role in maintaining agricultural productivity and plant health, offering promising avenues for sustainable disease management using microbial biocontrol agents. This chapter explores the potential of beneficial microorganisms, such as *Bacillus thuringiensis*, *Trichoderma* spp., and microbial volatile organic compounds (MVOCs), in controlling plant pathogens. These biocontrol agents operate through various mechanisms, including antibiosis, competition, parasitism, and the induction of systemic resistance, providing an environmentally friendly alternative to chemical pesticides. We review the current understanding of the interactions between soil microorganisms and plant pathogens, highlighting the efficacy and application of microbial biocontrol agents in diverse agricultural settings. The study also addresses the challenges associated with the consistency, formulation, and field application of these agents, proposing future research directions to optimize their use. By integrating microbial biocontrol agents into sustainable agricultural practices, it is possible to reduce reliance on chemical inputs, enhance soil health, and improve crop resilience. This article underscores the need for interdisciplinary collaboration, regulatory support, and farmer education to facilitate the adoption of microbial biocontrol strategies, ultimately contributing to sustainable agriculture and food security.

Keywords Soil microbiome · *Bacillus thuringiensis* · *Trichoderma* spp. · Microbial volatile organic compounds (MVOCs), etc.

1 Introduction

Soil microbiomes, the diverse community of microorganisms living in the soil, play a crucial role in maintaining soil health and supporting plant growth. Soil ecological function is maintained by soil microbes, which confer stability to the soil

S. B. Bastipati · D. S. Vaishnavi · S. Prasad · P G. Deepu · K. V. Reddy (✉)
University College of Engineering Science and Technology Hyderabad, Jawaharlal Nehru Technological University Kukatpally, Hyderabad, Telangana, India

S. A. Aransiola et al. (eds.), *Soil Microbiome in Green Technology Sustainability*,
https://doi.org/10.1007/978-3-031-71844-1_13

environment and stability during disturbance. The increased demand for agricultural products worldwide has been a consequence of population growth. Increased production leads to topsoil depletion, reduced organic matter content, and compromised soil ecological function. The need for sustainable and healthy food is also reflected in a population that is sensitive to environmental problems. Therefore, eco-friendly strategies for sustainable agriculture are becoming increasingly popular. The diversity of the soil microbiota, which plays a crucial role in nutrient recycling and soil formation, is therefore a key issue in sustainable agriculture. This chapter explores the composition and functions of soil microbiomes, the types and mechanisms of soil-borne pathogens, and how beneficial microbes can be used as biocontrol agents to manage these pathogens sustainably.

1.1 What Is Soil Microbiome

A wide range of microorganisms make up the soil microbiome, which supports vital ecosystem functions like disease suppression, nutrient recycling, and soil structure protection. The most complex habitat is soil, which is home to an enormous diversity of microbial life—roughly 4–5 × 10^{30} microbial cells. There are many kinds of microorganisms in this ecosystem, such as viruses, fungus, bacteria, and other microbes. These microbes are essential to many soil activities, including the breakdown of organic matter, the cycling of nutrients, and the biocontrol of plant diseases (Mendes et al., 2013).

1.2 Diversity of Microbiomes

The four basic groups of soil microorganisms are actinomycetes, protozoans, fungus, and bacteria. Each of these categories is essential to plant development, nutrient cycling, and soil health. Figure 1 contains diversity of soil microbiomes.

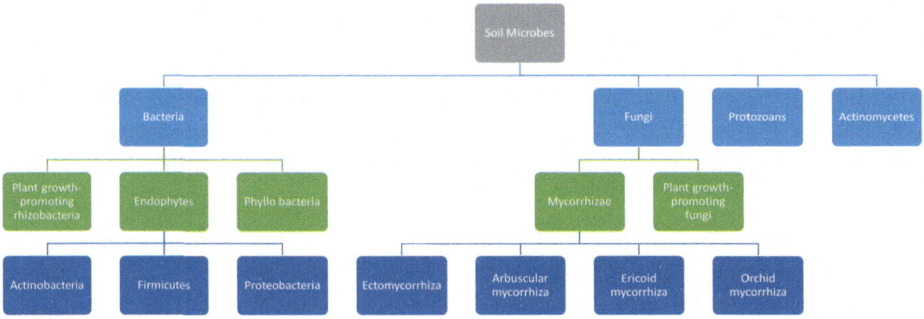

Fig. 1 Diversity of soil microbiome

There are three types of bacteria found in soil: phyllobacteria, endophytes, and plant growth-promoting rhizobacteria (PGPR). It is well recognized that PGPR can promote plant growth through several processes, such as nitrogen fixation, growth hormone synthesis, and disease prevention. Endophytes are microorganisms that inhabit plant tissues. They aid in plant growth and provide defense against diseases. On the surfaces of plant leaves, phyllobacteria are microorganisms that help shield plants against pathogens. Certain varieties of bacteria exist, including *actinobacteria*, *firmicutes*, and *proteobacteria*. To break down complex organic materials and increase soil fertility, *actinobacteria* are essential. *Firmicutes* and *proteobacteria* play a crucial role in the cycling of nutrients and the upkeep of plant health by aiding in the breakdown of organic matter and increasing nutrient availability (Chauhan et al., 2023).

Fungi, which can be divided into mycorrhizae and fungi that promote plant growth, are another essential part of soil microbial communities. Mycorrhizae and plant roots establish symbiotic partnerships that greatly improve nutrient and water intake. Subclassifications of mycorrhizae include ectomycorrhiza, arbuscular mycorrhiza, ericoid mycorrhiza, and orchid mycorrhiza. Ectomycorrhiza mostly live in trees and aid in their uptake of nutrients from the soil. Most plant species create relationships with arbuscular mycorrhiza, which enhance the plants' capacity to absorb nutrients. Typically present in plants belonging to the Ericaceae family, ericoid mycorrhiza facilitates the uptake of nutrients in soil deficient in nutrients. Because they provide the required nutrients, orchid mycorrhiza is vital to the germination and growth of orchids (Chauhan et al., 2023).

2 Functions of the Soil Microbiome

In agricultural environments, the soil microbiome is essential for preserving soil health and sustainability. Soil microorganisms carry out vital tasks such as maintaining soil structure, cycling nutrients, controlling biological populations, and managing carbon dynamics. Through the breakdown of organic matter, transformation of nutrients, and control of soil pH and nutrient availability, they play the role of "chemical engineers". Furthermore, bacteria serve as "biological regulators," guarding against pathogens, managing pests and illnesses, and assisting in the development of soil structure (Nadeu et al., 2023).

The rhizosphere is home to a wide variety of soil microorganisms, including bacteria, viruses, actinomycetes, blue-green algae, protozoa, and fungi. These bacteria perform a variety of tasks, such as maintaining soil structure, biodegrading pollutants, and moving biogenic components that give plants nourishment. The area of soil surrounding a plant's roots where the root's activities influence the qualities of the soil is known as the rhizosphere. Microbial diversity has a major role in the preservation of ecosystems and biological diversity. Listed below are some of the important functions of microbiome present in the soil (Nadeu et al., 2023).

2.1 Nitrogen Fixation

Bacteria that fix nitrogen are microorganisms that can convert atmospheric nitrogen into fixed nitrogen, which is an inorganic chemical that plants can use. These organisms play a crucial role in the nitrogen cycle since they are responsible for more than 90% of all nitrogen fixing.

There are two known species of nitrogen-fixing bacteria. The first type of bacteria is the free-living, nonsymbiotic ones, which include genera like *Azotobacter*, *Beijerinckia*, *Clostridium* as well as blue-green algae like *cyanobacteria*, *Anabaena* and *Nostoc*. The second category consists of mutualistic (symbiotic) bacteria. Some examples of these are certain species of *Azospirillum*, which is linked to cereal grasses, *Frankia*, which is linked to certain dicotyledonous species (actinorhizal plants), and *Rhizobium*, which is linked to leguminous plants.

The nitrogen-fixing bacteria live in symbiotic relationships with their hosts. They enter the root hairs of the plants, multiply, and promote the production of root nodules, which are enlargements of plant cells surrounded by bacteria. The bacteria in the nodules change free nitrogen into ammonia, which the host plant uses to grow (*Nitrogen-fixing Bacteria | Definition & Types*, 2024).

2.2 Nutrient Recycling

If there is a carbon source for energy, soil microbes can be found in vast numbers in the soil. Although there are many bacteria in the soil, their biomass is low due to their small size. Actinomycetes resemble bacteria in biomass, although being ten times fewer in number and greater in size. When the soil is not disturbed, fungi predominate in the biomass of the soil while having a lesser population. In tilled soils, fungal and nematode populations tend to predominate, while bacteria, actinomycetes, and protozoa are more resilient and can withstand greater soil disturbance than fungal populations (Kalakoutskii & Agre, 1976).

2.3 Plant Health

Beneficial microbes can facilitate the availability of essential nutrients such as nitrogen, phosphorus, and potassium by breaking down organic matter and converting these elements into forms that plants can absorb. Additionally, the soil microbiome contributes to plant health by producing growth hormones and suppressing harmful pathogens through the production of antimicrobial compounds and competitive exclusion. For instance, mycorrhizal fungi form networks with plant roots, extending their reach and improving water and nutrient absorption. Similarly, rhizobacteria can induce systemic resistance in plants, enhancing their ability to withstand biotic and abiotic stresses. These interactions underline the importance of maintaining a healthy and diverse soil microbiome to ensure sustainable agricultural practices and optimal plant health (Van Der Heijden et al., 2007; Berg, 2014; Berg et al., 2017).

3 Factors Effecting Soil Microbiome

Soil microbiomes are influenced by a multitude of biological and environmental variables that operate in both space and time. These comprise both Biotic and Abiotic, such as soil structure, moisture content, pH, and texture, as well as Biotic elements like plant species and microbial interactions. Although there is no one element that determines everything, some factors, such as soil texture, pH, and moisture content, have a consistent effect on microbial populations all over the world. For example, the total microbial biomass and the makeup of protists are greatly influenced by soil moisture. For instance, when soil is contaminated with fuel oil, some of the helpful microbes that usually help clean up the soil might not work as well. This happens because the oil contains tough contents that these microbes struggle to break down. So, there aren't as many backup helpers to do the job (Van Der Heijden et al., 2007).

Microbial communities are extremely dynamic and context sensitive. They can form symbiotic partnerships with plant roots in nutrient-rich environments like as the rhizosphere. Deeper soil strata do contain some microbes; however, their activity is lower. The delicate balance of soil microbiomes is further complicated by soil pollution. Long-term soil persistence of heavy metals can cause disruptions to microbial populations and biochemical processes, as they come from sources such as industrial emissions and fertilizer application. In a similar vein, the buildup of microplastics in soil may be harmful to the structure and function of microorganisms. Even though pesticides are widely used in agriculture, their effects on soil microbiomes are frequently disregarded. Pesticide application rates range from 30 to 50 percent, which means that microbial populations and soil health may be impacted. All things considered, knowing the intricate relationships between biotic and abiotic elements is essential to understanding the composition and roles of soil microbiomes as well as maintaining the health and productivity of the soil.The resilience and health of the soil are largely dependent on functional variety and redundancy. Redundancy guarantees that tasks can be performed by various species, enhancing soil resistance to disturbances, while functional diversity permits soils to execute a wide range of biological activities. Research has demonstrated a correlation between increasing multifunctionality in several ecosystems and a higher diversity of soil microbes. Linking microbial taxa to soil activities is still difficult, though, especially because most research has been done primarily on bacterial communities, leaving other microbial groups mostly unexplored (Van Der Heijden et al., 2007).

4 Rhizosphere: Hub of Microbial Activity

Plant-microbe interactions are universal and essential to the well-being of both the plant and the soil. Plants engage in a variety of complicated interactions with heterotrophic microbes, from parasitism to symbiosis. Important interactions between important components of the soil, plants, and microbes occur in the rhizosphere, which is the region next to plant roots. These exchanges may be advantageous, detrimental, or neutral. With up to 10–100 times more bacteria than bulk soil, rhizosphere is an ecosystem with perhaps the highest levels of biodiversity on Earth, richer in microbes than the surrounding soil (Chauhan et al., 2023).

The rhizosphere is an extraordinarily dense microbial home, with up to 10^11 cells of microbes per gram of root. Microbes linked with plants are called the plant's second genome. In the rhizosphere, interactions constitute more than 80% of the total, suggesting a strong possibility of mutually beneficial interactions. The rhizosphere is the primary site of activities such as respiration, gas exchange, and uptake of nutrients and water. It is also the location of labile carbon and other nutrients. All these efforts underscore the significance of the rhizosphere for the well-being of plants and soil (Chauhan et al., 2023).

5 Plant Microbial Interactions

Microbes and plants always interact, and this relationship is essential to the wellbeing of the soil and both partners. The intricate process involves the plants interacting with a variety of heterotrophic microbes, with which they may have close relationships ranging from symbiosis to parasitism.This interaction is beneficial most of the times but also harmful to the plants.

5.1 Legume-Rhizobia Symbiosis

Leguminous plants that generate specialized root nodules containing rhizobia, Gram-negative alpha-proteobacteria, are subject to endosymbiotic interactions with their counterparts. Nitrogenase activity in this symbiotic relationship fixes atmospheric nitrogen.Genera *Rhizobium, Azorhizobium, Mesorhizobium, Bradyrhizobium*, and *Ensifer* are examples of Rhizobia. Complementary soil microorganisms identify plant signals and generate Nod factors, which set off a gene cascade that results in bacterial infection and cell proliferation and the creation of nodules. Via an infection thread, bacteria go from the nodule primordium to the root hair cells. They mature into Bacteroides that fix nitrogen. The bacteroids are surrounded by a symbiosome membrane, which produces a microaerobic environment ideal for nitrogen fixation. Additionally, Rhizobia and Parasponia, a non-leguminous plant, create endosymbiotic relationships (Chauhan et al., 2023).

5.2 *Cyanobacteria-Plant Symbiosis*

Cyanobacteria, although widespread, rarely form symbiotic relationships with eukaryotes. The genus Nostoc is the most common cyanobacterial partner in such associations with land plants. These cyanobacteria colonize plant organs either intracellularly in angiosperms or extracellularly in gymnosperms, pteridophytes, and bryophytes. Bryophytes, such as mosses, hornworts, and liverworts, host heterocystous cyanobacteria, with mosses having epiphytic associations and hornworts maintaining endophytic relationships. Most of these symbioses are facultative, but in Azolla (a pteridophyte), the cyanobacteria are consistently transmitted through generations. Gymnosperms like cycads have coralloid roots infected by cyanobacteria, primarily Nostoc. The only known angiosperm-cyanobacteria symbiosis is the Gunnera-Nostoc partnership, where cyanobacteria enter through specialized stem glands and form an endosymbiotic interface for nutrient exchange (Chauhan et al., 2023).

5.3 *Plant-Pathogen Interaction*

Plant pathogens cover bacteria, viruses, fungus, and nematodes, with fungi and nematodes causing the greatest harm. Chemical triggers are used by pathogens to identify and infect hosts. Using effector-triggered immunity (ETI), PAMP-triggered immunity (PTI), and phytohormone signalling, plants fight pathogens. Necrotrophs, which destroy host cells to obtain nutrition, and biotrophs, which preserve host life, are the two categories of phytopathogens. The most dangerous organisms are fungi, which are mostly found as necrotrophs in the form of sclerotia and chlamydospores. Toxic metabolites and enzymes that break down cell walls are released by them to aid in infection. In contrast to necrotrophic soil fungi, biotrophic soil fungi are uncommon (Chauhan et al., 2023).

5.4 *Antimicrobial Root Exudates*

Exudates from plant roots function as antimicrobials in the rhizosphere, attracting good bacteria and inhibiting bad ones. The defence against microbial infections is aided by secondary metabolites found in root exudates, including Iso flavonoids, terpenoids, flavonoids, phenolics, and indole. Plant defensive responses are regulated by key signalling networks involving salicylic acid, abscisic acid, jasmonates, and jasmonic acid. As antimicrobial compounds build up, pathogenic interactions change the protein composition of the rhizosphere. Through competition between pathogenic and helpful bacteria for resources, the rhizobiome helps reduce disease. All things considered, antimicrobial root exudates are essential for keeping the soil microbiota healthy and shielding plants from soil-borne diseases (Chauhan et al., 2023).

6 Soil Borne Pathogens

The rhizosphere is described as both a playground and a battlefield where soilborne pathogens and beneficial microorganisms interact. The rhizosphere's complex microbial ecosystem, where interactions between plant roots, pathogens, and beneficial microorganisms shape plant health and soil fertility the rhizosphere is a critical zone where plants release root exudates that attract a diverse array of microorganisms, creating a hotspot for microbial activity (Raaijmakers et al., 2009).

Conversely, soil pathogens including bacteria, fungi, viruses, and nematodes, pose significant threats to agricultural productivity by causing diseases that affect plant health. Key soilborne pathogens such as *Phytophthora*, *Fusarium*, and *Rhizoctonia* are notorious for causing root rots, wilts, and damping-off diseases, leading to substantial crop losses (Mendes et al., 2013). Figure 2 contains types of soil pathogens and examples for each type.

6.1 *Types of Soil Pathogens*

6.1.1 Fungal Pathogens

Fungal pathogens are the most prevalent soil pathogens, including species such as *Phytophthora*, *Pythium*, *Fusarium*, and *Rhizoctonia*. These fungi produce spores that can survive in the soil for long periods, making them particularly resilient. They typically infect plants through natural openings like stomata or wounds, and some can directly penetrate plant surfaces using specialized structures called appressoria. Once inside, they secrete enzymes such as cellulases and pectinases that degrade plant cell walls, facilitating their spread. Fungal diseases manifest as root rots, wilts, and damping-off, leading to symptoms like yellowing leaves, stunted growth, and plant death (Agrios, 2005).

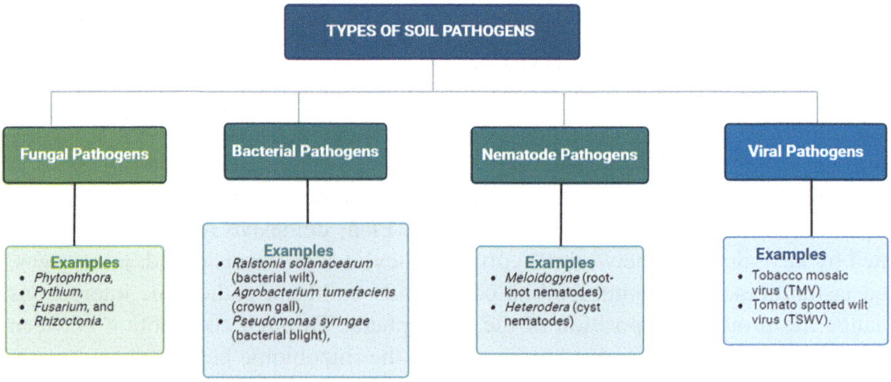

Fig. 2 Types of soil pathogens

6.1.2 Bacterial Pathogens

Soil-borne bacterial pathogens, such as *Ralstonia solanacearum* (bacterial wilt), *Agrobacterium tumefaciens* (crown gall), and *Pseudomonas syringae* (bacterial blight), thrive in moist soil conditions. Bacteria enter plants through natural openings, wounds, or via vectors like nematodes and insects. They produce a variety of toxins and enzymes that degrade plant tissues and disrupt normal plant functions. Bacterial diseases result in wilts, blights, galls, and soft rots, with symptoms including chlorosis, necrosis, cankers, and abnormal growths (Schaad, 1981).

6.1.3 Nematode Pathogens

Nematodes are microscopic roundworms, with pathogenic species including *Meloidogyne* (root-knot nematodes) and *Heterodera* (cyst nematodes). These nematodes can be free-living or parasitic. Parasitic nematodes infect plants by penetrating root tissues, where they establish feeding sites that disrupt normal root function. They release enzymes and other compounds that facilitate root penetration and feeding. Nematode infections lead to root galls, cysts, and lesions, impairing water and nutrient uptake, and causing symptoms like stunted growth, yellowing, wilting, and reduced yields (Perry & Moens, 2006).

6.1.4 Viral Pathogens

Plant viruses, though not always soil-borne, can be transmitted through soil by vectors such as nematodes and fungi. Notable examples include Tobacco mosaic virus (TMV) and Tomato spotted wilt virus (TSWV). Viruses rely on vectors to enter plants, where they hijack the host's cellular machinery to replicate. Infected cells produce new virus particles that spread throughout the plant. Viral infections lead to symptoms such as mosaic patterns on leaves, stunted growth, leaf curling, and fruit deformities, severely affecting plant health and productivity. Pressure and maintaining plant health (Matthews, 2012).

6.2 Mechanism of Infection

6.2.1 Penetration and Colonization

Through a variety of techniques, pathogens physically penetrate plant surfaces before moving on to invade and colonies plant tissues. Typically, fungal infections break through plant cuticle and epidermal cell barriers by applying mechanical pressure through specialized structures called appressoria. After entering, the pathogen grows hyphae between and within plant cells to spread its infection. On the other hand, bacterial infections typically enter through wounds or stomata, which are natural holes. Once inside, they multiply in the gaps between cells and start colonizing the tissue by releasing effector proteins and enzymes that promote infection and reduce plant defenses (Freeman, 2008).

6.2.2 Enzymatic Degradation of Plant Tissues

Plant cell walls are made up of complex polysaccharides including cellulose, hemicellulose, and pectin, which are broken down by a range of enzymes produced by pathogens. Pectinases, hemicellulases, and cellulases are some of these enzymes. Pectinases, for example, degrade pectin, which is a key constituent of the middle lamella of plant cell walls and helps the pathogen proliferate throughout plant tissues. This enzymatic breakdown provides nutrients that the pathogen can use for growth and reproduction in addition to facilitating the pathogen's penetration and colonization of plant cells (Jones & Dangl, 2006).

6.2.3 Toxin Production

To harm plants directly or obstruct their physiological functions, several plant diseases release poisons. Host-specific or non-host-specific toxins can be distinguished from one another. Selective plant species are the target of host-specific toxins, including victorin, which is generated by the Cochliobolus victoriae bacterium. These toxins frequently combine with genes particular to the host to induce illness. When they obstruct vital metabolic functions like protein synthesis or photosynthesis, non-host-specific toxins like tabtoxin, which is generated by Pseudomonas syringae, can harm a variety of plants. A plant's ability to withstand disease invasion is enhanced by toxins, which can also damage membrane integrity, kill cells, and obstruct signaling pathways (Jones & Dangl, 2006).

6.2.4 Impact on Plants

Pathogen infections have significant impacts on plant health, affecting growth, productivity, and survival. Pathogens such as bacteria, fungi, viruses, and oomycetes employ various strategies to infect plants, including the secretion of toxins, enzymes, and immune suppressors. These infections can lead to substantial yield losses, compromising food security and economic stability globally.When pathogens invade, they disrupt normal plant physiological processes, often causing symptoms like wilting, chlorosis, necrosis, and stunted growth. For example, fungal pathogens like *Fusarium oxysporum* can alter the root microbiome, weakening the plant's defense mechanisms and leading to disease progression (Mendes et al., 2023).

7 Biocontrol of Pathogens

Biocontrol, or biological control, refers to the use of living organisms to manage or suppress the population of pests, including insects, weeds, and plant diseases. This method leverages natural mechanisms and predator-prey relationships to maintain pest populations at levels that are not economically harmful. It is an eco-friendly and sustainable alternative to chemical pest control methods. Plant diseases are a major threat to global agriculture and cause significant crop losses annually. They can range in size from microscopic viroids to more sophisticated organisms like viruses, bacteria, fungus, oomycetes, and nematodes. The economic impact is huge, with plant diseases estimated to cause $40 billion in losses worldwide annually. These losses can be held responsible for between 20 and 40 percent of the overall drop in crop productivity brought on by infectious illnesses (Lahlali et al., 2022).

With the development of chemical fertilizers and pesticides during the twentieth century, modern agriculture has shifted toward more environmentally friendly methods. Nowadays, the intent is to produce more food with a minimum of negative effects on the environment. Crop protection techniques, which historically depended on chemical inputs, are changing to become more environmentally friendly. The idea of using natural enemies or antagonists to control plant infections is known as biocontrol, and it is becoming more and more popular in agricultural sectors. With this strategy, sustainable agriculture with lower ecological costs is the goal (Lahlali et al., 2022).

7.1 Types of Biocontrol Agents

Since the 1970s, research on the role of soil microbiota in reducing soil-borne diseases has been conducted. The discovery that suppressive soils can regulate pathogens has prompted more research into the many kinds of soils that can suppress illness. According to recent research, the microbiota in some soils can produce unfavourable environmental circumstances that hinder the growth of plant pathogens, producing extra suppressiveness on top of the soil's inherent suppressiveness. The soil microbial community and soil organic matter (SOM), which also improve plant nutrition and vigor, are the main forces behind this suppression.

Phytophthora nicotianae, Phytophthora capsici, Phytophthora cinnamomi, Rhizoctonia solani, Pythium ultimum, Pythium irregulare, Pythium aphanidermatum, and Fusarium oxysporum are just a few examples of the oomycetes and fungi against which natural capacity to suppress pathogens has been studied in many disease-suppressive soils. Comprehending the mechanisms underlying illness suppression is crucial in order to optimize the suppressive impact through soil microbiota manipulation. The suppressive qualities can be accounted for by the antagonistic mechanisms between microorganisms and pathogens or by the combined antimicrobial activities of molecules and microbes. Increased microbial activity and fungistasis, improved soil structure, the release of mineral nutrients during SOM decomposition, competition for space and nutrients, the induction of hyperparasitism and microbiostasis, the release of diffusible antibiotic-like compounds, and the activation of systemic disease resistance in the host plant are among the biological factors that typically contribute to disease suppression.

The two main types of soil suppression are "specific suppression," which is more readily transferred between soils, and "general suppression," which is linked to a small number of microbial taxonomic groupings actively combating a single pathogen. Specific suppression processes involve predation, parasitism, and the activation of disease resistance, while general suppression mechanisms involve competition and the synthesis of antibiotics. These two groups frequently coexist.

Microbiota based on biocontrol consists of filamentous fungus (like Trichoderma, Fusarium, Gliocladium, Aspergillus, and Penicillium), actinomycetes (like Streptomyces), and bacteria (like Bacillus, Paenibacillus, and Pseudomonas). These may set off processes linked to the suppression of illness. Plant Growth-Promoting Rhizobacteria (PGPR) and non-pathogenic species of Fusarium and Pythium are found in the microbiota of disease-suppressive compost. These species can enhance plant growth and vigor, making the host more resilient or tolerant to disease. By encouraging the growth of plants and the availability of nutrients, certain soils can increase crop output. This is achieved through the action of microbiota, which includes species of Bacillus, Paenibacillus, actinomycetes, and arbuscular mycorrhizal fungus (AMF) (De Corato, 2020).

7.1.1 Bacteria

Bacillus spp Bacillus species produce a variety of antibiotics (e.g., iturins, fengycins, and surfactins) that inhibit the growth of pathogenic fungi and bacteria. They also produce lytic enzymes such as chitinases and glucanases, which degrade the cell walls of pathogens. These bacteria can be applied to soil as a suspension or incorporated into seed coatings to protect plants from soil-borne diseases such as Fusarium wilt and Pythium damping-off (Compant et al., 2005).

Pseudomonas spp Pseudomonas species produce secondary metabolites like phenazines, pyoluteorin, and 2,4-diacetylphloroglucinol, which have strong antifungal properties. They also enhance plant growth by solubilizing phosphorus and producing plant growth hormones such as indoleacetic acid.These bacteria are often used as soil inoculants or seed treatments to control diseases like take-all in wheat and root rot in various crops (Compant et al., 2005).

7.1.2 Fungi

Trichoderma spp Trichoderma species are known for their mycoparasitic activity, where they directly attack and parasitize fungal pathogens. They also compete for space and nutrients, produce antifungal compounds, and induce plant resistance mechanisms. These fungi can be applied to soil as granules, powders, or liquid formulations to combat a variety of pathogens including Rhizoctonia solani, Sclerotium rolfsii, and Fusarium spp. (Compant et al., 2005).

Gliocladium spp Gliocladium species produce antibiotics such as gliotoxin and viridin, which inhibit the growth of many fungal pathogens. They also demonstrate mycoparasitism and competition for nutrients and space. These fungi are used similarly to Trichoderma and are effective against pathogens like Sclerotinia sclerotiorum and Rhizoctonia solani (Compant et al., 2005).

7.1.3 Nematodes

Entomopathogenic Nematodes (EPNs) EPNs, such as those from the genera Steinernema and Heterorhabditis, infect and kill insect larvae in the soil. The bacteria symbiotically associated with these nematodes produce toxins that kill the insect host, indirectly reducing soil pathogens associated with insect damage. EPNs are applied to the soil as aqueous suspensions to control insect pests like root weevils and cutworms, which can also reduce pathogen levels by eliminating vectors and entry points for pathogens (Kerry, 2000).

7.1.4 Viruses

Mycoviruses Mycoviruses infect fungal pathogens and can attenuate their virulence or cause lysis of the fungal cells. This biocontrol method targets the pathogen directly, reducing its ability to cause disease. Although still largely experimental, mycoviruses have potential as biocontrol agents against fungi like Botrytis cinerea and Cryphonectria parasitica (Gao et al., 2004).

7.1.5 Actinomycetes

Streptomyces spp Streptomyces species produce a wide range of antibiotics (e.g., streptomycin, actinomycin) and hydrolytic enzymes that degrade fungal cell walls. They also play a role in decomposing organic matter, which enhances soil health and nutrient availability. These bacteria can be incorporated into the soil or used as seed treatments to control pathogens like Verticillium dahliae and Fusarium oxysporum (Schloter et al., 2003).

7.1.6 Protozoa

Amoebae and Ciliates Protozoa feed on soil bacteria and fungi, including pathogens. This predatory activity helps control the populations of harmful microbes in the soil ecosystem. Although not as commonly used in commercial biocontrol products, promoting the activity of beneficial protozoa through soil health practices can contribute to pathogen suppression (Fravel., 2005).

7.2 Mechanisms of Biocotrol

7.2.1 Direct Antagonism

Direct antagonistic relationships arise from direct contact and/or a high level of pathogen selectivity mediated by the mechanism(s) exhibited by the BCA(s). As no other organism's actions would be necessary to have a suppressive effect, hyperparasitism by obligate parasites of a plant pathogen would be regarded as the most direct form of antagonism under such a scheme. Under hyperparasitism, a particular BCA targets and kills the pathogen or its propagules causing direct attack. Predators, facultative parasites, hypoviruses, and obligatory bacterial pathogens are the four main categories of hyperparasites in general. A BCA that has been utilised is Pasteuria penetrans, an obligatory bacterial pathogen of root-knot nematodes. Hypoparasites are hypoviruses. An instance that is well-known is the virus that causes hypovirulence, or a decrease in the pathogen's ability to produce illness, in the fungus Cryphonectria parasitica, which causes chestnut blight. In numerous locations, the phenomena have managed to suppress chestnut blight (Milgroom & Cortesi, 2004).

7.2.2 Mixed- Path Antagonism

Mixed-path antagonism is a biocontrol strategy where multiple biological control agents or mechanisms are employed simultaneously to combat soil pathogens. This approach leverages the synergistic or additive effects of different agents, leading to more effective and sustainable disease management. The rationale behind mixed-path antagonism includes enhancing biocontrol efficacy, reducing the likelihood of pathogen resistance, and providing broader-spectrum disease control. Antibiotics, produced by many microbes, can suppress plant pathogens and their diseases when produced in sufficient quantities near the pathogen. Detection and measurement of effective in situ antibiotic quantities are challenging due to their small production amounts relative to other organic compounds in the phytosphere. Studies manipulating genes responsible for antibiotic biosynthesis in biocontrol bacteria (e.g., phenazines and phloroglucinols) have shown that these antibiotics are crucial for effective pathogen suppression. Some biocontrol strains produce multiple antibiotics, enhancing their ability to suppress diverse pathogens. For instance, Bacillus cereus strain UW85 produces zwittermycin and kanosamine, which helps suppress microbial competitors, including plant pathogens. Genetically engineered Pseudomonas putida strains have also demonstrated improved disease suppression in field-grown crops. (Thomashow et al., 1997).

In situ production of antibiotics by biocontrol agents.) Various bacteria release and secrete additional metabolites that may impede the growth and/or activities of pathogens. Numerous microbes generate and secrete lytic enzymes that can hydrolyze a broad range of polymeric substances, such as proteins, chitin, cellulose, hemicellulose, and DNA. Plant pathogen activity can occasionally be directly suppressed by the expression and secretion of these enzymes by certain microorganisms. For instance, b-1,3-glucanase plays a major role in the biocontrol activities of Lysobacter enzymogenes strain C3 (Palumbo et al., 2005).

7.2.3 Indirect Antagonism

Indirect antagonism refers to the suppression of pathogens through mechanisms other than direct microbial interaction. This can include competition for nutrients and space, alteration of environmental conditions, and the production of antimicrobial metabolites. Microbial competition in nutrient-limited environments, such as soils and plant surfaces, is crucial for successful colonization and biocontrol. Microbes must compete for nutrients from plant exudates, leachates, senesced tissue, and waste products like aphid honeydew. Effective colonization by nonpathogenic microbes depletes nutrients, preventing pathogen growth. For example, Pseudomonas putida's production of agglutinin is linked to its root colonization ability and disease suppression. Additionally, microbes like Enterobacter cloacae suppress pathogens through nutrient catabolism and metabolite production, targeting nutrient-rich sites on plants (Kageyama & Nelson, 2003). Enterobacter cloacae and Pythium ultimum suppression.

Plants respond to various environmental and chemical stimuli, including those produced by soil- and plant-associated microbes, which can induce host defenses against pathogens. These defenses can be local or systemic and are mediated through different biochemical pathways. Systemic Acquired Resistance (SAR) is mediated by salicylic acid (SA), leading to the production of pathogenesis-related (PR) proteins that lyse invading cells, reinforce cell walls, or induce cell death. Induced Systemic Resistance (ISR) is mediated by jasmonic acid (JA) and/or ethylene, triggered by non-pathogenic rhizobacteria. The SA and JA pathways can be mutually antagonistic, with some pathogens, like Pseudomonas syringae, exploiting this by producing JA-like compounds to overcome SA-mediated defenses. Host resistance can be enhanced if multiple stimuli synergize, suggesting that the magnitude and duration of induced defenses vary (Ma & Ma, 2016).

7.3 Application of Biocontrol in Agriculture

7.3.1 Case Study: Exploring the Potential of Microbial Antagonists for Biocontrol of Phytopathogens in Agricultural Soils

The use of microbial antagonists as a biocontrol method to manage phytopathogens in agricultural soils. Phytopathogens, which are responsible for a range of plant diseases, pose significant challenges to agricultural productivity and sustainability. The key findings from the study were:

Effectiveness of Microbial Antagonists

Several microbial antagonists, including bacteria, fungi, and actinomycetes, that showed significant potential in suppressing phytopathogens. These antagonists demonstrated the ability to reduce the incidence and severity of plant diseases in various crops.

Mechanisms of Action

The effectiveness of microbial antagonists is attributed to several mechanisms:

- **Antibiosis:** Production of antimicrobial compounds that inhibit pathogen growth.
- **Competition:** Competing with pathogens for nutrients and space, thereby reducing their ability to establish and proliferate.
- **Parasitism:** Directly attacking and consuming pathogenic organisms.
- **Induced Resistance:** Enhancing the plant's own defense mechanisms against pathogens.

Application in Sustainable Agriculture

The integration of microbial antagonists into sustainable agricultural practices. The use of these biocontrol agents can lead to:

- **Reduced Chemical Pesticide Use:** Lower dependency on synthetic chemicals, which is beneficial for the environment and human health.
- **Enhanced Soil Health:** Promoting a diverse and balanced soil microbiome, which contributes to overall soil fertility and plant health.
- **Sustainable Crop Production:** Providing a reliable method for disease management that aligns with organic and sustainable farming principles (Suprapta, 2012).

7.3.2 Case Study: Harnessing Microbial Volatile Organic Compounds for Biocontrol of Plant Pathogens

The review by Srikamwang et al. (2023) investigates the potential of microbial volatile organic compounds (MVOCs) for biocontrol of plant pathogens. MVOCs are small molecules produced by microorganisms that can affect the growth and development of plants and pathogens. This review provides insights into how MVOCs function, their role in plant-microbe interactions, and their application in sustainable agriculture.

Key Findings:

Mechanisms of Action

MVOCs control plant pathogens through several mechanisms:

- **Antimicrobial Activity:** MVOCs can directly inhibit the growth of pathogens by disrupting cellular processes.
- **Induced Systemic Resistance (ISR):** MVOCs can enhance plant defenses by triggering ISR, making plants more resistant to subsequent pathogen attacks.
- **Allelopathy:** MVOCs can influence the growth and activity of neighboring plants and microorganisms, creating an environment less favorable for pathogens.
- **Modulation of Plant Hormones:** MVOCs can alter plant hormone levels, affecting plant growth and defense responses.

Plant-Microbe Interactions

MVOCs play a significant role in the complex interactions between plants and microorganisms:

- **Communication:** MVOCs act as signaling molecules that facilitate communication between plants and beneficial microbes, enhancing mutualistic relationships.
- **Microbiome Modulation:** MVOCs can influence the composition and activity of the plant microbiome, promoting beneficial microbes and suppressing pathogens.

Agricultural Applications

The review highlights the potential of MVOCs in sustainable agriculture:

- **Biocontrol Agents:** MVOCs can be developed as biocontrol agents to reduce the reliance on chemical pesticides.
- **Crop Protection:** Incorporating MVOC-producing microorganisms into farming practices can enhance crop protection against a wide range of pathogens.
- **Sustainable Practices:** MVOCs align with sustainable agriculture principles, offering a natural and eco-friendly approach to disease management (Thomas et al., 2020).

7.4 *Techniques for Enhancing Biocontrol*

7.4.1 Optimized Application Methods

Precision targeting of biocontrol agents to pest hotspots can improve efficacy while minimizing environmental impact. Techniques such as aerial spraying and drip irrigation application are discussed for their potential to enhance biocontrol effectiveness (Majumdar, 2023).

7.4.2 Improved Formulations

Encapsulation of biocontrol agents in biodegradable polymers can improve stability and prolong activity in the field. Nanoencapsulation and microencapsulation techniques are explored for their ability to protect biocontrol agents from environmental stressors (Majumdar, 2023).

7.5 Advanced Techniques Used for Improving the Efficacy of MBCAS

7.5.1 Protoplast Fusion

Microbial biocontrol agents (MBCAs) have been shown to have improved biocontrol capabilities by the application of protoplast fusion procedures. As an example, different strains of Trichoderma have been successfully used for protoplast fusion, isolation, and regeneration. The synthesis of enzymes that break down fungal cell walls, such as chitinase and endoglucanases, which are essential to Trichoderma's biocontrol mechanisms, has greatly increased thanks to this technique. Trichoderma HF9, a highly active fusant, was created via protoplast fusion of T. viride and T. harzianum, while fusant FU3, which was developed from endophytic T. harzianum strains, shown increased antagonistic activity against a variety of diseases. Furthermore, T. harzianum gamma radiation mutants showed enhanced enzyme synthesis and pathogen inhibition. T. viride and T. hazianum protoplast fusion generated fusants with enhanced antagonistic properties against a range of soilborne diseases.

Enhancing the biocontrol capacity of bacteria against plant harmful nematodes and fungus has also been accomplished by the application of protoplast fusion. Following protoplast fusion, Bacillus amyloliquefaciens and Lysinibacillus sphaericus demonstrated enhanced nematicidal activity against Meloidogyne incognita. Fusants that were sourced from Pseudomonas aeruginosa and Bacillus cereus shown increased antagonistic potential against nematodes known to cause knots in roots and prevented the germination of weed seeds. Additionally, virulent fungal strains have had hypoviruses introduced by protoplast fusion to lessen their virulence. For example, the introduction of the hypovirus FgV11-DK21 into Cryphonectria parasitica and other Fusarium spp. resulted in decreased pathogen virulence, whereas the apple valsa canker pathogen Valsa mali was infected and attenuated by Cryphonectria hypovirus 1 and Mycoreovirus 1. All of these findings point to protoplast fusion as a useful method for improving MBCAs' capacity for biocontrol in a variety of species (Majumdar, 2023).

7.5.2 Transposon Mutagenesis

When transposable elements like Tn5 or Tn3 are randomly inserted into an organism's genome to enhance its biocontrol capabilities, a variety of mutants have been created. For example, the 2,4-diacetylphloroglucinol (2,4-DAPG) biosynthesis locus phlACBDE was effectively integrated into a mini-Tn5 transposon after being cloned from Pseudomonas strain CPF-10. The strain Pseudomonas fluorescens P32, which does not produce 2,4-DAPG, had this construct put into its chromosome. The resulting strains greatly enhanced the tomato's resistance to bacterial wilt produced by Ralstonia solanacearum and the wheat's resistance to Gaumannomyces graminis var. tritici's take-all disease. Comparably, in a different investigation, a set of randomly selected transposon (Tn) mutants from Pseudomonas parafulva JBCS1880 were employed, and they shown strong antagonistic activity against Burkholderia glumae and Xanthomonas axonopodis pv. Glycines (Majumdar, 2023).

7.5.3 Genetic Recombination and Transformation

A microbial biocontrol agent (MBCA) can have its biocontrol activity increased by introducing foreign DNA into its genome that specifically codes for biocontrol metabolites. In particular strains, long-term biocontrol action can be improved by blocking or eliminating specific genes. The crown gall disease caused by Agrobacterium tumefaciens has been effectively treated with these techniques. Through the transfer of plasmid genes, a modified strain of Agrobacterium radiobacter K84 was utilized to stop the formation of Agrobacterium tumefaciens strains that were resistant to agrocin 84. In a different investigation, Chaetomium globosum CG10 was exposed to an endochitinase gene that was transferred from the genomic DNA of Trichoderma viride. About one-third of the changed strains showed a significant increase in endochitinase activity as a result of this transformation (Majumdar, 2023).

Rhizoctonia solani on bean seedlings was efficiently managed by inserting a chitinase gene from Serratia marcescens into an endophytic strain of Pseudomonas fluorescens. By expressing this gene, the P. fluorescens Rif1 strain was able to inhibit pathogens through increased chitinase activity. Furthermore, the competitive colonization ability of P. fluorescens strains on tomato root tips was enhanced by the introduction of the sss colonization gene. This shows that acolonization ability can be improved by genetic engineering. By inserting pyrrolnitrin-coding genes from Pseudomonas protegens, recombinant Pseudomonas synxantha strains with enhanced biocontrol efficacy against many wheat diseases have been created recently. When compared to the parent strain, these recombinant strains demonstrated increased biocontrol efficacy (Majumdar, 2023).

8 Conclusion and Future Directions

8.1 Conclusion

The productivity and health of agricultural ecosystems depend on the soil microbiome, which offers a long-term solution for managing disease with microbial biocontrol agents. Beneficial microbes have shown significant promise in the management of a variety of plant diseases, including Bacillus thuringiensis, Trichoderma spp., and microbial volatile organic compounds (MVOCs). These agents provide an environmentally acceptable substitute for chemical pesticides by working through a variety of processes, such as antibiosis, competition, parasitism, and the generation of systemic resistance in plants.

Microbial biocontrol agents improve soil health and support sustainable crop production in addition to lowering reliance on chemical inputs in agricultural practices. Future studies should concentrate on clarifying these biocontrol agents' mechanisms of action, creating efficient formulations and delivery methods, and carrying out comprehensive field research in order to fully realise the benefits of these biocontrol agents.

8.2 Future Directions

- Perform comprehensive investigations to clarify the molecular and metabolic pathways through which microbial biocontrol agents function. Gaining an understanding of these mechanisms will help to maximise their use and increase their efficacy.
- Investigate the possibility of manipulating the soil microbiome to increase the quantity and activity of beneficial microbes. For this, methods like prebiotics, microbial inoculants, and habitat modification can be researched.
- Provide microbial biocontrol agents with stable and efficient formulations that guarantee their viability and activity both during application and storage. It is important to investigate advanced delivery methods including encapsulation and controlled-release formulations.
- Confirm the effectiveness and uniformity of microbial biocontrol agents by conducting large-scale field experiments in a variety of crops, soil types, and climates. Case studies from the real world will offer insightful information and statistics for widespread application.
- Examine how employing microbial consortia—combinations of various beneficial microorganisms—can improve biocontrol efficacy and increase the range of disease control.
- Make use of metagenomics and genomics techniques to find new biocontrol agents and comprehend the intricate relationships found in the soil microbiome.

The identification of novel microbial strains and their functional roles may be aided by these technologies.

- Collaborate with regulatory agencies to expedite the microbial biocontrol agent approval procedure. Create regulations that uphold safety and efficacy requirements while encouraging the use of biocontrol techniques.
- By addressing these future directions, researchers and practitioners can enhance the effectiveness and adoption of microbial biocontrol agents. Continued interdisciplinary collaboration and innovation will be key to unlocking the full potential of the soil microbiome in the biocontrol of plant pathogens.

References

Agrios, G. N. (2005). *Plant pathology*. Elsevier. https://shop.elsevier.com/books/plant-pathology/agrios/978-0-08-047378-9

Berg, G. (2014). The plant microbiome and its importance for plant and human health. *Frontiers in Microbiology, 5*. https://doi.org/10.3389/fmicb.2014.00491

Berg, F., Almeland, O. W., Skadal, J., Slotte, A., Andersson, L., & Folkvord, A. (2017, November 30). *Biological and otolith shape data for a parental generation and their subsequent F1-generation of Atlantic/Baltic herring*. https://doi.org/10.1594/pangaea.883661.

Chauhan, P., Sharma, N., Tapwal, A., Kumar, A., Verma, G. S., Meena, M., Seth, C. S., & Swapnil, P. (2023). Soil microbiome: Diversity. *Benefits and Interactions with Plants. Sustainability, 15*(19), 14643. https://doi.org/10.3390/su151914643

Compant, S., Duffy, B., Nowak, J., CléMent, C., & Barka, E. A. (2005). Use of plant growth-promoting bacteria for biocontrol of plant diseases: Principles, mechanisms of action, and future prospects. *Applied and Environmental Microbiology, 71*(9), 4951–4959. https://doi.org/10.1128/aem.71.9.4951-4959.2005

De Corato, U. (2020). Soil microbiota manipulation and its role in suppressing soil-borne plant pathogens in organic farming systems under the light of microbiome-assisted strategies. *Chemical and Biological Technologies in Agriculture, 7*(1). https://doi.org/10.1186/s40538-020-00183-7

Fravel. (2005). Commercialization and implementation of biocontrol. *Annual Review of Phytopathology, 43*(1), 337–359. https://doi.org/10.1146/annurev.phyto.43.032904.092924

Freeman, N. (2008). An overview of plant defenses against pathogens and herbivores. *The Plant Health Instructor Index*. https://doi.org/10.1094/phi-i-2008-0226-01.

Gao, X., Jackson, T. A., & Hartley, C. (2004). Control of subterranean termites, Coptotermes spp., using a virulent strain of Metarhizium anisopliae and fungal metabolites. *Biological Control, 29*(1), 108–116.

Jones, J. D. G., & Dangl, J. L. (2006). The plant immune system. *Nature, 444*(7117), 323–329. https://doi.org/10.1038/nature05286

Kageyama, K., & Nelson, E. B. (2003). Differential inactivation of seed exudate stimulation ofPythium ultimumSporangium germination byEnterobacter cloacaeInfluences biological control efficacy on different plant species. *Applied and Environmental Microbiology, 69*(2), 1114–1120. https://doi.org/10.1128/aem.69.2.1114-1120.2003

Kalakoutskii, L. V., & Agre, N. S. (1976). Comparative aspects of development and differentiation in actinomycetes. *Bacteriological Reviews, 40*(2), 469–524. https://doi.org/10.1128/br.40.2.469-524.1976

Kerry, B. R. (2000). Rhizosphere interactions and the exploitation of microbial agents for the biological control of plant-parasitic nematodes. *Annual Review of Phytopathology, 38*(1), 423–441. https://doi.org/10.1146/annurev.phyto.38.1.423

Lahlali, R., Ezrari, S., Radouane, N., Kenfaoui, J., Esmaeel, Q., Hamss, H. E., Belabess, Z., & Barka, E. A. (2022). Biological control of plant pathogens: A global perspective. *Microorganisms, 10*(3), 596. https://doi.org/10.3390/microorganisms10030596

Ma, K. W., & Ma, W. (2016). Phytohormone pathways as targets of pathogens to facilitate infection. *Plant Molecular Biology, 91*(6), 713–725. https://doi.org/10.1007/s11103-016-0452-0

Majumdar, A. (2023). Molecular techniques for the improvement of microbial biocontrol agents against plant pathogens. *Egyptian Journal of Biological Pest Control, 33*(1). https://doi.org/10.1186/s41938-023-00746-4

Matthews, R. C. (2012). *Plant virology*. Elsevier. https://shop.elsevier.com/books/plant-virology/matthews/978-0-12-480553-8

Mendes, R., Garbeva, P., & Raaijmakers, J. M. (2013). The rhizosphere microbiome: Significance of plant beneficial, plant pathogenic, and human pathogenic microorganisms. *FEMS Microbiology Reviews, 37*(5), 634–663. https://doi.org/10.1111/1574-6976.12028

Mendes, L. W., Raaijmakers, J. M., De Hollander, M., Sepo, E., Expósito, R. G., Chiorato, A. F., Mendes, R., Tsai, S. M., & Carrión, V. J. (2023). Impact of the fungal pathogen Fusarium oxysporum on the taxonomic and functional diversity of the common bean root microbiome. *Environmental Microbiomes, 18*(1). https://doi.org/10.1186/s40793-023-00524-7

Milgroom, M. G., & Cortesi, P. (2004). Biological control of chestnut blight with hypovirulence: A critical analysis. *Annual Review of Phytopathology, 42*(1), 311–338. https://doi.org/10.1146/annurev.phyto.42.040803.140325

Nadeu, E., van Dijk, R., & Hiller, N. (2023). *The soil microbiome: Its contribution to soil health and one health*. Institute for European Environmental Policy. https://ieep.eu/wp-content/uploads/2023/12/The-Soil-Microbiome-ESAD-IEEP-2023.pdf

Palumbo, J. S., Talmage, K. E., Massari, J. V., La Jeunesse, C. M., Flick, M. J., Kombrinck, K. W., Jirouskova, M., & Degen, J. L. (2005). Platelets and fibrin(ogen) increase metastatic potential by impeding natural killer cell–mediated elimination of tumor cells. *Blood, 105*(1), 178–185. https://doi.org/10.1182/blood-2004-06-2272

Perry, R. N., & Moens, M. (2006). *Plant nematology*. CABI.

Raaijmakers, J. M., Paulitz, T. C., Steinberg, C., Alabouvette, C., & Moenne-Loccoz, Y. (2009). The rhizosphere: A playground and battlefield for soilborne pathogens and beneficial microorganisms. *Plant and Soil, 321*(1–2), 341–361. https://doi.org/10.1007/s11104-008-9568-6

Schaad, N. W. (Ed.). (1981). *Laboratory guide for identification of plant pathogenic bacteria* (pp. 72). https://www.cabidigitallibrary.org/doi/full/10.5555/19811370571

Schloter, M., Dilly, O., & Munch, J. (2003). Indicators for evaluating soil quality. *Agriculture, Ecosystems & Environment, 98*(1–3), 255–262. https://doi.org/10.1016/s0167-8809(03)00085-9

Srikamwang, C., Onsa, N. E., Sunanta, P., Sangta, J., Chanway, C. P., Thanakkasaranee, S., & Sommano, S. R. (2023). Role of microbial volatile organic compounds in promoting plant growth and disease resistance in horticultural production. *Plant Signaling & Behavior, 18*(1), 2227440. https://doi.org/10.1080/15592324.2023.2227440

Suprapta, D. (2012). Potential of microbial antagonists as biocontrol agents against plant fungal pathogens. *Journal of the International Society for Southeast Asian Agricultural Sciences, 18*, 1–8.

Thomas, G., Withall, D., & Birkett, M. (2020). Harnessing microbial volatiles to replace pesticides and fertilizers. *Microbial Biotechnology, 13*(5), 1366–1376. https://doi.org/10.1111/1751-7915.13645

Thomashow, L. S., Bonsall, R. F., & Weller, D. M.. (1997). *Antibiotic production by soil and rhizosphere microbes In Situ* (pp. 493–499). https://www.cabidigitallibrary.org/doi/full/10.5555/19981907905

Van Der Heijden, M. G. A., Bardgett, R. D., & Van Straalen, N. M. (2007). The unseen majority: Soil microbes as drivers of plant diversity and productivity in terrestrial ecosystems. *Ecology Letters, 11*(3), 296–310. https://doi.org/10.1111/j.1461-0248.2007.01139.x

Soil Regeneration Through Soil Microbiome: Role in Food Security

Katherine Georgina Menon, Pariki Nithin, B. H. Boje Gowd, and Kondakindi Venkateswar Reddy

Abstract Soil regeneration is a critical component in addressing global food security, with the soil microbiome playing a pivotal role in this process. The soil microbiome, consisting of diverse communities of bacteria, fungi, archaea, and other microorganisms, is fundamental to soil health and fertility. These microorganisms contribute to nutrient cycling, organic matter decomposition, and the suppression of soil-borne pathogens, thus enhancing plant growth and crop yields. Soil regeneration through the enhancement of the soil microbiome involves practices such as crop rotation, cover cropping, reduced tillage, and organic amendments. These practices promote biodiversity and create favorable conditions for beneficial microorganisms, leading to improved soil structure, increased organic matter content, and enhanced nutrient availability.

Microbial processes are essential for nitrogen fixation, phosphorus solubilization, and the breakdown of complex organic materials into plant-accessible forms. By enhancing these microbial processes, farmers can reduce the reliance on chemical fertilizers, which often lead to soil degradation and environmental pollution. Emerging research highlights the potential of specific microbial inoculants and biofertilizers to further boost soil regeneration efforts. These products, tailored to local soil conditions and crop requirements, can significantly enhance microbial activity and diversity. The soil microbiome is integral to soil regeneration and sustainable agricultural practices. By leveraging the natural processes mediated by soil microorganisms, it is possible to improve soil health, enhance crop productivity, and secure food supplies for the growing global population. Therefore, fostering a healthy soil microbiome should be a key focus in efforts to achieve long-term food security and environmental sustainability.

K. G. Menon
Department of Microbiology, School of Allied Health Sciences, Malla Reddy University, Hyderabad, Telangana, India

P. Nithin
Department of Clinical Nutrition Dietetics and Food Science, School of Allied Health Sciences, Malla Reddy University, Hyderabad, Telangana, India

B. H. Boje Gowd · K. V. Reddy (✉)
University College of Engineering Science and Technology Hyderabad JNTUH Kukatpally, Hyderabad, Telangana, India

S. A. Aransiola et al. (eds.), *Soil Microbiome in Green Technology Sustainability*,
https://doi.org/10.1007/978-3-031-71844-1_14

Keywords Food security · Soil regeneration · Soil microbiome · Environmental sustainability

1 Introduction

For both the long-term sustainability of the global community and the alleviation of poverty in less developed nations, food security has proven essential. "When all people, at all times, have physical and economic access to sufficient, safe, and nutritious food to meet their dietary needs and food preferences for an active and healthy life," is how the United Nations Food and Agriculture Organization (FAO) defines food security. By using this definition, it suggests that in addition to providing people with adequate food, the food distribution system should ensure that the food has the right amount of nutrients and is free of harmful substances to meet people's needs for health (Rosegrant & Cline, 2003).

Qualified expert assessments state that demand for agricultural output is expected to rise by at least 70% by 2050 due to factors including the exponential growth of the global population (expected to reach 9.6 billion by the end of 2050) and the depletion of agricultural productivity is necessary to supply the growing global population's food needs using non-renewable natural resources in a sustainable manner (Bandyopadhyay et al., 2017). People are starting to realize that intensive farming methods are essential to supplying the world's agricultural needs in the future. Because of this, research on agricultural productivity is becoming more and more important globally. It aims to address concerns about eco-friendly sustainability by reducing the use of non-renewable natural resources and eliminating greenhouse gas emissions because they have an impact on climate conditions (Altieri, 2004).

In order to support ecologically friendly agro-technological practices, a number of low-input biotechnology strategies for better using beneficial microbial resources have been and are being developed (Barea, 2015). Utilizing rhizosphere microorganisms to improve plant nutrient uptake and strengthen resilience to biotic and abiotic stressors targets a set of underutilized reservoirs of agricultural opportunity for sustainability (Zavala-Gonzalez et al., 2017). In this regard, various strategies are being employed to improve the soil quality by introducing soil microorganisms which have beneficial effects over the soil and plant growth.

2 Soil and Soil Microbes

2.1 Environmental Elements that Impact the Rhizospheric Microbiome of Plants

The microbial organization in plant organs is influenced by a range of biotic and abiotic stressors. Changes in the external environment, including environment, phyto-pathogen and nutrient competitions, etc., are much more relevant to subsurface plant components. These variables may include influences like water content, soil pH, salinity, type of soil and texture as well as structure, organic matter in soil, and root exudates (Bulgarelli et al., 2015).

Through their effects on plant signaling (i.e., acquired or induced systemic resistance) and root exudate composition, various host-influence factors, including plant development and early stage, health, fitness, and age, are also known to affect the bacterial composition of the plant roots microbiota (Singh et al., 2020).

Climate variables that affect plants and rhizospheric microbiomes include temperature, precipitation, salinity, and drought (Fierer et al., 2007; Dastogeer et al., 2020). In another study it was observed that the ratio of bacteria to fungi may increase as mean annual precipitation (MAP) rises and more organic chemicals accumulate. It is often observed that a variety of soil parameters, including pH, the C:N ratio, soil type, soil structure, soil moisture, and the amount of accessible N, P, and K, influence plant development and immunity to phytopathogens, which in turn determines the rhizospheric microbial populations. Numerous studies have shown that the rhizosperic microbial composition may be influenced by environmental variability, such as soil pH, soil carbon, soil nitrogen, C:N ratio, soil moisture, and biogeography (Pfeiffer et al., 2017).

By regulating the chemical forms of the soil components, soil pH directly affects the availability of vital nutrients for plants. Additionally, it has been proposed that these could be indirect constraints on the rhizospheric microbial communities (Santoyo et al., 2017). Similarly, pH was not always determined to be the most significant factor influencing the composition and diversity of rhizospheric microbial communities in a different study conducted over a mountain range in China that included tropical rainforests, boreal coniferous forests, and a variety of altitudes, climates, and soils. Instead, multiple factors, such as temperature, altitude, and nutrient availability, which vary depending on the region, were linked to these communities (Singh et al., 2013; Angel et al., 2009).

Therefore, in order to maximize the benefits of the rhizosphere microbiome for agricultural output, it is imperative to comprehend the interactions between biotic and abiotic factors. Similarly, knowledge of the ongoing changes in the environment and the consequences of human activity on the climate and environment must be included in the development of agricultural methods in the future.

The ability of a chosen microbial inoculum to coexist and function with the abiotic elements and autochthonous bacteria in that environment is crucial to its effectiveness (Finkel et al., 2017). The way a microbe interacts with other biotic elements in the ecosystem determines its survival and persistence in the soil; frequently, plant interactions with microbial consortia are more successful than interactions with individual microorganisms (Verbruggen et al., 2012; Nemergut et al., 2013). Thus, microbial diversity and its growth-promoting properties are definitely important for soil production (Cordero & Polz, 2014).

2.2 Role of Soil Microbiome in Soil Regeneration

An average gram of soil is thought to contain between 90 and 100 million bacteria and 200,000 fungus, the majority of which are found near the roots of plants. This localization is a result of the fact that, in most cases, a significant portion of the carbon fixed by photosynthesis is released by plant roots, and soil bacteria feed on this carbon. Plants and soil bacteria can interact in ways that are advantageous, detrimental, or neutral to the plant. Plant growth-promoting bacteria (PGPB) are bacteria that are good for a plant's growth and development (Glick, 2004).

Recent years have also seen a significant increase in interest in the significance of preserving the biological characteristics of the soil for food security. The most amazing ecosystem on Earth that supports a diverse range of organisms is undoubtedly soil. The biota of soil includes accountable for the decomposition of organic matter and the conversion of nutrients, including sulfur and nitrogen, making it a crucial component of soil quality. It is now commonly understood that soil biota influences plant growth and, consequently, productivity (Wardle et al., 2004).

Excessive planting of crops have the potential to reduce soil organic matter and nutrient stores, which in time will become insufficient to support plant growth and development in the natural environment (Gu et al., 2015). Furthermore, the degradation of soil, encompassing acidification, hardness, and salinization, has emerged as a significant risk to the advancement of sustainable global agriculture development (Coban et al., 2022).

Microorganisms control fertility and soil characteristics in a variety of ways.

- Soil nutrients can be activated and promoted by microorganisms
- The extracellular secretions of microbes can improve the stability of soil aggregates
- Nitrogen-fixing bacteria improve soil fertility by converting the nitrogen elements; and
- They increase soil organic matter content by decomposing plant and animal remains (Majeed et al., 2018)

By stimulating soil nutrients, some beneficial microorganisms can also increase fertility. For example, *Paenibacillus, Agrobacterium, Acinetobacter, and Bacillus can solubilize potassium, while Enterobacter, Brevibacillus, Mortierella, Trichoderma, and Phyllobacterium* can solubilize phosphate (Sarikhani et al., 2018; Sun et al., 2020).

Conversely, through secretions and hyphae networks, bacteria support the stability of soil aggregates (Coban et al., 2022). Clay particles can be bound together to form soil aggregates by extracellular polymeric substances (EPSs), such as polysaccharides, polyuronic, and amino acids with adhesive qualities from various bacterial species (Rashid et al., 2016; Stevens et al., 2014). This increases interparticle cohesiveness. For instance, the soil proteins associated with glomalin act as particle glue agents to improve soil aggregate stability, supporting water-holding capacity, carbon storage, and erosion resistance (Wu et al., 2012; Henao & Mazeau,

2009). In the meantime, actinomycete and fungal hyphae entangle particles to create a network that further stabilizes the soil structure. Microbes assist in increasing the amount of organic matter in the soil by breaking down plant and animal wastes, which can enhance the soil's ability to retain fertility and function as a buffer (Kallenbach et al., 2016; Six et al., 2004).

2.3 Employing Biofertilizers or Organic Additives

Using biofertilizers or organic additions, such as animal and green manure, organic wastes, composts, or biochar, is a proven method for restoring soil fertility. Nutrient-rich composites can provide nutrients, such as nitrogen, phosphorus, and carbon, to soils in a more gradual, targeted, and sustainable manner than chemical fertilizers. Using different biofertilizers or additions can affect soil fertility and health by promoting or enhancing specific microbial communities (Cesarano et al., 2017; Zhao et al., 2011; Melero et al., 2006; Bronick & Lal, 2005).

Biofertilizers and organic amendments improve crop yields by enhancing soil microbial structure, composition, and diversity. Adding organic additions like manure and sewage sludge over time might boost microbial biomass and alter community structure (Marschner et al., 2003; LeBlanc, 2023). The grade of organics and the type of biofertilizer used can impact the response to soilborne diseases in different crops (Bonanomi et al., 2010). Soil microbes play important roles in soil fertility, plant nourishment, and natural pest prevention. Adding biofertilizers has been linked to an increase in Actinobacteria in soils comparable to what was observed in conventional soils versus organic farms (Ramirez et al., 2012).

The process of composting is a structured biological process that breaks down organic materials into humic compounds through the action of microorganisms (Mehta et al., 2014). Composting has been used in farming for centuries to increase soil fertility, nitrogen levels, and crop health. It also reduces soil illnesses and pathogen attacks by boosting microbial population growth and activity (Cheuk et al., 2005; Joshi et al., 2009). Pseudomonas, Streptomyces, and Trichoderma strains are renowned for their ability to promote plant growth and combat soil diseases (Mavrodi et al., 2012). Compost contains a diverse range of bacteria, including *Bacillus, Burkholderia, Enterobacter, Lysobacter, Pantoea,* and the *fungus Gliocladium* (Castaño et al., 2011).

Composts made from agricultural leftovers, agro-industrial co/by-products, and plant green waste have a complex microbiome structure that suppresses growth. The circular economy is projected to benefit agriculture by promoting the use of compost from agro-wastes, agro-industrial leftovers, and bio-energy co/by-products (De Corato et al., 2019).

2.4 Crop Rotation

Crop rotation is a widely employed tactic to mitigate, and occasionally even rectify, soil erosion and nutrient deficiencies. Rather than using monoculture systems, it is possible to prevent soil fatigue and manage the variety of microbes in the soil in a more sustainable way by varying the crops grown in a given terrain over time. Several crop associations have tried in various parts of the world, with some plants said to be more prolific in those environments. Different agricultural management techniques can impact the microbiological community composition of bulk soil and the rhizosphere. Crop rotation is one of the most widely used conventional strategies because of its productivity and lower costs. (Maarastawi et al., 2018).

The native soil taxa may progressively displace some genera and species, and this transition may take place in just one cropping season. Certain taxa may be more important for preserving the microbial networks in the rhizosphere. Crop rotation applied to eroded perennial fodder-planted cropland resulted in improved soil conditions, nitrogen cycling, and soil organic matter (Lin et al., 2020; Franzluebbers, 2007).

It is common to find reports of increased plant productivity when crop rotation techniques are used. In Europe, the rotation of wheat and oilseeds is widely used (Kerdraon et al., 2019). In temperate climates, crop-pasture rotation is a common practice. It has been linked to increases in soil carbon sequestration, structural stability, organic matter, and biological diversity, all of which support more sustainable agricultural practices in these regions. Crop rotation seems to primarily affect bacterial populations because there aren't many noticeable changes in the fungus community composition in either the bulk or rhizosphere (Granzow et al., 2017).

2.5 Microorganisms Function in Soil Bioremediation

The recovery of the ecosystems depends heavily on the soil biota, and the steady growth of plant populations is mostly fueled by biological regulation driven by the microbiome. To remove and break down the external harmful substances, microbial metabolic activities are essential to the bioremediation of polluted soil. Additionally, a variety of beneficial impacts of plant growth-promoting microbes can assist plants in cleaning up soil pollutants by enhancing plant growth and improving the bioavailability and absorbability of toxins, hence enhancing the effectiveness of phytoremediation.

Pesticides (insecticides, fungicides, molluscicides, rodenticides, and nematocides) have long been used extensively around the world to prevent and manage plant diseases and insect pests. Regarding the pesticide residues in the soil, plants absorb and progressively accumulate them because they cannot break down quickly over time. Through the food chain, the pesticide-contaminated crops eventually reach consumers, posing a risk to their health. Furthermore, the pesticides that have accumulated in the soil travel with rain and irrigation water, contaminating and degrading the natural environment and water supplies (Kaur et al., 2021; Ahmad et al., 2020).

Due to their slow rate of breakdown, chemical pesticides and other organic pollutants including PCBs, TPHs, and PAHs are introduced into farmlands and end up building up in the soil. Thankfully, bacteria can break down these dangerous substances and significantly improve human health, environmental preservation, and agricultural output (Wu et al., 2021).

The ability of plants to remediate organic xenobiotics, such as volatile organic compounds, polycyclic aromatic hydrocarbons, and radionuclides, as well as heavy metals can be improved by plant-rhizospheric microorganisms. Using plants and the rhizospheric microorganisms that are associated with them, phyto-remediation techniques can shift, reorganize, and stabilize heavy metals as well as break down environmental pollutants (Azcón et al., 2013).

Microorganisms degrade soil contaminants using gene expression and numerous enzymes. Cytochrome P450 genes (CYPs) encode heme-thiolate proteins that modify and oxidize both foreign and endogenous chemicals (Renault et al., 2014). Soil microorganisms use enzymes such as dehalogenase, dehydrogenase, dehydrochlorinase, esterase, phosphatase, salicylate hydroxylase, paraoxonase, and dioxygenase to remove pesticides and high-molecular-weight organics (Kumar et al., 2018a, b).

For phyto-remediation to be effective, plants must be able to thrive in contaminated environments since the microorganisms that are associated to them must be adapted to the presence of pollutants (Pongrac et al., 2013). Heavy metal (HM) removal and the xenobiotic degradation of organic substances are two of the phyto-remediation strategies that use rhizosphere microorganisms (Germaine et al., 2013). These mechanisms include the development of low molecular weight siderophores, the acquisition of essential nutrients by plants, the production of phytohormones, the regulation of ACC deaminase activity by stress hormone (ethylene), and the degradation of polycyclic aromatic hydrocarbons (Afzal et al., 2013).

2.6 Elimination of Heavy Metals in Soil

Heavy metals, which are primarily As, Cd, Zn, Pb, Mn, Cr, Cu, and Hg, have the potential to cause phytotoxicity to MPs through oxidative damage, interference with enzyme activities, membrane damage, stomatal closure, reduction in photosynthesis, and retardation of carbon metabolism. These effects are thought to be harmful to plant growth and seed germination (Zulfiqar et al., 2019; Redha et al., 2021; Huang et al., 2019). The use of soil microorganisms in bioremediation is suggested as an economical and sustainable way to restore heavy metal-contaminated land, hastening the restoration of ecosystems and biodiversity (Islam et al., 2021).

Aeromonas, Bacillus, and *Pseudomonas,* meantime, play a significant role in the restoration of soil contaminated with heavy metals (Fakhar et al., 2022). By absorbing heavy metals from the soil, microbes can accumulate them. Certain ions found in functional groups on the surface of cells, like phosphorus, nitrogen, sulfur, and oxygen, can form coordination atoms with metal ions (Jin et al., 2018). AMF also produces phytochelatins, glutathione, metallothionein, and glomalin, which can immobilize metals and help host plants adapt to harsh conditions.

Furthermore, by reducing the phytotoxicity of heavy metals and encouraging plant uptake, soil microbes improve the phytoremediation of contaminated soil. Organic acids and other rhizosphere exudates speed up the solubilization of metals to improve bioavailability. According to Kumar et al. [164], the isolate Pseudomonas lurida EOO26 exhibited drought resistance, multi-metal tolerance, and plant growth-promoting characteristics. Additionally, the inoculation of EOO26 increased Cu uptake in Helianthus annuus roots by 8.6 times and in leaves by 1.9 times compared to uninoculated individuals. In order to alter the valence of heavy metals and lessen their biological toxicity, microorganisms can also initiate redox processes (Jin et al., 2018).

2.7 Remedial Bioengineering of Microorganisms to Restore Soil Health

Genetically modified microorganisms could be used for enhanced performance because native bacteria have difficulty adapting to their new surroundings and efficiently degrading contaminants (Yuanfan et al., 2010). Most of the pollutants that are incompatible with regular native microbes can be effectively removed by these synthetic microorganisms. For the purpose of creating genetically modified organisms (GMOs), a variety of molecular techniques are available, including conjugation, electroporation, horizontal transfer of bacterial DNA, biolistic transformation, and protoplast transformation.

The researchers have proposed four distinct approaches for the development of bioengineered microbes: (a) adjusting the specificity and affinity of the enzymes; (b) creating new genes and regulatory pathways; (c) developing, managing, and overseeing the bioremediation process; and (d) using sensor-based bio affinity reporters to detect contaminants, lessen toxicity, and forecast the endpoints (Kumar et al., 2018a, b). The potential for a single bacterium to break down a variety of xenobiotics is increased by its capacity to integrate numerous genes involved in xenobiotic degradation. (Pant et al., 2021).

Many bacteria can induce signs of sickness by generating photosynthetic pigments and phyto-toxic proteins. For instance, *Pseudomonas syringae* is a well-known plant disease pathogen that affects a wide variety of hosts, such as green beans, tobacco, strawberries, and olives. Another pathogen known to cause firebird illnesses in ornamental plants and fruit trees is *Erwinia amylovora*. Numerous important crop diseases, including those that affect potatoes and bananas, are also linked to *Xanthomonas sp., Ralstonia solanacearum, and Xylella fastidiosa*

(Vrancken et al., 2013). The impact of plant-pathogen interactions is determined by a combination of factors including the magnitude of the pathogen population, host vulnerability, favourable biotic, and environmental stressors (such as plant microbiome) (Lareen et al., 2016).

2.8 The Role of Plant Rhizospheric Microbiota as Metabolic Niches and Their Possible Advantages

An essential framework for examining plant-microbial interactions for various metabolic processes is the rhizosperic microbiome niche idea. In order to illustrate how metabolic activities impact the idea of rhizospheric microbial ecological niches, a series of examples illustrating various consequences, including competition, codependency, and coexistence, can be shown.

Cross-feeding is one way that rhizospheric microbial niches are created. When a single rhizospheric microbial strain excretes a novel secondary metabolite derived from microbes that was not previously present in the root metabolite pools, a new niche is created that can be occupied by other cross-feeding strains (Ponomarova et al., 2017). Consequently, there is a good chance that using novel approaches and tools to investigate these biochemical pathways will provide fresh perspectives on the metabolic mechanisms influencing the makeup of rhizospheric microbiomes. Over the past ten years, studies have identified extremely complex microbial assemblies connected to various plants and plant-specific organs (Reinhold-Hurek et al., 2015).

Members of the plant microbiome might be harmful, neutral, or helpful microorganisms. Plant Growth Promoting Bacteria (PGPBs) can promote plant growth in a number of ways, both directly and indirectly. Certain phytohormone-producing bacteria (PGPBs) generate gibberellin, cytokines, and auxin hormones, which influence plant growth by regulating the amounts of plant auxin hormone and metabolic processes (Gupta & Pandey, 2019).

Furthermore, some PGPBs have the ability to secrete the enzyme 1-aminocyclopropane-1-carboxylate (ACC) deaminase, which hydrolyzes ACC into α-ketobutyrate and ammonia for plant nutrition, hence lowering the amount of the stress hormone ethylene (Glick, 2004; De Souza et al., 2015). It has been shown that certain strains of *Pseudomonas, Arthrobacter, Bacillus,* and other bacteria can promote plant growth by producing multifunctional PGP characteristics and ACC deaminase. *A. thaliana* exhibits enhanced resistance to cold by virtue of the induction of stronger cell walls and more photosynthetic pigments by *Paraburkholderia phytofirmans* (Su et al., 2015).

Numerous molecular pathways that result in advantageous outcomes have been discovered. These pathways include chemically boosting nutrient concentration and accessibility (nitrogen fixation, phosphate or potassium solubilization, iron uptake), as well as host physiological modulation through signalling molecules (Olanrewaju et al., 2017). Constant usage of agricultural soils can lead to the growth of pathogens that contain microorganisms that mediate disease inhibition and can also generate pathogen pressure. By adjusting the amount of plant growth regulators and eliciting defensive reactions against the phyto-pathogens, these advantageous microbial communities shield the plant from phyto-pathogens. Sources of disease-fighting antibiosis activities and invasive pathogens by the synthesis of lytic enzymes, antibiotics, siderophores, and volatile chemicals that inhibit pathogens (Hopkins et al., 2017).

2.8.1 The Role of Soil Health in Food Security

A key component of food security and a prerequisite for guaranteeing the welfare and subsistence of the world's population is soil health. In addition to serving as a growing medium for plants, healthy soil is an evolving, living system that is essential to agricultural systems' sustainability and production. Increased agricultural yields, improved food nutrition, environmental sustainability, and economic stability are all impacted by soil health.

2.8.2 Enhancing Agricultural Productivity

Because soil health directly affects agricultural output, it plays a major role in ensuring food security. Rich in nutrients, organic matter, and microbiological activity, soils can promote strong plant development because they are more fertile. Rich soils allow for enhanced root growth, increased water retention, and improved crop availability of vital nutrients. Higher agricultural yields are the result of all these variables together. The Food and Agriculture Organisation (FAO) states that activities that enhance soil health have the potential to raise crop yields by as much as 58% (FAO, 2015). Meeting the increasing world population's demand for food depends on this yield enhancement.

2.8.3 Improving Nutritional Quality

An important factor in determining the nutritional value of the food produced is the health of the soil. Crops that grow in deficient soils—such as those lacking in zinc, iron, or iodine—also grow in deficient crops. For communities who depend on these crops as their main sources of food, this increases the risk of malnutrition and associated health problems. According to research by Lal (2009), improving soil nutrient shortages with suitable soil management techniques can greatly increase crop micronutrient content. Global public health outcomes and the reduction of malnutrition depend on this increase in food quality.

2.8.4 Environmental Sustainability

Sustainable environments depend on healthy soils. They play a vital role in reducing climate change because they operate as carbon sinks, removing carbon dioxide from the atmosphere. In healthy soils, the natural processes that preserve soil fertility and structure are supported by the presence of organic matter and a varied microbial community. The breakdown of organic matter, nitrogen fixation, and the prevention of soil-borne infections are some of these processes.

Furthermore, healthy soils contribute to biodiversity. They provide habitat for a myriad of organisms, from microorganisms to larger soil fauna, which are essential for maintaining ecological balance. These organisms help in pest control, pollination, and nutrient cycling, all of which are critical services for sustainable agriculture. The United Nations Convention to Combat Desertification (UNCCD) emphasizes that restoring degraded soils can help combat desertification and land degradation, ensuring the availability of arable land for future generations (UNCCD, 2017).

Significant financial gains might also result from investments in soil health. Farmers' production expenses are reduced when healthier soils eliminate the need for expensive inputs like pesticides and fertilisers. Because they frequently have narrow profit margins, smallholder farmers in underdeveloped nations would benefit greatly from this decrease in input costs. Furthermore, robust soils increase the ability of agricultural systems to withstand climatic change. Crops grown on well-structured and organic-rich soils are more resilient to harsh weather events like droughts and floods because they can better absorb and hold onto water. In the face of increasingly uncertain climate circumstances, this resilience is essential for preserving secure food supply.

Governments, agricultural organisations, and farmers must collaborate to adopt and disseminate practices that support sustainable land management, such as conservation tillage, crop rotation, cover crops, organic farming, and the use of compost and other organic amendments. Education and awareness programmes are also necessary to educate farmers about the advantages of soil health and to train them in sustainable farming methods.

3 Conclusion

In conclusion, the pivotal role of soil microbiomes in soil regeneration cannot be overstated, especially in the context of ensuring global food security. Through their intricate interactions and functions, soil microbes contribute to vital processes such as nutrient cycling, soil structure improvement, and plant health promotion. As we navigate the challenges posed by climate change, land degradation, and diminishing resources, understanding and harnessing the power of soil microbiomes becomes increasingly imperative.

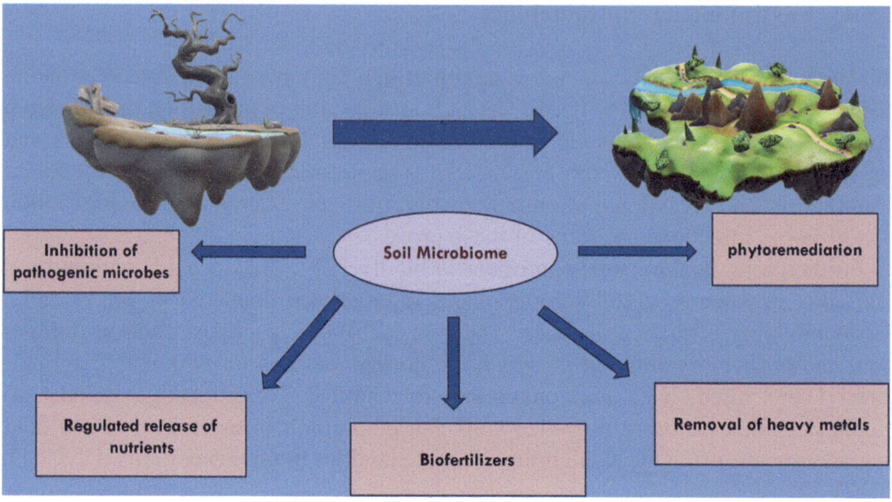

Fig. 1 Role of soil microorganisms in soil fertility

Efforts to promote soil regeneration through microbiome management offer promising solutions to enhance agricultural productivity sustainably. By employing practices such as cover cropping, crop rotation, reduced tillage, and organic amendments, we can cultivate diverse microbial communities that foster soil health and resilience. Moreover, advancements in microbiome research and technology, including microbiome sequencing and precision agriculture, provide valuable tools for optimizing soil management strategies.

Furthermore, fostering interdisciplinary collaborations between scientists, policymakers, farmers, and other stakeholders is essential for implementing effective soil regeneration initiatives on a global scale. By prioritizing soil health and microbiome conservation in agricultural policies and practices, we can mitigate the adverse effects of soil degradation, preserve natural ecosystems, and secure food production for future generations.

In essence, recognizing the critical role of soil microbiomes in soil regeneration underscores the importance of adopting holistic and sustainable approaches to agriculture. By nurturing the intricate web of life beneath our feet, we can cultivate fertile soils, resilient crops, and thriving ecosystems, thereby safeguarding food security and environmental sustainability for generations to come (Fig. 1).

References

Afzal, M., Yousaf, S., Reichenauer, T. G., & Sessitsch, A. (2013). Ecology of alkane-degrading bacteria and their interaction with the plant. *Molecular Microbial Ecology of the Rhizosphere, 1*, 975–989. https://doi.org/10.1002/9781118297674.ch92

Ahmad, K. S., Gul, P., & Gul, M. M. (2020). Efficient fungal and bacterial facilitated remediation of thiencarbazone methyl in the environment. *Environmental Research, 188*, 109811. https://doi.org/10.1016/j.envres.2020.109811

Altieri, M. A. (2004). Linking ecologists and traditional farmers in the search for sustainable agriculture. *Frontiers in Ecology and the Environment, 2*(1), 35–42. https://doi.org/10.1890/1540-9295(2004)002[0035:LEATFI]2.0.CO;2

Angel, R., Soares, M. I. M., Ungar, E. D., & Gillor, O. (2009). Biogeography of soil archaea and bacteria along a steep precipitation gradient. *The ISME Journal, 4*(4), 553–563. https://doi.org/10.1038/ismej.2009.136

Azcón, R., Medina, A., Aroca, R., & Ruiz-Lozano, J. M. (2013). Abiotic stress remediation by the arbuscular mycorrhizal symbiosis and rhizosphere bacteria/yeast interactions. *Molecular Microbial Ecology of the Rhizosphere, 1*, 991–1002. https://doi.org/10.1002/9781118297674.ch93

Bandyopadhyay, P., Bhuyan, S. K., Yadava, P. K., Varma, A., & Tuteja, N. (2017). Emergence of plant and rhizospheric microbiota as stable interactomes. *Protoplasma, 254*, 617–626. https://doi.org/10.1007/s00709-016-1003-x

Barea, J. M. (2015). Future challenges and perspectives for applying microbial biotechnology in sustainable agriculture based on a better understanding of plant-microbiome interactions. *Journal of Soil Science and Plant Nutrition, 15*(2), 261–282. https://doi.org/10.4067/S0718-95162015005000021

Bonanomi, G., Antignani, V., Capodilupo, M., & Scala, F. (2010). Identifying the characteristics of organic soil amendments that suppress soilborne plant diseases. *Soil Biology and Biochemistry, 42*(2), 136–144. https://doi.org/10.1016/j.soilbio.2009.10.012

Bronick, C. J., & Lal, R. (2005). Soil structure and management: A review. *Geoderma, 124*(1–2), 3–22. https://doi.org/10.1016/j.geoderma.2004.03.005

Bulgarelli, D., Garrido-Oter, R., Münch, P. C., Weiman, A., Dröge, J., Pan, Y., et al. (2015). Structure and function of the bacterial root microbiota in wild and domesticated barley. *Cell Host & Microbe, 17*(3), 392–403. https://doi.org/10.1016/j.chom.2015.01.011

Castaño, R., Borrero, C., & Avilés, M. (2011). Organic matter fractions by SP-MAS 13C NMR and microbial communities involved in the suppression of Fusarium wilt in organic growth media. *Biological Control, 58*(3), 286–293. https://doi.org/10.1016/j.biocontrol.2011.05.011

Cesarano, G., De Filippis, F., La Storia, A., Scala, F., & Bonanomi, G. (2017). Organic amendment type and application frequency affect crop yields, soil fertility and microbiome composition. *Applied Soil Ecology, 120*, 254–264. https://doi.org/10.1016/j.apsoil.2017.08.017

Cheuk, W., Lo, K. V., Copeman, R., Joliffe, P., & Fraser, B. S. (2005). Disease suppression on greenhouse tomatoes using plant waste compost. *Journal of Environmental Science and Health, Part B, 40*(3), 449–461. https://doi.org/10.1081/PFC-200047582

Coban, O., De Deyn, G. B., & van der Ploeg, M. (2022). Soil microbiota as game-changers in restoration of degraded lands. *Science, 375*(6584), abe0725. https://doi.org/10.1126/science.abe0725

Cordero, O. X., & Polz, M. F. (2014). Explaining microbial genomic diversity in light of evolutionary ecology. *Nature Rev Microbiol., 12*(4), 263–273. https://doi.org/10.1038/nrmicro3218

Dastogeer, K. M., Tumpa, F. H., Sultana, A., Akter, M. A., & Chakraborty, A. (2020). Plant microbiome–an account of the factors that shape community composition and diversity. *Current Plant Biology, 23*, 100161. https://doi.org/10.1016/j.cpb.2020.100161

De Corato, U., Patruno, L., Avella, N., Lacolla, G., & Cucci, G. (2019). Composts from green sources show an increased suppressiveness to soilborne plant pathogenic fungi: Relationships between physicochemical properties, disease suppression, and the microbiome. *Crop Protection, 124*, 104870. https://doi.org/10.1016/j.cropro.2019.104870

De Souza, R., Ambrosini, A., & Passaglia, L. M. (2015). Plant growth-promoting bacteria as inoculants in agricultural soils. *Genetics and Molecular Biology, 38*(4), 401–419. https://doi.org/10.1590/s1415-475738420150053

Fakhar, A., Gul, B., Gurmani, A. R., Khan, S. M., Ali, S., Sultan, T., et al. (2022). Heavy metal remediation and resistance mechanism of Aeromonas, bacillus, and pseudomonas: A review. *Critical Reviews in Environmental Science and Technology, 52*(11), 1868–1914. https://doi.org/10.1080/10643389.2020.1863112

Fierer, N., Bradford, M. A., & Jackson, R. B. (2007). Toward an ecological classification of soil bacteria. *Ecology, 88*(6), 1354–1364. https://doi.org/10.1890/05-1839

Finkel, O. M., Castrillo, G., Paredes, S. H., González, I. S., & Dangl, J. L. (2017). Understanding and exploiting plant beneficial microbes. *Current Opinion in Plant Biology, 38*, 155–163. https://doi.org/10.1016/j.pbi.2017.04.018

Food and Agriculture Organization. (2015). *Soil health and agricultural sustainability*. FAO. https://openknowledge.fao.org/server/api/core/bitstreams/57388c14-f84f-4bd5-b372-0640b8b882eb/content

Franzluebbers, A. J. (2007). *Integrated crop–livestock systems in the southeastern USA*. https://doi.org/10.2134/agronj2006.0076

Germaine, K. J., McGuinness, M., & Dowling, D. N. (2013). Improving phytoremediation through plant-associated bacteria. *Molecular Microbial Ecology of the Rhizosphere, 1*, 961–973. https://doi.org/10.1002/9781118297674.ch91

Glick, B. R. (2004). Bacterial ACC deaminase and the alleviation of plant stress. *Advances in Applied Microbiology, 56*, 291–312.

Granzow, S., Kaiser, K., Wemheuer, B., Pfeiffer, B., Daniel, R., Vidal, S., & Wemheuer, F. (2017). The effects of cropping regimes on fungal and bacterial communities of wheat and faba bean in a greenhouse pot experiment differ between plant species and compartment. *Frontiers in Microbiology, 8*, 216435. https://doi.org/10.3389/fmicb.2017.00902

Gu, B., Ju, X., Chang, J., Ge, Y., & Vitousek, P. M. (2015). Integrated reactive nitrogen budgets and future trends in China. *Proceedings of the National Academy of Sciences, 112*(28), 8792–8797. https://doi.org/10.1073/pnas.1510211112

Gupta, S., & Pandey, S. (2019). ACC deaminase producing bacteria with multifarious plant growth promoting traits alleviates salinity stress in French bean (Phaseolus vulgaris) plants. *Frontiers in Microbiology, 10*, 442191. https://doi.org/10.3389/fmicb.2019.01506

Henao, L. J., & Mazeau, K. (2009). Molecular modelling studies of clay–exopolysaccharide complexes: Soil aggregation and water retention phenomena. *Materials Science and Engineering: C, 29*(8), 2326–2332. https://doi.org/10.1016/j.msec.2009.06.001

Hopkins, S. R., Wojdak, J. M., & Belden, L. K. (2017). Defensive symbionts mediate host–parasite interactions at multiple scales. *Trends in Parasitology, 33*(1), 53–64. https://doi.org/10.1016/j.pt.2016.10.003

Huang, X. H., Zhu, F., Yan, W. D., Chen, X. Y., Wang, G. J., & Wang, R. J. (2019). Effects of Pb and Zn toxicity on chlorophyll fluorescence and biomass production of Koelreuteria paniculata and Zelkova schneideriana young plants. *Photosynthetica, 57*(2), 688–697. https://doi.org/10.32615/ps.2019.050

Islam, M. S., Kormoker, T., Idris, A. M., Proshad, R., & Kabir, M. H. (2021). Plant-microbe-metal interactions for heavy metal bioremediation: A review. *Crop & Pasture Science, 73*, 181–201. https://doi.org/10.1071/CP21322

Jin, Y., Luan, Y., Ning, Y., & Wang, L. (2018). Effects and mechanisms of microbial remediation of heavy metals in soil: A critical review. *Applied Sciences, 8*(8), 1336. https://doi.org/10.3390/app8081336

Joshi, D., Hooda, K. S., Bhatt, J. C., Mina, B. L., & Gupta, H. S. (2009). Suppressive effects of composts on soil-borne and foliar diseases of French bean in the field in the western Indian Himalayas. *Crop Protection, 28*(7), 608–615. https://doi.org/10.1016/j.cropro.2009.03.009

Kallenbach, C. M., Frey, S. D., & Grandy, A. S. (2016). Direct evidence for microbial-derived soil organic matter formation and its ecophysiological controls. *Nature Communications, 7*(1), 13630. https://doi.org/10.1038/ncomms13630

Kaur, I., Gaur, V. K., Regar, R. K., Roy, A., Srivastava, P. K., Gaur, R., et al. (2021). Plants exert beneficial influence on soil microbiome in a HCH contaminated soil revealing advantage of microbe-assisted plant-based HCH remediation of a dumpsite. *Chemosphere, 280*, 130690. https://doi.org/10.1016/j.chemosphere.2021.130690

Kerdraon, L., Balesdent, M. H., Barret, M., Laval, V., & Suffert, F. (2019). Crop residues in wheat-oilseed rape rotation system: A pivotal, shifting platform for microbial meetings. *Microbial Ecology, 77*, 931–945. https://doi.org/10.1007/s00248-019-01340-8

Kumar, N. M., Muthukumaran, C., Sharmila, G., & Gurunathan, B. (2018a). Genetically modified organisms and its impact on the enhancement of bioremediation. In *Bioremediation: Applications for environmental protection and management* (pp. 53–76). Springer. https://doi.org/10.1007/978-981-10-7485-1_4

Kumar, S., Kaushik, G., Dar, M. A., Nimesh, S., Lopez-Chuken, U. J., & Villarreal-Chiu, J. F. (2018b). Microbial degradation of organophosphate pesticides: A review. *Pedosphere, 28*(2), 190–208. https://doi.org/10.1016/S1002-0160(18)60017-7

Lal, R. (2009). Soil degradation as a reason for inadequate human nutrition. *Food Security, 1*, 45–57. https://doi.org/10.1007/s12571-009-0009-z

Lareen, A., Burton, F., & Schäfer, P. (2016). Plant root-microbe communication in shaping root microbiomes. *Plant Molecular Biology, 90*, 575–587. https://doi.org/10.1007/s11103-015-0417-8

LeBlanc, N. (2023). Green manures alter taxonomic and functional characteristics of soil bacterial communities. *Microbial Ecology, 85*(2), 684–697. https://doi.org/10.1007/s00248-022-01975-0

Lin, D., McCulley, R. L., Nelson, J. A., Jacobsen, K. L., & Zhang, D. (2020). Time in pasture rotation alters soil microbial community composition and function and increases carbon sequestration potential in a temperate agroecosystem. *Science of the Total Environment, 698*, 134233. https://doi.org/10.1016/j.scitotenv.2019.13423

Maarastawi, S. A., Frindte, K., Linnartz, M., & Knief, C. (2018). Crop rotation and straw application impact microbial communities in Italian and Philippine soils and the rhizosphere of Zea mays. *Frontiers in Microbiology, 9*, 342847. https://doi.org/10.3389/fmicb.2018.01295

Majeed, A., Muhammad, Z., & Ahmad, H. (2018). Plant growth promoting bacteria: Role in soil improvement, abiotic and biotic stress management of crops. *Plant Cell Reports, 37*(12), 1599–1609. https://doi.org/10.1007/s00299-018-2341-2

Marschner, P., Kandeler, E., & Marschner, B. (2003). Structure and function of the soil microbial community in a long-term fertilizer experiment. *Soil Biology and Biochemistry, 35*(3), 453–461. https://doi.org/10.1016/S0038-0717(02)00297-3

Mavrodi, O. V., Walter, N., Elateek, S., Taylor, C. G., & Okubara, P. A. (2012). Suppression of Rhizoctonia and Pythium root rot of wheat by new strains of Pseudomonas. *Biological Control, 62*(2), 93–102. https://doi.org/10.1016/j.biocontrol.2012.03.013

Mehta, C. M., Palni, U., Franke-Whittle, I. H., & Sharma, A. K. (2014). Compost: Its role, mechanism and impact on reducing soil-borne plant diseases. *Waste Management, 34*(3), 607–622. https://doi.org/10.1016/j.wasman.2013.11.012

Melero, S., Porras, J. C. R., Herencia, J. F., & Madejon, E. (2006). Chemical and biochemical properties in a silty loam soil under conventional and organic management. *Soil and Tillage Research, 90*(1–2), 162–170. https://doi.org/10.1016/j.still.2005.08.016

Nemergut, D. R., Schmidt, S. K., Fukami, T., O'Neill, S. P., Bilinski, T. M., Stanish, L. F., et al. (2013). Patterns and processes of microbial community assembly. *Microbiology and Molecular Biology Reviews, 77*(3), 342–356. https://doi.org/10.1128/mmbr.00051-12

Olanrewaju, O. S., Glick, B. R., & Babalola, O. O. (2017). Mechanisms of action of plant growth promoting bacteria. *World Journal of Microbiology and Biotechnology, 33*, 1–16. https://doi.org/10.1007/s11274-017-2364-9

Pant, G., Garlapati, D., Agrawal, U., Prasuna, R. G., Mathimani, T., & Pugazhendhi, A. (2021). Biological approaches practised using genetically engineered microbes for a sustainable environment: A review. *Journal of Hazardous Materials, 405*, 124631. https://doi.org/10.1016/j.jhazmat.2020.124631

Pfeiffer, S., Mitter, B., Oswald, A., Schloter-Hai, B., Schloter, M., Declerck, S., & Sessitsch, A. (2017). Rhizosphere microbiomes of potato cultivated in the High Andes show stable and

dynamic core microbiomes with different responses to plant development. *FEMS Microbiology Ecology, 93*(2), fiw242. https://doi.org/10.1093/femsec/fiw242
Pongrac, P., Vogel-Mikuš, K., Poschenrieder, C., Barcelo, J., Tolra, R., & Regvar, M. (2013). Arbuscular mycorrhiza in glucosinolate-containing plants: The story of the metal hyperaccumulator Noccaea (Thlaspi) praecox (Brassicaceae). *Molecular Microbial Ecology of the Rhizosphere, 1*, 1023–1032. https://doi.org/10.1002/9781118297674.ch96
Ponomarova, O., Gabrielli, N., Sévin, D. C., Mülleder, M., Zirngibl, K., Bulyha, K., et al. (2017). Yeast creates a niche for symbiotic lactic acid bacteria through nitrogen overflow. *Cell Systems, 5*(4), 345–357. https://doi.org/10.1016/j.cels.2017.09.002
Ramirez, K. S., Craine, J. M., & Fierer, N. (2012). Consistent effects of nitrogen amendments on soil microbial communities and processes across biomes. *Global Change Biology, 18*(6), 1918–1927. https://doi.org/10.1111/j.1365-2486.2012.02639.x
Rashid, M. I., Mujawar, L. H., Shahzad, T., Almeelbi, T., Ismail, I. M., & Oves, M. (2016). Bacteria and fungi can contribute to nutrients bioavailability and aggregate formation in degraded soils. *Microbiological Research, 183*, 26–41. https://doi.org/10.1016/j.micres.2015.11.007
Redha, A., Al-Hasan, R., & Afzal, M. (2021). Synergistic and concentration-dependent toxicity of multiple heavy metals compared with single heavy metals in Conocarpus lancifolius. *Environmental Science and Pollution Research, 28*(18), 23258–23272. https://doi.org/10.1007/s11356-020-12271-0
Reinhold-Hurek, B., Bünger, W., Burbano, C. S., Sabale, M., & Hurek, T. (2015). Roots shaping their microbiome: Global hotspots for microbial activity. *Annual Review of Phytopathology, 53*, 403–424. https://doi.org/10.1146/annurev-phyto-082712-102342
Renault, H., Bassard, J. E., Hamberger, B., & Werck-Reichhart, D. (2014). Cytochrome P450-mediated metabolic engineering: Current progress and future challenges. *Current Opinion in Plant Biology, 19*, 27–34. https://doi.org/10.1016/j.pbi.2014.03.004
Rosegrant, M. W., & Cline, S. A. (2003). Global food security: Challenges and policies. *Science, 302*(5652), 1917–1919. https://doi.org/10.1126/science.1092958
Santoyo, G., Pacheco, C. H., Salmerón, J. H., & León, R. H. (2017). The role of abiotic factors modulating the plant-microbe-soil interactions: Toward sustainable agriculture. A review. *Spanish Journal of Agricultural Research, 15*(1), 13. ISSN 1695-971X.
Sarikhani, M. R., Oustan, S., Ebrahimi, M., & Aliasgharzad, N. (2018). Isolation and identification of potassium-releasing bacteria in soil and assessment of their ability to release potassium for plants. *European Journal of Soil Science, 69*(6), 1078–1086. https://doi.org/10.1111/ejss.12708
Singh, D., Shi, L., & Adams, J. M. (2013). Bacterial diversity in the mountains of South-West China: Climate dominates over soil parameters. *Journal of Microbiology, 51*, 439–447. https://doi.org/10.1007/s12275-013-2446-9
Singh, S., Kumar, V., Singh, S., Dhanjal, D. S., Datta, S., & Singh, J. (2020). Global scenario of plant–microbiome for sustainable agriculture: Current advancements and future challenges. In *Plant microbiomes for sustainable agriculture* (pp. 425–443). https://doi.org/10.1007/978-3-030-38453-1_14
Six, J., Bossuyt, H., Degryze, S., & Denef, K. (2004). A history of research on the link between (micro) aggregates, soil biota, and soil organic matter dynamics. *Soil and Tillage Research, 79*(1), 7–31. https://doi.org/10.1016/j.still.2004.03.008
Stevens, W. B., Sainju, U. M., Caesar, A. J., West, M., & Gaskin, J. F. (2014). Soil-aggregating bacterial community as affected by irrigation, tillage, and cropping system in the northern great plains. *Soil Science, 179*(1), 11–20. https://doi.org/10.1097/SS.0000000000000036
Su, F., Jacquard, C., Villaume, S., Michel, J., Rabenoelina, F., Clément, C., et al. (2015). Burkholderia phytofirmans PsJN reduces impact of freezing temperatures on photosynthesis in Arabidopsis thaliana. *Frontiers in Plant Science, 6*, 810. https://doi.org/10.3389/fpls.2015.00810
Sun, F., Ou, Q., Wang, N., xuan Guo, Z., Ou, Y., Li, N., & Peng, C. (2020). Isolation and identification of potassium-solubilizing bacteria from Mikania micrantha rhizospheric soil and

their effect on M. micrantha plants. *Global Ecology and Conservation, 23*, e01141. https://doi.org/10.1016/j.gecco.2020.e01141

United Nations Convention to Combat Desertification. (2017). *The global land outlook*. UNCCD. https://www.unccd.int/sites/default/files/documents/2017-09/GLO_Full_Report_low_res.pdf

Verbruggen, E., Van Der Heijden, M. G., Weedon, J. T., Kowalchuk, G. A., & Röling, W. F. (2012). Community assembly, species richness and nestedness of arbuscular mycorrhizal fungi in agricultural soils. *Molecular Ecology, 21*(10), 2341–2353. https://doi.org/10.1111/j.1365-294X.2012.05534.x

Vrancken, K., Holtappels, M., Schoofs, H., Deckers, T., & Valcke, R. (2013). Pathogenicity and infection strategies of the fire blight pathogen Erwinia amylovora in Rosaceae: State of the art. *Microbiology, 159*(Pt_5), 823–832.

Wardle, D. A., Bardgett, R. D., Klironomos, J. N., Setala, H., Van Der Putten, W. H., & Wall, D. H. (2004). Ecological linkages between aboveground and belowground biota. *Science, 304*(5677), 1629–1633. https://doi.org/10.1126/science.1094875

Wu, Q. S., He, X. H., Zou, Y. N., He, K. P., Sun, Y. H., & Cao, M. Q. (2012). Spatial distribution of glomalin-related soil protein and its relationships with root mycorrhization, soil aggregates, carbohydrates, activity of protease and β-glucosidase in the rhizosphere of Citrus unshiu. *Soil Biology and Biochemistry, 45*, 181–183. https://doi.org/10.1016/j.soilbio.2011.10.002

Wu, T., Liu, Y., Yang, K., Zhu, L., White, J. C., & Lin, D. (2021). Synergistic remediation of PCB-contaminated soil with nanoparticulate zero-valent iron and alfalfa: Targeted changes in the root metabolite-dependent microbial community. *Environmental Science: Nano, 8*(4), 986–999. https://doi.org/10.1039/D1EN00077B

Yuanfan, H., Jin, Z., Qing, H., Qian, W., Jiandong, J., & Shunpeng, L. (2010). Characterization of a fenpropathrin-degrading strain and construction of a genetically engineered microorganism for simultaneous degradation of methyl parathion and fenpropathrin. *Journal of Environmental Management, 91*(11), 2295–2300. https://doi.org/10.1016/j.jenvman.2010.06.010

Zavala-Gonzalez, E. A., Rodríguez-Cazorla, E., Escudero, N., Aranda-Martinez, A., Martínez-Laborda, A., Ramírez-Lepe, M., et al. (2017). Arabidopsis thaliana root colonization by the nematophagous fungus Pochonia chlamydosporia is modulated by jasmonate signaling and leads to accelerated flowering and improved yield. *New Phytologist, 213*(1), 351–364. https://doi.org/10.1111/nph.14106

Zhao, Q., Dong, C., Yang, X., Mei, X., Ran, W., Shen, Q., & Xu, Y. (2011). Biocontrol of Fusarium wilt disease for Cucumis melo melon using bio-organic fertilizer. *Applied Soil Ecology, 47*(1), 67–75. https://doi.org/10.1016/j.apsoil.2010.09.010

Zulfiqar, U., Farooq, M., Hussain, S., Maqsood, M., Hussain, M., Ishfaq, M., et al. (2019). Lead toxicity in plants: Impacts and remediation. *Journal of Environmental Management, 250*, 109557. https://doi.org/10.1016/j.jenvman.2019.109557

The Impact of Heavy Metal Contamination in Soils on Soil Microbial Communities and Its Potential Health Risks for Humans

Olaniyi JohnPaul Popoola, Olusola David Ogundele, Emmanuel Abayomi Ladapo, and Samson Senbore

Abstract Heavy metal contamination poses significant threats to soil ecosystems, human health, and biodiversity due to its toxicity and persistence in the environment. Metal pollution, primarily from human activities is a significant contributor to soil levels. The most common metals include lead, cadmium, mercury, chromium, zinc, and nickel, each with unique health and environmental impacts. Understanding the impact of heavy metal pollution on soil microbial communities is crucial as they play a vital role in nutrient cycling, organic matter breakdown, and soil structure maintenance. Changes in microbial diversity can serve as early indicators of soil deterioration and provide insights for developing bioremediation approaches. These metals exhibit varying degrees of toxicity to soil microorganisms, affecting microbial biomass, diversity, and metabolic activities. These metals interfere with important physiological processes in soil microbes, such as enzymatic activities and nutrient cycling, posing threats to soil health and ecosystem functioning. Advancements in analytical methods and technologies have improved our ability to identify and measure traces of heavy metals in soil, facilitating the tracking and simulation of heavy metal pollution dynamics. Multidisciplinary research, and technological advancements are essential for developing efficient strategies to mitigate the negative effects of heavy metals on soil health. The interactions between metals and microbial diversity play a critical role in ecosystem health, biogeochemical cycles, and human welfare. Understanding microbial responses to metals, modifications to microbial communities, metal transformation and cycling, antagonistic and

O. J. Popoola (✉)
Department of Geological Sciences, Achievers University, Owo, Ondo State, Nigeria
e-mail: adewumiadeniyi@achievers.edu.ng

O. D. Ogundele
Department of Chemical Sciences, Achievers University, Owo, Ondo State, Nigeria

E. A. Ladapo
School of Earth, Environment and Society, Bowling Green, OH, USA

S. Senbore
Department of Geology, University of Pretoria, Pretoria, South Africa

S. A. Aransiola et al. (eds.), *Soil Microbiome in Green Technology Sustainability*,
https://doi.org/10.1007/978-3-031-71844-1_15

synergistic interactions, and disruptions to microbial processes and functions is crucial for controlling environmental pollution and maintaining ecosystem sustainability.

Keywords Contamination · Health risks · Heavy metals · Microbial communities · Soils

1 Introduction

1.1 Heavy Metal Contamination in Soils

Soil contamination by heavy metals poses a serious threat to ecosystems, human health, and biodiversity (Aransiola et al., 2024). The increase in heavy metal levels in soil is a result of anthropogenic activities such as mining, agriculture, industrial operations, and inappropriate waste disposal (Ahmed et al., 2022; Adewumi & Ogundele, 2024). Heavy metals that are commonly found in the environment include nickel, copper, zinc, arsenic, lead, cadmium, and mercury (Adewumi & Ogundele, 2024; Adewumi et al., 2022; Adewumi & Laniyan, 2020) (Fig. 1). Through the food chain, these activities can degrade the soil, which can result in decreased agricultural yields, changed nutritional content, and possible health risks (Adewumi & Lawal, 2022; Ferreira et al., 2022). Potable water sources and aquatic habitats may potentially be under jeopardy due to groundwater contamination (Sharma et al., 2023). Children are especially susceptible to the long-term neurological, developmental, respiratory, and carcinogenic effects of heavy metal exposure (Anyanwu et al., 2018; Adewumi, 2022).

1.2 Significance of Studying the Impact on Soil Microbial Communities

Heavy metal contamination significantly affects soil microbial populations due to their significance in soil structure, organic matter degradation, and nutrient cycling (Olaniran et al., 2013). Early warning indicators of soil degradation can be microorganisms that have changed in terms of variety, number, and activity (Bhaduri et al., 2022). Certain microbial species may flourish in the presence of high quantities of heavy metals, providing direction for bioremediation methods (Saeed et al., 2022). Symbiotic connections between plants and soil microbes are common and promote plant development and well-being (Sati et al., 2020). Pollution from heavy metals may disrupt these connections by influencing the availability and uptake of nutrients by plants (Mani & Kumar, 2014; Gonzalez-Henao & Ghneim-Herrera, 2021; Bhaduri et al., 2022; Yaashikaa et al., 2022). Changes in soil microbiology have been connected to potential threats to human health. There are a number of factors

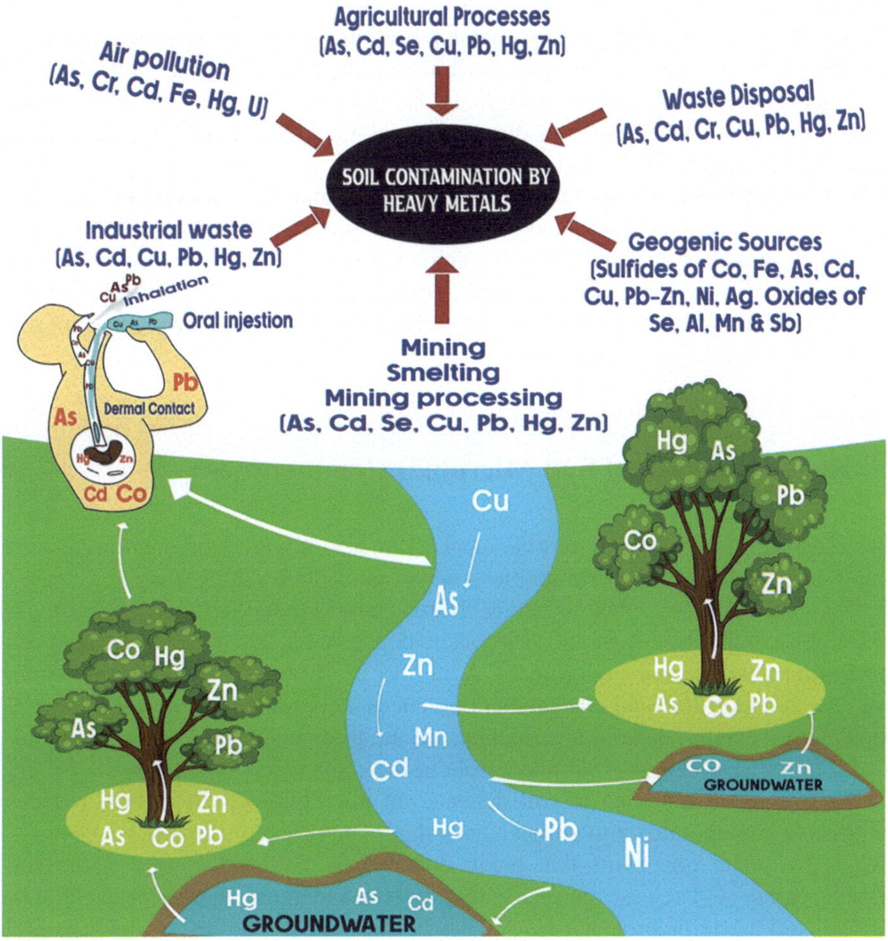

Fig. 1 Pathways of heavy metals in the environment

that can affect soil microbial populations, such as land use practices, climate change, and human activity. A decrease in the variety of microorganisms may lead to the emergence of harmful species (Hoque et al., 2021).

Urbanization and the construction of infrastructure can disturb the soil and liberate heavy metals that are held within the soil matrix (Głodowska & Wozniak, 2019; Brevik et al., 2020; Zhou et al., 2023). Leaded gasoline and lead-based paint in urban areas, inappropriate home, business, and electronic waste disposal, air deposition from industrial processes, volcanic eruptions, and wildfires are some of the past key culprits (Laniyan & Adewumi, 2019). Our understanding of the consequences of heavy metal contamination on soil ecosystems has been shaped by scientific advancements, environmental awareness, and the recognition of the ecological repercussions of human activities (Laidlaw et al., 2017; Nieder et al., 2018;

Aransiola et al., 2019). With the growing prevalence of thorough assessments of ecological issues related to heavy metal pollution in the late twentieth century, knowledge of the bioaccumulation and biomagnification of heavy metals in soil ecosystems became increasingly essential. Our ability to detect and quantify heavy metal traces in soil has improved due to technological developments in analytical techniques (Adewumi & Ogundele, 2024).

With international treaties and laws like the Minamata Convention on Mercury addressing transboundary flow and consequences on ecosystems, attitudes on heavy metal pollution have changed globally (Selin & Selin, 2020). Technological breakthroughs, multidisciplinary research, and international cooperation are necessary for the development of effective strategies to mitigate the effects of heavy metals on soil health and ecological sustainability.

2 Heavy Metals and Soil Microbial Communities

The ecosystem's health and fertility depend on the diversity of microorganisms found in soil microbial communities, which include bacteria, fungus, viruses, archaea, and protozoa (Nkongolo & Narendrula-Kotha, 2020; Punetha et al., 2022). The entire health of the ecosystem depends on the interactions between these groups and the soil environment. A number of variables, including vegetation, soil type, climate, and land management practices, have an impact on soil microbial populations. Fungi, bacteria, and archaea are the three major types of microorganisms found in soil, and they all play a role in the ecosystem (Punetha et al., 2022).

The cycling of nutrients in soil depends on microbial activity, which reduces complicated compounds into simpler ones that plants can consume (Chaves et al., 2020). The main organisms that break down dead plant matter into stable soil organic matter and release nutrients are bacteria and fungi (Gupta et al., 2022). Certain soil bacteria protect plants against illnesses by acting as their natural adversaries, maintaining plant health and productivity.

The composition and texture of the soil have an impact on microbial diversity as well; microbial communities in sandy soils differ from those in clayey soils. Microbial communities can be changed by agricultural practices like as tillage, irrigation, fertilizers, and pesticides (Wang et al., 2021a). Farmers and researchers utilize microbial markers, like diversity, microbial biomass, and enzyme activity, to assess the health of their soil (Wang et al., 2021b).

3 Heavy Metals and Their Effects on Soil Microbes

When present in excess, heavy metals—metals with large atomic weights—can have a detrimental effect on the populations of microorganisms in soil (Alengebawy et al., 2021). The causes of these metals are frequently human activities including mining,

Table 1 Effect of heavy metals on soil microbial communities

S/N	Metals	Effects on microbial communities	Reference
1.	Cadmium (Cd)	Inhibits microbial growth and enzyme activity. Reduces microbial diversity Decreases the abundance of beneficial bacteria such as Nitrosomonas and Nitrobacter, which play critical roles in nitrogen cycling Long-term exposure to cadmium could lead to the development of cadmium-resistant microbial strains, altering community composition	Giller et al. (1998) Xu et al. (2021)
2.	Lead (Pb)	Reduce microbial biomass and alters community structure Impacts nitrogen cycle by causing a significant decline in the population of nitrogen-fixing bacteria, thereby impacting soil fertility and plant growth	Giller et al. (1998) Ghosh and Singh (2005) Li et al. (2020)
3.	Mercury (Hg)	Highly toxic Significantly decreases in microbial biomass and diversity Disruption of microbial metabolic pathways, particularly those involved in carbon and sulfur cycling Increase in horizontal gene transfer, contributing to the spread of mercury resistance genes among microbial populations	Barkay et al. (2003) Wang et al. (2021a, b)
4.	Arsenic (As)	Alters microbial community composition Some microbes develop resistance to this metal It increases the number of arsenic-resistant bacteria such as Acidithiobacillus and Desulfosporosinus which possess genes that enable them to metabolize arsenic, thereby altering the biogeochemical cycling of arsenic in the environment	Mukhopadhyay et al. (2002) Zhang et al. (2022b)
5.	Zinc (Zn)	Stimulate or inhibit microbial activity depending on concentration High levels of this metal reduce microbial diversity Low concentration enhances microbial activity and biomass Negatively impacted soil enzyme activities, particularly those involved in organic matter decomposition and nutrient cycling	Baath (1989) Sun et al. (2020)
6.	Copper (Cu)	Inhibits microbial growth and enzyme activity Affects soil microbial community structure	Chen et al. (2019)

manufacturing, and farming. The diversity, quantity, and overall activity of soil microorganisms are impacted by the varying levels of toxicity of various metals (Bai et al., 2021).

The effect of heavy metals on soil microbial communities are stated out in Table 1. It is well known that lead (Pb) inhibits the development and metabolic activity of fungus and bacteria in soil, which lowers microbial diversity and biomass (Abdu et al., 2017; Liu et al., 2023). For soil microorganisms, especially bacteria, cadmium (Cd) is extremely hazardous. It damages microbial cell membranes, stifles enzyme activity, and impairs microbial respiration (Raza et al., 2020; Yang et al.,

2024). Microorganisms and plants both require zinc (Zn), but excessive concentrations can be hazardous (Jagaba et al., 2022). The growth of bacteria, fungi, and archaea in the soil is impacted by copper (Cu) toxicity, which also affects the activity of the enzymes involved in nitrification and denitrification and disrupts the nitrogen cycle (Shabbir et al., 2020; Guo et al., 2021). For soil microorganisms, chromium (Cr) is toxic, causing microbial cell disruption and inhibition of metabolic processes (Monib et al., 2024). Certain soil fungi and bacteria are prevented from growing by nickel toxicity, which also alters the composition and reduces the variety of microbial communities (Zhang et al., 2022a). High levels of mercury toxicity have a substantial effect on the growth and activity of soil microorganisms, which may cause biomagnification and interfere with the nitrogen cycle in the soil (Wang et al., 2020).

4 Interactions Between Heavy Metals and Microbial Diversity

Lead, mercury, cadmium, and arsenic are examples of heavy metals that can have a substantial negative influence on human wellbeing, biogeochemical cycles, and ecosystem health. Human activities including mining, manufacturing, and farming can introduce these components. It is essential to comprehend how these metals impact microbial populations in order to manage pollution and preserve the sustainability of ecosystems (Rizvi et al., 2020; Pal et al., 2022).

Different bacteria react differently to heavy metals; some develop defense mechanisms against high metal concentrations (Verma & Kuila, 2019; Tuomainen & Candolin, 2011). Certain microorganisms may grow more slowly or be inhibited by heavy metals, whereas other germs may be stimulated or inhibited (Roane et al., 2015; Li et al., 2021). Additionally, the presence of heavy metals affects the species diversity of microbial communities, which in turn affects the flow of pollutants from soil and sediments into water bodies (Muhammad et al., 2020).

5 Bioavailability of Heavy Metals in Soils

In toxicological and environmental research, the bioavailability of heavy metals in soils is vital. It speaks of the quantity of metals that plants, microbes, and other living things are capable of absorbing (Alengebawy et al., 2021; Kumar et al., 2021). The pH of the soil, the amount of organic matter there is, the mineral content and texture of the soil, the redox conditions, microbial activity, competition and interaction, plant uptake, and the hazards to human health and the environment all have an impact on the bioavailability of heavy metals.

Since soil pH has an impact on metal speciation and plant absorption, it is a major factor influencing the bioavailability of heavy metals (Kim et al., 2015). Soil organic matter contributes to metal complexation, which can increase or decrease a metal's bioavailability (Pontoni et al., 2022). The adsorption and desorption of heavy metals are also influenced by the texture of the soil; metals are bound by clay minerals and soil particles, which lowers their bioavailability (Li et al., 2022; Lin et al., 2021; Aransiola et al., 2023). The solubility and speciation of heavy metals are influenced by redox circumstances; iron and manganese become more bioavailable in reducing settings, while less bioavailable metal forms precipitate in oxidizing ones (Caporale & Violante, 2016; Abdu et al., 2017; Pardue & Patrick, 2018).

6 Cellular and Physiological Responses of Soil Microbes to Heavy Metals

Heavy metals are enduring environmental pollutants that build up in soils as a result of manmade or natural processes (Adewumi & Ogundele, 2024). Examples of these pollutants include zinc, lead, mercury, and cadmium. Since soil microbes are more sensitive to changes in their microenvironment, they respond dynamically to heavy metal stress (Philippot et al., 2024). This includes bacteria, fungi, and archaea. It is essential to comprehend soil microbes' cellular and physiological reactions to heavy metals in order to comprehend their adaptation strategies (Loutet et al., 2021).

Microorganisms in the soil employ a variety of tactics, including mechanisms of absorption and accumulation, to control their exposure to heavy metals. Metals can be absorbed and stored as precipitates with the help of membrane transporters. Heavy metals have the potential to disrupt microbial metabolism, which can have an effect on nutrition cycling, energy production, and cellular function (Philippot et al., 2024). Enzyme activities linked to respiration, nitrogen fixation, and sulfur metabolism may be hindered or altered when exposed to heavy metal stress (Sharma et al., 2022).

Genetic alterations in soil microorganisms can result in the selection of resistant populations. Comprehending the genetic foundation of metal resistance can facilitate the creation of microbially based approaches for metal detoxification and bioremediation (Gonzalez-Henao & Ghneim-Herrera, 2021). Larger ecological processes involving plant-microbe interactions, nitrogen cycle, and soil health are impacted by heavy metals on soil bacteria.

7 Disruption of Microbial Processes and Functions

A disturbance in the activities and functions of microorganisms can have profound effects on a range of scientific fields, such as ecology, environmental science, microbiology, agriculture, and medicine. Ecosystems, diverse industrial uses, and the health of living creatures all depend on these processes. Natural or man-made disturbances to these processes include pollution, habitat loss, climate change, and modifications to land use (Gonzalez-Henao & Ghneim-Herrera, 2021).

Another effect of disruptions is the loss of biodiversity since microbial diversity is essential to the stability of ecosystems. The depletion of habitat, pollution, and invasive species are the main causes of biodiversity loss (Bissett et al., 2013). Disturbances in microbial communities can also lead to disease and pathogen emergence by fostering an environment that is conducive to the emergence and spread of diseases (Baker et al., 2022).

Microbes are essential to agricultural environments because they cycle nutrients, keep the soil healthy, and interact with plants. Microbial population changes have the potential to reduce agricultural yield, hence requiring the restoration of microbial diversity and ecosystem services.

The economy and ecology may suffer if biotechnological applications—such as wastewater treatment, biofuel production, bioremediation, and pharmaceuticals—are disrupted. Understanding the factors that obstruct microbial operations and functions—such as ecosystem restoration, sustainable land management, enhanced waste management strategies, and cutting-edge technology for observing and managing microbial communities—is essential to mitigating and managing these problems.

8 Transfer of Heavy Metals from Soil to Humans

8.1 Routes of Exposure for Humans

Anthropogenic activities such as mining, industrial processes, agricultural practices, and waste disposal can result in the accumulation of heavy metals in soil, including lead, cadmium, mercury, arsenic, and chromium. Ingestion of these metals through food crops, water, or inhalation is possible. The two main routes of exposure are through bioaccumulation in contaminated food crops such as vegetables, fruits, cereals, and rice, and through water contamination, where pollutants can reach surface water sources such as lakes, rivers, and streams. Exposure to heavy metals can also result via inhaling dust and aerosol particles contaminated with heavy metals, especially in high-pollution industrial and urban environments. Direct contact with contaminated dust or soil can also lead to heavy metal exposure, especially for young infants. It is essential to comprehend these pathways in order to lessen the risks related to exposure to heavy metals. Through targeting these pathways, we can endeavor to mitigate the hazards linked to exposure to heavy metals.

8.2 Bioaccumulation and Biomagnification of Heavy Metals

Heavy metal bioaccumulation and biomagnification are ecological processes that contribute to the gradual rise in concentration of harmful compounds within living things and across food chains (Adewumi & Lawal, 2022). These processes occur when organisms absorb materials faster than they can be metabolized or eliminated, leading to the accumulation of heavy metals in biological tissues. These pollutants can be passed to higher trophic levels through the consumption of microbes, plants, and algae, which absorb the metals from the air, water, or soil. The accumulation of heavy metals in organisms, particularly in organs like the liver, kidneys, and muscles, has significant consequences on ecosystems, animals, and human health at all trophic levels (Adewumi & Laniyan, 2020).

Biomagnification occurs when predators ingest prey with accumulated concentrations of heavy metals, leading to increased concentrations in their tissues. This poses a significant challenge for apex predators like humans, as they may be exposed to high concentrations of heavy metals like lead and mercury through their consumption of tainted meat, fish, and dairy products. Long-term exposure to these metals can lead to developmental delays, neurological diseases, and reproductive difficulties.

8.3 Health Implications of Chronic Exposure to Contaminated Environments

Long-term exposure to polluted environments, including heavy metals, pesticides, industrial chemicals, and infections, poses serious health hazards worldwide (Abioye et al., 2021). These contaminants can linger in the air, water, soil, and food, causing respiratory disorders such as lung cancer, asthma, and chronic bronchitis. In metropolitan areas with poor air quality, indoor air pollution from sources like cooking stoves, tobacco smoke, vehicles, and industries exacerbate respiratory difficulties, particularly in areas with poor air quality (Maung et al., 2022). Reproductive difficulties are also a significant health concern.

Exposure to neurotoxic compounds like lead, mercury, arsenic, and pesticides can lead to neurological problems, cognitive impairment, and neurodegenerative illnesses like Alzheimer's and Parkinson's (Nabi & Tabassum, 2022). Long-term exposure to air pollution, industrial chemicals, heavy metals, and other carcinogenic contaminants increases the risk of developing cancer (Adewumi & Laniyan, 2020; Adewumi, 2022). Endocrine-disrupting chemicals (EDCs) can cause reproductive and developmental disorders, leading to low birth weight, premature birth, birth abnormalities, and decreased fertility (Ghosh et al., 2022; Margiana et al., 2023).

Cardiovascular diseases, such as heart attacks, strokes, and hypertension, are associated with air pollution, particularly fine particulate matter (PM2.5) and nitrogen dioxide (NO_2), which can promote inflammation, oxidative stress, endothelial dysfunction, and atherosclerosis, contributing to cardiovascular morbidity and mortality (de Bont et al., 2022). Chronic exposure to these pollutants can also compromise immune function, increasing susceptibility to infections and autoimmune diseases. Persistent organic pollutants (POPs), heavy metals, and microbial pathogens can disrupt immune responses, leading to chronic inflammation, allergic reactions, and autoimmune disorders like rheumatoid arthritis and lupus (Agache et al., 2024).

Mental health effects of living in contaminated environments include stress, anxiety, depression, and reduced quality of life. Environmental degradation, pollution-related health concerns, and socioeconomic disparities contribute to mental health burdens in affected communities (Legg et al., 2023). Therefore, it is crucial to be aware of the potential health risks associated with exposure to these pollutants.

9 Health Risks Associated with Altered Soil Microbial Communities

9.1 *Impact of Soil Microbial Changes on Nutrient Cycling*

Changes in soil microbiology have a major effect on the cycling of nutrients, which affects plant development, soil fertility, and ecosystem output. The mediating role of microorganisms such as bacteria, fungi, and archaea in the nutrient cycle is vital for processes like plant absorption, mineralization, immobilization, and decomposition. These processes are highly susceptible to changes in microbial populations, which can modify cycling dynamics and nutrient availability.Soil bacteria engage in vital processes called decomposition and organic matter breakdown, which convert organic molecules into simpler forms and liberate nutrients like carbon, nitrogen, phosphorus, and sulfur (Aransiola et al., 2013). Changes in the makeup and activity of microbes can affect the rate and efficacy of this process. The process of converting organic nutrients into inorganic forms that plants can absorb is known as "nutrient mineralization." Microbial populations have the power to change how these processes are balanced, which can affect how readily available nutrients are in the soil (Mahala et al., 2020; Čapek et al., 2021).

In order for soil microorganisms to convert atmospheric nitrogen gas (N_2) into ammonium (NH_4^+) for plant use, nitrogen fixation and cycling are also essential activities. Changes in microbial population can have an impact on the dynamics of the nitrogen cycle and the rates at which nitrogen is fixed, which can impact soil fertility and plant uptake of nitrogen (Shafreen et al., 2021).

The cycles of phosphorus and sulfur are ones in which soil microorganisms take part. Microbial enzymes help liberate phosphorus from organic compounds and change sulfur into forms that plants can absorb (Tian et al., 2021). Microbial population changes have the potential to reduce ecosystem productivity and restrict plant nutrient availability. Because plant-microbe interactions result in symbiotic partnerships that impact ecosystem dynamics, plant health, and nutrient absorption, they are essential to the functioning of ecosystems. Plant development and productivity may be impacted by the disruption of these symbiotic relationships caused by shifts in microbial populations.

10 Potential for Pathogenic Microorganisms in Contaminated Soils

Pathogenic bacteria can be found in contaminated soils and present a serious risk to human health. Numerous sources, such as inappropriate waste management, animal excrement, industrial runoff, and home wastewater, can introduce these germs into the soil (Adewumi & Ogundele, 2024). Humans can contract bacterial infections from contaminated water and food crops, as well as by direct touch, ingestion, inhalation, or exposure to Salmonella, Listeria, Clostridium, Mycobacterium, and Escherichia coli (*E. coli*).

Giardia, Toxoplasma gondii, and Cryptosporidium are examples of protozoan pathogens that can contaminate soil through animal or human waste. This can lead to contaminated water supplies and foodborne transmission through crops grown in contaminated soil. Humans can get lung infections, systemic mycoses, and skin infections from fungi such as Aspergillus, Candida, Histoplasma, and Coccidioides.

When sick people evacuate their feces, viruses that cause diseases like hepatitis and enteric viruses can remain in soil and sediments for a long time. They can enter soil by improper waste disposal methods, runoff from agriculture, or sewage discharge. Exposure to soil-borne diseases is higher in those who work in agriculture, live near polluted areas, or are young.

Skin infections, respiratory infections, gastrointestinal problems, and systemic ailments are among the health hazards linked to coming into touch with dangerous microorganisms in polluted soils. Little ones, the elderly, expectant mothers, and people with compromised immune systems are among the vulnerable demographics.

Strategies including improving sanitary standards, correctly treating sewage, putting waste management systems into place, supporting safe farming practices, and routinely testing and monitoring soil and water quality are required to reduce the risks associated with pathogenic microorganisms in polluted soils.

10.1 Human Health Effects Linked to Disrupted Soil Microbial Communities

In order to sustain soil fertility, nutrient cycling, plant growth, and ecosystem functioning, soil microbial populations are essential. Because disturbances to these communities alter the composition and diversity of soil microorganisms, they can have a substantial effect on human health. Reduced nutrition and food security, elevated exposure to soil-borne infections, intensified airborne allergens, compromised immunological regulation, contaminated soil microbial communities, and effects on mental health and general well-being are some of the major repercussions (Kumar et al., 2022).

The cycle of nutrients and their availability to plants is facilitated by soil microbes, and their absence can result in malnutrition and deficiencies in important micronutrients such as iron, zinc, and vitamin A. Reduced food yields and nutritional quality due to unhealthy soil can have major negative health implications, especially for those who are already susceptible.

In addition to raising the possibility of human exposure to soil-borne pathogens, disturbances in soil microbial communities can also raise the risk of respiratory, gastrointestinal, and skin illnesses. Changes in soil microbial populations can worsen airborne allergen release, which is influenced by the decomposition of organic matter and the cycling of nutrients.

Disturbed soil microbial populations can disrupt immune homeostasis, resulting in immunological dysregulation and heightened susceptibility to allergies, autoimmune disorders, and inflammatory illnesses. Immune-related diseases are on the rise among urbanized populations with little access to natural settings because of reduced contact to healthy soil microbes.

Because organic pollutants, pesticides, heavy metals, and other environmental toxins can influence the fate and migration of soil microbial populations, soil microbial populations can also be associated with contaminated exposure and health hazards. Variations in the microbiological activity of soil might prolong the pesticides' environmental durability, raising the possibility of human exposure to contaminated food, water, and air.

11 Case Studies and Research Findings

11.1 Overview of Key Studies Investigating Heavy Metal Effects on Soil Microbes

Understanding the impact of heavy metals on soil microbial communities is essential for assessing the ecological consequences of soil contamination and developing strategies for soil remediation and ecosystem restoration. Over the years, numerous studies have investigated the effects of heavy metals on soil microbes, focusing on their diversity, abundance, metabolic activity, and functional diversity. Here's an overview of some key studies in this field:

1. Giller et al. (1998): This groundbreaking study used a long-term field experiment to investigate the impact of heavy metal pollution on soil microbial populations. The researchers discovered that heavy metal contamination impeded important soil processes including nutrient cycling and organic matter breakdown, decreased microbial biomass and activity, and changed the organization of microbial communities.
2. Cao et al. (2008): This study used molecular methods, including DNA sequencing and microbial functional gene analysis, to examine how soil microbial communities were affected by cadmium exposure. The findings showed that the makeup of the microbial community had changed, with some bacterial species becoming more prevalent in soils polluted with cadmium. The significance of

microbial community adaptability and resistance to heavy metal stress was emphasized by the study.

3. Chen et al. (2019): In a polluted mining region, the effect of lead pollution on soil microbial diversity and enzyme activity was evaluated by researchers. When compared to uncontaminated soils, the results indicated a considerable drop in both enzyme activity and microbial diversity in lead-contaminated soils. The study emphasized how important soil microbial communities are to the health of soil ecosystems and how sensitive they are to heavy metal contamination.
4. Ma et al. (2016): The results of several investigations on the impact of heavy metals on soil microbial communities were combined and analyzed in this meta-analysis. Consistent trends were identified by the analysis across research, such as decreased microbial biomass and diversity, modifications in the makeup of microbial communities, and adjustments in soil enzyme activity in response to heavy metal pollution. Important information on the cumulative impact of heavy metals on soil microbial communities in various ecosystems and geographical areas was obtained from the meta-analysis.
5. Sun et al. (2020): This study examined the impact of several heavy metals on soil bacterial and fungal communities in polluted mine soils using high-throughput sequencing methods. According to the findings, there were noticeable changes in the diversity and composition of the microbial community in response to heavy metal stress, with certain metal-tolerant microbial taxa becoming more prevalent in polluted soils. The study emphasised how soil ecosystem resilience may be affected by microbial communities' capacity to respond to heavy metal pollution.
6. Xiao et al. (2020): In contaminated mine soils, this study examined the effects of various heavy metals on soil bacterial and fungal communities using high-throughput sequencing techniques. In response to heavy metal stress, the results demonstrated clear changes in the diversity and composition of the microbial community, with some metal-tolerant microbial taxa becoming dominant in contaminated soils. The study emphasized the relevance of microbial community adaptation to heavy metal pollution for soil ecosystem resilience.

All things considered, this significant studies provide valuable information about the impact of heavy metals on soil microbial communities and the implications for environmental health and soil ecosystem function. By elucidating the mechanisms underpinning microbial responses to heavy metal contamination, these studies contribute to the development of workable strategies for pollution control, soil remediation, and sustainable land management techniques. More research in this area is needed to comprehend the long-term repercussions of heavy metal contamination on soil microbial populations and to develop fresh approaches to lessen its detrimental effects.

11.2 Insights from Field Experiments and Long-Term Monitoring

Experiments in the field and long-term observation studies have yielded important information about how soil bacteria are affected by heavy metal pollution. According to these studies, soil microbial communities can be affected by heavy metal pollution in terms of richness and composition, which can modify the structure of the community and repress vulnerable species. Even if metal inputs are halted, these changes might endure in the long run.

According to field research, even in the face of heavy metal stress, soil microbial communities are able to maintain ecosystem services because they still have functional redundancy. The robustness of these systems can be impacted by factors such as microbial diversity, soil properties, and metal content.

The functioning of soil ecosystems can be disrupted by heavy metal contamination, which can limit soil enzyme activity involved in nutrient cycling, organic matter breakdown, and pollutant degradation. On plant growth, soil fertility, and total ecosystem production, this may have cascade effects. When assessing the effects of heavy metals on soil microorganisms, field research and long-term monitoring have shown how crucial it is to take into account environmental parameters such soil pH, moisture, organic matter content, and land use practices.

The management and repair of soil in metal-contaminated areas are affected by these findings. Determining bioindicators of soil health, developing targeted remediation techniques, and optimizing the use of microbial-based technologies for metal detoxification and bioremediation can all be aided by knowledge of how microbial populations respond to metal contamination.

Essential empirical data on the intricate interactions between heavy metals and soil microbial populations in natural ecosystems can be obtained from field research and long-term monitoring initiatives. By implementing these findings into soil management methods, policymakers, land managers, and environmental practitioners can endeavor to limit the detrimental impacts of metal pollution on soil health and encourage the regeneration of contaminated soils to support ecosystem resilience and sustainability.

11.3 Variability in Findings and Gaps in Existing Research

Experiments in the field and long-term observation studies have yielded important information about how soil bacteria are affected by heavy metal pollution. According to these studies, soil microbial communities can be affected by heavy metal pollution in terms of richness and composition, which can modify the structure of the community and repress vulnerable species.

Even if metal inputs are halted, these changes might endure in the long run.

According to field research, even in the face of heavy metal stress, soil microbial communities are able to maintain ecosystem services because they still have functional redundancy. The robustness of these systems can be impacted by factors such as microbial diversity, soil properties, and metal content.

The functioning of soil ecosystems can be disrupted by heavy metal contamination, which can limit soil enzyme activity involved in nutrient cycling, organic matter breakdown, and pollutant degradation. On plant growth, soil fertility, and total ecosystem production, this may have cascade effects. When assessing the effects of heavy metals on soil microorganisms, field research and long-term monitoring have shown how crucial it is to take into account environmental parameters such soil pH, moisture, organic matter content, and land use practices.

The management and repair of soil in metal-contaminated areas are affected by these findings. Determining bioindicators of soil health, developing targeted remediation techniques, and optimizing the use of microbial-based technologies for metal detoxification and bioremediation can all be aided by knowledge of how microbial populations respond to metal contamination.

Essential empirical data on the intricate interactions between heavy metals and soil microbial populations in natural ecosystems can be obtained from field research and long-term monitoring initiatives. By implementing these findings into soil management methods, policymakers, land managers, and environmental practitioners can endeavor to limit the detrimental impacts of metal pollution on soil health and encourage the regeneration of contaminated soils to support ecosystem resilience and sustainability.

12 Challenges and Limitations in Existing Literature

12.1 Methodological Challenges in Studying Heavy Metal Effects on Soil Microbes

Numerous methodological issues arise in the investigation of the effects of heavy metals on soil microorganisms. These include spatial variability and sampling, which can be difficult because soil microbial populations vary geographically. Heterogeneity in contaminants and bioavailability are important concerns as well because heavy metal-contaminated soil frequently demonstrates regional variation. Sampling strategies that guarantee replication and take into consideration regional heterogeneity are crucial for obtaining trustworthy data.

Assessing the effects of heavy metals on soil microorganisms requires careful consideration of experimental design and control conditions. Experiments can have different outcomes depending on how variables like metal content, exposure duration, soil properties, and microbial inoculum are controlled. Dose-response testing, appropriate controls, and confounding variables must all be taken into account when interpreting findings.

Sample preparation, data analysis, sequencing, and DNA extraction are among the obstacles posed by microbial community analysis. The accuracy and resolution of microbial community profiles can be impacted by the use of molecular techniques, such as amplicon sequencing or metagenomics. Reducing methodological biases, standardizing procedures, and interpreting data using bioinformatics tools are all necessary for reproducibility and cross-study comparability.

Comprehending the functional implications of heavy metal impacts on soil microorganisms necessitates an ecological relevance and functional assessment. Overcoming these challenges can be aided by multidisciplinary methods such as microbial activity assays, stable isotope tracking, and functional gene analysis.

To find out how long-term exposure to heavy metals affects soil microbes and ecosystem function, field research and long-term monitoring are required. On the other hand, obstacles include environmental changes, resource scarcity, and logistical problems. Through collaborative networks, ongoing monitoring initiatives, and state-of-the-art monitoring technologies, addressing these issues can improve our comprehension of heavy metal-microbe interactions in soil ecosystems.

12.2 Variability in Study Designs and Outcomes

The variability in studies investigating the effects of heavy metals on soil microorganisms is a significant challenge. This variability can be attributed to variations in metal concentrations, microbial populations, ambient conditions, and experimental designs. To ensure the accuracy of study results, strict statistical analysis and standardization of experimental procedures are necessary.

Environmental conditions, such as soil pH, texture, organic matter concentration, moisture levels, and microbial community composition, greatly influence the way microorganisms react to heavy metal pollution. Studies conducted under disparate environmental settings may provide inconsistent results due to differences in microbial community composition and metal bioavailability. Therefore, it is crucial to incorporate environmental data and accurately characterize soil features in study designs.

Metal characteristics and speciation also play a role in the impacts of heavy metals on soil microbial communities. Differences in toxicity, solubility, and chemical forms can cause varied results in different investigations. Additionally, changes in metal speciation and interactions with soil constituents can influence microbial community dynamics and microbial-metal interactions. Understanding these concepts is crucial for accurately evaluating their impact on soil microorganisms.

Microbial community dynamics are dynamic and varied, displaying intricate relationships with environmental elements and heavy metals. Variations in microbial composition, abundance, and functional diversity can impact the sensitivity and resilience of microbial communities to heavy metal pollution. Molecular methods like functional gene analysis and DNA sequencing can shed light on these dynamics.

Geographical and temporal scales, such as experimental plot size and duration, sample intervals, and geographic locations, can also contribute to variability in research findings. Multi-site comparisons and long-term monitoring studies are needed to capture temporal and geographical heterogeneity in microbial responses and uncover underlying patterns.Data interpretation and synthesis can be challenging due to variability in research results. Integrated modelling techniques, systematic reviews, and meta-analyses can help synthesize and combine available data, identify patterns, and clarify underlying causes of variability.

12.3 Areas Requiring Further Investigation and Clarification

The understanding of heavy metals' impact on soil microorganisms has significantly improved, but there are still unanswered questions that need further research. These include the interactive effects of multiple heavy metals, metal speciation and bioavailability, microbial community resilience and adaptation, long-term monitoring and field studies, ecosystem-level responses and functional implications, and integrated risk assessment and management.

The majority of studies on the impact of heavy metals on soil microorganisms have focused on specific metals, which is essential to understand how co-contamination affects microbial responses, metal bioavailability, and ecosystem functioning. Future studies should examine how different metal concentrations and environmental circumstances affect soil bacteria in terms of additive, antagonistic, and synergistic effects.

Metal speciation and bioavailability are crucial for determining the toxicological impact of heavy metals on soil microorganisms. Advanced analytical methods like isotope tracking and synchrotron-based spectroscopy can help clarify the dynamics of metal bioavailability and speciation in soil ecosystems.

Microbial community resilience and adaptation are also essential for understanding soil microbial communities' long-term viability and ecosystem functioning. Systems biology, microbial ecology, and omics methods can provide insights into the adaptation mechanisms and functional capabilities of metal-tolerant microbial species.

Long-term monitoring and field studies are needed to evaluate the cumulative impacts of heavy metals on soil microbial communities, soil health, and ecosystem resilience over time. Combining microbial ecology and ecosystem ecology can improve the relationship between microbial responses and the consequences of heavy metal pollution on ecosystems. Addressing the complex issues of heavy metal contamination and preserving soil health and ecosystem integrity requires bridging the gap between scientific research, policy-making, and community engagement.

13 Integration of Soil Microbial and Human Health Studies

13.1 Bridging the Gap Between Soil Science and Human Health Research

"Bridging the gap between soil science and human health research" aims to bridge the gap between these two fields. Soil science focuses on soil composition, structure, and function, while human health studies focus on health-related variables. Understanding the relationship between soil quality and human health is crucial. Soil's impact on the food chain, crop nutrient content, and potential health risks like herbicides and heavy metals can be improved. This approach can lead to better environmental management, agricultural practices, and soil biodiversity. By bridging the gap, a comprehensive understanding of soil, food, and health can be achieved.

13.2 Emerging Trends and Future Directions in Research

With emerging trends highlighting the intricate connection between soil ecosystems and human health, the interaction between soil microbes and human health is changing quickly. Understanding how many facets of human health are impacted by soil microbial populations requires the application of multidisciplinary approaches. By preserving soil health, the soil microbiome affects crop quality and the cycling of nutrients. To lower the risk of diseases connected to soil, researchers are also concentrating on the transmission of diseases through the soil. They are detecting pathogens and comprehending how they propagate.

Soil plays a crucial role in climate change and human health by retaining carbon and reducing greenhouse gas emissions. It serves as a carbon sink, impacting climate resilience, public health, and food security. Future research aims to explore sustainable soil management methods to improve health outcomes and climate adaptation. Advanced technologies like metagenomics and bioinformatics are being used to study soil microbial communities, providing insights into their roles and interactions.

14 Conclusion

In summary, heavy metal contamination poses significant threats to soil ecosystems, human health, and biodiversity due to its toxicity and persistence. Anthropogenic activities contribute to escalating levels of heavy metals in soil, including lead, cadmium, mercury, arsenic, chromium, copper, zinc, and nickel. Understanding the impact on soil microbial communities is vital for nutrient cycling and soil health, as heavy metals can disrupt symbiotic plant-microbe associations, affecting agricultural sustainability.

Interdisciplinary research is crucial for addressing the complex challenges of heavy metal contamination and its impacts on ecosystems and human well-being, including adverse health effects and disruptions to soil microbial communities. Areas requiring further investigation include understanding the mechanisms of heavy metal toxicity, evaluating the long-term impacts on microbial diversity and functionality, investigating the interactions between heavy metals and other environmental factors, and studying the bioavailability and mobility of heavy metals in soil matrix.

Developing efficient bioremediation strategies is also essential, as it involves creating microbial strains with improved metal tolerance and remediation capacities and investigating the possibilities of native microbial communities for metal remediation. The wider effects of heavy metal pollution on ecosystem services that soil bacteria perform are also crucial, such as evaluating the effects of alterations in microbial populations and functions on plant production, soil fertility, carbon sequestration, and clean air and water availability.

Translating research findings into practical applications is crucial for closing the knowledge gap between scientific research and real-world applications in agriculture and environmental management. Future research should focus on converting scientific discoveries into workable plans for pollution repair and sustainable soil management.

References

Abdu, N., Abdullahi, A. A., & Abdulkadir, A. (2017). Heavy metals and soil microbes. *Environmental Chemistry Letters, 15*(1), 65–84. https://doi.org/10.1007/s10311-016-0587-x

Abioye, O. P., Ijah, U. J. J., Aransiola, S. A., Auta, S. H., & Ojeba, M. I. (2021). Bioremediation of toxic pesticides in soil using microbial products. In R. Prasad, S. C. Nayak, R. N. Kharwar, & N. K. Dubey (Eds.), *Mycoremediation and environmental sustainability. Fungal biology*. Springer. https://doi.org/10.1007/978-3-030-54422-5_1

Adewumi, A. J. (2022). Heavy metals in soils and road dust in Akure City, Southwest Nigeria: Pollution, sources, and ecological and health risks. *Exposure and Health, 14*(2), 375–392. https://doi.org/10.1007/s12403-021-00456-y

Adewumi, A. J., & Laniyan, T. A. (2020). Contamination, sources and risk assessments of metals in media from Anka artisanal gold mining area, Northwest Nigeria. *Science of the Total Environment, 718*, 137235. https://doi.org/10.1016/j.scitotenv.2020.137235

Adewumi, A. J., & Lawal, A. E. (2022). Heavy metals in paddy soils and their uptake in rice plants collected along Ogbese River, Southwest Nigeria: Implications for contamination and health risk. *Ife Journal of Science, 24*(3), 569–582. https://doi.org/10.1080/27658511.2023.2293239

Adewumi, A. J., & Ogundele, O. D. (2024). Hidden hazards in urban soils: A meta-analysis review of global heavy metal contamination (2010–2022), sources and its ecological and health consequences. *Sustainable Environment, 10*(1), 2293239. https://doi.org/10.1080/27658511.2023.2293239

Adewumi, A. J., Akinleye, C., & Akinleye, R. O. (2022). Ecological and human health implication of heavy metals in soils of urban areas of Nigeria: A systematic review. *Achievers Journal of Scientific Research, 4*(1), 95–119.

Agache, I., Akdis, C., Akdis, M., Al-Hemoud, A., Annesi-Maesano, I., Balmes, J., & Nadeau, K. C. (2024). Immune-mediated disease caused by climate change-associated environmental hazards: Mitigation and adaptation. *Frontiers in Science, 2*, 1279192. https://doi.org/10.3389/fsci.2024.1279192

Ahmed, S. F., Kumar, P. S., Rozbu, M. R., Chowdhury, A. T., Nuzhat, S., Rafa, N., & Mofijur, M. (2022). Heavy metal toxicity, sources, and remediation techniques for contaminated water and soil. *Environmental Technology and Innovation, 25*, 102114. https://doi.org/10.1016/j.eti.2021.102114

Alengebawy, A., Abdelkhalek, S. T., Qureshi, S. R., & Wang, M. Q. (2021). Heavy metals and pesticides toxicity in agricultural soil and plants: Ecological risks and human health implications. *Toxics, 9*(3), 42. https://doi.org/10.3390/toxics9030042

Anyanwu, B. O., Ezejiofor, A. N., Igweze, Z. N., & Orisakwe, O. E. (2018). Heavy metal mixture exposure and effects in developing nations: An update. *Toxics, 6*(4), 65. https://doi.org/10.3390/toxics6040065

Aransiola, S. A., Ijah, U. J. J., & Abioye, O. P. (2013). Phytoremediation of Lead Polluted Soil by *Glycine max L. Applied and Environmental Soil Science, 2013*, Article ID 631619. https://doi.org/10.1155/2013/631619. https://www.hindawi.com/journals/aess/2013/631619/abs/

Aransiola, S. A., Ijah, U. J. J., Abioye, O. P., & Bala, J. D. (2019). Microbial-aided phytoremediation of heavy metals contaminated soil: A review. *European Journal of Biological Research, 9*(2), 104–125. http://www.journals.tmkarpinski.com/index.php/ejbr/article/view/157

Aransiola, S. A., Ijah, U. J. J., Abioye, O. P., & Bala, J. D. (2023). Vermicompost-assisted phytoremediation of toxic trace element-contaminated soil in Madaka, Nigeria using *Melissa officinalis* L and *Sida acuta. International Journal of Environmental Science and Technology*. https://doi.org/10.1007/s13762-022-04105-y

Aransiola, S. A., Ijah, U. J. J., Abioye, O. P., Bala, J. D., Rivadeneira-Mendoza, R. F., Luque, R., Rodríguez-Díaz, J. M., & Maddela, N. R. (2024). Micro and vermicompost assisted remediation of heavy metal contaminated soils using phytoextractors. *Case Studies in Chemical and Environmental Engineering, 9*(2024), 100755. https://doi.org/10.1016/j.cscee.2024.100755

Baath, E. (1989). Effects of heavy metals in soil on microbial processes and populations (a review). *Water, Air, and Soil Pollution, 47*, 335–379. https://doi.org/10.1007/BF00279331

Bai, X. T., Wang, J., Dong, H., Chen, J. M., & Ge, Y. (2021). Relative importance of soil properties and heavy metals/metalloids to modulate microbial community and activity at a smelting site. *Journal of Soils and Sediments, 21*, 1–12. https://doi.org/10.1007/s11368-020-02743-8

Baker, R. E., Mahmud, A. S., Miller, I. F., Rajeev, M., Rasambainarivo, F., Rice, B. L., & Metcalf, C. J. E. (2022). Infectious disease in an era of global change. *Nature Reviews Microbiology, 20*(4), 193–205. https://doi.org/10.1038/s41579-021-00639-z

Barkay, T., Miller, S. M., & Summers, A. O. (2003). Bacterial mercury resistance from atoms to ecosystems. *FEMS Microbiology Reviews, 27*(2–3), 355–384. https://doi.org/10.1016/S0168-6445(03)00046-9

Bhaduri, D., Sihi, D., Bhowmik, A., Verma, B. C., Munda, S., & Dari, B. (2022). A review on effective soil health bio-indicators for ecosystem restoration and sustainability. *Frontiers in Microbiology, 13*, 938481. https://doi.org/10.3389/fmicb.2022.938481

Bissett, A., Brown, M. V., Siciliano, S. D., & Thrall, P. H. (2013). Microbial community responses to anthropogenically induced environmental change: Towards a systems approach. *Ecology Letters, 16*, 128–139. https://doi.org/10.1111/ele.12109

Brevik, E. C., Slaughter, L., Singh, B. R., Steffan, J. J., Collier, D., Barnhart, P., & Pereira, P. (2020). Soil and human health: Current status and future needs. *Air, Soil and Water Research, 13*, 1178622120934441. https://doi.org/10.1177/117862212093444

Cao, L., Shen, G., & Lu, Y. (2008). Combined effects of heavy metal and polycyclic aromatic hydrocarbon on soil microorganism communities. *Environmental Geology, 54*, 1531–1536. https://doi.org/10.1016/j.ese.2022.100169

Čapek, P., Choma, M., Tahovská, K., Kaňa, J., Kopáček, J., & Šantrůčková, H. (2021). Coupling the resource stoichiometry and microbial biomass turnover to predict nutrient mineralization and immobilization in soil. *Geoderma, 385*, 114884. https://doi.org/10.1016/j.geoderma.2020.114884

Caporale, A. G., & Violante, A. (2016). Chemical processes affecting the mobility of heavy metals and metalloids in soil environments. *Current Pollution Reports, 2*, 15–27. https://doi.org/10.1007/s40726-015-0024-y

Chaves, P. J. L., Cepeda, J. T., Álvarez, F. J. G., & Orzáez, M. J. H. (2020). Influence of moisture, temperature and microbial activity in biomass sustainable storage. Special focus on olive biomasses. International journal of environmental science and natural. *Resources, 25*(3), 556165. https://doi.org/10.19080/IJESNR.2020.25.556165

Chen, J., Zhang, Y., & Wei, H. (2019). Effects of copper pollution on soil microbial community and enzyme activity in orchard soils. *Journal of Soils and Sediments, 19*(2), 707–716.

de Bont, J., Jaganathan, S., Dahlquist, M., Persson, Å., Stafoggia, M., & Ljungman, P. (2022). Ambient air pollution and cardiovascular diseases: An umbrella review of systematic reviews and meta-analyses. *Journal of Internal Medicine, 291*(6), 779–800. https://doi.org/10.1111/joim.13467

Ferreira, C. S., Seifollahi-Aghmiuni, S., Destouni, G., Ghajarnia, N., & Kalantari, Z. (2022). Soil degradation in the European Mediterranean region: Processes, status and consequences. *Science of the Total Environment, 805*, 150106. https://doi.org/10.1016/j.scitotenv.2021.150106

Ghosh, M., & Singh, S. P. (2005). A review on phytoremediation of heavy metals and utilization of its byproducts. *Applied Ecology and Environmental Research, 3*(1), 1–18. https://doi.org/10.15666/aeer/0301_001018

Ghosh, A., Tripathy, A., & Ghosh, D. (2022). Impact of endocrine disrupting chemicals (EDCs) on reproductive health of human. *Proceedings of the Zoological Society, 75*(1), 16–30. New Delhi: Springer India. https://doi.org/10.1007/s12595-021-00412-3

Giller, K. E., Witter, E., & Mcgrath, S. P. (1998). Toxicity of heavy metals to microorganisms and microbial processes in agricultural soils: A review. *Soil Biology and Biochemistry, 30*(10–11), 1389–1414. https://doi.org/10.1016/S0038-0717(97)00270-8

Głodowska, M., & Wozniak, M. (2019). Changes in soil microbial activity and community composition as a result of selected agricultural practices. *Agricultural Sciences, 10*(3), 330–351. https://doi.org/10.4236/as.2019.103028

Gonzalez-Henao, S., & Ghneim-Herrera, T. (2021). Heavy metals in soils and the remediation potential of bacteria associated with the plant microbiome. *Frontiers in Environmental Science, 9*, 604216. https://doi.org/10.3389/fenvs.2021.604216

Guo, H., Gu, J., Wang, X., Song, Z., Qian, X., Sun, W., & Yu, J. (2021). Negative effects of oxytetracycline and copper on nitrogen metabolism in an aerobic fermentation system: Characteristics and mechanisms. *Journal of Hazardous Materials., 403*, 123890. https://doi.org/10.1016/j.jhazmat.2020.123890

Gupta, A., Singh, U. B., Sahu, P. K., Paul, S., Kumar, A., Malviya, D., & Saxena, A. K. (2022). Linking soil microbial diversity to modern agriculture practices: A review. *International Journal of Environmental Research and Public Health, 19*(5), 3141. https://doi.org/10.3390/ijerph19053141

Hoque, M. N., Tahjib-Ul-Arif, M., Hannan, A., Sultana, N., Akhter, S., Hasanuzzaman, M., & Brestič, M. (2021). Melatonin modulates plant tolerance to heavy metal stress: Morphological responses to molecular mechanisms. *International Journal of Molecular Sciences, 22*(21), 11445. https://doi.org/10.3390/ijms222111445

Jagaba, A. H., Kutty, S. R. M., Isa, M. H., Ghaleb, A. A. S., Lawal, I. M., Usman, A. K., & Soja, U. B. (2022). Toxic effects of xenobiotic compounds on the microbial community of activated sludge. *ChemBioEng Reviews, 9*(5), 497–535. https://doi.org/10.1002/cben.202100055

Kim, R. Y., Yoon, J. K., Kim, T. S., Yang, J. E., Owens, G., & Kim, K. R. (2015). Bioavailability of heavy metals in soils: Definitions and practical implementation—A critical review. *Environmental Geochemistry and Health, 37*, 1041–1061. https://doi.org/10.1007/s10653-015-9695-y

Kumar, V., Pandita, S., Sidhu, G. P. S., Sharma, A., Khanna, K., Kaur, P., & Setia, R. (2021). Copper bioavailability, uptake, toxicity and tolerance in plants: A comprehensive review. *Chemosphere, 262*, 127810. https://doi.org/10.1016/j.chemosphere.2020.127810

Kumar, S., Bamboriya, S. D., Rani, K., Meena, R. S., Sheoran, S., Loyal, A., & Jhariya, M. K. (2022). Grain legumes: A diversified diet for sustainable livelihood, food, and nutritional security. In *Advances in legumes for sustainable intensification* (pp. 157–178). Academic.

Laidlaw, M. A., Filippelli, G. M., Brown, S., Paz-Ferreiro, J., Reichman, S. M., Netherway, P., & Mielke, H. W. (2017). Case studies and evidence-based approaches to addressing urban soil lead contamination. *Applied Geochemistry, 83*, 14–30. https://doi.org/10.1016/j.apgeochem.2017.02.015

Laniyan, T. A., & Adewumi, A. J. (2019). Health risk assessment of heavy metal pollution in groundwater around an exposed dumpsite in southwestern Nigeria. *Journal of Health and Pollution., 9*(24), 191–210. https://doi.org/10.5696/2156-9614-9.24.191210

Legg, R., Prior, J., Adams, J., & McIntyre, E. (2023). The relations between mental health and psychological wellbeing and living with environmental contamination: A systematic review and conceptual framework. *Journal of Environmental Psychology, 87*, 101994. https://doi.org/10.1016/j.jenvp.2023.101994

Li, B., He, X., & Song, Y. (2020). Effects of lead pollution on soil microbial activity and community structure. *Environmental Pollution, 256*, 113443. https://doi.org/10.1007/s42832-022-0134-6

Li, Y., Zhang, M., Xu, R., Lin, H., Sun, X., Xu, F., & Sun, W. (2021). Arsenic and antimony co-contamination influences on soil microbial community composition and functions: Relevance to arsenic resistance and carbon, nitrogen, and sulfur cycling. *Environment International, 153*, 106522. https://doi.org/10.1016/j.envint.2021.106522

Li, Q., Wang, Y., Li, Y., Li, L., Tang, M., Hu, W., & Ai, S. (2022). Speciation of heavy metals in soils and their immobilization at micro-scale interfaces among diverse soil components. *Science of the Total Environment, 825*, 153862. https://doi.org/10.1016/j.scitotenv.2022.153862

Lin, H. Y., Chuang, T. J., Yang, P. T., Guo, L. Y., & Wang, S. L. (2021). Adsorption and desorption of Thallium (I) in soils: The predominant contribution by clay minerals. *Applied Clay Science., 205*, 106063. https://doi.org/10.3390/agronomy14040718

Liu, X., Ju, Y., Mandzhieva, S., Pinskii, D., Minkina, T., Rajput, V. D., & Rensing, C. (2023). Sporadic Pb accumulation by plants: Influence of soil biogeochemistry, microbial community and physiological mechanisms. *Journal of Hazardous Materials, 444*, 130391. https://doi.org/10.1016/j.jhazmat.2022.130391

Loutet, S. A., Chan, A. C., Kobylarz, M. J., Verstraete, M. M., Pfaffen, S., Ye, B., & Murphy, M. E. (2021). The fate of intracellular metal ions in microbes. In *Trace metals and infectious diseases [Internet]*. MIT Press. 2015. Chapter 4.

Ma, Y., Oliveira, R. S., Freitas, H., & Zhang, C. (2016). Biochemical and molecular mechanisms of plant-microbe-metal interactions: Relevance for phytoremediation. *Frontiers in Plant Science, 7*, 918. https://doi.org/10.3389/fpls.2016.00918

Mahala, D. M., Maheshwari, H. S., Yadav, R. K., Prabina, B. J., Bharti, A., Reddy, K. K., & Ramesh, A. (2020). Microbial transformation of nutrients in soil: An overview. In *Rhizosphere microbes: Soil and plant functions* (pp. 175–211). https://doi.org/10.1007/978-981-15-9154-9_7

Mani, D., & Kumar, C. (2014). Biotechnological advances in bioremediation of heavy metals contaminated ecosystems: An overview with special reference to phytoremediation. *International Journal of Environmental Science and Technology, 11*, 843–872. https://doi.org/10.1007/s13762-013-0299-8

Margiana, R., Yousefi, H., Afra, A., Agustinus, A., Abdelbasset, W. K., Kuznetsova, M., & Mohammadi, M. J. (2023). The effect of toxic air pollutants on fertility men and women, fetus and birth rate. *Reviews on Environmental Health, 38*(3), 565–576. https://doi.org/10.1515/reveh-2022-0032

Maung, T. Z., Bishop, J. E., Holt, E., Turner, A. M., & Pfrang, C. (2022). Indoor air pollution and the health of vulnerable groups: A systematic review focused on particulate matter (PM), volatile organic compounds (VOCs) and their effects on children and people with pre-existing lung disease. *International Journal of Environmental Research and Public Health, 19*(14), 8752. https://doi.org/10.3390/ijerph19148752

Monib, A. W., Niazi, P., Barai, S. M., Sawicka, B., Baseer, A. Q., Nikpay, A., & Thea, B. (2024). Nitrogen cycling dynamics: Investigating volatilization and its interplay with N2 fixation. *Journal for Research in Applied Sciences and Biotechnology, 3*(1), 17–31. https://doi.org/10.55544/jrasb.3.1.4

Muhammad, M. H., Idris, A. L., Fan, X., Guo, Y., Yu, Y., Jin, X., & Huang, T. (2020). Beyond risk: Bacterial biofilms and their regulating approaches. *Frontiers in Microbiology, 11*, 928. https://doi.org/10.3389/fmicb.2020.00928

Mukhopadhyay, R., Rosen, B. P., Phung, L. T., & Silver, S. (2002). Microbial arsenic: From geocycles to genes and enzymes. *FEMS Microbiology Reviews, 26*(3), 311–325. https://doi.org/10.1111/j.1574-6976.2002.tb00617.x

Nabi, M., & Tabassum, N. (2022). Role of environmental toxicants on neurodegenerative disorders. *Frontiers in Toxicology, 4*, 837579. https://doi.org/10.3389/ftox.2022.837579

Nieder, R., Benbi, D. K., & Reichl, F. X. (2018). Soil-borne particles and their impact on environment and human health. In *Soil components and human health*. Springer. https://doi.org/10.1007/978-94-024-1222-2_3

Nkongolo, K. K., & Narendrula-Kotha, R. (2020). Advances in monitoring soil microbial community dynamic and function. *Journal of Applied Genetics, 61*(2), 249–263. https://doi.org/10.1007/s13353-020-00549-5

Olaniran, A. O., Balgobind, A., & Pillay, B. (2013). Bioavailability of heavy metals in soil: Impact on microbial biodegradation of organic compounds and possible improvement strategies. *International Journal of Molecular Sciences, 14*(5), 10197–10228. https://doi.org/10.3390/ijms140510197

Pal, A., Bhattacharjee, S., Saha, J., Sarkar, M., & Mandal, P. (2022). Bacterial survival strategies and responses under heavy metal stress: A comprehensive overview. *Critical Reviews in Microbiology, 48*(3), 327–355. https://doi.org/10.1080/1040841X.2021.1970512

Pardue, J. H., & Patrick, W. H. (2018). Changes in metal speciation following alteration of sediment redox status. In *Metal contaminated aquatic sediments* (pp. 169–185). Routledge.

Philippot, L., Chenu, C., Kappler, A., Rillig, M. C., & Fierer, N. (2024). The interplay between microbial communities and soil properties. *Nature Reviews Microbiology, 22*(4), 226–239. https://doi.org/10.1038/s41579-023-00980-5

Pontoni, L., La Vecchia, C., Boguta, P., Sirakov, M., D'Aniello, E., Fabbricino, M., & Locascio, A. (2022). Natural organic matter controls metal speciation and toxicity for marine organisms: A review. *Environmental Chemistry Letters, 20*(1), 797–812. https://doi.org/10.1007/s10311-021-01310-y

Punetha, A., Punetha, S., & Khan, A. (2022). Soil community composition and ecosystem processes. In *Agriculture, environment and sustainable development: Experiences and case studies* (pp. 217–236). Springer International Publishing. https://doi.org/10.1023/A:1006299100678

Raza, A., Habib, M., Kakavand, S. N., Zahid, Z., Zahra, N., Sharif, R., & Hasanuzzaman, M. (2020). Phytoremediation of cadmium: Physiological, biochemical, and molecular mechanisms. *Biology, 9*(7), 177. https://doi.org/10.3390/biology9070177

Rizvi, A., Zaidi, A., Ameen, F., Ahmed, B., AlKahtani, M. D., & Khan, M. S. (2020). Heavy metal induced stress on wheat: Phytotoxicity and microbiological management. *RSC Advances, 10*(63), 38379–38403. https://doi.org/10.1039/D0RA05610C

Roane, T. M., Pepper, I. L., & Gentry, T. J. (2015). Microorganisms and metal pollutants. In *Environmental microbiology* (3rd ed., pp. 415–439). Elsevier. https://doi.org/10.1016/B978-0-12-394626-3.00018-1

Saeed, M. U., Hussain, N., Sumrin, A., Shahbaz, A., Noor, S., Bilal, M., & Iqbal, H. M. (2022). Microbial bioremediation strategies with wastewater treatment potentialities–A review. *Science of the Total Environment, 818*, 151754. https://doi.org/10.1016/j.scitotenv.2021.151754

Sati, D., Pandey, S. C., Pande, V., Upreti, S., Gouri, V., Joshi, T., & Samant, M. (2020). Toward an enhanced understanding of plant growth promoting microbes for sustainable agriculture. In *Recent advancements in microbial diversity* (pp. 87–112). Academic. https://doi.org/10.1016/B978-0-12-821265-3.00005-0

Selin, H., & Selin, N. E. (2020). *Mercury stories: Understanding sustainability through a volatile element.* MIT Press. https://doi.org/10.7551/mitpress/11856.001.0001

Shabbir, Z., Sardar, A., Shabbir, A., Abbas, G., Shamshad, S., Khalid, S., & Shahid, M. (2020). Copper uptake, essentiality, toxicity, detoxification and risk assessment in soil-plant environment. *Chemosphere, 259*, 127436. https://doi.org/10.1016/j.chemosphere.2020.127436

Shafreen, M., Vishwakarma, K., Shrivastava, N., & Kumar, N. (2021). Physiology and distribution of nitrogen in soils. In *Soil nitrogen ecology* (pp. 3–31). Springer International Publishing. https://doi.org/10.1007/978-3-030-71206-8_1

Sharma, P., Chouhan, R., Bakshi, P., Gandhi, S. G., Kaur, R., Sharma, A., & Bhardwaj, R. (2022). Amelioration of chromium-induced oxidative stress by combined treatment of selected plant-growth-promoting rhizobacteria and earthworms via modulating the expression of genes related to reactive oxygen species metabolism in *Brassica juncea*. *Frontiers in Microbiology, 13*, 802512. https://doi.org/10.3389/fmicb.2022.802512

Sharma, A. K., Sharma, M., Sharma, A. K., & Sharma, M. (2023). Mapping the impact of environmental pollutants on human health and environment: A systematic review and meta-analysis. *Journal of Geochemical Exploration, 255*, 107325. https://doi.org/10.1016/j.gexplo.2023.107325

Sun, R., Zhang, X., & Guo, X. (2020). Influence of zinc contamination on soil microbial communities and enzyme activities in agricultural soils. *Ecotoxicology and Environmental Safety, 189*, 109966.

Tian, J., Ge, F., Zhang, D., Deng, S., & Liu, X. (2021). Roles of phosphate solubilizing microorganisms from managing soil phosphorus deficiency to mediating biogeochemical cycle. *Biology, 10*(2), 158. https://doi.org/10.3390/biology10020158

Tuomainen, U., & Candolin, U. (2011). Behavioural responses to human-induced environmental change. *Biological Reviews, 86*(3), 640–657. https://doi.org/10.1111/j.1469-185X.2010.00164.x

Verma, S., & Kuila, A. (2019). Bioremediation of heavy metals by microbial process. *Environmental Technology and Innovation, 14*, 100369. https://doi.org/10.1016/j.eti.2019.100369

Wang, L., Wang, L. A., Zhan, X., Huang, Y., Wang, J., & Wang, X. (2020). Response mechanism of microbial community to the environmental stress caused by the different mercury concentration in soils. *Ecotoxicology and Environmental Safety., 188*, 109906. https://doi.org/10.1016/j.ecoenv.2019.109906

Wang, Y., Feng, J., & Lin, Q. (2021a). Impact of mercury contamination on microbial communities and function in freshwater sediments. *Science of the Total Environment, 753*, 141826.

Wang, L., Kaur, M., Zhang, P., Li, J., & Xu, M. (2021b). Effect of different agricultural farming practices on microbial biomass and enzyme activities of celery growing field soil. *International Journal of Environmental Research and Public Health, 18*(23), 12862. https://doi.org/10.3390/ijerph182312862

Xiao, W., Ye, X., Liu, C., Cheng, X., Zhu, L., & Zhou, H. (2020). Effects of heavy metals on microbial communities in soils under different flooding conditions: A high-throughput sequencing analysis. *Journal of Environmental Management, 264*, 110454. https://doi.org/10.1016/j.jenvman.2020.110454

Xu, M., Zhang, H., & Wang, Q. (2021). Impact of cadmium pollution on soil microbial community structure and diversity in different land use patterns. *Journal of Hazardous Materials, 407*, 124382.

Yaashikaa, P. R., Kumar, P. S., Jeevanantham, S., & Saravanan, R. (2022). A review on bioremediation approach for heavy metal detoxification and accumulation in plants. *Environmental Pollution, 301*, 119035. https://doi.org/10.1016/j.envpol.2022.119035

Yang, R., Hou, B., & Zhang, L. (2024). Dissolved organic compounds in shale gas extraction flowback water as principal disturbance factors of soil nitrogen dynamics. *Science of the Total Environment, 908*, 168197. https://doi.org/10.1016/j.scitotenv.2023.168197

Zhang, X., Chen, B., Yin, R., Xing, S., Fu, W., Wu, H., & Zhang, X. (2022a). Long-term nickel contamination increased soil fungal diversity and altered fungal community structure and co-occurrence patterns in agricultural soils. *Journal of Hazardous Materials, 436*, 129113.

Zhang, L., Zhou, S., & Yang, H. (2022b). Responses of microbial communities to arsenic contamination in paddy soils. *Environmental Science and Technology, 56*(4), 2378–2388. https://doi.org/10.3390/microorganisms11061392

Zhou, X., Zhou, X., Wang, C., & Zhou, H. (2023). Environmental and human health impacts of volatile organic compounds: A perspective review. *Chemosphere, 313*, 137489. https://doi.org/10.1016/j.jhazmat.2022.129113

Part III
Soil Microbiomes and Roles in Green Technology

Application of Nanotechnology in the Remediation of Heavy Metals and Endocrine Disrupting Chemicals in Soil Environment

Kingsley Erhons Enerijiofi, Thankgod Onuoha, Taiwo Stephen Okanlawon, and Cyril Chinedu Otali

Abstract The indiscriminate release of heavy metals and endocrine disrupting chemicals are continuously on the increase due to the rapid growth and urbanization of society. They accumulate in soils and pose threat to food security, human health, and the environment. The adverse effects of their indiscriminate discharge on environmental quality and public health require priority attention on a global scale. This review will provide insight into the concept of heavy metals and endocrine disrupting chemicals. It will also take a holistic overview on the use of nanotechnology, a current technology to remediate heavy metals and endocrine disrupting chemicals. The effects of remediating contaminated soil environments using nanotechnology on public health including effect on crops and soil enzymes were presented as well as current studies on Nano technological clean-up of the aforementioned environmental wastes. Finally, the paper presented the prospects of using nanomaterials research in the remediation of heavy metals and endocrine disrupting chemicals contaminated soils.

Keywords Bioremediation · Nanotechnological application · Sustainable environment · Public health · Soil biodiversity

1 Introduction

There is industrial revolution as a result of the need to meet the population' demand for quality and affordable goods and services. Some selected sites and unrestricted industrial and mining areas are more likely to have soil impurities of ecological concerns. The contamination from strong metals damage crop productivities and

K. E. Enerijiofi (✉) · T. S. Okanlawon
Department of Biological Sciences, College of Basic and Applied Sciences, Glorious Vision University, Ogwa, Edo State, Nigeria

T. Onuoha · C. C. Otali
Department of Biological Sciences, Novena University, Ogume, Delta State, Nigeria

S. A. Aransiola et al. (eds.), *Soil Microbiome in Green Technology Sustainability*,
https://doi.org/10.1007/978-3-031-71844-1_16

quality, influence food chain, and subsequently the human body (Enerijiofi & Ekhaise, 2019). As a result of these, the public health is exposed to risks by posed by mutagens, teratogens, and carcinogens (Bolan et al., 2014). Significantly, nutrients in the soil are utilized by plants which aid the foundation of agricultural production (Parikh & James, 2012). Though, agriculture disrupts natural environment but it still defines and sustains human existence. However, earlier studies by Pretty and Bharucha (2014); Struik and Kuyper (2017) investigated the conflict of interest concerning the importance of managing agricultural land and its maintenance. Land contamination has served as a threat to Sustainable development, quality of life, and livelihoods (Bhowmik et al., 2023). With the continuous increase in population and reduction in the available arable land, tracked to different technological activities, priority of soil conservation is therefore necessary in the modern period (Martin et al., 2016).

The emergence of nanotechnology as an effective method of treatment or regeneration of contaminated soil has been gained prominence (Medina-Pérez et al., 2019; Rajput et al., 2021). Nanotechnology has a lot of possibilities as a more sustainable approach, particularly by reducing the adverse effects of specific metals, corresponding to various sources (Kumari et al., 2021; Raffa et al., 2021). Nanotechnology has demonstrated high effectiveness in the delineation of contaminants in various surroundings (Rajput et al., 2021). Cai et al. (2019) and Fajardo et al. (2020) explained that when suitable, such process could be referred to as environmental nano-remediation where by the contaminants found in the soil, water, and air are removed using nanomaterials. The need to dredge and transfer soil that are polluted could be minimized by applying Nano-remediation techniques since the treatment process typically occurs in-situ. Mehta et al. (2015) juxtaposed that a variety of techniques were utilized to reclaiming nanomaterials in the contamination applications for therapy (such as the iron nanoparticle magnetic separation and metal recovery from used nano-sorbents. The different nanomaterials may be divided into three distinct types according to their specifics; membranes (pressure-driven technique used in the treatment of wastewater and seawater), catalysts (for the transformation of contaminants), and adsorbents (Anjum et al., 2019).

Through the aid of nanoparticles (NPs), hyper-accumulators and natural soil bacteria may be able to accelerate biodegradation, thereby extending the range of mending. This process is also referred to as nano-phytoremediation and microbial-mediated nano-remediation. Nanoparticles can be employed with bioremediation procedures and may offer a lot of benefits due to their significant dimension, tiny size (1–100 nm), and reactivity (Alazaiza et al., 2021). The utilization of NPs typically lessens the toxicity of metal contaminants and encourage plant growth, which causes a considerable increase in elemental toxic content of plant tissues (Ghani et al., 2022). The extent of pollution, bioavailability, and metal buildup by the plants from affected regions are some elements which influence the efficacy of nano-phytoremediation at removing heavy metals (Rajput et al., 2021). Though receiving a large audience lately, NPs in form of metal-organic framework (MOF) were used periodically to disinfect wastewater (Younis et al., 2021). So, there is a sizable window of opportunities for investigators to think about conceivable uses for MOF in soil remediation. Considering the fact that fewer investigations are currently done, there are problems surrounding the secure application of NPs for fixing damaged soils and their discharge into ecosystems (Moameri & Khalaki, 2021).

Lead and some other heavy metals, are antagonistic to living things and depending on the magnitude of their exposure, they may lead to immediate or over time damage. (Roy & Bharadvaja, 2021; Aransiola et al., 2023). Flotation and soaking represent a few of the tried-and-true techniques for clean-up. These, nevertheless, pose certain drawbacks, including needs for more energy needs, and the creation of compounds that cannot be recycled. However, according to Roy et al. (2021), Nanotechnology appears to offer a developing answer to these issues. The majority of synthetic chemicals affect the quality of soils that are momentously growing into a public menace due to the presence of industrial facilities. These frequently increase their risks, which can be traced to accidents, poor waste management, uncontrolled large scale production, and widespread utilization while disregarding the role of environmental regulations (Enerijiofi & Ekhaise, 2022). In addition, atmospheric emissions and transits widely distribute the pollutants with the soil ecosystems acting as their primary sinks. The adage "pollutants have no borders", with the advent of analytical methodologies. Concurrently, it has become more accurate with high resolution system which enable recognizing and gauging contaminants from heterogeneous media like soils (Enerijiofi et al., 2022; Aransiola et al., 2024). This chapter posits the use of nanotechnology to remediate heavy metals and endocrine disrupting chemicals contaminated soils, it also took a look at the public health implications of nanotechnology on soil, soil microorganisms, human health and other forms of soil biodiversity. Finally, the current studies so far on the utilization of nanotechnology to cleanup of heavy metals and endocrine disrupting chemicals contaminated soils were also documented. The various nanotechnological applications for heavy metal clean-up are as in Fig. 1 below.

2 Methodologies and Tools that Fall Under the Umbrella of Nano-bioremediation

The following nanosorbents, nanobiosensors, nano-phytoremediation, and nanobiosurfactants, are some of the diverse methodologies and tools that fall under the umbrella of nano-bioremediation and are discussed below.

2.1 Nano-biosensor

In order to create nano-biosensors, living components are modified biosensors that are coupled to an atomic-level transducer to perform and review data (Andrade-Zavaleta et al., 2022). A physical and chemical smaller and connected to active

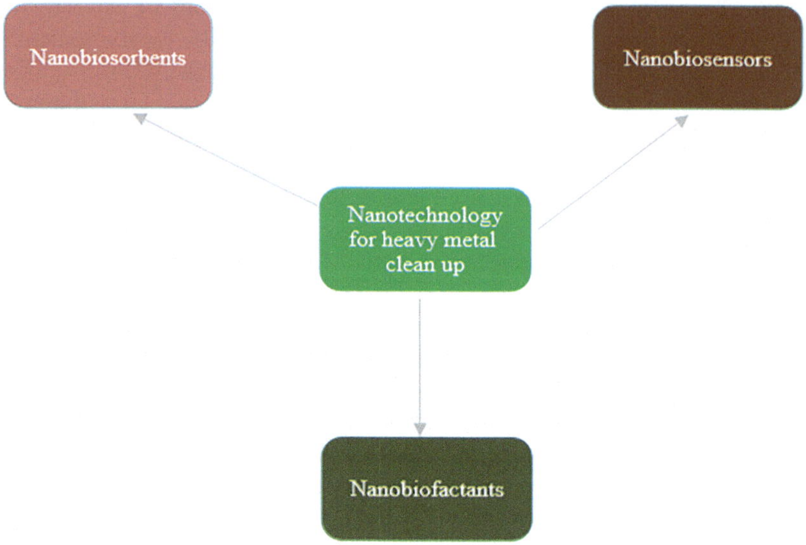

Fig. 1 The various nanotechnological applications for heavy metal clean up

components transducer reduces and detects the products at various concentrations (Arduini et al., 2016). Nano biosensors feature a number of impressive properties that improve the ability of NMs to integrate biological elements into massive structures, such as preventing particle movement, miniaturization, and fictionalization. In comparison to biosensors of more traditional origin, nano-biosensors have demonstrated tremendous value since they exhibit an increase in the detection of specific compounds and are used in many different industries, such as soil monitoring, fertilizer, environmental management, agriculture, and pharmaceutical bioprocessing (Ashraf et al., 2019). By using fertilizers, land, pesticides and water in exchange, through early detection and quick decision-making, nano-biosensor technology can improve crop output. Heavy metals like lead, cadmium, arsenic, and mercury are one of the most harmful materials in the environment. They may accumulate in crops, which would worsen the situation for both plants and the people who consume them. They can be identified utilizing nanomaterials created by biochemical and environmentally friendly methods of synthesis (Briffa et al., 2020). Numerous applications for the detection of heavy metals in soil using nanomaterials (nanoclusters, antibodies, and aptamers) and nanodevices (FRET, or fluorescence resonance energy transfer, quantum dots, and cantilever biosensors), are now being developed (Awasthi et al., 2022). By monitoring the amount of chlorophyll, nanobiosensors can be used to identify the phytotoxicity of heavy metals. One of the heavy metals that seriously endangers the health of ecosystem's inhabitants is Arsenic. In a study on the plant, Pteris cretica's innate ability to accumulate arsenic, Lew and his colleagues employed nanobionic sensors that were implanted which made the plant's internal dynamics to be studied by nanobionic sensors in real time (Borghese et al., 2014).

2.2 *Nano-biosorbents*

According to Bakshi and Abhilash (2020), biosorption is the process of using biologically produced materials that can be combined with organic and inorganic stuff. Nano-biosorbents are materials that break down pollutants using nanotechnology and biosorption. Nanoadsorbents, which are environmentally benign, reasonably priced, more environmentally favorable, and made of renewable materials are frequently utilized for specific pollutant (Biswas et al., 2019). For a biosorbent to effectively bind to radionuclides, dyes, and heavy metals, it must have both carboxyl (-COOH) and hydroxyl (-OH) functional groups (Ballen et al., 2021). Waste materials including bagasse, fly ash, used up microorganisms, and rice husk are excellent candidates that serve as biosorbents or nanosorbents. In an earlier study, Ballen et al. (2021) created rice husk that was employed as a biosorbent with graphene Quantum Dots to remove Pb (II). Starch, cellulosic material, gummies, pectins, and other types of alginates like chitin can be harmful and all contribute to the development of powerful nanobiosorbents.

2.3 *Nano-biofactants*

The chemicals known as biosurfactants are produced by microbes on life surfaces or secreted extracellularly. The surface tension is decreased and these compounds exhibit hydrophilic and lipophilic characteristics. Biofactants have emerged as an environmentally friendly and promising raw material for nanoparticle synthesis in sustainable chemistry. Heavy metal remediation, such as zinc, copper, and nickel, occurs as biofactants produced from microbes. Benjamin et al., 2019 studied the relationship between the two biofactants. The study revealed a high electronegative Nano Zirconia surfaces which interact with one another. The biofactant's capacity to minimize toxicity, operate as a dispersion and foaming agent, and reduce heavy metal stress made it a potent nano-biofactant. Heavy metals and hydrocarbon-contaminated soils were successfully remediated using lipopeptide biosurfactant made from *Bacillus subtilis* (Cao et al., 2020a).

3 Heavy Metals

Heavy metals are naturally found in undisturbed soils and little concentrations are vital for the healthy living by plants (Aransiola et al., 2019). Heavy metals are a group of metals that cannot be biodegraded by microorganisms or any other agent of degradation which makes them to accumulate in the environment. There are lots of human activities that generate significant quantities of heavy metals into the soil environment. When rain increases, the heavy metals leach into the soil from where translocated eventually occurs and gets to plants which get to man when taken either as food for energy or medicine for treating ailments (Enerijiofi & Ekhaise, 2019). However, two types of heavy metals are found in soil. They are the inactive fraction, which is known as the non-toxic and the labile fraction, known to be potentially lethal (Wuana & Okieimen, 2011).

3.1 Utilization of Nanotechnology for Heavy Metal Cleanup

According to Abhilash et al. (2009), nanobioremediation is a technique for cleaning up contaminants utilizing processed green nanoparticles or nanomaterials. The techniques for reducing pollution and improving soil quality by using microorganisms and plants to break down hazardous substances are affordable. By degrading down pollutants, the procedure can either eliminate all dangerous compounds from the soil or reduce their concentration (Alabresm et al., 2018). The effectiveness of bioremediation was examined and enhanced by biotechnology methods Ajuzie et al. (2015) and nanotechnology has, however, given the procedure a new dimension (Alabresm et al., 2018; Bhowmik et al., 2023). By using nanoparticles to clean up organic matter tainted with pollutants and soil that harbors heavy metals, for example, phytoremediation based on nanotechnology is more effective. Because NPs have favorable physical and chemical properties, they can work as a catalyst to reduce the amount of energy required to activate molecules, so inducing the degradation or annihilation of the pollutants that are being targeted. The main ingredients used are nanoparticles made of metal and carbon (Ameen et al., 2021). However, as HM are difficult to biodegrade and prone to ingestion by food chains, it is important to understand their biological mechanisms to aid their complete elimination (Enerijiofi & Ajuzie, 2012; Amin et al., 2021).

3.2 The Degradation of Heavy Metals by Nanoparticles and Microbes

In contrast to their sole use, nanomaterials are more effective when combined with microorganisms to increase heavy metals reduction (Abdi & Kazemi, 2015). Nanomaterials interact with microbes based on their chemical properties, shape, size, coating characteristics, metabolism, crystalline phase of the material, extent of contamination, resistance of the material to hazards (Enerijiofi et al., 2024a). Nanocomposites increase the absorption and adsorption of HMs, and increase the number of chemical pathways for reducing microbial biological biostimulation and bioaccumulation as well as biotransformation of HMs. The removal of Pb (II) and Cd (II) using iron oxide nano-nanomaterials coated in polyvinyl was proven to include *Halomonas* sp. As a green synthesis device, the microbe can produce nanoparticles in contrast (Al-Gheethi et al., 2015). Despite the fact that separating or recovering heavy metals from nanomaterials require time and effort, magnetic nanoparticles have gained a lot of attention recently. Iron or iron oxide coating, encapsulation, and surface modification have all been designed to make it easier to separate or retrieve heavy metals (Ali et al., 2013).

4 Endocrine Disrupting Chemicals (EDCs)

An endocrine disrupting chemicals (EDC) is "an exogenous substance or mixture that can alter normal hormonal functions in humans and animals, and subsequently impact the hormonal machinery of living organisms," based on the World Health Organization. In accordance with their application, place of origin, and consequences, these pollutants fall under various classes. According to Li et al. (2013), the origin of EDCs are medications, personal care items, and pesticides. Among the three fundamental processing systems in humans, the endocrine system is decisive in handling a range of bodily practices and preserving the relative stability of the internal environment and diseases may be brought on by imbalances within the system (Marty et al., 2018). The chemicals which include a variety of organic and synthetic substances can be deleterious to the normal regulation of the endocrine system by blocking or attaching to the receptors and affecting activities within the system (Yang et al., 2015). These terms "EDCs" are a significant class of new pollutants that are frequently referred to as "endocrine disruptors" or "environmental hormones" and are discovered in aquatic habitats and process effluents at low quantities (Sarangapani et al., 2017).

4.1 Application of Nanotechnology in Remediation of Endocrine Disrupting Chemicals (EDCs)

Currently, biodegradation is one of the effective methods for treating endocrine disrupting chemicals. It offers several benefits for both economic purposes, and to safeguard the integrity of the ecosystem, including minimal initial cost, a broad range of application, a lengthy time required, and straightforward equipment and space needs. Biodegradation is a process that poses less damage to the environment and does not create secondary pollution (Enerijiofi et al., 2020; Gao et al., 2020). At this point, semiconductor materials of the TiO_2 and ZnO types are the most often employed photocatalysts. Based on this, an increasing number of nano-scale composite materials, including nano-rods and nano-composites, have been created and experimental evidence has shown that they all have outstanding photocatalytic effects. Then, as an urgent challenge to be tackled, it is important to comprehend the pathways produced by photocatalytic degradation (Gao et al., 2020). The bulk of photocatalysts currently under investigation are heterogeneous oxide semiconductor materials, with compounds made of clean and free TiO_2 being the most common. The most researched photocatalyst is TiO_2. For instance, suspended TiO_2 is linked to the recovery of precious metals, and photocatalysts in addition to its unmatched benefits, such as its safe, low cost, and its availability (Dolat et al., 2012).

However, some studies Sornalingam et al. (2017) and Zatloukalová et al. (2017) have discovered that utilizing TiO_2 as a photocatalyst might efficiently degrade endocrine disrupting chemicals. It was demonstrated that progesterone and all forms of estradiol could be easily broken down by the powder photocatalyst NnF Ceram TiO_2 made of nanofibers (Zatloukalová et al., 2017). Two commercial TiO_2 nano-powders were used as photocatalysts to remove six known pesticides (Vela et al., 2017). Benzophenone-3 (BP3), one of the most often used ultraviolet (UV) filters, is an ingredient in a variety of lotions and sunscreens. BP3 has a negative impact on the endocrine systems of various organisms. The photocatalytic degradation of BP3 by particles of TiO_2 showed that organic carbon can be removed after a certain period of treatment (Zúñiga-Benítez et al., 2016). This led to the confirmation that BP3 may be removed from water via photocatalysis utilizing TiO_2. It was created using a straightforward sol-gel technique and shown improved UV and visible light photocatalytic performance. With strong crystallinity, permeable, and UV loving areas, the produced photocatalyst was nano-scale and shown good reusability. The reduction level of some pesticides increased more than 90% when a nano-rod ZnO/SiC nano-composite was used as a photocatalyst (Meenakshi & Sivasamy, 2018).

The application of nanoparticles in soil has detoxified numerous organic, inorganic, and metal contaminants. Organic contaminants in soil were quickly destroyed with the application of nano-zerovalent iron and magnetite nanoparticles (Madhura et al., 2019). According to Li et al. (2016) and Kalidhasan et al. (2017), nanoparticles of polyethylenimine (PEI)-copper and TiO_2 decreased the level of common pesticides. Specific bacteria can exploit heavy metal contaminants in the bio-organic nanoparticle synthesis process, which results in eliminating them from the environs, and the creation of wealth from materials (Enerijiofi et al., 2024a). *Enterococcus faecelis* was employed in this method to remove and recover lead. Some bacteria produce lead nanoparticles of about 10 nm in size both intracellularly and extracellularly. According to Cao et al. (2020a), these particles had a high binding activity and lessens Chromium within few hours. The health of people and the environment is harmed by organic pollutants, especially persistent organic compounds. Soil remediation is crucial in order to remove these chemicals. Dibenzothiophene (DBT) was desulfurized as a result of magnetic nanoparticles and *Rhodococcus engthropolis* (Ansari et al., 2009).

The reduction of the impact of traditional sewage treatment techniques on environmental endocrine disrupting chemicals is thought to be minimal, but it is hoped that the two biodegradation processes of EDCs will compensate for this.

EDCs' estrogenic activity can be successfully decreased by it. The chemicals that are deleterious in the natural world can be eliminated by the catalytic activity of photocatalysts, which is characterized by the reduction of persistent organic matter pollution. Naturally, it is assumed that nanotechnology will become more reliable and efficient (Gao et al., 2020).

4.2 *Mechanism of Nanoparticles-Microbes Interaction in Endocrine Disrupting Chemicals*

The biodegradation and biotransformation of Endocrine Disrupting Chemicals endocrine disrupting chemicals by bacteria involves enzymatic reactions. The bioadsorption and bioaccumulation of EDCs, on the other hand, are carried out by the aid of the enzymes, whereas only biotransformation is identical to some microbes (Wojcieszynska et al., 2020). Additionally, the organism's mode of distribution and the carrier substance had a significant impact on biodegradation (Roccuzzo et al., 2021). Enzymes play different roles in maintaining different metabolic processes of organisms as a fundamental biocatalyst. Environmental EDCs that degrade through the action of microbial products such as enzymes have been used. The enzymes that microbes release provide them the ability to break down EDCs. Manganese peroxidase (MnP) and laccase have both been widely studied for their roles in the biodegradation of endocrine disrupting chemicals (Gao et al., 2020).

Laccases participate in developmental processes, spore production, and the removal of harmful compounds produced by the plant immune system in phytopathogenic organisms. Bacterial laccases, for instance, can aid in shielding the spore from UV light and hydrogen peroxide. Plant laccases take part in the polymerization of lignin, among other things (Wang et al., 2021). Conversely, laccases convert the pigment in it's structure into the equivalent materials that are essential for cuticle production in insects (Yang et al., 2022). Additionally, the protein molecule's structure varies between laccases. Laccases found in fungi and plants are glycoproteins. Copper ions are affected by the glycosylation process, giving the enzyme thermal stability and defense against proteolytic destruction. According to Arregui et al. (2019), intracellular bacterial enzymes are frequently not glycosylated.

5 Application of Some Widely Used Nanomaterials for Removing some Soil Pollutants

Nanotechnology is increasingly gaining grounds in the Nano-bioremediation technology. The table below shows the summary of observations when a specific nanoparticle is applied to remediate an individual pollutant with the observations recorded in Table 1 as reported by Hemalatha et al. (2022).

Table 1 Some widely used nanomaterials for removing pollutants

S/no	Nanoparticles used	Pollutants	Observations	References
1	Activated carbon	Copper and sulfates	The hydrophobicity of activated carbon nanoparticles and enhanced the removal efficiencies.	Hosseini et al. (2018)
2	Aluminum	Arsenate	Arsenate was competently cleaned up at optimal pH and temperatures using aluminum nanoparticles.	Prabhakar and Samadder (2018)
3	Polystyrene	Estrone	Nanoparticles of polystyrene were low; about 40% clean-up efficacy in comparison with nanofilteration.	Akanyeti et al. (2017)
4	Titanium dioxide	Cadmium	Titanium dioxide nanoparticles totally removed cadmium.	Zand et al. (2020)
5	Zinc oxide	Benzophenones	Zinc oxide nanoparticles efficiently breakdown endocrine disrupting hormones.	Rajesha et al. (2017)
6	Magnesium oxide	Estradiol	Estrogens in the soil had a removal efficiency of 88% with magnesium oxide nanoparticles; a reduction in the speed of inserting nanoparticles into the polluted soils improved the degradation prowess.	Han et al. (2017)
7	Palladium	Pentachlorobiphenyl (PCB)	Palladium nanoparticles fixed with carbon (iv) oxide fluid at a pressure of 200 atm was able to remediate PCB in soils	Wang and Chiu (2009)
8	MWCNTs	Crude oil	The use of MWCNTs can improve the breakdown of hydrocarbons through increase in the total microbial population as revealed in the results.	Abbasian et al. (2016)
9	PS-Z/nZVI	Trichloroethane (TCE)	The results showed that Z/nZVI had high potential for PS stimulation and superb clean up efficacy (98.8%) of TCE at a recorded pH of 7 within 2 hours.	Huang et al. (2019a, b)
10	nZVI combined with compost from organic waste	Hydrocarbons (TPH, PAH and heavy metals)	The results showed that the amalgamation of nZVI pooled with compost from organic waste could reduce the concentration of aliphatic hydrocarbons by up to 60% at even below wild conditions. In addition, there was a significant reduction in ecotoxicity in the biopile of the soil	Galdames et al. (2017)

6 Public Health Effect of Nanotechnological Clean-Up of Contaminated Soil Environment

Although nanotechnology has advanced and played various roles, there are still concerns relating to its effects on public health (Enerijiofi et al., 2024b). The sub-topic will provide insights into to the fate and environmental issues caused by the intentional release of nanomaterials and nanoparticles, primarily during remediation processes (Lew et al., 2021). In 2010, the amount of engineered nanomaterials produced worldwide was predicted to be between 260,000 and 307,000 tons with about 9–17% in soil, 2–5% in water and 1–5% in the atmosphere. Ecotoxicological concerns are posed by nanomaterials because of their persistence in the environment, biodegradability, accumulation in trophic levels and the spread of species throughout food chains. In order for nanomaterials to enter the food chain, it is necessary for plants to act as both entry points and sinks. More recent research has focused on the general implications of the uptake of plants by nano-materials. The significance of natural minerals (NMs) in phytoremediation is discussed, along with how they impact plant responses and biomass output. While there is no reason to believe that nanotechnology cannot solve problems, one of the biggest challenges is keeping it from being used and banned. There are serious problems with NPs for agriculture, according to recent research. Nanoparticles (NPs) can change the properties of soil by purposefully building up or increasing its concentration (Laslo et al., 2022). It has been discovered that pH is affected, and that this alteration has a significant influence on soil microbiology, plant growth and development, nutrient availability, and soil health. A research by Cullen found that change of soil at the nanoscale by zero-valent iron (nZVI) can cause a large increase in pH (Lal, 2015). This method has been used to study the effects of CuO and CuO nano-nodes on soil pH, and it's been proven to work better in acidic soils. Basically, these nano-nodes (NPs) made use of the H^+ to produce Cu ions and other Cu (OH) complexes in the soil. It's been shown that many nano-nodes, like Silver, Gold and Titanium can change the pH of soil and have negative effects on soil microorganisms. Some of the primary elements that influence the detrimental impact of nano-nodes in soil include how many and what kind of nanoparticle they are, and what kind of soil they're in (Konate et al., 2017). Increased nanoparticle concentration can lower dehydrogenase activity, which can affect the soil's fertility and nutritional balance (Enerijiofi et al., 2024b). Furthermore, ingesting and absorbing these nanoparticles may adversely impact the regular functioning of bacteria and significantly impair their mycelium (Ibrahim et al., 2013). The three primary processes that bond, shift surface complexation, hydrophobic partitioning, and ion exchange are the three NP processes in soil. Jaafar et al. (2016) claimed that these routes also play a leading role in the coupling of contaminants to colloid surfaces. Nanoparticles emissions become normal when there is a constant flow into the environment. At different trophic levels, NPs are probably going to enter the food chain. Numerous aquatic and terrestrial animals, as well as people, are susceptible to their toxicological effects (Jośko et al., 2014). Nutrients (NPs) turn become pollutants when they get into water bodies and are eaten by

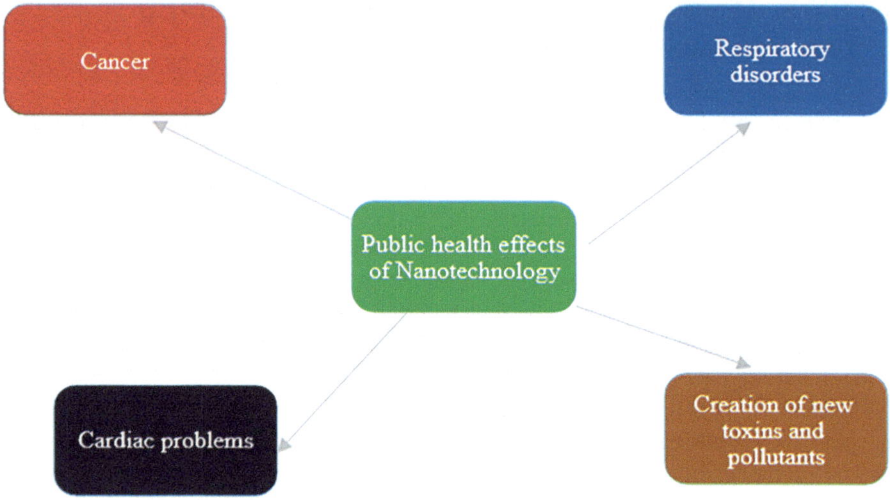

Fig. 2 The public health effects of nanotechnological applications

smaller aquatic creatures. Different trophic levels of animals, like the daphnia and zebra fish Danio rero. The toxicity of metal NPs was investigated in Daphnia pulex and the microalgae *Pseudokirchneriella subcapitata*. According to Khan et al. (2022), daphnia and microalgae are more susceptible to nitrogen pollution than zebra fish because they feed via particle filters. NPs have an impact on biochemical processes at the molecular and tissue levels as soon as they reach a biological system (Kandiah & Chandrasekaran, 2021). According to Kasture et al. (2008), NPs have been linked to a multitude of toxicological consequences, such effects of oxidative stress on DNA due to the presence of reactive oxygen species (ROS) which can lead to DNA damage, alteration of protein folding, and cell death. So, these environmental issues should come first before whether or not NPs are acceptable for nanobioremediation. Also, this would make it possible to design NPs in a way that promotes long-term uses. It's a huge danger to people's health and the planet due to the fast discharge of toxins into the atmosphere. Figure 2.

7 Current Studies on Nano-technological Cleanup of Heavy Metals and Endocrine Disrupting Chemicals

Zhang et al. (2019) reported the potential of micro-nano-hydroxyapatite to increase the pH of soil while reducing the amount of readily available heavy metals. Micron hydroxyapatite enriches and diversifies fungal communities, whereas nanometer hydroxyapatite does not. Five percent (5%) nanoscale zero-valent iron (nZVI) is potent at removing lead and cadmium from polluted areas (Fajardo et al. (2019). Although nanometer zero-valent iron (NZVI) reacts quickly, its long-term effectiveness is still unknown. According to Arif et al. (2020), carbon-based nanomaterials like carbon nanotubes, metal oxides (including ferric oxide and titanium oxide), and other nanocomposites are some of the manmade nanomaterials that have been utilized to immobilize soil pollutants. Ferric oxide nanoparticles can effectively absorb and immobilize heavy metals like Cadmium and Arsenic from a variety of media samples. Using gold and silver nanoparticles, the traditional two-step method for dye degradation involves concentrating electrons on the particle surface in step one and diffusing the dye for a reduction reaction in step two. Using gold and silver nanoparticles, the traditional two-step method for dye degradation involves concentrating electrons on the particle surface in step one and diffusing the dye for a reduction reaction in step two.

According to Zheng (2020), polyacrylamide-treated magnetite nanoparticles were effective in reducing soil erosion and arsenate leaking. Magnetic nanoparticles have several applications in the treatment of contaminated environment (Ajmal et al., 2018). In their investigation, Usman et al. (2020) used sodium carboxymethyl cellulose-stabilized nZVI and it successfully eliminated 80% of metals in the soil. Some pollutants can be remedied using reductive processes and immobilization techniques. These procedures have a direct impact on the mobility and toxicity of the contaminants. For instance, adding nZVI to soil contaminated with pyrene led to a significant reducing conditions that enabled pyrene reduction. Nanomaterials are also used in some processes such as oxidation, which use a range of compounds to degrade impurities in damaged soil. Blundell and Owen (2021) were able to lower a particular concentration of DDT and chromium (VI) to 25% and 62%, respectively, over three days by employing nZVI nanomaterials. Research by Tian et al. (2020) found that tri-chloroethylene (TCE) was reduced to 84.73% in the soil dechlorination of TCE. In order to improve the bioremediation of soil that had been tainted with Pb and Cd using a Gram-negative bacterium known as *Halomonas* sp., Alabresm et al. (2018) used nanoparticles of FeO coated with polyvinylpyrrolidone (PVP). This approach effectively removed lead and all metals than bacteria biodegradation or using NPs.

Hung et al. (2021) employed a range of nanomaterials to treat soil affected with heavy metals. Carbon nanotubes (CNTs) NPs, TiO_2 NPs, and Ag NPs are well studied for their importance in enhancing the structure of soils. How well heavy metals are removed by CNTs depends on several factors, including the pH and temperature of the surrounding environment, contact time, and CNT dose. Mitzia et al. (2020) reported that the combination of nZVI and biochar greatly improved the immobilization of Zinc, Lead, and Cadmium in soils. In a related research, Shi et al. (2018) posited that CuO NPs modify soil pH by utilizing ambient H^+ to create Cu ions and the ionic complex $Cu(OH)^+$. It's also said that acidic soil accelerates this process. In addition to iron oxides, other NPs can be used to remediate soils that have been tainted with various metals and metalloids (Martínez-Fernández et al., 2017). However, the detrimental presence of NPs in the soils may depend on the kind, concentration, and enzymatic activities of the soil. The uptake and absorption of such NPs by bacteria also has a detrimental effect on mycelium and their normal cell activity (Ameen et al., 2021).

8 Conclusion and Future Research Direction

The use of nanotechnology as an evolving field in science has found application in almost all facets of life including in the remediation of emerging environmental pollutants of public health concern such endocrine disrupting chemicals and heavy metals as discussed in this chapter. However, factors such as type, concentration, and enzymatic activity of the soil play leading roles in the clean-up of the aforementioned environmental pollutants. Considering the public health implications of nano-technological applications in clean-up. Some of the nanomaterials and the pollutants they remediate were presented in the course of this write-up. However, there will be need to extensively study them in other to allow for better understanding of their roles and impacts in the removal of soil pollutants. The fact that emerging contaminants like heavy metals and endocrine disrupting chemicals will continue to be associated with environmental and public health issues that arising from developmental strides, calls for the need to look critically into a sustained effort to ensure their clean-up. This requires a collaborative effort to ensure a robust remediation and sustainable environment for all biodiversities. There is also the need for improved instantaneous facilities that will facilitate the easy detection and quantification of the aforementioned pollutants so as to effective monitor them to prevent transfer through food chain and the attendant public health consequences.

References

Abbasian, F., Lockington, R., Palanisami, T., Megharaj, M., & Naidu, R. (2016). Multiwall carbon nanotubes increase the microbial community in crude oil contaminated fresh water sediments. *Science of the Total Environment, 539*, 370–380.

Abdi, O., & Kazemi, M. (2015). A review study of biosorption of heavy metals and comparison between different biosorbents. *Journal of Materials and Environmental Science, 6*, 1386–1399.

Abhilash, P. C., Jamil, S., & Singh, N. (2009). Transgenic plants for enhanced biodegradation and phytoremediation of organic xenobiotics. *Biotechnology Advances, 27*, 474–488. https://doi.org/10.1016/j.biotechadv.2009.04.002

Ajmal, Z., Muhmood, A. M., & Usman, E. T. (2018). Phosphate removal from aqueous solution using iron oxides: Adsorption, desorption and regeneration characteristics. *Journal of Colloid and Interface Science, 528*, 145–155.

Ajuzie, C. U., Atuanya, E. I., & Enerijiofi, K. E. (2015). Biodegradation potentials of microorganisms isolated from Eleme petrochemical industrial effluent. *Nigerian Society for Experimental Biology Journal, 15*(4), 128–136.

Akanyeti, I., Kraft, A., & Ferrati, M. C. (2017). Hybrid polystyrene nanoparticle ultrafiltration system for hormone removal from water. *Journal of Water Process Engineering, 17*, 102–109.

Alabresm, A., Chen, P. Y., Decho, A. W., & Lead, J. (2018). A novel method for the synergistic remediation of oil-water mixtures using nanoparticles and oil-degrading bacteria. *Science of the Total Environment, 630*, 1292–1297. https://doi.org/10.1016/j.scitotenv.2018.02.277

Alazaiza, M. Y. D., Albahnasawi, A., Ali, G. A. M., Bashir, M. J. K., Copty, K., Amr, S. S. A., Abushammala, M. F. M., & Al Maskari, T. (2021). Recent advances of nanoremediation technologies for soil and groundwater remediation: A review. *Water, 13*, 2186.

Al-Gheethi, A. A. S., Lalung, J., Noman, E. A., Bala, J. D., & Norli, I. (2015). Removal of heavy metals and antibiotics from treated sewage effluent by bacteria. *Clean Technologies and Environmental Policy, 17*, 2101–2123. https://doi.org/10.1007/s10098-015-0968-z

Ali, H., Khan, E., & Sajad, M. A. (2013). Phytoremediation of heavy metals—Concepts and applications. *Chemosphere, 91*, 869–881. https://doi.org/10.1016/j.chemosphere.2013.01.075

Ameen, F., Alsamhary, K., Alabdullatif, J. A., & ALNadhari, S. (2021). A review on metal-based nanoparticles and their toxicity to beneficial soil bacteria and fungi. *Ecotoxicology and Environmental Safety, 213*, 112027.

Amin, I., Nazir, R., & Rather, M. A. (2021). Nano-bioremediation: An innovative approach for remedying heavy metals using fungi. *Journal of Bioremediation & Biodegradation, 12*, 2.

Andrade-Zavaleta, K., Chacon-Laiza, Y., Asmat-Campos, D., & Raquel-Checca, N. (2022). Green synthesis of superparamagnetic iron oxide nanoparticles with eucalyptus globulus extract and their application in the removal of heavy metals from agricultural soil. *Molecules, 27*, 1367. https://doi.org/10.3390/molecules27041367

Anjum, M., Miandad, R., Waqas, M., Gehany, F., & Barakat, M. A. (2019). Remediation of wastewater using various nano-materials. *Arabian Journal of Chemistry, 12*(8), 4897–4919.

Ansari, F., Grigoriev, P., Libor, S., Tothill, I. E., & Ramsden, J. J. (2009). DBT degradation enhancement by decorating *Rhodococcus erythropolis* IGST8 with magnetic Fe3O4 nanoparticles. *Biotechnology and Bioengineering, 102*, 1505–1512. https://doi.org/10.1002/bit.22161

Aransiola, S. A., Ijah, U. J. J., Abioye, O. P., & Bala, J. D. (2019). Microbial-aided phytoremediation of heavy metals contaminated soil: A review. *European Journal of Biological Research, 9*(2), 104–125. http://www.journals.tmkarpinski.com/index.php/ejbr/article/view/157

Aransiola, S. A., Ijah, U. J. J., Abioye, O. P., & Bala, J. D. (2023). Vermicompost-assisted phytoremediation of toxic trace element-contaminated soil in Madaka, Nigeria using Melissa officinalis L and Sida acuta. *International Journal of Environmental Science and Technology*. https://doi.org/10.1007/s13762-022-04105-y

Aransiola, S. A., Ijah, U. J. J., Abioye, O. P., Bala, J. D., Rivadeneira-Mendoza, R. F., Luque, R., Rodríguez-Díaz, J. M., & Maddela, N. R. (2024). Micro and vermicompost assisted remediation of heavy metal contaminated soils using phytoextractors. *Case Studies in Chemical and Environmental Engineering, 9*(2024), 100755. https://doi.org/10.1016/j.cscee.2024.100755

Arduini, F., Micheli, L., Moscone, D., Palleschi, G., Piermarini, S., & Ricci, F. (2016). Electrochemical biosensors based on nanomodified screen-printed electrodes: Recent applications in clinical analysis. *Trends in Analytical Chemistry, 79*, 114–126. https://doi.org/10.1016/j.trac.2016.01.032

Arif, Z., Sethy, N. K., Kumari, L., Mishra, P. K., & Upadhyay, S. N. (2020). Recent advances in functionalized poly-mer-based composite photocatalysts for wastewater treatment. In *Nanomaterials as Photocatalysts for degradation of environmental pollutants* (pp. 39–64). Elsevier.

Arregui, L., Ayala, M., Gómez-Gil, X., Gutiérrez-Soto, G., Hernández-Luna, C. E., Herrera deLos Santos, M., Levin, L., Rojo-Domínguez, A., Romero-Martínez, D., & Saparrat, M. C. N. (2019). Laccases: Structure, function, and potential application in water bioremediation. *Microbial Cell Factories, 18*, 200.

Ashraf, S., Siddiqa, A., Shahida, S., & Qaisar, S. (2019). Titanium-based nanocomposite materials for arsenic removal from water: A review. *Heliyon, 5*, e01577. https://doi.org/10.1016/j.heliyon.2019.e01577

Awasthi, G., Nagar, V., Mandzhieva, S., Minkina, T., Sankhla, M. S., & Pandit, P. P. (2022). Sustainable amelioration of heavy metals in soil ecosystem: Existing developments to emerging trends. *Minerals, 12*, 85. https://doi.org/10.3390/min12010085

Bakshi, M., & Abhilash, P. C. (2020). Nanotechnology for soil remediation: Revitalizing the tarnished resource. In *Nano-materials as Photocatalysts for degradation of environmental pollutants* (pp. 345–370). Elsevier. https://doi.org/10.1016/B978-0-12-818598-8.00017-1

Ballen, S. C., Steffens, J., & Steffens, C. (2021). Stability characteristics of cantilever nanobiosensors with simple and complex molecules for determination of cadmium. *Sensors and Actuators A: Physical, 324*, 112686. https://doi.org/10.1016/j.sna.2021.112686

Benjamin, S. R., Lima, F. D., Florean, E. O. P. T., & Guedes, M. I. F. (2019). Current trends in nanotechnology for bioremediation. *International Journal of Environment and Pollution, 66*, 19–40. https://doi.org/10.1504/IJEP.2019.104526

Bhowmik, D., Chetri, S., Enerijiofi, K. E., Naha, A., Kanungo, T. S., Shah, M. P., & Nath, S. (2023). Multitudinous approaches, challenges and opportuinities of bio-electrochemical systems in conversion of wastes to energy from wastewater treatment plants. *Journal of Cleaner and Circular Bioeconomy – Elsevier, 4*, 100040.

Biswas, B., Warr, L. N., Hilder, E. F., Goswami, N., Rahman, M. M., & Churchman, J. G. (2019). Biocompatible functionalisation of nanoclays for improved environmental remediation. *Chemical Society Reviews, 48*, 3740–3770. https://doi.org/10.1039/C8CS01019F

Blundell, S. P., & Owen, G. (2021). Evaluation of enhancement techniques for the dichlorination of DDT by nanoscale zero-valent iron. *Chemosphere, 264*(1), 123824. https://doi.org/10.1016/j.chemosphere.2020.128324

Bolan, N., Kunhikrishnan, A., & Thangarajan, R. (2014). Remediation of heavy metal (loid) s contaminated soil-to mobilize or to immobilize? *Journal of Hazardous Materials, 266*, 141–166.

Borghese, R., Baccolini, C., Francia, F., Sabatino, P., Turner, R. J., & Zannoni, D. (2014). Reduction of chalcogen oxyanions and generation of nanoprecipitates by the photosynthetic bacterium *Rhodobacter capsulatus*. *Journal of Hazardous Materials, 269*, 24–30. https://doi.org/10.1016/j.jhazmat.2013.12.028

Briffa, J., Sinagra, E., & Blundell, R. (2020). Heavy metal pollution in the environment and their toxicological effects on humans. *Heliyon, 6*, e04691. https://doi.org/10.1016/j.heliyon.2020.e04691

Cai, C., Zhao, M., Yu, Z., Rong, H., & Zhang, C. (2019). Utilization of nanomaterials for in-situ remediation 1of heavy metal (loid) contaminated sediments: A review. *Science of the Total Environment, 662*, 205–217.

Cao, X., Alabresm, A., Chen, Y. P., Decho, A. W., & Lead, J. (2020a). Improved metal remediation using a combined bacterial and nanoscience approach. *Science of the Total Environment, 704*, 135378. https://doi.org/10.1016/j.scitotenv.2019.135378

Cao, F., Dai, H., Hao, P. F., & Wu, F. (2020b). Silicon regulates the expression of vacuolar H+-pyrophosphatase 1 and decreases cadmium accumulation in rice (*Oryza sativa* L.). *Chemosphere, 240*, 124907. https://doi.org/10.1016/j.chemosphere.2019.124907

Dolat, D., Quici, N., Kusiak-Nejman, E., Morawski, A. W., & Puma, G. L. (2012). One-step, hydrothermal synthesis of nitrogen, carbon co-doped titanium dioxide (N,C TiO_2) photocatalysts. Effect of alcohol degree and chain length as carbon dopant precursors on photocatalytic activity and catalyst deactivation. *Applied Catalysis B: Environmental, 115–116*, 81–89.

Enerijiofi, K. E., & Ajuzie, C. U. (2012). Health impact of some heavy metals resistance in microbes isolated from Ikhueniro refuse dumpsite, Benin City. *Nigerian Journal of Microbiology, 26*, 2593–2599.

Enerijiofi, K. E., & Ekhaise, F. O. (2019). Physicochemical and microbiological qualities of government approved solid waste dumpsites in Benin City. *Dutse Journal of Pure and Applied Sciences, 5*(2a), 12–22.

Enerijiofi, K. E., & Ekhaise, F. O. (2022). Achieving ecofriendly environment through sustainable management of solid wastes in the ecosystem. In T. Aftab (Ed.), *Sustainable management of environmental contaminants: Eco-friendly remediation approaches* (pp. 451–470). Springer. https://doi.org/10.1007/978-3-031-08446-1_17

Enerijiofi, K. E., Ahonsi, C. O., & Ajao, E. K. (2020). Biodegradation potentials of waste engine oil by three bacterial isolates. *Journal of Applied Sciences and Environmental Management, 24*(3), 483–487. https://doi.org/10.4314/jasem.v24i34

Enerijiofi, K. E., Okuguni, N. N., & Ajayi, A. V. (2022). Physicochemical composition and heavy metals tolerance of bacterial isolates in leachate from solid waste dumpsites in Delta North Senatorial District, Delta State. *Journal of Earth and Environmental Sciences – Elsevier., 13*(1), 48–52.

Enerijiofi, K. E., Musa, S. I., Igiebor, F. A., Lawani, M., Chuka, N. E., Odozi, E. B., & Ikhajiagbe, B. (2024a). Nanotechnological approaches for enhanced microbial activities, sustainable agriculture and the environment. In V. D. Rajput, A. Singh, K. Ghazaryan, & T. M. Minkina (Eds.), *Sustainable agriculture nanotechnology, biotechnology and management* (pp. 153–183). Walter de Gruyter. https://doi.org/10.1515/9783111123494-009

Enerijiofi, K. E., Maduka, N., Osaro-Matthew, R. C., & Ahonsi, C. O. (2024b). Nanoparticles in food: Sources, application, health impact and regulations. In V. D. Rajput, A. Singh, K. Ghazaryan, A. Alexiou, & A. R. M. Said Al-Tawaha (Eds.), *Harnessing NanoOmics and nanoenzymes for sustainable agriculture*. IGI Global. https://doi.org/10.4018/979-8-3693-1890-4.ch018

Fajardo, C., Costa, G., Nande, M., Martin, C., Martín, M., & Sánchez-Fortún, S. (2019). Heavy metals immobilization capability of two iron-based nanoparticles (nZVI and Fe3O4): Soil and freshwater bioassays to assess ecotoxicological impact. *Science of the Total Environment, 656*, 421–432.

Fajardo, C., Sánchez-Fortún, S., Costa, G., Nande, M., Botías, P., García-Cantalejo, J., Mengs, G., & Martín, M. (2020). Evaluation of nanoremediation strategy in a Pb, Zn and Cd contaminated soil. *Science of the Total Environment, 706*, 136041.

Galdames, A., Mendoza, A., Orueta, M., de Soto Garcia, I. S., Sanchez, M., Virto, I., & Vilas, J. L. (2017). Development of new remediation technologies for contaminated soils based on the application of zero-valent iron nanoparticles and bioremediation with compost. *Resource-Efficient Technologies, 3*, 166–176.

Gao, X., Kang, S., Xiong, R., & Chen, M. (2020). Environment-friendly removal methods for endocrine disrupting chemicals. *Sustainability, 12*(18), 7615. https://doi.org/10.3390/su12187615

Ghani, M. I., Saleem, S., Rather, S. A., Rehmani, M. S., Alamri, S., Rajput, V. D., Kalaji, H. M., Saleem, N., Sial, T., & Liu, M. (2022). Foliar application of zinc oxide nanoparticles: An effective strategy to mitigate drought stress in cucumber seedling by modulating antioxidant defense system and osmolytes accumulation. *Chemosphere, 289*, 133202.

Han, B., Zhang, M., & Zhao, D. (2017). In-situ degradation of soil-sorbed 17ß-estradiol using carboxymethyl cellulose stabilized manganese oxide nanoparticles: Column studies. *Environmental Pollution, 223*, 238–246.

Hemalatha, I., Harika, D., & Karnena, M. J. (2022). Sustainable nano-bioremediation approaches for the treatment of polluted soils. *Nature Environment and Pollution Technology, 21*, 10.46488/NEPT.2022.v21i04.036.

Hosseini, S. M., Amini, S. H., Khodabakhshi, A. R., Bagheripour, E., & Van der Bruggen, B. (2018). Activated carbon nanoparticles entrapped mixed matrix polyethersulfone-based nanofiltration membrane for sulfate and copper removal from water. *Journal of the Taiwan Institute of Chemical Engineers, 82*, 169–178.

Huang, H., Ullah, F., Zhou, D. X., Yi, M., & Zhao, Y. (2019a). Mechanisms of ROS regulation of plant development and stress responses. *Frontiers in Plant Science, 10*, 800. https://doi.org/10.3389/fpls.2019.00800

Huang, J., Yi, S., Zheng, C., & Lo, I. M. C. (2019b). Persulfate activation by natural zeolite supported nanoscale zero-valent iron for trichloroethylene degradation in groundwater. *Science of the Total Environment, 684*, 351–359.

Hung, C. M., Huang, C. P., Chen, C. W., & Dong, C. D. (2021). Degradation of organic contaminants in marine sediments by peroxymono-sulfate over $LaFeO_3$ nanoparticles supported on water caltrop shell-derived biochar and the associated microbial community responses. *Journal of Hazardous Materials, 420*, 126553.

Ibrahim, M. L., Ijah, U. J. J., Manga, S. B., Bilbis, L. S., & Umar, S. (2013). Production and partial characterization of biosurfactant produced by crude oil degrading bacteria. *International Biodeterioration & Biodegradation, 81*, 28–34. https://doi.org/10.1016/j.ibiod.2012.11.012

Jaafar, R., Al-Sulami, A., & Al-Taee, A. (2016). The biosorption ability of shewanella oneidensis for cadmium and lead isolated from soil in Basra Governorate, Iraq. *African Journal of Microbiology Research, 38*, 267–270.

Jośko, I., Oleszczuk, P., & Futa, B. (2014). The effect of inorganic nanoparticles (ZnO, Cr2O3, CuO and Ni) and their bulk counterparts on enzyme activities in different soils. *Geoderma, 232*, 528–537. https://doi.org/10.1016/j.geoderma.2014.06.012

Kalidhasan, S., Dror, I., & Berkowitz, B. (2017). Atrazine degradation through PEI-copper nanoparticles deposited onto montmorillonite and sand. *Scientific Reports, 7*, 1–13. https://doi.org/10.1038/s41598-017-01429-5

Kandiah, M., & Chandrasekaran, K. N. (2021). Green synthesis of silver nanoparticles using *Catharanthus roseus* flower extracts and the determination of their antioxidant, antimicrobial, and photocatalytic activity. *Journal of Nanotechnology, 2021*, 5512786. https://doi.org/10.1155/2021/5512786

Kasture, M. B., Patel, P., Prabhune, A. A., Ramana, C. V., Kulkarni, A. A., & Prasad, B. L. V. (2008). Synthesis of silver nanoparticles by sophorolipids: Effect of temperature and sophorolipid structure on the size of particles. *Journal of Chemical Sciences, 120*, 515–520. https://doi.org/10.1007/s12039-008-0080-6

Khan, A., Malik, S., Ali, N., Bilal, M., Yang, Y., & Akhter, M. S. (2022). Nanobiosorbents: basic principles, synthesis, and application for contaminants removal. In *Nano-Biosorbents for Decontamination of Water, Air, and Soil Pollution* (pp. 45–59). Elsevier. https://doi.org/10.1016/B978-0-323-90912-9.00003-4

Konate, A., He, X., Zhang, Z., Ma, Y., Zhang, P., & Alugongo, G. M. (2017). Magnetic (Fe3O4) nanoparticles reduce heavy metals uptake and mitigate their toxicity in wheat seedling. *Sustainability, 9*, 790. https://doi.org/10.3390/su9050790

Kumari, A., Kumari, P., Rajput, V. D., Sushkova, S. N., & Minkina, T. (2021). Metal(loid) nanosorbents in restoration of polluted soils: Geochemical, ecotoxicological, and remediation perspectives. *Environ. Geochem. Health, 44*, 235–246.

Lal, R. (2015). Restoring soil quality to mitigate soil degradation. *Sustainability, 7*, 5875–5895. https://doi.org/10.3390/su7055875

Laslo, V., Pinzaru, S. C., Zaguła, G., Kluz, M., Vicas, S. I., & Cavalu, S. (2022). Synergic effect of selenium nanoparticles and lactic acid bacteria in reduction cadmium toxicity. *Journal of Molecular Structure, 1247*, 131325. https://doi.org/10.1016/j.molstruc.2021.131325

Lew, T. T. S., Park, M., Cui, J., & Strano, M. S. (2021). Plant nanobionic sensors for arsenic detection. *Advanced Materials, 33*, 2005683. https://doi.org/10.1002/adma.202005683

Li, J., Jiang, L., Liu, X., & Lv, J. (2013). Adsorption and aerobic biodegradation of four selected endocrine disrupting chemicals in soil–water system. *International Biodeterioration & Biodegradation, 76*, 3–7.

Li, Q., Chen, X., Zhuang, J., & Chen, X. (2016). Decontaminating soil organic pollutants with manufactured nanoparticles. *Environmental Science and Pollution Research, 23*, 11533–11548. https://doi.org/10.1007/s11356-016-6255-7

Madhura, L., Singh, S., Kanchi, S., Sabela, M., & Bisetty, K. (2019). Nanotechnology-based water quality management for wastewater treatment. *Environmental Chemistry Letters, 17*, 65–121. https://doi.org/10.1007/s10311-018-0778-8

Martin, J. L., Maris, V., & Simberloff, D. S. (2016). The need to respect nature and its limits challenges society and conservation science. *Proceedings of the National Academy of Sciences of the United States of America, 113*, 6105.

Martínez-Fernández, D., Vítková, M., Michálková, Z., & Komárek, M. (2017). Engineered nanomaterials for phytoremediation of metal/metalloid-contaminated soils: Implications for plant physiology. In A. A. Ansari, S. S. Gill, R. Gill, G. R. Lanza, & L. Newman (Eds.), *Phytoremediation: Management of environmental contaminants* (pp. 369–403). Springer.

Marty, M. S., Borgert, C., Coady, K., Green, R., Levine, S. L., Mihaich, E., Ortego, L., Wheeler, J. R., Yi, K. D., & Zorrilla, L. M. (2018). Distinguishing between endocrine disruption and non-specific effects on endocrine systems. *Regulatory Toxicology and Pharmacology, 99*, 142–158.

Medina-Pérez, G., Fernández-Luqueño, F., Vazquez-Nuñez, E., López-Valdez, F., Prieto-Mendez, J., Madariaga-Navarrete, A., & Miranda-Arámbula, M. (2019). Remediating polluted soils using nanotechnologies: Environmental benefits and risks. *Polish Journal of Environmental Studies, 28*, 1013–1030.

Meenakshi, G., & Sivasamy, A. (2018). Nanorod ZnO/SiC nanocomposite: An efficient catalyst for the degradation of an endocrine disruptor under UV and visible light irradiations. *Journal of Environmental Chemical Engineering, 6*, 3757–3769.

Mehta, D., Mazumdar, S., & Singh, S. K. (2015). Magnetic adsorbents for the treatment of water/wastewater—A review. *Journal of Water Process Engineering, 7*, 244–265.

Mitzia, A., Vítková, M., & Komárek, M. (2020). Assessment of biochar and/or nano zero-valent iron for the stabilisation of Zn, Pb and Cd: A temporal study of solid phase geochemistry under changing soil conditions. *Chemosphere, 242*, 125248.

Moameri, M., & Khalaki, M. A. (2021). Chapter 5–Toxicity/risk assessment of nanomaterials when used in soil treatment. In A. Amrane, D. Mohan, T. A. Nguyen, A. A. Assadi, & G. Yasin (Eds.), *Nanomaterials for Soil Remediation* (pp. 87–100). Elsevier.

Parikh, S. J., & James, B. R. (2012). Soil: The Foundation of Agriculture. *National Knowledge Network, 3*, 2.

Prabhakar, R., & Samadder, S. R. (2018). Low cost and easy synthesis of aluminum oxide nanoparticles for arsenite removal from groundwater. *Journal of Molecular Liquids, 250*, 192–201.

Pretty, J., & Bharucha, Z. P. (2014). Sustainable intensification in agricultural systems. *Annals of Botany, 4*(114), 1571–1596.

Raffa, C. M., Chiampo, F., & Shanthakumar, S. (2021). Remediation of metal/metalloid-polluted soils: A short review. *Applied Sciences, 11*, 4134.

Rajesha, J. B., Ramasami, A. K., Nagaraju, G., & Balakrishna, G. R. (2017). Photochemical elimination of endocrine disrupting chemical (EDC) by ZnO nanoparticles, synthesized by gel combustion. *Water Environment Research, 89*(5), 396–405.

Rajput, V., Minkina, T., Kumari, A., Shende, S., Ranjan, A., Faizan, M., Barakvov, A., Gromovik, A., Gorbunova, N., & Rajput, P. (2021). A review on nanobioremediation approaches for restoration of contaminated soil. *Eurasian Journal of Soil Science, 11*, 43–60.

Roccuzzo, S., Beckerman, A. P., & Trögl, J. (2021). New perspectives on the bioremediation of endocrine disrupting compounds from wastewater using algae-, bacteria- and fungi-based technologies. *International Journal of Environmental Science and Technology, 18*, 89–106. https://doi.org/10.1007/s13762-020-02691-3

Roy, A., & Bharadvaja, N. (2021). Efficient removal of heavy metals from artificial wastewater using biochar. *Environmental Nanotechnology, Monitoring & Management, 16*, 100602.

Roy, A., Elzaki, A., & Tirth, V. (2021). Biological synthesis of nanocatalysts and their applications. *Catalysts, 11*(12), 1494.

Sarangapani, C., Danaher, M., Tiwari, B., Lu, P., Bourke, P., & Cullen, P. J. (2017). Efficacy and mechanistic insights into endocrine disruptor degradation using atmospheric air plasma. *Chemical Engineering Journal, 326*, 700–714.

Shi, J., Ye, J., Fang, H., Zhang, S., & Xu, C. (2018). Effects of copper oxide nanoparticles on paddy soil properties and components. *Nanomaterials, 8*, 839.

Sornalingam, K., Mcdonagh, A., Zhou, J. L., Mah, J., & Ahmed, M. B. (2017). Photocatalysis of estrone in water and wastewater: Comparison between Au-TiO2 nanocomposite and TiO2, and degradation by-products. *Science of the Total Environment, 610–611*, 521.

Struik, P. C., & Kuyper, T. W. (2017). Sustainable intensification in agriculture: The richer shade of green. A review. *Agronomy for Sustainable Development, 37*, 39.

Tian, H., Liang, Y., Yang, D., & Sun, Y. (2020). Characteristics of PVP–stabilised NZVI and application to dechlorination of soil–sorbed TCE with ionic surfactant. *Chemosphere, 239*, 124807. https://doi.org/10.1016/j.chemosphere.2019.124807

Usman, M., Farooq, M., & Wakeel, A. (2020). Nanotechnology in agriculture: Current status, challenges and future opportunities. *The Science of the Total Environment, 721*, 137778.

Vela, N., Calín, M., Yáñez-Gascón, M. J., Garrido, I., Pérez-Lucas, G., Fenoll, J., & Navarro, S. (2017). Photocatalytic oxidation of six pesticides listed as endocrine disruptor chemicals from wastewater using two different TiO2 samples at pilot plant scale under sunlight irradiation. *Journal of Photochemistry and Photobiology A: Chemistry, 353*(271), 278.

Wang, J. S., & Chiu, K. (2009). Destruction of pentachlorobiphenyl in soil by supercritical CO2 extraction coupled with polymer-stabilized palladium nanoparticles. *Chemosphere, 75*(5), 629–633.

Wang, L., Tan, L., Hu, L., Wang, X., Koppolu, R., Tirri, T., van Bochove, B., Ihalainen, P., Seleenmary Sobhanadhas, L. S., & Seppälä, J. V. (2021). On laccase-catalyzed polymerization of biorefinery lignin fractions and alignment of lignin nanoparticles on the nanocellulose surfacevia one-pot water-phase synthesis. *ACS Sustainable Chemistry & Engineering, 9*, 8770–8782.

Wojcieszynska, D., Marchlewicz, A., & Guzik, U. (2020). Suitability of immobilized systems for microbiological degradation of endocrine-disrupting compounds. *Molecules, 25*(19), 4473. https://doi.org/10.3390/molecules25194473

Wuana, R. A., & Okieimen, F. E. (2011). *Heavy metals in contaminated soils: A review of sources, chemistry, risks and best available strategies for remediation* (pp. 1–20). International Scholarly Research Network.

Yang, O., Kim, H. L., Weon, J. Z., & Seo, Y. R. (2015). Endocrine-disrupting chemicals: Review of toxicological mechanisms using molecular pathway analysis. *Journal of Cancer Prevention, 20*, 12–24.

Yang, C. H., Zhang, Q., Zhu, W. Q., Shi, Y., Cao, H. H., Guo, L., Chu, D., Lu, Z., & Liu, T. X. (2022). Involvement of Laccase2 in cuticle sclerotization of the whitefly, Bemisia tabaci Middle East–Asia Minor 1. *Insect, 13*, 471.

Younis, S. A., Serp, P., & Nassar, H. N. (2021). Photocatalytic and biocidal activities of $ZnTiO_2$ oxynitride heterojunction with MOF-5 and g-C3N4: A case study for textile wastewater treatment under direct sunlight. *Journal of Hazardous Materials, 410*, 124562.

Zand, A. D., Mikaeili-Tabrizi, A., & Vaezi-Heir, A. (2020). Application if Titinium dioxide nanoparticles to promote phytoremediation Cd-polluted soil: Contribution of PGPR inoculation. *Bioremediation Journal, 24*(2-3), 171–189.

Zatloukalová, K., Obalová, L., Koči, K., Čapek, L., Matěj, Z., Šnajdhaufová, H., Ryczkowski, J., Słowik, G., Zatloukalová, K., & Obalová, L. (2017). Photocatalytic degradation of endocrine disruptor compounds in water over immobilized TiO_2 photocatalysts. *Iranian Journal of Chemistry and Chemical Engineering, 36*, 29–38.

Zhang, E. H., Sun, R. B., Xu, L., Liang, J. N., Wu, T. Y., & Zhou, J. (2019). Effects of micro-/nano-hydroxyapatite and phytoremediation on fungal community structure in copper contaminated soil. *Ecotoxicology and Environmental Safety, 174*, 100–109.

Zheng, M. (2020). Simultaneous control of soil erosion and arsenic leaching at disturbed land using polyacrylamide modified magnetite nanoparticles. *Science of the Total Environment, 702*, 134997.

Zúñiga-Benítez, H., Aristizábal-Ciro, C., & Peñuela, G. A. (2016). Heterogeneous photocatalytic degradation of the endocrine-disrupting chemical Benzophenone-3: Parameters optimization and by-products identification. *Journal of Environmental Management, 167*, 246–258.

Soil Microorganisms and Mycotoxins

Suresh Babu Bastipati, Saloni Prasad, D. Sai Vaishnavi, Kathuroju Harikrishna, and Kondakindi Venkateswar Reddy

Abstract Soil microbiomes, consisting of a wide array of microorganisms, are vital for soil health and ecosystem functionality. These microbial communities are integral to nutrient cycling, organic matter breakdown, and supporting plant growth. However, they are also involved in both the production and degradation of mycotoxins—hazardous secondary metabolites generated by certain fungi. Mycotoxins, which are toxic secondary metabolites produced by fungi, pose a significant risk to food and feed safety. Biological methods, including the use of microorganisms for mycotoxin biodegradation, are promising due to their efficiency, cost-effectiveness, and simplicity compared to physical and chemical methods. Advances in metagenomics and bioinformatics have allowed for detailed analysis of soil microbial communities and their capabilities. Research has shown that some soil bacteria and fungi can break down mycotoxins, suggesting potential bioremediation methods. Environmental conditions, such as soil pH, moisture levels, and organic matter content, can affect both microbial community structures and mycotoxin levels. Combining soil microbiome management with sustainable agricultural practices can improve soil resilience and lower mycotoxin risks. Continued research is necessary to fully understand the complex interactions between soil microorganisms and mycotoxins, leading to innovative approaches in soil management and ensuring food safety.

Keywords Soil microorganisms · Biodegradation · Mycotoxins · Metagenomics · Bioinformatics · Sustainable agriculture

S. B. Bastipati · S. Prasad · D. S. Vaishnavi · K. Harikrishna · K. V. Reddy (✉)
University College of Engineering Science and Technology Hyderabad, Jawaharlal Nehru Technological University Kukatpally, Hyderabad, Telangana, India

S. A. Aransiola et al. (eds.), *Soil Microbiome in Green Technology Sustainability*,
https://doi.org/10.1007/978-3-031-71844-1_17

1 Soil Microbiome

Soil microbiomes are intricate communities of microorganisms, including bacteria, fungi, archaea, viruses, and protozoa, that inhabit the soil environment. These microorganisms play vital roles in nutrient cycling, organic matter decomposition, and the promotion of plant health. Bacteria, for instance, are key players in nitrogen fixation and organic matter breakdown, while fungi, particularly mycorrhizal fungi, enhance nutrient uptake and contribute to soil structure through their hyphal networks. The diversity and function of soil microbiomes are influenced by various factors such as soil type, pH, moisture, and organic content. Understanding these microbial communities is crucial for sustainable agriculture, soil health management, and ecosystem stability, as they are essential for processes like nutrient acquisition, disease suppression, and stress resilience in plants. The rhizosphere is a hotspot for microbial activity due to root exudates that provide nutrients for microorganisms. The interactions in the rhizosphere include competitive, exploitative, commensal, and mutualistic relationships, all of which impact plant health and soil microbial community structure. Mycorrhizal fungi and nitrogen-fixing bacteria form symbiotic relationships with plants, facilitating nutrient exchange. In return for carbohydrates from the plant, mycorrhizal fungi enhance the uptake of water and nutrients, particularly phosphorus.

1.1 Composition of Soil Microbiome

The soil microbiome is a diverse and complex community of microorganisms that play critical roles in soil health and ecosystem functioning. Figure 1 depicts the composition of soil microbiome. Its composition varies depending on factors such as soil type, climate, vegetation, and land use practices. Here is a detailed overview of the main groups of microorganisms that make up the soil microbiome (Wei et al., 2019).

1.1.1 Bacteria

Gram-Positive Bacteria: This group includes Actinobacteria and Firmicutes, which are known for their ability to decompose complex organic matter and produce antibiotics that can suppress plant pathogens.

Gram-Negative Bacteria: Groups such as Proteobacteria, Bacteroidetes, and Acidobacteria are important for nutrient cycling, nitrogen fixation, and maintaining soil pH balance.

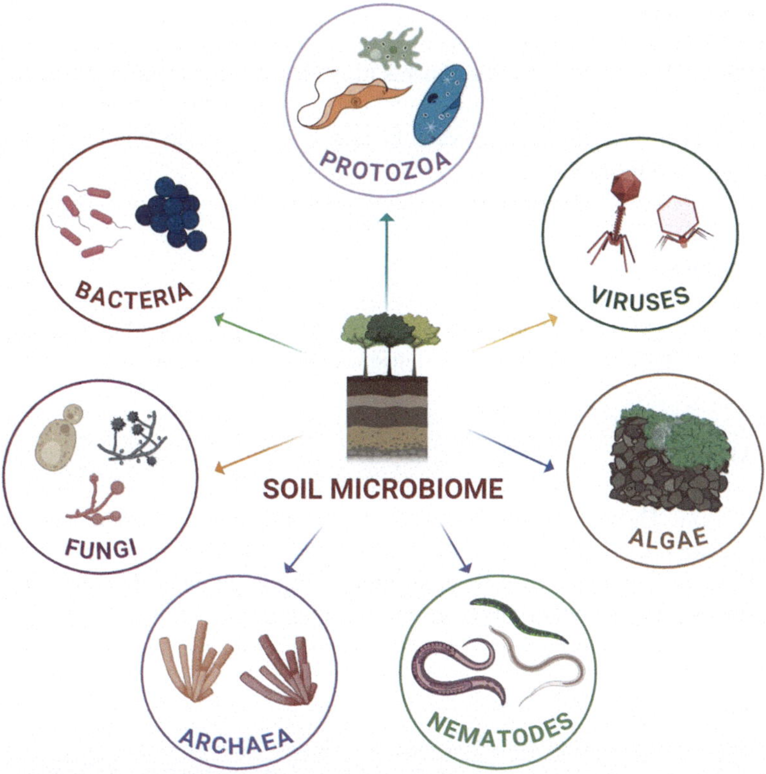

Fig. 1 Composition of soil microbiome

1.1.2 Fungi

Mycorrhizal Fungi: These fungi form symbiotic relationships with plant roots, aiding in nutrient uptake, especially phosphorus and nitrogen. Arbuscular mycorrhizal fungi (AMF) and ectomycorrhizal fungi are common types.

Saprotrophic Fungi: These fungi decompose dead organic matter, contributing to nutrient cycling and soil structure. Common examples include Ascomycetes and Basidiomycetes.

1.1.3 Archaea

Ammonia-Oxidizing Archaea: These microorganisms play a role in the nitrogen cycle by converting ammonia to nitrite, a crucial step in nitrification.

Methanogenic Archaea: Found in anaerobic environments, these archaea produce methane as a byproductof organic matter decomposition.

1.1.4 Protozoa

Flagellates: These single-celled organisms move using flagella and feed on bacteria, fungi, and other microorganisms, contributing to nutrient cycling.
Ciliates: Ciliates use cilia for movement and play a role in regulating bacterial populations in the soil.

1.1.5 Viruses

Soil viruses, also known as bacteriophages, infect and replicate within bacterial cells, influencing bacterial populations and nutrient cycling in the soil.

1.1.6 Algae

Cyanobacteria: These photosynthetic bacteria fix atmospheric nitrogen and produce organic carbon, contributing to soil fertility and carbon cycling.
Green Algae: Green algae in soils contribute to organic matter decomposition and nutrient cycling, especially in moist environments.

1.1.7 Nematodes

Bacterial-Feeding Nematodes: These microfauna feed on bacteria, influencing bacterial populations and nutrient cycling.
Predatory Nematodes: These nematodes feed on other soil organisms, helping regulate microbial populations.

1.2 Importance of Soil Microbiome

The soil microbiome plays a crucial role in maintaining soil health and ecosystem functioning. It is a complex and diverse community of microorganisms, including bacteria, fungi, archaea, protozoa, and viruses, that interact with each other and with plants in the soil environment. Here are some key aspects highlighting the importance of the soil microbiome (Thakur & Geisen, 2019).

1.2.1 Nutrient Cycling

Soil microorganisms are essential for decomposing organic matter and recycling nutrients in the soil. They break down complex organic compounds into simpler forms that can be taken up by plants, contributing to the nutrient cycle in ecosystems.

1.2.2 Plant Health

Many soil microorganisms form symbiotic relationships with plants, providing them with essential nutrients, such as nitrogen and phosphorus. Mycorrhizal fungi, for example, form associations with plant roots, increasing the plants' ability to absorb water and nutrients from the soil.

1.2.3 Disease Suppression

Some beneficial soil microorganisms can help suppress soil-borne plant diseases by outcompeting or inhibiting the growth of pathogenic microbes. This can contribute to overall plant health and reduce the need for chemical pesticides.

1.2.4 Soil Structure and Aggregation

Soil microorganisms play a role in soil structure by producing organic compounds that help bind soil particles together, forming aggregates. This improves soil porosity, water infiltration, and root penetration, which are crucial for plant growth.

1.2.5 Carbon Sequestration

Soil microorganisms are involved in the decomposition of organic matter, which contributes to the cycling of carbon in ecosystems. Some soil microbes can also help sequester carbon in the soil, reducing carbon dioxide levels in the atmosphere and mitigating climate change.

1.2.6 Bioremediation

Certain soil microorganisms have the ability to degrade or detoxify pollutants in the soil, a process known as bioremediation. Microorganisms can break down contaminants such as pesticides, heavy metals, and hydrocarbons, helping to clean up polluted sites.

1.2.7 Resilience to Environmental Stress

The diversity and activities of soil microorganisms contribute to the resilience of ecosystems to environmental stressors such as drought, pollution, and climate change. A diverse soil microbiome can enhance ecosystem stability and productivity under changing conditions.

1.2.8 Sustainable Agriculture

Understanding and harnessing the beneficial functions of the soil microbiome is key to promoting sustainable agricultural practices. Practices such as crop rotation, cover cropping, and reduced tillage can help maintain a healthy soil microbiome and improve overall soil fertility.

2 Fungi and Mycotoxin

Throughout many ecosystems, fungi—a diverse collection of eukaryotic organisms—play vital roles. From the top of the troposphere to deep-sea vents, they may be found in practically every region on Earth. From tiny single-celled yeasts to massive multicellular mushrooms, fungi come in a variety of shapes and sizes. These microbes are necessary for the cycling of nutrients, their breakdown, and their symbiotic connections with other species and plants.

Fungi secrete various substances into the soil as part of their ecological roles and interactions with their environment. These secretions play crucial roles in nutrient cycling, decomposition, and overall soil health. Here are some of the key substances that fungi secrete into the soil:

Enzymes: Fungi release enzymes into the soil to break down complex organic matter such as cellulose, lignin, and other recalcitrant compounds. These enzymes help in the decomposition process by converting organic residues into simpler forms that can be absorbed by the fungi for nutrition.

Organic Acids: Fungi also secrete organic acids like oxalic acid, citric acid, and others into the soil. These acids help in solubilizing minerals and nutrients from the soil matrix, making them more available for both the fungi and other organisms in the ecosystem.

Antibiotics: Some fungi produce antibiotics as a defense mechanism against competing microorganisms in the soil. These antibiotics can inhibit the growth of pathogenic bacteria or other fungi, helping to maintain a balance in the microbial community.

Growth Factors: Fungi may release growth factors or signaling molecules that can influence plant growth and development. These factors can stimulate root growth, enhance nutrient uptake efficiency, or trigger specific responses in plants that benefit both the fungi and their host plants.

Mycotoxins: While not all fungi produce mycotoxins, some species do secrete toxic compounds into the soil. Mycotoxins can have detrimental effects on other organisms in the ecosystem and are often associated with fungal pathogens that cause diseases in plants or animals (Pitt, 2000).

2.1 Mycotoxins

Mycotoxins are low molecular weight secondary metabolites that are produced by filamentous fungus species. They are very hazardous to people and animals in varying degrees, and their chemical structures are significantly varied. Although they can be generated by one or more fungal species, mycotoxins are frequently genotypically specific; a single species can create many types of mycotoxins. More than 200 different types of mycotoxins can be found in the environment. While some of them have the potential to be neurotoxic, teratogenic, carcinogenic, or mutagenic, others have the ability to be ant tumorous, cytotoxic, or antibacterial.

These toxins are typically produced when mold growth occurs before or during storage, especially under warm and humid conditions. Mycotoxins are known to be chemically stable, allowing them to survive food processing. There are hundreds of mycotoxins identified, with some of the most common ones being aflatoxins, ochratoxin A, patulin, fumonisins, zearalenone, and nivalenol.

2.2 Impact of Mycotoxin on Human Health

Mycotoxins are toxic compounds produced by certain types of molds that can have significant impacts on human health. These toxins can contaminate various foodstuffs such as cereals, dried fruits, nuts, and spices, leading to adverse health effects when consumed. Exposure to mycotoxins can occur directly through the ingestion of contaminated food or indirectly through animals that have consumed contaminated feed. The most commonly observed mycotoxins that pose a concern to human health include aflatoxins, ochratoxin A, patulin, fumonisins, zearalenone, and nivalenol/deoxynivalenol. Mycotoxins present a significant risk to human health through contaminated food sources. They can lead to acute symptoms or have long-term effects such as cancer or immune deficiency. Proper food storage practices and vigilance in inspecting for moldy or contaminated foods are essential steps to reduce the risk of mycotoxin exposure.

2.3 Types of Mycotoxins

The below are the wide range of mycotoxin (Elkenany & Awad, 2021). Figure 2 depicts the types of mycotoxins. The most common fungal species that produces various types of mycotoxins are included in Table 1 (Kępińska-Pacelik & Biel, 2021).

Aflatoxins

- Aflatoxins are potent mycotoxins produced by *Aspergillus* species such as *A. flavus, A. parasiticus, A. nomius, A. pseudotamarii* and *A. ochraceoroseus.*

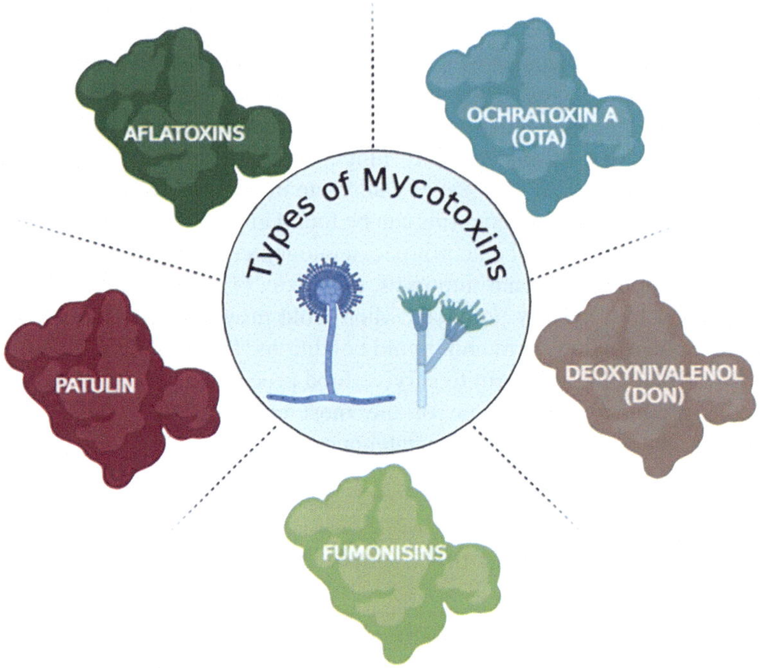

Fig. 2 Types of mycotoxins

Table 1 Most common fungal species that produces various types of Mycotoxins

S.no	Type of mycotoxin	Fungal species
1	Aflatoxins	*A. flavus* *A. parasiticus* *A. nomius* *A. pseudotamarii* *A. ochraceoroseus.*
2	Ochratoxin A (OTA)	*A. ochraceus,* *A. carbonarius* *Penicillium verrucosum.*
3	Patulin	*Penicillium expansum, Aspergillus* species *Byssochlamys.*
4	Fumonisins	*Aspergillus niger* *Fusarium oxysporum* *Fusarium Verticullioides* *Fusarium proliferatum*
5	Deoxynivalenol (DON)	*F. gramine arum* *F. culmorum* *F. pseudograminearum*

- Not all toxins produced by different strains are toxigenic, only 35% of *A. flavus* strains produce aflatoxin whereas the majority of strains of *A. parasiticus* are toxigenic to humans and animals.
- Aflatoxins are of four main types: B1, B2, G1 and G2.
- Aflatoxin B1 is the most toxic compound that causes adverse effects like liver cancer, reproductive problem, suppression of the immune system and death.
- Aflatoxin M1 is associated with nursing animals that may pass the toxin through milk to their babies.
- The maximum limit of Aflatoxin B1 and M1 in food, feed and milk should be less than 2 μg/kg and 0.05 μg/kg according to European Union.

Ochratoxin A

- Ochratoxin A (OTA) is a toxin produced by *A. ochraceus, A. carbonarius* and *Penicillium verrucosum.*
- OTA is mainly associated with barley, coffee, cocoa, vine fruits, spices and beverages.
- Ochratoxin A targets the kidney organ and forms a urinary tract tumor and is considered carcinogenic to humans.
- This toxin remains for a long time in the animal body even after slaughtering but it is reported to date that OTA is not produced in plants in the field.

Patulin

- Patulin is the smallest mycotoxin molecule produced by the molds such as *Penicillium expansum, Aspergillus* species and *Byssochlamys.*
- This smallest mycotoxin contaminates fruit and vegetable-based products, especially apples.
- Fresh apple juice is the major contaminating source of patulin in humans who consume it on a daily basis.
- Acute toxicity occurs when 9 to 55 mg/kg of patulin concentration is consumed which causes agitation, convulsions, dyspnea, pulmonary congestion and ulceration.

Fumonisins

- Common soil fungus, Fusarium species generate a variety of toxins, including fumonisins, zearalenone (ZEN), and trichothecenes such deoxynivalenol (DON), nivalenol (NIV), and T-2 and HT-2 toxins.
- Numerous distinct grain crops are affected by the development of poisons and molds. Certain varieties of grain are linked to distinct fusarium poisons. For instance, fumonisins are frequently linked to maize (corn), T-2 and HT-2 toxins to oats, and both DON and ZEN to wheat. Trichothecenes have the potential to be extremely hazardous to humans, resulting in diarrhea and quick skin or intestinal mucosal irritation.
- Immune system suppression is one of the chronic consequences in animals that have been documented. It has been demonstrated that ZEN has estrogenic, hormonal effects and, at high doses, can result in infertility, especially in pigs.

- Animal toxicity to the liver and kidneys as well as human esophageal cancer have both been linked to fumonisins.

  Zearalenone: Zearalenone is a mycotoxin associated with hormonal estrogenic effects that can lead to infertility.

  Trichothecenes: Trichothecenes are toxic mycotoxins that can cause irritation and diarrhea in acute cases while chronic exposure may lead to immune suppression.

Deoxynivalenol
- Deoxynivalenol (DON) toxin is produced by *Fusarium* species such as *F. graminearum. F. culmorum* and *F. pseudograminearum.*
- These species contaminate corn, wheat, oats, barley, rice and other grains during storage and have been detected in buckwheat, flour, bread, breakfast cereals, noodles and infant foods.
- Deoxynivalenol toxin causes acute health effects such as nausea, vomiting, diarrhea, abdominal pain, throat irritation and fever.
- The maximum limit of DON toxin is 1.1 μg/kg (Kępińska-Pacelik & Biel, 2021).

3 Interaction Between Soil Microbiome and Mycotoxin

The interaction between the soil microbiome and mycotoxins involves a complex web of ecological and biochemical processes that can significantly affect soil health and plant productivity.

3.1 Mycotoxin Production and Soil Microbiome Composition

Fungal Producers
Certain soil-dwelling fungi are prolific producers of mycotoxins. For example:

- Aspergillus species produce aflatoxins.
- Fusarium species produce fumonisins and trichothecenes.
- Penicillium species produce ochratoxins.

These fungi can influence the composition of the soil microbiome by outcompeting other microorganisms for nutrients and space. The production of mycotoxins can further affect microbial community dynamics by inhibiting the growth of susceptible microorganisms (Vinale et al., 2008).

Bacterial Interactions
Soil bacteria can interact with mycotoxin-producing fungi in various ways:

- *Competition*: Bacteria compete with fungi for nutrients, potentially reducing fungal growth and mycotoxin production.
- *Antagonism:* Some bacteria produce antifungal compounds that inhibit fungal growth. For instance, Pseudomonas species produce phenazine antibiotics that inhibit Fusarium spp. (Raaijmakers et al., 2002).

3.2 Degradation of Mycotoxins

Microbial Degradation
Certain soil microorganisms have the ability to degrade mycotoxins, thereby reducing their toxicity:

- Nocardia and Mycobacterium species can degrade aflatoxins.
- Pseudomonas and Rhodococcus species can degrade ochratoxins and zearalenone.

These bacteria use enzymatic pathways to break down mycotoxins into less harmful compounds (Alberts et al., 2006).

3.3 Impact on Soil Health and Plant Growth

Negative Effects
Mycotoxins can adversely affect soil health by:

- Inhibiting the growth of beneficial soil microorganisms.
- Reducing microbial diversity and activity, which can impair nutrient cycling and soil structure.

These effects can lead to reduced soil fertility and impaired plant growth (Abbas et al., 2013).

Biocontrol Potential
Beneficial microorganisms can be employed as biocontrol agents to reduce the impact of mycotoxins:

- Trichoderma species are known to antagonize mycotoxin-producing fungi and can be used in biocontrol strategies.
- Beneficial bacteria like Bacillus and Pseudomonas species can inhibit fungal growth and mycotoxin production through the production of antifungal compounds and competition for nutrients (Harman et al., 2004).

3.4 Environmental Factors

Soil pH and Moisture

The composition of the soil microbiome and the activity of mycotoxin-producing fungi are influenced by environmental factors such as soil pH and moisture:

- High moisture levels can promote fungal growth and mycotoxin production.
- Soil pH can influence the growth of specific microbial communities, with certain pH levels favoring mycotoxin-degrading bacteria (Kumar & Pal, 2015).

Crop Residues and Organic Matter

The presence of crop residues and organic matter can impact mycotoxin dynamics:

- Crop residues can provide substrates for fungal growth and mycotoxin production.
- Organic matter can support the growth of mycotoxin-degrading microorganisms, enhancing the biodegradation process (Yates et al., 1999).

3.5 Research and Application

Bioremediation

Bioremediation strategies aim to exploit the natural mycotoxin-degrading capabilities of soil microorganisms to detoxify contaminated soils. Enhancing the microbial population capable of degrading mycotoxins can be achieved through:

- Adding specific microbial strains known for their mycotoxin-degrading abilities.
- Amending soil with organic matter to support microbial growth (Zain, 2011).

Sustainable Agriculture

Integrating beneficial microbial communities into soil management practices can be a sustainable approach to managing mycotoxin contamination. This includes:

- Using microbial inoculants as part of integrated pest management (IPM) strategies.
- Promoting soil health through practices that enhance microbial diversity and activity (Saghafi et al., 2018).

4 Detection and Analysis of Mycotoxins in Soil

- The detection and analysis are as follows: (Pestka, 1994).

4.1 Sampling and Preparation

Soil Sampling

- *Representative Sampling:* Collect soil samples from various locations within the area of interest to obtain a representative sample. Ensure samples are taken from different depths if necessary.
- *Avoiding Contamination:* Use clean tools and containers to avoid contamination of samples.

Sample Preparation

- *Drying and Sieving:* Air-dry the soil samples and pass them through a sieve to remove large debris and obtain a uniform sample.
- *Homogenization:* Homogenize the sample to ensure consistency.

4.2 Extraction Methods

Solvent Extraction

- *Common Solvents:* Use organic solvents such as methanol, acetonitrile, or a mixture of water and acetonitrile to extract mycotoxins from soil samples.
- *Procedure:* Mix the soil sample with the solvent, shake or sonicate the mixture, and then filter or centrifuge to separate the liquid extract.

Solid Phase Extraction (SPE)

- *Clean-up Step:* Use SPE columns to clean up and concentrate the mycotoxin extract. This helps in removing interfering substances from the soil matrix.
- *Elution:* Elute the mycotoxins from the SPE column using appropriate solvents.

4.3 Detection and Quantification Methods

Enzyme-Linked Immunosorbent Assay (ELISA)

- *Principle:* Use antibodies specific to the mycotoxins of interest to detect and quantify their presence in the soil extract.
- *Advantages:* ELISA is sensitive, relatively quick, and can be used for screening large numbers of samples.
- *Limitations:* May require confirmation by more specific methods due to possible cross-reactivity.

High-Performance Liquid Chromatography (HPLC)

- *Principle:* Separate mycotoxins based on their chemical properties using a chromatographic column, followed by detection with UV, fluorescence, or mass spectrometry.
- *Advantages:* High sensitivity and specificity; can quantify multiple mycotoxins simultaneously.
- *Limitations:* Requires expensive equipment and technical expertise (Misihairabgwi et al., 2019).

Liquid Chromatography-Mass Spectrometry (LC-MS/MS)

- *Principle:* Combines liquid chromatography with mass spectrometry to provide highly sensitive and specific detection and quantification of mycotoxins.
- *Advantages:* High sensitivity, specificity, and ability to analyze multiple mycotoxins in a single run.
- *Limitations:* High cost and complexity of operation (Berthiller et al., 2013).

4.4 Validation and Quality Control

Method Validation

- *Accuracy and Precision:* Validate the analytical methods for accuracy (recovery studies) and precision (repeatability and reproducibility).
- *Limit of Detection (LOD) and Limit of Quantification (LOQ):* Determine the LOD and LOQ for each mycotoxin in the soil matrix.

Quality Control

- *Use of Standards*: Employ certified reference materials and standards to ensure the reliability of the results.
- *Control Samples:* Include blank samples, spiked samples, and replicates in the analysis to monitor for potential contamination and ensure method consistency.

4.5 Data Interpretation and Reporting

Data Analysis

- *Quantification*: Calculate the concentration of mycotoxins in soil samples using calibration curves generated from standards.
- *Statistical Analysis:* Perform statistical analysis to interpret the data and understand the distribution and levels of mycotoxins in the soil samples.

Reporting

- *Detailed Reporting:* Provide a detailed report including the methodology, results, quality control data, and any deviations from standard procedures.
- *Regulatory Compliance:* Ensure that the reporting meets regulatory requirements and guidelines for mycotoxin analysis.

5 Strategies to Mitigate Mycotoxin Effects on Soil Microbiome

Mitigating the effects of mycotoxins on the soil microbiome involves several strategies that aim to reduce mycotoxin levels, enhance soil health, and promote the growth of beneficial microorganisms. Here is a detailed examination of these strategies with references:

5.1 Bioremediation

Microbial Bioremediation

- *Utilization of Mycotoxin-Degrading Microorganisms:* Introducing or promoting microorganisms that can degrade mycotoxins can help detoxify contaminated soils. Bacteria such as Pseudomonas, Rhodococcus, and fungi such as Trichoderma spp. have shown the ability to degrade mycotoxins.
- *Enzymatic Degradation:* Microorganisms produce enzymes like laccases, peroxidases, and dehydrogenases that can break down mycotoxins into less toxic compounds (Alberts et al., 2006).

Phytoremediation

- *Use of Mycotoxin-Tolerant Plants:* Certain plants can tolerate or even accumulate mycotoxins, reducing their levels in soil. Planting such species can help mitigate mycotoxin contamination.
- *Rhizoremediation:* Plant roots can promote the growth of mycotoxin-degrading microorganisms in the rhizosphere, enhancing the breakdown of mycotoxins (Kuiper et al., 2004).

5.2 Soil Amendments

Organic Amendments

- *Compost and Manure:* Adding organic matter like compost and manure can improve soil structure and microbial activity, which can enhance the degradation of mycotoxins.

- *Biochar:* Biochar can adsorb mycotoxins, reducing their bioavailability and toxicity to soil microorganisms (Park et al., 2011).

Inorganic Amendments

- *Clay Minerals:* Bentonite and other clay minerals can adsorb mycotoxins, reducing their impact on the soil microbiome.
- *Zeolites:* Natural zeolites can also adsorb mycotoxins and improve soil health (Bennett & Klich, 2003).

5.3 Cultural Practices

Crop Rotation

- *Diverse Crop Rotation:* Rotating crops, especially with non-host plants, can reduce the buildup of mycotoxin-producing fungi in soil.
- *Cover Crops:* Planting cover crops can improve soil health, reduce erosion, and suppress mycotoxin-producing fungi (Abbas et al., 2006).

Tillage Practices

- *Reduced Tillage:* Minimizing soil disturbance can help maintain a stable soil microbiome and reduce the spread of mycotoxin-producing fungi.
- *Incorporation of Residues*: Properly incorporating crop residues can enhance microbial activity and mycotoxin degradation (Kohl et al., 2009).

5.4 Biological Control

Beneficial Microorganisms

- *Antagonistic Fungi:* Trichoderma spp. can inhibit the growth of mycotoxin-producing fungi and degrade mycotoxins.
- *Probiotic Bacteria:* Introducing beneficial bacteria like Bacillus and Pseudomonas can suppress pathogenic fungi and degrade mycotoxins (Harman et al., 2004).

5.5 Chemical Methods

Mycotoxin Binders

- *Chemical Adsorbents:* Adding chemical adsorbents such as activated charcoal or aluminosilicates to soil can bind mycotoxins, reducing their bioavailability.
- *Detoxifying Agents:* Chemical agents that break down mycotoxins into less harmful compounds can be applied to contaminated soils (Kabak et al., 2006).

5.6 *Integrated Pest Management (IPM)*

Combining Strategies

- *Holistic Approach:* Integrating biological, cultural, physical, and chemical methods to manage mycotoxin contamination can be more effective than relying on a single strategy.
- *Monitoring and Assessment:* Regular monitoring of soil and crops for mycotoxin levels and adjusting management practices accordingly (Stinson et al., 1981).

6 Bioremediation Strategies for Mycotoxin-Contaminated Soil

Bioremediation strategies for mycotoxin-contaminated soil involve the use of biological agents and processes to detoxify and degrade mycotoxins, thereby restoring soil health.

6.1 *Microbial Bioremediation*

Mycotoxin-Degrading Microorganisms

Bacteria: Several bacterial species have been identified with the capability to degrade mycotoxins. For example:

- Rhodococcus erythropolis can degrade aflatoxins.
- Pseudomonas and Rhodococcus species can degrade ochratoxin A and zearalenone.
- Nocardia and Mycobacterium species can also break down various mycotoxins.

Mechanisms: These bacteria use specific enzymes to degrade mycotoxins. For instance, aflatoxins can be broken down by enzymes like laccases, peroxidases, and oxidoreductases (Alberts et al., 2006).

6.2 *Combined Approaches*

Fungi: Fungal species such as Trichoderma spp. and Pleurotus spp. can degrade mycotoxins through enzymatic processes.

- Trichoderma spp. produce enzymes that can degrade aflatoxins, ochratoxins, and trichothecenes.
- Pleurotusostreatus (oyster mushroom) can degrade aflatoxins through enzymatic activity (Vanhoutte et al., 2016).

6.3 Phytoremediation

Mycotoxin-Tolerant Plants

Accumulation and Tolerance: Certain plants can tolerate mycotoxins or even accumulate them in their tissues, thereby reducing soil contamination. Plants such as alfalfa and maize have been used in studies for their ability to tolerate mycotoxins (Kuiper et al., 2004).

Rhizoremediation

Rhizosphere Enhancement: Plant roots can enhance the microbial activity in the rhizosphere (root zone), promoting the growth of mycotoxin-degrading microorganisms. The root exudates can provide nutrients and stimulate microbial populations capable of degrading mycotoxins (Zhuang et al., 2007).

6.4 Biostimulation

Enhancing Indigenous Microbial Activity

Nutrient Addition: Adding nutrients such as nitrogen, phosphorus, and carbon sources can stimulate the indigenous microbial populations in soil, enhancing their ability to degrade mycotoxins. Organic amendments like compost and manure can provide these nutrients (Kuiper et al., 2004).

6.5 Bioaugmentation

Introducing Degrading Microorganisms

Inoculation with Specific Strains: Introducing specific strains of bacteria or fungi that are known to degrade mycotoxins can enhance the biodegradation process. For example, inoculating soil with Rhodococcuserythropolis or Trichoderma spp. can significantly reduce mycotoxin levels (Alberts et al., 2006).

6.6 Combined Approaches

Integrated Bioremediation Strategies

Combination of Methods: Combining bioremediation strategies such as phytoremediation, bioaugmentation, and bio stimulation can be more effective than using a single method. For instance, planting mycotoxin-tolerant crops along with inoculating the soil with mycotoxin-degrading microorganisms and adding organic amendments can provide a synergistic effect (Vanhoutte et al., 2016).

6.7 Monitoring and Optimization

Continuous Monitoring

Regular Soil Testing: Regular monitoring of soil mycotoxin levels and microbial populations is essential to assess the effectiveness of bioremediation strategies and make necessary adjustments.

Optimization of Conditions: Optimizing environmental conditions such as soil pH, moisture, and temperature can enhance the effectiveness of bioremediation processes (Kuiper et al., 2004).

7 Emerging Technologies for Studying Soil Microbiome and Mycotoxins

Emerging technologies for studying the soil microbiome and mycotoxins are revolutionizing our understanding and ability to manage soil health and contamination. These technologies provide detailed insights into microbial communities, their functions, and interactions with mycotoxins. These technologies are enhancing our ability to study, monitor, and mitigate mycotoxin contamination, promoting soil health and sustainable agricultural practices. These are some technologies:

7.1 Next-Generation Sequencing (NGS)

Metagenomics

- *Description:* Metagenomics involves the direct sequencing of DNA extracted from soil samples, allowing for the identification and quantification of microbial communities without the need for culturing.
- *Applications*: This technology can identify the diversity and abundance of microorganisms in soil, including those involved in mycotoxin degradation.
- *Advantages:* Provides comprehensive insights into the entire microbial community, including rare and unculturable species (Fierer & Jackson, 2006).

Meta Transcriptomics

- *Description:* Meta transcriptomics involves sequencing RNA extracted from soil to study the active metabolic processes and gene expression of the microbial community.
- *Applications:* Can reveal which genes are actively involved in the degradation of mycotoxins and other soil processes.
- *Advantages:* Provides information on functional activity and metabolic pathways in the soil microbiome (Mendes et al., 2011).

7.2 Single-Cell Genomics

- *Description:* Single-cell genomics isolates and sequences the genome of individual microbial cells, providing insights into the functions and capabilities of specific microorganisms.
- *Applications:* Allows for the study of rare or unculturable microbes and their roles in mycotoxin degradation and soil health.
- *Advantages:* Provides high-resolution data on individual microorganisms, overcoming the limitations of bulk metagenomics (Woyke et al., 2010).

7.3 Metabolomics

- *Description:* Metabolomics involves the comprehensive analysis of metabolites in soil, providing insights into the biochemical activities and interactions within the soil microbiome.
- *Applications:* Can identify metabolic changes due to mycotoxin presence and the metabolic pathways involved in mycotoxin degradation.
- *Advantages:* Provides a functional perspective on microbial activity and interactions in soil (Bhattacharyya & Jha, 2012).

7.4 Stable Isotope Probing (Sip)

- *Description:* SIP involves the incorporation of stable isotopes (e.g., ^13C or ^15N) into microbial DNA, RNA, or proteins to track the flow of nutrients and identify active microorganisms involved in specific processes.
- *Applications:* Can be used to trace the degradation pathways of mycotoxins and identify active mycotoxin-degrading microorganisms in soil.
- *Advantages:* Provides direct evidence of microbial activity and functional roles in situ (Dumont & Murrell, 2005).

7.5 CRISPR-Based Technologies

- *Description:* CRISPR-based techniques can be used to edit microbial genomes to study gene function, including genes involved in mycotoxin degradation.
- *Applications:* Can create knockout or overexpression strains to understand the genetic basis of mycotoxin degradation and microbial interactions in soil.
- *Advantages:* Provides precise genetic manipulation capabilities to study microbial functions and interactions (Barrangou & Doudna, 2016).

7.6 *Bioinformatics and Computational Tools*

- *Description:* Advanced bioinformatics and computational tools are used to analyze large datasets generated from sequencing, metabolomics, and other high-throughput technologies.
- *Applications:* Can integrate multi-omics data to provide comprehensive insights into microbial community structure, function, and interactions with mycotoxins.
- *Advantages:* Enhances data interpretation and provides predictive models for microbial behavior and mycotoxin dynamics in soil (Segata et al., 2011).

7.7 *Microfluidics and Lab-on-a-Chip*

- *Description:* Microfluidics involves the manipulation of small volumes of fluids in micro-scale channels, enabling high-throughput and precise analysis of microbial interactions and metabolic activities.
- *Applications:* Can be used to study the effects of mycotoxins on microbial communities and to screen for mycotoxin-degrading microorganisms.
- *Advantages:* Allows for high-throughput and real-time analysis of microbial processes in controlled environments (Whitesides, 2006).

7.8 *Nanotechnology*

- *Description:* Nanotechnology involves the use of nanoparticles and nanomaterials to detect and remediate mycotoxins in soil.
- *Applications:* Can be used for the development of sensitive and selective sensors for mycotoxin detection, as well as for delivering remediation agents.
- *Advantages:* Provides high sensitivity and specificity in detection and targeted delivery of remediation agents (Ingle et al., 2020).

8 Applications of Soil Microbiome Research in Agriculture and Environmental Protection

Research into the soil microbiome has numerous potential applications in agriculture and environmental protection. These applications leverage the complex interactions within soil microbial communities to enhance crop production, manage soil health, and mitigate environmental pollutants. By leveraging the soil microbiome, these applications can promote sustainable agricultural practices, improve crop productivity, enhance soil health, and protect the environment. The integration of advanced microbiome research techniques with traditional agricultural practices holds significant promise for future innovations in both agriculture and environmental management.

8.1 Enhancing Crop Production

Plant Growth-Promoting Rhizobacteria (PGPR)

- Description: PGPR are beneficial bacteria that colonize plant roots and enhance plant growth by various mechanisms such as nitrogen fixation, phosphate solubilization, and production of growth hormones.
- Applications: Inoculating crops with PGPR can increase nutrient availability, enhance root growth, and improve overall plant health and yield.
- Examples: Strains of Bacillus, Pseudomonas, and Azospirillum are commonly used as PGPR (Bhattacharyya & Jha, 2012).

Biofertilizers

- Description: Biofertilizers are formulations containing living microorganisms which, when applied to seeds, plant surfaces, or soil, promote plant growth by increasing the supply or availability of primary nutrients to the host plant.
- Applications: Biofertilizers can reduce the need for chemical fertilizers, promote sustainable agriculture, and improve soil fertility.
- Examples: Rhizobium for leguminous crops, Azotobacter for non-leguminous crops, and mycorrhizal fungi for various crops (Malusa & Vassilev, 2014).

8.2 Biocontrol of Plant Diseases

Antagonistic Microorganisms

- *Description:* Certain soil microorganisms can inhibit the growth of plant pathogens through the production of antibiotics, competition for nutrients, or induction of plant defense mechanisms.
- *Applications:* Using biocontrol agents such as Trichoderma, Bacillus, and Pseudomonas can reduce the reliance on chemical pesticides and manage soil-borne diseases.
- *Examples:* Trichoderma spp. are effective against fungal pathogens like Fusarium and Rhizoctonia (Harman et al., 2004).

8.3 *Improving Soil Health and Fertility*

Organic Matter Decomposition

- *Description:* Soil microorganisms play a crucial role in decomposing organic matter, cycling nutrients, and maintaining soil structure.
- *Applications:* Promoting the activity of decomposer microbes can enhance soil organic matter content, improve soil structure, and increase nutrient availability.
- *Examples:* Fungi like Penicillium and bacteria like Bacillus are involved in the decomposition of organic residues (Kuzyakov & Blagodatskaya, 2015).

Biochar and Soil Amendments

- *Description:* Adding biochar or organic amendments to soil can enhance microbial activity and diversity, improving soil health and fertility.
- *Applications:* Biochar can increase soil carbon content, enhance nutrient retention, and provide habitat for beneficial microbes.
- *Examples:* Biochar applications have been shown to increase microbial biomass and activity in various soil types (Lehmann & Joseph, 2015).

8.4 *Mitigating Environmental Pollution*

Bioremediation

- *Description:* Bioremediation uses microorganisms to degrade or detoxify pollutants such as heavy metals, pesticides, and hydrocarbons in soil.
- *Applications:* Microorganisms can be employed to clean up contaminated soils, reducing environmental pollution and restoring soil health.
- *Examples:* Pseudomonas putida and Mycobacterium spp. are known for their ability to degrade petroleum hydrocarbons (Juwarkar et al., 2010).

Phytoremediation

- *Description:* Phytoremediation involves the use of plants and their associated soil microbes to remove, degrade, or contain soil contaminants.
- *Applications:* Plants can be used to extract heavy metals, degrade organic pollutants, and stabilize contaminated soils.
- *Examples:* Plants like Brassica juncea (Indian mustard) and Populus spp. (poplars) are used for phytoremediation of heavy metals and organic pollutants (Pilon-Smits, 2005).

8.5 *Sustainable Agricultural Practices*

Conservation Agriculture

- *Description:* Conservation agriculture practices, such as minimal soil disturbance, cover cropping, and crop rotation, enhance soil microbial diversity and function.
- *Applications:* These practices can improve soil health, increase carbon sequestration, and enhance resilience to climate change.
- *Examples:* No-till farming and the use of diverse cover crops improve soil structure and microbial activity (Hobbs et al., 2008).

8.6 *Monitoring Soil Health*

Microbial Indicators

- *Description:* Soil microbial communities and their activities can serve as indicators of soil health and fertility.
- *Applications:* Monitoring microbial biomass, diversity, and enzymatic activities can provide insights into soil health and guide management practices.
- *Examples:* Enzymatic activities like dehydrogenase, phosphatase, and urease are commonly used as indicators of soil health (Bastida et al., 2008).

8.7 *Climate Change Mitigation*

Carbon Sequestration

- *Description:* Soil microorganisms play a crucial role in carbon cycling and sequestration, helping to mitigate climate change by storing carbon in soil organic matter.
- *Applications*: Practices that promote microbial activity and organic matter accumulation can enhance carbon sequestration in soils.
- *Examples:* Agroforestry and organic farming practices can increase soil carbon stocks and microbial activity (Lal, 2004).

9 Future Directions in Soil Microbiome and Mycotoxin Research

Research on soil microbiomes and mycotoxins is rapidly evolving, and several future directions promise to further our understanding and management of soil health and contamination.

9.1 Integrative Multi-Omics Approaches

Comprehensive Profiling

- *Description:* Integrating genomics, transcriptomics, proteomics, metabolomics, and metagenomics to provide a holistic view of soil microbial communities and their functions.
- *Applications:* Understanding the complex interactions between soil microbes and their environment, identifying key microbial functions involved in mycotoxin degradation.
- *Advancements Needed:* Improved computational tools and data integration methods to handle and analyze large datasets (Singh et al., 2019).

9.2 Advanced Bioinformatics and Machine Learning

Data Analysis and Prediction

- *Description:* Utilizing machine learning and artificial intelligence to analyze complex soil microbiome data, predict microbial interactions, and identify key functional genes involved in mycotoxin degradation.
- *Applications:* Developing predictive models for soil health, mycotoxin presence, and microbial community dynamics under different environmental conditions.
- *Advancements Needed:* Development of robust machine learning models and databases to support predictive analytics in soil microbiome research (Noecker et al., 2016).

9.3 Synthetic Biology and Microbial Engineering

Engineered Microorganisms

- *Description:* Using synthetic biology to design and construct microbial strains with enhanced capabilities for mycotoxin degradation, nutrient cycling, and plant growth promotion.
- *Applications:* Developing bioengineered microorganisms that can be introduced into soil to degrade specific mycotoxins or improve soil fertility.
- *Advancements Needed:* Ensuring the stability, safety, and ecological compatibility of engineered microorganisms in the soil environment (Karlovsky, 2011).

9.4 Microbiome-Plant Interactions

Rhizosphere Engineering

- *Description:* Exploring the interactions between plant roots and soil microbiomes to enhance nutrient uptake, stress tolerance, and resistance to pathogens.
- *Applications:* Developing microbial inoculants and soil amendments that enhance beneficial plant-microbe interactions, promote healthy crop growth, and suppress mycotoxin-producing fungi.
- *Advancements Needed:* Understanding the molecular mechanisms underlying plant-microbiome interactions and identifying key microbial species that confer benefits to plants (Philippot et al., 2013).

9.5 Climate Change and Soil Microbiomes

Impact Assessment and Adaptation

- *Description:* Investigating how climate change affects soil microbiomes and their functions, including mycotoxin production and degradation.
- *Applications:* Developing strategies to mitigate the negative impacts of climate change on soil health and microbial diversity, enhancing the resilience of soil ecosystems.
- *Advancements Needed:* Long-term field studies and predictive models to assess the impact of climate change on soil microbiomes and their ecological functions (Delgado-Baquerizo et al., 2016).

9.6 Bioremediation and Environmental Biotechnology

Innovative Remediation Strategies

- *Description*: Developing novel bioremediation techniques using microbial consortia, bioaugmentation, and bio stimulation to degrade mycotoxins and other soil pollutants.
- *Applications:* Effective remediation of contaminated soils, restoration of soil health, and sustainable management of agricultural lands.
- *Advancements Needed:* Field-scale validation of bioremediation technologies and understanding the ecological impacts of introduced microbial consortia (Das & Chandran, 2011).

9.7 Soil Health Monitoring and Diagnostics

Real-Time Monitoring Tools

Description: Developing advanced sensors and diagnostic tools for real-time monitoring of soil health, microbial activity, and mycotoxin levels.

Applications: Early detection of soil contamination, monitoring the effectiveness of bioremediation strategies, and guiding soil management practices.

Advancements Needed: Miniaturization of sensors, integration with mobile and IoT technologies, and validation under field conditions (Bhardwaj et al., 2020).

9.8 Policy and Sustainable Agriculture

Regulatory Frameworks

- *Description:* Developing policies and regulatory frameworks that promote the use of microbial technologies in agriculture and environmental protection.
- *Applications:* Encouraging sustainable agricultural practices, reducing reliance on chemical inputs, and promoting the use of microbial inoculants and bioremediation strategies.
- *Advancements Needed:* Collaborations between scientists, policymakers, and stakeholders to ensure the safe and effective implementation of microbial technologies (Malusá et al., 2012).

10 Conclusion

The interaction between soil microbiomes and mycotoxins is a dynamic and complex field that holds significant implications for agriculture, environmental health, and food safety. Mycotoxins, toxic secondary metabolites produced by fungi, pose a substantial threat to crops, soil health, and human health. Understanding and managing these interactions is critical for developing sustainable agricultural practices and effective environmental protection strategies.

The study of soil microbiome and mycotoxin interactions is essential for advancing sustainable agriculture and environmental protection. By harnessing the capabilities of soil microorganisms to degrade mycotoxins, promote plant health, and enhance soil fertility, we can develop innovative solutions to mitigate the risks associated with mycotoxin contamination. Future research and technological advancements will continue to uncover the intricate dynamics of these interactions, enabling more effective and sustainable management practices to safeguard soil health and food security.

References

Abbas, H. K., Zablotowicz, R. M., Bruns, H. A., & Abel, C. A. (2006). Biocontrol of aflatoxin in corn by inoculation with non-aflatoxigenic *Aspergillus flavus* isolates. *Biocontrol Science and Technology, 16*(5), 437–449. https://doi.org/10.1080/09583150500532477

Abbas, H. K., Zablotowicz, R. M., Weaver, M. A., Shier, W. T., Bruns, H. A., Bellaloui, N., et al. (2013). Implications of Bt traits on mycotoxin contamination in maize: Overview and recent experimental results in southern United States. *Journal of Agricultural and Food Chemistry, 61*(48), 11759–11770. https://doi.org/10.1021/jf400754g

Alberts, J. F., Engelbrecht, Y., Steyn, P. S., Holzapfel, W. H., & Van Zyl, W. H. (2006). Biological degradation of aflatoxin B1 by Rhodococcus erythropolis cultures. *International Journal of Food Microbiology, 109*(1–2), 121–126. https://doi.org/10.1016/j.ijfoodmicro.2006.01.019

Barrangou, R., & Doudna, J. A. (2016). Applications of CRISPR technologies in research and beyond. *Nature Biotechnology, 34*(9), 933–941. https://doi.org/10.1038/nbt.3659

Bastida, F., Kandeler, E., Moreno, J. L., Ros, M., García, C., & Hernández, T. (2008). Application of fresh and composted organic wastes modifies structure, size and activity of soil microbial community under semiarid climate. *Applied Soil Ecology, 40*(2), 318–329. https://doi.org/10.1016/j.apsoil.2008.05.007

Bennett, J. W., & Klich, M. (2003). Clinical microbiology reviews. *Mycotoxins, 16*(1), 497–516.

Berthiller, F., Crews, C., Dall'Asta, C., Saeger, S. D., Haesaert, G., Karlovsky, P., et al. (2013). Masked mycotoxins: A review. *Molecular Nutrition & Food Research, 57*(1), 165–186. https://doi.org/10.1002/mnfr.201100764

Bhardwaj, A. K., Garg, A., Ram, S., Gajpal, Y., & Zheng, C. (2020). Research trends in green product for environment: A bibliometric perspective. *International Journal of Environmental Research and Public Health, 17*(22), 8469. https://doi.org/10.3390/ijerph17228469

Bhattacharyya, P. N., & Jha, D. K. (2012). Plant growth-promoting rhizobacteria (PGPR): Emergence in agriculture. *World Journal of Microbiology and Biotechnology, 28*, 1327–1350. https://doi.org/10.1007/s11274-011-0979-9

Das, N., & Chandran, P. (2011). Microbial degradation of petroleum hydrocarbon contaminants: An overview. *Biotechnology Research International, 2011*, 1–13. https://doi.org/10.4061/2011/941810

Delgado-Baquerizo, M., Maestre, F. T., Reich, P. B., Jeffries, T. C., Gaitan, J. J., Encinar, D., Berdugo, M., Campbell, C. D., & Singh, B. K. (2016). Microbial diversity drives multifunctionality in terrestrial ecosystems. *Nature Communications, 7*(1). https://doi.org/10.1038/ncomms10541

Dumont, M. G., & Murrell, J. C. (2005). Stable isotope probing—Linking microbial identity to function. *Nature Reviews Microbiology, 3*(6), 499–504. https://doi.org/10.4061/2011/941810

Elkenany, R., & Awad, A. (2021). Types of mycotoxins and different approaches used for their detection in foodstuffs. *Mansoura Veterinary Medical Journal, 22*(1), 25–32. https://doi.org/10.21608/mvmj.2021.161191

Fierer, N., & Jackson, R. B. (2006). The diversity and biogeography of soil bacterial communities. *Proceedings of the National Academy of Sciences, 103*(3), 626–631. https://doi.org/10.1073/pnas.0507535103

Harman, G. E., Howell, C. R., Viterbo, A., Chet, I., & Lorito, M. (2004). Trichoderma species—Opportunistic, avirulent plant symbionts. *Nature Reviews Microbiology, 2*(1), 43–56. https://doi.org/10.1038/nrmicro797

Hobbs, P. R., Sayre, K., & Gupta, R. (2008). The role of conservation agriculture in sustainable agriculture. *Philosophical Transactions of the Royal Society B: Biological Sciences, 363*(1491), 543–555. https://doi.org/10.1098/rstb.2007.2169

Ingle, A. P., Gupta, I., Jogee, P., & Rai, M. (2020). Role of nanotechnology in the detection of mycotoxins: A smart approach. In *Nanomycotoxicology* (pp. 11–33). Academic. https://doi.org/10.1016/B978-0-12-817998-7.00002-1

Juwarkar, A. A., Singh, S. K., & Mudhoo, A. (2010). A comprehensive overview of elements in bioremediation. *Reviews in Environmental Science and Bio/Technology, 9*, 215–288. https://doi.org/10.1007/s11157-010-9215-6

Kabak, B., Dobson, A. D., & Var, I. I. L. (2006). Strategies to prevent mycotoxin contamination of food and animal feed: A review. *Critical Reviews in Food Science and Nutrition, 46*(8), 593–619. https://doi.org/10.1080/10408390500436185

Karlovsky, P. (2011). Biological detoxification of the mycotoxin deoxynivalenol and its use in genetically engineered crops and feed additives. *Applied Microbiology and Biotechnology, 91*, 491–504. https://doi.org/10.1007/s00253-011-3401-5

Kępińska-Pacelik, J., & Biel, W. (2021). Alimentary risk of mycotoxins for humans and animals. *Toxins, 13*(11), 822. https://doi.org/10.3390/toxins13110822

Kohl, T., Van de Vondel, P., Stressig, R., Wartenberg, H. C., Heep, A., Keiner, S., et al. (2009). Percutaneous fetoscopic laser decompression of congenital high airway obstruction syndrome (CHAOS) from laryngeal atresia via a single trocar–current technical constraints and potential solutions for future interventions. *Fetal Diagnosis and Therapy, 25*(1), 67–71. https://doi.org/10.1159/000200017

Kuiper, I., Lagendijk, E. L., Bloemberg, G. V., & Lugtenberg, B. J. (2004). Rhizoremediation: A beneficial plant-microbe interaction. *Molecular Plant-Microbe Interactions, 17*(1), 6–15. https://doi.org/10.1094/MPMI.2004.17.1.6

Kumar, R., & Pal, P. (2015). Assessing the feasibility of N and P recovery by struvite precipitation from nutrient-rich wastewater: A review. *Environmental Science and Pollution Research, 22*, 17453–17464. https://doi.org/10.1007/s11356-015-5450-2

Kuzyakov, Y., & Blagodatskaya, E. (2015). Microbial hotspots and hot moments in soil: Concept & review. *Soil Biology and Biochemistry, 83*, 184–199. https://doi.org/10.1016/j.soilbio.2015.01.025

Lal, R. (2004). Soil carbon sequestration to mitigate climate change. *Geoderma, 123*(1–2), 1–22. https://doi.org/10.1016/j.geoderma.2004.01.032

Lehmann, J., & Joseph, S. (2015). Biochar for environmental management: An introduction. In *Biochar for environmental management* (pp. 1–13). Routledge. ISBN9780203762264.

Malusa, E., & Vassilev, N. (2014). A contribution to set a legal framework for biofertilisers. *Applied Microbiology and Biotechnology, 98*, 6599–6607. https://doi.org/10.1007/s00253-014-5828-y

Malusá, E., Sas-Paszt, L., & Ciesielska, J. J. T. S. W. J. (2012). Technologies for beneficial microorganisms inocula used as biofertilizers. *The Scientific World Journal, 2012*, 1–12. https://doi.org/10.1100/2012/491206

Mendes, R., Kruijt, M., De Bruijn, I., Dekkers, E., Van Der Voort, M., Schneider, J. H., et al. (2011). Deciphering the rhizosphere microbiome for disease-suppressive bacteria. *Science, 332*(6033), 1097–1100. https://doi.org/10.1126/science.1203980

Misihairabgwi, J. M., Ezekiel, C. N., Sulyok, M., Shephard, G. S., & Krska, R. (2019). Mycotoxin contamination of foods in southern Africa: A 10-year review (2007–2016). *Critical Reviews in Food Science and Nutrition, 59*(1), 43–58. https://doi.org/10.1080/10408398.2017.1357003

Noecker, C., Eng, A., Srinivasan, S., Theriot, C. M., Young, V. B., Jansson, J. K., et al. (2016). Metabolic model-based integration of microbiome taxonomic and metabolomic profiles elucidates mechanistic links between ecological and metabolic variation. *MSystems, 1*(1), 10–1128. https://doi.org/10.1128/msystems.00013-15

Park, J. H., Choppala, G. K., Bolan, N. S., Chung, J. W., & Chuasavathi, T. (2011). Biochar reduces the bioavailability and phytotoxicity of heavy metals. *Plant and Soil, 348*, 439–451. https://doi.org/10.1007/s11104-011-0948-y

Pestka, J. J. (1994). Application of immunology to the analysis and toxicity assessment of mycotoxins. *Food and Agricultural Immunology, 6*(3), 219–233. https://doi.org/10.1080/09540109409354833

Philippot, L., Raaijmakers, J. M., Lemanceau, P., & Van Der Putten, W. H. (2013). Going back to the roots: The microbial ecology of the rhizosphere. *Nature Reviews Microbiology, 11*(11), 789–799.

Pilon-Smits, E. (2005). Phytoremediation. *Annual Review of Plant Biology, 56*, 15–39. https://doi.org/10.1146/annurev.arplant.56.032604.144214

Pitt, J. I. (2000). Toxigenic fungi and mycotoxins. *British Medical Bulletin, 56*(1), 184–192. https://doi.org/10.1258/0007142001902888

Raaijmakers, J. M., Vlami, M., & De Souza, J. T. (2002). Antibiotic production by bacterial biocontrol agents. *Antonie Van Leeuwenhoek, 81*, 537–547. https://doi.org/10.1023/A:1020501420831

Saghafi, D., Ghorbanpour, M., & Lajayer, B. A. (2018). Efficiency of Rhizobium strains as plant growth promoting rhizobacteria on morpho-physiological properties of Brassica napus L. under salinity stress. *Journal of Soil Science and Plant Nutrition, 18*(1), 253–268. https://doi.org/10.4067/S0718-95162018005000903

Segata, N., Izard, J., Waldron, L., Gevers, D., Miropolsky, L., Garrett, W. S., & Huttenhower, C. (2011). Metagenomic biomarker discovery and explanation. *Genome Biology, 12*, 1–18. https://doi.org/10.1186/gb-2011-12-6-r60

Singh, A., Shannon, C. P., Gautier, B., Rohart, F., Vacher, M., Tebbutt, S. J., & Lê Cao, K. A. (2019). DIABLO: An integrative approach for identifying key molecular drivers from multi-omics assays. *Bioinformatics, 35*(17), 3055–3062. https://doi.org/10.1093/bioinformatics/bty1054

Stinson, E. E., Osman, S. F., Heisler, E. G., Siciliano, J., & Bills, D. D. (1981). Mycotoxin production in whole tomatoes, apples, oranges, and lemons. *Journal of Agricultural and Food Chemistry, 29*(4), 790–792. https://doi.org/10.1021/jf00106a025

Thakur, M. P., & Geisen, S. (2019). Trophic regulations of the soil microbiome. *Trends in Microbiology, 27*(9), 771–780. https://doi.org/10.1016/j.tim.2019.04.008

Vanhoutte, I., Audenaert, K., & De Gelder, L. (2016). Biodegradation of mycotoxins: Tales from known and unexplored worlds. *Frontiers in Microbiology, 7*, 187744. https://doi.org/10.3389/fmicb.2016.00561

Vinale, F., Sivasithamparam, K., Ghisalberti, E. L., Marra, R., Woo, S. L., & Lorito, M. (2008). Trichoderma–plant–pathogen interactions. *Soil Biology and Biochemistry, 40*(1), 1–10. https://doi.org/10.1016/j.soilbio.2007.07.002

Wei, Z., Gu, Y., Friman, V. P., Kowalchuk, G. A., Xu, Y., Shen, Q., & Jousset, A. (2019). Initial soil microbiome composition and functioning predetermine future plant health. *Science Advances, 5*(9), eaaw0759. https://doi.org/10.1126/sciadv.aaw0759

Whitesides, G. M. (2006). The origins and the future of microfluidics. *Nature, 442*(7101), 368–373. https://doi.org/10.1038/nature05058

Woyke, T., Tighe, D., Mavromatis, K., Clum, A., Copeland, A., Schackwitz, W., et al. (2010). One bacterial cell, one complete genome. *PLoS One, 5*(4), e10314. https://doi.org/10.1371/journal.pone.0010314

Yates, I. E., Meredith, F., Smart, W., Bacon, C. W., & Jaworski, A. J. (1999). Trichoderma viride suppresses fumonisin B1 production by Fusarium moniliforme. *Journal of Food Protection, 62*(11), 1326–1332. https://doi.org/10.4315/0362-028X-62.11.1326

Zain, M. E. (2011). Impact of mycotoxins on humans and animals. *Journal of Saudi Chemical Society, 15*(2), 129–144. https://doi.org/10.1016/j.jscs.2010.06.006

Zhuang, X., Chen, J., Shim, H., & Bai, Z. (2007). New advances in plant growth-promoting rhizobacteria for bioremediation. *Environment International, 33*(3), 406–413. https://doi.org/10.1016/j.envint.2006.12.005

Soil Microbial Enzymes and Applications

Abdullahi Dabban Idris, Aisha Bisola Bello, Ibrahim Mohammed Hussaini, Umar Aliyu Umar, and Umar Abdulrahim

Abstract A plethora of enzymes produced by microorganisms found in the soil play significant pivotal roles as they mediate several biochemical processes vital to soil fertility, organic matter breakdown, and nutrient cycling, which is why they are so important to soil health and overall ecosystem. These enzymes are synthesized by a variety of soil microbes including bacteria, fungi, algae, protozoa, actinomycetes, archaea and viruses help break down complex organic substances into simpler ones so that plants can absorb the nutrients. Cellulases, proteases, phosphatases, and catalases are some important soil enzymes that target particular substrates and support important ecological processes. Applications of soil microbial enzymes include environmental biotechnology, agriculture, medicine and industries. They

A. D. Idris (✉)
Department of Biological Sciences, School of Applied and Natural Sciences, The Federal Polytechnic Bida, Niger State, Nigeria

Department of Microbiology, School of Life Sciences, Federal University of Technology Minna, Niger State, Nigeria
e-mail: abdullahi.idris@fedpolybida.edu.ng

A. B. Bello
Department of Biological Sciences, School of Applied and Natural Sciences, The Federal Polytechnic Bida, Niger State, Nigeria

Department of Pharmaceutical Microbiology, Faculty of Pharmaceutical Sciences, Ahmadu Bello University, Zaria, Kaduna State, Nigeria

I. M. Hussaini
Department of Microbiology, Faculty of Life Sciences, Ahmadu Bello University, Zaria, Kaduna State, Nigeria

U. A. Umar
Department of Biochemistry, Faculty of Life Sciences, Ahmadu Bello University, Zaria, Kaduna State, Nigeria

U. Abdulrahim
School of Basic Medical Sciences, Department of Medical Microbiology, Central South University, Changsha, Hunan Province, China

Department of Pharmaceutical Microbiology and Biotechnology, Faculty of Pharmacy, University of Maiduguri, Borno State, Nigeria

S. A. Aransiola et al. (eds.), *Soil Microbiome in Green Technology Sustainability*,
https://doi.org/10.1007/978-3-031-71844-1_18

have been applied extensively in bioremediation to break down pollutants and purify contaminated soils. In agriculture, these microbial biocatalysts have been used to improve soil fertility and overall plant development. The creation of biofertilizers and soil amendments based on enzymes has been made possible by developments in enzyme biotechnology, supporting sustainable agricultural methods. Furthermore, soil enzymes provide information on microbial activity and the effects of environmental changes, acting as bio-indicators/biosensors of soil health and quality. This chapter therefore highlights the types of microbes found in the soil, classes of soil microbial enzymes, applications of these soil microbial biocatalysts in agriculture, environment, medicine and industries. Some challenges associated with the applications of these enzymes were also elucidated.

Keywords Microorganisms · Soil · Enzymes · Bacteria · Applications · Environment

1 Introduction

Soil is a very complex system in which plants and microorganisms collaborate to obtain sources of water and nutrients. It is typically regarded as the cornerstone of agricultural production and widely acknowledged as a crucial component for ecological processes (Gossner et al., 2016), environmental sustainability, and global climate change mitigation (Le Quéré et al., 2018; Mir et al., 2023). These forms of life within the soil cooperate with one another together with the soil to continually create changing conditions to improve and maintain soil quality (Stamenov et al., 2018; Holík et al., 2019). Soil is colonized by a vast variety of diverse microbial communities, making up around 95% of total soil biomass, playing a pivotal role in soil ecological processes (van Leeuwen et al., 2017). The microorganisms, when in abundance, a balanced ecological system is achieved. They act in synergy to support plants growth and development, by fostering a healthy rhizosphere that can help fend off attack by pathogens. Majority of microorganisms reside in the upper portion of the soil profile because there are more organic deposits, air, and water available there (Fierer et al., 2003). However, studies have indicated that some anthropogenic activities such as application of fertilizers, tillage, and crop rotation exert a notable influence on soil microbial communities (Ding et al., 2017), which subsequently impairs soil ecosystem functioning (Kracmarova et al., 2020; Ullah et al., 2023).

Multiple studies also showed that the microbial population decreases with soil depth, because deeper soil strata have less substrate (Piotrowska-Długosz et al., 2022; Stone et al., 2014). Additionally, because a wide range of metabolic pathways can be used, microbial diversity is linked to improvement of soil quality and stress resistance (Kracmarova et al., 2020). These microbes are essential to the cycling of important elements such as carbon (C), nitrogen (N), phosphorous (P), and sulfur (S), because they play a pivotal role in sustaining soil fertility and quality (Wei et al., 2019; Song et al., 2020).

Soil extracellular enzymes, sometimes referred to as "biological fingerprints" of soil, are synthesized and released by soil microorganisms, and are associated with the synthesis and decomposition of soil organic matter (Burns et al., 2013; Holík et al., 2019; Aransiola et al., 2022a). Cells of both living and dead microorganisms release these extracellular microbial enzymes into the cell and invariably, there is a strong correlation between these microbial enzymes and substrate availability, microbial biomass, the size of soil particles, community structure and environmental conditions (Knight & Dick, 2004; Marx et al., 2005; Kracmarova et al., 2020). They are highly catalytic and influence a number of significant biochemical processes in soils, which are vital for the microbial cycles, soil structure stabilization, soil organic matter (SOM) creation, organic pollutant biodegradation, nutrient cycling, and control of plant diseases (Erdel et al., 2023; Baldoni et al., 2020; Badzinski et al., 2021; Sobucki et al., 2021). Since the enzymes offer prompt and precise information on slight alterations in the soil environment, they are regarded as reliable markers of biological and biochemical changes (Li et al., 2019; Erdel et al., 2023). In addition, soil enzymes can be derived from plants or animals, but microbes are the primary source. They also catalyze the release of inorganic nutrients from organic matter that are necessary for plant growth (Adetunji et al., 2017; Aransiola et al., 2022b).

The synthesis of enzymes such as β-1,4-glucosidase (BG), β-1,4-N-acetyl-glucosaminidase (NAG), leucine aminopeptidase (LAP), and acid/alkaline phosphatases (AP) indicates that soil microbial activities match the metabolic and stoichiometric requirements of microorganisms. The ecological stoichiometry theory and ecological metabolic theory are linked by these extracellular enzyme activities (EEA), indicating a crucial connection between microbial metabolism and nutrient availability in the environment (Allen & Gillooly, 2009; Sinsabaugh et al., 2009; Daunoras et al., 2024). Furthermore, it has been suggested that the BG:NAG, BG:AP, and NAG:AP relative activities serve as markers for the relative activities of the acquisition enzymes for carbon against nitrogen, carbon versus phosphorus, and nitrogen versus phosphorus (Sinsabaugh et al., 2008; Waring et al., 2014). Additional soil enzymes include oxidoreductases, which aid in the breakdown of organic compounds, and hydrolases, which aid in the acquisition of carbon, nitrogen, and phosphorus to support primary metabolism (Tiemann & Billings, 2011; Holík et al., 2019).

The principal source of energy for soil microbes, glycosidases are a class of carbon (C) cycling enzymes that are essential in the breakdown of low molecular weight carbohydrates (Nannipieri et al., 2011; Ndabankulu et al., 2022). The hydrolysis of the disaccharides, a-D-galatopyranosides, in soils is catalyzed by the glycosidase, a-galactosidase, also referred to as cellobiose (Zhang et al., 2020). An essential enzyme in the hydrolysis of N-acetyl-b-D-glucosamine residues from the terminal non-reducing ends of chito oligosaccharides is β-glucosaminidase (Turner, 2010; Acosta-Martínez et al., 2019). It is believed that this hydrolysis is essential to the cycling of carbon and nitrogen in soils. Measured activities of such enzymes can be used to evaluate microbial nutrient demands and to indicate the ecosystem's response to environmental changes. They also reflect the direction and intensity of different biochemical processes occurring in the soil environment (Wang et al., 2015; Li et al., 2017). The pH of the soil has an impact on the activity of soil enzymes, including β-D-Cellobioside (Delgado-Baquerizo et al., 2016), peroxidase (Bhuyan et al., 2019), N-acetyl-β-D-Glucosaminidase (Acosta-Martínez et al., 2019), and laccase (Olajuyigbe & Fatokun, 2017). On the other hand, some microorganisms and plants react to soil acidity and Phosphorous deficiency by secreting and activating more enzymes, such as phosphatases (Nannipieri et al., 2011; Lacava et al., 2021). Many microbial enzymes found in soil promote soil fertility and other soil activities. Therefore, for a better understanding of the many soil microbial enzymes and their application, this overview is crucial.

2 Soil Microorganisms

The soil is one of the most complex and dynamic natural resource that forms the foundation of all terrestrial ecosystems (Surendran et al., 2023). It consists of a mixture of mineral particles, organic matter, water, air, and living organisms (Kaviya et al., 2019). Soil plays critical roles in supporting plant growth, regulating water and nutrient cycles as well as providing habitat for a diverse array of macro and microorganisms (De Deyn and Kooistra, 2021). The soil harbors a vast assortment of microorganisms referred to as soil microbiome (Banerjee & Van Der Heijden, 2023). This complex community of microorganisms found in the soil including viruses, bacteria, actinomycetes fungi, archaea, protozoa, and algae (Fig. 1) affects composition of the soil, overall plant health, decomposition of organic matter, nutrient cycling, and stability of soil ecosystems as a whole (Adewara et al., 2024). The soil is a reservoir for a plethora of nutrients ranging from simple sugars to the most complex substrates such as humus, which are abundantly present and utilized by soil microbiome for growth and subsequent multiplication (Minhas et al., 2024). The nature of the soil as well as the type and concentration of nutrients found the soil makes the occurrence, population and distribution of microorganism differ from soil to soil (Costantini & Mocali, 2022). However, the rhizosphere, a small area of soil directly affected by plant roots, is home to more microorganisms than other soil regions (Wendel et al., 2022).

2.1 Bacteria

In the complex microbial ecosystems, bacteria stand out as the most prevalent group, with millions of individual bacterial cells per gram of soil. This dominance renders bacteria to be the most abundant and diverse group from the array of microorganisms present in the soil (Timofeeva et al., 2023). This class of microorganisms carries out vital processes such carbon breakdown, nitrification, denitrification, and nitrogen

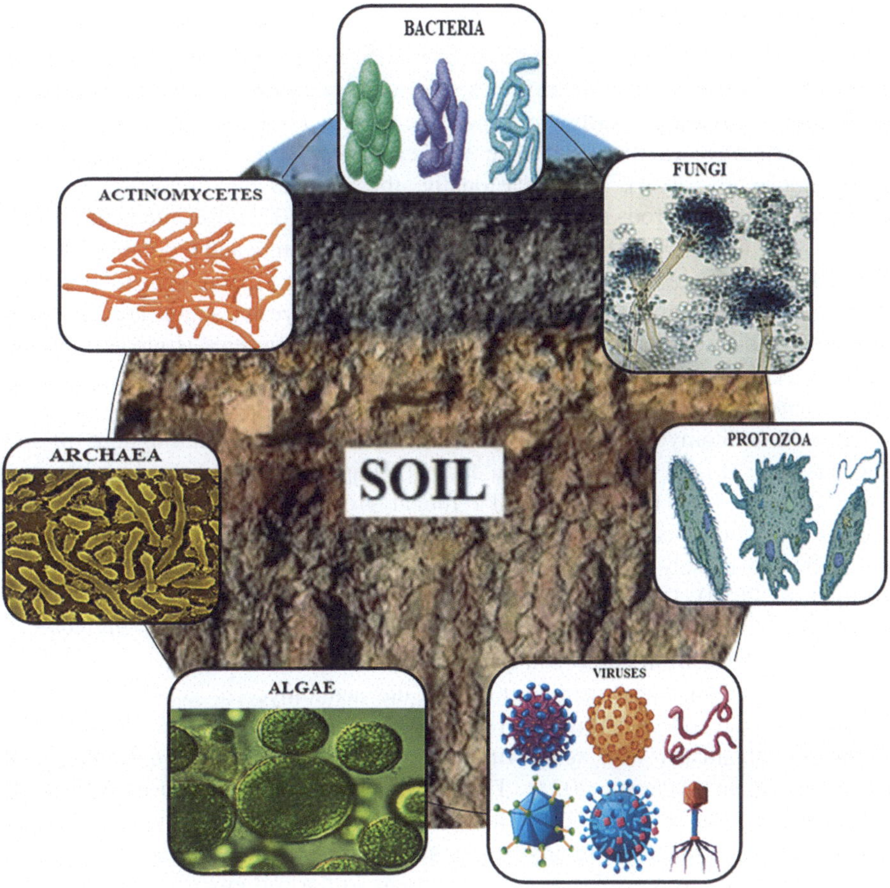

Fig. 1 Microorganisms present in the soil

fixation among other biogeochemical recycling of soil nutrients (Singh, 2024). Some important bacterial taxa in this group include Firmicutes, Acidobacteria, Proteobacteria, and Actinobacteria (Zhu et al., 2023). Many bacterial species frequently isolated from the soil belong to the genera *Clostridium*, *Rhizobium*, *Arthrobacter*, *Pseudomonas*, *Azospirillum*, *Agrobacterium*, *Bacillus*, *Klebsiella, Micrococcus*, *Xanthomonas*, and *Mycobacterium* (Bansal et al., 2024). A major significance of soil bacteria is the enzymatic breakdown and transformation of complex organic soil compounds into nutrients, as well as the subsequent release of these nutrients and trace enzymes from the cells into the soil fraction (Neemisha & Sharma, 2022; Kandeler, 2024). Consequently, most bacteria isolated from the soil usually have potentials for production of an array of biotechnologically important enzymes (Nayak et al., 2020).

2.2 Actinomycetes

Filamentous bacteria, or actinomycetes, are more commonly found in neutral to alkaline soils and are primarily Gram-positive bacteria (Adeoyo, 2023). The majorities of actinomycetes are anaerobic and can grow large mycelia or colonies (Aluvihara et al., 2023). But occasionally, the mycelia may separate, forming structures resembling rods or coccids (Aluvihara et al., 2023). Despite being bacteria, they were classified differently due to their unique feature and biomass (Adeoyo, 2023). The size range of actinomycetes varies from 0.5 to 1.5 μm, depending on the species (Kontro & Yaradoddi, 2022). Individual actinomycetes strains are present in all layers of the soil, although the population of actinomycetes is greatest on the soil's surface layer and gradually diminishes with soil depth (Aluvihara et al., 2023). An estimated population ranging from 10^4 to 10^8 cells per gram of soil has been reported for actinomycetes, they are therefore widely distributed throughout the soil (Meenakshi et al., 2024). They have a distinct damp newly over-turned healthy soil fragrance caused by their hyphae-like fungal growth (Nagendran et al., 2021). Species of actinomycetes such as *Streptomyces* are the most predominant in the soil, followed closely by *Micromonospora*, *Thermoactinomycetes* and *Nocardia* (Katti et al., 2022). Furthermore, a few actinomycetes species from genus such as *Microtetraspora*, *Saccharothrix*, *Microbispora*, *Amycolaptosis* and *Actinomadura* have been isolated from desert soil (Katti et al., 2022).

Greater than ninety percent (>90%) of the actinomycetes genera have been isolated from the soil (Janatiningrum et al., 2024). However, they are known to be mesophilic organisms with sensitivity to low pH (acidic soil) and waterlogged soil environment (Adeoyo, 2023). They posses an ideal pH range of 6.5 to 8.0 (Ngamcharungchit et al., 2023). Their ability to produce hydrolytic and other essential enzymes makes actinomycetes a predominant class of soil microorganisms that are important in recycling organic materials in the soil environment (Shanthi, 2021). Actinomycetes are well known for their ability to synthesize a wide variety of enzymes, many of which are essential for soil ecology, nitrogen cycling, and the breakdown of organic materials (Sağlam et al., 2024). Numerous factors however affect the production of enzymes by these soil microbes including nutrient availability, moisture content, pH, temperature, oxygen availability, substrate complexity, microbial interactions and production of secondary metabolite (Sağlam et al., 2024).

2.3 Fungi

A notable and important group of microorganisms present in soil is fungi (Mason et al., 2023). They aid in the breakdown of complex organic matter by secreting various enzymes (Babaniyi et al., 2023). They are also involved in the recycling of soil nutrients (Babaniyi et al., 2023). Fungi are also known to collaborate with plants roots to create symbiotic relationships known as mycorrhizae, which help plants absorb nutrients (Taylor & Bhatnagar, 2024). The Ascomycota, Basidiomycota and Zygomycota are the most predominant fungal groups found in the soil (Taylor & Bhatnagar, 2024). Species of *Funneliformis*, *Boletus*, *Amanita*, *Glomus*, *Tricholoma*, *Laccaria*, *Rhizophagus*, *Torrendia*, *Rhizophagus*, *Descomyces*, *Rhizophagus*, *Saccharomyces*, *Thelephora*, *Candida*, *Verticillium*, *Penicillium*, *Geotricum*, *Phytophthora*, *Aspergillus*, *Rhizoctonia*, *Trichoderma*, *Pythium*, *Coprinus*, *Pleurotus* and *Agaricus* are a few examples of often occurring fungi in the soil (Adeleke et al., 2019; Lugo & Pagano, 2022; Fan et al., 2022).

Like bacteria, soil fungi have also been reported to be good producers of biotechnologically important enzymes (Kour et al., 2019). In order to effectively break down complex organic substrates found in the soil environment, soil fungi must produce enzymes through a carefully regulated process (Chia et al., 2024) which aids in the decomposition of organic matter, the cycling of nutrients, and the health of ecosystems (Bhardwaj et al., 2024). Numerous factors, such as genetic regulation, substrate availability, and environmental conditions, affect the regulation of enzyme production (Bhardwaj et al., 2024).

2.4 Archaea

Archaea, another important microbial group found in the soil (Bandopadhyay & Shade, 2024) are less common than bacteria and fungi, they are nonetheless very important for carbon metabolism, methane oxidation, and nitrogen cycle (Robertson & Groffman, 2024). They are frequently found in extreme soil environments, like anaerobic, high salinity, extreme pH and temperatures conditions (Zhang et al., 2024). Notable examples are the methane-producing methanogenic archaea and the ammonia-oxidizing archaea (Zhang et al., 2024). Species of *Nitrososphaera*, *Methanosarcina*, *Halobacterium*, *Candidatus*, *Methanococcus*, *Pyrolobus* and *Picrophilus* are some examples of archaea isolated from the soil (Gupta et al., 2021; Gordon, 2023).

Production of enzymes by soil archaea is a complicated process controlled by a variety of genetic, physiological, and environmental factors such as genetic regulation, nutrient availability, substrate utilization, physiological adaptation, microbial interactions, ecosystem functioning and environmental stressors (Hartmann & Six, 2023; Bandopadhyay & Shade, 2024). Despite receiving less research attention than bacteria and fungi, soil archaea and the enzymes they produce are crucial for breakdown of organic matter, cycling of nutrients, and overall health of the soil (Mishra et al., 2023).

2.5 *Protozoa*

Protozoa are unicellular eukaryotes and another important group of soil microorganisms that feed on bacteria, fungus, and other microorganisms (Neher, 2023). They aid in the recycling of nutrients and the regulation of other microorganisms (Zwair, 2023). They are also known to take part in the breakdown of organic substances and control the populations of bacteria. Amoebae, ciliates, flagellates, and testate amoebae are examples of common soil protozoa (Adewara et al., 2024). Protozoa are also known to produce various enzymes to help them break down nutrients, defend against pathogens, and carry out other vital functions necessary for their survival and reproduction (Shukla et al., 2024). These enzymes play critical roles in the protozoan's physiology and are integral to their overall biology (Shukla et al., 2024).

2.6 *Algae*

Photosynthetic organisms also known as algae live on the soil and are found in the upper layers of the soil where light can penetrate easily (Garcia-Pichel & Belnap, 2021). They enhance soil fertility by fixing atmospheric carbon dioxide and generate organic matter through photosynthesis (Ramakrishnan et al., 2023). Among the algae found in soil habitats are diatoms, cyanobacteria, brown, red and green algae. Algae species are predominantly aquatic, but some can also thrive in terrestrial environments (Wehr & van Vuuren, 2024). Some examples include species of *Trentepohlia*, *Nostoc*, *Stigonema*, *Desmococcus*, *Chroococcidiopsis* and *Klebsormidium* (Sanders & Masumoto, 2021). These microorganisms play essential roles in soil by fixing nitrogen and carbon by the synthesis of exopolysaccharides that increase soil fertility and water retention capacity (Ramakrishnan et al., 2023). Enzymes are essential for various metabolic processes within algae, including photosynthesis, respiration, and nutrient uptake (Parveen et al., 2023). Soil algae therefore produce enzymes to break down organic matter, facilitate biochemical reactions, and regulate metabolic pathways (Shahi et al., 2023). These enzymes play crucial roles in the algae's ability to obtain nutrients, grow and respond to environmental changes (Shahi et al., 2023).

2.7 *Virus*

Soil viruses are another important group of soil microbes; they include the bacteriophages that infect and influence the dynamics of bacterial and archaeal populations (Bi et al., 2022). They play roles in controlling microbial abundance, genetic exchange, and nutrient cycling (Albright et al., 2022). Soil viral communities are diverse and dynamic, with complex interactions with other soil microorganisms (Bi et al., 2022). The soil typically contains approximately 10^9 virus particles per gram of dry weight (Durham et al., 2022). Tailed bacteriophages, which are the predominant type of soil viruses, exhibit a preference for wetland forest soil compared to drier agricultural soils (Coclet et al., 2023).

Among the viruses commonly found in soil are small spherical particles that resemble single-stranded (ss) RNA-containing bacteriophages of the Leviviridae family or certain plant viruses (Novik et al., 2018). Additionally, larger spherical viruses akin to double-stranded (ds) DNA-containing viruses from families such as Partitiviridae, Chrysoviridae, and Totiviridae are also prevalent (Kumar et al., 2023). The Sahara desert surface sands harbor a diverse range of viruses belonging to the Myoviridae, Siphoviridae, and Podoviridae families, with genome sizes ranging from 45 to 270 kb (Zablocki et al., 2016). While soils serve as reservoirs for viruses, they are likely not entirely static repositories, as some viruses appear to move readily between different environments (Reavy et al., 2014).

By and large, the soil microbiome is an intricate web of relationships between many microbial taxa and other abiotic factors, all of which support the resilience and efficiency of soil ecosystems (Hartmann & Six, 2023). For sustainable soil management techniques, agricultural productivity and environmental preservation, it is essential to comprehend the dynamics, composition, and ecological, industrial, medical and biotechnological roles of these microbial communities (Iquebal et al., 2022).

3 Classes of Soil Microbial Enzymes

Microorganisms in the soil produce different types of enzymes depending on the substrate available as carbon, nitrogen and phosphorus sources (Daunoras et al., 2024). They facilitate the breakdown of organic compounds in the soil and cycling of nutrients in the ecosystem. Enzymes produced by soil microorganisms can be largely classified into two main categories namely: hydrolases (hydrolytic enzymes) and oxidoreductases (enzymes involved in oxidation-reaction reactions).

3.1 Hydrolases

These enzymes are involved in the hydrolytic break down of complex organic compounds into simple molecules. Examples of hydrolases produced by soil microorganisms include:

3.1.1 Cellulases

Cellulases are microbial enzymes involved in the hydrolysis of cellulose into D-glucose as well as decomposition of organic matter present in the soil (Daunoras et al., 2024).

3.1.2 Invertases

Invertases catalyze the hydrolysis of sucrose to produce equal molar proportion of D-glucose and D-fructose (Lincoln & More, 2017; Osiebe et al., 2023). It act by hydrolyzing the α-1,2-glycosidic link in sucrose (Lincoln & More, 2017).

3.1.3 Glucosidases

Glucosidases are one of the most common and predominant class of microbial enzymes in the soil. They play a vital role in the hydrolysis of different derivatives of glucose in the soil (Daunoras et al., 2024). The most prominent glucosidases produced by soil microorganisms are the β-Glucosidases (Utobo & Tewari, 2015). β-Glucosidase catalyze the rate limiting step in the production of glucose from cellulose (Adetunji et al., 2017).

3.1.4 N-Acetyl-Glucosaminidases

N-Acetyl-Glucosaminidases are another class of soil microbial enzymes. They catalyze the conversion of chitin oligosaccharides and chitin dimers into a metabolically useful monomeric amino sugar called N-acetylglucosamine (Krolicka et al., 2018). Amino sugars serve as a major source of carbon and nitrogen in soil (Luo et al., 2017; Daunoras et al., 2024).

3.1.5 Peptidases/Proteases

These are proteolytic enzymes involved in the selective hydrolysis of amino acid residues or dipeptides at the N-termini of proteins (Carroll et al., 2013). They are one of various enzymes produced by soil microorganisms to absorb nitrogen (Liew et al., 2013; Daunoras et al., 2024).

3.1.6 Lipases

Lipases are serine hydrolases that catalyze the hydrolytic cleavage of ester bond of triglycerides form fatty acids and glycerol. They degrade microbial, plant and animal derived lipids found in the soil (Bhandari et al., 2021).

3.1.7 Phosphatases

Microbial phosphatases play a vital role in organic phosphorus mineralization in the soil (Stone & Plante, 2014). They catalyze phosphomonoesters hydrolysis as well as hydrolysis of phosphodiesters in some instances which results in the release of phosphate (Daunoras et al., 2024).

3.2 *Oxidoreductases*

Oxidoreductases are involved in the reduction / oxidation (redox) of organic compounds. Examples of oxidoreductases produced by soil microorganisms include:

3.2.1 Dehydrogenases

Dehydrogenases are intracellular oxidoreductase enzymes that catalyze the oxidation of organic matter in the soil through the transfer of protons and electrons from acceptors to substrates (Daunoras et al., 2024).

3.2.2 Oxygenases

Oxygenases are oxidoreductase enzymes involved in the oxidation of substrate through oxygen molecule incorporation using FAD, NADH or NADPH as co-factors. There are two types of oxygenase namely: monooxygenases and dioxygenases (Arora et al., 2009).

3.2.3 Laccases

Laccases are multiple copper containing microbial enzymes that catalyze the oxidation of phenol (and phenolic compounds), aromatic amine as well as their substituted compounds. The reaction is characterized by the formation of two molecules of water with concomitant loss of electron in a single oxygen molecule (Bhandari et al., 2021).

3.2.4 Peroxidases

Peroxidases are oxidoreductases that catalyze the reduction of peroxides (e.g. H_2O_2) and the oxidation of several organic and inorganic compounds. Their activity involves the donation of electrons that bind to other substrates (such as ferricyanides and ascorbate) thereby breaking them into harmless compounds (Bansal & Kanwar, 2013) (Fig. 2).

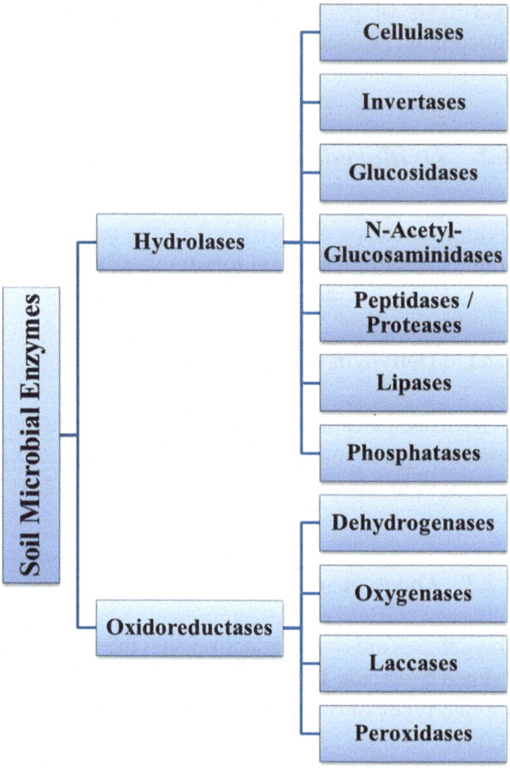

Fig. 2 Classes of soil microbial enzymes

4 Applications of Soil Microbial Enzymes in Agriculture, Medicine, Industries and Environment

Soil-borne microorganisms have been demonstrated to play significant role in promoting soil health and sustainable agriculture through solubilizing nutrients to increase their availability to plants and endure nutrient cycle (Zungu et al., 2020). Soil microbial communities in general and their enzyme activities are influenced by different agricultural activities including tillage, fertilizer and pesticide applications, irrigation and cropping system (Jat et al., 2021). Microorganisms in the soils produce numerous enzymes that could be isolated for many applications. There are wide range of soil microbial enzymes derived from diverse microorganisms with dynamic functions and applications such as agriculture, environment, medicines and industries (Aransiola et al., 2023).

Indeed soil enzymatic activities have been used as an important determinant of soil quality, due to their important roles in decomposition of organic matter and biogeochemical cycling of soil nutrients (Ranjith et al., 2015; Santos et al., 2018; Gorde et al., 2022). Furthermore, soil microbial enzymes provide sensitive biochemical information on the effect of various environmental management practices on soil health and proper balance of ecosystem. Hence, they have been regarded as important biochemical determinants of soil fertility which is essential for sustainable agriculture planning and development (Jat et al., 2021). Table 1 provides a summary of some of the soil microbial enzymes and their applications in agriculture and environment. Microbial enzymes are cost effective industrial inputs for many applications as alternative to chemical products and chemical-based technologies that are often associated with drawbacks such as high cost, high energy demands and negative environmental impact. However, enzymes are energy-efficient, produce less toxic waste and can be recovered easily at low cost. Some of the applications of soil microbial enzymes in medicine and industries are summarized in Table 2. Indeed, soil microorganisms can be readily isolated and grown in low-cost culture media for the production of enzyme of interest. Additionally, with advent of recombinant DNA technology, the microbial genome can be manipulated to produce enzymes that can perform optimally at different conditions and for various applications (Okpara, 2022).

4.1 Lipase

Soil microbial lipases are produced from vast species of microorganism such as *Acinetobacter sp*., *Rhodococcus sp*., *Pseudomonas Sp* and *Mycobacterium S*p., *Trichoderma* sp., and *Penicillium* sp. (Chandra et al., 2020). *Pseudomanas sp.* has been reported to be the major lipase producing soil bacteria (Willerding et al., 2011; Mobarak-Qamsari et al., 2011; Ilesanmi et al., 2020). The enzyme has commercial applications in many fields such as bioremediation of oil-contaminated soils, detergent and paper industries (Mobarak-Qamsari et al., 2011). Lipase is the most important commercially available enzymes used in the bioremediation of soil contaminated with oil residues including petroleum (Hwang et al., 2018; Bhandari et al., 2021). It also has wide applications in many industries including pulp and paper, cosmetics, as well as research and therapeutics (Gurung et al., 2013; Arora et al., 2020). In the dairy industry for instance, lipases are commonly used to enhance flavour of cheese (Chandra et al., 2020).

4.2 Protease

Bacillus and *Pseudomonas sp.* sp. are the main bacterial strains reported for protease production while *Aspergillus* sp. are the main fungal sources (Mageswari et al., 2017). Soil microbial proteases are extracellular enzymes belonging to hydrolase family. The products of catalytic activities of protease are short chain Polypeptides or amino acids derived from the hydrolytic breakdown of peptides bonds in proteins (Okpara, 2022). Soil microbial protease catalyzes the breakdown of peptide bonds in organic matter present in the soil. This makes the enzyme important for the

Table 1 Applications of soil microbial enzymes in agriculture and environment

Enzyme	Source	Chemical Reaction	Application(s)	Reference(s)
Urease	Plants and microorganisms	Hydrolysis of urea to CO_2 and NH_3	Nitrogen cycling; biological indicator of soil fertility	Ranjith et al. (2015); Santos et al. (2018)
Phosphatases	Plants and microorganisms	Hydrolyze phosphate esters to inorganic phosphorus	Phosphorus cycling; soil quality indicator	Ranjith et al. (2015); Santos et al. (2018); Kenarova and Boteva (2023)
Arylsulphatase	Mainly bacteria, but also fungi, plants and animals	Hydrolysis of organic sulphate (sulphate esters)	Sulphur cycling; indicator of soil acidification and heavy metals contamination	Santos et al. (2018)
Dehydrogenase	Microorganisms	Catalyze oxidation of organic matter in the soil	Indicator of microbial activity in the soil; biological indicator of soil fertility; involved in carbon cycling	Ranjith et al. (2015); Santos et al. (2018); Gorde et al. (2022)
Cellulase	Microorganisms	Hydrolysis of cellulotic materials to glucose	Indicator of soil fertility; involved in carbon cycling; useful in assessing the effect of soil management practices; bioremediation of organic pollutants	Aziz et al. (2018); Chia et al. (2024)
β-Glucosidase	Mainly microorganisms, but also plants and animals	Hydrolysis of glycosidic bonds of cellulose residues in the soil	Indicator for the effect of agricultural practice on soil biomass (soil quality) as well as soil acidification; involved in carbon cycling	Santos et al. (2018); Zungu et al. (2020); Kenarova and Boteva (2023)
Invertase	Microorganisms, plants and some animals	Hydrolyze sucrose into glucose and fructose	Decomposition of plant biomass; indicator for efficiency of soil microbial metabolism	Kenarova and Boteva (2023)
Catalase	Microorganisms	Reduces H_2O_2 into H_2O and O_2	Indicator of soil oxidative stress tolerance (soil contamination)	Zungu et al. (2020)
L-Asparaginase	Mainly bacteria, yeast and fungi, but also plants and animals	Hydrolysis of asparagine into aspartic acid and ammonia	Involved in nitrogen cycling and mineralization	Zungu et al. (2020)

β-D-glucosaminidase	Mainly microorganisms	Catalyze breakdown of chito-oligosaccharides	Involved in nitrogen cycling and mineralization	Zungu et al. (2020)
Lipases	Animals, plants and microorganisms	Hydrolysis of mono-, di- and triglycerides into fatty acids and glycerol	Bioremediation of organic pollutants and oil spills; treatment of industrial effluent	Gurung et al. (2013); Bhandari et al. (2021); Narayanan et al. (2023); Chia et al. (2024)
Laccase	Microorganisms	Uses molecular oxygen to catalyze oxidation of aromatic compounds	Bioremediation of pesticide-contaminated soils, textile dyes and waste water	Bhandari et al. (2021); Ayilara and Babalola (2023); Chia et al. (2024)
Protease	Microorganisms	Catalyze breakdown of peptide bonds in proteins	Bioremediation of organic pollutants (feathers, nail, hair); waste water treatment	Bhandari et al. (2021); Chia et al. (2024)
Peroxidase	Microorganisms		Bioremediation of pesticide and other xenobiotics in the soil	Chia et al. (2024)
Glutathione-S-transferase	Microorganisms	Phase II conjugation reaction of xenobiotics	Bioremediation of pesticide and other xenobiotics in the soil	Chia et al. (2024)
Sulfotransferase	Microorganisms	Phase II conjugation reaction of xenobiotics	Bioremediation of pesticide-contaminated soils	Chia et al. (2024)
Methyltransferase	Microorganisms	Phase II conjugation reaction of xenobiotics	Bioremediation of pesticide-contaminated soils	Chia et al. (2024)
Glucuronyltransferase	Microorganisms	Phase II conjugation reaction of xenobiotics	Bioremediation of pesticide-contaminated soils	Chia et al. (2024)
Cytochrome P450	Microorganisms	Catalyze oxidation of pesticides and other xenobiotics in the soil	Bioremediation of pesticide-contaminated soils	Chia et al. (2024)

Table 2 Application of soil microbial enzymes in medicine and industries

Enzyme	Application	Reference(s)
α-Amylase	Baking industry (dough making); beverage industry (production of corn syrup); cosmetic industry (detergent production); paper industry; used in design, discovery and development of drugs	Muruvanda et al. (2023)
β-Amylase	Beverage industry (production of high maltose syrup); cosmetic industry (detergent production)	Gurung et al. (2013); Muruvanda et al. (2023)
Cellulase	Production of ethanol and single cell protein; textile processing (biocarbonization and wool scouring as well as biostone washing); pulp and paper production	Gurung et al. (2013); Naqvi Syed et al. (2018); Muruvanda et al. (2023); Korsa et al. (2023)
Laccase	Breakdown of lignin for paper and pulp production; remove phenolic substances and enhance shelf-life of beverages	Muruvanda et al. (2023)
Protease	Baking industry (bread dough improvement); meat tenderization	Kuddus (2019)
Xylanase	Baking industry (dough rising); wine and beverage production; paper industry (breakdown of lignocellulosic materials as well as used as bleaching agent); animal feed industry (for breakdown of non-starch polysaccharides)	Walsh et al. (1993); Naqvi Syed et al. (2018); Kuddus (2019); Muruvanda et al. (2023)
Invertase	Beverage industry	Muruvanda et al. (2023)
Lipase	Food industry (production of specialty fats); cheese production; animal feed (to improve digestibility); cosmetic industry (use in detergent production and for hair waving preparation); used in drug design, discovery and development	Gurung et al. (2013); Muruvanda et al. (2023)
L-Asparaginase	Treatment of leukemia; food industry (flavor enhancing agent)	Gurung et al. (2013); Orabi et al. (2019)
L-Glutaminase	Treatment of leukemia; food industry (flavor enhancing agent as alternative to monosodium glutamate)	Williams and Woessner (2009); Gurung et al. (2013); Orabi et al. (2019)
Collagenase	Treatment of skin ulcer; wound healing	Sabu (2003); Alipour et al. (2016)

control of environmental pollution apart from its other wide applications in food and leather industries, as well as for treatment of wastewater (Mazotto et al., 2011; Kumar & Sharma, 2019; Bhandari et al., 2021). The enzymes have tremendous application in various industries as well as environment (Alam et al., 2017). In dairy industry, protease are used to prevent the formation and accumulation of casein micelles in curds during cheese production (Okpara, 2022). In Leather Industry, the enzyme is widely used for dehairing processes (Alam et al., 2017). Keratinase, a protease enzyme for instance, is used in leather industry for dehairing processes. It is also useful in bioremediation of poultry wastes such as chicken feathers for recycling into useful byproducts (Bhange et al., 2016; Akhter et al., 2020). Proteases are also been used to hydrolyze gluten protein in wheat flour thereby promoting dough formation during bread and biscuit production as well as other baking products. They are also useful in improving meat tenderization and their additions in animal feed also enhance amino acid availability from the protein content (Okpara, 2022).

4.3 Cytochrome P450

Cytochrome P450 is a group of heme-containing enzymes that catalyze the biotransformation of xenobiotics and other toxic chemicals. The enzyme derives from soil microorganisms is used for metabolism of pesticide, which makes it useful in enzymatic bioremediation of pesticide-contaminated environments (Bhandari et al., 2021). Several microorganism including *Rhodococcus*, *Mycobacterium* and *Pseudomonas* have been reported to produce this enzyme for bioremediation of organic pollutants and crude oil spills from the environment (Bhandari et al., 2021).

4.4 Laccases

They are multi-copper-containing oxidoreductases produced by plants, bacteria and fungi. The major microbial sources of laccases are actinomycetes by species including *Streptomyces bikinesis*, *S. cyaneaus* and *S. ipomoea* (Bhandari et al., 2021). Laccase is used in beverage industry to enhance shelf life through removal of oxygen from wine and beer (Okpara, 2022). Laccase catalyze oxidation of diverse array of compounds including phenolics, and aromatic amines, which makes it a versatile enzyme for degradation of pollutants such as synthetic dyes and crude oil in the environment (Bhandari et al., 2021).

4.5 Catalase

This is an oxidoreductase enzyme that reduces hydrogen peroxide (H_2O_2) into water (H_2O) and oxygen (O_2). Thus, it is biological indicator of oxidative stress tolerance in soil (Zungu et al., 2020). Soil catalase is produced by plant, animals and microorganism. Major microbial sources of catalase include *Bacillus* sp and *Aspergillus* sp. Catalase is been used in dairy industry for prevention of rancidity in milk due to its catalytic ability to remove peroxide, as well as in wine making to increase shelf life through removal of oxygen from the wine (Okpara, 2022).

4.6 Dehydrogenase

Dehydrogenase is an intracellular enzyme belonging to the oxidoreductase family and catalyzes the oxidation of organic matter in the soil (Aziz et al., 2018). The enzyme is an important biological indicator of microbial activity in the soil, and also useful for assessing the effectiveness of agricultural practices (Partoazar et al., 2011; Jat et al., 2021). The activity of the enzyme has been demonstrated to be increased (stimulated) or decreased (indicted) by environmental management practices and pesticide application (Kenarova & Boteva, 2023). For instance, different fungicides affect soil dehydrogenase activity in a different manner based on the type and exposure period of the fungicide use. Hence measuring the enzyme activity provides insights on the environmental impact of pesticide, which has implication on sustainable agriculture (Bacimaga et al., 2015; Sun et al., 2020; Kenarova & Boteva, 2023).

4.7 Phosphatases

Acid and Alkaline phosphatases catalyzed the hydrolytic breakdown of phosphate ester (organic phosphates) depending on low and high environmental pH, respectively. They are extracellular enzymes produced mainly by bacteria, fungi and plants for acid phosphate while alkaline phosphatase from soil microorganisms and invertebrate animals (Kenarova & Boteva, 2023). Phosphomonoestarases are the most extensively studied soil phosphatases and are commonly used for assessment of soil quality for sustainable agriculture due to its role in phosphorus cycling (Reid & Wilson, 1971; Zungu et al., 2020; Kenarova & Boteva, 2023).

4.8 Arylsulphatase

Sulfur exist predominantly in an organic form in the soil, accounting for over 90% of the total soil sulfur, with sulfate esters constituting large proportion (30–75%) of the soil organic sulfur (Chen et al., 2019; Daunoras et al., 2024). Arylsulphatases are important members of hydrolase class of enzymes that catalyze the hydrolysis of sulfate esters (Daunoras et al., 2024). Soil microbial arylsulphatase catalyzes the hydrolysis of sulfates esters in the soil, and therefore are involved in sulfur cycling and mineralization, (Kertesz & Mirlean, 2004). Soil arylsulphatase is an extracellular enzyme produced mainly by soil bacteria in response to sulfur deficiency in the environment (Kertesz & Mirlean, 2004; Kunito et al., 2022; Daunoras et al., 2024).

However, plants have the ability for the uptake of inorganic sulphate only but not organic sulphur, including sulfate esters. Hence, arylsulphatase plays crucial role in plant sulphur nutrition (Sherene, 2017). The enzyme is been used for assessing soil acidification and heavy metals contamination, and it is considered an important biochemical indicator of soil sulfur cycling and soil mineralization (Acosta-Martinez et al., 2007; Chen et al., 2019; Daunoras et al., 2024).

4.9 Urease

Soil urease is an important biochemical indicator of soil quality, involved in nitrogen cycling and mineralization. The enzyme catalyzes the hydrolytic breakdown of urea into CO_2 and NH_3 (Yang et al., 2006; Daunoras et al., 2024). Ammonium ions (NH_4^+) generated from ammonia (NH_3) following enzymatic activity of urease causes increase in soil pH leading to volatilization of NH_3 with resultant loss of nitrogen to the atmosphere (Sherene, 2017; Adetunji et al., 2017; Daunoras et al., 2024). It is also used to assess the impact of agricultural practices on soil health, due to its sensitivity for increased activity by organic fertilization and decreased activity by tillage practices (Utobo & Tewari, 2015; Daunoras et al., 2024). The main sources of the enzyme in soil are bacteria, actinomycetes and fungi (Santos et al., 2018; Kenarova & Boteva, 2023).

4.10 Amylase

Soil α—amylase is an extracellular enzyme produced by plants, animals and microorganisms for the breakdown of α—1, 4—glycosidic bonds in starch. Hence, it is useful in decomposition of organic matter and soil microbial energy metabolism, as well as carbon cycle (Ryan et al., 2006; Kenarova & Boteva, 2023). Several soil bacteria and fungi including species of *Bacillus, Proteus, Streptomyces, Aspergillus* are good sources of soil amylase (Luang-in et al., 2019; Yassin et al., 2021; Abo-Kamer et al., 2023). Microbial amylases are widely used across many industries such as food manufacturing, beverage, textile, pharmaceutical and paper (Singh & Kumari, 2016; Yassin et al., 2021).

4.11 Invertase

It is an extracellular hydrolytic enzyme that is secreted by soil microorganisms as well as plants and animals for the catalytic breakdown of α-1,2-glycosidic bond in sucrose, releasing glucose and fructose units. Hence it plays significant role in carbon cycling, and serve as an important indicator of efficiency in soil microbial energy metabolism and decomposition of organic pollutants (Belcarz et al., 2002; Jin et al., 2009; Kenarova & Boteva, 2023; Daunoras et al., 2024). The enzyme activity can be stimulated by low concentration of pesticides and inhibited by higher concentration, hence providing insights on soil health. (Ramudu et al., 2011; Walia et al., 2014; Kenarova & Boteva, 2023).

4.12 β—Glucosidase

The major source of soil β—glucosidase is microorganisms, but is also produced by plants and animals. It belongs to cellulolytic enzyme complex that catalyze the breakdown of glycosidic bond in cellulose residues yielding glucose units (Kenarova & Boteva, 2023). The enzyme has been shown to be essential in decomposition of organic matter as energy source for microorganisms in the soil. Hence, an important determinant of the effect of agricultural practices on soil fertility (Turner et al., 2002; de la Horra et al., 2003; Kenarova & Boteva, 2023).

4.13 Cellulase

Cellulase is an extracellular hydrolytic soil enzyme produced mainly by microorganisms such as *Bacillus* sp. and *Streptomyces* sp. and to a lesser extent by plants and certain invertebrates (Sadhu & Maiti, 2013; Okpara, 2022). It catalyzes the hydrolytic breakdown of cellulose into D-glucose units. Soil microbial cellulase is composed of at least three enzymes that include endo—1, 4-β—glucosidase, exo—1, 4-β—glucosidase and β—D—glucosidase which randomly breakdown inner glycosidic bonds, releases glucose from non-reducing end of cellulose and hydrolyzes cellobiose to glucose, respectively (Makoi & Ndakidemi, 2008; Kenarova & Boteva, 2023). In beverage industry, cellulase is used as biochemical fruit-softening agent in fruit and vegetable juice, thereby hydrolyzing cellulose and preventing cloudiness in the juice. They also have important application in animal husbandry, as they are used to improve digestibility of animal feed for ruminant animals (Okpara, 2022).

4.14 Collagenase

Soil microbial collagenases belong to protease class of enzyme that catalyze hydrolytic degradation of both native and denatured collagen. *Bacillus* sp. and *Clostridium* sp., are among the major bacterial strains that have been demonstrated to produce collagenases (Lima et al., 2015). Species of *Aspergillus, Penicillium, Cladosporium and Candida* are the main fungal sources of collagenases in the soil (Wanderley et al., 2016). Collagenases have important applications in medicine and industries as well as environment. In medicine, they have been reported to exhibit therapeutic potentials for prevention of osteoporosis and gastric ulcer among many other ailments (Lima et al., 2015). Collagenase is been used in medical research as well as for cosmetics formulation (skin moisturizer) and wound healing (Pal & Suresh, 2016; Gautam & Azmi, 2017). Collagenase is also used in meat industry for tenderizing and flavour enhancement (Pal & Suresh, 2016; Chauhan & Prabha, 2017).

4.15 Xylanase

Xylanases are group of hydrolase enzymes that catalyze the breakdown of β-1,4-glycosidic bonds in xylan to produce xylose sugars (Okpara, 2022). In baking industry, xylanase is been used to improve bread dough formation as well as in fruit juice industry to improve juice extraction and quality (Bhardwaj et al., 2019). The enzyme is used for brightening of pulp for paper production (Muruvanda et al., 2023). They are also used in feed formulation for ruminant animals to improve digestibility of complex carbohydrates such as arabinoxylan (Okpara, 2022).

4.16 L-Asparaginase

Produced by several soil bacteria including *Bacillus* sp., Asparaginase catalyzes the hydrolysis of amino acid Asparagine, yielding Aspartic acid and ammonia. It is an important microbial therapeutic enzyme widely recognized for its anticancer activity against leukemia (Sabu, 2003).

4.17 L-Glutaminase

Glutaminase belongs to hydrolase class of enzymes, catalyzing the hydrolytic deamination of Glutamine into Glutamate and ammonia. It is used in medical industry for treatment of leukemia (Sabu, 2003).

5 Challenges/Limitations in the Use of Soil Microbial Enzymes

The following challenges limit the use of enzymes produced by microorganisms in the soil (Chia et al., 2024):

(i) Soil microbial enzymes tend to become inactivated following their interaction with substrates.
(ii) In their free states, microbial enzymes are usually unstable and tend to become degraded within short time under unfavourable condition.
(iii) In the absence of cofactors, the activity of microbial enzymes might be impeded.
(iv) Enzymes are non-reusable since their recovery from medium after reaction is difficult.
(v) Enzyme extraction and purification is time-consuming and is sometimes costly.

6 Conclusion and Future Prospects

This chapter focused on the complex nature of the soil and the array of microbial groups found in the soil environment, their characteristics, specific examples, as well as the roles they play in the overall soil health. It also covers the classes of soil microbial enzymes, their applications in various sectors such as medicine, agriculture, industries and biotechnology. The challenges encountered in study, production and application of these important biocatalysts is also highlighted. Soil microbial enzymes hold the potential to revolutionize agriculture by promoting sustainable farming practices (Goyal et al., 2023). They can enhance soil health, increase nutrient availability to plants, and improve crop yields, reducing the reliance on chemical fertilizers and pesticides (Khan et al., 2023). In addition, the mechanisms of carbon sequestration and nutrient cycling, which are critical for reducing climate change, depend heavily on the enzymes produced by soil microorganisms (Mason et al., 2023). Subsequent investigations could reveal innovative enzyme-oriented approaches to improve soil carbon sequestration and reduce greenhouse gas emissions (Bhandari et al., 2023). By dissolving toxins and pollutants in soil, soil microbial enzymes also present viable options for environmental repair (Kour et al., 2021). The development of enzyme-based bioremediation technologies to clean up polluted locations and restore ecosystem health is one of the near future expectations (Kumar & Bharadvaja, 2019). Enzymes produced by soil microbes can also be utilized in bioenergy production processes, such as the conversion of biomass into biofuels (Binod et al., 2019). Advancements in this field of study may lead to the optimization of enzyme cocktails for efficient biomass degradation and biofuel production (Singh et al., 2022). The utilization of soil microbial enzymes in precision agriculture has tremendous promise for optimizing nutrient management and boosting soil fertility (Bargaz et al., 2018). The creation of soil testing kits based on enzymes and precision farming methods to improve crop yields and optimize fertilizer delivery is another possible advancement in soil-microbial-enzyme technology (Goyal et al., 2023). Biotechnological uses of soil microbial enzymes are numerous and include the production and processing of food and beverages as well as pharmaceuticals. In the near future, designing and discovery of novel enzymes with special functions for a range of industrial uses are some foreseeable possibilities

(Akinsemolu et al., 2024). The development of microbiome engineering may make it possible to modify the microbial communities in soil to improve overall soil health and enzyme production (Jansson et al., 2023). Subsequent investigations could reveal tactics for adjusting soil microbiomes to enhance agricultural output and sustainability (French et al., 2021). In a nutshell, soil microbial enzymes have a great deal of promise to help with global issues pertaining to agriculture, pollution, health and sustainable development (Ahirwar et al., 2020). Future advancements in this area of study and research should present fresh possibilities for using microbial enzymes' beneficial effects on the environment and humankind (Brevik et al., 2020).

References

Abo-Kamer, A. M., Abd-El-Salam, I. S., Mostafa, F. A., Mustafa, A. A., & Al-Madboly, L. A. (2023). A promising microbial a-amylase production, and purification from *Bacillus cereus* and its assessment as antibiofilm agent against *Pseudomonas aeruginosa* pathogen. *Microbial Cell Factories, 22*, 141.

Acosta-Martinez, V., Cruz, L., Sotomayor-Ramirez, D., & Perez-Alegria, L. (2007). Enzyme activities as affected by soil properties and land use in a tropical watershed. *Applied Soil Ecology, 35*, 35–45.

Acosta-Martínez, V., Pérez-Guzmán, L., & Johnson, J. M. (2019). Simultaneous determination of β-glucosidase, β-glucosaminidase, acid phosphomonoesterase, and arylsulfatase activities in a soil sample for a biogeochemical cycling index. *Applied Soil Ecology, 142*, 72–80. https://doi.org/10.1016/j.apsoil.2019.05.001

Adeleke, R. A., Nunthkumar, B., Roopnarain, A., & Obi, L. (2019). Applications of plant–microbe interactions in agro-ecosystems. In *Microbiome in plant health and disease: Challenges and opportunities* (pp. 1–34). Springer.

Adeoyo, O. R. (2023). Actinomycetes: Dependable sources for industrially useful secondary metabolites. *International Journal of Recent Research in Life Sciences (IJRRLS), 10*(2), 37–42.

Adetunji, A. T., Lewu, F. B., Mulidzi, R., & Ncube, B. (2017). The biological activities of β-glucosidase, phosphatase and urease as soil quality indicators: A review. *Journal of Soil Science and Plant Nutrition, 17*(3), 794–807. https://doi.org/10.4067/S0718-95162017000300018

Adewara, O. A., Adebayo-Olajide, T. C., Ayedun, J. S., Kotun, B. C., Adeleke, A. J., Brown, A. D., Alabi, O. J., & Ogunbanwo, S. T. (2024). Soil formation, soil health and soil biodiversity. In *Prospects for soil regeneration and its impact on environmental protection* (pp. 95–121). Springer.

Ahirwar, N. K., Singh, R., Chaurasia, S., Chandra, R., & Ramana, S. (2020). Effective role of beneficial microbes in achieving the sustainable agriculture and eco-friendly environment development goals: A review. *Frontiers in Microbiology, 5*, 111–123.

Akhter, M., Wal, M. L., Akter, Y., & Shimizu, K. (2020). *Microbial bioremediation of feather waste for keratinase production: An outstanding solution for leather dehairing in tanneries* (Vol. 13, p. 117863612091328). Microbiology Insights.

Akinsemolu, A. A., Onyeaka, H., Odion, S., & Adebanjo, I. (2024). Exploring Bacillus subtilis: Ecology, biotechnological applications, and future prospects. *Journal of Basic Microbiology, 64*, e202300614.

Alam, M. G., Ekhlas Uddin, M., Rahman, S., Ahmad, T., Reazul Karim, M., Mahmood, M. S., et al. (2017). Protease activity of extracellular enzyme produced by *B. Subtilis* isolated from soil. *International Journal of Environment, Agriculture and Biotechnology, 2*(1), 382–388.

Albright, M. B., Gallegos-Graves, L. V., Feeser, K. L., Montoya, K., Emerson, J. B., Shakya, M., & Dunbar, J. (2022). Experimental evidence for the impact of soil viruses on carbon cycling during surface plant litter decomposition. *ISME Communications, 2*(1), 24.

Alipour, H., Raz, A., Zakeri, S., & Djadid, N. D. (2016). Therapeutic applications of collagenase (metalloproteases): A review. *Asian Pacific Journal of Tropical Biomedicine, 6*(11), 975–981.

Allen, A. P., & Gillooly, J. F. (2009). Towards an integration of ecological stoichiometry and the metabolic theory of ecology to better understand nutrient cycling. *Ecology Letters, 12*(5), 369–384. https://doi.org/10.1111/j.1461-0248.2009.01302.x

Aluvihara, S., Weerawardena, T. E., Gunaratne, H. D. A. K., Deniyapahala, D. V. R. I., Dissanayake, S. S., & Gunawardana, S. I. (2023). Advanced review on soil microbiology and fertility of soils. *Mechanization in agriculture & Conserving of the resources, 67*(1), 27–31.

Aransiola, S. A., Joseph, F., Oyedele, O. J., & Maddela, N. R. (2022a). Editorial – Ecological interplays in microbial enzymology. In N. R. Maddela, S. A. Aransiola, & R. Prasad (Eds.), *Ecological interplays in microbial enzymology* (1st ed., pp. 2662–1681). Springer Nature Singapore Pte Ltd. https://www.springer.com/series/16324

Aransiola, S. A., Victor-Ekwebelem, M. O., & Maddela, N. R. (2022b). Microbial enzymes for sustainable development – Future guidelines. In N. R. Maddela, S. A. Aransiola, & R. Prasad (Eds.), *Ecological interplays in microbial enzymology* (1st ed.). Springer Nature Singapore Pte Ltd. ISSN: 2662-1681. https://www.springer.com/series/16324

Aransiola, S. A., Afolabi, F., Joseph, F., & Maddela, N. R. (2023). Soil enzymes: Distribution, interactions and influencing factors. In P. Jeschke & E. B. Starikov (Eds.), *Agricultural biocatalysis: Enzymes in agriculture and industry* (1st ed., pp. 303–333). Jenny Stanford Publishing. ISBN 9789814968478. https://www.routledge.com/Agricultural-Biocatalysis-Enzymes-in-Agriculture-and-Industry/Jeschke-Starikov/p/book/9789814968478

Arora, P. K., Kumar, M., Gajendra, C. A., Raghava, P. S., & Jain, R. K. (2009). OxDBase: A database of oxygenases involved in biodegradation. *BMC Research Notes, 2*, 67. https://doi.org/10.1186/1756-0500-2-67

Arora, N. K., Mishra, J., & Mishra, V. (2020). Microbial enzymes: Roles and applications in industries. In *Microorganisms for sustainability*. Springer.

Ayilara, M. S., & Babalola, O. O. (2023). Bioremediation of environmental wastes: The role of microorganisms. *Frontiers in Agronomy, 5*, 1183691.

Aziz, M. A., Hazra, F., Salma, S., & Nursyamsi, D. (2018). Soil enzyme activities and their relationship to Total soil bacteria, soil microbial biomass and soil chemical characteristics of organic and conventional farming. *Journal of Tropical Soils, 23*(3), 133–141.

Babaniyi, G. G., Olagoke, O. V., & Aransiola, S. A. (2023). Extracellular enzymatic activity of bacteria in aquatic ecosystems. In *Microbiology for cleaner production and environmental sustainability* (pp. 277–300). CRC Press.

Bacimaga, M., Kucharski, J., & Wyszkowska, J. (2015). Microbial and enzymatic activity of soil contaminated with azoxystrobin. *Environmental Monitoring and Assessment, 187*, 615.

Badzinski, C., Ramos, R. F., Godoi, B., & Daroit, D. J. (2021). Soil acidification and impacts over microbial indicators during attenuation of soybean biodiesel (B100) as compared to a diesel-biodiesel blend (B8). *Fuel, 289*, 119989. https://doi.org/10.1016/j.fuel.2020.119989

Baldoni, D. B., Antoniolli, Z. I., Mazutti, M. A., Jacques, R. J. S., Dotto, A. C., de Oliveira Silveira, A., et al. (2020). Chitinase production by Trichoderma koningiopsis UFSMQ40 using solid state fermentation. *Brazilian Journal of Microbiology, 51*, 1897–1908. https://doi.org/10.1007/s42770-020-00334-w

Bandopadhyay, S., & Shade, A. (2024). Soil bacteria and archaea. In *Soil microbiology, ecology and biochemistry* (pp. 41–74). Elsevier.

Banerjee, S., & Van Der Heijden, M. G. (2023). Soil microbiomes and one health. *Nature Reviews Microbiology, 21*(1), 6–20.

Bansal, N., & Kanwar, S. S. (2013). Peroxidase(s) in environment protection. *The Scientific World Journal, 2013*, 1–9. https://doi.org/10.1155/2013/714639

Bansal, K., Raturi, A., Katiyar, U., Mishra, A., & Tewari, S. (2024). Microbial rhizoremediation as a strategy for decontaminating polluted sites and augmenting plant growth. In *Microbiome drivers of ecosystem function* (pp. 181–227). Academic.

Bargaz, A., Lyamlouli, K., Chtouki, M., Zeroual, Y., & Dhiba, D. (2018). Soil microbial resources for improving fertilizers efficiency in an integrated plant nutrient management system. *Frontiers in Microbiology, 9*, 1606.

Belcarz, A., Ginalska, G., Lobarzewski, J., & Penel, C. (2002). The novel non-glycosylated invertase from *Candida utilis* (the properties and the conditions of production and purification). *Biochimica et Biophysica Acta, 1594*(1), 40–53.

Bhandari, S., Poudel, D. K., Marahatha, R., Dawadi, S., Khadayat, K., Phuyal, S., Shrestha, S., Gaire, S., Basnet, K., Khadka, U., & Parajuli, N. (2021). Microbial enzymes used in bioremediation. *Journal of Chemistry, 2021*, 1–17. https://doi.org/10.1155/2021/8849512

Bhandari, Y., Sajwan, H., Pandita, P., & Koteswara Rao, V. (2023). Chloroperoxidase applications in chemical synthesis of industrial relevance. *Biocatalysis and Biotransformation, 41*(6), 403–420.

Bhange, K., Chaturvedi, V., & Bhatt, R. (2016). Feather degradation potential of *Stenotrophomonas maltophia* KB13 and feather protein hydrolysate (FPH) mediated reduction of hexavalent chromium. *Biotech, 6*(1), 42.

Bhardwaj, N., Kumar, B., & Verma, P. (2019). A detailed overview of Xylanases: An emerging biomolecule for current and future prospective. *Bioresources and Bioprocessing, 6*, 40.

Bhardwaj, G., Chauhan, A., Walia, A., & Brar, P. S. (2024). Microbial enzymes for soil health. In *Advancements in microbial biotechnology for soil health* (pp. 97–117). Springer Nature Singapore.

Bhuyan, M. B., Hasanuzzaman, M., Mahmud, J. A., Hossain, M. S., Alam, M. U., & Fujita, M. (2019). Explicating physiological and biochemical responses of wheat cultivars under acidity stress: Insight into the antioxidant defense and glyoxalase systems. *Physiology and Molecular Biology of Plants, 25*, 865–879. https://doi.org/10.1007/s12298-019-00678-0

Bi, L., Yu, D. T., Han, L. L., Du, S., Yuan, C. Y., He, J. Z., & Hu, H. W. (2022). Unravelling the ecological complexity of soil viromes: Challenges and opportunities. *Science of the Total Environment, 812*, 152217.

Binod, P., Gnansounou, E., Sindhu, R., & Pandey, A. (2019). Enzymes for second generation biofuels: Recent developments and future perspectives. *Bioresource Technology Reports, 5*, 317–325.

Brevik, E. C., Slaughter, L., Singh, B. R., Steffan, J. J., Collier, D., Barnhart, P., & Pereira, P. (2020). Soil and human health: Current status and future needs. *Air, Soil and Water Research, 13*, 1178622120934441.

Burns, R. G., DeForest, J. L., Marxsen, J., Sinsabaugh, R. L., Stromberger, M. E., Wallenstein, M. D., et al. (2013). Soil enzymes in a changing environment: Current knowledge and future directions. *Soil Biology and Biochemistry, 58*, 216–234. https://doi.org/10.1016/j.soilbio.2012.11.009

Carroll, R. K., Veillard, F., Gagne, D. T., Lindenmuth, J. M., Poreba, M., Drag, M., Potempa, J., & Shaw, L. N. (2013). The *Staphylococcus aureus* leucine aminopeptidase is localized to the bacterial cytosol and demonstrates a broad substrate range that extends beyond leucine. *Biological Chemistry, 394*, 791–803.

Chandra, P., Enespa, S. R., & Arora, P. K. (2020). Microbial lipases and their industrial applications: A comprehensive review. *Microbial Cell Factories, 19*, 169.

Chauhan, A., & Prabha, V. (2017). Production and partial purification of collagenase from *Bacillus sp.* isolated from soil sample. *International Journal of Advanced Research and Development, 2*(5), 60–65.

Chen, H., Liu, J., Li, D., Xiao, K., & Wang, K. (2019). Controls on soil arylsulfatase activity at a regional scale. *European Journal of Soil Biology, 90*, 9–14.

Chia, X. K., Hadibarata, T., Kristanti, R., Jusoh, M. N. H., Tan, I. S., & Foo, H. C. Y. (2024). The function of microbial enzymes in breaking down soil contaminated with pesticides: A review. *Bioprocess and Biosystems Engineering, 47*, 597–620. https://doi.org/10.1007/s00449-024-02978-6

Coclet, C., Sorensen, P. O., Karaoz, U., Wang, S., Brodie, E. L., Eloe-Fadrosh, E. A., & Roux, S. (2023). Virus diversity and activity is driven by snowmelt and host dynamics in a high-altitude watershed soil ecosystem. *Microbiome, 11*(1), 237.

Costantini, E. A., & Mocali, S. (2022). Soil health, soil genetic horizons and biodiversity. *Journal of Plant Nutrition and Soil Science, 185*(1), 24–34.

Daunoras, J., Kačergius, A., & Gudiukaitė, R. (2024). Role of soil microbiota enzymes in soil health and activity changes depending on climate change and the type of soil ecosystem. *Biology, 13*(2), 85. https://doi.org/10.3390/biology13020085

De Deyn, G. B., & Kooistra, L. (2021). The role of soils in habitat creation, maintenance and restoration. *Philosophical Transactions of the Royal Society B, 376*(1834), 20200170.

de la Horra, A. M., Conti, M. E., & Palma, R. M. (2003). β– Glucosidase and proteases activities as affected by long-term management practices in a typical arigiudoll soil. *Communications in Soil Science and Plant Analysis, 24*, 2395–2404.

Delgado-Baquerizo, M., Grinyer, J., Reich, P. B., & Singh, B. K. (2016). Relative importance of soil properties and microbial community for soil functionality: Insights from a microbial swap experiment. *Functional Ecology, 30*(11), 1862–1873. https://doi.org/10.1111/1365-2435.12674

Ding, J., Jiang, X., Guan, D., Zhao, B., Ma, M., Zhou, B., et al. (2017). Influence of inorganic fertilizer and organic manure application on fungal communities in a long-term field experiment of Chinese Mollisols. *Applied Soil Ecology, 111*, 114–122. https://doi.org/10.1016/j.apsoil.2016.12.003

Durham, D. M., Sieradzki, E. T., Ter Horst, A. M., Santos-Medellín, C., Bess, C. W. A., Geonczy, S. E., & Emerson, J. B. (2022). Substantial differences in soil viral community composition within and among four Northern California habitats. *ISME Communications, 2*(1), 100.

Erdel, E., Şimşek, U., & Kesimci, T. G. (2023). Effects of fungi on soil organic carbon and soil enzyme activity under agricultural and pasture land of Eastern Türkiye. *Sustainability, 15*(3), 1765. https://www.mdpi.com/2071-1050/15/3/1765#

Fan, X., Pan, H., Ping, Y., Jin, G., & Song, F. (2022). The underlying mechanism of soil aggregate stability by fungi and related multiple factor: A review. *Eurasian Soil Science, 55*(2), 242–250.

Fierer, N., Allen, A. S., Schimel, J. P., & Holden, P. A. (2003). Controls on microbial CO2 production: A comparison of surface and subsurface soil horizons. *Global Change Biology, 9*(9), 1322–1332. https://doi.org/10.1046/j.1365-2486.2003.00663.x

French, E., Kaplan, I., Iyer-Pascuzzi, A., Nakatsu, C. H., & Enders, L. (2021). Emerging strategies for precision microbiome management in diverse agroecosystems. *Nature Plants, 7*(3), 256–267.

Garcia-Pichel, F., & Belnap, J. (2021). Cyanobacteria and algae. In *Principles and applications of soil microbiology* (pp. 171–189). Elsevier.

Gautam, M., & Azmi, W. (2017). Screening and isolation of collagenase producing microorganism from proteins waste found in Himalayan Region. *Journal Of Applied Biotechnology Reports, 4*(1), 558–565.

Gorde, N. B., Zade, S. P., Bodke, V. S., & Nakhate, P. S. (2022). Effect of organic farming on enzymatic activity and microbial populations of soybean. *The Pharma Innovation Journal, 11*(12), 2182–2186.

Gordon, R. (2023). When we were triangles: Shape in the origin of life via abiotic, shaped droplets to living, polygonal archaea during the Abiocene. In *Conflicting models for the origin of life* (pp. 213–261). Wiley.

Gossner, M. M., Lewinsohn, T. M., Kahl, T., Grassein, F., Boch, S., Prati, D., et al. (2016). Land-use intensification causes multitrophic homogenization of grassland communities. *Nature, 540*(7632), 266–269. https://doi.org/10.1038/nature20575

Goyal, V., Rani, D., Ritika, Mehrotra, S., Deng, C., & Wang, Y. (2023). Unlocking the potential of nano-enabled precision agriculture for efficient and sustainable farming. *Plants, 12*(21), 3744.

Gupta, R., Gupta, N., Gupta, R., Gupta, N., & Saini, M. (2021). Diversity of prokaryotes. In *Fundamentals of bacterial physiology and metabolism* (pp. 3–42).

Gurung, N., Ray, S., Bose, S., & Rai, V. (2013). A broader view: Microbial enzymes and their relevance in industries, medicine and beyond. *BioMed Research International*. Article ID 329121.

Hartmann, M., & Six, J. (2023). Soil structure and microbiome functions in agroecosystems. *Nature Reviews Earth & Environment, 4*(1), 4–18.

Holík, L., Hlisnikovský, L., Honzík, R., Trögl, J., Burdová, H., & Popelka, J. (2019). Soil microbial communities and enzyme activities after long-term application of inorganic and organic fertilizers at different depths of the soil profile. *Sustainability, 11*(12), 3251. https://doi.org/10.3390/su11123251

Hwang, S. W. M., Hala, H., & Ali, S. M. (2018). Production of cold-active lipase by free and immobilized marine *Bacillus cereus* HSS: Application in wastewater treatment. *Frontiers in Microbiology, 9*, 2377.

Ilesanmi, O. I., Adekunle, A. E., Omolaiye, J. A., Olorode, E. M., & Ogunkanmi, A. L. (2020). Isolation, optimization, and molecular characterization of Lipase producing bacteria from contaminated soil. *Scientific African, 8*, e00279.

Iquebal, M. A., Jagannadham, J., Jaiswal, S., Prabha, R., Rai, A., & Kumar, D. (2022). Potential use of microbial community genomes in various dimensions of agriculture productivity and its management: A review. *Frontiers in Microbiology, 13*, 708335.

Janatiningrum, I., Zahra, A., Fitriyanti, F., & Anggia, V. (2024). Rhizosphere actinobacteria isolated from Pometia pinnata and its antimicrobial activity. *Biodiversitas Journal of Biological Diversity, 25*(3).

Jansson, J. K., McClure, R., & Egbert, R. G. (2023). Soil microbiome engineering for sustainability in a changing environment. *Nature Biotechnology, 41*(12), 1716–1728.

Jat, H. S., Datta, A., Choudhary, M., Sharma, P. C., Dixit, B., & Jat, M. L. (2021). Soil enzymes activity: Effect of climate smart agriculture on rhizosphere and bulk soil under cereal based systems of north-west India. *European Journal of Soil Biology, 103*, 103292.

Jin, K., Sleutel, S., Buchan, D., De Neve, S., Cai, D. X., Gabriels, D., & Jin, J. Y. (2009). Changes of soil enzyme activities under different tillage practices in the Chinese Loess Plateau. *Soil and Tillage Research, 104*(1), 115–120.

Kandeler, E. (2024). Physiological and biochemical methods for studying soil biota and their functions. In *Soil microbiology, ecology and biochemistry* (pp. 193–227). Elsevier.

Katti, A. K. S., AK, S., & Mudgulkar, S. B. (2022). Diversity and classification of rare Actinomycetes. In *Actinobacteria: Ecology, diversity, classification and extensive applications* (pp. 117–142). Springer Nature Singapore.

Kaviya, N., Upadhayay, V. K., Singh, J., Khan, A., Panwar, M., & Singh, A. V. (2019). Role of microorganisms in soil genesis and functions. In *Mycorrhizosphere and pedogenesis* (pp. 25–52). Springer.

Kenarova, A., & Boteva, S. (2023). Fungicides in agriculture and their side effects on soil enzyme activities: A review. *Bulgarian Journal of Agricultural Science, 29*(1), 33–42.

Kertesz, M. A., & Mirlean, P. (2004). The role of soil microbes in plant sulphur nutrition. *Journal of Experimental Botany, 55*(404), 1939–1945.

Khan, A., Panthari, D., Sharma, R. S., Punetha, A., Singh, A. V., & Upadhayay, V. K. (2023). Biofertilizers: A microbial-assisted strategy to improve plant growth and soil health. In *Advanced microbial techniques in agriculture, environment, and health management* (pp. 97–118). Academic.

Knight, T. R., & Dick, R. P. (2004). Differentiating microbial and stabilized β-glucosidase activity relative to soil quality. *Soil Biology and Biochemistry, 36*(12), 2089–2096. https://doi.org/10.1016/j.soilbio.2004.06.007

Kontro, M. H., & Yaradoddi, J. S. (2022). Actinomycetes in environmental applications. In *Actinobacteria: Ecology, diversity, classification and extensive applications* (pp. 247–270). Springer Nature Singapore.

Korsa, G., Konwarh, R., Masi, C., & Haile, S. (2023). Microbial cellulose production and its potential application for textile industries. *Annals of Microbiology, 73*, 13.

Kour, D., Rana, K. L., Yadav, N., Yadav, A. N., Singh, J., Rastegari, A. A., & Saxena, A. K. (2019). Agriculturally and industrially important fungi: Current developments and potential biotechnological applications. In *Recent advancement in white biotechnology through fungi: Volume 2: Perspective for value-added products and environments* (pp. 1–64). Springer.

Kour, D., Kaur, T., Devi, R., Yadav, A., Singh, M., Joshi, D., Singh, J., Suyah, D. C., Kumar, A., Rajput, V. D., Yadav, A. N., Singh, K., Singh, J., Sayyed, R. Z., Arora, N. K., & Saxena, A. K. (2021). Beneficial microbiomes for bioremediation of diverse contaminated environments for environmental sustainability: Present status and future challenges. *Environmental Science and Pollution Research, 28*, 24917–24939.

Kracmarova, M., Kratochvilova, H., Uhlik, O., Strejcek, M., Szakova, J., Cerny, J., et al. (2020). Response of soil microbes and soil enzymatic activity to 20 years of fertilization. *Agronomy, 10*(10), 1542. https://doi.org/10.3390/agronomy10101542

Krolicka, M., Hinz, S. W. A., Koetsier, M. J., Eggink, G., Van Den Broek, L. A. M., & Boeriu, C. G. (2018). β-N-Acetylglucosaminidase MthNAG from Myceliophthorathermophila C1, a thermostable enzyme for production of N-Acetylglucosamine from chitin. *Applied Microbiology and Biotechnology, 102*, 7441–7454.

Kuddus, M. (2019). *Enzymes in food biotechnology. Production, applications and future prospects*. Elsevier Inc.

Kumar, L., & Bharadvaja, N. (2019). Enzymatic bioremediation: A smart tool to fight environmental pollutants. In *Smart bioremediation technologies* (pp. 99–118). Academic.

Kumar, A., & Sharma, S. (2019). Microbes and enzymes in soil health and bioremediation. In *Microorganisms for sustainability*. Springer.

Kumar, A., Kumar, S., Pal, D., & Sharma, J. (2023). Soil microbiota effect for sustainable agriculture. *Integrated Publications TM New Delhi*, 29.

Kunito, T., Kurita, H., Kumori, M., Sakaguchi, K., Nishizawa, S., Fujita, K., Moro, H., Sawada, K., Miyabara, Y., Toda, H., et al. (2022). Microbial synthesis of arylsulfatase depends on the soluble and adsorbed sulfate concentration in soils. *European Journal of Soil Biology, 111*, 103418.

Lacava, P. T., Machado, P. C., & de Andrade, P. H. M. (2021). Phosphate solubilization by endophytes from the tropical plants. *Endophytes: Mineral Nutrient Management, 3*, 207–226. https://doi.org/10.1007/978-3-030-65447-4_9

Le Quéré, C., Andrew, R. M., Friedlingstein, P., Sitch, S., Pongratz, J., Manning, A. C., et al. (2018). Global carbon budget 2017. *Earth System Science Data, 10*(1), 405–448. https://doi.org/10.5194/essd-10-405-2018

Li, W. T., Liu, M., Jiang, C. Y., Wu, M., Chen, X. F., Ma, X. Y., & Li, Z. P. (2017). Changes in soil aggregate-associated enzyme activities and nutrients under long-term chemical fertilizer applications in a phosphorus-limited paddy soil. *Soil Use and Management, 33*(1), 25–33. https://doi.org/10.1111/sum.12322

Li, Y., Nie, C., Liu, Y., Du, W., & He, P. (2019). Soil microbial community composition closely associates with specific enzyme activities and soil carbon chemistry in a long-term nitrogen fertilized grassland. *Science of the Total Environment, 654*, 264–274. https://doi.org/10.1016/j.scitotenv.2018.11.031

Liew, S. M., Tay, S. T., & Puthucheary, S. D. (2013). Enzymatic and molecular characterisation of leucine aminopeptidase of *Burkholderia pseudomallei*. *BMC Microbiology, 13*, 110.

Lima, L. A., Filho, R. F. C., dos Santos, J. G., & Silva, W. C. (2015). Protease with collagenolytic activity produced by *Bacillus sp.* DPUA 1728 from Amazonian soil. *Brazilian Journal of Microbiology, 46*(4), 1217–1223.

Lincoln, L., & More, S. S. (2017). Bacterial invertases: Occurrence, production, biochemical characterization, and significance of Transfructosylation. *Journal Basic Microbiology, 57*, 803–813.

Luang-in, V., Yotchaisara, M., Saengha, W., Undomwing, P., Deeseenthum, S., & Maneewan, K. (2019). Isolation and identification of amylase-producing bacteria from soil in Nasinuan community Forest, Maha Sarakham, Thailand. *Biomedical and Pharmacology Journal, 12*(3), 1061–1068.

Lugo, M. A., & Pagano, M. C. (2022). Overview of the biodiversity, conservation, and sustainable food production with Mycorrhizal fungi in South America. In *Mycorrhizal fungi in South America: Biodiversity, conservation, and sustainable food production* (pp. 1–23). Springer International Publishing.

Luo, L., Meng, H., & Gu, J.-D. (2017). Microbial extracellular enzymes in biogeochemical cycling of ecosystems. *Journal Environmental Management, 197*, 539–549.

Mageswari, A., Subramanian, P., Chandrasekaran, S., Karthikeyan, S., & Gothandam, K. M. (2017). Systematic functional analysis and application of a cold-active serine protease from a novel *Chryseobacterium sp. Food Chemistry, 217*, 18–27.

Makoi, J. H. J. R., & Ndakidemi, P. A. (2008). Selected soil enzymes: Examples of their potential roles in the ecosystem. *African Journal of Biotechnology, 7*(3), 181–191.

Marx, M. C., Kandeler, E., Wood, M., Wermbter, N., & Jarvis, S. C. (2005). Exploring the enzymatic landscape: Distribution and kinetics of hydrolytic enzymes in soil particle-size fractions. *Soil Biology and Biochemistry, 37*(1), 35–48. https://doi.org/10.1016/j.soilbio.2004.05.024

Mason, A. R. G., Salomon, M. J., Lowe, A. J., & Cavagnaro, T. R. (2023). Microbial solutions to soil carbon sequestration. *Journal of Cleaner Production, 417*, 137993.

Mazotto, A. M., De Melo, A. C. N., & Macrae, A. (2011). Biodegradation of feather waste by extracellular keratinases and gelatinases from *Bacillus spp. Word Journal of Microbiology and Biotechnology, 27*(6), 1355–1365.

Meenakshi, S., Hiremath, J., Meenakshi, M. H., & Shivaveerakumar, S. (2024). Actinomycetes: Isolation, cultivation and its active biomolecules. *Journal of Pure & Applied Microbiology, 18*(1), 118–143.

Minhas, B., Katoch, P., Singha, A., Minhas, N., & Kaushik, N. (2024). Microbial metabolites and their role to improve soil health. In *Advancements in microbial biotechnology for soil health* (pp. 67–95). Springer Nature Singapore.

Mir, Y. H., Ganie, M. A., Shah, T. I., Bangroo, S. A., Mir, S. A., Shah, A. M., et al. (2023). Soil microbial and enzyme activities in different land use systems of the Northwestern Himalayas. *PeerJ, 11*, e15993. https://doi.org/10.7717/peerj.15993

Mishra, A., Singh, L., & Singh, D. (2023). Unboxing the black box—One step forward to understand the soil microbiome: A systematic review. *Microbial Ecology, 85*(2), 669–683.

Mobarak-Qamsari, E., Kasra-Kermanshaki, R., & Moosari-nejad, Z. (2011). Isolation and identification of a novel, lipase-producing bacterium, *Pseudomonas aeruginosa* KM110. *Iranian Journal of Microbiology, 3*(2), 92–98.

Muruvanda, B. U., Prakash, A. S., Kalapriya, K., & Sampath, K. (2023). Green enzymes from plants and microbial sources used in industries: An overview. *Rasayan Journal of Chemistry, 16*(1), 483–493.

Nagendran, S., Agrawal, S. S., & Patwardhan, A. G. (2021). Eco-friendly association of plants and actinomycetes. In *Symbiotic soil microorganisms: Biology and applications* (pp. 99–116). Springer.

Nannipieri, P., Giagnoni, L., Landi, L., & Renella, G. (2011). Role of phosphatase enzymes in soil. In *Phosphorus in action: Biological processes in soil phosphorus cycling* (pp. 215–243). https://doi.org/10.1007/978-3-642-15271-9_9

Naqvi Syed, Z. H., Akram, W., Atta, A., & Mahboob, L. (2018). Fungal enzymes and their applications. *International Journal of Advanced Biotechnology and Research, 9*(2), 123–132.

Narayanan, M., Ali, S. S., & El-Sheekh, M. (2023). A comprehensive review on the potential of microbial enzymes in multipollutant bioremediation: Mechanisms, challenges and future prospects. *Journal of Environmental Management, 334*, 117532.

Nayak, S. K., Dash, B., Nayak, S., Mohanty, S., & Mishra, B. B. (2020). Chitinase producing soil bacteria: Prospects and applications. In *Frontiers in soil and environmental microbiology* (pp. 289–298). CRC Press.

Ndabankulu, K., Egbewale, S. O., Tsvuura, Z., & Magadlela, A. (2022). Soil microbes and associated extracellular enzymes largely impact nutrient bioavailability in acidic and nutrient poor grassland ecosystem soils. *Scientific Reports, 12*(1), 12601. https://doi.org/10.1038/s41598-022-16949-y

Neemisha, S., & Sharma, S. (2022). Soil enzymes and their role in nutrient cycling. In *Structure and functions of pedosphere* (pp. 173–188). Springer Nature Singapore.

Neher, D. A. (2023). Moving up within the soil food web: Protists, nematodes, and other microfauna. In *Biological approaches to regenerative soil systems* (pp. 157–168). CRC Press.

Ngamcharungchit, C., Chaimusik, N., Panbangred, W., Euanorasetr, J., & Intra, B. (2023). Bioactive metabolites from terrestrial and marine actinomycetes. *Molecules, 28*(15), 5915.

Novik, G., Savich, V., & Kiseleva, E. (2018). Biodiversity conservation of phages and microbial populations. In *Microbial resource conservation: Conventional to modern approaches* (pp. 261–301). Springer.

Okpara, M. O. (2022). Microbial enzymes and their applications in food industry: A mini-review. *Advances in Enzyme Research, 10*, 23–47.

Olajuyigbe, F. M., & Fatokun, C. O. (2017). Biochemical characterization of an extremely stable pH-versatile laccase from Sporothrix carnis CPF-05. *International Journal of Biological Macromolecules, 94*, 535–543. https://doi.org/10.1016/j.ijbiomac.2016.10.037

Orabi, H. M., El-Fakharany, E. M., Abdelkhalek, E. S., & Sidkey, N. M. (2019). L-Asparaginase and L-Glutaminase: Sources, production and applications in medicine and industry. *Journal of Microbiology, Biotechnology and Food Sciences, 9*(2), 179–190.

Osiebe, O., Adewale, I. O., & Omafuvbe, B. O. (2023). Production and characterization of intracellular Invertase from *Saccharomyces cerevisiae* (OL629078.1), using cassava-soybean as a cost-effective substrate. *Scientific Reports, 13*, 16295.

Pal, G. K., & Suresh, P. V. (2016). Microbial collagenase: Challenges and prospects in production and potential applications in food and nutrition. *RSC Advances, 6*, 33763–33780.

Partoazar, M., Hoodaji, M., & Tahmaourespour, A. (2011). The effect of glyphosate application on soil microbial activities in agricultural land. *African Journal of Biotechnology, 10*(83), 19419–19424.

Parveen, A., Bhatnagar, P., Bisht, B., Kumar, S., Joshi, S., Gautam, P., Kumar, V., Adarchenko, I., & Anna, K. (2023). Algal nanobionics to enhance value added products–A review. *Current Research in Biotechnology, 5*, 100124.

Piotrowska-Długosz, A., Długosz, J., Frąc, M., Gryta, A., & Breza-Boruta, B. (2022). Enzymatic activity and functional diversity of soil microorganisms along the soil profile–A matter of soil depth and soil-forming processes. *Geoderma, 416*, 115779. https://doi.org/10.1016/j.geoderma.2022.115779

Ramakrishnan, B., Maddela, N. R., Venkateswarlu, K., & Megharaj, M. (2023). Potential of microalgae and cyanobacteria to improve soil health and agricultural productivity: A critical view. *Environmental Science: Advances, 2*(4), 586–611.

Ramudu, A. C., Mohiddin, G. J., Srinivasulu, M., Madakka, M., & Rangaswamy, V. (2011). *Impact of fungicides chlorothalonil and propiconazole on microbial activities in groundnut (Arachis hypogaea L.) soils* (pp. 1–7). ISRN Microbiology.

Ranjith, M., Sridevi, S., Venkata, R. M., & Chandrasekhar, R. P. (2015). Soil enzyme activity, soil microbial communities, microbial biomass Arbon changes and seed cotton yield under different nutrient management practices in Bt cotton. *The Bioscan, 10*(4), 1499–1504.

Reavy, B., Swanson, M. M., & Taliansky, M. (2014). Viruses in soil. In *Interactions in soil: Promoting plant growth* (pp. 163–180). Springer.

Reid, T. W., & Wilson, I. B. (1971). *E. coli* alkaline phosphatase. In P. D. Boyer (Ed.), *The enzymes* (Vol. 4). Academic Press.

Robertson, G. P., & Groffman, P. M. (2024). Nitrogen transformations. In *Soil microbiology, ecology and biochemistry* (pp. 407–438). Elsevier.

Ryan, S. M., Fitzgerald, G. F., & van Sindeven, D. (2006). Screening for and identification of starch-, amylopectin-, and pullulan-degrading activities in bifidobacterial strains. *Applied and Environmental Microbiology, 27*(8), 5289–5296.

Sabu, A. (2003). Sources, properties and applications of microbial therapeutic enzymes. *Indian Journal of Biotechnology, 2*, 334–341.

Sadhu, S., & Maiti, T. K. (2013). Cellulase production by bacteria: A review. *British Microbiology Research Journal, 3*(3), 235–258.

Sağlam, N. G., Rachid, N. A., & Güngör, N. D. (2024). Secondary metabolites and biological compounds of actinomycetes and their applications. In *Bacterial secondary metabolites* (pp. 123–145). Elsevier.

Sanders, W. B., & Masumoto, H. (2021). Lichen algae: The photosynthetic partners in lichen symbioses. *The Lichenologist, 53*(5), 347–393.

Santos, M. S., Villela, E. S. D., Pazianotto, R. A. A., Reynaldo, E. F., Silva, E. H. F. M., Bueno, A. C. S. O., & May, A. (2018). Valuation of microbial and enzymatic communities in soil and Rizosphere from soybean plants. *African Crop Science Journal, 26*(1), 79–92.

Shahi, K. A. B., Kavusi, E., Dehghanian, Z., Pandey, J., Asgari Lajayer, B., Price, G. W., & Astatkie, T. (2023). Removal of organic and inorganic contaminants from the air, soil, and water by algae. *Environmental Science and Pollution Research, 30*(55), 116538–116566.

Shanthi, V. (2021). Actinomycetes: Implications and prospects in sustainable agriculture. In *Biofertilizers: Study and impact* (pp. 335–370). Wiley.

Sherene, T. (2017). Role of soil enzymes in nutrient transformation: A review. *Bio Bulletin, 3*(1), 109–131.

Shukla, R., Soni, J., Kumar, A., & Pandey, R. (2024). Uncovering the diversity of pathogenic invaders: Insights into protozoa, fungi, and worm infections. *Frontiers in Microbiology, 15*, 1374438.

Singh, V. (2024). Biogeochemical cycles. In *Textbook of environment and ecology* (pp. 75–94). Springer.

Singh, P., & Kumari, P. (2016). Isolation and characterization of amylase producing *Bacillus sp.* from selected soil sample. *International Journal for Research in Bioscience, 2*, 24–29.

Singh, R., Langyan, S., Rohtagi, B., Darjee, S., Khandelwal, A., Shrivastava, M., Kothari, R., Mohan, H., Raina, S., Kaur, J., & Singh, A. (2022). Production of biofuels options by contribution of effective and suitable enzymes: Technological developments and challenges. *Materials Science for Energy Technologies, 5*, 294–310.

Sinsabaugh, R. L., Lauber, C. L., Weintraub, M. N., Ahmed, B., Allison, S. D., Crenshaw, C., & Zeglin, L. H. (2008). Stoichiometry of soil enzyme activity at global scale. *Ecology Letters, 11*(11), 1252–1264. https://doi.org/10.1111/j.1461-0248.2008.01245.x

Sinsabaugh, R. L., Hill, B. H., & Follstad Shah, J. J. (2009). Ecoenzymatic stoichiometry of microbial organic nutrient acquisition in soil and sediment. *Nature, 462*(7274), 795–798. https://doi.org/10.1038/nature08632

Sobucki, L., Ramos, R. F., Meireles, L. A., Antoniolli, Z. I., & Jacques, R. J. S. (2021). Contribution of enzymes to soil quality and the evolution of research in Brazil. *Revista Brasileira de Ciência do Solo, 45*, e0210109. https://doi.org/10.36783/18069657rbcs20210109

Song, D., Chen, L., Zhang, S., Zheng, Q., Ullah, S., Zhou, W., & Wang, X. (2020). Combined biochar and nitrogen fertilizer change soil enzyme and microbial activities in a 2-year field trial. *European Journal of Soil Biology, 99*, 103212. https://doi.org/10.1016/j.ejsobi.2020.103212

Stamenov, D., Đurić, S., Jafari, T. H., Ćirić, V., & Manojlović, M. (2018). Microbiological activity in the soil of various agricultural crops in organic production. *Contemporary Agriculture, 67*(1), 34–39. https://doi.org/10.2478/contagri-2018-0005

Stone, M. M., & Plante, A. F. (2014). Changes in phosphatase kinetics with soil depth across a variable tropical landscape. *Soil Biology and Biochemistry, 71*, 61–67.

Stone, M. M., DeForest, J. L., & Plante, A. F. (2014). Changes in extracellular enzyme activity and microbial community structure with soil depth at the Luquillo Critical Zone Observatory. *Soil Biology and Biochemistry, 75*, 237–247. https://doi.org/10.1016/j.soilbio.2014.04.017

Sun, T., Miao, J., Saleem, M., Zhang, H., Yang, Y., & Zhang, Q. (2020). Bacterial compatibility and immobilization with biochar improved tebuconazole degradation, soil microbiome composition and functioning. *Journal of Hazardous Materials, 398*, 122941.

Surendran, U., Jayakumar, M., Raja, P., Gopinath, G., & Chellam, P. V. (2023). Microplastics in terrestrial ecosystem: Sources and migration in soil environment. *Chemosphere, 318*, 137946.

Taylor, D. L., & Bhatnagar, J. M. (2024). Fungi in soil: A rich community with diverse functions. In *Soil microbiology, ecology and biochemistry* (pp. 75–129). Elsevier.

Tiemann, L. K., & Billings, S. A. (2011). Indirect effects of nitrogen amendments on organic substrate quality increase enzymatic activity driving decomposition in a Mesic grassland. *Ecosystems, 14*, 234–247. https://doi.org/10.1007/s10021-010-9406-6

Timofeeva, A. M., Galyamova, M. R., & Sedykh, S. E. (2023). Plant growth-promoting bacteria of soil: Designing of consortia beneficial for crop production. *Microorganisms, 11*(12), 2864.

Turner, B. L. (2010). Variation in pH optima of hydrolytic enzyme activities in tropical rain forest soils. *Applied and Environmental Microbiology, 76*(19), 6485–6493. https://doi.org/10.1128/AEM.00560-10

Turner, B. L., Hopkins, D. W., Haygarth, P. M., & Ostle, N. J. (2002). β–Glucosidase activity in pasture soils. *Applied Soil Ecology, 20*, 157–162.

Ullah, S., Raza, M. M., Abbas, T., Guan, X., Zhou, W., & He, P. (2023). Responses of soil microbial communities and enzyme activities under nitrogen addition in fluvo-aquic and black soil of North China. *Frontiers in Microbiology, 14*, 1249471. https://doi.org/10.3389/fmicb.2023.1249471

Utobo, E. B., & Tewari, L. (2015). Soil enzymes as bioindicators of soil ecosystem status. *Applied Ecology and Environmental Research, 13*(1), 147–169.

Van Leeuwen, J. P., Djukic, I., Bloem, J., Lehtinen, T., Hemerik, L., De Ruiter, P. C., & Lair, G. J. (2017). Effects of land use on soil microbial biomass, activity and community structure at different soil depths in the Danube floodplain. *European Journal of Soil Biology, 79*, 14–20. https://doi.org/10.1016/j.ejsobi.2017.02.001

Walia, A., Mehta, P., Guleria, S., Chauhan, A., & Shirkot, C. K. (2014). Impact of fungicide mancozeb at different application rates on soil microbial population, soil biological processes and enzyme activities in soil. *The Scientific World Journal, 2014*, 702909.

Walsh, G. A., Power, R. F., & Headon, D. R. (1993). Enzymes in animal feed industry. *Trends in Biotechnology, 11*, 424–430.

Wanderley, M. C. A., Duarte Neto, J. M. W., Lima, C. A., Silverio, S. I., Filho, J. L., Teixeira, J. A. C., & Porto, A. L. F. (2016). Production and characterization of collagenase by *Penicillium sp.* UCP 1286 isolated from Caatinga soil. *Journal of Applied Biology and Biotechnology, 4*(4). https://doi.org/10.7324/JABB.2016.40401

Wang, R. Z., Dorodnikov, M., Yang, S., Zhang, Y. Y., Filley, T. R., Turco, R. F., Zhang, Y. G., Xu, Z. W., Li, H., & Jiang, Y. (2015). Responses of enzymatic activities within soil aggregates to 9-year nitrogen and water addition in a semi-arid grassland. *Soil Biology and Biochemistry, 81*, 159–167. https://doi.org/10.1016/j.soilbio.2014.11.015

Waring, B. G., Weintraub, S. R., & Sinsabaugh, R. L. (2014). Ecoenzymatic stoichiometry of microbial nutrient acquisition in tropical soils. *Biogeochemistry, 117*, 101–113. https://doi.org/10.1007/s10533-013-9849-x

Wehr, J., & van Vuuren, S. J. (2024). Algae and cyanobacteria communities. In *Wetzel's limnology* (pp. 463–510). Academic.

Wei, X., Zhu, Z., Wei, L., Wu, J., & Ge, T. (2019). Biogeochemical cycles of key elements in the paddy-rice rhizosphere: Microbial mechanisms and coupling processes. *Rhizosphere, 10*, 100145. https://doi.org/10.1016/j.rhisph.2019.100145

Wendel, A. S., Bauke, S. L., Amelung, W., & Knief, C. (2022). Root-rhizosphere-soil interactions in biopores. *Plant and Soil, 475*(1), 253–277.

Willerding, A., de Oliveira, L. A., Moreira, F. W., Germano, M. G., & Chagas, A. F., Jr. (2011). *Lipase activity among bacteria isolated from Amazonian soils*. Enzyme Research. Article ID 720194.

Williams, A. N., & Woessner, K. M. (2009). Monosodium glutamate 'allergy': Menace or myth? *Clinical and Experimental Allergy, 39*(5), 640–646.

Yang, Z., Liu, S., Zheng, D., & Feng, S. (2006). Effects of cadmium, zinc and lead on soil enzyme activities. *Journal of Environmental Sciences, 18*, 1135–1141.

Yassin, S. N., Jiru, T. M., & Indracanti, M. (2021). Screening and characterization of thermostable amylase-producing bacteria isolated from soil samples of Afdera, Afar region, and molecular detection of amylase-coding gene. *International Journal of Microbiology*. Article ID5592885.

Zablocki, O., Adriaenssens, E. M., & Cowan, D. (2016). Diversity and ecology of viruses in hyperarid desert soils. *Applied and Environmental Microbiology, 82*(3), 770–777.

Zhang, L., Chen, X., Xu, Y., Jin, M., Ye, X., Gao, H., et al. (2020). Soil labile organic carbon fractions and soil enzyme activities after 10 years of continuous fertilization and wheat residue incorporation. *Scientific Reports, 10*(1), 11318. https://doi.org/10.1038/s41598-020-68163-3

Zhang, M., Huang, W., Zhang, L., Feng, Z., Zuo, Y., Xie, Z., & Xing, W. (2024). Nitrite-dependent anaerobic methane oxidation (N-DAMO) in global aquatic environments: A review. *Science of the Total Environment, 921*, 171081.

Zhu, Y. L., Huang, Y. J., Nuerhamanti, N., Bai, X. Y., Wang, H. N., Zhu, X. Y., & Zhang, W. (2023). Composition and distribution characteristics of rhizosphere bacterial community of Ammodendron bifolium growing in Takeermohuer desert are different from those in non-rhizosphere. *Microbial Ecology, 86*(4), 2461–2476.

Zungu, N. S., Egbewale, S. O., Olaniran, A. O., & Perez-Fernandez, M. (2020). Soil nutrition, microbial composition and associated soil enzyme activities in Kwazulu-Natal grasslands and Savannah ecosystems soils. *Applied Soil Ecology, 155*, 103663.

Zwair, H. (2023). Protista: Protozoa, characteristics. *Scientific Signification. Academia Repository, 4*(09), 67–79.

Characterization of Microbial Communities from the Rhizosphere of the Royal Palm (*Roystonea oleracea*) in a RAMSAR Wetland in Ecuador

Jean Carlo Andrade, Santiago Mafla, Emily Avila, and José Herándes

Abstract The interaction between the rhizosphere and the microbial communities that develop in this area is essential to understanding plant-microorganism processes such as nutrient availability or the development and suppression of plant diseases. Sequencing approaches have provided relevant insights into the microbiomes of plant rhizospheres such as the ecological interactions that are essential for the regulation of biogeochemical cycles inside wetlands. This research characterized some microbial communities present around an area close to the rhizosphere of the *Roystonea oleracea* palm. In this sense, ions present in four soil samples (M1, M2, M3 & M4) were quantified by spectrophotometry techniques to get an approximation of soil nutrients variation in this area. Later, the metabolic diversity of the bacterial communities cultivated from the samples was carried out using the BIOLOG EcoPlate™ kit. In addition, some isolated strains were analyzed through DGGE analysis of the 16 s rDNA gene amplified by PCR, founding that the most representative genera of the microorganisms obtained were *Bacillus* and *Bovista*. The results of a multidimensional scaling analysis of the banding pattern showed that M1 and M4 are 80% similar to each other. Finally, an antagonism test was performed between the microbial communities resulting in no antagonistic relationships observed.

Keywords Rhizosphere · Wetland · RAMSAR · Microbial communities

J. C. Andrade (✉)
Universidad Tecnológica Indoamérica, Quito, Ecuador
e-mail: jeanandrade@uti.edu.ec

S. Mafla
Universidad Católica del Ecuador Ibarra, Ibarra, Ecuador

E. Avila
Independent Researcher, Guayaquil, Ecuador

J. Herándes
Universidad Tecnológica ECOTEC, Guayaquil, Ecuador

Universidad Central de Venezuela, Caracas, Venezuela

S. A. Aransiola et al. (eds.), *Soil Microbiome in Green Technology Sustainability*,
https://doi.org/10.1007/978-3-031-71844-1_19

1 Introduction

Kadykalo and Findlay (2016) describe wetlands as multifunctional systems that serve as nature-based solutions, acting as transition zones between terrestrial and aquatic environments. They enhance groundwater recharge, retain nutrients and chemicals, and influence flow regimes, particularly in flood attenuation and low flow augmentation by controlling erosion and sedimentation (Newall & Fisk, 2023). Wetlands also modulate peak flows by storing runoff and releasing it slowly, benefiting soil moisture and overall water balance through evapotranspiration, infiltration, and groundwater recharge (Abarca, 2001). These ecosystems vary along a hydrologic continuum from low to high hydrologic throughput, connecting with rivers, estuaries, or lakes (Dodds & Whiles, 2020). They are highly productive ecosystems, contributing to biodiversity conservation because their basins and their high seasonal water level variations create diverse niches that supports dynamic and diverse communities of plants and animals (Mitsch et al., 2015). Additionally, they play a crucial role in the hydrological cycle by storing and releasing water, regulating nutrient cycling, and storing carbon. Additionally, wetlands help mitigate climate change by retaining atmospheric carbon (Finlayson & Dam, 2023). Peatlands, covering just 3–4% of the Earth's surface, store approximately 540 gigatons of carbon. This accounts for 1.5% of the global carbon storage and 25–30% of the carbon contained in terrestrial vegetation and soils. At the same time, peatlands and wet grasslands in river basins act as natural sponges, absorbing rainfall, creating surface pools, and reducing flooding of streams and rivers (Fennessy et al., 2023). This same storage capacity also protects against drought (Romero et al., 2018; Yepes, 2015).

Wetlands offer prime examples of plant-microorganism interactions, particularly through the rhizosphere. Kertesz and Frossard (2015) emphasize the significance of the rhizosphere in understanding these interactions. The rhizosphere, the biologically active zone surrounding plant roots, facilitates interactions between plants and various microorganisms depending on factors like arrangement, intensity, concentration, and characteristics. High microbial activity in the rhizosphere is fueled by substances exuded by plant roots, such as carbohydrates, derived from photosynthesis. These substances stimulate microbial populations and serve as nutrients or energy sources for microorganisms (Rivero et al., 2009). Ecuador has been a member of the Convention since 1990 and has 18 sites designated as Wetlands of International Importance (Ramsar Sites), with an area of 1.064,483 hectares (Cunalata García & López Pumalema, 2020).

2 Characterization of Microbial Communities from the Rhizosphere of the Royal Palm (*Roystonea oleracea*) in a Ramsar Wetland in Ecuador

Invasive exotic plants pose a significant threat to global biodiversity conservation, serving as drivers of ecosystem change. These species, settling beyond their natural distribution areas, disrupt local environments and native communities through processes like hybridization and antagonism. The *Roystonea oleracea* is classified as an invasive exotic species (Herrera et al., 2017). In Ecuador, the presence of the palm has been recently detected on Santay Island (Bermeo et al., 2022).

The objective in this investigation was to explore the microbial communities present in the Santay wetland in Ecuador, with particular emphasis on their functional attributes in association with ions in the soil. Additionally, the physical and chemical characteristics of the soil where *Roystonea oleracea* was found were studied. The microorganisms present in the rhizosphere where isolated to later carry out antagonism tests and analyze the interaction between species. Ultimately, this chapter explains the soil and microbial ecology where *Roystonea oleracea* is found.

Santay Island is a RAMSAR site where the generally flat land areas are permanently or intermittently flooded and regularly covered with water, where the soil is saturated and devoid of oxygen, resulting in a hybrid ecosystem between purely aquatic and terrestrial (Mendoza, 2008). For this reason, its water table is almost superficial due to the accumulation of groundwater that is found at a relatively shallow depth below ground level. In that sense, when digging between 40 cm to 60 cm deep in the soil of Santay we can observe the presence of water. On the other hand, according to the morphology of the palms, their roots do not have mechanisms to tolerate anoxia, which is the total absence of dissolved oxygen in the water (Roblero et al., 2020), thus it is explained why the root grow shallowly at the phreatic level and not at depth due to the development and structure of root systems are generally controlled by genetic characters and influenced by the environment around them (Malekpoor Mansoorkhani et al., 2014; Młodzińska-Michta, 2023).

2.1 Methodology

A total of four (4) soil samples were collected inside Santay Island located at the following UTM coordinates: Latitude 6°27′45.1″ E/ Longitude: 97°56′04.5" N. This area was selected due to the high abundance of the *Roystonea oleracea* palms around Santay Island.

The samples were taken at depths of twenty centimeters (20 cm) after removing vegetable material, due to the roots intolerance to anoxia as it was explained in the previous section, however, it is important to keep in mind that palms respond favorably to this ecosystem because they have an adventitious root system, that is, they are plant roots that form from any non-root tissue and are produced both during normal development and in response to stress conditions such as flooding, nutrient deprivation and wounding (Steffens & Rasmussen, 2016). The samples were collected at distances of 5 cm (M1), 50 cm (M2), and 100 cm (M3). The last sample (M4) was collected in an area away from the presence of *Roystonea oleracea*.

A total of 2 kg was collected in the extraction process, all the samples were packed inside Ziploc® bags previously disinfected and hermetically sealed, labeled and finally transported in a conditionate cooler at 4 °C until they were taken and stored in the laboratories to perform preliminary analysis.

2.2 Measurement of Ions in Soil by Colorimetry

For this process, 15 g of each soil sample was mixed with 100 ml of sterilized distilled purified water and left to filter for 24 hours. The quantification was performed by colorimetry, using the specific kit for each element. The readings were obtained with a photometric measurement range ± 3.0 Abs and a wavelength range of 400 nm and 900 nm.

In the case of Iron, Phosphate and Ammonium, a NOVA 60 A Spectroquant© photometer was used to determine the concentration in mg/L, while a La Motte® colorimeter was used to measure Nitrates, Nitrites and Potassium. All the Merck protocols for the spectrophotometer were followed as depicted by the manufacturer instructions and can be found online on the official website of Merck. The rest of the protocols followed by La Motte® were depicted inside the user's manual which also is available online. For pH measurement, the potentiometer used was a LAQUA—Horiba Scientific model F-71.

2.3 Characterization of the Physical and Chemical Properties of the Soil

2.3.1 Moisture

Initially, the soil samples were weighed and later placed in an oven for 24 hours at 105 °C. The difference between the initial weight and the final weight after drying was calculated to obtain the percentage of moisture loss following the next formula:

$$M\% = \frac{(P1 - P2)}{P1} \times 100 \tag{1}$$

$$M\% = \%$$

Where:

- P1: wet weight (105 °C)
- P2: dried weight after desiccator (2 hours)

Also, the porcelain capsules used were calibrated in the oven at the same temperature.

2.3.2 Texture

It was used the gravimetric method to separate the constituent particles of the soil aggregates according to their sizes by weighting 25 g of soil from each sample and leaving them to dry at 25 °C for 48 hours. Once this process was completed, the samples were placed in a mortar, homogenizing, and crushing until the particle size was reduced. Subsequently, the homogenized sample was placed in a mechanical shaker in 40 mm, 60 mm and 100 mm sieves, resulting in the separation of the gravel, sand, silt, and clay coatings.

2.3.3 Organic Matter

The quantification of organic matter (OM) was carried out by the calcination method. The calculation of OM was made by weight difference at different temperatures, using the following formula:

$$OM\% = \frac{(P1 - P2) * 100}{P2} \tag{2}$$

$$OM\% = \%$$

Where:

- P1: weight before calcination (105 °C)
- P2: weight after calcination (360 °C)

The difference in weight obtained was equivalent to the amount of organic matter that was lost from the sample due to calcination.

2.4 DNA Purification and Extraction

The PureLink® Genomic kits for DNA purification and extraction from Invitrogen™ were used in the process of extraction of bacterial DNA from the isolates cultured in Nutrient Broth, following the protocols of the manufacturer.

2.4.1 Polymerase Chain Reaction (PCR) and DNA Amplification

The PCR of the 16 s rDNA sequence was performed using the reagents specified in Table 1.

The primers used for DNA amplification are detailed in Table 2.

DNA amplification was performed using a Labnet Multigene Optimax thermal cycler. The basic amplification protocol in the thermal cycler consisted of an initial denaturation at 94 °C for 5 min; then 35 cycles of denaturation at 94 °C for 30 s, followed by alignment at 55.6 °C for 45 s and elongation at 72 °C for 90 s. Subsequently, another denaturation process was performed at 94 °C for 30 s followed by an alignment at 55.6 °C for 45 s. As a last step a final extension was performed at 72 °C for 90 s.

The samples (M1, M2, M3 & M4) and four isolates (C1, C2, C3, C4) were sent for sequencing to Macrogen® and the sequences obtained were subsequently uploaded to the NCBI (National Center for Biotechnology Information) website. The reagents used as a denaturant stock solution are described in Table 3.

2.4.2 Denaturing Gradient Gel Electrophoresis (DGGE)

The samples were characterized by DGGE with a GBS Scientifc Equipement© dual cassette Cipher. The DGGE was performed with a denaturing gradient of 10 to 60% for band extraction. For the acrylamide polymerization process, a solution of 50 mg of ammonium persulfate (Promega©) was prepared in 500 μL of sterile Mili-Q water, of which 130 μL were used along with 13 μL N′-Tetramethylethylenediamine [TEMED]. Next, 15 μL of the PCR products of the primers (907r-P3f) were loaded

Table 1 Reagents used for PCR

Reagent	Quantity(μL)
Sterile distilled water	4.5
Primers	1
MM dNTPs (deoxynucleotide triphosphates)	12.5
DNA	5

Table 2 Partition sequence and expected size of the 16S rDNA gene amplification products

Reagent	Quantity(μL)	Expected size
907r	5′-AGT TTG ATC MTG GCT CAG-3’	926 bp
P3	5’-CCG TCA ATT CMT TTG AGT TT-3	

Table 3 Reagents of the denaturant stock solution (7.5% gel)

Percentage	10%	50%
40% acrylamide/Bis (ml)	18.8	18.8
50% TAE buffer (ml)	2	2
Formamide (ml)	4	20
Urea (g)	4.2	21

onto the gel, after being previously mixed with 3 μL of loading buffer (6X Load Dye Promega©) in a 3:1 ratio. Finally, the electrophoresis was carried out under a 85 V and 500A for 660 minutes at 60 °C.

2.4.3 Identification of Microorganisms in the Rhizosphere by Sanger PCR Method

For all microbiological analyses, asepsis procedures were followed to prevent contamination of samples. As a culture medium it was used solid Agar and nutrient media. Solid media was distributed among Petri dishes and the liquid media was distributed in test tubes. It was used Himedia® Nutrient Broth and Seaweed Solution Laboratories® Nutrient Agar sterilized by autoclave.

For this and the following sections the words "sterile or sterilized" shall be used interchangeably for all substances, laboratory instruments, or similar materials that have been autoclaved at 121 °C for 15 minutes and 1.2 atmospheres of pressure.

2.5 Metabolic Characterization by Ecoplate™

Ecoplate microplates assist to acquire the physiological profiles at the community level. For carrying out this test, we used serial dilutions. The plates contain a total of 96 wells each, of which, they are divided into 3 replicates of 31 carbon sources, whose categories are 6 groups: carboxylic acids, complex carbon sources, carbohydrates, carbon phosphates, amino acids, amines (Chazarenc et al., 2010).

A spectrophotometer for microplate Epoch® with monochromatic-based absorbance and with a wavelength of 200 nm to 999 nm for UV-Vis applications was used to perform the measurements. The results were analyzed with the Gen 5® software (Microplate Reader and Imager Software). These results were obtained in a period of 5 days recorded at 24 hours, 48 hours, 72 hours, and 120 hours to later analyze the Average Well Color Development (AWCD) index, which shows the overall potential metabolic activity of the microbial community based on the patterns of substrates use.

2.6 Antagonism Tests

The antagonism tests were carried out with the cultivated strains. It was used the cross-streak method described by Tóth et al. (2013), which emphasizes in the detection of antagonism between microbes using cross-streak experiments and measuring the antimicrobial effect by observing the inhibition distance.

3 Results and Discussion

3.1 *Measurement of Ions by Colorimetry*

The results of colorimetry measurements are presented in Table 4 below. In a general view, sample M2 had more iron ions than the other samples, being M3 the sample with the lowest value (0,26). Moreover, the results of the potassium ion in the M3 sample have the highest value (7,40). About the phosphate ion we can see that M2 obtained the highest result with a value of 6,08. Considering that the area of study in Santay Island is Ramsar wetland, Sutton-Grier et al. (2011) explained that to understand the role of wetlands, it is necessary to understand the differences in the availability of the influences of electron donors and electron final acceptors on anaerobic mineralization of soil organic matter. In their study with wetland soils, they examined how organic carbon mineralization rates changed resulting in sulfate predominating together with this process. These results indicated that electron donors and acceptors play an important role in the regulation of organic mineralization by anaerobic microorganisms. This provides an approximation to analyze M1, as a site where there could be a higher rate of mineralization. On the other side, Gu et al. (2021) indicate that there is a marked effect of climate on the availability of phosphorus in the soil, with the wetter areas presenting the greatest problems of phosphorus deficiency, thus, this could result in an interesting approach to further analyze phosphate in Santay at different locations. Organic phosphorus in the soil comes mainly from animal and plant residues, and acts as fertilizer when broken down by microorganisms. It can even become insoluble when microorganisms release phosphates from the compounds they've transformed (Acevedo et al., 2014). Thus, it's likely that the area near the palm had a high organic matter content, primarily due to the surrounding plant residue (Yarwood, 2018). In this sense, is also possible to consider that phosphate levels, decrease as the sampling distance increases. However, this is not a definitive conclusion about this relationship.

On the other side, excessive levels of Potassium (K) can have adverse impacts on both plants and the environment. Elevated concentrations of this nutrient in the soil can hinder seed germination and plant growth due to the saline effect. Additionally, it can lead to antagonism in magnesium absorption, heightened leaching losses, and constraints on enzymatic activity linked to nitrogen absorption through the plant-microorganism relationship. However, potassium gain in soil can also occur by

Table 4 Ions measured by colorimetry

Samples	pH	Phosphates (ppm)	Potassium (ppm)	Iron (ppm)
Sample 1 (M1)	6979	3,80	>0,10	4,21
Sample 2 (M2)	7237	6,08	0,00	>5
Sample 3 (M3)	7367	<1,0	7,40	0,26
Sample 4 (M4)	7332	1,93	0,00	0,56

weathering, fertilization or by biological action, for example, in the release of ions from microorganisms (Meneses et al., 2017). The registered values for K did not show a clear understanding of the ion. As can be seen in Table 4, values in M2 and M4 could not be noted. This leads to think that the concentration of the ion was too high for the detection range of the used equipment, however it can be seen also as a malfunctioning. Thus, conclusions are not clear about the ion.

The allocation of iron by biological means has been recognized as one of the oldest forms of microbial metabolism on Earth. Moreover, the diverse metabolic capabilities of the microorganisms engaged in these processes have led to the creation of biotechnological approaches aimed at remediating polluted environments and harnessing energy (Weber et al., 2006a). Another study by Weber et al. (2006b) depicts that microbially catalyzed nitrate-dependent Fe (II) oxidation has the potential to contribute to a dynamic anaerobic Fe redox cycle in freshwater sediments. Similar results were described by Sutariya et al. (2024) as they tested the impact of iron (Fe) nanoparticles on paddy rhizospheric and rhizoplanar microbes. This can be slightly related to the results observed for the iron ion, which showed a higher concentration in M1 than the other sampled sites.

The total nitrogen corresponds to the ratio between the sum of all forms of nitrogen present in the soil. The Table 5 indicates that in M1 (8,17) and M4 (8,96) there is a higher concentration, while M3 (1,26) showed the lowest value. All organic forms are transient in the soil, therefore the amounts of inorganic N in the soil are extremely variable (Crumpton et al., 2020).

Wetland microbes are crucial to the biogeochemical cycles of nutrients (Islam et al., 2020). The study of Zhang et al. (2020) used a similar approach to determine seasonal changes by testing pH, carbon, total nitrogen, and the microbial diversity, richness and structure. They found that these are key factors structuring the microbial communities from wetland soil samples obtained from the Suaeda wetland in China. Another study found that some microorganisms might allocate their resources to enzymes synthesis in response to the N supply rate rather than the size of the easily available N pools. Hence, it can be said that the rhizosphere is a critical interface that supports the exchange of resources between plants and their soil-associated environment, therefore the microbial diversity of the rhizosphere is influenced by the physical and chemical properties of this area, some of which are determined by the host plant genetics (Fujita et al., 2018).

Table 5 Total Nitrogen measured by spectrophotometry

Samples	Ammonium (ppm)	Nitrate (ppm)	Nitrites (ppm)	Total Nitrogen (ppm)
Sample 1 (M1)	1,18	6,59	0,40	8,17
Sample 2 (M2)	1,56	>3	1,04	5,6
Sample 3 (M3)	0,44	0,62	0,20	1,26
Sample 4 (M4)	0,19	7,97	>0,80	8,96

3.2 Soil Physical Properties

The analysis of the physical properties of the soil was performed by evaluating texture, moisture content, and organic matter (OM).

In this sense, a composite sample was made with M1, M2, and M3. The result was that 37% of soil moisture was contained in the composite sample, where the initial weight of the sample (10,298 g) and the final weight of the sample after calcination (6488 g) were calculated the Eq. (1).

$$MC\% = \frac{(10{,}298 - 6{,}488)}{10{,}298} \times 100$$

$$MC\% = 37\%$$

For OM, the sieved soil fraction from 2 mm was placed and weighted (23,9934 g). Then, the capsule was removed and placed in the desiccator to be reweighed, obtaining a value of 23,3571 g after calcining. The difference in weight between the measurements before and after calcination was calculated to determine the amount of organic matter that was lost from due to calcination.

$$O_m = \frac{(23,9934 - 23,3571)}{23,3571} \times 100$$

$$O_m = 2,72\%$$

The solution for the equation (Eq. 2) results into a total of 2.72% organic matter calculated.

On the other side, the weight of each soil fraction obtained by gravimetry can be seen in the Table 6.

Table 6 represents the proportions of soil texture, with 44,62% gravels and sands, 9,92% silts, and 45,46% of clays. This means that this is a sandy clay type soil according to the soil texture triangle (Aarthi & Sivakumar, 2020).

Surface soils are the most fertile and have the highest levels of organic matter. Contrasting the results with the investigation of Castiblanco Duran and Zapata Herazo (2020) were they found silt loam, sandy loam and silty; a higher percentage of organic matter was obtained. This can be linked to the fact of wetlands stational variability and to the amount of phosphate present in the soil. The more organic matter, the greater the availability of phosphorus in the soil. Additionally, wetland

Table 6 Soil fractions of composed sample

Sample	Sieves (mm)	Type of soil	Weight (g)	Percent (%)
Composite sample	40	Gravel+sand	26,3796	44,62
	60	Silt	58,621	9,92
	100	Clay	26,8723	45,46
Total				100

soils have unique chemistry due to the deficiency of oxygen, where microorganisms utilize electron acceptors other than oxygen in respiration. It has been demonstrated that the microorganisms found in wetlands can allow the removal or reduce the number of organic and inorganic pollutants for example pesticides, heavy metals, hydrocarbons, and others (Huyssteen, 2023). The surface layers of sandy clay soils have a moderate ability to retain nutrients, and leaching is less severe. Low levels of phosphorus, calcium, potassium, sulfur, nitrogen, and trace elements of copper and molybdenum have been recorded (Swamy & Veere Gowda, 2006). This could be linked with the results obtained from measuring several ions in the previous section. Reduced cultivation and stubble retention techniques will help prevent the loss of surface soil and nutrients through wind and water erosion. Productivity is greatly reduced with the loss of surface soil; sedimentation of dams, waterways, and low-lying areas may also be a cause for concern (Palta et al., 2016).

3.3 Metabolic Profiles Obtained from EcoPlate Analysis

After the readings were obtained from the Ecoplate Biolog® Kit, it was used ANOVA and TUKEY statistical analyses to observe the variance of the means and compare the significant differences between values, considering a significance level of 5%.

As can be observed in Fig. 1, the ANOVA statistical analysis of the carboxylic acid consumption of the 4 soil samples shows that all the comparisons between the M2, M3, and M4 samples have non-significant differences, however, M1 showed a significant difference with respect to the rest of the samples. Referring to the substrate of complex carbon sources where the representative absorbance average of each sample analysis is that all comparisons between the samples have no

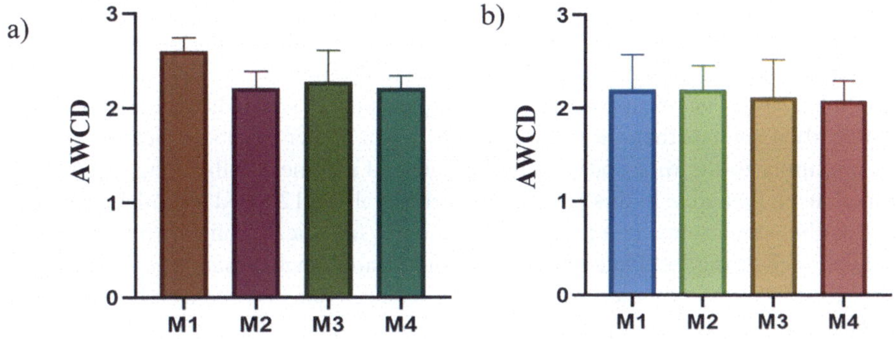

Fig. 1 Metabolic profile analysis based on carbon sources consumption measured at 24, 48, 72, and 120 hours. (**a**) Carboxylic acid consumption; (**b**) Consumption of complex carbon sources. (Note. AWCD stands for Average Well Color Development)

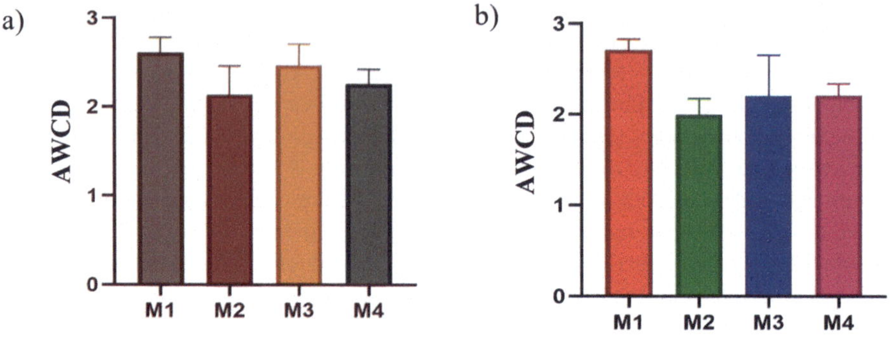

Fig. 2 Metabolic profile analysis based on carbon sources consumption measured at 24, 48, 72, and 120 hours. (**a**) Carbohydrate consumption; (**b**) Carbon-Phosphate Consumption. (Note. AWCD stands for Average Well Color Development)

significant differences, that means that the samples have similarities between them. The Tukey test indicated that there is no difference between the samples.

Regarding the carbohydrate consumption of the 4 soil samples, the ANOVA indicated that the samples have a significant difference ($p < 0.05$). However, the comparison between M1 vs M3 and M2 vs M4 showed a non-significant difference. The carbon-phosphate analysis showed that M1 vs M2, M1 vs M3 and M1 vs M4 had a significant difference ($p < 0.05$). In addition to that, the comparison between M2 vs M3, M2 vs M4 and M3 vs M4 showed a non-significant difference (Fig. 2).

Regarding the amino acids consumption in the 4-sample, most of the samples have a highly significant difference $p < 0.05$, except for the comparison of M1 vs M3 and M2 vs M4, since comparisons between the two showed that they have non-significant differences. In other hand, the amines analysis determined that there are highly significant differences ($p < 0.05$) between the M1 vs M2 samples. However, it was observed that between M3 and M4 there are not a significant difference. The Tukey analysis showed that the M1 and M4 samples had a greater difference compared to the other samples (Fig. 3).

The Ecoplate method used in this study enabled to define the consumption of various carbon sources by the microbial communities found in the four samples (M1, M2, M3 & M4). The results showed that samples closer to the palms' base consumed more carbon sources than those further out, thus a higher metabolic activity due to the proximity to the rhizosphere may be deduced summed to the concentration of some ions in that area. In this context Janniche et al. (2012) used a similar approach to describe, at microbial community-level the physiological profiles with 7 types of polluted soils using EcoPlate analysis. It was detailed that communities respond to induced stress from pollution by consuming in accelerated rates the carbon sources. In the same way, Nakamura et al. (2022), experimented with 33 types of soils and their properties were analyzed parallelly with a physiological profile essay using EcoPlate ® to determine in a multifunctional way the soil microbial communities,

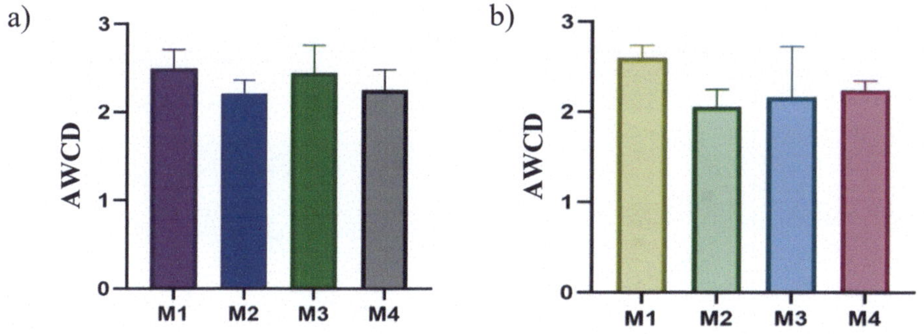

Fig. 3 Metabolic profile analysis based on carbon sources consumption measured at 24, 48, 72, and 120 hours. (**a**) Amino Acid consumption; (**b**) Amines Consumption. (Note. AWCD stands for Average Well Color Development)

due to the metabolic activity associated with the utilization of diverse external Carbon sources.

3.4 DGGE and Identification of the Microorganisms by the Sanger PCR Method

The DGGE analysis helped to determine the correlation of the ecological index between samples. It was found by means of the multidimensional analysis that there is an 80% similarity between the M1 and M4 samples, likewise, indicating that there is a 60% similarity between M2 and M3. Also, a 40% similarity between all samples can be observed in the Fig. 4.

The bands obtained from the DGGE analysis from the soil samples (M1, M2, M3, M4) and the 4 isolated strains (C1, C2, C3, C4) were sent to MacroGen DNA Sequencing® and the sequences obtained were uploaded in the NCBI (National Center for Biotechnology Information) website.

The genera of the microorganisms found in the most representative bands of each sample are indicated. In M1 and M2 the most representative gender is *Bacillus*, having an identity percentage of 80% and 86% respectively. It is important to note that the sequencing of sample 3 was not successful, as the sequences obtained from the PCR were no sufficient. This could be because 16 s rRNA gene amplification was not completely successful, due to several process externalities and possible errors in the technique protocol. Therefore, it was not possible to determine which type of microorganisms were at that specific sampling point (M3). Finally, M4 found the genus *Bovista* with 98% identity.

Villarreal-Delgado et al. (2018), indicate that the genus *Bacillus* is widely distributed worldwide due to its ability to form endospores, a characteristic that gives them resistance and enhances their insolation in various habitats, both aquatic and terrestrial ecosystems, and even in environments under extreme conditions. This

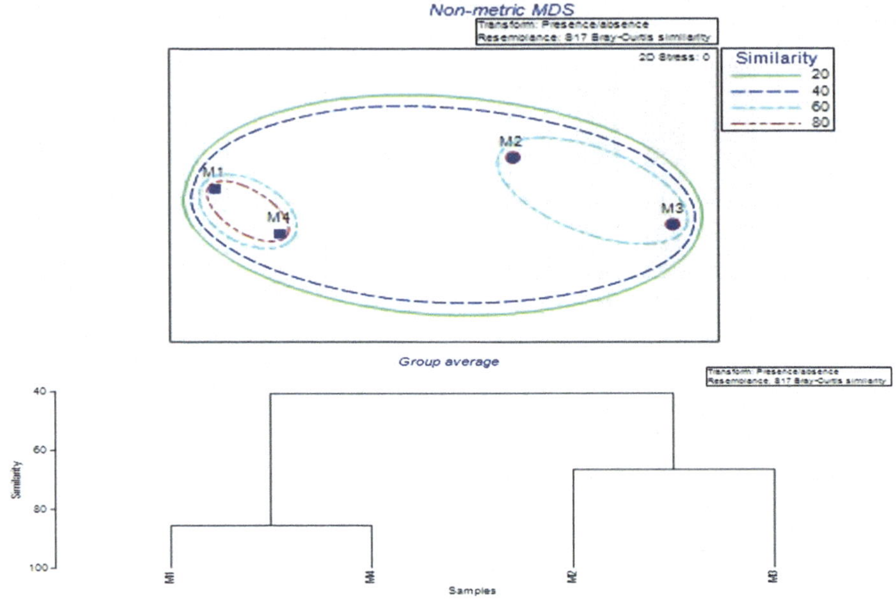

Fig. 4 Multidimensional Scaling (MDS) of the band patterns obtained by DGGE

research managed to understand the diversity and population dynamics of *Bacillus* in rhizospheric soils, where the bacterial communities they inhabit respond particularly to soil fertility and plant root exudates, but these vary according to plant phenology and genotype. Therefore, microorganisms that interact with plants and present associative, endophytic or symbiotic processes to adapt to rhizospheric conditions are recognized as potential microbial inoculants. On the other hand, it should be noted that no specific test was performed with the genus found and therefore it cannot be stated that these bacteria develop as a microbial inoculant in the rhizosphere of *Roystonea oleracea.* Lastly, the genre *Bovista* is commonly found growing on rich soils of tropical forests but a lack of information about this kind of microorganisms was found around their relationship within wetlands (Calonge et al., 2004).

3.5 Antagonistic Interactions

Through this technique, it was possible to observe and analyze the growth of the strains while they were in close contact with each other to witness whether inhibitory effects. As a result of the respective antagonism tests to determine the ecological interactions between microbial communities, no antagonistic relationships were found. The microbial community structures vary depending on plant species and can differ among specific plant root exudates and soil type (Djebaili et al., 2021;

Schoustra et al., 2012). Antagonism in microbial communities involves the inhibition, deterioration, or death of certain species by the action of others, highlighting the complex dynamics within these communities (Giometto et al., 2021; Wytock et al., 2020). These interactions occur in many ecological niches such as the rhizosphere (Kent & Triplett, 2002). In this context, we used the same approach described in a study carried out in China were combining molecular and morphological characteristics, to analyze the rhizospheric fungal diversity. A total of 16,130 paired-end reads of the nuclear ribosomal internal transcribed spacer 2 were generated and clustered into 860 operational taxonomic units. In total, 113 fungal isolates were isolated from rhizosphere soil. They were assigned to five classes, eight orders, 26 genera, and 43 species based on morphological characteristics and phylogenetic analysis of the interval transcribed spacer. *Furasium* was the most isolated genus with six species. Their results suggest that diverse fungi including potential pathogenic ones exist in the rhizosphere soil to *P. notoginseng* and that antagonistic isolates may be useful for biological control of pathogens (Miao et al., 2015). Summarizing, the results showed that most microorganisms isolated from the rhizosphere of the palm, live in a stable community and could be also useful for further investigation related to microbial consortiums.

4 Conclusions

The measurement of the ions showed that although there is a variation of data for each of the analyzed ions, M1 was the one that presented the greatest accumulation of these. Only similar values were observed in the total N concentration, taking in M1 (8,17 ppm) and in M4 (8,96 ppm) possibly by the amount of organic matter present in the soil.

The characterization of the physical properties of the soil indicated that is a sandy clay type soil, with a moisture content of 37% and a percentage of organic matter of 2,72%.

The metabolic profiles analysis showed a higher metabolic activity next to the palm inside the rhizosphere zone. The genera of the most representative microorganisms identified in this research were *Bacillus* and *Bovista* with an 86% and 98% of similarity respectively.

By means of the MDS analysis an 80% similarity between the M1 and M4 samples was observed, a 60% similarity between M2 and M3 and finally a 40% similarity between all strains isolated from the four samples.

No antagonistic relationships were found after performing the tests. This suggests that microbial communities were cohabiting in a functional ecological structure, probably result of the rhizosphere microhabitat.

References

Aarthi, R., & Sivakumar, D. (2020). An enhanced agricultural data mining technique for dynamic soil texture prediction. *Procedia Computer Science, 171*, 2770–2778. https://doi.org/10.1016/j.procs.2020.04.301

Abarca, F. (2001). Definition and importance of wetlands. *Revista de Ciencias Ambiental (Tropical Journal of Environment Sciences), 21*(1), 4–8. https://doi.org/10.15359/rca.21-1.1

Acevedo, E., Galindo-Castañeda, T., Prada, F., Navla, M., & Romero, H. (2014). Phosphate-solubilizing microorganisms associated with the rhizosphere of oil palm (Elaeis guineensis Jacq.) in Colombia. *Applied Soil Ecology, 80*, 26–33. https://doi.org/10.1016/j.apsoil.2014.03.011

Bermeo, K., Ayala, C., & Hernandez-Rosas, J. (2022). Mycorrhizal activity in roots of climbing plants of the Ramsar. In *Third student congress on food science, agriculture and Enviroment* (p. 170). Agrarian University of Ecuador.

Calonge, F. D., Kreisel, H., & Guzmán, G. (2004). Bovista sclerocystis, a new species from Mexico. *Mycologia, 96*(5), 1152–1154. https://doi.org/10.2307/3762097

Castiblanco Duran, M., & Zapata Herazo, W. (2020). *Evaluación de la biodiversidad fúngica en suelos de la reserva natural banco totumo – Bijibana (municipio de Repelón) y el humedal el Limón (corregimiento de las Compuertas-municipio de Manatí) en el departamento del Atlántico*. Universidad de la Costa, Departamento de Ingeniería Civil y Ambiental, Barranquilla. Retrieved from https://repositorio.cuc.edu.co/handle/11323/6346

Chazarenc, F., Brisson, J., & Merlin, G. (2010). Seasonal and spatial changes of microorganism communities in constructed wetlands: A community level physiological profiling analysis. *International Journal of Chemical Engineering, 2010*, 490240. https://doi.org/10.1155/2010/490240

Crumpton, W. G., Stenback, G. A., Fisher, S. W., Stenback, J. Z., & Green, D. I. (2020). Water quality performance of wetlands receiving nonpoint-source nitrogen loads: Nitrate and total nitrogen removal efficiency and controlling factors. *Journal of Environmental Quality, 49*(3), 735–744. https://doi.org/10.1002/jeq2.20061

Cunalata García, A., & López Pumalema, J. (2020). Turismo de humedales en Ecuador: Análisis a los sitios RAMSAR. *Green World Journal, 03*, 1–12. https://www.greenworldjournal.com/doi-017-ea-2020

Djebaili, R., Pellegrini, M., Ercole, C., Farda, B., Kitouni, M., & Gallo, M. D. (2021). Biocontrol of soil-borne pathogens of *Solanum lycopersicum* L. and *Daucus carota* L. by plant growth-promoting actinomycetes: In vitro and in planta antagonistic activity. *Pathogens, 10*, 1305. https://doi.org/10.3390/pathogens10101305

Dodds, W. K., & Whiles, M. R. (2020). Freshwater ecosystems. In W. K. Doods (Ed.), *Acuatic ecology, freshwater ecology* (3rd ed., pp. 723–764). Academic. https://doi.org/10.1016/B978-0-12-813255-5.00024-7

Fennessy, M. S., Moon, J. B., & Finlayson, C. M. (2023). Linking wetland ecological processes with the delivery of ecosystem services. In P. Gell, M. Finlayson, & N. Davidson (Eds.), *Ramsar wetlands. Values, assessment, management* (pp. 153–178). Elsevier. https://doi.org/10.1016/B978-0-12-817803-4.00011-5

Finlayson, C. M., & Dam, A. A. (2023). Climate change and wetlands: Vulnerability, adaptation, mitigation, resolutions, and scientific societies. In P. Gell, N. Davidson, & M. Finlayson (Eds.), *Ramsar wetlands. Values, assessment, management* (pp. 495–524). Elsevier. https://doi.org/10.1016/B978-0-12-817803-4.00016-4

Fujita, K., Kunito, T., Matsushita, J., Nakamura, K., Moro, H., Yoshida, S., Toda, H., Otsuka, S., & Nagaoka, K. (2018). Nitrogen supply rate regulates microbial resource allocation for synthesis of nitrogen-acquiring enzymes. *PLoS One, 13*(8), 1–18. https://doi.org/10.1371/journal.pone.0202086

Giometto, A., Nelson, D. R., & Murray, A. W. (2021). Antagonism between killer yeast strains as an experimental model for biological nucleation dynamics. *eLife, 10*, e62932. https://doi.org/10.7554/eLife.62932

Gu, C., Zhou, Q.-H., Cusick, R., & Margenot, A. (2021). Evaluating agronomic soil phosphorus tests for soils amended with struvite. *Geoderma, 399*, 115093. https://doi.org/10.1016/J.GEODERMA.2021.115093

Herrera, I., Hernández-Rosas, J. I., Suárez, C. F., Cornejo, X., Amaya, A., Goncalves, E., & Ayala, C. (2017). New record and potential distribution of an exotic ornamental palm. *Rodriguésia, 68*(2), 759–769. https://doi.org/10.1590/2175-7860201768226

Huyssteen, C. W. (2023). Wetland soils. In *Encyclopedia of soils in the environment* (Vol. 4, 2nd ed., pp. 436–446). https://doi.org/10.1016/B978-0-12-822974-3.00064-1

Islam, W., Noman, A., Naveed, H., Huang, Z., & Chen, H. Y. (2020). Role of environmental factors in shaping the soil microbiome. *Environmental Science and Pollution Research, 27*, 41225–41247. https://doi.org/10.1007/s11356-020-10471-2

Janniche, G. S., Spliid, H., & Albrechtsen, H.-J. (2012). Microbial community-level physiological profiles (CLPP) and herbicide mineralization potential in groundwater affected by agricultural land use. *Journal of Contaminant Hydrology, 140–141*, 45–55. https://doi.org/10.1016/j.jconhyd.2012.08.008

Kadykalo, A. N., & Findlay, C. S. (2016). The flow regulation services of wetlands. *Ecosystem Services, 20*, 91–103. https://doi.org/10.1016/j.ecoser.2016.06.005

Kent, A., & Triplett, E. (2002). Microbial communities and their interactions in soil and rhizosphere ecosystems. *Annual Review of Microbiology, 56*, 211–236. https://doi.org/10.1146/annurev.micro.56.012302.161120

Kertesz, M. A., & Frossard, E. (2015). Biological cycling of inorganic nutrients and metals in soils and their role in soil biogeochemistry. In E. A. Paul (Ed.), *Soil microbiology, ecology and biochemistry* (4th ed., pp. 471–503). Academic. https://doi.org/10.1016/B978-0-12-415955-6.00016-5

Malekpoor Mansoorkhani, F., Seymour, G. B., Swarup, R., Moeiniyan Bagheri, H., Ramsey, R. J., & Thompson, A. J. (2014). Environmental, developmental, and genetic factors controlling root system architecture. *Biotechnology and Genetic Engineering Reviews, 30*(2), 95–112. https://doi.org/10.1080/02648725.2014.995912

Mendoza, J. D. (2008, Febrero 01). *Los amigos de Santay*. Retrieved from https://www.islasantay.info/2008/02/.

Meneses, N., Mendoza-Cortez, J. W., & Ceciliofilho, A. B. (2017). Fertilización potásica del maíz dulce en suelo con alta disponibilidad de potasio. *Agrociencia (Uruguay), 21*(2), 54–58. Obtenido de http://www.scielo.edu.uy/scielo.php?script=sci_arttext&pid=S2301-15482017000200054&lng=es&nrm=iso

Miao, C.-P., Mi, Q.-L., Qiao, X.-G., Zheng, Y.-K., Chen, Y.-W., Xu, L.-H., et al. (2015). Rhizospheric fungi of Panax notoginseng: Diversity and antagonism to host phytopathogens. *Journal of Ginseng Research, 40*(2), 127–134. https://doi.org/10.1016/j.jgr.2015.06.004

Mitsch, W. J., Bernal, B., & Hernandez, M. E. (2015). Ecosystem services of wetlands. *International Journal of Biodiversity Science, Ecosystem Services & Management, 11*(1), 1–4. https://doi.org/10.1080/21513732.2015.1006250

Młodzińska-Michta, E. (2023). Abiotic factors determine the root system architecture-review and update. *Acta Societatis Botanicorum Poloniae, 92*(1), 1–18. https://doi.org/10.5586/asbp/168700

Nakamura, M., Terada, C., Ito, K., Matsui, K., Niwa, S., Ishihara, M., Kenta, T., Yoshikawa, T., Kadoya, T., Hiura, T., Muraoka, H., Ishida, K., Agetsuma, N., Nakamura, R., Sakio, H., Takagi, M., Mori, A. S., Kimura, M. K., Kurokawa, H., Enoki, T., Seino, T., Takashima, A., Kobayashi, H., Matsumoto, K., Takahashi, K., Tateno, R., Yoshida, T., Nakaji, T., Maki, M., Kobayashi, K., Fukuzawa, K., Hoshizaki, K., Ohta, K., Kobayashi, K., Hasegawa, M., Suzuki, S. N., Sakimoto, M., Kitagawa, Y., Sakai, A., Kondo, H., Ichie, T., Kageyama, K., Hieno, A., Kato, S., Otani, T., Utsumi, Y., Kume, T., Homma, K., Kishimoto, K., Masaka, K., Watanabe, K.,

Toda, M., Nagamatsu, D., Miyazaki, Y., Yamashita, T., Tokuchi, N. (2022). Evaluating the soil microbe community-level physiological profile using EcoPlate and soil properties at 33 forest sites across Japan (Abstract). *Ecological Research, 37*(3), 432–445. https://doi.org/10.1111/1440-1703.12293

Newall, P. R., & Fisk, G. (2023). Trajectories in wetland condition: Setting limits of acceptable change. In *Ramsar wetlands* (pp. 335–355). https://doi.org/10.1016/B978-0-12-817803-4.00010-3

Palta, M. M., Ehrenfeld, J. G., Giménez, D., Groffman, P. M., & Subroy, V. (2016). Soil texture and water retention as spatial predictors of denitrification in urban wetlands. *Soil Biology & Biochemistry, 101*, 237–250. https://doi.org/10.1016/j.soilbio.2016.06.011

Rivero, S. T., Moorillón, V. N., & Borunda, E. O. (2009). Growth, yield, and nutrient status of pecans fertilized with biosolids and inoculated with rizosphere fungi. *Bioresource Technology, 100*(6), 1992–1998. https://doi.org/10.1016/j.biortech.2007.12.078

Roblero, M. D., Pineda, J. P., León, M. T., & Castellanos, J. S. (2020). El oxígeno en la zona radical y su efecto en las plantas. *Revista Mexicana de Ciencias Agrícolas, 11*(4), 931–943. Retrieved from https://dialnet.unirioja.es/descarga/articulo/7498408.pdf

Romero, A. L., Batista, R. G., Aguirre, P. U., & Jaramillo, F. V. (2018). Economic importance of ecosystem services of wetlands. *Revista Científica Agroecosistemas, 6*(3), 40–48. Retrieved from http://aes.ucf.edu.cu/index.php/aes/index

Schoustra, S., Dench, J., Dali, R., Aaron, S., & Kassen, R. (2012). Antagonistic interactions peak at intermediate genetic distance in clinical and laboratory strains of Pseudomonas aeruginosa. *BMC Microbiology, 12*, 1–9. https://doi.org/10.1186/1471-2180-12-40

Steffens, B., & Rasmussen, A. (2016). The physiology of adventitious roots. *Plant Physiology, 170*(2), 603–617. https://doi.org/10.1104/pp.15.01360

Sutariya, B. P., Vyas, T. K., & Vala, A. K. (2024). Simulation study on the effect of iron nanoparticles on Rhizospheric and rhizoplane microbes of Paddy (Oryza sativa) plant. *Communications in Soil Science & Plant Analysis, 55*(8), 1136–1147. https://doi.org/10.1080/00103624.2023.2296539

Sutton-Grier, A., Keller, J. K., Koch, R., Gilmour, C., & Megonigal, J. P. (2011). Electron donors and acceptors influence anaerobic soil organic matter mineralization in tidal marshes. *Soil Biology and Biochemistry, 43*(7), 1576–1583. https://doi.org/10.1016/j.soilbio.2011.04.008

Swamy, K. R. M., & Veere Gowda, R. (2006). 22 – Leek and shallot. In K. V. Peter (Ed.), *Handbook of herbs and spices* (pp. 365–389). Woodhead Publishing. https://doi.org/10.1533/9781845691717.3.365

Tóth, E. M., Borsodi, A. K., Felföldi, T., Vajna, B., Sipos, R., & Márialigeti, K. (2013). Practical microbiology based on the Hungarian practical not esentitled "Mikrobiológiai Laboratóriumi Gyakorlatok". In E. L. University (Ed.), *Mikrobiológiai Laboratóriumi Gyakorlatok* (p. 89). Hungría. Retrieved from https://ttk.elte.hu/dstore/document/893/book.pdf

Villarreal-Delgado, M. F., Villa-Rodríguez, E. D., Cira-Chávez, L. A., Estrada-Alvarado, M. I., Parra-Cota, F. I., & Santos-Villalobos, S. D. (2018). The genus Bacillus as a biological control agent and its implications in the agricultural biosecurity. *Mexican Journal of Phytopathology, 36*(1), 95–130. https://doi.org/10.18781/R.MEX.FIT.1706-5

Weber, K. A., Achenbach, L. A., & Coates, J. D. (2006a). Microorganisms pumping iron: Anaerobic microbial iron oxidation and reduction. *Nature Reviews Microbiology, 4*(10), 752–764. https://doi.org/10.1038/nrmicro1490

Weber, K. A., Urrutia, M. M., Churchill, P. F., Kukkadapu, R. K., & Roden, E. E. (2006b). Anaerobic redox cycling of iron by freshwater sediment microorganisms. *Environmental Microbiology, 8*(1), 100–113. https://doi.org/10.1111/j.1462-2920.2005.00873.x

Wytock, T. P., Zhang, M., Jinich, A., Fiebig, A., Crosson, S., & Motter, A. E. (2020). Extreme antagonism arising from gene-environment interactions. *Biophysical Journal, 199*, 2074–2086. https://doi.org/10.1016/j.bpj.2020.09.038

Yarwood, S. A. (2018). The role of wetland microorganisms in plant-litter decomposition and soil organic matter formation: A critical review. *FEMS Microbiology Ecology, 94*(11), 175. https://doi.org/10.1093/femsec/fiy175

Yepes, G. Y. (2015). Servicios ecosistémicos y variables sociambientales determinantes en ecosistemas de humedales altoandinos. Sector el ocho y paramo de letras Manizales Colombia. *Revista Mexicana de Ciencias Agrícolas, 1*, 173–179. Retrieved from http://www.redalyc.org/articulo.oa?id=263139243024

Zhang, X., Ji, Z., Shao, Y., Guo, C., Zhou, H., Liu, L., & Qu, Y. (2020). Seasonal variations of soil bacterial communities in Suaeda wetland of Shuangtaizi River estuary, Northeast China. *Journal of Environmental Sciences (Elsevier), 97*, 45–53. https://doi.org/10.1016/j.jes.2020.04.012

The Impact of Gold Mining on Soil Biogeochemistry and Environmental Health

Olaniyi JohnPaul Popoola, Olusola David Ogundele, Emmanuel Abayomi Ladapo, and Samson Senbore

Abstract Gold is one the precious minerals that have contributed to economic growth of many nations in the world. Gold mining activities have different effects on the environment in which the disruption of biogechemical processes is a major impact. The depletion of biogeochemical processes may lead to nutrient cycling, microbial communities, and ecosystem health. Depletion of biogeochemical processes by gold mining activities may affect microbial communities which are essential to soil fertility and indirectly to human wellbeing. The use of toxic metals like cyanide and mercury in gold processing pose dangers to both the ecosystem and human health of people living within the surroundings. Pathways for the presence of toxic elements in the environment include: surficial water contamination, sod and sediments tainting, bioaccumulation and biomagnification, and occupational exposure. Toxic elements in the environment may also contribute to many health issues experienced in human beings. Remediation procedures can be employed to mitigate the effect of gold mining activities on the environment. It is also important to carry out detailed environmental impact assessment (EIA) before the commencement of gold exploitation. Governmental and professional regulations for gold exploitation should be strictly adhered to by players in the auriferous ore industry. Also, there must be cordial relationships between the exploitation companies and the community who are important stakeholders in this matter through corporate social responsibilities. The community must also be involved in decision making to reduce the impact of the exploitation processes.

O. J. Popoola (✉)
Department of Geological Sciences, Achievers University, Owo, Ondo State, Nigeria
e-mail: adewumiadeniyi@achievers.edu.ng

O. D. Ogundele
Department of Chemical Sciences, Achievers University, Owo, Ondo State, Nigeria

E. A. Ladapo
School of Earth, Environment and Society, Bowling Green, OH, USA

S. Senbore
Department of Geology, University of Pretoria, Pretoria, South Africa

S. A. Aransiola et al. (eds.), *Soil Microbiome in Green Technology Sustainability*,
https://doi.org/10.1007/978-3-031-71844-1_20

Keywords Environmental health · Gold exploitation · Soil biogeochemistry · Remediation · Toxic elements

1 Introduction

1.1 Background on Gold Mining

Gold, valued for its rarity, beauty, and malleability since ancient times, has been used for currency, jewelry, religious artifacts, and as an emblem of affluence and influence (De Witte & Oosterbaan, 2013). The gold surges of the nineteenth century in California, Australia, and South Africa significantly shaped these regions' demographics and economies (Wilson, 2001; Walker, 2001; Verbrugge & Geenen, 2019). The life cycle of a mine is divided into seven stages as shown in Fig. 1. Various extraction methods have evolved over the centuries, including panning, sluice boxes, and placer mining (Mathioudakis et al., 2023). Modern gold mining know-how include open-pit excavation, underground mining, and cyanide heap leaching (Manning & Kappes, 2016). Each method has its inherent group of environmental and social impacts. Open-pit mining and cyanide heap leaching cause habitat destruction, sod erosion, and toxic substances release into water sources (Adewumi & Laniyan, 2020). Mercury adoption in minuscule-prorate gold mining also poses additional environmental risks (Schwartz et al., 2023). Gold mining has been a catalyst for economic development, providing employment opportunities and contributing to national revenues (Ericsson & Löf, 2019). However, it often comes with social challenges, such as land disputes, displacement of local communities, and issues related to worker safety and labor practices (Ikhumetse et al., 2019).

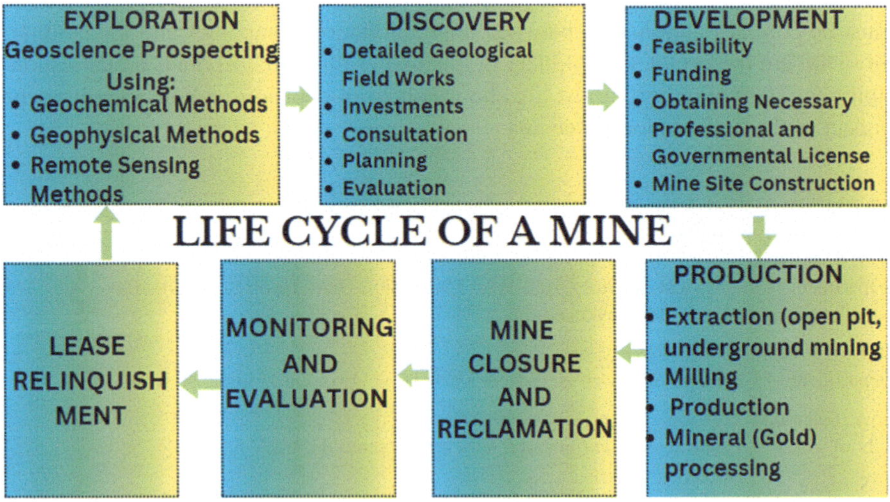

Fig. 1 The life cycle of a mine

Balancing economic benefits with social and environmental responsibility remains a complex challenge for the industry.

1.2 Significance of Studying Gold Mining's Impact on Sod Biogeochemistry

The study of gold mining's impact on sod biogeochemistry is crucial for understanding the interplay between human activities, environmental health, and ecosystem sustainability. Gold mining, a global industry, can alter sod biogeochemical processes, affecting nutrient cycling, microbial communities, and ecosystem health. The use of chemicals like cyanide and mercury, along with extensive land disturbance, can lead to alterations in sod composition, nutrient availability, and microbial diversity (Artiola et al., 2019). Understanding these changes is essential for predicting and managing the consequences on vegetation and ecosystem productivity.

Sod biogeochemistry is also linked to water quality, as sod serves as a buffer and filtration system for water. Mining activities can release contaminants into nearby water sources, affecting aquatic ecosystems and potentially posing risks to human populations. Investigating the impact of gold mining on sod biogeochemistry provides a holistic understanding of the pathways through which contaminants may enter water systems (Moreno-Brush et al., 2020).

Studies on sod biogeochemistry can help develop informed strategies for sustainable environmental management, such as reclamation and remediation plans to restore sod health, promote ecosystem resilience, and minimize the long-term environmental footprint of gold mining activities (Bhaduri et al., 2022). This knowledge is crucial for designing effective reclamation and remediation plans that aim to restore sod health, promote ecosystem resilience, and minimize the long-term environmental footprint of gold mining activities.

1.3 Relevance to Environmental Health Concerns

Gold mining, despite its economic significance, poses significant environmental health risks due to its potential impact on ecosystems, water quality, and human populations. The use of toxic chemicals like cyanide and mercury in ore extraction and processing can contaminate sod and water, posing serious risks to both terrestrial and aquatic ecosystems (Talukder et al., 2023). Understanding the pathways and consequences of chemical contamination is crucial for preventing long-term environmental degradation and safeguarding biodiversity.

Water quality degradation is another concern, as mining activities can lead to the release of pollutants into nearby water bodies, affecting aquatic life and potentially posing risks to human health (Akhtar et al., 2021). Assessing and addressing water quality degradation is vital for protecting aquatic ecosystems and local communities' health.

Open-pit mining and other extraction methods can result in extensive land disturbance, sod erosion, and habitat destruction, disrupting ecosystems and reducing biodiversity (Wang et al., 2022). Monitoring and mitigating air quality concerns is essential to prevent respiratory issues and other health problems. Proximity to gold mining activities can affect the health of local communities, with exposure to mining-related pollutants potentially leading to respiratory and neurological problems (Addo et al., 2023). Studying and addressing potential health impacts on nearby communities is vital for promoting social responsibility within the mining industry. The environmental impacts of gold mining can persist long after operations cease, necessitating effective reclamation and remediation strategies.

2 Gold Mining Processes and Sources of Contamination

2.1 Overview of Gold Extraction Methods

Gold extraction methods vary depending on the ore type and deposit location. Gravity concentration is a method used for free-milling ores where gold is easily liberated and has a high density (Agorhom & Owusu, 2022). Cyanidation, such as heap leaching and carbon-in-pulp (CIP) and carbon-in-leach (CIL), dissolves auriferous mineral from the native mineral using a weak cyanide solution (Ilyas, 2018). Flotation uses the hydrophobic nature of gold particles to selectively attach and float gold particles, which are then skimmed off (Pawlik, 2022). Amalgamation involves using mercury to form an amalgam with gold, which is heated, vaporizing the mercury and leaving behind the gold (Tibau & Grube, 2019).

Bioleaching employs microorganisms to enhance gold extraction from refractory ores by oxidizing sulfide minerals, making them amenable to conventional extraction methods (Roberto & Schippers, 2022). Pressure oxidation treats sulfide ores with oxygen under high pressure and temperature, making gold accessible for cyanidation (Saim et al., 2022). Electrowinning and electrowinning are common methods used to recover gold from solution after leaching processes (Murali et al., 2022). Carbon-in-Column (CIC) and Resin-in-Column (RIC) methods involve passing a gold-laden solution through columns filled with activated carbon or resin, respectively, to adsorb and concentrate the gold (Fedyukevich & Vorob'ev-Desyatovskii, 2016).

Placer mining involves extracting gold from alluvial deposits, usually through the use of water to separate gold particles from gravel and sediment (Mathioudakis et al., 2023). This method is applied in riverbeds and areas with loose sediments where auriferous mineral has huddled over season. The choice of method depends on factors such as ore type, geological characteristics, environmental considerations, and economic viability. Modern gold extraction practices often involve a combination of these methods to optimize recovery and minimize environmental impact.

2.1.1 Identification of Potential Sources of Contamination

Gold contamination in the environment can occur through various sources, including mining operations, smelting and refining, industrial processes, waste disposal, landfills, tailings and mine waste, urbanization and construction, abandoned or historical gold processing sites, atmospheric deposition, minuscule-prorate gold mining, and wastewater from jewelry manufacturing (Yang et al., 2016). Environmental impact of gold mining and processing in different countries are shown in Table 1. Mining operations involve crushing, grounding, and treating ores with chemicals like cyanide, leading to the release of gold into surrounding sods and water bodies. Smelting and refining of gold ores can result in the emission of gold particles and vapors into the air, which may settle on nearby surfaces, contributing to sod contamination. Industrial processes like metal plating, electronics manufacturing, and the production of gold-containing products can also release gold particles and compounds into the environment, with wastewater discharges from these industries potentially containing elevated levels of gold.

Improper disposal of gold-containing products can lead to the leaching of gold into the sod, while landfills containing gold-bearing waste may contribute to local contamination. Mine tailings, waste materials generated during the mining process, can also lead to gold contamination in nearby areas if not properly managed. Urbanization and construction activities can disturb sods, potentially releasing gold particles that may have accumulated over time (Adewumi & Ogundele, 2024). Abandoned or historical gold processing sites may still contain residual gold and associated contaminants, if not remediated. Atmospheric deposition, natural processes like weathering of gold-bearing rocks and volcanic activity, and unregulated minuscule-prorate gold mining can result in the release of gold and mercury into the environment (Clackett, 2017). Wastewater from jewelry manufacturing, which is widely used in goldsmithing and jewelry-making processes, may contain gold particles if not properly treated (UNICEF, 2021).

2.1.2 Introduction to Sod Biogeochemical Changes Induced by Auriferous Mineral Mining

Gold contamination in the environment can occur through various sources, including mining operations, smelting and refining, industrial processes, waste disposal, landfills, tailings and mine waste, urbanization and construction, abandoned or historical gold processing sites, atmospheric deposition, minuscule-prorate gold mining, and wastewater from jewelry manufacturing (Agboola et al., 2020; Akinleye et al., 2022). Mining operations involve crushing, grounding, and treating ores with chemicals like cyanide, leading to the release of gold into surrounding sods and

Table 1 Location, estimated amount, major companies and environmental impact associated with gold mining across the world

S/n	Location	Estimated amount (million tons)	Major gold processing companies	Environmental impact	References
1.	China	370	China Gold International Resources Shandong Gold Zijin Mining Group Green Mine Construction	Hg pollution Acid Mine Drainage Groundwater pollution Soil pollution Health challenges	Belder (2024) Lin et al. (1997) Li et al. (2022)
2.	Australia	310	Newmont's mine Boddington mine	Land contamination Water pollution Air pollution Reduction in biodiversity Climate change	Belder (2024) ATSE (2017)
3.	Russia	310	Olimpiada Mine Natalka Mine Blagodatnoye Mine Kupol and Dvoinoye Mine Verninskoye Mine	Flooding Water contamination Annihilation of fish communities Destruction of plants	Belder (2024) Kedr.media (2023) Glotov et al. (2018)
4.	Canada	200	Newmont's Brucejack gold mine Red Chris copper-gold mine Goliath Resources	Lake sediment pollution Acid Mine Drainage Destruction of plant communities Destruction of watersheds	Belder (2024) Clark et al. (2021) Wong et al. (2002)
5.	United States of America	170	Newmont Corporation Alcoa Corporation Anglogold Ashanti Hecla Mining Calibre Mining Corps	Displacement of communities Contamination of drinking water Acid Mine Drainage Destruction of biodversity	Belder (2024) Asamoah et al. (2018)
6.	Kazakhstan	130	Altyntau Kokshetau mine Anglo-Russian company Polymetal International	Destruction of habitat and biodiversity Release of mine waste Air pollution Acid drainages Groundwater contamination	Belder (2024) Djenchuraev (1999)

(continued)

Table 1 (continued)

S/n	Location	Estimated amount (million tons)	Major gold processing companies	Environmental impact	References
				Noise pollution Increased radiation Deterioration of public health	
7.	Mexico	120	Herradura mine	Reduced quality of life Depletion of groundwater quality Destruction of plant and animal biodiversity	Belder (2024) Duran (2021)
8.	Indonesia	110	Grasberg Mining District Indonesia Asahan Aluminium	Destruction of ecosystems, Loss of biodiversity, Disruption of water sources, Production of hazardous waste High level pollution Threat to public health	Belder (2024) Meutia et al. (2023)
9.	South Africa	100	Gold fields Sibanye Stillwater Harmony gold mining	Soil degradation Water contamination Destruction of plant biodiversity Air pollution	Globaldata (2024) Tibane and Mamba (2022)
10.	Uzbekistan	100	Navoi Mining Almalyk Mining	Water pollution Air pollution Biodiversity loss Land degradation	Globaldata (2024) UNDRR (2023)

water bodies (Candeias et al., 2018). Smelting and refining of gold ores can result in the emission of gold particles and vapors into the air, which may settle on nearby surfaces, contributing to sod contamination. Industrial processes like metal plating, electronics manufacturing, and the production of gold-containing products can also release gold particles and compounds into the environment, with wastewater discharges from these industries potentially containing elevated levels of gold. (Do et al., 2023).

Improper disposal of electronic waste and other gold-containing products can lead to the leaching of gold into the sod, while landfills containing electronic components and gold-bearing waste may contribute to local contamination. Mine tailings, waste materials generated during the mining process, can also lead to gold contamination in nearby areas if not properly managed. Urbanization and construction activities can disturb sods, potentially releasing gold particles that may have accumulated over time. Abandoned or historical gold processing sites may still contain residual gold and associated contaminants, if not remediated. Atmospheric deposition, natural processes like weathering of gold-bearing rocks and volcanic activity, and unregulated minuscule-prorate gold mining can result in the release of gold and mercury into the environment (Ramzey et al., 2021). Wastewater from jewelry manufacturing, which is widely used in goldsmithing and jewelry-making processes, may contain gold particles if not properly treated.

3 Sod Biogeochemical Alterations

3.1 Changes in Sod Composition and Structure

Sod biogeochemical changes are influenced by various factors, including human activities, climate variability, and natural processes. Land use change, climate change, erosion and sedimentation, pollution and contamination, and sod management practices all play a role in affecting sod biogeochemistry (de Lacerda et al., 2022). Land use change, such as the conversion of natural ecosystems to agricultural or urban land, can deplete sod organic matter, disrupt sod structure, and decrease microbial diversity, leading to nutrient imbalances and reduced sod fertility. Climate change, on the other hand, can alter temperature and precipitation patterns, leading to increased nutrient mineralization and turnover (Das et al., 2024). Sod erosion due to water or wind can remove topsoil rich in organic matter and nutrients, altering sod composition and fertility. Pollution and contamination from industrial activities, mining, and improper waste disposal can introduce pollutants and contaminants into sods, altering microbial communities and nutrient cycling (Kumari et al., 2024). Sod management practices, such as tillage, irrigation, fertilization, and crop rotation, can also influence sod composition and structure. Conservation practices like no-till agriculture, cover cropping, and organic amendments can enhance sod structure, increase organic matter content, and promote microbial diversity, improving sod health and productivity. Natural disturbances, such as wildfires, floods, and volcanic eruptions, can also alter sod properties and biogeochemical cycles. Understanding these dynamics is crucial for sustainable land management and ecosystem health.

3.2 Impact on Nutrient Cycles and Sod Fertility

Nutrient cycles and sod fertility exhibit a momentous task in shaping sod biogeochemical alterations, influencing the availability of essential nutrients, microbial activity, and overall sod health (Daunoras et al., 2024). Changes in nutrient cycles can disrupt sod biogeochemistry, leading to nutrient imbalances, which affect microbial communities, nutrient uptake by plants, and sod primal matter disintegration paces (Leuschner et al., 2017).

Sod fertility influences the disintegration of primal matter, a crucial process in nutrient cycling. High sod fertility facilitates the breakdown of organic residues into simpler compounds, releasing nutrients like nitrogen, phosphorus and carbon back into the sod (Gerke, 2022). Conversely, low sod fertility can slow down decomposition rates, leading to nutrient immobilization and reduced nutrient availability for plant uptake (D'Angioli et al., 2022).

Microbial activity is also affected by sod fertility. Changes in sod fertility can alter microbial community composition and activity, impacting nutrient cycling processes (Bogati & Walczak, 2022). For example, nutrient-rich sods may support diverse microbial populations capable of efficient nutrient mineralization, while nutrient-poor sods may exhibit reduced diversity and activity.

Sod pH strongly influences nutrient availability and microbial activity. Acidic sods limit the availability of essential nutrients, impairing plant growth and microbial activity. Alkaline sods may lead to nutrient imbalances and toxicities, affecting sod biogeochemistry. Sustainable sod management practices, such as liming or acidification, can help adjust sod pH levels and improve nutrient availability (Pattnaik et al., 2021).

Supplement leaking and drainage are likewise influenced by sod virility. Disproportionate supplement addition can upsurge the risk of supplement leaking into subsurface water or superficial water, leading to ecological defilement and eutrophication (Craswell, 2021). Sod virility administration blueprints that enhance supplement detainment and downplay drainage can allay these repercussions.

3.2.1 Ecological Wellness Repercussions

Pathways of Hazards to Toxin from Auriferous Ore Mining

Auriferous ore extirpation ventures can introduce inimical items into the surrounding, as well as chemical substances such as mercury (Hg), lead (Pb), and arsenic (As), as well as cyanide and other chemical pollutants (Adewumi & Ogundele, 2024). Understanding these pathways is crucial for assessing ecological and wellbeing dangers related to auriferous ore extirpation and implementing effective mitigation controls.

1. **Surficial water contamination**: Runoff from extraction sites can carry contaminants into nearby surficial water makeups, contaminating rivers, streams, and lakes. Chemicals used in auriferous element extirpation, such as cyanide and mercury, may leach into surface water, posing risks to aquatic creatures and downstream communities (Timsina et al., 2022). Contaminants can also infiltrate groundwater aquifers through leaching and percolation, affecting drinking water

sources and potentially exposing communities to health hazards (Abanyie et al., 2023).
2. **Sod and sediment tainting**: Extraction operations can lead to the deposition of adulterant in sod and sediment, posing risks to terrestrial and water surroundings (Adewumi & Laniyan, 2023; Laniyan & Adewumi, 2023). Dust dispersion during extraction operations can transport contaminants over short and long distances, affecting soil quality and mortal wellbeing (Onemu et al., 2024). Inhalation of airborne particles containing chemical elements and other precarious materials can induce respiratory problems and other health issues (Adewumi & Laniyan, 2020).
3. **Bioaccumulation and biomagnification**: Adulterant released into the surrounding can be absorbed up by plants, animals, and microbes, leading to buildup in tissues (Adewumi et al., 2020). This process can amplify the risks of vulnerability for creature at the pinnacle of the food web, including humans who consume contaminated fish or wildlife.
4. **Occupational exposure**: Individuals directly involved in gold mining operations may face occupational vulnerability to adulterant through gasping, skin interaction, or eating. Communities living near mining sites may also experience exposure to contaminants through various pathways, including consumption of contaminated water or food, inhalation of dust, and direct contact with contaminated sod or sediment (Gyamfi et al., 2021).

Wellbeing Consequences on Mortal Communities

Auriferous ore extraction present significant wellness dangers to both excavator and nearby neighborhoods. Dangerous chemicals like mercury and cyanide are often used in the extirpation process, leading to Hg defilement in humans. This vulnerability can occur through inhalation, ingestion, or skin contact. Respiratory issues, such as chronic bronchitis, silicosis, and pulmonary fibrosis, can be caused by dust and particulate matter generated during mining operations (Vanka et al., 2022). Water contamination from cyanide and other elements can cause gastrointestinal illnesses, skin rashes, and long-term health effects like cancer (Karri et al., 2021).

Physical injuries from accidents, cave-ins, and equipment malfunctions can permit long-lasting bodily and psychological impacts on miners. Cerebral wellbeing outcomes, such as abasement, nervousness, and substance abuse, can also arise due to harsh working conditions, long hours, and isolation experienced by miners (Reuben et al., 2022). The uncertainty of employment in the mining industry adds to the anxiety and cognitive burden on miners and their families (Carrington & McIntosh, 2013).

Substantial auriferous ore extraction activities repeatedly result in the displacement of neighborhoods and disruption of traditional livelihoods, leading to societal tensions, loss of ethnic status, and increased vulnerability to poverty and exploitation. These wellness indications have important connotations for both cerebral and bodily health.

Ecological Consequences for Territorial Ecological Community

Auriferous ore mining has momentous ecological burdens on territorial ecological communities, including abode annihilation, detriment of biodiversity, water and sod tainting. This carnage occurs when large breadth of acreages of ecological communities are cleaned for exploration, extirpation, and facility development, championing the detriment of strategic ecological communities like thickets, wetlands, and riparian zones. Deforestation is also a notable issue, as thickets are often unloaded for open-pit mines, roads, and other excavation facilities (Grbeš et al., 2024).

Sod debasement is another pivotal issue, with unearthing exercises eliminating sod structure and fertility, resulting in despoliation, compaction, and impairment of topsoil. Chemical elements and other hazardous contents from excavation activities can impair ecological community productivity and resilience. Water tainting is another concern, as toxic substances like cyanide and mercury leach into neighboring water bodies, tainting surficial water and groundwater sources. This poses risks to water-related ecological communities, affecting water-related creatures, as well as neighborhoods that depend on them for food and livelihoods.

Sedimentation is another crucial issue, as excavation processes can escalate sedimentation in rivers, streams, and other water frames due to erosion and runoff. This can smother aquatic abodes, degrade water quality, and disrupt aquatic food chains, causing downturn in biodiversity and ecological communities wellbeing.

The social-ecological onus of auriferous ore excavation are intertwined with social dynamics, as neighborhoods rely on ecological communities for their livelihoods and cultural identity. The degeneration of pure reserves can undermine the flexibility and adaptive capacity of communities, exacerbating poverty, food insecurity, and communal conflict.

To address these environmental consequences, stringent environmental rules, comprehensive burden assessments, and effective mitigation measures are necessary. Viable extraction conventions, reclamation of degraded landscapes, and collaboration between governments, extraction corporations, conservation organizations, and indigenous neighborhood are crucial for achieving viable management of natural resources and mitigating the environmental footprint of auriferous ore extraction.

4 Options for Mitigation and Remediation

4.1 Recap of New Techniques to Alleviate Ecological Impacts

With regard to detrimental environmental conditions and rising temperatures, living beings' tasks—like mining—have become increasingly detrimental on the surroundings. An array of sectors has begun using strategies for lowering this detrimental effect, including the adoption of environmentally friendly innovations, assets optimized performance, Environmental Impact Assessments (EIAs), Best Management

Practices (BMPs), rehabilitation of ecosystems and sustainability, and efforts to conserve water.

Clean automation includes appliances that are current-capable, protocols that lessen emissions and released greenhouse gases, and renewable resource causes like solar and wind power. By reusing, recycling, and recovering resources, and simplifying industrial operations, resource enhancement and dirt elimination strive to augment applications of resources and lower generated dirt (Zhou et al., 2022).

Before starting any significant projects, environmental impact assessments, or EIAs, are necessary to evaluate any negative effects on the environment and determine mitigating strategies. BMPs include actions to stop pollution, appropriately handle trash, save biodiversity, and advance sustainable resource management. Restoration of damaged habitats, preservation of important ecosystems, and assistance with conservation projects in collaboration with regional communities and conservation organizations are all parts of ecosystem restoration and conservation activities. Water-saving technology are being used, wastewater is being recycled and treated, and water sources are not being contaminated.

To abate the sway on the surroundings, novel tools, procedures, and blends must be advanced via research and vicissitude. Productions may lessen their ecological burden and help safeguard and rehabilitate habitats for eras to come by using these steps.

4.2 Difficulties and Restrictions in Lessening the Adverse Consequences of Auriferous Ore Extraction

Lessening the ecosystem and adverse consequences of auriferous ore extraction is fraught with difficulties and constraints. A multifaceted strategy that incorporates technical expertise, regulatory reform, stakeholder engagement, and sustainable development principles is necessary. These include the complexity of the surroundings, the legacy of historical exploitation procedures, the lack of technological solutions, financial constraints, insufficient regulatory enforcement and compliance, conflicts with indigenous rights and land tenure, limited stakeholder engagement, global supply chain complexity, changing climate and uncertainty, and the need for a holistic approach.

The environment are intricately linked systems that make it challenging to anticipate and minimize every possible effect. Resolving the legacy of prior mining operations and subpar preservation of the surroundings may be expensive and technically difficult.

Limited technological solutions, such as the treatment and disposal of mine tailings, present ongoing technical and financial challenges. Financial constraints may make it prohibitive for smaller mining companies or operations in economically disadvantaged regions to implement robust environmental mitigation measures.

Regulatory enforcement and compliance may be inadequate due to limited administration capacity, corruption, or political influence, which can undermine efforts to mitigate ecological influence and hold extraction syndicate answerable. Conflicts with indigenous rights and land tenure are also significant, leading to conflict over land rights, resource utilization, and sustainability. Limited stakeholder engagement is crucial for identifying ecological concerns, building trust, and fostering societal acceptance of extraction projects.

It is challenging to track the origin of gold and hold all parties responsible for upholding principled and surrounding standards due to the worldwide nature of the gold supply chain. In addition to rising water shortages and harsh climatic proceedings, weather variability puts extra pressure on gold mining operations to minimize carbon emissions.

To sum up, tackling these issues calls for an all-encompassing strategy that incorporates technological know-how, legislative change, stakeholder involvement, and sustainable development principles. Encouraging responsible mining practices that put social justice, the environment, and economic growth first requires cooperation between governments, mining firms, civil society organizations, and local people.

4.3 Case Studies Highlighting Successful Remediation Efforts

Gold mining has shown potential for mitigating environmental impacts and restoring ecosystems through successful remediation efforts, with notable case studies highlighting successful examples.

1. **Omai Gold Mine, Guyana:** The Omai Gold Mine in Guyana, operated by Omai Gold Mines Limited, faced environmental challenges after its closure in 2005 (David, 2023). Acid mine drainage and heavy metal contamination in water bodies posed risks to human health and ecosystems. The government of Guyana, in collaboration with international partners and mining companies, implemented a comprehensive remediation plan, including water treatment facilities, reclamation of mine sites, and rehabilitation of affected landscapes. Over time, water standard improved and surroundings began to recover.
2. **Porgera Gold Mine, Papua New Guinea:** Barrick Gold Corporation, the operator of the Porgera gold mine in Papua New Guinea, has committed to a remediation and compensation program following allegations of ecological damage and mortal rights abuses. The program aims to tackle water contamination, restore land, and assist sustainable livelihoods for affected neighborhoods. By interacting with parties involved and implementing targeted remediation efforts, Barrick Gold demonstrated accountable mining processes and contributed to ecological revitalization (Alphonse, 2011).

3. **Revegetation Efforts in Western Australia:** Western Australia's gold mining regions have seen successful revegetation efforts by companies like Newmont Corporation. Their Boddington gold mine implemented a program using native plant species to reclaim disturbed areas and promote biodiversity. This demonstrates the potential for ecological rehabilitation in gold mining regions through careful planning, soil stabilization, and ongoing monitoring (De Sousa & Amoah, 2012).
4. **Mercury-Free auriferous ore exploitation in Colombia:** The Alliance for Responsible Mining (ARM) has launched initiatives to promote mercury-free alternatives and sustainable livelihoods in artisanal and diminutive auriferous ore exploitation. In Colombia, ARM collaborated with miners in the Chocó region to implement mercury-free gold extraction techniques and promote accountable operations (Martinez, 2022). This involved enabling training, technical assistance, and access to alternative technologies, thereby reducing environmental impact and improving health and safety.

5 Monitoring and Assessment

5.1 Techniques for Assessing Soil Contamination Levels

Soil contamination levels in gold mining areas are crucial for understanding environmental degradation and informing remediation efforts. Various techniques and methods are used to assess soil contamination, including soil sampling, chemical analysis, field portable analytical instruments, geospatial analysis, bioassays and ecotoxicological tests, and soil quality guidelines and standards.

Soil sampling involves collecting specimen from copious localities inside the mining site, such as mine tailings, waste dumps, and surrounding soils. These samples are typically analyzed for concentrations of impurities like chemical elements, cyanide, and other pollutants associated with gold mining. Chemical inspection approaches include atomic absorption spectroscopy (AAS), inductively coupled plasma mass spectrometry (ICP-MS), and X-ray fluorescence (XRF) (Adewumi & Ogundele, 2024).

Field portable analytical instruments provide rapid, real-time measurements of specific contaminants without the need for laboratory scrutiny (Lemiere, 2018). Geospatial inquiry uses geographic information systems (GIS) and remote sensing to appraise sod tainting levels over large spatial scales (Shi et al., 2018). Satellite generated imagery can identify land adjustment changes, vegetation stress, and other indicators of sod tainting in auriferous ore exploitation sites. GIS-based analysis allows for the integration of spatial data layers to assess contamination risks and prioritize remediation efforts (Shi et al., 2018).

Bioassays and ecotoxicological tests expose organisms to contaminated sod specimens to appraise the toxicity and ecological impacts of contaminants. Sod quality guidelines and standards, established by regulatory agencies and environmental authorities, serve as benchmarks for assessing soil contamination levels in gold mining areas. The degree of tainting and the necessity of reconditioning strategies are assessed by comparing sod tainting levels to these requirements. Environmental scientists and regulators can create efficient remediation solutions to reduce environmental hazards and safeguard ecological and mortal wellbeing by combining various tactics and strategies.

5.2 *Long-Term Monitoring of Environmental Health Post-mining*

Inspecting ecological health after gold mining is pivotal for tracking changes in habitat health, evaluating the success of remediation initiatives, and guaranteeing the long-term viability of impacted landscapes. Threshold evidence gathering, the creation of inspection protocols, important habitat indicators, dimensional and time-based gradation, data handling and analysis, stakeholder engagement and communication, adaptive handling, administrative submissiveness, and legacy considerations are important components of long-term inspection.

Data on biodiversity, ecosystem dynamics, soil and water quality, and the region around the mining site are all included in the baseline data. In order to guarantee consistency and comparability of data acquired over time, inspecting procedures describe parameters to be monitored, sample prevalence, sampling sites, and analysis techniques. Key habitat indicators include sod attributes, water variable, vegetation cover and biodiversity, wildlife populations, and ecosystem functions.

Spatial and temporal scales are used to capture variability and shift in habitat wellbeing post gold mining. Evidences gathered through long-term inspection of efforts are managed, scrutinized, and elucidated to identify sequences, inclination, and potential ecological dangers. Advanced statistical and modeling techniques may be employed to analyze complex datasets and appraise the interactions between several habitat variables.

Effective stakeholder engagement and communication are essential for long-term monitoring initiatives, involving collaboration with local communities, indigenous groups, regulatory agencies, mining companies, and other associates. Long-term inspection enables adaptive handling approaches, allowing for adjustments to remediation strategies and handling practices based on monitoring findings.

Obedience to administrative specifications, ecological standards, and commitments made by excavation companies is also promoted through regular divulging of inspection outcomes. By executing robust long-term inspection programs, associates can constructively evaluate the habitat wellbeing of gold exploitation areas post excavation, identify emerging threats, and applying adaptive handling strategies to safeguard habitats and safeguard mortal wellbeing for progeny.

5.3 The Relevance of Swift Detection and Prophylactic Actions

For a number of reasons, swift diagnosis and preventative measures are fundamental in the gold mining industry. They guarantee sustainable resource management, lessen negative effects on the surroundings and society, and protect the general wellness. By detecting any hazardous substances early on and taking preventative steps like appropriate waste management, erosion prevention, and reclamation planning, we can safeguard the habitat. The proactive preservation of limited resources such as current, water, and minerals is another way that resource conservation is accomplished. Early recognition of possible wellbeing and protection dangers, such as openness to dangerous substances and dangers to the body, protects the community's wellbeing and safety.

Maintaining a social license to operate in gold mining areas requires proactive engagement with local communities and stakeholders. Addressing community concerns, respecting indigenous rights, and promoting transparent communication foster trust and collaboration, minimizing societal conflicts and regulatory challenges. Early detection of habitat and regulatory concern issues allows excavation corporations to take corrective actions and avoid potential fines, penalties, and licit liabilities.

Monetary viability is ensured by early detection of operational inefficiencies, environmental liabilities, and market risks. Investing in preventive measures, such as risk assessments, contingency planning, and environmental insurance, helps protect against unexpected costs and liabilities, ensuring the continual financial viability of excavations.

In the auriferous exploitation enterprises durability and formal accountability are equally essential. Exploitation businesses may increase their corporate citizenship reputation and minimize negative consequences while maximizing good contributions by incorporating Environmental Social and Governance (ESG) aspects into decision-making processes. Therefore, proactive administration of habitat societal and financial hazards throughout the mining lifetime is made possible by early identification and prevention which are crucial elements of charged with auriferous ore excavation techniques.

6 Regulatory Scheme and Policies

6.1 Overview of Extant Rules Analogous to Auriferous Ore Excavation

Auriferous ore exploitation vary across regions and jurisdictions but they generally aim to assure culpable standards protect the environment safeguard public health and promote socio-economic expansion these regulations amongst others are excavation permits and licenses societal impact assessments (EIAs) water and air condition administration, debris administration and tailings disposal reclamation and closure requirements health and safety regulations indigenous rights and community engagement Corporate Social Responsibility (CSR) initiatives and global norms and guidelines.

Most nations require excavation enterprises acquire permits or licenses before conducting inspection or abstraction extortion these guaranteed actions. These authorizations outline the terms and conditions exploitation actions including EIA, reclamation requirements and financial assurances. EIA evaluate potential habitat and societal wellbeing effect of excavation actions and propose measures to forestall adverse effects. Conformity with these rules typically monitored and enforced by environmental agencies.

Debris administration and tailings disposal are also regulated requiring associations to implement measures to minimize waste generation treat contaminated water and safely store or dispose of mine waste. Reclamation and closure requirements are often included to restore disturbed landscapes and minimize long-term environmental impacts. Health and safety regulations are crucial for protecting workers and communities from hazards associated with auriferous ore excavation actions.

Indigenous rights and community engagement are also regulated aiming to respect indigenous rights protect cultural heritage and contribute to tenable progress many excavation enterprise voluntarily adopt CSR initiatives to resolve societal and habitat concerns related to auriferous ore excavation global best guidelines established by organizations like the cyanide management code the world auriferous ore councils accountable for auriferous ore excavation and the extraction sector clarity idea promote openness accountability and liable extraction action on a worldwide scale.

6.2 Recommendations for Strengthening Regulatory Measures

The text emphasizes the importance of strengthening regulatory measures in auriferous extraction to guarantee accountable actions, preserve the habitat, protect community wellbeing and promote viable growth. It suggests several recommendations for enhancing these measures:

1. conduct comprehensive environmental impact assessments (EIAs) before commencing operations including assessments of potential habitat, societal and wellbeing burdens. Guarantee that EIAs are conducted by independent experts and involve consultation with affected communities and stakeholders
2. establish and enforce stringent water and atmospheric attribute standards to limit tainting from excavation actions. Set clear limits on the discharge of pollutants into water bodies and emission of particulate matter and throttle into the air.

Implement regular monitoring and reporting requirements to guarantee compliance with ecological thresholds.
3. Impose standards for the safe administration and dumping of excavation refuse including tails, refuse rocks and other by-products of auriferous ore excavation.
4. Reinforce regulations recovery and ore pit shutdown to secure the rehabilitation of disturbed landscapes and durability of excavation holes
5. enhance occupational wellbeing and protection regulations to better protect workers and neighborhoods from burdens associated with auriferous ore excavation actions. Incorporate provisions for indigenous rights community involvement and benefit-sharing agreements into extraction regulations
6. enhance clearness and liability mechanisms in the regulation of gold mining through measures such as public disclosure of mining permits habitat burden assessments and compliance reports. Institute systems for independent monitoring and oversight of mining activities by civil society consortiums, administrative agents and affected communities
7. provide coaching, capacity enhancement and technical assistance to regulation agent, law officials and interest groups involved in auriferous ore excavation actions, foster global cooperation and collaboration on gold mining regulation by adopting and implementing international standards and guidelines
8. conduct regular reviews and updates of gold extraction rules to control emerging habitat, societal and high-tech difficulties.

By carrying out these recommendations, rules by authorities, administrative agents, relevant participants and civil society organizations collaborate together to strengthen set measures in auriferous ore extraction, control habitat and societal dangers and foster accountable and viable risks for present and future generations.

6.3 Steps for Fostering Partnership Between Participants

Partnership between participants in excavation sector is crucial for endurable growth, functional, efficiency and equitable distribution of gains. Schemes to nurture partnership include establishing clear communication channels, building trust and mutual respect through community engagement programs, and implementing collaborative frameworks and agreements like partnership agreements and public-private partnerships.

Common vision and goals should be established, focusing viable growth, ecological safety and financial boom shared value creation projects should be determined and developed for example infrastructure progress training programs and wellbeing initiatives. Instructional and teaching programs should be implemented to strengthen domestic groups awareness of excavation actions and their capacity to efficiently engage in discussions and negotiations economic empowerment can be fostered through procurement from local suppliers and capacity-building workshops.

Viable exercises should be committed to play down habitat impact and fostering societal welfare. Corporate Social Responsibility (CSR) programs should tackle the specific needs and priorities of native settlements designed in consultation with community leaders and stakeholders. Continuous monitoring systems should be established including regular assessments and independent auditors to ensure conformity with agreed standards. Feedback mechanisms should be created to allow participants to provide input on mining operations and community projects which should be taken seriously and used to make necessary adjustments and improvements.

Conflict resolution and mediation should be established with mediation services being impartial and aiming to find mutually acceptable solutions. Grievance mechanisms should be developed transparent and timely responses to raise issues and seek redress. Overall, collaboration between extraction companies, domestic settlements, governments, non-governmental organizations (NGOs) and other relevant parties is fundamental for attaining common aspirations and foster viable growth.

6.4 Community Empowerment for Viable Excavation Activities

Ore extraction is a critical sector for economic prosperity and technical advancement, but conventional exercises often cause habitat degradation, societal interruption and financial inequalities. Communal empowerment is a key strategy to mitigate these effects and promote viable activities by ensuring domestic settlements' voices are heard, their rights are protected, and their livelihoods are improved.

6.4.1 The Role of Community

Societal empowerment is a varied procedure that involves empowering domestic settlements play a pivotal role in decision-making resource management and benefit-sharing mechanisms. This involves participation in decision-making processes such as extraction, project approvals, and observing actions. Teaching and development exercises are also fundamental to equip settlement members with essential awareness and competence to engage effectively with extraction enterprise and administrative members. This approach encompasses societal financial and habitat aspects.

6.4.2 Fiscal Empowerment

This writeup emphasizes the significance of fair benefit distribution and viable livelihoods in extraction activities. It suggests mechanisms such as communal advancement agreements, commissions, domestic job creations to ensure fair distribution of financial gains. It also suggests that excavation companies should support local trades by sourcing goods and services locally. This approach will reduce dependency on ore excavation and promote viable sustenance.

6.4.3 Ecological Empowerment

Settlements should be empowered to protect and manage their native assets through instruction on viable land use water conservation and biodiversity preservation engaging. Settlements in environmental monitoring should ensures that excavation activities comply with environmental standards post-exploitation land rehabilitation is fundamental for habitat viability empowering. Settlements to participate in reclamation processes supports future land uses and biodiversity.

6.4.4 Strategies for Efficient Communal Empowerment

To empower communities open and transparent communication between exploitation enterprise, administrative and settlement is crucial institutional assistance from administration, and NGOs can provide legal frameworks financial means and technical assistance. Collective frameworks including multi-stakeholder forums and community-company partnerships can lead to shared decision-making and responsibilities monitoring and accountability mechanisms such as independent audits community-led monitoring and grievance redress mechanisms are important for upholding commitments to viable actions and community benefits these strategies help build trust and ensure the availability of information about mining activities.

6.4.5 Case Studies

I. **Canada's first nations and extraction corporations**: Canada's extraction agreements with First Nations settlements have enhanced financial possibilities and sustainability actions often incorporating local job, revenue, sharing and habitat safe guarding.
II. **Peru's Yanacocha Gold Mine**: the Yanacocha gold mine in Peru has implemented community engagement programs focusing on health, education, and infrastructure development, fostering stronger connections between operators and local communities
III. **Ghana's small-scale extraction sector**: Ghana is fostering communal empowerment through technical and legitimate for small-scale extraction resulting in viable action and decreased habitat destruction.

7 Future Research Directions

7.1 Identified Gaps in Understanding Gold Mining Impacts

Gold mining despite its socio-economic benefits poses significant habitat and social conflicts. Notwithstanding extensive research, gaps exist in the understanding of its full consequences, which are crucial for developing viable extraction actions particularly, in key areas 1

7.1.1 Ecological Consequences

Ecological disturbance and biodiversity loss: Detailed data on how extraction affects local flora and fauna is often incomplete. Lasting studies on the recovery of habitats post-mining are scarce, leaving uncertainties about the resilience of these environments. The extent to which excavation actions fragment habitats and interrupt wildlife corridors needs further study. This fragmentation can have cascading effects on ecosystems that are not well understood.

Water Resources: There is a lack of comprehensive understanding of how mining alters local and regional hydrology the cumulative effects of multiple excavation actions on groundwater tables and stream flows remain poorly quantified.

Soil Debasement: The effects of excavation on soil health, comprising nutrient impoverishment and contamination are not fully understood. Studies on soil recovery post-mining are limited particularly in different climatic and geological settings while the immediate consequences of acid mine drainage are well-documented, the lasting effects on water trait, including the persistence of heavy metals amongst other contaminants require further investigation.

7.1.2 Social and Economic Impacts

Community Health and Safety

The long-term health impacts of mining-related pollutants on local inhabitants, including respiratory issues, cancers, and chronic conditions, are not well-documented. Understanding societal dynamics contributing to unsafe practices, particularly in artisanal and small-scale mining, is crucial, as there is a gap in knowledge about current safety regulations and enforcement mechanisms.

Economic Dependency and Resilience

Research on transitioning communities from mining-dependent economies to more resilient ones are limited, requiring more comprehensive study of successful case studies and adaptable strategies. Additionally, more comprehensive study is needed on the socio-economic consequences of displacement due to excavation actions.

7.1.3 Climatic Variability and Carbon Emission

Gold excavation actions need a detailed life cycle analysis to understand their total carbon emission including direct and indirect exudation. Research is needed to explore the potential of integrating sustainable power origins into excavation actions to lessen carbon exudations. The effects of weather variability on actions, such as precipitation patterns and extreme weather events, require further study for developing adaptive strategies. Current environmental regulations need to be assessed and possibly updated to ensure they remain robust under future climate scenarios.

7.1.4 Technological and Operational Improvements

Investigation into viable actions in ore excavation is limited, with progresses for instance bio-mining and debris administration needing further exploration. Remote sensing methods are promising but underutilized for real-time monitoring of mining impacts, while big data and artificial intelligence are emerging fields that can be harnessed for habitat and societal monitoring.

The dangers of auriferous ore excavation are multifaceted encompassing ecological, societal and fiscal dimensions. Future investigations should concentrate on advanced habitat monitoring and mitigation, ecosystem restoration, and biodiversity conservation, sustainable water

References

Abanyie, S. K., Apea, O. B., Abagale, S. A., Amuah, E. E. Y., & Sunkari, E. D. (2023). Sources and factors influencing groundwater quality and associated health implications: A review. *Emerging Contaminants, 9*(2), 100207. https://doi.org/10.1016/j.emcon.2023.100207

Addo, A. M., Nyantakyi, E. K., Appiah-Adjei, E., Ackerson, N. O. B., Yeboah, S. I. I. K., Borkloe, J. K., & Owusu, M. (2023). Environmental and health impacts of mining: A case study in Kenyasi-Ahafo region, Ghana. *Arabian Journal of Geosciences, 16*(5), 334. https://doi.org/10.1007/s12517-023-11424-y

Adewumi, A. J., & Laniyan, T. A. (2020). Contamination, sources and risk assessments of metals in media from Anka artisanal gold mining area, Northwest Nigeria. *Science of the Total Environment, 718*, 137235. https://doi.org/10.1016/j.scitotenv.2020.137235

Adewumi, A. J., & Laniyan, T. A. (2023). Contamination, ecological, and human health risks of heavy metals in water from a Pb–Zn–F mining area, North Eastern Nigeria. *Journal of Water and Health, 21*(10), 1470–1488. https://doi.org/10.2166/wh.2023.132

Adewumi, A. J., & Ogundele, O. D. (2024). Hidden hazards in urban sods: A meta-analysis review of global heavy metal contamination (2010-2022), sources and its ecological and health consequences. *Sustainable Environment, 10*(1), 2293239. https://doi.org/10.1080/27658511.2023.2293239

Adewumi, A. J., Laniyan, T. A., Xiao, T., Liu, Y., & Ning, Z. (2020). Exposure of children to heavy metals from artisanal gold mining in Nigeria: Evidences from bio-monitoring of hairs and nails. *Acta Geochimica, 39*, 451–470. https://doi.org/10.1007/s11631-019-00371-9

Agboola, O., Babatunde, D. E., Fayomi, O. S. I., Sadiku, E. R., Popoola, P., Moropeng, L., & Mamudu, O. A. (2020). A review on the impact of mining operation: Monitoring, assessment and management. *Results in Engineering, 8*, 100181. https://doi.org/10.1016/j.rineng.2020.100181

Agorhom, E. A., & Owusu, C. (2022). The effects of pulp rheology on gravity gold recovery in free milling gold ore of the Tarkwaian systems of Ghana. *Mineral Processing and Extractive Metallurgy Review, 43*(1), 55–60. https://doi.org/10.1080/08827508.2020.1825955

Akhtar, N., Syakir Ishak, M. I., Bhawani, S. A., & Umar, K. (2021). Various natural and anthropogenic factors responsible for water quality degradation: A review. *Water, 13*(19), 2660. https://doi.org/10.3390/w13192660

Akinleye, C., Adewumi, A. J., & Akinleye, R. O. (2022). A systematic review of ecological and human health risk associated with metals in soils around mining areas in Nigeria. *Achievers Journal of Scientific Research, 4*(1), 120–141.

Alphonse, A. (2011, March–April). *Press review: Mining in the South Pacific,* Vol. 3, No. 2, 93 pp.

Artiola, J. F., Walworth, J. L., Musil, S. A., & Crimmins, M. A. (2019). Sod and land pollution. In *Environmental and pollution science* (pp. 219–235). Academic.

Asamoah, E. O., Xu, W., Huang, W., & Yang, W. (2018). Environmental impacts of artisanal gold mining: A case study of Nkaseim community-Ghana. *Journal of Environment and Earth Sciences, 8*(12), 116–130.

ATSE. (2017). *Addressing the environmental impacts of Australian mining's past and future.* Available at https://www.atse.org.au/wp-content/uploads/2019/01/AS-2019-01-30-APPLIED_Minerals-Addressing-the-environmental-impacts-01-D5-web.pdf

Belder, D. (2024). *10 Largest producers of Gold by country (Updated 2024).* Accessed at: https://investingnews.com/daily/resource-investing/precious-metals-investing/gold-investing/top-gold-producing-countries/

Bhaduri, D., Sihi, D., Bhowmik, A., Verma, B. C., Munda, S., & Dari, B. (2022). A review on effective sod health bio-indicators for ecosystem restoration and sustainability. *Frontiers in Microbiology, 13*, 938481. https://doi.org/10.3389/fmicb.2022.938481

Bogati, K., & Walczak, M. (2022). The impact of drought stress on soil microbial community, enzyme activities and plants. *Agronomy, 12*(1), 189. https://doi.org/10.3390/agronomy12010189

Candeias, C., Ávila, P., Coelho, P., & Teixeira, J. P. (2018). Mining activities: Health impacts. *Reference Module in Earth Systems and Environmental Sciences*, 1–21. https://doi.org/10.1016/B978-0-444-52272-6.00488-8

Carrington, K., & McIntosh, A. (2013). *A literature review of wellness, wellbeing and quality of life issues as they impact upon the Australian mining sector.* Crime and Justice Research Centre, Queensland University of Technology.

Clackett, S. (2017). *Long-term changes in global and local atmospheric mercury as recorded in white spruce tree-rings in Central Yukon territory.* University of Toronto (Canada).

Clark, A. J., Labaj, A. L., Smol, J. P., Campbell, L. M., & Kurek, J. (2021). Arsenic and mercury contamination and complex aquatic bioindicator responses to historical gold mining and modern watershed stressors in urban Nova Scotia. *Canada. Science of The Total Environment, 787*, 147374. https://doi.org/10.1016/j.scitotenv.2021.147374

Craswell, E. (2021). Fertilizers and nitrate pollution of surface and ground water: An increasingly pervasive global problem. *SN Applied Sciences, 3*(4), 518. https://doi.org/10.1007/s42452-021-04521-8

D'Angioli, A. M., Giles, A. L., Costa, P. B., Wolfsdorf, G., Pecoral, L. L., Verona, L., & Abrahao, A. (2022). Abandoned pastures and restored savannas have distinct patterns of plant–soil feedback and nutrient cycling compared with native Brazilian savannas. *Journal of Applied Ecology, 59*(7), 1863–1873. https://doi.org/10.1111/1365-2664.14193

Das, D., Ray, P., & Datta, S. P. (2024). Nutrient cycling and climate change. In *Climate change impacts on soil-plant-atmosphere continuum* (pp. 37–62). Springer Nature Singapore.

Daunoras, J., Kačergius, A., & Gudiukaitė, R. (2024). Role of soil microbiota enzymes in soil health and activity changes depending on climate change and the type of soil ecosystem. *Biology, 13*(2), 85. https://doi.org/10.3390/biology13020085

David, S. (2023). Reconciling green and oil state goals: Guyana as a case study. *GNOSI: An Interdisciplinary Journal of Human Theory and Praxis, 6*(1), 1–18.

de Lacerda, L. D., Ward, R. D., Borges, R., & Ferreira, A. C. (2022). Mangrove trace metal biogeochemistry response to global climate change. *Frontiers in Forests and Global Change, 5*, 817992. https://doi.org/10.3389/ffgc.2022.817992

De Sousa, K., & Amoah, N. (2012). Long term closure planning for an evolving mine site–a case study of Newmont Boddington gold mine in Western Australia. In *Mine closure 2012: Proceedings of the seventh international conference on mine closure* (pp. 251–264). Australian Centre for Geomechanics.

De Witte, M., & Oosterbaan, M. (2013). Introduction gold. *Etnofoor, 25*(1), 7–13.

Djenchuraev, N. (1999). *Current environmental issues associated with mining wastes in Kyrgyzstan* (Master of Science). Department of Environmental Sciences and Policy of Central European University.

Do, M. H., Nguyen, G. T., Thach, U. D., Lee, Y., & Bui, T. H. (2023). Advances in hydrometallurgical approaches for gold recovery from E-waste: A comprehensive review and perspectives. *Minerals Engineering, 191*, 107977. https://doi.org/10.1016/j.mineng.2022.107977

Duran, P. (2021). *Mining's true impact in Mexico*. Available at https://mexicobusiness.news/mining/news/minings-true-impact-mexico

Ericsson, M., & Löf, O. (2019). Mining's contribution to national economies between 1996 and 2016. *Mineral Economics, 32*(2), 223–250. https://doi.org/10.1007/s13563-019-00191-6

Fedyukevich, V. A., & Vorob'ev-Desyatovskii, N. V. (2016). Advantages and disadvantages of various kinds of adsorbents used in industrial extraction of [Au (CN) 2]–anions from cyanide solutions and pulps. *Russian Journal of Applied Chemistry, 89*(4), 577–582. https://doi.org/10.1134/S1070427216040091

Gerke, J. (2022). The central role of soil organic matter in soil fertility and carbon storage. *Soil Systems, 6*(2), 33. https://doi.org/10.3390/soilsystems6020033

Globaldata (2024). *The five largest gold mines in operation in South Africa*. Available at https://www.mining-technology.com/data-insights/five-largest-gold-mines-south-africa/?cf-view&cf-closed

Glotov, V. E., Chlachula, J., Glotova, L. P., & Little, E. (2018). Causes and environmental impact of the gold-tailings dam failure at Karamken, the Russian Far East. *Engineering Geology, 245*, 236–247. https://doi.org/10.1016/j.enggeo.2018.08.012

Grbeš, A., Dokoza, V., Lukač, G., & Bilić, Š. (2024). Potential for restoration of biodiversity in karstic mining regions with small-scale mining sites: Birds and pits. *Euro-Mediterranean Journal for Environmental Integration*, 1–15. https://doi.org/10.1007/s41207-024-00488-1

Gyamfi, O., Sørensen, P. B., Darko, G., Ansah, E., Vorkamp, K., & Bak, J. L. (2021). Contamination, exposure and risk assessment of mercury in the soils of an artisanal gold mining community in Ghana. *Chemosphere, 267*, 128910. https://doi.org/10.1016/j.chemosphere.2020.128910

Ikhumetse, A. A., Abioye O. P., & Aransiola, S. A. (2019). Biosorption potential of bacteria on lead and chromium in groundwater obtained from mining community. *Acta Scientific Microbiology, 2*(6), 123–137. https://doi.org/10.31080/ASMI.2019.02.0252

Ilyas, S. (2018). Cyanidation of gold-bearing ores. In *Gold metallurgy and the environment* (pp. 51–102). CRC Press.

Karri, R. R., Ravindran, G., & Dehghani, M. H. (2021). Wastewater—Sources, toxicity, and their consequences to human health. In *Soft computing techniques in solid waste and wastewater management* (pp. 3–33). Elsevier.

Kedr.Media. (2023). *A polluted river runs through it: How gold mining is destroying Eastern Siberia*. Available at https://www.rferl.org/a/krasnoyarsk-siberia-polluted-rivers-gold-mining-russia-catastrophe/32599027.html

Kumari, D., Thakur, A., Tiwari, A. K., & Singh, R. (2024). Types and sources of agricultural pollution: Identifying the different pollutants generated by farming practices and their origins. In *A comprehensive exploration of soil, water, and air pollution in agriculture* (p. 71). BFC Publications.

Laniyan, T. A., & Adewumi, A. J. (2023). Ecological and human health risk associated with potentially toxic metals in water from Ijero mining area, Southwest Nigeria. *International Journal of Environmental Analytical Chemistry, 103*(16), 4677–4701. https://doi.org/10.1080/03067319.2021.1931159

Lemiere, B. (2018). A review of pXRF (field portable X-ray fluorescence) applications for applied geochemistry. *Journal of Geochemical Exploration, 188*, 350–363. https://doi.org/10.1016/j.gexplo.2018.02.006

Leuschner, C., Ellenberg, H., Leuschner, C., & Ellenberg, H. (2017). The central European vegetation as the result of millennia of human activity. *Ecology of Central European Forests: Vegetation Ecology of Central Europe, 1*, 31–116. https://doi.org/10.1007/978-3-319-43042-3_3

Li, S., Yu, L., Jiang, W., Yu, H., & Wang, X. (2022). The recent progress China has made in green mine construction, Part I: Mining groundwater pollution and sustainable mining. *International Journal of Environmental Research and Public Health, 19*(9), 5673. https://doi.org/10.3390/ijerph19095673

Lin, Y., Guo, M., & Gan, W. (1997). Mercury pollution from small gold mines in China. *Water, Air, & Soil Pollution, 97*, 233–239. https://doi.org/10.1023/A:1018384822608

Manning, T. J., & Kappes, D. W. (2016). Heap leaching of gold and silver ores. In *Gold ore processing* (pp. 413–428). Elsevier.

Martinez, G. (2022). *Formalization, certification, and technology: The complex landscape of artisanal and small-scale gold mining interventions in Peru* (Doctoral dissertation). Colorado School of Mines.

Mathioudakis, S., Xiroudakis, G., Petrakis, E., & Manoutsoglou, E. (2023). Alluvial gold mining technologies from ancient times to the present. *Mining, 3*(4), 618–644. https://doi.org/10.3390/mining3040034

Meutia, A. A., Bachriadi, D., & Gafur, N. A. (2023). Environment degradation, health threats, and legality at the artisanal small-scale gold mining sites in Indonesia. *International Journal of Environmental Research and Public Health, 20*(18), 6774. https://doi.org/10.3390/ijerph20186774

Moreno-Brush, M., McLagan, D. S., & Biester, H. (2020). Fate of mercury from artisanal and small-scale gold mining in tropical rivers: Hydrological and biogeochemical controls. A critical review. *Critical Reviews in Environmental Science and Technology, 50*(5), 437–475. https://doi.org/10.1080/10643389.2019.1629793

Murali, A., Zhang, Z., Shine, A. E., Free, M. L., & Sarswat, P. K. (2022). E-wastes derived sustainable Cu recovery using solvent extraction and electrowinning followed by thiosulfate-based gold and silver extraction. *Journal of Hazardous Materials Advances, 8*, 100196. https://doi.org/10.1016/j.hazadv.2022.100196

Onemu, S. O., Obeagu, E. I., Popoola, O. J., Osuntuyi, M. A., & Isibor, C. N. (2024). An assessment of the immune status of some stone quarry workers in Ondo state, Nigeria. *Medicine, 103*(2), e36969. https://doi.org/10.1097/MD.0000000000036969

Pattnaik, S., Mohapatra, B., & Gupta, A. (2021). Plant growth-promoting microbe mediated uptake of essential nutrients (Fe, P, K) for crop stress management: Microbe–soil–plant continuum. *Frontiers in Agronomy, 3*, 689972. https://doi.org/10.3389/fagro.2021.689972

Pawlik, M. (2022). Fundamentals of froth flotation. *ChemTexts, 8*(4), 19. https://doi.org/10.1007/s40828-022-00170-5

Ramzey, M., Abd El-Rahman, Y., & Said, A. (2021). Weathering products at the historic Qulaan gold prospect, Eastern Desert, Egypt: Implication on the mobility and distribution of arsenic, gold and silver. *Journal of African Earth Sciences, 182*, 104276. https://doi.org/10.1016/j.jafrearsci.2021.104276

Reuben, A., Manczak, E. M., Cabrera, L. Y., Alegria, M., Bucher, M. L., Freeman, E. C., & Perry, M. J. (2022). The interplay of environmental exposures and mental health: Setting an agenda. *Environmental Health Perspectives, 130*(2), 025001.

Roberto, F. F., & Schippers, A. (2022). Progress in bioleaching: Part B, applications of microbial processes by the minerals industries. *Applied Microbiology and Biotechnology, 106*(18), 5913–5928. https://doi.org/10.1007/s00253-022-12085-9

Saim, A. K., Ofori-Sarpong, G., & Amankwah, R. K. (2022). Oxidation behaviour and bio-oxidation of gold-bearing sulphide ores: Oxygen capabilities and challenges. *Ghana Mining Journal., 22*(2), 15–25.

Schwartz, M., Smits, K., & Phelan, T. (2023). Quantifying mercury use in artisanal and small-scale gold mining for the Minamata Convention on Mercury's national action plans: Approaches and policy implications. *Environmental Science & Policy, 141*, 1–10. https://doi.org/10.1016/j.envsci.2022.12.002

Shi, T., Guo, L., Chen, Y., Wang, W., Shi, Z., Li, Q., & Wu, G. (2018). Proximal and remote sensing techniques for mapping of soil contamination with heavy metals. *Applied Spectroscopy Reviews, 53*(10), 783–805. https://doi.org/10.1080/05704928.2018.1442346

Talukder, P., Ray, R., Sarkar, M., Das, A., & Chakraborty, S. (2023). Adverse effects of mining pollutants on terrestrial and aquatic environment and its remediation. *Environmental Quality Management, 33*, 595–610. https://doi.org/10.1002/tqem.22121

Tibane, L. V., & Mamba, D. (2022). Ecological risk of trace metals in soil from gold mining region in South Africa. *Journal of Hazardous Materials Advances, 7*, 100118. https://doi.org/10.1016/j.hazadv.2022.100118

Tibau, A. V., & Grube, B. D. (2019). Mercury contamination from dental amalgam. *Journal of Health and Pollution, 9*(22), 190612. https://doi.org/10.5696/2156-9614-9.22.190612

Timsina, S., Hardy, N. G., Woodbury, D. J., Ashton, M. S., Cook-Patton, S. C., Pasternack, R., & Martin, M. P. (2022). Tropical surface gold mining: A review of ecological impacts and restoration strategies. *Land Degradation & Development, 33*(18), 3661–3674. https://doi.org/10.1002/ldr.4430

UNDRR. (2023). *Uzbekistan makes efforts to strengthen capacity on mine tailings and prevention of accidental water pollution, with UNECE's support*. Available at https://www.preventionweb.net/news/uzbekistan-makes-efforts-strengthen-capacity-mine-tailings-and-prevention-accidental-water

UNICEF. (2021). *Wastewater from jewelry manufacturing, which is widely used in goldsmithing and jewelry-making processes, may contain gold particles if not properly treated*. Available at https://ntnuopen.ntnu.no/ntnu-xmlui/handle/11250/2982963

Vanka, K. S., Shukla, S., Gomez, H. M., James, C., Palanisami, T., Williams, K., & Horvat, J. C. (2022). Understanding the pathogenesis of occupational coal and silica dust-associated lung disease. *European Respiratory Review, 31*(165), 210250. https://doi.org/10.1183/16000617.0250-2021

Verbrugge, B., & Geenen, S. (2019). The gold commodity frontier: A fresh perspective on change and diversity in the global gold mining economy. *The Extractive Industries and Society, 6*(2), 413–423. https://doi.org/10.1016/j.exis.2018.10.014

Walker, R. A. (2001). California's golden road to riches: Natural resources and regional capitalism, 1848–1940. *Annals of the Association of American Geographers, 91*(1), 167–199. https://doi.org/10.1111/0004-5608.00238

Wang, Z., Luo, K., Zhao, Y., Lechner, A. M., Wu, J., Zhu, Q., & Wang, Y. (2022). Modelling regional ecological security pattern and restoration priorities after long-term intensive open-pit coal mining. *Science of the Total Environment, 835*, 155491. https://doi.org/10.1016/j.scitotenv.2022.155491

Wilson, F. (2001). Minerals and migrants: How the mining industry has shaped South Africa. *Daedalus, 130*(1), 99–121.

Wong, H. K. T., Gauthier, A., Beauchamp, S., & Tordon, R. (2002). Impact of toxic metals and metalloids from the Caribou gold-mining areas in Nova Scotia, Canada. *Geochemistry: Exploration, Environment, Analysis, 2*(3), 235–241. https://doi.org/10.1144/1467-787302-027

Yang, M., Wang, S., Zhang, L., Wu, Q., Wang, F., Hui, M., & Hao, J. (2016). Mercury emission and speciation from industrial gold production using roasting process. *Journal of Geochemical Exploration, 170*, 72–77. https://doi.org/10.1016/j.gexplo.2016.08.014

Zhou, H., Bhattarai, R., Li, Y., Si, B., Dong, X., Wang, T., & Yao, Z. (2022). Towards sustainable coal industry: Turning coal bottom ash into wealth. *Science of the Total Environment., 804*, 149985. https://doi.org/10.1016/j.scitotenv.2021.149985

Arbuscular Mycorrhizal: Invisible Allies in Plant Development

Juan Carlos González Rojas and Marjorie Jazmin Salazar Orellana

Abstract Arbuscular mycorrhizal fungi (AMF) establish mutualistic associations with almost 80% of terrestrial plants, playing a fundamental role in soil vitality and plant development. They improve nutrient absorption, crop quality and contribute to sustainable agriculture. In addition, they help plants resist diseases and tolerate the presence of heavy metals and water scarcity, obtaining sugars and organic substances from the plants.

The AMF-plant association constitutes an important evolutionary adjustment that supports plant functionality. The mycorrhization process involves a complex interaction that improves nutrient absorption and plant development. AMF enhances crop production, nutrient and water intake. Strengthens stress tolerance, with variable benefits depending on the type of plant and environmental conditions.

AFM interact with other soil microorganisms, improving the development and resistance of plants to weather fluctuations. AFMs also improve soil structure and carbon storage, restoring soil health and promoting ecosystem stability. However, its application faces challenges, such as inconsistent results in the field and deficiencies

J. C. González Rojas (✉)
Centro de Investigación, Innovación y Transferencia de Tecnología, Universidad Católica de Cuenca, Cuenca, Ecuador

Laboratorio de Entomología Agrícola Fitosanitario, Universidad Católica de Cuenca, Jardín Botánico de Cuenca, Cuenca, Ecuador

Geociencias, Ambiente y Recursos Naturales - GARN - Grupo de Investigación, Universidad Católica de Cuenca, Cuenca, Ecuador
e-mail: jgonzalezr@ucacue.edu.ec

M. J. Salazar Orellana (✉)
Centro de Investigación, Innovación y Transferencia de Tecnología, Universidad Católica de Cuenca, Cuenca, Ecuador

Laboratorio de Ecología Microbiana y Principios Activos, Universidad Católica de Cuenca, Cuenca, Ecuador

Geociencias, Ambiente y Recursos Naturales - GARN - Grupo de Investigación, Universidad Católica de Cuenca, Cuenca, Ecuador
e-mail: jazmin.salazar@ucacue.edu.ec

S. A. Aransiola et al. (eds.), *Soil Microbiome in Green Technology Sustainability*,
https://doi.org/10.1007/978-3-031-71844-1_21

in understanding its stress tolerance mechanisms. Future research should focus on creating efficient AFM communities tailored to specific crops and circumstances, leveraging new technologies to improve the understanding and optimization of formulations for inoculation.

Keywords Arbuscular mycorrhizal fungi · Soil microbiology · Plants · Sustainable agriculture

1 Introduction

Mycorrhizal fungi form a relationship with almost 80% of terrestrial plants (Roy & Choudhury, 2022), this has been proven to be invaluable for trees and agriculture in general, guaranteeing soil health and plant growth. But what exactly is mycorrhiza and what are its effects?

Mycorrhizae are symbiotic fungi that establish a mutualistic relationship with plants, benefiting both parties (Yadav et al., 2023; Kalamulla et al., 2022). These fungi improve the absorption of nutrients, especially phosphorus and other elements found in small quantities (Khaliq et al., 2022), promote growth, improve crop quality and contribute to sustainable agricultural practices (Prasad, 2021; Soti et al., 2023). It has been suggested that probably the main role is related to defense against pathogens and tolerating adverse conditions such as the presence of heavy metals and water stress (Frew et al., 2024). On the other hand, fungi receive sugars and other organic compounds from the plant as energy sources (Fig. 1). This exchange of

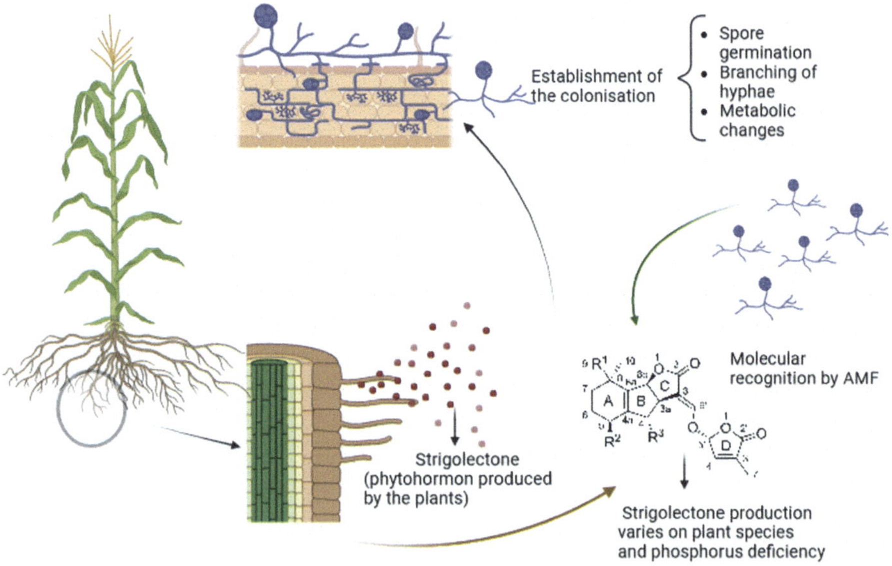

Fig. 1 Benefits of the symbiosis between arbuscular mycorrhizal fungi and plants

nutrients makes this relationship of great importance, it is estimated that almost the entire plant kingdom establishes some type of mycorrhiza and that 90% of these are arbuscular (Fitter & Moyersoen, 1996).

The diversity of AMF varies among plant species, influencing root colonization and spore number. There are two large groups of mycorrhizal fungi: ectomycorrhizal and endomycorrhizal. Ectomycorrhizal organisms are associated with species from the families *Fabaceae, Dipterocarpaceae, Gnetaceae, Polygonaceae and Nyctaginaceae* (Corrales et al., 2018; Vasco-Palacios et al., 2017), in this case the fungus does not depend on the plant to survive. In the case of endomycorrhizal fungi, they are obligate symbionts whose association is nonspecific, dominated by the presence of arbuscular mycorrhizal fungi and which require this to survive (Helgason & Fitter, 2009).

The association between AMF and plants represents an important evolutionary adaptation that favors sustainable agriculture and overall plant performance.

2 Mycorrhization Process

Although a lot has been done so far in the study of mycorrhiza, the interrogations continue regarding the mycorrhization process. Is there communication between fungi and plants? Do plants select with whom to establish this symbiotic relationship? Is it a general or specific relationship? These questions continue to generate interest in the scientific community.

The mycorrhizal process involves symbiotic associations with plant roots to enhance nutrient uptake and plant growth (Yang & Irman, 2023). AMF play a crucial role maintaining the nutritional balance of any ecosystem (Roy et al., 2018). Factors that influence the colonization process include soil ecology, genetic diversity of fungi and host plants, and agronomic practices (in 't Zandt et al., 2022).

During the process of root colonization by fungi, several structures are formed, including extraradical hyphae, vesicles, arbuscules, auxiliary cells, intracellular coiled hyphae, intercellular hyphae, appressoria, infection units, and different apoplastic compartments within the roots. The development of these structures is crucial for the bidirectional transfer of nutrients between symbionts and plays a fundamental role in the symbiosis success (Dreyer et al., 2014).

As a result of the association between AMF and radicular structures of the plants, arbuscules are formed within the root cortical cells, where an exchange of chemical signals takes place. During this process, secondary metabolites and hormones are released by the plant into the rhizosphere. These hormones are known as strigolactones (SL), which are phytohormones derived from carotenoid compounds, play a crucial role in root colonization (Lanfranco et al., 2018; Rochange et al., 2019), as they induce several responses in AMF, such as spore germination, hyphal branching and metabolic changes (Akiyama et al., 2005; Foo, 2013). The composition of strigolactones depends on the plant species and phosphorus deficiency in the soil (Akiyama & Hayashi, 2006), being this the first signal of communication

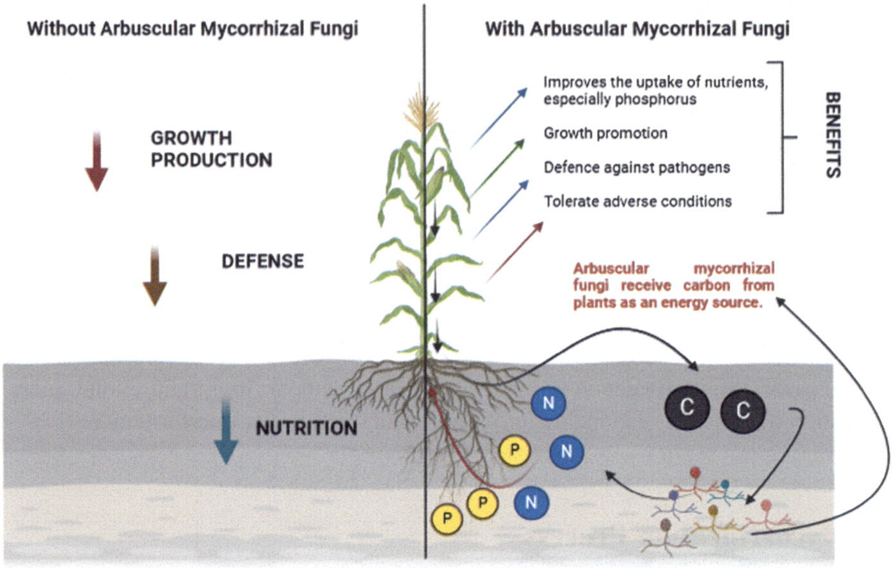

Fig. 2 Beginning of the root colonization process by AMF in the presence of the phytohormone strigolactone produced by plants

between the plant and its symbiotic partner and the start of cellular programming for colonization (Fig. 2).

After this process and if the fungi successfully recognize this signal, a molecular recognition is initiated in which mycorrhizal factors (Myc factors) are synthesized and secreted, the group of proteins that are synthesized come from the translation of mycorrhizal genes (Myc genes). This process makes the plant very specific and allows the establishment of arbuscular mycorrhizal symbiosis.

Once the fungi recognize this signal, it initiates a molecular recognition in which mycorrhization factors (Myc factors) are synthesixed and secreted. The group of proteins synthesized come from the translation of mycorrhizal genes (Myc Genes), this process gives specificity to the plant and allows the establishment of arbuscular mycorrhizal symbiosis.

In the process of contact and penetration (Choi et al., 2018), an attachment structure is generated on the surface of the root epidermis called hyphopodium (Nagahashi & Douds, 1997). Once it adheres, a pre-penetration apparatus (APP) is generated stimulated by Myc factors and the fungus itself. APP prevents it from having contact with the cytoplasm of the plant cell, preventing the host defense response (Genre et al., 2008).

Finally, when fungi are inside the plant cell, they undergo essential changes for their new function, basically morphological, such as: increase in cell size, reorganization and compaction of cellular organelles and the formation of the arbuscule.

These are the key structures for the exchange of nutrients between the plant species and the fungi (Saha et al., 2014).

Overall, the mycorrhizal process involves intricate interactions between AMF, plant roots, soil conditions to optimize nutrient uptake and promote plant health.

3 Benefits for Plants

AMF have a significant impact on the crop growth by improving nutrient, water absorption and tolerance to various stress factors (Kalamulla et al., 2022).

AMF facilitates biomass production, increases essential nutrient levels and enhances antioxidant activity in crops facing stressful conditions, this leads to improvements in both growth and yield outcomes (Khaliq et al., 2022). The ability of AMF to mitigate biotic and abiotic stress factors such as drought, salinity, high temperatures and exposure to heavy metals is also considered, significantly contributing to improved plant performance and levels of overall productivity (Alam et al., 2023). By establishing an intricate network of hyphae in the soil environment, AMF plays a crucial role in improving soil structure, increasing water holding capacity and increasing the availability of essential nutrients for plants (Chen et al., 2014). All of these effects combined result in a positive influence on overall plant growth and productivity, highlighting the intricate and multifaceted mechanisms through which AMF can have a positive impact on agricultural systems.

The benefits of AMF vary across plant species and environmental conditions; Nutrient uptake efficiency can vary depending on the specific AMF species, host plant type, and soil conditions (Ma et al., 2022; Karunanantham et al., 2022).

In vegetable crops, AMF reduces mortality and improves plant establishment, which is essential for transplant success and subsequent yield (Nogueira de Sousa, 2023). Studies have shown that AMF, such as *Rhizophagus clarus, R. irregularis* and *Paraglomus occultum*, increase root length, shoot mass and overall biomass, leading to a 28% increase in dry weight of shoots and a 20% increase in yield under drought conditions (Alam et al., 2023; Zargar & Balwan, 2022). Inoculation with *Claroideoglomus claroideum* produced an 84% increase in fresh weight and a 124.5% increase in lycopene concentration in tomatoes (Orine et al., 2022), and improved the dissolution of low-solubility fertilizers, which increases the absorption and formation of biomass by 22% (Felföldi et al., 2022).

In the context of cereals, inoculation with *Rhizophagus irregularis* and *Funneliformis mosseae* and *Claroideoglomus etunicatum* has been found to improve cereal yield (Basiru et al., 2021). AMF in wheat improves morphological parameters, chlorophyll content and soluble protein content, especially under water deficit conditions, thus optimizing production in arid and semi-arid regions (Ould Amer et al., 2023; Agbodjato et al., 2022). In the case of corn, it has been shown that species of the genus *Rhizophagus* and *Gigaspora margarita* have a significant positive impact on the crop (Carrara & Heller, 2022). Research has demonstrated that coinoculation of native strains of *Glomus caledonius*, *Rhizophagus intraradices*,

Funneliformis geosporum together with a medium dose of chemical fertilization increases yields (Aguégué et al., 2022).

In citrus, a mixed treatment with *Funneliformis mosseae*, *Diversispora versiformis*, *F. mosseae* and *Rhizoglomus intraradices* recorded an increase in yield (Qiang-Sheng et al., 2020). In olive nursery plants, native strains of *Funneliformis geosporum* and *Septoglomus constrictum*, give rise to more robust plants (Wu et al., 2022).

In apple cultivation, *Gigaspora albida* increases yield on micropropagated rootstocks, due to efficiency in the use of nutrients (Dalla Costa et al., 2021) and in crop fields, *Glomus versiforme*, *Rhizophagus intraradices* and *Claroideoglomus etunicatum* have additive effects on performance (Hosseini & Gharghani, 2015).

The positive effect on the increase in biomass and essential oils in thyme and lavender plants has been determined due to the presence of *Funneliformis mosseae* and *Rhizophagus irregularis* respectively (Akachoud et al., 2022), in turn *Glomus viscosum*, *Gigaspora margarita*, *Gigaspora rosea* and *F. mosseae* increase the essential oil yield of various species associated with the increase in trichomes (Israel et al., 2022).

4 Interaction Between Organisms and the Soil

Soil microorganisms maintain intra-specific relationships between bacteria-fungi and fungi-nematodes and interspecific relationships between fungi of the same genus (Selim & Zayed, 2017). These microbial interactions constitute the most relevant activities that occur in terrestrial ecosystems, including agroecosystems. Thus, depending on each other to fulfill essential functions in the soil, therefore the approach to study soil life must be comprehensive (Orozco-Mosqueda et al., 2022).

Interactions between mycorrhizal fungi and other microorganisms, such as phosphate-solubilizing bacteria (PSB), nitrogen-fixing bacteria and biocontrol agents, play a crucial role in regulating plant growth, development and resistance to climate change (Rahimzadeh & Pirzad, 2017; Sharma et al., 2023). Inoculation of bacteria can also stimulate mycorrhization processes, therefore the inoculation of both bacteria and AMF is recommended, which is directly reflected in the improvement of soil structure, nutrient mobilization and moisture retention (Noceto et al., 2021).

Among other benefits, the improvement in the nutritional and organoleptic properties of foods is considered, also known as biofortification, it is the natural enrichment of the components of vegetables and fruits through the application of microorganisms such as mycorrhizal fungi and plant growth promoting bacteria (PGPR) (Gío-Trujillo et al., 2022).

The most successful and profitable co-inoculation is HMA with *Azospirillum*, as it benefits both the microorganisms and the host plant. Overall, these three-way interactions between AMF, microorganisms and plants are essential for sustainable agriculture and crop productivity (Domínguez-Núñez et al., 2015).

On the other hand, the role of AMF in the formation of soil aggregates is fundamental, since it influences the microbial environment of the rhizosphere and improves soil structure. AMF improve soil fertility through biochemical, biophysical and biological processes, as demonstrated by their association with organic carbon, phosphorus, cation exchange capacity, orientation of soil particles, production and release of binding substances that intervene. in the soil structure forming agglomerates or microaggregates (Ebbisa, 2023). Microorganisms also influence pH changes in the rhizosphere, mobilize and release nutrients, retain moisture, reduce the presence of diseases and even bioremediation processes (Philippot et al., 2024; Sudha Samuel & Veeramani, 2021).

Additionally, AMF offer benefits such as carbon sequestration, tolerance to water stress and protection against pests, making them valuable fertilizers for sustainable agricultural practices. In general, AMF contribute directly to improving soil health.

Carbon sequestration and the priming effect improve soil fertility, contribute to the stabilization of organic matter and the cycling of nutrients, thus enhancing carbon sequestration in the soil (Fall et al., 2022). AMF promote decomposition through the priming effect, facilitating nitrogen acquisition and potentially reducing soil carbon storage. This process is influenced by factors such as the C:N ratio and absolute latitude (Choreño-Parra & Treseder, 2024).

AMF also accelerate soil carbon turnover, increase the rhizosphere priming effect, and improves the stabilization of rhizodeposits, significantly contributing to carbon accumulation in soil-plant systems (Zhen et al., 2022). It is estimated that global plant communities annually allocate 13.12 Gt of CO_2 to the underground mycelium of mycorrhizal fungi (Hawkins et al., 2023).

Overall, the symbiotic relationship between plants and AMF influences soil biogeochemical cycling, microbial activities, and soil structure, highlighting the importance of AMF in sustainable soil management and carbon sequestration efforts.

5 Practical Applications

AMF represent a feasible, economical and ecological, biotechnological tool for the remediation of contaminated soils, particularly those with heavy metal contamination. They achieve this through the secretion of glomalin, a glycoprotein that binds heavy metals and enhances soil properties (Trocio & Paguntalan, 2023). Arbuscular mycorrhizae contribute to phytoremediation by facilitating the absorption and compartmentalization of heavy metals within the roots of plants, thus preventing their translocation to the aerial parts and the food chain (Khan & Mohammad, 2022; Adetunji et al., 2022). In addition, it favors the remediation of contaminated soils, since it increases the tolerance of plants to contaminants, facilitates the degradation, stabilization and extraction of contaminants (Sun et al., 2023). Despite the potential of AMF in bioremediation, its effectiveness may vary depending on factors such as AMF species, crop type, and soil conditions (Kumar & Suprasanna, 2021).

The AMF application on degraded soils holds great promise for improving soil quality and plant growth, making it a vital tool for ecological restoration and sustainable agriculture. Studies have shown that AMF inoculation, combined with soil improving agents such as biochar (Alotaibi et al., 2023), fly ash and surface vegetation, significantly increases microbial biomass and plant growth, as seen in soil restoration of land tailings from mining (Adetunji et al., 2022; Bekti et al., 2022).

AMF have demostrated significant potential in enhacing plant tolerance to various types of abiotic stress, such as drought, salinity, and extreme temperatures. They achieve this by upregulating antioxidant systems, improving photosynthesis, maintaining membrane integrity and supporting hydric relationships that contribute the adequate metabolic functioning of plants (Pratibha et al., 2023; Yang & Irman, 2023). In drought stress situations, AMF improve water use efficiency, osmolyte accumulation and reduce reactive oxygen species by enhancing antioxidant activities and gene expression, thereby protecting the photosynthetic apparatus and improving the general plant growth (Tang et al., 2022; Karunanantham et al., 2022). Under saline conditions, AMF inoculation modulates the expression of stress-responsive genes, improves nutrient status and nitrogen balance index, thereby increasing growth and stress tolerance in plants (Singh et al., 2022b; Liang et al., 2022). Furthermore, AMF can alleviate the negative effects of waterlogging and drought by increasing root colonization, improving dry weight and relative growth of plants (Meena et al., 2022). The use of AMF as biofertilizers not only strengthens the adaptability of plants to challenging environments but also improves soil health and ecosystem stability (Abbaspour et al., 2021).

It has been argued that AMF do not provide tolerance to biotic stress in plant development (Ortas et al., 2021), however, several studies consider that they definitely play a primary role. Since they improve the tolerance of host plants against bacteria, fungi, viruses, phytopathogenic nematodes and herbivores through competition for nutrients, space and host photosynthates, alteration of the rhizosphere and induction of host defenses (Dowarah et al., 2022). This contributes to plant health (Branco et al., 2022), and promotes environmentally sustainable practices (Suchitra et al., 2020).

AMF contributes to the stabilization of soil organic matter by transporting substantial amounts of plant-derived C into the soil. This carbon serves as a substrate for other organisms (Singh et al., 2022a), influencing soil microbial dynamics and increasing both its activity and diversity (Hnini et al., 2024). This favors the decomposition of organic matter and the formation of stable carbon reserves (Son et al., 2021). Consequently, this process is facilitated by the production of glomalin -related soil protein, which is known for its high C content and its ability to stabilize soil aggregates, thus protecting soil organic C (Peduru et al., 2021). Furthermore, AMF hyphae interact with soil minerals such as goethite, which enhances the stabilization of rhizodeposits and promotes the formation of aggregates, which helps to further sequester C (Cheng et al., 2021).

6 Challenges and Future Research

AMF represent significant potential for sustainable agriculture, but their application faces several challenges and requires further research. One of the main challenges are the inconsistent results seen when translating success from controlled environments to real-world agricultural fields, which is due in part to the complex interactions between AMF and indigenous soil microbes. Furthermore, there are gaps in the understanding of the cellular and molecular mechanisms by which AMF confers stress tolerance. It is also necessary to carefully evaluate the possible ecotoxicological impacts of phytosanitary products on AMF to ensure that these fungi can be safely integrated into agricultural practices.

The lack of robust methodologies for large-scale production and quality control of inoculants complicates their widespread adoption.

Future research should focus on developing effective AMF communities adapted to specific crops and environmental conditions, taking advantage of new technologies, such as next-generation sequencing, to better understand AMF interactions and optimize inoculant formulations; Additionally, field studies are essential to validate the benefits of AMF in various agricultural settings and to refine application techniques for maximum effectiveness.

7 Conclusions

The complex and mutually beneficial association that exists between arbuscular mycorrhizal fungi (AMF) and plants encompasses a wide range of substantial benefits, from improvements in soil quality and nutrient uptake to increased plant resilience against various stress factors, both abiotic and biotic. This distinctive interconnection between AMF and plants plays a critical role in the realm of sustainable agriculture and overall plant performance. Despite the many advantages it offers, there are obstacles that require further investigation to achieve a comprehensive understanding of its practicality and optimize its effectiveness in authentic agricultural environments. By focusing on cultivating effective AMF communities adapted to specific crops and environmental circumstances, along with refining application methodologies, notable progress can be made towards the successful integration of AMF into sustainable agricultural methodologies.

References

Abbaspour, H., Pour, F. S. N., & Abdel-Wahhab, M. A. (2021). Arbuscular mycorrhizal symbiosis regulates the physiological responses, ion distribution and relevant gene expression to trigger salt stress tolerance in pistachio. *Physiology and Molecular Biology of Plants, 27*, 1765–1778. https://doi.org/10.1007/s12298-021-01043-w

Adetunji, C., Anani, O., Thangadurai, D., & Islam, S. (2022). Arbuscular mycorrhizas: Applications in organic agriculture and beyond. In J. Sangeetha, K. Soytong, D. Thangadurai, & A. Al-Tawaha (Eds.), *Organic farming for sustainable development* (1st ed., p. 27). Apple Academic Press.

Agbodjato, A., Assogba, A., Babalola, O., Koda, D., Aguégué, M., Sina, H., Dagbénonbakin, D., Adjanohoun, A., & Baba-Moussa, L. (2022). Formulation of biostimulants based on Arbuscular Mycorrhizal fungi for maize growth and yield. *Frontiers in Agronomy, 4*. https://doi.org/10.3389/fagro.2022.894489

Aguégué, M. R., Ahoyo Adjovi, N. R., Agbodjato, N. A., Noumavo, P., Assogba, S., Salami, H., Salako, V., Ramón, R., Baba-Mousa, F., Adjanohoun, A., Kakai, R., & Baba-Mousa, L. (2022). Efficacy of native strains of Arbuscular Mycorrhizal fungi on maize productivity on Ferralitic soil in Benin. *Agricultural Research, 11*, 627–641. https://doi.org/10.1007/s40003-021-00602-7

Akachoud, O., Bouamama, H., Facon, N., Laruelle, F., Zoubi, B., Benkebboura, A., Ghoulam, C., Qaddoury, A., & Lounès-Hadj Sahraoui, A. (2022). Mycorrhizal inoculation improves the quality and productivity of essential oil distilled from three aromatic and medicinal plants: Thymus satureioides, Thymus pallidus, and Lavandula dentata. *Agronomy, 12*, 2223. https://doi.org/10.3390/agronomy12092223

Akiyama, K., & Hayashi, H. (2006). Strigolactones: Chemical signals for fungal symbionts and parasitic weeds in plant roots. *Annals of Botany, 97*, 925–931. https://doi.org/10.1093/aob/mcl063

Akiyama, K., Matsuzaki, K. I., & Hayashi, H. (2005). Plant sesquiterpenes induce hyphal branching in arbuscular mycorrhizal fungi. *Nature, 435*, 824–827. https://doi.org/10.1038/nature03608

Alam, M., Choudhury, T., & Mridha, M. (2023). Arbuscular Mycorrhizal fungi enhance biomass growth, mineral content, and antioxidant activity in tomato plants under drought stress. *Journal of Food Quality, 2023*, 14–14. https://doi.org/10.1155/2023/2581608

Alotaibi, M., Ikram, M., Alotaibi, N., Hussain, G. S., Younis, A. M., Naz, N., & Danish, S. (2023). Examining the role of AMF-biochar in the regulation of spinach growth attributes, nutrients concentrations, and antioxidant enzymes in mitigating drought stress plant. *Stress, 10*, 100205. https://doi.org/10.1016/j.stress.2023.100205

Basiru, S., Mwanza, H. P., & Hijri, M. (2021). Analysis of Arbuscular Mycorrhizal fungal inoculant benchmarks. *Microorganisms, 9*, 81. https://doi.org/10.3390/microorganisms9010081

Bekti, H. S., Budi, S. W., & Wibowo, C. (2022). Arbuscular mycorrhizal fungi inoculum and soil ameliorants enhance the growth of Falcataria moluccana in revegetation of post-silica sand mining land in Sukabumi, Indonesia. *Biodiversitas, 23*, 1181–1187.

Branco, S., Schauster, A., Liao, H. L., & Ruytinx, J. (2022). Mechanisms of stress tolerance and their effects on the ecology and evolution of mycorrhizal fungi. *The New Phytologist, 235*, 2158–2175. https://doi.org/10.1111/nph.18308

Carrara, J., & Heller, W. (2022). Arbuscular mycorrhizal species vary in their impact on nutrient uptake in sweet corn (Zea mays) and butternut squash (Cucurbita moschata). *Frontiers in Agronomy, 4*. https://doi.org/10.3389/fagro.2022.1040054

Chen, X., Song, F., Liu, F., Tian, C., Liu, C., Xu, H., & Zhu, X. (2014). Effect of different Arbuscular Mycorrhizal fungi on growth and physiology of maize at ambient and low temperature regimes. *The Scientific World Journal, 2014*, 7. https://doi.org/10.1155/2014/956141

Cheng, H., Wang, J., Tu, C., Lin, S., Xing, D., Hill, P., Chadwick, D., & Jones, D. L. (2021). Arbuscular mycorrhizal fungi and biochar influence simazine decomposition and leaching. *Global Change Biology Bioenergy, 13*, 708–718. https://doi.org/10.1111/gcbb.12802

Choi, J., Summers, W., & Paszkowski, U. (2018). Mechanisms underlying establishment of Arbuscular Mycorrhizal symbioses. *Annual Review of Phytopathology, 56*, 135–160. https://doi.org/10.1146/annurev-phyto-080516-035521

Choreño-Parra, E. M., & Treseder, K. K. (2024). Mycorrhizal fungi modify decomposition: A meta-analysis. *The New Phytologist, 242*, 2763–2774. https://doi.org/10.1111/nph.19748

Corrales, A., Henkel, T. W., & Smith, M. E. (2018). Ectomycorrhizal associations in the tropics—Biogeography, diversity patterns and ecosystem roles. *The New Phytologist, 220*, 1076–1091. https://doi.org/10.1111/nph.15151

Dalla Costa, M., Rech, T. D., Primieri, S., Pigozzi, B., Werner, S., & Stürmer, S. (2021). Inoculation with isolates of arbuscular mycorrhizal fungi influences growth, nutrient use efficiency and gas exchange traits in micropropagated apple rootstock 'Marubakaido'. *Plant Cell, Tissue and Organ Culture, 145*, 89–99. https://doi.org/10.1007/s11240-020-01994-0

Domínguez-Núñez, J. A., Berrocal-Lobo, M., & Albanesi, A. S. (2015). Interaction of Azospirillum and mycorrhiza. In F. Cassán, Y. Okon, & C. Creus (Eds.), *Handbook for Azospirillum* (pp. 419–432). Springer. https://doi.org/10.1007/978-3-319-06542-7_23

Dowarah, B., Gill, S. S., & Agarwala, N. (2022). Arbuscular Mycorrhizal fungi in conferring tolerance to biotic stresses in plants. *Journal of Plant Growth Regulation, 41*, 1429–1444. https://doi.org/10.1007/s00344-021-10392-5

Dreyer, B., Honrubia, M., & Morte, A. (2014). How root structure defines the Arbuscular Mycorrhizal Symbiosis and what we can learn from it? In A. Morte & A. Varma (Eds.), *Root engineering, soil biology* (pp. 145–169). Springer. https://doi.org/10.1007/978-3-642-54276-3_7

Ebbisa, A. (2023). Arbuscular Mycorrhizal Fungi (AMF) in optimizing nutrient bioavailability and reducing agrochemicals for maintaining sustainable agroecosystems. In N. de Sousa (Ed.), *Arbuscular mycorrhizal fungi in agriculture – New insights*. IntechOpen. https://doi.org/10.5772/intechopen.104271

Fall, A. F., Nakabonge, G., Ssekandi, J., Founoune-Mboup, H., Apori, S. O., Ndiaye, A., Badji, A., & Ngom, K. (2022). Roles of Arbuscular Mycorrhizal fungi on soil fertility: Contribution in the improvement of physical, chemical, and biological properties of the soil. *Frontiers in Fungal Biology, 3*, 723892. https://doi.org/10.3389/ffunb.2022.723892

Felföldi, Z., Vidican, R., Stoian, V., Roman, I. A., Sestras, A. F., Rusu, T., & Sestras, R. E. (2022). Arbuscular Mycorrhizal fungi and fertilization influence yield, growth and root colonization of different tomato genotype. *Plants, 11*, 1743. https://doi.org/10.3390/plants11131743

Fitter, A. H., & Moyersoen, B. (1996). Evolutionary trends in root-microbe symbioses. *Philosophical Transactions of the Royal Society of London B, 351*, 1367–1375. https://doi.org/10.1098/rstb.1996.0120

Foo, E. (2013). Something old, something new: Auxin and strigolactone interact in the ancient mycorrhizal symbiosis. *Plant Signaling and Behavior, 8*, 4. https://doi.org/10.4161/psb.23656

Frew, A., Weinberger, N., Powell, J. R., Watts-Williams, S. J., & Aguilar-Trigueros, C. A. (2024). Community assembly of root-colonizing arbuscular mycorrhizal fungi: Beyond carbon and into defense? *The ISME Journal, 18*, 1. https://doi.org/10.1093/ismejo/wrae007

Genre, A., Chabaud, M., Faccio, A., Barker, D. G., & Bonfante, P. (2008). Prepenetration apparatus assembly precedes and predicts the colonization patterns of arbuscular mycorrhizal fungi within the root cortex of both Medicago truncatula and Daucus carota. *Plant Cell, 20*, 1407–1420. https://doi.org/10.1105/tpc.108.059014

Gío-Trujillo, J. A., Alvarado-López, C. J., Pacheco-López, N. A., Cristóbal-Alejo, J., & Reyes-Ramírez, A. (2022). Future perspectives of food biofortification: The association with soil microorganisms. *Ra Ximhai, 18*, 175–199. https://doi.org/10.35197/rx.18.04.2022.08.jg

Hawkins, H. J., Cargill, R. I. M., Van Nuland, M. E., Hagen, S. C., Field, K. J., Sheldrake, M., Soudzilovskaia, N. A., & Kiers, E. T. (2023). Mycorrhizal mycelium as a global carbon pool. *Current Biology, 33*, R560–R573. https://doi.org/10.1016/j.cub.2023.02.027

Helgason, T., & Fitter, A. H. (2009). Natural selection and the evolutionary ecology of the arbuscular mycorrhizal fungi (Phylum Glomeromycota). *Journal of Experimental Botany, 60*, 2465–2480. https://doi.org/10.1093/jxb/erp144

Hnini, M., Rabeh, K., & Oubohssaine, M. (2024). Interactions between beneficial soil microorganisms (PGPR and AMF) and host plants for environmental restoration: A systematic review. *Plant Stress, 11*, 100391. https://doi.org/10.1016/j.stress.2024.100391

Hosseini, A., & Gharghani, A. (2015). Effects of Arbuscular Mycorrhizal fungi on growth and nutrient uptake of apple rootstocks in calcareous soil. International. *Journal of Horticultural Science and Technology, 2*, 173–185. https://doi.org/10.22059/ijhst.2015.56434

in 't Zandt, D., Hoekstra, N. J., de Caluwe, H., Cruijsen, P. M. J. M., Visser, E. J. W., & de Kroon, H. (2022). Plant life-history traits rather than soil legacies determine colonization of soil patches in a multi-species grassland. *Journal of Ecology, 110*, 889–901. https://doi.org/10.1111/1365-2745.13850

Israel, A., Langrand, J., Fontaine, J., & Lounès-Hadj Sahraoui, A. (2022). Significance of Arbuscular Mycorrhizal fungi in mitigating abiotic environmental stress in medicinal and aromatic plants: A review. *Food, 11*, 2591. https://doi.org/10.3390/foods11172591

Kalamulla, R., Karunarathna, S. C., Tibpromma, S., Galappaththi, M. C. A., Suwannarach, N., Stephenson, S. L., Asad, S., Salem, Z. S., & Yapa, N. (2022). Arbuscular Mycorrhizal Fungi in sustainable agriculture. *Sustainability, 14*, 12250. https://doi.org/10.3390/su141912250

Karunanantham, K., Lakshminarayanan, S., Ganesamurthi, A., Ramasamy, K., & Rajamony, V. (2022). Arbuscular mycorrhiza—A health engineer for abiotic stress alleviation. In R. C. Dubey & P. Kumar (Eds.), *Rhizosphere engineering* (pp. 171–198). Academic. https://doi.org/10.1016/B978-0-323-89973-4.00019-3

Khaliq, A., Perveen, S., Alamer, K. H., Ul Haq, M. Z., Rafique, Z., Alsudays, I. M., Althobaiti, A. T., Saleh, M. A., Hussain, S., & Attia, H. (2022). Arbuscular Mycorrhizal Fungi symbiosis to enhance plant–soil interaction. *Sustainability, 14*, 13. https://doi.org/10.3390/su14137840

Khan, A., & Mohammad, A. (2022). Mass production of Arbuscular Mycorrhizal fungus inoculum and its use for enhancing biomass yield of crops for food, for in situ Nano-Phyto-Mycorrhizo remediation of contaminated soils and water, and for sustainable bioenergy production. In F. Policarpo, R. Bhat, & G. Dar (Eds.), *Nanotechnology for environmental pollution decontamination* (p. 16). Apple Academic Press.

Kumar, V., & Suprasanna, P. (2021). Mycorrhizoremediation: A novel tool for bioremediation. In V. Kumar (Ed.), *Rhizomicrobiome dynamics in bioremediation* (p. p12). CRC Press.

Lanfranco, L., Fiorilli, V., & Gutjahr, C. (2018). Partner communication and role of nutrients in the arbuscular mycorrhizal symbiosis. *New Phytologist, 220*, 1031–1046. https://doi.org/10.1111/nph.15230

Liang, S. M., Li, Q. S., Liu, M. Y., Hashem, A., Al-Arjani, A. B. F., Alenazi, M. M., Abdallah, E. F., Muthuramalingam, P., & Wu, Q. S. (2022). Mycorrhizal effects on growth and expressions of stress-responsive genes (aquaporins and SOSs) of tomato under salt stress. *Journal of Fungi, 8*, 1305. https://doi.org/10.3390/jof8121305

Ma, Q., Chadwick, D., Wu, L., & Jones, D. (2022). Arbuscular mycorrhiza fungi colonisation stimulates uptake of inorganic nitrogen and sulphur but reduces utilisation of organic forms in tomato. *Soil Biology and Biochemistry, 172*, 108719. https://doi.org/10.1016/j.soilbio.2022.108719

Meena, K. K., Kumar, P., Sorty, A. M., Bitla, U., & Pathak, H. (2022). Ecology of Arbuscular mycorrhizae and influence on drought tolerance in crop plants. In N. K. Arora & B. Bouizgarne (Eds.), *Microbial BioTechnology for sustainable agriculture Volume 1. Microorganisms for sustainability*. Springer. https://doi.org/10.1007/978-981-16-4843-4_8

Nagahashi, G., & Douds, D. D. (1997). Appresorium formation by AM fungi on isolated cell walls of carrot roots. *New Phytologist, 136*, 299–304. https://doi.org/10.1046/j.1469-8137.1997.00739.x

Noceto, P. A., Bettenfeld, P., Boussageon, R., Hériché, M., Sportes, A., van Tuinen, D., Courty, P. E., & Wipf, D. (2021). Arbuscular mycorrhizal fungi, a key symbiosis in the development of quality traits in crop production, alone or combined with plant growth-promoting bacteria. *Mycorrhiza, 31*, 655–669. https://doi.org/10.1007/s00572-021-01054-1

Nogueira de Sousa, R. (2023). Introductory chapter: Mycorrhizal fungi—A current overview on agricultural productivity and soil health. *IntechOpen*. https://doi.org/10.5772/intechopen.109021

Orine, D., Defossez, E., Vergara, F., Uthe, H., van Dam, N., & Rasmann, S. (2022). Arbuscular mycorrhizal fungi prevent the negative effect of drought and modulate the growth-defence trade-off in tomato plants. *Journal of Sustainable Agriculture and Environment, 1*, 177–190. https://doi.org/10.1002/sae2.12018

Orozco-Mosqueda, M. C., Fadiji, A. E., Babalola, O. O., Glick, B. R., & Santoyo, G. (2022). Rhizobiome engineering: Unveiling complex rhizosphere interactions to enhance plant growth and health. *Microbiological Research, 263*, 127137. https://doi.org/10.1016/j.micres.2022.127137

Ortas, I., Rafique, M., & Çekiç, F. Ö. (2021). Do Mycorrhizal fungi enable plants to cope with abiotic stresses by overcoming the detrimental effects of salinity and improving drought tolerance? In N. Shrivastava, S. Mahajan, & A. Varma (Eds.), *Symbiotic soil microorganisms*. Springer. https://doi.org/10.1007/978-3-030-51916-2_23

Ould Amer, S., Aliat, T., Kucher, D. E., Bensaci, O. A., & Rebouh, N. Y. (2023). Investigating the potential of Arbuscular Mycorrhizal fungi in mitigating water deficit effects on durum wheat (Triticum durum Desf.). *Agriculture, 13*, 552. https://doi.org/10.3390/agriculture13030552

Peduru, J., Luo, Y., Yu, G., Fu, Y., He, X., Van Zwieten, L., Liang, C., Kumar, A., He, Y., Kuzyakov, Y., Qin, H., Guggenberger, G., & Xu, J. (2021). Arbuscular mycorrhizal fungi and goethite promote carbon sequestration via hyphal-aggregate mineral interactions. *Soil Biology and Biochemistry, 162*, 108417. https://doi.org/10.1016/j.soilbio.2021.108417

Philippot, L., Chenu, C., Kappler, A., Rillig, M. C., & Fierer, N. (2024). The interplay between microbial communities and soil properties. *Nature Reviews Microbiology, 22*, 226–239. https://doi.org/10.1038/s41579-023-00980-5

Prasad, K. (2021). Potential impression of Arbuscular Mycorrhizal fungi on agricultural growth, productivity, and environment toward global sustainable development for green technology. *Fungal Biology*. https://doi.org/10.1007/978-3-030-90649-85

Pratibha, Y., Bindu, Y., Chhaya Deepi, D., & Om Prakash, N. (2023). Role of arbuscular mycorrhizas in regulating physiological and molecular aspects of plants under abiotic stress. In S. Anket, P. Sangeeta, B. Renu, Z. Bingsong, & T. Durgesh (Eds.), *The role of growth regulators and phytohormones in overcoming environmental stress* (pp. 399–433). Academic Press. https://doi.org/10.1016/B978-0-323-98332-7.00015-9

Qiang-Sheng, W., Wei-Qin, G., Srivastava, A., Zhang, F., & Ying-Ning, Z. (2020). Nutrient acquisition and fruit quality of Ponkan mandarin in response to AMF inoculation. *Indian Journal of Agricultural Sciences, 90*, 8.

Rahimzadeh, S., & Pirzad, A. R. (2017). Microorganisms (AMF and PSB) interaction on linseed productivity under water-deficit conditions. *GUASNR International Journal of Plant Production, 11*, 2.

Rochange, S., Goormachtig, S., Lopez-Raez, J. A., & Gutjahr, C. (2019). The role of strigolactones in plant–microbe interactions. In *Strigolactones – Biology and applications* (pp. 121–142). https://doi.org/10.1007/978-3-030-12153-2_4

Roy, S., & Choudhury, L. (2022). *The symbiotic relationship between fungi and plants* (pp. 52–74). https://doi.org/10.2174/9789815040340122020009.

Roy, S., Vigyan Kendra, K., Kumari, P., Kumar, U., Yadav, M., Prasad Scientist, R., Sinha, D., Kumar, P., Roy, S., & Prasad, R. (2018). Vesicular arbuscular mycorrhiza as potential biocontrol agent for nutrition and management of soil-borne diseases: A review. *International Journal of Chemical Studies, 6*, 1.

Saha, R., Mondal, B., & Naskar, B. (2014). AMF inoculation changes, the root development pattern of plants at early stage of colonization. *International Journal of Bioresource Science, 1*, 1.

Selim, S. M., & Zayed, M. S. (2017). Microbial interactions and plant growth. In D. Singh, H. Singh, & R. Prabha (Eds.), *Plant-microbe interactions in agro-ecological perspectives* (pp. 1–15). Springer. https://doi.org/10.1007/978-981-10-5813-4_1

Sharma, S., Bhuvaneswari, V., Saikia, B., Karthik, R., Rajeshwaran, B., Shree Naveena, P., & Gayithri, M. (2023). Multitrophic reciprocity of AMF with plants and other soil microbes in relation to biotic stress. In P. Mathur, R. Kapoor, & S. Roy (Eds.), *Microbial symbionts and plant health: Trends and applications for changing climate. Rhizosphere biology* (pp. 329–366). Springer. https://doi.org/10.1007/978-981-99-0030-5_13

Singh, H. V., Singh, U. B., Sahu, P. K., Malviya, D., Singh, S., & Saxena, A. K. (2022a). Arbuscular Mycorrhizal fungal symbiosis for mutual benefit: More than expectation. In U. B. Singh, J. P. Rai, & A. K. Sharma (Eds.), *Re-visiting the rhizosphere eco-system for agricultural sustainability. Rhizosphere biology*. Springer. https://doi.org/10.1007/978-981-19-4101-6_6

Singh, M., Bisht, S., Sharma, J. G., & Singh, S. (2022b). Implications of abiotic stress tolerance in arbuscular mycorrhiza colonized plants: Importance in plant growth and regulation. *Journal of Applied Biology & Biotechnology, 10*, 1–11. https://doi.org/10.7324/JABB.2022.100601

Son, Y., Stott, K., Manning, D. A. C., & Cooper, J. M. (2021). Carbon sequestration in artificial silicate soils facilitated by arbuscular mycorrhizal fungi and glomalin-related soil protein. *European Journal of Soil Science, 72*, 863–870. https://doi.org/10.1111/ejss.13058

Soti, P., Kariyat, R., & Racelis, A. (2023). Effective farm management promotes native AMF and benefit organic farming systems. *Agriculture, Ecosystems & Environment, 342*, 108240. https://doi.org/10.1016/j.agee.2022.108240

Suchitra, R., Rajaram, K., Arunkumar, N., & Kumar, D. S. S. (2020). Contribution of beneficial fungi for maintaining sustainable plant growth and soil fertility. In A. Varma, S. Tripathi, & R. Prasad (Eds.), *Plant microbe symbiosis*. Springer. https://doi.org/10.1007/978-3-030-36248-5_6

Sudha Samuel, S., & Veeramani, A. (2021). Advantages of Arbuscular Mycorrhizal fungi (AMF) production for the profitability of agriculture and biofertilizer industry. In R. Radhakrishnan (Ed.), *Mycorrhizal fungi – Utilization in agriculture and forestry*. IntechOpen. https://doi.org/10.5772/intechopen.95458

Sun, D., Kotianová, M., Rozmoš, M., Hršelová, H., Bukovská, P., & Jansa, J. (2023). Arbuscular mycorrhizal hyphae selectively suppress soil ammonia oxidizers—But probably not by production of biological nitrification inhibitors. *Plant and Soil, 491*, 627–643. https://doi.org/10.1007/s11104-023-06144-x

Tang, H., Hassan, U., Feng, L., Nawaz, M., Shah, N., Qari, H., Liu, Y., & Miao, J. (2022). The critical role of Arbuscular Mycorrhizal fungi to improve drought tolerance and nitrogen use efficiency in crops. *Frontiers in Plant Science, 13*. https://doi.org/10.3389/fpls.2022.919166

Trocio, D. Y., & Paguntalan, D. (2023). Review on the use of Arbuscular Mycorrhizal fungi in bioremediation of heavy metal contaminated soils in The Philippines. *The Philippine Journal of Science, 152*, 1139–1159. https://doi.org/10.56899/152.03.31

Vasco-Palacios, A. M., Hernandez, J., Peñuela-Mora, M. C., Franco-Molano Ana, E., & Teun, B. (2017). Ectomycorrhizal fungi diversity in a white sand forest in western Amazonia. *Fungal Ecology, 31*, 9–18.

Wu, T., Pan, L., Zipori, I., Mao, J., Li, R., Li, Y., Li, Y., Jing, Y., & Chen, H. (2022). Arbuscular mycorrhizal fungi enhanced the growth, phosphorus uptake and Pht expression of olive (Olea europaea L.) plantlets. *PeerJ, 10*, e13813. https://doi.org/10.7717/peerj.13813

Yadav, A., Batra, D., Kaushik, P., & Mohanta, T. K. (2023). Abundance and distribution of arbuscular mycorrhizal fungi associated with oil-yielding plants. *Journal of Basec Microbiology, 63*, 814–827. https://doi.org/10.1002/jobm.202300012

Yang, S., & Irman, O. I. (2023). Impact of mycorrhiza on plant nutrition and food security. *Journal of Plant Nutrition, 46*, 3247–3272. https://doi.org/10.1080/01904167.2023.2192780

Zargar, J. I., & Balwan, W. K. (2022). Effect of vesicular arbuscular mycorrhiza (VAM) inoculants on growth of tomato. *International Journal of Biological Innovations, 4*, 375–378. https://doi.org/10.46505/IJBI.2022.4213

Zhen, L., Songlin, W., Longbin, H., & Yuanfang, H. (2022). *Arbuscular mycorrhizal colonization enhanced organic carbon and nitrogen sequestration in eco-engineered technosols from Fe ore tailings with different plant biomass residues*. EGU General Assembly. https://doi.org/10.5194/egusphere-egu22-6667

Harnessing the Bacterial Rhizobiome in the Green Technologies of Biofertilisation and Phytoremediation Towards Environmental Sustainability

Yahaya Yunusa Riko, Zubairu Umar Darma, and Sirajo Salisu Jibia

Abstract Population explosion and industrialization threaten environmental sustainability today, as chemical fertilisers, organic pollutants, heavy metals, etc. contaminate and exert deleterious effects on the biosphere. Green technologies, including biofertilisation and phytoremediation, can mitigate those problems and promote sustainability. The rhizobiome harbours a complex array of microorganisms, including bacteria (the bacterial rhizobiome/rhizobacteriome); and can be utilized in those green technologies to substitute for synthetic fertilisers/physicochemical remediation approaches. The rhizobacteriome also aid plants growth and degradation/bioremoval of contaminants. This chapter overviewed the urgency for those green technologies globally, and detailed the key players in the rhizobacteriome involved in biofertilisation; the mechanisms employed; plant-microbiome interactions and signalling mechanisms; the application of genetic engineering and microbiome manipulation techniques for enhanced biofertilisation; and the critical roles played by the rhizobacteriome in facilitating plant growth; degradation of organic pollutants and heavy metals bioremoval. Recent case studies of successful applications of the rhizobacteriome in biofertilisation and phytoremediation were successfully reviewed as well. The chapter concludes by highlighting major challenges preventing the widespread and successful use of these green technologies, and proffering solutions to them; before giving a panoramic sweep of the future prospects in these research fields by offering recommendations for further researches.

Keywords Biofertilisation · Phytoremediation · Rhizobacteriome · Rhizosphere · Sustainability

---

Y. Y. Riko (✉) · Z. U. Darma (✉)
Department of Microbiology, Faculty of Natural and Applied Sciences, Umaru Musa 'Yar'adua University, Katsina, Nigeria
e-mail: yahaya.yunusa@umyu.edu.ng; zubairu.umar@umyu.edu.ng

S. S. Jibia
Department of Plant and Soil Sciences, Faculty of Agriculture, Chiang Mai University, Chiang Mai, Thailand
e-mail: sirajo_salisu@cmu.ac.th

S. A. Aransiola et al. (eds.), *Soil Microbiome in Green Technology Sustainability*,
https://doi.org/10.1007/978-3-031-71844-1_22

1 Introduction

The consistent increase in human population and industrial activity are recognised as posing significant threats to environmental sustainability. Green technologies are employed to find sustainable solutions to these and other environmental challenges. The bacterial rhizobiome is a community of microorganisms living in association with plant roots and the soil microenvironment, that is interfacing with both plants and the soil, and plays a crucial role in biofertilization which may supplant the use of synthetic fertilizers. Furthermore, the rhizobiome can be utilized in microbially-assisted phytoremediation, by combining plants and their associated microbes in the clean-up of contaminated environments, and the promotion of environmental sustainability through innovative green technologies for a healthier planet (Rashid et al., 2024).

Specifically, the field of rhizobiome engineering is a nascent, burgeoning field that has applications in numerous green technologies such as biofertilisation and phytoremediation, as it explores the pivotal roles played by beneficial bacteria in enhancing plant growth and nutrient uptake, thereby facilitating sustainable agriculture (Kumar et al., 2021a), and highlights the use of plants in contaminated soils remediation for environmental conservation (Alotaibi et al., 2021).

This research field is receiving a recent increase in relevant publications (Benaissa, 2023; González-Reguero et al., 2023; Ibáñez et al., 2023), therefore, there is the need for a comprehensive overview of key concepts undergirding rhizobiome-plant interactions and their applications in/implications for soil health and ecosystem sustainability, towards sustainable agriculture, environmental sustainability, conservation and sustainable development, which is what this chapter aims to achieve.

1.1 An Overview of the Soil Microbiome

The term microbiome generally connotes the totality of microorganisms associated with a particular ecological niche. In the context of the soil, the microbiome refers to the diverse community of microorganisms present therein, known for their crucial functions in the soil eosystem, including nutrient cycling, enhancement of plant health, biodegradation and pathogen suppression. The microbiome is dominated mainly by the bacteria, which form the soil bacteriome, including members of the alpha-proteobacteria, *Lysobacter*, *Dietzia*, *Pseudomonas*, and *Extingomonas* commonly known for facilitating soil health and biodegradation. Additionally, plant growth-promoting bacteria like *Rhizobium*, *Azospirillum*, *Bacillus*, and *Pseudomonas* enhance biofertilisation and promote plant growth. Other microbes, typically fungi, form an important component of the soil microbiome as well, as part of the mycobiome, typically mycorrhizal fungi, superparasites, and generalists, and they enhance plant health, soil stability, and biogeochemical cycles by facilitating nutrient transfer, bioremediation, and maintaining a balance in the soil microbiome in complex associations and networks with the bacteria and other soil microorganisms (archaea, protozoa, viruses, etc.) towards enhanced plant growth and development (Islam et al., 2020; Bhat et al., 2022; Frąc et al., 2022; Nabi, 2023).

1.2 The Rhizobiome as a Micro-hotspot in the Soil

The soil microbiome is also further subdivided based on specific areas. For instance, the soil environment can be classified into the following microenvironments: bulk/free soil, the rhizospheric soil (having a 5–15 cm proximity to plant roots), and the rhizoplane soil (directly attached to roots), each harbouring distinct microbiomes. Additionally, the endosphere also has a unique microbiota as a constituent of the soil rhizobacteriome (Riko et al., 2023, 2024).

Among these microenvironments, the rhizobiome is considered a melting pot of microbial, and more importantly bacterial diversity (Liang et al., 2021; Camargo et al., 2023). These bacteria cohabiting the roots region are essential for the health and development of plants, and encompass members of the Plant Growth Promoting Rhizobacteria (PGPR) family, such as *Pseudomonas*, *Bacillus*, *Azosprillum*, and *Rhizobium*, which help in nutrient solubilisation, secretion of phytohormones and conferring of stress resistance to plants. Nitrogen-fixing symbionts are also classified here, together with mycorrhizal fungi, both ectomycorrhizae and arbuscular mycorrhizae, which gacilitate nutrient absorption and water uptake in crops. These components of the rhizobiome are constantly interacting through plant-microbe and microbe-microbe interactions to improve plant nutrition and confer resilience to environmental stresses on plants (Tian et al., 2019; Richardson et al., 2020; Prakash et al., 2022).

1.3 Bacterial Inhabitants of the Rhizobiome: Diversity and Influencing Factors

Numerous previous studied had illuminated that the rhizobacteriome exhibits significant, crop-type dependent diversity. Proteobacteria, Acidobacteria, and Actinobacteria had been identified as the dominant phyla, as evidenced by their higher operational taxonomic units (OTUs) numbers, and the sheer diversity of the bacteriobiome as evaluated by other diversity indices. Specifically, the Bradyrhizobiaceae were found to be the most dominant, with 1% representation. Other studies had shown the dominance of Firmicutes and Bacteroidetes members as abundant taxa, in addition to the three mentioned earlier, with *Pseudomonas* revealed as the prevalent genus; in contrast to other studies revealing *Bacillus*, *Stenotrophomonas*, *Delftia* and *Herbaspirillum* as the prevalent species, in ddition to *Pseudomonas,* in the rhizobacteriome. Other studies reveal the presence of bacteria from the genera: *Cupriavidus*, *Acinetobacter* and *Enterobacter* as the dominant species, in addition to the previously mentioned species. In other comparative studies in different soils, Proteobacteria was identified as the dominant phylum, within limits of regional-geographical variations. PGPR, including *Rhizobium leguminosarum* and *Luteibacter rhizovicinus* ealso evidenced their dominance in the rhizobacteriome. Another quasi-microbial legacy effect study identified some 6202 bacteria-specific OTUs, with a core microbiome shared among different species, indicating a diverse rhizosphere microbial community with potential for crop improvement through breeding strategies (Acuña et al., 2020; Vermeulen et al., 2020; Ambardar et al., 2021; Rahal & Chekireb, 2021; Malviya et al., 2022).

Regarding the influencing factors, soil type, pH and salinity were identified as the major determining factors (Vermeulen et al., 2020). In general, genetic variation, soil properties, plant species, environmental conditions, and interactions between bacterial strains are the key factors shaping this diversity in the rhizobacteriome. Specifically, genetic diversity among bacterial strains influences facilitative and competitive interactions, which in turn affect community composition; likewise soil physicochemical parameter, such as pH, nutrient content (nitrogen, phosphorus), and cation exchange capacity. The type of plant species and the properties of its roots, specifically its foraging traits, also shape the composition and diversity of the rhizobacteriome. Climatic factors, including seasons of the year and the presence of environmental stressors, further affect the bacterial diversity of the rhizobacteriome, as does metabolic versatility of the species involved, which enable them to flourish and function in the rhizosphere microenvironment. These factors play crucial roles in governing bacterial diversity, improving agricultural practices and ecological restoration, and facilitating beneficial plant-microbe interactions in the soil (Zhou et al., 2020; Akinola et al., 2021; Fields et al., 2021; Thomas-Barry et al., 2021).

1.4 Green Technologies in the Context of Environmental Sustainability

Green technologies are defined as vital techniques, spanning solar photovoltaic, wind energy, biofuel, biogas, biomass, organic farming, integrated pest management, phytoremediation, biofertilisation, etc. critical to promoting environmental sustainability by offering solutions to mitigate the adverse effects of conventional farming practices and industrial processes on the environment. These technologies contribute to reducing environmental degradation, global warming, and biodiversity loss while enhancing profitability and sustainability in production processes, towards the overall adoption of green technologies for achieving sustainable growth and ensuring a healthier environment for future generations (Anagah, 2023; Chovancová et al., 2023; Oinam & Chakraborty, 2023).

1.5 Biofertilisation: An Outline

Biofertilisation, the use of biological organisms to enhance soil fertility and plant growth, has emerged as a sustainable alternative to chemical fertilisers. This approach leverages the natural capabilities of microorganisms to improve nutrient availability, enhance plant resilience to stress, and promote overall crop productivity. *Bacillus, Rhizobium* and *Pseudomonas* are the major bacterial genera involved in biofertilisation nitrogen fixation, phosphate solubilisation, and production of growth-promoting substances, even under stressful conditions; cumulatively enhancing the nutrient content of plants/biofortification, innovative design for new biofertiisation and improving environmental sustainability (Jung & Kim, 2020; Raimi et al., 2020; Soumare et al., 2020; Sun et al., 2020; Khanghahi et al., 2022; Mumtaz et al., 2022).

1.6 Phytoremediation: An Outline

Phytoremediation is a promising bioremediation technique that utilizes plants and their associated microorganisms to extract, sequester, or reduce the toxicity of soil pollutants, especially heavy metals (HMs) and petroleum hydrocarbons (Abioye et al., 2017; Aransiola et al., 2023). Microorganisms play crucial roles in facilitating this process, with the rhizobacteriome in particular—consisting of rhizosphere bacteria interacting with plant roots—playing a crucial role in enhancing phytoremediation efficiency by facilitating the availability of metals to plants, promoting plant growth, and aiding in the degradation of pollutants like petroleum hydrocarbons, by influencing soil physicochemical parameters, metal translocation factors, and the synthesis of proteins related to metal transport, ultimately impacting the phytoremediation capacity of plants for heavy metals (Aransiola et al., 2013; Karmakar et al., 2022; Rabani et al., 2022; Xiao, et al., 2023; Aransiola et al., 2024).

1.7 Need for Biofertilisation and Phytoremediation in the Twenty-First Century

The need for phytoremediation and biofertilization assumes a status of urgency in the twenty-first century due to escalating environmental contamination arising from intensive farming, industrialization, and urbanization, leading to reduced yields, and soil and groundwater pollution with harmful chemical compounds and heavy metals, ultimately exerting deleterious effects on environmental quality and human well-being. Through phytoremediation approaches, organic contaminants and heavy metals can be cost-effectively removed in an eco-friendly technique. Incorporating microbes in this endeavor further enhances soil fertility, reduces pollution, and promotes sustainability for future generations. On the other hand, the use of biofertilizers enhances soil fertility, promotes plant growth, and contributes to environmental sustainability by reducing the reliance on synthetic fertilizers, ultimately improving crop yield and food safety but also aiding in bioremediation, making them eco-friendly, cost-effective, and essential for ensuring global food security in a world witnessing incremental population expansion (Wijegunawardana et al., 2022; Khan et al., 2023; Verma & Pandey, 2023; Yapa et al., 2022).

1.8 An Overview of the Potential of Harnessing the Bacterial Rhizobiome in Biofertilisation and Phytoremediation

The bacterial rhizobiome, the community of bacteria associated with plant roots, plays a crucial role in biofertilization and phytoremediation by enhancing plant growth through improving nutrient availability and helping in the degradation of pollutants, making them valuable for sustainable agriculture and environmental restoration. From the biofertilisation perspective, the bacterial rhizobiome holds enormous potential for nitrogen fixation, siderophore production, phosphate solubilisation and hormonal secretion. These activities, chiefly orchestrated by members of genera such as *Bacillus, Brevibacillus, Burkholderia, Enterobacter*, etc., increasing nutrient availability, uptake, plant growth and biomass. From the phytoremediation perspective, the rhizobacteriome plays critical roles in pollutant degradation, heavy metal tolerance and remediation. Genetic modification leading to overexpression of enzymes can further enhance the phytoremediation efficacy (Agarwal et al., 2020; Nor, 2020; Yuan et al., 2021; Jach et al., 2022; Rosariastuti & Yudha, 2022).

2 Bacterial Rhizobiomes for Biofertilization

Bacterial rhizobiomes consist of diverse and specific bacterial communities that inhabit the rhizosphere and endo-rhizosphere of plants (Shahrajabian et al., 2021). These rhizobiomes are primarily composed of Proteobacteria, Acidobacteria, and Bacteroidetes, harbouring well-known genera such as *Rhizobium*, *Pseudomonas*, *Bacillus*, *Azospirillum*, and *Azotobacter* (Bhat et al., 2022). Their presence significantly contributes to plant health and growth through various mechanisms, including enhanced nutrient availability, root growth stimulation, and root disease control. Additionally, bacterial rhizobiomes produce phytohormones, exopolysaccharides, antibiotics, and play important roles in nitrogen fixation, promoting overall plant development (Shahrajabian et al., 2021; Bhat et al., 2022). These bacterial communities serve as critical connectors between soil and plant microbiomes, improving soil texture and modulating plant stress responses (Shahrajabian et al., 2021). Furthermore, their potential application as biofertilizers and biopesticides contributes to sustainable food production by reducing reliance on chemical fertilizers and pesticides.

2.1 *Biofertilization Mechanisms Employed by the Bacterial Rhizobiome*

Plant growth-promoting rhizobacteria (PGPR) bring about biofertilisation by engaging in processes like biological nitrogen fixation (BNF), phosphate, potassium, and iron solubilization. Some bacterial species are characterized as P-solubilizing microorganisms, while others enhance the solubility of micronutrients, including siderophores that chelate iron (Bhat et al., 2022). For instance, *Pseudomonas* species contribute to plant growth by producing water-soluble vitamins like niacin (Riaz et al., 2021).

PGPR also enhance soil nutrient status through mechanisms such as root surface area expansion, nitrogen fixation, and phosphate solubilisation (Riaz et al., 2021). They also act as phytostimulators by producing phytohormones like indole acetic acid (IAA), cytokinin, gibberellins, and ethylene, influencing plant growth and development (Riaz et al., 2021; Rani et al., 2023). When plants are inoculated with PGPR, their physiology and molecular processes are significantly impacted, potentially making them more efficient in nutrient uptake from the soil.

Biological nitrogen fixation, where specific bacteria convert atmospheric nitrogen into ammonia, is essential for sustainable agriculture, as it reduces reliance on synthetic fertilizers (Ramírez-Cariño et al., 2023). The symbiotic relationship between *Rhizobium* and leguminous plants facilitates access to atmospheric nitrogen, improves soil fertility, and reduces the need for chemical fertilizers. Rhizobacteria can also solubilize inorganic phosphates by producing organic acids, thereby increasing plant-available phosphorus (Parewa et al., 2018). However, the efficacy of biofertilization by bacterial rhizobiomes is influenced by environmental factors such as pH, water availability, heat, and essential mineral nutrients. The molecular mechanisms underlying nitrogen fixation and phosphorus solubilisation involve the conversion of atmospheric nitrogen into ammonia, symbiotic relationships between bacteria and plants, and the solubilization of inorganic phosphate (Riaz et al., 2021; Rani et al., 2023; Ramírez-Cariño et al., 2023; Parewa et al., 2018).

2.2 *Beneficial Rhizobacteria Elicit Plant Defence Mechanisms and Induce Systemic Resistance*

PGPR can directly and indirectly enhance plant health. They have the capacity to instigate systemic resistance (ISR) in host plants against pathogenicity, through a multistep process that includes the perception of elicitors produced by beneficial rhizobacteria, signal transduction, and the expression of defence mechanisms that limit further pathogen penetration into host tissues. Additionally, beneficial rhizobacteria can confer systemic tolerance to abiotic stresses, such as temperature, drought, and salinity, through the enhancement of antioxidative responses and the accumulation of osmolytes and phenolics (Chakraborty & Chakraborty, 2022).

The key signalling pathways involved in the interaction between beneficial rhizobacteria, and plants include the release of diverse signal molecules by plants to draw beneficial bacteria, mediating the activities of plant genetics and biochemistry (Zhu et al., 2022). Studies have demonstrated that both ethylene and jasmonic acid are the main signalling molecules for ISR mediated by rhizobacteria, while salicylic acid is the primary activator in locally and systemically induced resistance response (Seth et al., 2023). In essence, the mechanisms by which beneficial rhizobacteria elicit plant defence responses and induce systemic resistance involve a sophisticated interplay of signalling pathways and molecular interactions (Chakraborty & Chakraborty, 2022; Zhu et al., 2022; Seth et al., 2023).

2.3 Plant-Microbe Interactions Between the Bacterial Rhizobiome and Crop Plants

In sustainable agriculture, interactions between plants and microbes within the rhizosphere play a crucial role, impacting crop yield, quality, and overall plant health. These interactions stimulate plant growth, suppress harmful pathogens, and mitigate abiotic stresses through various mechanisms, thereby exerting positive effects on plant growth and soil health through nitrogen fixation, nutrient solubilization, hormonal stimulation, and improved root growth (Kumar et al., 2021b). The intricate communication systems between plant holobionts and microorganisms result in different strategies and interplays, shaping bacterial communities and plant fitness simultaneously (Borrero de Acuña & Bernal, 2021).

Thus, the importance of this interaction calls for more research into the dynamic interactions between crop plants, the rhizosphere microbiome, and the environment, to better guide harnessing the microbiome to increase crop yield and quality (Zhang et al., 2021) as these interactions have significant implications for sustainable agriculture, impacting crop yield, quality, and overall plant health.

2.4 Communication and Signalling Between Plants and Rhizobacteria

Plants engage in intricate communication with rhizobacteria through a complex exchange of signals. This process involves the secretion of low molecular weight molecules by plants and the release of specific signalling molecules by rhizobacteria (Ahmed & Tariq, 2023). The interaction begins with flavonoids released from plant roots, triggering changes in rhizobial gene expression. Consequently, rhizobia secrete Nod factors (NFs) recognized by host receptors, ultimately initiating symbiosis (Ghantasala & Roy Choudhury, 2022).

Ahmed and Tariq (2023) identified the key mechanisms and molecules involved in this communication, as including quorum-sensing and signalling molecules employed to coordinate behaviours among the rhizobiome members, and between the bacteria and plants; plant-to-microorganism signalling, through the release of low molecular weight molecules, such as flavonoids, to initiate interactions with rhizobacteria, leading to the secretion of *Nod* factors from rhizobia; and microorganism-to-plant signalling, via the secretion of signalling molecules like acyl-homoserine lactone by the rhizobacteria, facilitating the establishment of beneficial interactions with plants. These interactions jointly and severally influence nutrient uptake, growth hormone production, and stress tolerance.

2.5 Biofertilization Mechanisms Employed by the Rhizobacteriome Upon the Inoculation as Biofertilisers

Biofertilizers, as described by Dineshkumar et al. (2017), consist of living or dormant microbes, including bacteria, actinomycetes, fungi, and algae, either individually or in combination. Rhizobacteria-based biofertilizers offer several mechanisms to benefit plants, as summarized from de Andrade et al. (2023), key among which include:

Nitrogen Fixation Biological nitrogen fixation (BNF) provides an alternative to synthetic nitrogen fertilizers, as it accounts for nearly 60% of the Earth's total fixed nitrogen. Rhizobacterial inoculation optimizes BNF for sustainable food production.

Phosphorus Solubilization, Mineralization, and Siderophore Production Phosphorus-solubilizing members of the rhizobacteriomd excrete low molecular weight organic acids that dissolve inorganic phosphorus, modifying soil characteristics and indirectly impacting the rhizosphere microbial population. Rhizobacteria also produce siderophores, which dissolve iron from minerals or organic compounds, making it available to plants and influencing various metabolic processes.

Production of Phytohormones to Reduce Pesticide Usage PGPR synthesize phytohormones (e.g., indole-3-acetic acid, cytokinin and gibberellin) that impact plant hormonal balance, and enhance root architecture and overall plant performance.

2.6 Genetic Engineering of Rhizobacteria for Enhanced Plant-Microbe Interactions

Genetic engineering techniques have been widely employed to improve the potentials of rhizobacteria for enhanced mutualistic relationship with plants. These techniques target improvements in nitrogen fixation, chitinases activity, phosphate solubility (transfer of *Nap A* gene responsible for acid phosphatase), potassium solubilization, PGR's chitinase activity for disease resistance, indole acetic acid production for stress response among others (Kaur et al., 2021). Improvement of N-fixation potentials of rhizobacteria provide an avenue for crop productivity through the transfer of nitrogen-fixing (*nif*) genes from fixer to non-fixer bacterial strains. For example, a previous research reported a successful transfer of functional *nif* genes to *Escherichia coli* and *Pseudomonas putida* (non-symbiotic) from *Klebsiella pneumoniae* (symbiotic) which led to higher nitrogenase activity in the resultant mutants (Kaur et al., 2021).

2.7 Rhizospheric Microbiome Engineering for Enhanced Biofertilisation

Numerous strategies of rhizosphere engineering harnessing the potentials of indigenous and exogenous microbial consortia for enhanced biofertilization continue to emerge. These usually entail the selection and maintenance of microbial population through rhizodeposits, genetically modifying microbes to encourage selective rhizodeposits, good agricultural management, modification of rhizosphere through fertilization, irrigation and tillage, inoculation of suitable and persistent PGR, etc. (Babalola et al., 2021).

Babalola et al. (2021) outlined various strategies for engineering the plant rhizospheric microbiome to enhance biofertilisation, including:

Beneficial Microbiota The utilization of plant growth-promoting microbes (PGPM) and endophytes potentially enhances plant nutrient and water uptake, phytohormone biosynthesis, and nitrogen fixation. However, competition with native microbes for resources is a potential drawback.

Assemblage-Inspired Plant Breeding and Genetic Transformation Plants can be engineered to instigate enhanced root exudate production, thereby recruiting specific beneficial memebrs of the rhizobacteriome to improve soil microclimate and plant growth. It also has the advantage of not requiring any change in agronomic practices, however, potential drawbacks include rapid degradation of plant exudates and the enrichment of undesired microbial groups.

Biochemical Pathway Modification The enhanced biosynthesis of volatile organic compounds and other chemicals of microbial origin can suppress pathogens and enhance stress resistance. However, modification of certain critical pathways may mimic quorum sensing and disrupt communication among beneficial rhizobacterial communities.

Molecular/Genetic Engineering Approaches Plasmids can be utilised in inserting desired genetic traits and pathways, such as toluene degradation, to establish a high-functioning rhizosphere with specific microbial communities. The major challenge is difficulty in achieving stability in the dynamic soil environment.

Agronomic Approaches Crop rotation and the use of organic fertility enhancements can improve nutrient cycling, increase soil organic carbon content, and enhance biological activity. However, potential drawbacks include soil acidification and disruption of the indigenous soil microbiome.

2.8 *Recent Case Studies of Successful Applications of Biofertilisation (Table 1)*

Table 1 Key insights from recent case studies that successfully utilised the rhizobacteriome in biofertilisation

Crop	PGPR	Main effects reported	Source
Tomato (*Solanum lycopersicon* L.)	*Bacillus subtilis* and *Bacillus amyloliquefaciens*	Improved growth, fruit yield and quality	Gashash et al. (2022)
Strawberry (*Fragaria x ananassa* Duch.)	*Variovorax paradoxus*, *Pseudomonas* sp., *Bacillus velezensis*, *Pseudarthrobacter oxydans*, and *Bacillus zhangzhouensis*	Despite phosphorus limitations, treated plants exhibited increased biomass, water use efficiency, net carbon assimilation rate as well as improved N: P balance.	Valle-Romero et al. (2023)
Chili (*Capsicum annuum* L.)	*Bacillus* sp., *Pseudomonas* sp., *Bacillus amyloliquefaciens*, and *Sinorhizobium meliloti*	Increased the general performance of the plant, soil nitrogen, enzymatic activities as well as rhizobacterial diversity.	Gou et al. (2020)
Moringa (*Moringa oleifera* L.)	*Bacillus paralicheniformis, Acinetobacter guillouiae, Aeromonas caviae, Pseudomonas lini*	Improved *Moringa* forage yield and quality attributes	Mendoza et al. (2020)
Wheat (*Triticum aestivum*)	*Bacillus subtilis* and *Pseudomonas mandelii*	Increased root abscisic acid concentration, lowered chlorophyll degradation, and increased leaf area/gas exchange.	Arkhipova et al. (2020)
Maize (*Zea mays*)	*Pseudomonas aeruginosa* (+ foliar application of Salicylic Acid)	Reported alleviation of salinity stress on maize and a concomitant increase in chlorophyll a and b figments, biomass, grain yield among others.	Ali et al. (2023)
Onions (*Allium cepa*)	*Bacillus licheniformis* and *Pseudomonas fluorescens* (+ Kelpak®, a seaweed extract)	Improved mineral contents in leaves and bulbs, and enhanced secretion of phytohormones.	Gupta et al. (2021)

3 Microbially-Assisted Phytoremediation

Microbially-assisted phytoremediation is an encouraging approach for tackling environmental pollution anchored on utilizing the synergistic relationship between plants and microbes in the removal of soil pollutants (Aransiola et al., 2019). The key roles of microbes, particularly bacteria and fungi, in such relationships include growth promotion, secretion of beneficial compounds, stress alleviation, degradation of contaminants and metal tolerance enhancement, which aid plants in surviving high pollutant concentrations. They also mobilize the pollutants in the rhizosphere, facilitating their uptake by plants, which, concomitantly, release exudates and enzymes that enhance microbial activities, supporting bioremediation. This technique has been effectively used to treat many pollutants, including metals, pesticides, and hydrocarbons, contributing to land restoration, carbon sequestration, and biodiversity maintenance in an eco-friendly manner capable of mitigating environmental contamination challenges (Ismail et al., 2021; Tiodar et al., 2021; Kumar et al., 2022; Bhatt et al., 2023).

3.1 Phytoremediation for the Degradation of Organic Pollutants

Phytoremediation is an environmentally friendly and cost-effective technology that uses plants to remove, degrade, or stabilize organic pollutants from contaminated environments. It involves various mechanisms such as phytoextraction, phytostabilization, phytovolatilization, rhizofiltration, and phytotransformation to remove or degrade organic pollutants (Abioye et al., 2021). Plant growth promoting bacteria (PGPB) had been shown to assist in this degradation process. Genetically modified plants have been developed to enhance the degradation potential of organic pollutants. These transgenic plants express specific genes that improve the uptake, metabolism, and transportation of such contaminants. Recent methods focusing on combined tecniques, such as integrating biochar with plants and PGPB can overcome several barriers in phytoremediation, such as high concentrations of secondary toxins and nutrient limitations. Biochar acts as a bacterial carrier and soil health enhancer, improving the overall efficiency of the process. Moreover, phytoremediation in natural and constructed wetland settings effectively degrades organic pollutants by utilizing the natural ability of certain plants to absorb and degrade contaminants, making them a cost-effective and sustainable alternative. Other novel approaches, such as the development of single-atom catalysts (SACs) from hyperaccumulator plants, have shown high efficiency in degrading specific organic pollutants like rhodamine B, demonstrating the potential for innovative phytoremediation (Ezaz et al., 2020; Mishra et al., 2020; Tripathi et al., 2020; Cui et al., 2022; Xiang et al., 2022).

3.2 Phytoremediation for Heavy Metals Bioremoval

Various phytoremediation methods such as phytoextraction, phytovolatilization, phytostabilization, and phytofiltration have been developed to address heavy metal pollution. Numerous plants, like *Catharanthus roseus*, Indian mustard, and sunflower, etc., have shown promise in accumulating heavy metals like lead, nickel, zinc, and chromium from soil. Factors influencing phytoremediation efficiency include plant species, environmental conditions, and post-treatment disposal methods (Babaniyi et al., 2023). Additionally, the use of metal-binding proteins in plants and genetic modifications to enhance phytoremediation capacity have been explored. The success of phytoremediation has been linked to the diversity and metabolic functionality of the bacterial rhizobiome of the metal accumulating plants (Li et al., 2020; Khan, 2023; Naeem et al., 2023; Priya et al., 2023).

3.3 Roles of Rhizobacteria in Phytoremediation

Rhizobacteria—the bacteria that colonize the rhizosphere (the region of soil around plant roots), play a crucial role in enhancing the efficiency of phytoremediation processes. They can enhance the degradation of pollutants such as pharmaceuticals and the bioremoval of heavy metals by co-metabolism and metabolism, often under the influence of root exudates and changes in the rhizosphere environment. They may also participate indirectly by improving plant growth and tolerance to metal stress, thereby increasing the efficiency of phytoremediation. Additionally, they can produce plant growth-promoting substances and enhance nutrient availability. Their activities can also increase the bioavailability of heavy metals (Ni, As), through mechanisms such as chelation and redox reactions, facilitating their uptake by plants. The rhizobacteriome also mediate coping with abiotic stresses such as salinity and heavy metal toxicity by producing stress-alleviating compounds and enhancing antioxidant capacities in plants. As the composition and diversity of rhizobacterial communities can change dynamically in response to the presence of contaminants and plant developmental stages, keystone species within these communities can play significant roles in maintaining ecosystem functions and promoting phytoremediation. Co-inoculation with multiple rhizobacterial strains or with other beneficial microbes like *Rhizobium* can lead to synergistic effects, further enhancing phytoremediation efficiency and soil health (Ju et al., 2020; Li et al., 2020; Yuan et al., 2021).

3.4 Bacterial Biodegradation of Organic Contaminants during Phytoremediation

Bacterial biodegradation plays a crucial role in the context of phytoremediation, aiding in the removal of organic contaminants from soil (Umar et al., 2020).

Bacteria, both endophytic and rhizospheric, contribute significantly to the degradation of toxic organic compounds by plants. These bacteria possess metabolic genes encoding biosurfactants, specific enzymes, and detoxification pathways that facilitate the breakdown of organic pollutants in the soil (Umar et al., 2018). Genetic engineering of these bacteria shows promise in enhancing their biodegradation capabilities, with transgenic plants potentially offering an efficient and environmentally friendly solution for cleaning up polluted soils (Umar et al., 2022). The synergistic relationship between plants and microbes in the rhizosphere region further supports the bioremediation process, as microbes assist in plant growth and survival, and the subsequent degradation of organic contaminants through various mechanisms, including nutrient acquisition and production of growth-promoting metabolites (Agarwal et al., 2020; Umar et al., 2022; Girolkar et al., 2021).

3.5 Heavy Metals Sequestration and Immobilisation

Plants can remove, detoxify, or immobilize heavy metals through natural biological processes, such as uptake through roots and subsequent sequestration in tissues. The rhizobacteriome can also enhance the uptake of heavy metals. Recently, combining plants and biochar facilitate heavy metal immobilisation, as plants synergistically interact with biochar to enhance the effectiveness of heavy metal remediation in contaminated soils. Biochar immobilizes metals, making them less bioavailable and thus reducing their toxicity. Even though herbs and shrubs are commonly used, but woody trees are recommended for long-term metal sequestration; similarly, non-edible plants are preferred to prevent heavy metals from entering the food chain (Haque et al., 2021; Ghosh & Maiti, 2020). Recent Case Studies Highlighting Roles of Rhizobacteria in Phytoremediation (Table 2)

3.6 Challenges and Potential Solutions in Rhizobacteriome Research for Biofertilisation and Phytoremediation

An in-depth analysis of recent literature in the research fields of biofertilisation and phytoremediation using rhizobacteria enables the surmising of the following major challenges and the advancing of potential solutions to them, as follows (Fig. 1).

3.7 Conclusion/Summary

Changing realities of the twenty-first century, majorly population expansion and industrialization, against the backdrop of sustainability, amplify the need for biofertilisation and phytoremediation as alternatives for addressing food security

Table 2 Summary of methodological approaches and results highlights from recent case studies of utilising the rhizobacteriome in phytoremediation

S/No	Study approach/methodology	Results highlights	Reference
1	Investigating the rhizobacterial dynamics of *Typha angustifolia* during its lifecycle (seedling, vegetative, maturity, senescence stages, etc.) in constructed wetlands while treating ibuprofen (IBP)-polluted wastewater	At the mature stage, removal efficiencies of 92% organics, 52% ammonia, 60% phosphorus and 76% IBP were obtained, however, rhizobacterial diversity was adversely affected by IBP, as shown by the reduced evenness. Community dynamics changed significantly at the different stages of growth, compared to the control wetlands (free of IBP). Actinobacteria, with a final relative abundance of 0.79, dominated the rhizosphere.	Li et al. (2020)
2	Isolating PGPR from the rhizosphere of naturally-growing wild grasses in salt-affected areas, Lucknow, Uttar Pradesh, India, and evaluating their salt-tolerance and plant growth-promoting (PGP) abilities	*Pseudomonas plecoglossicida* (KM233646), *Acinetobacter calcoaceticus* (KM233647), *Bacillus flexus* (KM233648), and *Bacillus safensis* (KM233652) applied as bio-inoculants significantly ($P < 0.05$) increased the growth and yield of *Bacopa monnieri* (L.), a medicinal plant, under salt condition. Phytoremediation was indicated by substantial increases in shoot Na^+:K^+ ratio of bio-inoculant-treated plants, as well improved soil physico-chemical properties, compare to the control. The shoot and root biomass (fresh and dry weights), soil enzymes, and soil nutrient parameters also showed significant positive correlations with the shoot Na^+:K^+ ratio.	Pankaj et al. (2020)
3	Effect of planting (P) and not planting (NP) balsam poplars (*Populus balsamifera*) in agricultural and forest soils; and contamination with 50 mg kg^{-1} phenanthrene (PHE), on the bacterial communities of the rhizosphere (P) and the bulk soil (NP) measured using 16S Rrna; and the abundance of PAH ring-hydroxylating dioxygenase alpha subunits (PAH-RHDα), measured by real-time quantitative PCR	Soil type significantly affected the bacterial communities, including the dominant PAH-degrading community. Planting poplar was found to increase phenanthrene degradation in the forest (but not the agricultural) soil. Both total bacteria and specific PAH-degrading bacteria were limited by soil type, leading to different degradation patterns between soils.	Correa-García et al. (2021)

(continued)

Table 2 (continued)

S/No	Study approach/methodology	Results highlights	Reference
4	Twenty-nine heavy metals and salt-tolerant PGPR were evaluated for soil salinity and heavy metal stress alleviation properties in plants, for further use in phytoremediation, based on growth assays in increasing concentrations of NaCl, Zn, Pb, Cu, and Cd.	Seventeen rhizobacteria exhibited high heavy metals and salt tolerance (up to 100 g/L NaCl, 5 mM Cd, 6 mM Cu, 9 mM Pb and 10 mM Zn. PGP traits remain stable even under 10% NaCl and polymetal stresses. Seedling bioassay using *Peganum harmala* L. and *Lactuca sativa* L. seeds inoculated with a bacterial consortium harbouring *Mesorhizobium tamadayense, Enterobacter xiangfangensis, Pseudomonas azotifigens*, and *Streptomyces caelestis* showed the beneficial effects exerted by the consortia as evidenced by root and shoot elongation.	Madline et al. (2021)
5	Screening for hydrocarbon degradation and plant growth promotion (PGP) abilities, such as siderophore production, phosphate solubilization, and IAA production, in the presence of crude oil led to the identification of *Pseudomonas aeruginosa* BB-BE3, *P. aeruginosa* BBBJ, and *Gordonia amicalis* BB-DAC as fficient strains. Their rhizoremediation ability was confirmed via pot trials, and these isolates promoted the growth of *Azadirachta indica* in crude oil-contaminated soils.	Combining *A. indica* with the before-named bacteria led to 95.71%, 93.28%, and 89.88% removal of TPHs, respectively, however, the plant itself removed only 13.44% of TPHs. The control setup (sterilised bulk soil + crude oil) led to the removal of 5.87% of the hydrocarbons. Plants augmented with bacterial strains had significantly low catalase and peroxidase activities compared to the uninoculated control.	Bhuyan and Pandey (2022)
6	Changes in the activity of the rhizospheric microbiome of common sunflower (*Helianthus annuus* L.) and crested wheatgrass (*Agropyron cristatum* L.) used for the phytoremediation of oil polluted soils, in the presence of biopreparations of different spectrums of action. Both "Lenoil", a hydrocarbon-oxidizing preparation, and "Elena" and "Azolen", two non-specific action growth promoting preparations were used, at an oil pollutant concentration of 4%.	The total microbial count in the rhizosphere of plants increased under the pollution, amino-autotrophs and cellulolytics showed a decreased count. The non-specific biopreparations stimulated the growth and stabilised the rhizospheric bacterial population. Microbial activity was also increased. *Agropyron cristatum* L. rhizosphere, however, favoured the growth of amino-autotrophs, cellulolytics, micromycetes, and hydrocarbon-oxidizing bacteria.	Grigoriadi et al. (2023)

(continued)

Table 2 (continued)

S/No	Study approach/methodology	Results highlights	Reference
7	The ability of *Alternanthera* spp. to uptake and hence phytoremediate methylparaben (MP) in a hydroponic setting was evaluated. Similarly, rhizobacteria that are highly tolerant to methylparaben were isolated and characterized.	*Alternanthera* spp. was found to be efficient in removal of methylparaben from water, as assisted by rhizobacteria, such as *Cronobacter sakazakii*, which showed high methylparaben removal efficiency, and expressed the gyrB gene. The bioremoval rates were significantly higher in hydroponic reactors (66.6 ± 3%) compared to controls (<40%). The MP reduction data was fitted well in a first-order kinetics model ($R^2 > 0.9$), with the plants shortening the half-life ($T_{1/2}$) of MP in aqueous solutions to 3.41 days compared to 34.8 days in the control. The bioconcentration factors (BCFs) were found to be higher in roots (1.9–5.2 L/kg) compared to stems (0–0.5 L/kg).	Mutar et al. (2023)
8	Native soil bacteria were isolated and identified based on their fatty acid profiles, as analysed by the MIDI Sherlock system. Additionally, their biofertilisation, biofilm formation and heavy meyals phytoremediation properties were assessed for potential application in phytoremediation of mine waste deposits.	Most of the 48 identified bacterial strains, out of the 99 isolated, from 34 species and 23 genera tolerated arsenic, and or mercury. *Pseudomonas fluorescens* TL97 produced the most IAA, while *Herbaspirillum huttiense* TL36 solubilised the most phosphate and exhibited the highest siderophore production; *Bacillus atrophaeus* TL76 exhibited the biofilm producton ability. *Herbispirillum huttiense* TL36, *Klebsiella oxytoca* TL49 and *Rhizobium radiobacter* TL52 were also found to be tolerant to high mercury chloride concentrations.	Rojas-Solis et al. (2023)
9	The ability of *Echinacea purpurea* to phytoremediate two types of soil: DW, which was an aged excavation pit soil, and OS, obtained from an oil spill area, was evaluated in six test systems (non-inoculation, inoculation with B1/B2 microbial consortium, inoculation with B1 + γ-poly glutamic acid (γ-PGA), inoculation with B2 + γ-PGA and inoculation with γ-PGA solution, over 6 months. Remediation effectiveness was evaluated through	Phytoremediation plus bioaugmentation with B2 microbial consortium + γ-PGA was the most effective bioremediation method for soils contaminated with petroleum hydrocarbons, as it reduced the total petroleum hydrocarbons (TPHs) and polycyclic aromatic hydrocarbons (PAHs) contents in the tested soils by 53.98% and 49.54% (DW-5) and 60.47% and 37.55% (OS-5), respectively. However, the lowest bioremediation efficiency	Wojtowicz et al. (2023)

(continued)

Table 2 (continued)

S/No	Study approach/methodology	Results highlights	Reference
	chromatography of soil, plant roots/shoots and toxicological analyses using four kits.	was recorded in non-inoculated systems, with 18.40 and 16.14% (DW-1) and 21.87% and 18.20% (OS-1) decreases in TPH and PAH contents, respectively.	
10	The abilities of *Artemisia argyi* (LA), *Miscanthus floridulus* (LM), and *Boehmeria nivea* (LZ)—which were the dominant plants, in a lead-zinc mining area in Huayuan County (China), to phytoextract/phytostabilize various heavy metals was evaluated.	Based on the bioconcentration (BCF) and translocation (TF) factors, LA, LZ and LM preferred accumulating Cd, Cr/Sb and Cr/Ni, respectively. Significant ($p < 0.05$) differences were found among the microbes in their rhizosphere. LA had *Truepera* and *Anderseniella*; LM had *Paracoccus* and *Erythrobacter*, and LZ had *Novosphingobium*, as the dominant rhizobacterial genera. Correlation analysis delineated links between certain physicochemical parameters (e.g., organic matter and pH) and specific rhizosphere bacterial taxa (e.g., Actinomarinicola, Bacillariophyta and Oscillochloris), which as affected the rhizosphere soil and enhanced the TF of metals as well. Functional prediction analysis of soil bacterial communities showed that the relative abundances of genes for manganese/zinc-transporting P-type ATPase C, nickel transport protein and 1-aminocyclopropane-1-carboxylate deaminase, etc. were positively correlated with the plants' heavy metals phytoextraction/phytostabilization capacity.	Xiao et al. (2023)

and environmental sustainability. The bacterial rhizobiome plays crucial roles in these techniques. In biofertilisation, the bacteria adequately supplant chemical-based/synthetic fertilisers by facilitating nitrogen fixation, increasing nutrient uptake, facilitating stress tolerance and secreting plant hormones, among others. In phytoremediation, the rhizobacteriome help plants in the degradation of routine organic/emerging pollutants (such as pharmaceuticals), enhance plants' tolerance to heavy metals and other pollutants, and mobilise pollutants in the rhizosphere for uptake, sequestration or degradation. Understanding the evolving dynamics of plant-rhizobacteriome interactions can help revolutionize agricultural practices and environmental remediation towards environmental sustainability and ultimately a

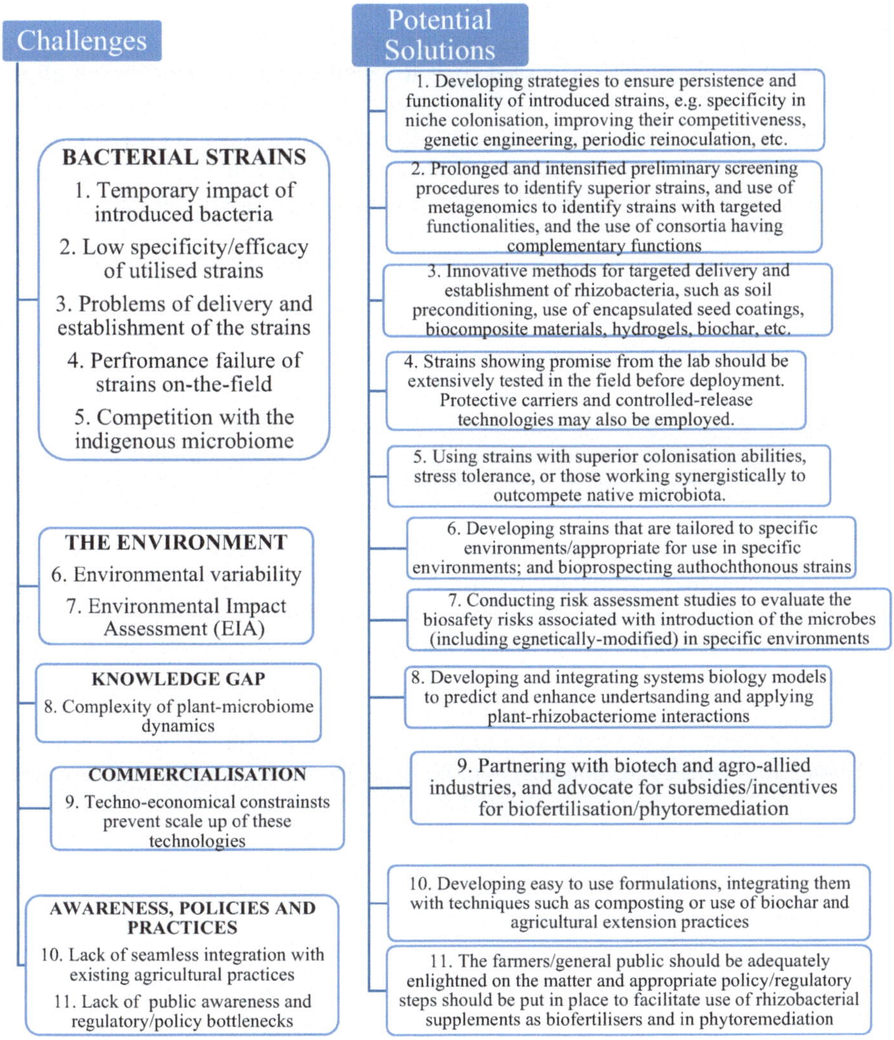

Fig. 1 Challenges to the application of the rhizobacteriome in biofertilisation and phytoremediation, and their potential solutions

healthier planet by leveraging on the green technologies of biofertilization and phytoremediation.

3.8 *Future Prospects/Recommendations*

It is recommended that the following action points be enshrined in the nearest future, to facilitate the realization of the enormous potentials that abound in the field of rhizobacteriome engineering for biofertilisation and phytoremediation applications:

- Engineering next generation biofertilisers with enhanced potency, specificity to various plants and soils and genes from an array of bacteria.
- Incorporating rhizobacteria as seed coats to facilitate early entrenching of plant-microbe symbiosis.
- Development and deployment of genetically engineered consortia with enhanced biofertilisation and phytoremeidation-enhancement capabilities.
- Use of nanotechnology-based carrier materials for precise delivery of biofertilisers.
- Utilisation of bacteria to alleviate stresses such as drought, salinity and temperature, to confer resilience to the plants.
- Employing potent bacteria in the remediation of emerging pollutants such as microplastics and pharmaceuticals.
- The advent of precision-agriculture, with crop- and soil-specific microbiomes employed to boost productivity and phytoremediation ability.
- Enhanced plant-microbe interactions by developing plant varieties that attract more rhizobacteria, and rhizobacteria that promote increased root size and metabolism.
- Combining microbially-assisted phytoremediation with other techniques such as bioaugmentation, biostimulation, cover cropping and organic matter amendments to further enhance cleanup capabilities and soil health.
- Development of microbial biosensors and biomarkers for real-time monitoring of phytoremediation.
- Exploiting untapped microbial diversity through metagenomics/culture-independent approaches for the characterisation of novel strains with enhanced potentials.
- Harnessing OMICs techniques, remote sensing, data analytics and artificial intelligence to develop further breakthroughs.

References

Abioye, O. P., Ijah, U. J. J., & Aransiola, S. A. (2017). Phytoremediation of soil contaminants by biodiesel plant *Jatropha curcas*, Chapter 4. In K. Bauddh et al. (Eds.), *Phytoremediation potential of bioenergy plants*. Springer Nature Singapore Pte Ltd. https://doi.org/10.1007/978-981-10-3084-0_4. https://link.springer.com/book/10.1007%2F978-981-10-3084-0

Abioye, O. P., Ijah, U. J. J., Aransiola, S. A., Auta, S. H., & Ojeba, M. I. (2021). Bioremediation of toxic pesticides in soil using microbial products. In R. Prasad, S. C. Nayak, R. N. Kharwar, & N. K. Dubey (Eds.), *Mycoremediation and environmental sustainability. Fungal biology*. Springer. https://doi.org/10.1007/978-3-030-54422-5_1

Acuña, J. J., Marileo, L. G., Araya, M. A., Rilling, J. I., Larama, G. A., Mora, M. L., Epstein, S., & Jorquera, M. A. (2020). In situ cultivation approach to increase the culturable bacterial diversity in the rhizobiome of plants. *Journal of Soil Science and Plant Nutrition, 20*, 1411–1426. https://doi.org/10.1007/S42729-020-00222-0

Agarwal, P., Giri, B. S., & Rani, R. (2020). Unravelling the role of rhizospheric plant-microbe synergy in phytoremediation: A genomic perspective. *Current Genomics, 21*(5), 334–342. https://doi.org/10.2174/1389202921999200623133240

Ahmed, A., & Tariq, A. (2023). Genetics basis of PGPR interactions. In T. A. Bhat & K. R. Hakeem (Eds.), *Biotechnologies and genetics in plant mutation breeding* (pp. 219–238). Apple Academic Press. https://doi.org/10.1201/9781003305101

Akinola, S., Ayangbenro, A., & Babalola, O. (2021). Metagenomic insight into the community structure of maize-rhizosphere bacteria as predicted by different environmental factors and their functioning within plant proximity. *Microorganisms, 9*(7), 1419. https://doi.org/10.3390/microorganisms9071419

Ali, Q., Ahmad, M., Kamran, M., Ashraf, S., Shabaan, M., Babar, B. H., Zulfiqar, U., Haider, F. U., Ali, M. A., & Elshikh, M. S. (2023). Synergistic effects of rhizobacteria and salicylic acid on maize salt-stress tolerance. *Plants, 12*(13), 2519. https://doi.org/10.3390/plants12132519

Alotaibi, F., Lee, S., St-Arnaud, M., & Hijri, M. (2021). *Salix purpurea* and *Eleocharis obtusa* rhizospheres harbor a diverse rhizospheric bacterial community characterized by hydrocarbons degradation potentials and plant growth-promoting properties. *Plants, 10*(10), 1987. https://doi.org/10.3390/plants10101987

Ambardar, S., Bhagat, N., Vakhlu, J., & Gowda, M. (2021). Diversity of rhizo-bacteriome of *Crocus sativus* grown at various geographical locations and cataloging of putative PGPRs. *Frontiers in Sustainable Food Systems, 5*, 644230. https://doi.org/10.3389/fsufs.2021.644230

Anagah, F. I. (2023). Review of green technologies use by farmers: Implications for environmental sustainability in Nigeria. *International Journal of Environment and Climate Change, 13*(9), 75–84. https://doi.org/10.9734/ijecc/2023/v13i92206

Aransiola, S. A., Ijah, U. J. J., & Abioye, O. P. (2013). Phytoremediation of lead polluted soil by *Glycine max L. Applied and Environmental Soil Science, 2013*, Article ID 631619. https://doi.org/10.1155/2013/631619. https://www.hindawi.com/journals/aess/2013/631619/abs/

Aransiola, S. A., Ijah, U. J. J., Abioye, O. P., & Bala, J. D. (2019). Microbial-aided phytoremediation of heavy metals contaminated soil: A review. *European Journal of Biological Research, 9*(2), 104–125. http://www.journals.tmkarpinski.com/index.php/ejbr/article/view/157

Aransiola, S. A., Ijah, U. J. J., Abioye, O. P., & Bala, J. D. (2023). Vermicompost-assisted phytoremediation of toxic trace element-contaminated soil in Madaka, Nigeria using *Melissa officinalis* L and *Sida acuta*. *International Journal of Environmental Science and Technology*. https://doi.org/10.1007/s13762-022-04105-y

Aransiola, S. A., Ijah, U. J. J., Abioye, O. P., Bala, J. D., Rivadeneira-Mendoza, R. F., Luque, R., Rodríguez-Díaz, J. M., & Maddela, N. R. (2024). Micro and vermicompost assisted remediation of heavy metal contaminated soils using phytoextractors. *Case Studies in Chemical and Environmental Engineering, 9*(2024), 100755. https://doi.org/10.1016/j.cscee.2024.100755

Arkhipova, T., Martynenko, E., Sharipova, G., Kuzmina, L., Ivanov, I., Garipova, M., & Kudoyarova, G. (2020). Effects of plant growth promoting rhizobacteria on the content of abscisic acid and salt resistance of wheat plants. *Plants, 9*(11), 1429. https://doi.org/10.3390/plants9111429

Babalola, O. O., Emmanuel, O. C., Adeleke, B. S., Odelade, K. A., Nwachukwu, B. C., Ayiti, O. E., Adegboyega, T. T., & Igiehon, N. O. (2021). Rhizosphere microbiome cooperations: Strategies for sustainable crop production. *Current Microbiology, 78*(4), 1069–1085. https://doi.org/10.1007/s00284-021-02375-2

Babaniyi, B. R., Ogundele, O. D., Bisi-omotosho, A., Babaniyi, E. E., & Aransiola, S. A. (2023). Remediation approaches in environmental sustainability. In N. R. Maddela, L. K. W. Eller, & R. Prasad (Eds.), *Microbiology for cleaner production and environmental sustainability*. CRC Press. ISBN: 9781032496061. https://www.routledge.com/Microbiology-for-Cleaner-Production-and-Environmental-Sustainability/Maddela-Eller-Prasad/p/book/9781032496061

Benaissa, A. (2023). Rhizosphere: Role of bacteria to manage plant diseases and sustainable agriculture—A review. *Journal of Basic Microbiology, 64*(3), 2300361. https://doi.org/10.1002/jobm.202300361

Bhat, B. A., Tariq, L., Nissar, S., Islam, S. T., Islam, S. U., Mangral, Z., Ilyas, N., Sayyed, R. Z., Muthusamy, G., Kim, W., & Dar, T. U. H. (2022). The role of plant-associated rhizobacteria in plant growth, biocontrol and abiotic stress management. *Journal of Applied Microbiology, 133*(5), 2717–2741. https://doi.org/10.1111/jam.15796

Bhatt, P., Chaudhary, P., Ahmad, S., Bhatt, K., Chandra, D., & Chen, S. (2023). Recent advances in the application of microbial inoculants in the phytoremediation of xenobiotic compounds. In D. Chandra & P. Bhatt (Eds.), *Unravelling plant-microbe synergy developments in applied microbiology and biotechnology* (pp. 37–48). Academia Press. https://doi.org/10.1016/b978-0-323-99896-3.00013-8

Bhuyan, B., & Pandey, P. (2022). Remediation of petroleum hydrocarbon contaminated soil using hydrocarbonoclastic rhizobacteria, applied through *Azadirachta indica* rhizosphere. *International Journal of Phytoremediation, 24*(13), 1444–1454. https://doi.org/10.1080/15226514.2022.2033689

Borrero de Acuña, J. M., & Bernal, P. (2021). Plant holobiont interactions mediated by the type VI secretion system and the membrane vesicles: Promising tools for a greener agriculture. *Environmental Microbiology, 23*(4), 1830–1836. https://doi.org/10.1111/1462-2920.15457

Camargo, A. P., de Souza, R. S. C., Jose, J., Gerhardt, I. R., Dante, R. A., Mukherjee, S., Huntemann, M., Kyrpides, N. C., Carazzolle, M. F., & Arruda, P. (2023). Plant microbiomes harbor potential to promote nutrient turnover in impoverished substrates of a Brazilian biodiversity hotspot. *The ISME Journal, 17*(3), 354–370. https://doi.org/10.1038/s41396-022-01345-1

Chakraborty, U., & Chakraborty, B. (2022). Utilization of plant growth promoting rhizobacteria with multiple beneficial traits in agricultural biotechnology for crop improvement. In P. Jeschke & E. V. Starikov (Eds.), *Agricultural biocatalysis: Biological and chemical applications* (pp. 131–172). Pan Stanford Publishing Pte. Ltd.. https://doi.org/10.1201/9781003313144

Chovancová, J., Majerník, M., Drábik, P., & Štofková, Z. (2023). Environmental technological innovations and the sustainability of their development. *Ecological Engineering & Environmental Technology, 24*(4), 245–252. https://doi.org/10.12912/27197050/162708

Correa-García, S., Rheault, K., Tremblay, J., Séguin, A., & Yergeau, E. (2021). Soil characteristics constrain the response of microbial communities and associated hydrocarbon degradation genes during phytoremediation. *Applied and Environmental Microbiology, 87*(2), e02170–e02120. https://doi.org/10.1128/aem.02170-20

Cui, P., Liu, C., Su, X., Yang, Q., Ge, L., Huang, M., Dang, F., Wu, T., & Wang, Y. (2022). Atomically dispersed manganese on biochar derived from a hyperaccumulator for photocatalysis in organic pollution remediation. *Environmental Science and Technology, 56*(12), 8034–8042. https://doi.org/10.1021/acs.est.2c00992

de Andrade, L. A., Santos, C. H. B., Frezarin, E. T., Sales, L. R., & Rigobelo, E. C. (2023). Plant growth-promoting rhizobacteria for sustainable agricultural production. *Microorganisms, 11*(4), 1088. https://doi.org/10.3390/microorganisms11041088

Dineshkumar, R., Kumaravel, R., Gopalsamy, J., Sikder, M. N. A., & Sampathkumar, P. (2017). Microalgae as bio-fertilizers for rice growth and seed yield productivity. *Waste and Biomass Valorization, 9*(5), 793–800. https://doi.org/10.1007/s12649-017-9873-5

Ezaz, Z., Azhar, R., Rana, A., Ashraf, S., Farid, M., Mansha, A., Naqvi, S., Zahoor, A., & Rasool, N. (2020). Current trends of phytoremediation in wetlands: Mechanisms and applications. In M. Hasanuzzaman (Ed.), *Plant ecophysiology and adaptation under climate change: Mechanisms and perspectives II* (pp. 747–765). Springer. https://doi.org/10.1007/978-981-15-2172-0_28

Fields, B., Moffat, E., Harrison, E., Andersen, S., Young, J., & Friman, V. (2021). Genetic variation is associated with differences in facilitative and competitive interactions in the *Rhizobium leguminosarum* species complex. *Environmental Microbiology, 24*(8), 3463–3485. https://doi.org/10.1111/1462-2920.15720

Frąc, M., Hannula, E. S., Bełka, M., Salles, J. F., & Jedryczka, M. (2022). Soil mycobiome in sustainable agriculture. *Frontiers in Microbiology, 13*, 1033824. https://doi.org/10.3389/fmicb.2022.1033824

Gashash, E. A., Osman, N. A., Alsahli, A. A., Hewait, H. M., Ashmawi, A. E., Alshallash, K. S., El-Taher, A. M., Azab, E. S., Abd El-Raouf, H. S., & Ibrahim, M. F. M. (2022). Effects of plant-

growth-promoting rhizobacteria (pgpr) and cyanobacteria on botanical characteristics of tomato (*Solanum lycopersicon* L.) plants. *Plants, 11*(20), 2732. https://doi.org/10.3390/plants11202732
Ghantasala, S., & Roy Choudhury, S. (2022). Nod factor perception: An integrative view of molecular communication during legume symbiosis. *Plant Molecular Biology, 110*(6), 485–509. https://doi.org/10.1007/s11103-022-01307-3
Ghosh, D., & Maiti, S. (2020). Biochar assisted phytoremediation and biomass disposal in heavy metal contaminated mine soils: a review. *International Journal of Phytoremediation, 23*(6), 559–576. https://doi.org/10.1080/15226514.2020.1840510
Girolkar, S., Thawale, P., & Juwarkar, A. (2021). Bacteria-assisted phytoremediation of heavy metals and organic pollutants: Challenges and future prospects. In G. Saxena & M. P. Shah (Eds.), *V. Kumar* (pp. 247–267). Elsevier. https://doi.org/10.1016/b978-0-12-820318-7.00012-5
González-Reguero, D., Robas-Mora, M., Lobo, A. P., Gómez, P., & a. J. (2023). Bioremediation of environments contaminated with mercury. Present and perspectives. *World Journal of Microbiology & Biotechnology Incorporating the MIRCEN Journal of Applied Microbiology and Biotechnology, 39*(9), 249. https://doi.org/10.1007/s11274-023-03686-1
Gou, J. Y., Suo, S. Z., Shao, K. Z., Zhao, Q., Yao, D., Li, H. P., Zhang, J. L., & Rensing, C. (2020). Biofertilizers with beneficial rhizobacteria improved plant growth and yield in chili (*Capsicum annuum* L.). *World Journal of Microbiology & Biotechnology, 36*(6), 86. https://doi.org/10.1007/s11274-020-02863-w
Grigoriadi, A., Zobkova, N., Sotnikova, Y., Yamaleeva, A., & Farkhutdinov, R. (2023). Response of the rhizosphere microbial community of phytoremediation plants to oil pollution and the application of biopreparations. *Proceedings of Universities. Applied Chemistry and Biotechnology, 13*(4), 523–531. https://doi.org/10.21285/2227-2925-2023-13-4-523-531
Gupta, S., Stirk, W. A., Plačková, L., Kulkarni, M. G., Doležal, K., & Van Staden, J. (2021). Interactive effects of plant growth-promoting rhizobacteria and a seaweed extract on the growth and physiology of *Allium cepa* L. (onion). *Journal of Plant Physiology, 262*, 153437. https://doi.org/10.1016/j.jplph.2021.153437
Haque, M., Biswas, K., & Sinha, S. (2021). Phytoremediation strategies of some plants under heavy metal stress. In A. Hossain (Ed.), *Plant stress physiology*. InTech Open. https://doi.org/10.5772/intechopen.94406
Ibáñez, A., Garrido-Chamorro, S., Vasco-Cárdenas, M., & Barreiro, C. (2023). From lab to field: Biofertilizers in the 21st century. *Horticulturae, 9*(12), 1306. https://doi.org/10.3390/horticulturae9121306
Islam, W., Noman, A., Naveed, H., Huang, Z., & Chen, H. Y. H. (2020). Role of environmental factors in shaping the soil microbiome. *Environmental Science and Pollution Research, 27*, 41225–41247. https://doi.org/10.1007/s11356-020-10471-2
Ismail, H. Y., Farouq, A. A., Rabah, A. B., Muhammad, A. B., Allamin, I. A., Ibrahim, U. B., & Bukar, U. A. (2021). Microbe-assisted phytoremediation of petroleum hydrocarbons. In J. A. Malik (Ed.), *Advances in environmental engineering and green technologies book series* (pp. 386–416). IGI Global. https://doi.org/10.4018/978-1-7998-7062-3.ch015
Jach, M., Sajnaga, E., & Ziaja, M. (2022). Utilization of legume-nodule bacterial symbiosis in phytoremediation of heavy metal-contaminated soils. *Biology, 11*(5), 676. https://doi.org/10.3390/biology11050676
Ju, W., Liu, L., Jin, X., Duan, C., Cui, Y., Wang, J., Ma, D., Zhao, W., Wang, Y., & Fang, L. (2020). Co-inoculation effect of plant-growth-promoting rhizobacteria and rhizobium on EDDS assisted phytoremediation of Cu contaminated soils. *Chemosphere, 254*, 126724. https://doi.org/10.1016/j.chemosphere.2020.126724
Jung, H., & Kim, J. (2020). Complete reutilisation of mixed mackerel and brown seaweed wastewater as a high-quality biofertiliser in open-flow lettuce hydroponics. *Journal of Cleaner Production, 247*, 119081. https://doi.org/10.1016/j.jclepro.2019.119081
Karmakar, S., Roy, R., Bhattacharyya, A., Kar, B., Kumar, S., Singh, R., Bauddh, K., & Kumar, N. (2022). Rhizobacteria assisted phytoremediation of oily sludge contaminated sites. In

K. Bauddh & Y. Ma (Eds.), *Advances in microbe-assisted phytoremediation of polluted sites* (pp. 403–418). Academic. https://doi.org/10.1016/b978-0-12-823443-3.00006-5

Kaur, R., Kaur, S., & Kaur, G. (2021). Molecular and physiological manipulations in rhizospheric bacteria. *Acta Physiologiae Plantarum, 43*(5), 77. https://doi.org/10.1007/s11738-021-03251-z

Khan, T. (2023). Phytoremediation of heavy metals from contaminated soil by *Catharanthus roseus*. *Indian Scientific Journal of Research in Engineering and Management, 07*(07), 1–7. https://doi.org/10.55041/ijsrem24535

Khan, S., Masoodi, T. H., Pala, N. A., Murtaza, S., Mugloo, J. A., Sofi, P. A., Zaman, M. U., Kumar, R., & Kumar, A. (2023). Phytoremediation prospects for restoration of contamination in the natural ecosystems. *Water, 15*(8), 1498. https://doi.org/10.3390/w15081498

Khanghahi, M., AbdElgawad, H., Verbruggen, E., Korany, S., Alsherif, E., Beemster, G., & Crecchio, C. (2022). Biofertilisation with a consortium of growth-promoting bacterial strains improves the nutritional status of wheat grain under control, drought and salinity stress conditions. *Physiologia Plantarum, 174*(6), e13800. https://doi.org/10.1111/ppl.13800

Kumar, D., Dutt, S., Raigond, P., Changan, S. S., Lal, M. K., Tiwari, R. K., Chourasia, K. N., & Singh, B. (2021a). Beneficial microorganisms in crop growth, soil health, and sustainable environmental management. In S. Tyagi, R. Kumar, B. Saharan, & A. K. Nadda (Eds.), *Plant-microbial interactions and smart agricultural biotechnology* (pp. 11–32). CRC Press. https://doi.org/10.1201/9781003213864-2

Kumar, A., Jha, M., Singh, D., Pathak, D., & Rajawat, M. (2021b). Prospecting catabolic diversity of microbial strains for developing microbial consortia and their synergistic effect on Lentil (Lensesculenta) growth, yield and iron biofortification. *Archives of Microbiology, 203*, 4913–4928. https://doi.org/10.1007/s00203-021-02446-9

Kumar, S., Thakur, N., Singh, A. K., Gudade, B. A., Ghimire, D., & Das, S. (2022). Microbes-assisted phytoremediation of contaminated environment: Global status, progress, challenges, and future prospects. In V. Kumar, M. P. Shah, & S. K. Shahi (Eds.), *Phytoremediation technology for the removal of heavy metals and other contaminants from soil and water* (pp. 555–570). Elsevier. https://doi.org/10.1016/b978-0-323-85763-5.00007-6

Li, Y., Lian, J., Wu, B., Zou, H., & Tan, S. (2020). Phytoremediation of pharmaceutical-contaminated wastewater: Insights into rhizobacterial dynamics related to pollutant degradation mechanisms during plant life cycle. *Chemosphere, 253*, 126681. https://doi.org/10.1016/j.chemosphere.2020.126681

Liang, D., Xiao, C., Song, F., Li, H., Liu, R., & Gao, J. (2021). Complete genome sequence and function gene identify of prometryne-degrading strain *Pseudomonas* sp. DY-1. *Microorganisms, 9*(6), 1261. https://doi.org/10.3390/microorganisms9061261

Madline, A., Benidire, L., & Boularbah, A. (2021). Alleviation of salinity and metal stress using plant growth-promoting rhizobacteria isolated from semiarid Moroccan copper-mine soils. *Environmental Science and Pollution Research International, 28*(47), 67185–67202. https://doi.org/10.1007/s11356-021-15168-8

Malviya, M. K., Li, C. N., Lakshmanan, P., Solanki, M. K., Wang, Z., Solanki, A. C., Nong, Q., Verma, K. K., Singh, R. K., Singh, P., Sharma, A., Guo, D. J., Dessoky, E. S., Song, X. P., & Li, Y. R. (2022). High-throughput sequencing-based analysis of rhizosphere and diazotrophic bacterial diversity among wild progenitor and closely related species of sugarcane (*Saccharum* spp. inter-specific hybrids). *Frontiers in Plant Science, 13*, 829337. https://doi.org/10.3389/fpls.2022.829337

Mendoza, V. G., Vázquez, A. E. H., Carrillo, J. L. R., Viramontes, U. F., Mata, J. S., Garza, H. M. Q., Sáenz, E. O., Ríos, P. C., & Martínez, J. E. G. (2020). Halophilic rhizobacteria maintain the forage quality of *Moringa oleifera* grown on a saline substrate. *Revista Mexicana de Ciencias Pecuarias, 11*(3), 718–737. https://doi.org/10.22319/rmcp.v11i3.5175

Mishra, S., Kumar, P., & Singh, R. (2020). Transgenic plants in phytoremediation of organic pollutants. In V. C. Pandey & V. Singh (Eds.), *Bioremediation of pollutants* (pp. 39–56). Elsevier. https://doi.org/10.1016/b978-0-12-819025-8.00003-x

Mumtaz, M., Ahmad, M., Zafar-ul-Hye, M., Saqib, M., Akhtar, M., & Zaheer, M. (2022). Seed-applied zinc-solubilising *Bacillus* biofertilisers improve antioxidant enzyme activities, crop productivity, and biofortification of maize. *Crop and Pasture Science, 73*, 503–514. https://doi.org/10.1071/CP21415

Mutar, Z. H., Al-Baldawi, I. A., Mohammed, A. A., Hameed, H. K., Abdullah, S. R. S., & Ismail, N. I. (2023). Efficacy of phytoremediation for methylparaben removal from contaminated water: A study by *Alternanthera* spp and associated rhizobacteria. *IOP Conference Series: Earth and Environmental Science, 1215*(1), 012002. https://doi.org/10.1088/1755-1315/1215/1/012002

Nabi, M. (2023). Role of microorganisms in plant nutrition and soil health. In T. Aftab & K. R. Hakeem (Eds.), *Sustainable plant nutrition molecular interventions and advancements for crop development* (pp. 263–282). Academic. https://doi.org/10.1016/b978-0-443-18675-2.00016-x

Naeem, M., Khadeeja, K., Salam, A., Maria, A., Iftikhar, U., Anwar, A., Saleem, M., Mazhar, M. W., & Tariq, I. (2023). Phytoremediation of heavy metals from irrigation water, Faisalabad, Pakistan. *Global Academic Journal of Pharmacy and Drug Research, 5*(01), 10–15. https://doi.org/10.36348/gajpdr.2023.v05i01.003

Nor, M. N. M. (2020). Isolation and characterization of effective microorganism from oil palm rhizhopheric soil and evaluation of their potential as biofertiliser. *IOP Conference Series: Earth and Environmental Science, 515*, 012040. https://doi.org/10.1088/1755-1315/515/1/012040

Oinam, S., & Chakraborty, C. (2023). Green IT for sustainable development. In *Lecture notes in networks and systems* (pp. 783–791). https://doi.org/10.1007/978-981-99-0769-4_68.

Pankaj, U., Singh, D. N., Mishra, P., Gaur, P., Babu, C. V., Shanker, K., & Verma, R. K. (2020). Autochthonous halotolerant plant growth-promoting rhizobacteria promote bacoside A yield of *Bacopa monnieri* (L.) Nash and phytoextraction of salt-affected soil. *Pedosphere, 30*(5), 671–683. https://doi.org/10.1016/s1002-0160(20)60029-7

Parewa, H. P., Meena, V. S., Jain, L. K., & Choudhary, A. (2018). Sustainable crop production and soil health management through plant growth-promoting Rhizobacteria. In V. Meena (Ed.), *Role of Rhizospheric microbes in soil: Stress management and agricultural sustainability* (pp. 299–329). Springer. https://doi.org/10.1007/978-981-10-8402-7_12

Prakash, R., Subramani, R., Krodi, A., Berde, C. V., Chandrasekhar, T., Prathyusha, A. M. V. N., Kariali, E., & Bramhachari, P. V. (2022). Rhizobacteriome: Plant growth-promoting traits and its functional mechanism in plant growth, development, and defenses. In V. P. Bramhachari (Ed.), *Understanding the microbiome interactions in agriculture and the environment* (pp. 315–344). Springer. https://doi.org/10.1007/978-981-19-3696-8_16

Priya, A. K., Muruganandam, M., Ali, S. S., & Kornaros, M. (2023). Clean-up of heavy metals from contaminated soil by phytoremediation: A multidisciplinary and eco-friendly approach. *Toxics, 11*(5), 422. https://doi.org/10.3390/toxics11050422

Rabani, M. S., Hameed, I., Mir, T. A., Wani, B. A., Gupta, M. K., Habib, A., Jan, M., Hussain, H., Tripathi, S., Pathak, A., Ahad, M. B., & Gupta, C. (2022). Microbial-assisted phytoremediation. In R. A. Bhat, F. M. P. Tonelli, G. H. Dar, & K. Hakeem (Eds.), *Phytoremediation, biotechnological strategies for promoting invigorating environs* (pp. 91–114). Academic. https://doi.org/10.1016/b978-0-323-89874-4.00006-6

Rahal, S., & Chekireb, D. (2021). Diversity of rhizobia and non-rhizobia endophytes isolated from root nodules of *Trifolium* sp. growing in lead and zinc mine site Guelma, Algeria. *Archives of Microbiology, 203*(7), 3839–3849. https://doi.org/10.1007/s00203-021-02362-y

Raimi, A., Roopnarain, A., Chirima, G., & Adeleke, R. (2020). Insights into the microbial composition and potential efficiency of selected commercial biofertilisers. *Heliyon, 6*, e04342. https://doi.org/10.1016/j.heliyon.2020.e04342

Ramírez-Cariño, H. F., Morales, I., Guadarrama-Mendoza, P. C., Tlilayatzi-Bello, A. A., & Valdez-Blanco, R. (2023). Beneficial traits of crop-associated rhizobacteria toward plant growth promotion, biofertilization, and nutrient enhancement of harvested agro-products. In K. A. Abd-E-Salam & H. I. Mohamed (Eds.), *Secondary metabolites: Synthesis and applications in agroecosystem—Nanotechnology for plant protection* (pp. 299–334). Elsevier. https://doi.org/10.1016/B978-0-323-95251-4.00005-3

Rani, A. A., Basha, S. M., Darsha, K. D., Christy, C. A., Nagaiah, H. P., Kasthuri, T., & Pandian, S. K. (2023). Plant growth promoting rhizobacteria and their biofilms in promoting sustainable agriculture and soil health. In S. Das & N. A. Kungwani (Eds.), *Understanding microbial biofilms: Fundamentals to applications* (pp. 629–647). Academic. https://doi.org/10.1016/b978-0-323-99977-9.00003-x

Rashid, A., Rasheed, R., & Ngah, A. (2024). Achieving sustainability through multifaceted green functions in manufacturing. *Journal of Global Operations and Strategic Sourcing, 17*(2), 402–428. https://doi.org/10.1108/jgoss-06-2023-0054

Riaz, U., Murtaza, G., Anum, W., Samreen, T., Sarfraz, M., & Nazir, M. Z. (2021). Plant Growth-Promoting Rhizobacteria (PGPR) as biofertilizers and biopesticides. In K. R. Hakeem, G. H. Dar, M. A. Mehmood, & R. A. Bhat (Eds.), *Microbiota and biofertilizers: A sustainable continuum for plant and soil health* (pp. 181–196). Springer International Publishing. https://doi.org/10.1007/978-3-030-48771-3_11

Richardson, A. E., Kawasaki, A., Condron, L. M., Ryan, P. R., & Gupta, V. V. S. R. (2020). Root microbiome structure and microbial succession in the rhizosphere. In V. V. S. R. Gupta & A. K. Sharma (Eds.), *Rhizosphere biology interactions between microbes and plants* (pp. 109–128). https://doi.org/10.1007/978-981-15-6125-2_5

Riko, Y. Y., Darma, Z. U., & Kabir, K. (2023). Temporal variations in bacterial population dynamics of the *Sorghum bicolor* rhizobiome reflect effective heavy metal phytoremediation. In E. I. Sahin & Z. Khalilli (Eds.), *Proceedings Book. 4. International Antalya Scientific Research and Innovative Studies Congress, May 9–10, 2023, Antalya, Türkiye* (pp. 742–743). IKSAD Publishing House.

Riko, Y. Y., Darma, Z. U., & Kabir, K. (2024). Metabolic diversity of the metallotolerant bacterial rhizobiome of *Sorghum bicolor* facilitates plant growth promotion and heavy metals phytoremediation. In Department of Microbiology, UMYUK (Ed.), *Book of Abstracts for the 2nd UMYU Conference on Microbiology and Related Sciences, Tuesday 27th – Thursday 29th February, 2024* (pp. 87–88). Darma-Kt Digital Prints.

Rojas-Solis, D., Larsen, J., & Lindig-Cisneros, R. (2023). Arsenic and mercury tolerant rhizobacteria that can improve phytoremediation of heavy metal contaminated soils. *PeerJ, 11*, e14697. https://doi.org/10.7717/peerj.14697

Rosariastuti, R., & Yudha, F. (2022). Potential of *Rhizobium* sp. I 3 and *Agrobacterium* sp. I 3 7 in formation of root nodules and stimulating growth of *Arachis hypogaea* L. *Engenharia Agricola, 42*(5), e20200209. https://doi.org/10.1590/1809-4430-eng.agric.v42n5e20200209/2022

Seth, K., Vyas, P., Deora, S., Gupta, A. K., Meena, M., Swapnil, P., & Harish, N. (2023). Understanding plant-plant growth-promoting rhizobacteria (PGPR) interactions for inducing plant defense. In P. Swapnil, M. Meena, H. A. Marwal, S. Vijayalakshmi, & A. Zehra (Eds.), *Plant-microbe interaction – Recent advances in molecular and biochemical approaches: Volume 2: Agricultural aspects of microbiome leading to plant Defence* (pp. 201–226). Academic. https://doi.org/10.1016/b978-0-323-91876-3.00010-5

Shahrajabian, M. H., Sun, W., & Cheng, Q. (2021). The importance of *Rhizobium, Agrobacterium, Bradyrhizobium, Herbaspirillum, Sinorhizobium* in sustainable agricultural production. *Notulae Botanicae Horti Agrobotanici Cluj-Napoca, 49*(3), 12183. https://doi.org/10.15835/nbha49312183

Soumare, A., Boubekri, K., Lyamlouli, K., Hafidi, M., Ouhdouch, Y., & Kouisni, L. (2020). From isolation of phosphate solubilizing microbes to their formulation and use as biofertilizers: Status and needs. *Frontiers in Bioengineering and Biotechnology, 7*, 425. https://doi.org/10.3389/fbioe.2019.00425

Sun, B., Bai, Z., Bao, L., Xue, L., Zhang, S., Wei, Y., Zhang, Z., Zhuang, G., & Zhuang, X. (2020). Bacillus subtilis biofertilizer mitigating agricultural ammonia emission and shifting soil nitrogen cycling microbiomes. *Environment International, 144*, 105989. https://doi.org/10.1016/j.envint.2020.105989

Thomas-Barry, G., Martin, C., Lynch, M., Ramsubhag, A., Rouse-Miller, J., & Charles, T. (2021). Driving factors influencing the rhizobacteriome community structure of plants adapted to multiple climatic stressors in edaphic savannas. *The Science of the Total Environment, 769*, 145214. https://doi.org/10.1016/j.scitotenv.2021.145214

Tian, T., Reverdy, A., She, Q., Sun, B., & Chai, Y. (2019). The role of rhizodeposits in shaping rhizomicrobiome. *Environmental Microbiology Reports, 12*(2), 160–172. https://doi.org/10.1111/1758-2229.12816

Tiodar, E. D., Văcar, C. L., & Podar, D. (2021). Phytoremediation and microorganisms-assisted phytoremediation of mercury-contaminated soils: Challenges and perspectives. *International Journal of Environmental Research and Public Health, 18*(5), 2435. https://doi.org/10.3390/ijerph18052435

Tripathi, S., Singh, V., Srivastava, P., Singh, R., Devi, R., Kumar, A., & Bhadouria, R. (2020). Phytoremediation of organic pollutants. In P. Singh, A. Kumar, & A. Borthakur (Eds.), *Abatement of environmental pollutants trends and strategies* (pp. 81–105). Elsevier. https://doi.org/10.1016/b978-0-12-818095-2.00004-7

Umar, Z. D., Aziz, N. A. A., Zulkifli, S. Z., & Muskhazli, M. (2018). Efficiency of effective polycyclic aromatic hydrocarbons (PAHs) degrading consortium in resisting toxic heavy metals during PAHs biodegradation. *International Journal of Environment, 7*(1), 14–27. https://doi.org/10.3126/ije.v7il.21291

Umar, Z. D., Aminu, M., & Yahaya, R. Y. (2020). Survival response of consortium isolates from diesel contaminated soil within Katsina state Nigeria. *International Journal of Environment, 9*(2), 51–66. https://doi.org/10.3126/ije.v9i2.32516

Umar, Z. D., Aminu, M., & Yahaya, Y. R. (2022). Optimization of diesel biodegrading conditions using response surface methodology based on central composite design. *Polycyclic Aromatic Compounds, 42*(5), 2002–2012. https://doi.org/10.1080/10406638.2020.1823859

Valle-Romero, P., García-López, J. V., Redondo-Gómez, S., Flores-Duarte, N. J., Rodríguez-Llorente, I. D., Idaszkin, Y. L., Pajuelo, E., & Mateos-Naranjo, E. (2023). Biofertilization with PGP bacteria improve strawberry plant performance under sub-optimum phosphorus fertilization. *Agronomy, 13*(2), 335. https://doi.org/10.3390/agronomy13020335

Verma, P., & Pandey, K. (2023). Biofertilizer: An ultimate solution for the sustainable development of agriculture. *Current Agriculture Research Journal, 10*(3), 193–203. https://doi.org/10.12944/carj.10.3.04

Vermeulen, M., Cason, E. D., & Swart, W. J. (2020). The rhizobiome of herbaceous plants in Clovelly and Sterkspruit soils of the Stevenson–Hamilton supersite. *Koedoe, 62*(2), a1596. https://doi.org/10.4102/koedoe.v62i2.1596

Wijegunawardana, N. D. A. D., Perera, E. G., & Ekanayake, M. S. (2022). Phytoremediation. In T. C. Bamunuarachchige & H. K. S. de Zoysa (Eds.), *Waste technology for emerging economies* (pp. 167–197). https://doi.org/10.1201/9781003132349-8

Wojtowicz, K., Steliga, T., & Kapusta, P. (2023). Evaluation of the effectiveness of bioaugmentation-assisted phytoremediation of soils contaminated with petroleum hydrocarbons using *Echinacea purpurea*. *Applied Sciences, 13*(24), 13077–13077. https://doi.org/10.3390/app132413077

Xiang, L., Harindintwali, J., Wang, F., Redmile-Gordon, M., Chang, S., Fu, Y., He, C., Muhoza, B., Brahushi, F., Bolan, N., Jiang, X., Ok, Y., Rinklebe, J., Schaeffer, A., Zhu, Y., Tiedje, J., & Xing, B. (2022). Integrating biochar, bacteria, and plants for sustainable remediation of soils contaminated with organic pollutants. *Environmental Science and Technology, 56*(23), 16546–16566. https://doi.org/10.1021/acs.est.2c02976

Xiao, Y., Chen, L., Li, C., Ma, J., Chen, R., Yang, B., Liu, G., Liu, S., & Fang, J. (2023). Role of the rhizosphere bacterial community in assisting phytoremediation in a lead-zinc area. *Frontiers in Plant Science, 13*, 1106985. https://doi.org/10.3389/fpls.2022.1106985

Yapa, N., Lakmali, D., De Zoysa, K. S., Silva, S., Manawadu, C., Herath, B. M., Madhushan, A., Perera, G., Ratnayakae, O., Kapilan, R., Rathnayake, A., Sirisena, A. I., Asad, S., Karunarathna, S. C., & Bamunuarachchige, C. (2022). Biofertilizers: An emerging trend in agricultural sustainability. *Chiang Mai Journal of Science, 49*(3), 608–640. https://doi.org/10.12982/cmjs.2022.050

Yuan, Q., Wang, P., Wang, X., Hu, B., & Tao, L. (2021). Phytoremediation of cadmium-contaminated sediment using *Hydrilla verticillata* and *Elodea canadensis* harbor two same keystone rhizobacteria Pedosphaeraceae and Parasegetibacter. *Chemosphere, 286*(Pt 1), 131648. https://doi.org/10.1016/j.chemosphere.2021.131648

Zhang, J., Cook, J., Nearing, J. T., Zhang, J., Raudonis, R., Glick, B. R., Langille, M. G. I., & Cheng, Z. (2021). Harnessing the plant microbiome to promote the growth of agricultural crops. *Microbiological Research, 245*, 126690. https://doi.org/10.1016/j.micres.2020.126690

Zhou, Z., Yu, M., Ding, G., Gao, G., & He, Y. (2020). Diversity and structural differences of bacterial microbial communities in rhizocompartments of desert leguminous plants. *PLoS One, 15*(12), e0241057. https://doi.org/10.1371/journal.pone.0241057

Zhu, L., Huang, J., Lu, X., & Zhou, C. (2022). Development of plant systemic resistance by beneficial rhizobacteria: Recognition, initiation, elicitation and regulation. *Frontiers in Plant Science, 13*, 952397. https://doi.org/10.3389/fpls.2022.952397

Harnessing Biosurfactants from Soil Bacteria: Advancing Green Technology for Environmental Sustainability

Mounia Arab, Wafa Malki, Nadjette Moumene, and Nihel Klouche

Abstract Green technologies are being revolutionized by surface active compounds produced by bacteria called biosurfactants that significantly lower surface tension between two phases. These biosurfactants are biodegradable, less toxicic and often obtained from renewable sources, making them highly eco-friendly. Soil bacteria stand out as excellent producers of biosurfactants because they can easily adapt to different environments due to their high metabolic diversity. This chapter explores the essential properties, biosynthesis, applications in green technologies, and environmental impact of biosurfactants. It aims to demonstrate how exploitation of soil bacteria can provide ecological solutions and contribute to a sustainable future. Progress requires interdisciplinary collaboration and continuous innovation.

Keywords Biosurfactants · Soil bacteria · Environmentail sustainability · Bioremediation

1 Introduction

Sustainable formulations are becoming increasingly important in a wide range of industrial sectors, including environmental applications. This need for sustainability is prompting the industry to move away from synthetic surfactants and adopt more eco-friendly alternatives (Eras-Muñoz et al., 2022). One route to achieve this goal is utiliziation of biosurfactants, which offer a sustainable and environmentally friendly

M. Arab (✉)
Laboratory of Applied Microbiology in Food, Biomedical and Environment, Department of Biology, Faculty of Nature Sciences and Life, University of Tlemcen, Tlemcen, Algeria

University of Sciences and Technology Houari Boumediene, Faculty of Biological Sciences, Algiers, Algeria
e-mail: marab1@usthb.dz

W. Malki · N. Moumene · N. Klouche
Laboratory of Applied Microbiology in Food, Biomedical and Environment, Department of Biology, Faculty of Nature Sciences and Life, University of Tlemcen, Tlemcen, Algeria

S. A. Aransiola et al. (eds.), *Soil Microbiome in Green Technology Sustainability*,
https://doi.org/10.1007/978-3-031-71844-1_23

solution. Biosurfactants can be produced by various microorganisms, for example, bacteria and yeasts.

Biosurfactants are amphiphilic molecules with both hydrophobic and hydrophilic components, allowing them to promote the formation of interfaces between liquids with different polarities. The hydrophilic part of biosurfactants typically consists of structures such as amino acids, anionic or cationic peptides, and carbohydrates. The hydrophobic tail is often made up of peptides, proteins, or fatty acids that can be saturated or unsaturated (Farias et al., 2021). The ability of biosurfactants to reduce interfacial tension varies depending on their chemical composition. Biosurfactants can be classified into two main groups based on their molecular weight: low molecular weight (glycolipids and lipopeptides) and high molecular weight (polysaccharides, lipopolysaccharides, proteins, and lipoproteins). Lower-molecular-weight biosurfactants are more effective at reducing surface tension, while higher-molecular-weight biosurfactants are better at stabilizing oil-water emulsions (Otzen, 2017; Karlapudi et al., 2018; Kumar & Ngueagni, 2021; Sánchez, 2022). The unique chemical properties of different biosurfactants are closely linked to their biological activities and applications (Morita et al., 2015). These molecules are among the strongest competitors to synthetic surfactants in terms of their safety, low toxicity, great biocompatibility, high biodegradability, environmentally benign character, and ability to be produced utilizing low-cost agro-industrial raw materials (Qamar & Pacifico, 2023).

Biosurfactants gained attention as hydrocarbon degradation agents in the 1960s, and their uses have significantly expanded in the past five decades as an improved alternative to chemical surfactants, particularly in the food, pharmaceutical, and oil industries. Over 10 million tons of chemical surfactants and microbial biosurfactants are annually produced. The synthesis of microbial surfactants is primarily influenced by the kind and quantity of the producing organism, as well as parameters such as nitrogen and carbon availability, temperature, air circulation, and trace elements (Banat et al., 2010).

Biosurfactants are extensively utilized in several industries including agriculture, food production, chemistry, cosmetics, and pharmaceutics due to their significant benefits. The various applications of biosurfactants are documented in numerous review papers (Muthusamy et al., 2008; Banat et al., 2010; Soberón-Chávez & Maier, 2011). This chapter primarily examines the use of biosurfactants produced by soil bacteria, emphasizing its crucial role in promoting environmental sustainability.

2 Soil Bacteria for Biosurfactant Production

Microorganismes isolates form soil play an important role in the synthesis of biosurfactants. Indeed, they have demonstated the ability to produce surface-active molecules in response to specific environmental conditions, such as the presence of pollutants or complex carbon sources (Kalvandi et al., 2022). Among the bacteria most studied for the production of biosurfactants, we find the genera *Pseudomonas*, *Bacillus*, and *Rhodococcus* (Xu et al., 2022).

Ibrahim et al. (2013) have identified *Achromobacter xylosoxidans*, *Bacillus licheniformis*, *Proteus vulgaris*, *Proteus mirabilis*, *Serratia marcescens*, *Sphingomonas paucimobilis*, and *Micrococcus kristinae* as biosurfactant-producing soil bacteria, and their biosurfactants can be used in MEOR under various environmental conditions.

Folasade and Hunter (2021) have isolated two aerobic endospore-forming bacteria, belonging to the genera *Brevibacillus* and *Bacillus*, from Mfabeni peatland in South Africa. The isolates produced surfactin isomers, plantazolicin precursor, fengycin and iturin.

Machado et al. (2022) have isolated four strains, *Rhodococcus erythropolis, Rhizobium radiobacter, Microbacterium liquefacien, Pseudomonas libanensis* and *Pseudomonas veronii*, from Antarctic soil contaminated by diesel. These bacteria were able to produce biosurfactants containing carbohydrates with potential applications.

Kalvandi et al. (2022) have isolated from saline-sodic soils contaminated with petroleum strains belonging to the genus *Bacillus*. These strains were able to produce biosurfactants that release TPH.

The structural variations confer specific physico-chemical properties to biosurfactants, suited for different applications. For instance, rhamnolipids produced by *Pseudomonas aeruginosa* are effective in hydrocarbon bioremediation, while surfactins from *Bacillus subtilis* show promising antimicrobial properties (Niamat et al., 2022).

3 Biosynthesis of Bacterial Biosurfactants

Biosurfactants are encoded by several genes and synthesized by a variety of microorganisms including *Pseudomonas* sp., *Acinetobacter* sp., *Bacillus* sp., *Candida* sp., *Cryptococcus* sp., *Penicillium* sp., *Aspergillus* sp., *Kurtzmanomyces* sp., *Rhodococcus* sp., *Sphingomonas* sp., *Arthrobacter* sp., *Lactococcus* sp., and *Pseudozyma* sp. (Borah et al., 2021).

Studies have indicated that microorganisms employ distinct pathways to produce the hydrophobic and hydrophilic components of biosurfactants, which are then merged together (Théatre et al., 2021). These biosurfactants are produced from a range of substrates and the production can be spontaneous or induced by factors like the presences of lipophilic compounds, pH variations, agitation speed, environmental stress and low nitrogen concentrations (Silva et al., 2018). This biosynthetic process varies also depending on the carbon source available to the microorganism during growth. The main carbon sources are carbohydrates and lipids or hydrocarbons (Sanches et al., 2021).

A study conducted by Batool et al. (2017), demonstrated that each of the chosen bacterial strains, isolated from petroleum-contaminated soil in different motor workshops, demonstrated increased production of biosurfactants when yeast extract was used as the nitrogen source and glucose as the carbon source. Datta et al. (2020) investigated how various carbon sources affected biosurfactant production. This was done by adding glucose, crude oil, glycerol, as well as light and heavy paraffin oils to the bacterial culture as carbon sources .The highest biosurfactant concentration was achieved using glycerol, a cost-effective carbon source that is also a by-product of trans-esterification processes (De Faria et al., 2011; Sharma et al., 2018). This finding suggests that utilizing glycerol can enhance the cost-effectiveness of biosurfactant production.

Microbial biosurfactants are synthesized intracellularly or extracellularly, requiring the activation of specific genes or enzymes in the presence of particular substrates.

3.1 Lipopeptides Biosurfactants

Cyclic lipopeptides exhibit an amphiphilic nature, featuring a hydrophilic portion composed of an oligopeptide ring linked to a fatty acid chain. The extensive structural diversity of these compounds arises from variations in oligopeptide composition, polar head sequence, and fatty acid tail diversity (Mnif & Ghribi, 2015). Within the realm of bacterial genera, *Bacillus* (Gram-positive) and *Pseudomonas* (Gram-negative) have emerged as focal points of study owing to their ability to yield a broad spectrum of efficacious lipopeptides biosurfactants (LPBSs) (Geissler et al., 2019).

Over recent years, there has been thorough scrutiny of the biosynthetic processes and genetic control mechanisms of LPBSs. These compounds are typically produced independently of ribosomes, utilizing large enzymes known as nonribosomal peptide synthetases (NRPSs). The generation of functional NRPSs necessitates more than just transcriptional activation and translation; it also involves post-translational modifications and assembly (Roongsawang et al., 2010). Among them, Surfactin stands out as the first and most well-known microbial surfactant within this category.

Surfactin is a robust biosurfactant .This biosurfactant is produced by most species belonging to the *Bacillus* genus, comprises a heptapeptide unit connected to a fatty acid chain. Its biosynthesis is catalyzed non-ribosomally by a sizable multienzyme peptide synthetase complex referred to as surfactin synthetase, which is composed of three proteins subunits (Das et al., 2008). Surfactin production by *Bacillus* species is facilitated by surfactin synthase, involving the assembly of amino acids into a peptide chain and subsequent formation of the lipopeptide (Jimoh et al., 2021).

The surfactin biosynthetic gene cluster contains a co-encoded regulatory gene known as *comS* (Hamoen et al., 1995). This gene is situated within the open reading frame of the *srfAB* gene (Liu & Zuber, 1998). *ComS* plays a role in regulating genetic competence and is also a component of the quorum sensing system (Stiegelmeyer & Giddings, 2013). In 2022, Yasmin et al. (2022) conducted a study focusing on the genomes of *Bacillus subtillis* strains MB415 and MB418. Annotation analysis uncovered non-ribosomal synthetase gene clusters responsible for producing lipopeptides, including surfactin and fengycin, via the *srf* and *fen* genes, respectively. Comparative genomic analysis highlighted differences in gene operons, particularly those related to surfactin biosynthesis.

Qiao et al. (2024) investigated in their study the non ribosomal synthesis of surfactin, a cyclic hexalipopeptide compound, by *Bacillus* species. The focus was on examining the regulatory pathways and mechanisms that affect the production and release of surfactin, with specific attention given to the genes regulating the transcription of the srfAA-AD operon.

3.2 Glycolipids Biosurfactants

Glycolipids biosurfactants consisted generally of a fatty acid linked to a carbohydrate by a glycosidic bond moiety, they are surface-active compounds synthesized by diverse microorganisms. They exhibit significant structural diversity of glycolipids arises from the varying nature of their carbohydrate portion and lipid backbones. This diversity leads to the categorization of glycolipids into sub-classes, such as rhamnolipids, sophorolipids, trehalolipids, succinoyl-trehalolipids, lipids derived from cellobiose or ustilagic acid, mannosyl-erythritol lipids or ustilipids, xylolipids, and lipids derived from oligosaccharides (Kitamoto et al., 2002).

Among these compounds, the rhamnolipids synthesized by *Pseudomonas sp.* stand out. The molecular biosynthetic regulation of rhamnolipids, a type of glycolipid biosurfactant produced by *P. aeruginosa*, was initially elucidated (Das et al., 2008). In these bacteria, the enzymes responsible for rhamnolipid production are encoded by the rhlAB operon and the *rhlC* gene (Wagner et al., 2003). Rhamnolipids can be produced during either the exponential or stationary phase of growth, depending on whether hydrophilic or hydrophobic substrates are employed (Mnif et al., 2018). *P. aeruginosa* produces two major types of rhamnolipid in liquid cultures: the monorhamnolipid and the dirhamnolipid (Déziel et al., 2000).

Sánchez (2022) has suggested for glycolipid biosynthesis that when carbohydrates serve as the primary carbon source, both lipogenic and glycolytic pathways contribute to the synthesis of lipid and hydrophilic portions, respectively.

Distinct biosynthetic pathways have been proposed for the synthesis of bioemulsifiers. For instance, in emulsan biosynthesis, fructose 6-phosphate undergoes transformations to form UDP-N-acetyl-D-galactosamine uronic acid, which, through sequential reactions, leads to the formation of emulsan (Singh et al., 1990; Nakar & Gutnick, 2001).

Understanding these intricate mechanisms is crucial for harnessing the potential of biosurfactants in various industrial and environmental applications.

4 Biosurfactants Applicaions Areas

Biosurfactants have a wide range of applications including bioremediation, oil recovery, emulsification stabilization, foaming and detergency, lubrication and drug delivery. Microbially-derived biosurfactants are related with several processes such as motility, biofilm contact, substrate consumption, and adhesion. The employment of biosurfactants in the biomedical sector is expanding due to their bioactive features, which include antifungal, antibacterial, anticancer, and antiviral activities (Camelia & Sumitra, 2023; Qamar & Pacifico, 2023).

4.1 Environmental Applications

Overall, biosurfactants used for environmental applications offer numerous benefits that make them suitable for use in highly contaminated environments. According to the literature, biosurfactants can be used in three forms: crude extract, pure biosurfactant, or biosurfactant-producing microorganisms (Eras-Muñoz et al., 2022).

Bioremediation techniques have demonstrated their efficacy as a promising and ecologically acceptable technology for restoring oil-contaminated soil and water to their original state (Borah & Yadav, 2017). Remediation refers to the process of converting harmful environmental pollutants into less or non-toxic substances through degradation and transformation techniques (Li et al., 2021). Bioremediation is seen as a promising approach due to its ability to harness microorganisms. The environment is unharmed by this technique because to the microbial capacity to break down dangerous chemical contaminants into benign substances including CO_2, CH_4, H_2O, and biomass (Ron & Rosenberg, 2014).

Microbial enhanced oil recovery (MEOR) can be carried out using two primary methods: *in-situ* and *ex-situ* procedures (Geetha et al., 2018). The *in-situ* technique entails inducing a production of biosurfactants within the reservoir by introducing nutrients or injecting native microorganisms that produce biosurfactants. On the other hand, the *ex-situ* approach entails manufacturing biosurfactants in a specialized container called a bioreactor and then introducing them into the reservoir using wells. The *ex-situ* approach is preferred due to its ability to overcome the difficulties presented by reservoir conditions on strains that produce biosurfactants. Biosurfactants employed in MEOR are generally chosen for their capacity to decrease the interfacial tension between crude oil and water. However, they also play a key role in modifying wettability and improving emulsification, as stated by Kryachko in 2018.

The biosurfactant produced by *Rhodococcus erythropolis* AQ5–07 on a medium containing 3% waste canola oil has shown a great ability to remove heavy metals, phenolic content and to degrade diesel fuel (Aransiola et al., 2019; Ibrahim et al., 2020).

Heavy metals pose serious health and environmental prolems due to their persistence, virulence, and toxicity, even at low concentrations. While various conventional techniques have been employed to decontaminate and remove these contaminants from the environment, their high operational costs, energy demands, and issues with post-disposal have rendered them unsuitable and environmentally unfriendly (Chiamaka et al., 2023). The mechanism for heavy metal removal using biosurfactants involves three main steps: binding and sorption of biosurfactants to the metal and soil particles, separation of metals from the soil, and association of heavy metals with micelles.

Biosurfactants are employed in the synthesis of nanoparticles, playing a crucial role in the rehabilitation of polluted marine ecosystems (Christopher et al., 2019) in several ways:

- Stabilization of metal nanoparticles: Surfactin, is used as a stabilizer in the formulation of gold and silver nanoparticles (Chiamaka et al., 2023).
- Production of Silver nanoplastics: The glycolipid derived from *Brevibacterium casei* is employed in the production of silver nanoplastics via a microemulsion process that utilizes reverse micelles (Chiamaka et al., 2023).
- Enhanced stability in nanoparticle synthesis: A biosurfactant produced by *Pseudomonas aeruginosa* enhances the stability of silver nanoparticles (Farias et al., 2014).
- Reduction and capping agents: Biosurfactants act as reducing and capping agents in nanoparticle synthesis (Chiamaka et al., 2023).
- Lipopeptide stabilization: *Bacillus subtilis* lipopeptides act as stabilizers, facilitating nanoparticle formation without the use of reducing agents (Rane et al., 2017).
- Optimization of polyethylene nanoparticles: Biosurfactants are used to optimize the synthesis of polyethylene nanoparticles (Chiamaka et al., 2023).
- Oil recovery mechanisms: A combination of biosurfactants and nanoparticles is employed in oil recovery processes, offering a competitive advantage by utilizing nanoparticles as chemical agents in remedial applications (Sircar et al., 2021).

4.2 Agriculture-Related Applications

Biosurfactants (BSs) offer significant advantages over chemically manufactured surfactants because of their dual hydrophobic and hydrophilic properties. These characteristics enable their use in various agricultural applications (Ranjan & Sow, 2021).

4.2.1 Improvement of Soil Quality

The presence of various organic and inorganic pollutants causes abiotic stress to crops, reducing agricultural productivity. Bioremediation is crucial for improving soil quality contaminated by hydrocarbons and heavy metals. Technologies such as soil washing and clean-up combination techniques utilize biosurfactants (BSs) to effectively remove these pollutants, as BSs enhance bioavailability and promote the biodegradation of hydrophobic substances (Kang et al., 2010).

BSs can also degrade chemical insecticides that accumulate in agricultural soil. Rhamnolipids, for instance, have been found effective in removing pentachlorophenol and polyaromatic hydrocarbons from soil, thus contributing to soil quality improvement (Moldes et al., 2011).

New foaming surfactant technologies have shown that BSs like surfactin and rhamnolipid can remove heavy metals such as Ni, Ba, Li, Mg, Mn, Ca, Cd, Cu, and Zn ions from soil. Therefore, utilizing BS overproducers could be more beneficial for bioremediation compared to harmful synthetic surfactants (Ranjan & Sow, 2021).

4.2.2 Plant Pathogen Elimination

Several biosurfactants derived from bacteria exhibit antimicrobial activity against various plant diseases, positioning them as promising biocontrol agents for long-term agricultural sustainability. Rhizobacteria-produced biosurfactants are known for their antagonistic properties (Nihorimbere et al., 2011).

Plant growth-promoting bacteria use various biocontrol mechanisms when biosurfactants are applied in agriculture, including parasitism, antibiosis, competition, induced systemic resistance, and hypovirulence. These surfactants enhance the antagonistic actions of microorganisms and microbial products when used in large quantities in agriculture. Biosurfactants have been demonstrated to enhance the insecticidal activities of other systems in numerous *in vitro* and *in situ* experiments. Additionally, these surfactants, in conjunction with fungi such as *Myrothecium verrucaria,* are used to control weed species that negatively impact land productivity. BSs have been found to inhibit aflatoxin production by *Aspergillus* species, which infect crops like peanuts, cottonseed, and corn during storage and in the field. Thus, these biomolecules play various roles in plant pathogen elimination and several agricultural processes (Shashank et al., 2023).

4.2.3 Assets for Beneficial Plant-Microbe Interactions

Biosurfactants influence microbial movement, signaling, differentiation, and biofilm creation (Ron & Rosenberg, 2011). These green surfactants are essential for microorganisms to form beneficial relationships with plant roots and promote plant growth. Additionally, rhizobacteria produce biosurfactants that increase the bioavailability of hydrophobic compounds, which can serve as nutrients. Soil bacteria produce biosurfactants that improve soil wettability and aid in the effective dispersion of chemical fertilizers, thereby enhancing plant growth (Dusane et al., 2010).

4.3 Industrial Applications

Biosurfactants have found widespread use across various industries due to their unique functional properties. These include applications in petroleum, metallurgy, petrochemicals, organic chemicals, mining, beverages, agrochemicals, pharmaceuticals, fertilizers, cosmetics, and more. Their ability to reduce surface tension enables their use as emulsifiers, demulsifiers, foaming agents, wetting agents, functional food components, spreading agents, and detergents, among other applications (Tirth et al., 2023).

Due to their inherent surface-active and emulsifying properties, biosurfactants are utilized as natural alternatives to chemical detergents (Van Renterghem et al., 2018). Their biodegradability makes them environmentally friendly upon disposal. Additionally, BSs are effective across a wide range of pH levels, temperatures, and salinities (Mukherjee & Das, 2010).

4.3.1 Biorefinery Processes

Despite the environmental pollution drawbacks, petroleum-based products and crude oil remain vital in modern society. Conventional oil extraction technologies have a limited ability to recover oil from reservoirs, typically only retrieving 10–40% of the available oil. This has led to the development of enhanced oil recovery techniques Microbially enhanced oil recovery (MEOR) is a process that utilizes the inherent capabilities of specific microbes to spread oil by use of biosurfactants. MEOR can be implemented by:

- Stimulating the reservoir by cultivating indigenous hydrocarbonoclastic bacteria and adding nitrogen sources;
- Inoculating the reservoir with selected consortia of targeted bacteria; or
- Adding *ex situ* products, such as biosurfactants, to the reservoirs.

Numerous microbial biosurfactants have been shown to enhance oil recovery (Khouloud & Osama, 2023).

Gas hydrates, often known as ice formations containing trapped gas molecules, offer an alternative approach for energy and carbon storage (Chong et al., 2016). Ongoing research is dedicated to the development of systems that can store and transport gas hydrates in a manner that is both efficient and secure. Research has discovered that biosurfactants have the ability to enhance the re-formation of methane hydrate, leading to an improvement in storage (Arora et al., 2014).

4.3.2 Food Processing

Biosurfactants play a crucial role in food processing due to their ability to reduce surface tension. They act as anti-adhesive agents, aiding in food formation and stabilization. Biosurfactants also help control fat globule clumping, stabilize aerated systems, enhance food texture, extend the shelf life of starch-containing products, modify the viscoelasticity of wheat dough, and improve the consistency and texture of fat-based products (Krishnaswamy et al., 2008). Incorporating microbial biosurfactant into food can enhance dough rheology, increase emulsification, and boost fat volume, benefiting the meat processing and bakery industries (Tripathi et al., 2018).

4.3.3 Cosmetic Industry

Chemically synthesized surfactants can cause skin allergies and irritations, driving consumer interest in natural cosmetic products. Biosurfactants are increasingly used as alternatives to synthetic surfactants in cosmetics to mitigate these harmful effects. Biosurfactants are utilized in a variety of applications, including foaming agents, emulsifiers, solubilizers, cleansers, wetting agents, antimicrobial agents, bath products, mediators of enzyme action, acne pads, anti-dandruff products, insect repellents, baby products, toothpaste, mascara, dentine cleansers, lipsticks, and contact lens solutions (Gharaei-Fathabad, 2011).

The choice of BS compounds for cosmetic formulations is determined by their capacity to emulsify and their surface properties, including hydrophilic-lipophilic balance (HLB) and critical micelle concentration (CMC). A higher HLB value implies a greater degree of hydrophilicity in the biosurfactant, whereas a lower HLB value suggests a higher degree of lipophilicity in the biosurfactant. Biosurfactantss possess attractive qualities in cosmetics due to their HLB values, making them suitable for use as emulsifiers, antifoaming agents, and wetting agents. The critical micelle concentration (CMC) refers to the lowest concentration of a surfactant needed to decrease the surface tension of water, resulting in the formation of micelles that effectively reduce both surface and interfacial tension. The surface-active characteristics of a biosurfactant are affected by the length of its side chain, the presence of unsaturated bonds, and the size of its hydrophilic groups. Higher hydrophobicity generally results in a lower CMC, meaning less biosurfactant is needed for micelle formation (Tripathi et al., 2018).

5 Sustainable Production of Biosurfactants

The main challenges for bringing biosurfactants to market include high manufacturing costs.

In recent years, there has been a growing interest in finding cost-effective materials that can be used as substrates for biosurfactant manufacturing. Promoting the utilization of renewable and cost-effective industrial and household wastes is a priority due to their potential to substitute expensive synthetic ingredients in the production of biosurfactants. Food residue and agro-industrial by-products can be used as organic waste to produce biosurfactants. Employing waste as an approach not only saves money but also successfully reduces the growing amount of environmental waste.

Moshtagh et al. (2018) reported the utilization of brewery waste as the exclusive carbon source for the biosurfactants production using *B. subtilis* N3–1 P. Under optimal conditions, the emulsification index reached a maximum of 63.11%, while the surface tension measured 27.31 mN/m. The maximum yield of biosurfactant obtained was 657 mg/L, which was achieved with a CMC value of 107 mg/L.

Pseudomonas aeruginosa used kitchen waste oil as a carbon source to produce rhamnolipid biosurfactant, which was found to be more effective compared to four other carbon sources. This biosurfactant reduced the interfacial tension to a value of 0.57 mN/m, while exhibiting a high optical density of 2.33 (OD600). The salinity (2–20%, w/v), pH (6.0–12.0), and temperature (20–100 °C) exhibited high stability, which was consistent with the CMC value of 55.87 mg/L (Chen et al., 2018).

Anionic glycolipids, consisting of 25% carbohydrates and 75% lipids, were produced from *Pseudomonas cepacia* CCT6659 by adding corn steep liquor and soybean waste frying oil in the production media. The results showed a substantial degradation (over 80%) of hydrophobic organic molecules within 10 days (Silva et al., 2014). Furthermore, a significant degradation was also observed when *Serratia marcescens* ZCF25 biosurfactant was used, with olive oil as sole carbon source (Huang et al., 2020).

De França et al. (2015) found that *Bacillus subtilis* ICA56 was able to produce efficient biosurfactant for heavy metal and hydrocarbon removal using cheese whey, cashew apple juice, glycerol and sunflower oil as carbon sources.

Pseudomonas aeruginosa SR17 cultivated on molasses and other waste substrates produced biosurfactants with aptitude to degrade >80% of total petroleum hydrocarbons and 65–100% for PAHs (Patowary et al., 2018).

Research perspectives include improving bacterial strains through genetic engineering techniques, optimizing fermentation processes, and developing new applications for biosurfactants (Banat et al., 2010).

6 Conclusion

Soil microorganisms play a crucial role in the mineralization of organic matter and the decontamination of decontaminating polluted ecosystems. They produce metabolites, such as biosurfactants, which are highly advantageous and have potential applcations in various sectors.

Biosurfactants offer strategic benefits for green technologies due to their biodegradability, low toxicity, and practical use in applications such as bioremediation, wastewater treatment, and enhanced oil recovery. However, the high cost of production restricts their widspread application in bioremediation procedures.Therefore, it is important to examine the culturing conditions that maximize their production, evaluate the economic utilization of new substrates derived from industrial waste, and assess strategies for separation and purification to enhance the economic feasibility of production.

Conflict of Interest Statement Conflict of interest statement: No conflicts declared.

References

Aransiola, S. A., Ayams, J. N., & Abioye, O. P. (2019). Poduction of biosurfactants using *Pseudomonas aeruginosa* for biodegradation of herbicide. *The International Journal of Biotechnology, 8*(1), 66–74. https://doi.org/10.18488/journal.57.2019.81.66.74. https://ideas.repec.org/a/pkp/tijobi/2019p66-74.html

Arora, A., Cameotra, S. S., Kumar, R., Kumar, P., & Balomajumder, C. (2014). Effects of biosurfactants on gas hydrates. *Journal of Petroleum & Environmental Biotechnology, 5*, 1–7.

Banat, I. M., Franzetti, A., Gandolfi, I., Bestetti, G., Martinotti, M. G., Fracchia, L., et al. (2010). Microbial biosurfactants production, applications and future potential. *Applied Microbiology and Biotechnology, 87*, 427–444.

Batool, R., Ayub, S., & Akbar, I. (2017). Isolation of biosurfactant producing bacteria from petroleum contaminated sites and their characterization. *Soil & Environment, 36*(1), 35–44.

Borah, D., & Yadav, R. N. S. (2017). Bioremediation of petroleum based contaminants with biosurfactant produced by a newly isolated petroleum oil degrading bacterial strain. *Egyptian Journal of Petroleum, 26*(1), 181–188.

Borah, D., Chaubey, A., Sonowal, A., Gogoi, B., & Kumar, R. (2021). Microbial biosurfactants and their potential applications: An overview. In *Microbial biosurfactants: Preparation, properties and applications* (pp. 91–116). Springer.

Camelia, B., & Sumitra, D. (2023). The exploitation of marine biosurfactants in India. In K. Se-Kwon & S. Kyung-Hoon (Eds.), *Marine surfactants: Preparations and applications* (pp. 213–128). CRC Press.

Chen, C., Sun, N., Li, D., Long, S., Tang, X., Xiao, G., & Wang, L. (2018). Optimization and characterization of biosurfactant production from kitchen waste oil using *Pseudomonas aeruginosa*. *Environmental Science and Pollution Research, 25*, 14934–14943.

Chiamaka, L.M., Xiaoyan, Z., Guiling, W., Edidiong, O.A, Lijia, J., & Chunfang, Z., (2023). Advances in biosurfactant production from marine waste and its potential application in the marine environment, in Se-Kwon, Kim, & Kyung-Hoon, S., Marine surfactants: Preparations and applications, , CRC Press, 224–245.

Chong, Z. R., Yang, S. H. B., Babu, P., Linga, P., & Li, X.-S. (2016). Review of natural gas hydrates as an energy resource: Prospects and challenges. *Applied Energy, 162*, 1633–1652.

Christopher, F. C., Ponnusamy, S. K., Ganesan, J. J., & Ramamurthy, R. (2019). Investigating the prospects of bacterial biosurfactants for metal nanoparticle synthesis—A comprehensive review. *IET Nanobiotechnology, 13*, 243–249.

Das, P., Mukherjee, S., & Sen, R. (2008). Genetic regulations of the biosynthesis of microbial surfactants: An overview. *Biotechnology and Genetic Engineering Reviews, 25*(1), 165–186.

Datta, P., Tiwari, P., & Pandey, L. M. (2020). Oil washing proficiency of biosurfactant produced by isolated *Bacillus tequilensis* MK 729017 from Assam reservoir soil. *Journal of Petroleum Science and Engineering, 195*, 107612.

De Faria, A. F., Teodoro-Martinez, D. S., de Oliveira Barbosa, G. N., Vaz, B. G., Silva, Í. S., Garcia, J. S., et al. (2011). Production and structural characterization of surfactin (C14/Leu7) produced by *Bacillus subtilis* isolate LSFM-05 grown on raw glycerol from the biodiesel industry. *Process Biochemistry, 46*(10), 1951–1957.

De França, Í. W. L., Lima, A. P., Lemos, J. A. M., Lemos, C. G. F., Melo, V. M. M., de Sant'ana, H. B., & Gonçalves, L. R. B. (2015). Production of a biosurfactant by *Bacillus subtilis* ICA56 aiming bioremediation of impacted soils. *Catalysis Today, 255*, 10–15.

Déziel, E., Lépine, F., Milot, S., & Villemur, R. (2000). Mass spectrometry monitoring of rhamnolipids from a growing culture of Pseudomonas aeruginosa strain 57RP. *Biochimica et Biophysica Acta (BBA)-Molecular and Cell Biology of Lipids, 1485*(2–3), 145–152.

Dusane, D., Rahman, P., Zinjarde, S., Venugopalan, V., McLean, R., & Weber, M. (2010). Quorum sensing; implication on rhamnolipid biosurfactant production. *Biotechnology and Genetic Engineering Reviews, 27*, 159–184.

Eras-Muñoz, E., Farré, A., Sánchez, A., Font, X., & Gea, T. (2022). Microbial biosurfactants: A review of recent environmental applications. *Bioengineered, 13*(5), 12365–12391.

Farias, C. B. B., Silva, A. F., Rufino, R. D., Luna, J. M., Souza, J. E. G., & Sarubbo, L. A. (2014). Synthesis of silver nanoparticles using a biosurfactant produced in low-cost medium as stabilizing agent. *Electronic Journal of Biotechnology, 17*, 122–125.

Farias, C. B. B., Almeida, F. C., Silva, I. A., Souza, T. C., Meira, H. M., Rita de Cássia, F., et al. (2021). Production of green surfactants: Market prospects. *Electronic Journal of Biotechnology, 51*, 28–39.

Folasade, A. A., & Hunter, C. H. (2021). Screening and identification of Lipopeptide biosurfactants produced by two aerobic endospore-forming bacteria isolated from Mfabeni peatland, South Africa. *Current Microbiology*. https://doi.org/10.1007/S00284-021-02516-7

Geetha, S. J., Banat, I. M., & Joshi, S. J. (2018). Biosurfactants: Production and potential applications in microbial enhanced oil recovery (MEOR). *Biocatalysis and Agricultural Biotechnology, 14*, 23–32.

Geissler, M., Heravi, K. M., Henkel, M., & Hausmann, R. (2019). Lipopeptide biosurfactants from Bacillus species. In *Biobased surfactants* (pp. 205–240). AOCS Press.

Gharaei-Fathabad, E. (2011). Biosurfactants in pharmaceutical industry: A mini-review. *American Journal of Drug Discovery and Development, 1*, 58–69.

Hamoen, L. W., Eshuis, H., Jongbloed, J., Venema, G., & van Sinderen, D. (1995). A small gene, designated *comS*, located within the coding region of the fourth amino acid activation domain of *srfA*, is required for competence development in *Bacillus subtilis*. *Molecular Microbiology, 15*, 55–63.

Huang, Y., Zhou, H., Zheng, G., Li, Y., Xie, Q., You, S., & Zhang, C. (2020). Isolation and characterization of biosurfactant-producing Serratia marcescens ZCF25 from oil sludge and application to bioremediation. *Environmental Science and Pollution Research, 27*, 27762–27772.

Ibrahim, M. L., Ijah, U. J. J., Manga, S. B., Bilbis, L. S., & Umar, S. (2013). Production and partial characterization of biosurfactant produced by crude oil degrading bacteria. *International Biodeterioration & Biodegradation, 81*, 28–34.

Ibrahim, S., Abdul Khalil, K., Zahri, K. N. M., Gomez-Fuentes, C., Convey, P., Zulkharnain, A., et al. (2020). Biosurfactant production and growth kinetics studies of the waste canola oil-degrading bacterium *Rhodococcus erythropolis* AQ5-07 from Antarctica. *Molecules, 25*(17), 3878.

Jimoh, A. A., Senbadejo, T. Y., Adeleke, R., & Lin, J. (2021). Development and genetic engineering of hyper-producing microbial strains for improved synthesis of biosurfactants. *Molecular Biotechnology, 63*, 267–288.

Kalvandi, S., Garousin, H., Pourbabaee, A. A., & Farahbakhsh, M. (2022). The release of petroleum hydrocarbons from a saline-sodic soil by the new biosurfactant-producing strain of bacillus sp. *Scientific Reports, 12*(1), 19770.

Kang, S. W., Kim, Y. B., Shin, J. D., & Kim, E. K. (2010). Enhanced biodegradation of hydrocarbons in soil by microbial biosurfactant, sophorolipid. *Applied Biochemistry and Biotechnology, 160*, 780–790.

Karlapudi, A. P., Venkateswarulu, T. C., Tammineedi, J., Kanumuri, L., Ravuru, B. K., Ramu Dirisala, V., & Kodali, V. P. (2018). Role of biosurfactants in bioremediation of oil pollution–A review. *Petroleum, 4*(3), 241–249.

Khouloud, M. B., & Osama, M. D. (2023). Marine Microbe Surfactants: Future Implementations. In K. Se-Kwon & S. Kyung-Hoon (Eds.), *Marine surfactants: Preparations and applications* (pp. 40–70). CRC Press.

Kitamoto, D., Isoda, H., & Nakahara, T. (2002). Functions and potential applications of glycolipid biosurfactants—From energy-saving materials to gene delivery carriers—. *Journal of Bioscience and Bioengineering, 94*(3), 187–201.

Krishnaswamy, M., Subbuchettiar, G., Ravi, T. K., & Panchaksharam, S. (2008). Biosurfactants properties, commercial production and application. *Current Science, 94*, 736–747.

Kryachko, Y. (2018). Novel approaches to microbial enhancement of oil recovery. *Journal of Biotechnology, 266*, 118–123.

Kumar, P. S., & Ngueagni, P. T. (2021). A review on new aspects of lipopeptide biosurfactant: Types, production, properties and its application in the bioremediation process. *Journal of Hazardous Materials, 407*, 124827.

Li, W., Qamar, S. A., Qamar, M., Basharat, A., Bilal, M., & Iqbal, H. M. (2021). Carrageenan-based nano-hybrid materials for the mitigation of hazardous environmental pollutants. *International Journal of Biological Macromolecules, 190*, 700–712.

Liu, J. J., & Zuber, P. (1998). A molecular switch controlling competence and motility: Competence regulatory factors ComS, MecA, and ComK control sigma(D)-dependent gene expression in *Bacillus subtilis*. *Journal of Bacteriology, 180*, 4243–4251.

Machado, F. F., Cordeiro, H. G., de Oliveira, M. A., Livio, D. F., Silva, T. N. L., Meira, H. G. R., Tarabal, V. S., Parreira, A. G., da Silva, J. A., Gonçalves, D. B., Cury, J. d. C., & Granjeiro, P. A. (2022). Production of biosurfactants by bacteria collected in soil contaminated by diesel oil on the Antarctic continent. *Research, Society and Development, 11*(13), e426111335593.

Mnif, I., & Ghribi, D. (2015). Review lipopeptides biosurfactants: Mean classes and new insights for industrial, biomedical, and environmental applications. *Peptide Science, 104*(3), 129–147.

Mnif, I., Ellouz-Chaabouni, S., & Ghribi, D. (2018). Glycolipid biosurfactants, main classes, functional properties and related potential applications in environmental biotechnology. *Journal of Polymers and the Environment, 26*, 2192–2206.

Moldes, A. B., Paradelo, R., Rubinos, D., Devesa-Rey, R., Cruz, J. M., & Barral, M. T. (2011). Ex situ treatment of hydrocarbon-contaminated soil using biosurfactants from Lactobacillus pentosus. *Journal of Agricultural and Food Chemistry, 59*, 9443–9447.

Morita, T., Fukuoka, T., Imura, T., & Kitamoto, D. (2015). Mannosylerythritol lipids: Production and applications. *Journal of Oleo Science, 64*(2), 133–141.

Moshtagh, B., Hawboldt, K., & Zhang, B. (2018). Optimization of biosurfactant production by *Bacillus subtilis* N3-1P using the brewery waste as the carbon source. *Environmental Technology, 40*(25), 3371–3380.

Mukherjee, A. K., & Das, K. (2010). Microbial surfactants and their potential applications: An overview. In R. Sen (Ed.), *Biosurfactants. Advances in experimental medicine and biology* (Vol. 672, pp. 54–64). Springer.

Muthusamy, K., Gopalakrishnan, S., Ravi, T. K., & Sivachidambaram, P. (2008). Biosurfactants: Properties, commercial production and application. *Current Science*, 736–747.

Nakar, D., & Gutnick, D. L. (2001). Analysis of the wee gene cluster responsible for the biosynthesis of the polymeric bioemulsifier from the oil-degrading strain Acinetobacter lwoffii RAG-1. *Microbiology, 147*(7), 1937–1946.

Niamat, H., Qureshi, A. W., Rafi, U., Khaliq, Z., & Bokhari, S. S. (2022). Efficacy of indigenous microbes for removal of oil-contaminated soil by producing biosurfactant. *Pakistan Biomedical Journal, 5*(2), 75–82.

Nihorimbere, V., Ongena, M. M., Smargiassi, M., & Thonart, P. (2011). Beneficial effect of the rhizosphere microbial community for plant growth and health. *Biotechnology, Agronomy, Society and Environment, 15*(2), 327–337.

Otzen, D. E. (2017). Biosurfactants and surfactants interacting with membranes and proteins: Same but different? *Biochimica et Biophysica Acta (BBA)-Biomembranes, 1859*(4), 639–649.

Patowary, R., Patowary, K., Kalita, M. C., & Deka, S. (2018). Application of biosurfactant for enhancement of bioremediation process of crude oil contaminated soil. *International Biodeterioration & Biodegradation, 129*, 50–60.

Qamar, S. A., & Pacifico, S. (2023). Cleaner production of biosurfactants via bio-waste valorization: A comprehensive review of characteristics, challenges, and opportunities in bio-sector applications. *Journal of Environmental Chemical Engineering, 11*, 111555.

Qiao, J., Borriss, R., Sun, K., Zhang, R., Chen, X., Liu, Y., & Liu, Y. (2024). Research advances in the identification of regulatory mechanisms of surfactin production by *Bacillus:* A review. *Microbial Cell Factories, 23*(1), 100.

Rane, A. N., Baikar, V. V., Ravi Kumar, V., & Deopurkar, R. L. (2017). Agro-industrial wastes for production of biosurfactant by Bacillus subtilis ANR 88 and its application in synthesis of silver and gold nanoparticles. *Frontiers in Microbiology, 8*, 492.

Ranjan, S., & Sow, S. (2021). Biosurfactants as a biological tool to increase agricultural productivity. *Agriculture Food and Newsletter, 3*(4), 274–277.

Ron, E. Z., & Rosenberg, E. (2011). Natural roles in biosurfactants. *Environmental Microbiology, 3*, 229–236.

Ron, E. Z., & Rosenberg, E. (2014). Enhanced bioremediation of oil spills in the sea. *Current Opinion in Biotechnology, 27*, 191–194.

Roongsawang, N., Washio, K., & Morikawa, M. (2010). Diversity of nonribosomal peptide synthetases involved in the biosynthesis of lipopeptide biosurfactants. *International Journal of Molecular Sciences, 12*(1), 141–172.

Sanches, M. A., Luzeiro, I. G., Cortez, A. C. A., de Souza, É. S., Albuquerque, P. M., Chopra, H. K., & de Souza, J. V. B. (2021). Production of biosurfactants by ascomycetes. *International Journal of Microbiology, 2021*, 6669263.

Sánchez, C. (2022). A review of the role of biosurfactants in the biodegradation of hydrophobic organopollutants: Production, mode of action, biosynthesis and applications. *World Journal of Microbiology and Biotechnology, 38*(11), 216.

Sharma, S., Saxena, V., Baranwal, A., Chandra, P., & Pandey, L. M. (2018). Engineered nanoporous materials mediated heterogeneous catalysts and their implications in biodiesel production. *Materials Science for Energy Technologies, 1*(1), 11–21.

Shashank, R., Vartika, V., & Nidhi, S. (2023). Marine biosurfactants applications in agriculture. In K. Se-Kwon & S. Kyung-Hoon (Eds.), *Marine surfactants: Preparations and applications* (pp. 255–265). CRC Press.

Silva, R. D. C. F., Almeida, D. G., Rufino, R. D., Luna, J. M., Santos, V. A., & Sarubbo, L. A. (2014). Applications of biosurfactants in the petroleum industry and the remediation of oil spills. *International Journal of Molecular Sciences, 15*(7), 12523–12542.

Silva, A. C. S. D., Santos, P. N. D., Silva, T. A. L. E., Andrade, R. F. S., & Campos-Takaki, G. M. (2018). Biosurfactant production by fungi as a sustainable alternative. *Arquivos do Instituto Biológico, 85*, e0502017.

Singh, S. H. R. I. K. R. I. S. H. N. A., Singh, U. S. H. A. S. I., Hogan, S. E., & Feingold, D. S. (1990). Formation of UDP-2-acetamido-2-deoxy-L-galactose and UDP-2-acetamido-2-deoxy-L-galacturonic acid by *Pseudomonas aeruginosa. Journal of Bacteriology, 172*(1), 299–304.

Sircar, A., Rayavarapu, K., Bist, N., Yadav, K., & Singh, S. (2021). Applications of nanoparticles in enhanced oil recovery. *Petroleum Research, 7*(1), 77–90.

Soberón-Chávez, G., & Maier, R. M. (2011). Biosurfactants: A general overview. In *Biosurfactants: From genes to applications* (pp. 1–11). Springer.

Stiegelmeyer, S. M., & Giddings, M. C. (2013). Agent-based modeling of competence phenotype switching in *Bacillus subtilis*. *Theoretical Biology & Medical Modelling, 10*, 23.

Théatre, A., Cano-Prieto, C., Bartolini, M., Laurin, Y., Deleu, M., Niehren, J., Fida, T., Gerbinet, S., Alanjary, M., Medema, M. H., Léonard, A., Lins, L., Arabolaza, A., Gramajo, H., Gross, H., & Jacques, P. (2021). The surfactin-like lipopeptides from *Bacillus* spp.: Natural biodiversity and synthetic biology for a broader application range. *Frontiers in Bioengineering and Biotechnology, 9*, 623701.

Tirth, B., Avani, B., Asmita, D., Dhruv, G., & Gaurav, S. (2023). Cosmetic application of surfactants from marine microbes. In K. Se-Kwon & S. Kyung-Hoon (Eds.), *Marine surfactants : Preparations and applications* (pp. 272–287). CRC Press.

Tripathi, L., Irorere, V. U., Marchant, R., & Banat, I. M. (2018). Marine derived biosurfactants: A vast potential future resource. *Biotechnology Letters, 40*, 1441–1457.

Van Renterghem, L., Roelants, S. L. K. W., Baccile, N., Uyttersprot, K., Taelman, M. C., Everaert, B., Mincke, S., Ledegen, S., Debrouwer, S., Scholtens, K., Stevens, C., & Soetaert, W. (2018). From lab to market: An integrated bioprocess design approach for new-to-nature biosurfactants produced by Starmerella bombicola. *Biotechnology and Bioengineering, 115*, 1195–1206.

Wagner, V. E., Bushnell, D., Passador, L., Brooks, A. I., & Iglewski, B. H. (2003). Microarray analysis of *Pseudomonas aeruginosa* quorum-sensing regulons: Effects of growth phase and environment. *Journal of Bacteriology, 185*(7), 2080–2095.

Xu, Y., Jing, Y., Zhang, Y., Liu, Q., Xiu, J., Zhang, K., & Wu, X.-L. (2022). Single-cell-based high-throughput cultivation and functional characterization of biosurfactant-producing bacteria from soil and oilfield-produced water. *Microorganisms, 10*(3), 450.

Yasmin, A., Aslam, F., & Fariq, A. (2022). Genetic evidences of biosurfactant production in two Bacillus subtilis strains MB415 and MB418 isolated from oil contaminated soil. *Frontiers in Bioengineering and Biotechnology, 10*, 855762.

Future Research of Soil Microbiomes and Green Technology Innovation for a Better Tomorrow

Mustapha Abdulsalam, Musa Ojeba Innocent, Miracle Uwa Livinus, Shehu-Alimi Elelu, Ganiyat Omotayo Ibrahim, Salami Olaitan Lateefat, Saheed Kayode Saheed, and Auwal Sagir Muhammad

Abstract Soil microbiome, a very complicated and dynamic ecosystem that is composed of various microbial communities, has high significance for sustainable development and innovation in green technology. The focus of this study is on the diversity of soil microbiota and how it can be used in a way that we can shape a better tomorrow. It elucidates the compositional structure, variability, and function, indicating the close connections between soil microbiomes and environmental factors. The recent developments in omics technologies allow us to unravel microbial communities as never before, giving a deeper understanding of their functions as soil ecosystems and resilience. In addition, this research looks at the nature of soil microbiomes as a tool for the conversion to green technology. Sustainable agriculture techniques, bioremediation strategies, and bioenergy production with soil microbiomes as the source are possible alternatives to environmental issues. The integration of soil microbiome study and precision agriculture as well as synthetic biology offers new opportunities for improving agricultural productivity and

M. Abdulsalam (✉) · M. O. Innocent
Department of Microbiology, Skyline University Nigeria, Kano, Nigeria
e-mail: mustapha.abdulsalam@sun.edu.ng

M. U. Livinus
Department of Biochemistry, School of Sciences and Information Technology, Skyline University Nigeria, Kano, Nigeria

S.-A. Elelu
Department of Chemistry, Howard University, Washington, DC, USA

G. O. Ibrahim
Department of Chemistry, Nottingham Trent University, Nottingham, UK

S. O. Lateefat
Department of Chemistry and Molecular Biology, University of Gothenburg, Gothenburg, Sweden

S. K. Saheed
Department of I.T and Computing, American University of Nigeria, Yola, Nigeria

A. S. Muhammad
School of Informatics, Xiamen University, Xiamen, China

S. A. Aransiola et al. (eds.), *Soil Microbiome in Green Technology Sustainability*,
https://doi.org/10.1007/978-3-031-71844-1_24

mitigating climate change besides restoring the ecosystem. However, translating soil microbiomes into full potential demands a solution for various issues, such as the deciphering of microbial functions, the scaling up of green technology innovations, and the crossing of regulatory barriers. Through interdisciplinary collaboration, cultivating innovation culture, and overcoming these challenges, researchers and relevant stakeholders can release the transformation potential of microbiomes of the soil for a sustainable future. This research emphasizes the role of microbes in the soul as a basis of ecological sustainability. This can trigger green technological development. Thus, we will make use of the existing resilience and diversity of soil microbial communities to create a future that's greener and probably more resilient for generations to come.

Keywords Soil microbiome · Green technology · Environmental sustainability · Omics technologies · Precision agriculture · Synthetic biology

1 Introduction

Soil microbiomes, that is the multitudinous families of microscopic organisms such as bacteria, fungi, archaea, and viruses that exist in the soil matrix, host a variety of other microorganisms as well. These microbial communities, being engaged with each other and plants, are contributing to the continuance of essential ecosystem processes like the circulation of nutrients, organic matter decomposition, and soil formation (Philippot et al., 2023). The scope of soil microbiome varies from the geophone (the region of soil which receives influence from plant roots) covering (Adeola et al., 2022). Soil microbiota is composed of an amazing diversity of microorganisms possessing multifunctional characteristics enabling their adaptation to, and the ability to survive, a diverse array of environmental conditions and human-associated disturbances. Soil microorganism composition and interaction must be understood to comprehend how microbial communities participate in soil quality assurance, ecosystem functioning, and sustainable functioning of the environment. Recent developments in molecular techniques, especially high throughput sequencing, and most genomics have changed our attitude toward the assessment of soil microbial biomass and their functional competencies (Abdulsalam et al., 2023).

1.1 Importance of Soil Microbiomes in Environmental Sustainability

As we delve into the intricacies of soil microbiomes, we discover that their roles go beyond just controlling soil fertility. They also significantly boost plant productivity by conserving nutrients in soils and reducing environmental pollution due to their ability to break down organic compounds. Microbial processes, for instance, nitrogen fixation and nutrient mineralization, are among the basic components that support agricultural productivity and help to decrease greenhouse gas emissions (Chataut et al., 2023).

More so, microbiomes in soil take part in soil erosion control, water filtration, and disease suppressing, and help in the stabilization of ecosystems and biodiversity (Hartmann & Six, 2023). The inherent problem-solving powers of microbes within the soil are invaluable for practical technologies that will facilitate land management and conserve ecosystem function.

1.2 Overview of Green Technology Innovation

In the context of sustainable development, green technological innovations include a variety of developed and applied strategies, technologies, and approaches that are geared towards minimizing the adverse impacts on the environment, as well as efforts to preserve nature and its resources (Haughton, 2021). In terms of microbiomes, bioremediation methods, biofertilizers, and the use of precision agriculture technologies that utilize microbial processes do the trick in boosting soil health and sustaining agricultural systems over time are green technology innovations (Pretty & Bharucha, 2014). Biotechnology, nanotechnology, and data sciences are at the forefront of driving discoveries in soil microbiome, hence technological developments in this field are fostering the development of new technologies for solving environmental challenges (Jaffe et al., 2005). Utilizing green technology together with traditional agriculture techniques promises a boost in productivity in the agricultural sector while minimizing the negative implications on the environment and resource input.

2 Understanding Soil Microbiomes

Understanding soil microbiomes is essential for comprehending their contributions to soil health, ecosystem functioning, and environmental sustainability. This section aims to explore microbiomes in soil biology by elaborating on their constitution, living organisms, ecological roles, and responses of microbiomes to environmental conditions.

2.1 Composition and Diversity of Soil Microbiomes

Soil microbiomes adopt a diverse family of microorganisms (bacteria, fungi, archaea, protozoa, and viruses) as members, each clear contributor to the community microbial structure (Geisen et al., 2019). The development and application of high-throughput sequencing technologies have led to the extensive identification of the diversity of microbial organisms within the soil, with millions of distinct microbial taxa of microorganisms existing within a single gram of soil (Stefanis et al., 2013).

The makeup of soil microbiomes differs in space and at different times, and it is controlled by factors such as soil type, the plants used, climate, and land use (Islam et al., 2020). A lot of times bacteria turn out to be the most prevalent and abundant group of soil microbes, and after them come fungi that are also important as they participate in metabolic cycles (nutrient cycling), organic substance decomposition, and plant-microorganism interactions (Jansson & Hofmockel, 2020). Analyzing the microbial population and diversity of the soil will help us to know their functional capacity, resulting in ecological benefits, and strategies for fertilization and preservation of the soil.

2.2 Ecological Functions and Roles of Soil Microorganisms

Moreover, these microbes in the soil carry out an immense scope of ecological functions necessary for the preservation of soil fertility, nutrients and cycling, and ecosystem resilience (Bhaduri et al., 2022). Microorganisms such as bacteria and fungi utilize organic material as they decompose it, contributing nutrients for plant uptake, while symbiotic nitrogen-fixing bacteria act on the conversion of atmospheric nitrogen into plant-available forms by assimilating nitrogen into organic matter (Sharma et al., 2023). The ectomycorrhizal fungi (don) are in a mutualistic relationship with plant roots, which facilitates nutrient exchange and supports tolerance to the surrounding environment by plants. The soil microorganisms also impede the plant pathogen by regulating soil structure, nutrient cycling, carbon reduction, and others (Hartmann & Six, 2023). Understanding the importance of soil microorganisms and their functionalities is a key aspect not to be disregarded in ecosystem management with, at the same time, the provision of increased agricultural productivity and reduced environmental pollution.

2.3 Impact of Environmental Factors on Soil Microbiomes

It is worth noting that some environmental variables including soil pH, moisture, temperature, land use practices, and pollution play significant roles in structuring the microbiome (Islam et al., 2020). The pH of soil for example, strongly dictates community composition in a way that species such as acidophils or alkalophils become the dominant where soils are acidic and alkaline respectively (Bandh et al., 2022). The extent to which land use change may occur, e g deforestation, agricultural intensification, and urbanization, is considered to have a significant effect on the soil microbial diversity and functionality that can lead to differences in ecosystem processes and services (Ortiz et al., 2021). More so, there are alterations in the characteristic thermal regime and precipitation pattern which mark another factor leading to soil microbiota modification which can affect soil carbon sequestration, methane emissions, and nutrient cycling (Mukhtar et al., 2023).

3 Advances in Soil Microbiome Research

Microbiome soil studies have driven a paradigm shift in our perception of soils; now we know how it's composed, what they do, and how they are transformed. This section will show new sectoral and technical approaches that are used in studying the biota of soil environment, like omics technologies, metagenomics, metatranscriptomics, and involvements with microbiota interactions and community ecology.

3.1 Omics Technologies for Studying Soil Microbiomes

The omics technologies e.g. genomics, proteomics, metabolomics, and transcriptomics, have enabled us to access the soil microbiomes at the molecular level increase (Sandrini et al., 2022). High-throughput techniques may be efficient for analyzing a thousand to millions of microbial genes, proteins, metabolites, and transcripts providing information about the community composition, its functional potentials, and actual metabolic activities (Kumar et al., 2021). Metagenomic investigation or analysis refers to the process of extracting DNA from the studied environmental sample and then using it for identifying microbial taxa and making functional estimates (Fadiji & Babalola, 2020). Omics technologies bring researchers the opportunity to reveal the metabolic capability of microbiomes in the soils, determine the important microbes involved in various ecosystem processes, and identify the reactions of microorganisms toward ecological variations.

3.2 Metagenomics and Metatranscriptomics Approaches

Metagenomics and metatranscriptomics are established approaches that can be used for studying the soil microbiome and they offer information about microbial diversity, functional potentials as well as the activities of microbes (Wagg et al., 2019). Metagenomic sequencing enables the gene content monitoring of microbial genomes present in environmental samples, resulting in metadata on metabolic pathways and potential interactions in microbial communities (Abdulsalam et al., 2024b). Metatranscriptomic analysis allows us to understand the active gene expression and pathways making it an instrumental complement to classical metagenomics data (Tripathi & Nailwal, 2020). Through gene expression profiling under various environmental conditions and management practices, the researchers read the microbial responses to the disturbances, the availability of nutrients, or climate change (Jansson & Hofmockel, 2020).

A combination of metagenomic and metatranscriptomic integrations for the soil microbial communities provides more comprehensive landscapes of the taxonomic composition and functional activities which are directly in association with ecosystem processes (Sokol et al., 2022). These methods not only connect gaps of knowledge of soil microbes, but also illuminate their ability to be resilient, and adaptable, and play an ecological role in terrestrial systems.

3.3 Microbial Interactions and Community Dynamics

Communities formed by microorganisms and their cooperation dictate the shaping of basic soil ecosystems (Waldrop et al., 2023). The fact that microbial communities in the soil are efficient is the basis for predicting ecosystem functions and soil health control (Bhaduri et al., 2022). Some of the latest conveniences in the field of network analysis, ecological modeling, and computational biology have given new life to the exploration of composition in the micro-bacterial communities in soil ecosystems (Kupolati et al., 2019). Networks display patterns of joint occurrences among microbial taxa (e.g. positive and negative associations), which are aimed at revealing the composition and structure of the communities, their stability, as well as resilience (Calabrese et al., 2022). In addition, longitudinal studies and time-series analyses permit inquiry into microbial community dynamics at different spatial and temporal scales discovering the peculiar succession, resilience, and reactions in case of environmental perturbations (Abdulsalam et al., 2024b).

4 Applications of Soil Microbiomes in Green Technology Innovation

The sophisticated arrangements in the soil microbial communities are a significant resource that can offer numerous green technology expansion possibilities. This is the part where the use of different methods to make the soil microbiome do useful agricultural work. The work involves mitigating air pollution and getting bioenergy and carbon sequestration.

4.1 Soil Microbiomes in Sustainable Agriculture

The biological section in the soil where microbiomes exist enables them to support sustainable agriculture through the improvement of soil fertility, recycling of nutrients, and adding productivity to crops (Suman et al., 2022). Advantageous microorganisms like nitrogen-fixing bacteria and mycorrhizal fungi can be considered partners of plants providing the essential nutrients back to the plants and reducing their stress levels (Ji et al., 2024). Some of the advanced agricultural practices are precision farming, cover cropping, and biofertilizer application which rely on the microbial community of soil to gain efficient nutrient use, lessen chemical inputs, and improve soil health (Tahat et al., 2020). The biologically-based microbial inoculants with increasing popularity as a synthetic fertilizer and pesticide alternatives are presented to be part of the sustainability and have an environmental impact reduction (Tyagi et al., 2022). With the help of microbe communities' functional capacity, future agriculture with enhanced food security and prevention of loss of natural resources while concurrently mitigating the impacts of climate change is promising for agriculture's sustainability (Fig. 1).

4.2 Bioremediation and Soil Health Management

Soil microbiomes are one of the biggest players in the sensorineural biodegradation, decontamination, and storage of pollutants in contaminated soil (Chaudhary et al., 2023). Big numbers of microbes are capable of biodegrading different organic pollutants, heavy metals, and agrochemicals by using diverse and enzymatic pathways (Abioye et al., 2021; Saravanan et al., 2021). Bioremediation remedies are one of the methods which include; phytoremediation (using plants), bioaugmentation (adding microorganisms), and biostimulation (stimulating microorganisms to cleanse contaminated sites). Genetically engineered microbe cohorts with enhanced pollutant-degrading potential will increase the precision of spot decontamination during contamination cleanup. In this context, ecotechnology will be a core unit of sustainable strategies for soil and environment remediation (Musa et al., 2024a, b). Soil microbiomes facilitate soil aggregation, structure, and fertility placing soil health on a high pedestal that is capable of handling environmental stresses (Livinus et al., 2024). Sustainable soil management techniques including conservation tillage, organic manurations such as cover crops, and agroecological approaches are adopted to improve soil microbiome diversity and functions which ultimately creates healthy and productive soils (Abdulsalam et al., 2024a).

4.3 Bioenergy Production and Carbon Sequestration

Microorganisms from the soil community have a profound influence on organic matter cycles in soils, including biomass degradation, mineralization, and stabilization (Bhattacharyya et al., 2022). Microbial methods, including anaerobic digestion, and fermentation, and applied for the generation of biofuel, biogas, and biochemicals are conducted using microorganisms (Musa et al., 2024a, b). Moreover, the

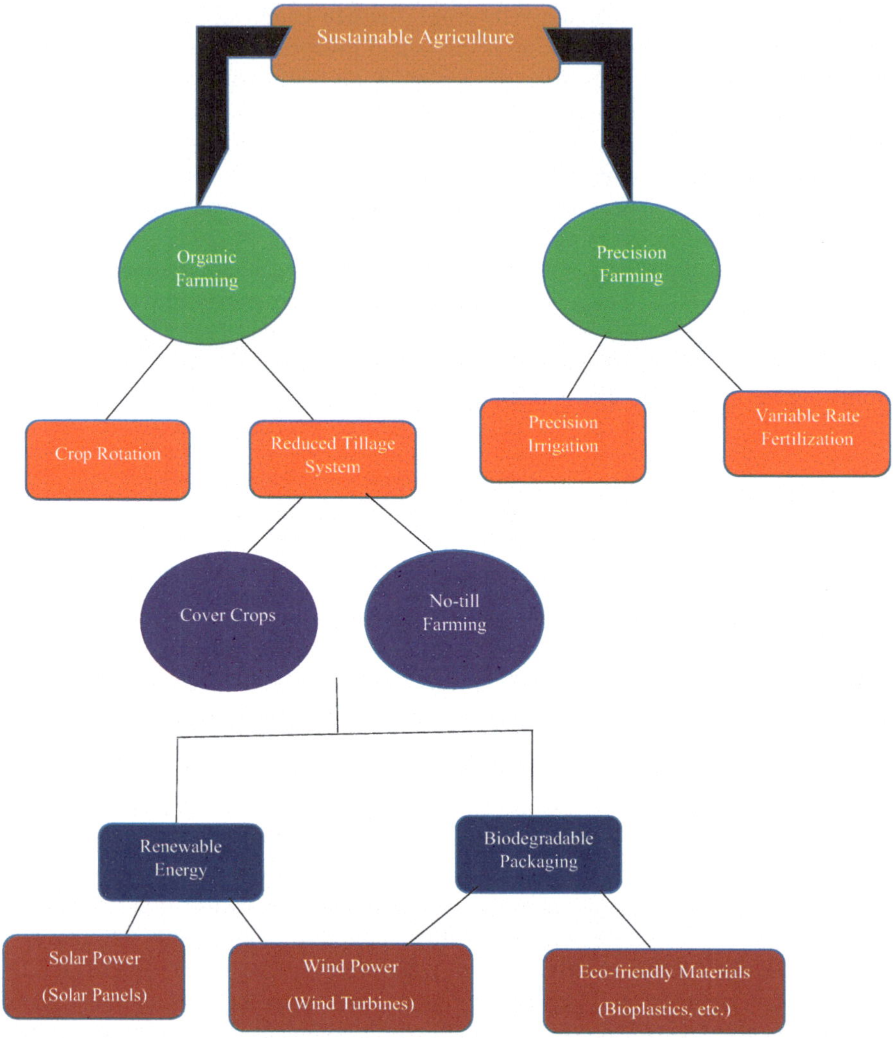

Fig. 1 Green technology applications in agricultural sustainability

microorganisms living in the soil delve into carbon sequestration in the form of organic carbon aggregation, humification, and microbial necromass (Yang et al., 2023). Converting towards conservation agriculture practices such as no-till farming, cover cropping, and agroforestry is the best way that enhance microbe-mediated procedures of organic matter stabilization in the soil thus making it store carbon for a longer (Lawal et al., 2023).

5 Emerging Trends and Future Directions

Given the nature of soil microbiome research, it is challenging, and it changes much with each time, owing to technology evolution, environmental challenges, and new paradigms of research. The last part of this chapter discusses the trending topics in soil microbiome studies and the direction it is heading after incorporating them into precision agriculture, mitigating climate change, and the use of synthetic biology and engineering to boost the functionality of soil microbiomes.

5.1 Harnessing Soil Microbiomes for Climate Change Mitigation

Microbiomes of soil may be used to some extent to mitigate climate change because they offer various mechanisms such as carbon sequestration, greenhouse gas mitigation, and soil fertility enhancement (Mona et al., 2021). Knowing the components of interactions occurring between soil microorganisms and organic matter cycling processes has been identified as one of the most important outcomes for planning interventions designed to enhance the store of carbon in soil and lower the production of greenhouse gases (Haj-Amor et al., 2022). New studies are currently intended for controlling microbial communities in soil and thus promote carbon sequestration by different techniques, such as biochar amendment or mix cropping (Livinus et al., 2024). Furthermore, microbially-controlled processes of denitrification, methane oxidation, and nitrous oxide reduction which reduce N-related greenhouse gas emissions from soil are some of the positive benefits of enhanced microbial activity (Li et al., 2022).

Future research agenda will be focused on microbial control mechanisms over greenhouse gas emission, as well as creating optimal methods for natural carbon sequestration, and exploring consequences of climate change for the soil microbiome functioning.

5.2 Integration of Soil Microbiome Research with Precision Agriculture

The precision agriculture method which involves the precision application of inputs such as soil, rainfall, evaporation, and runoff depending on spatial and temporal variation in the soil and the crops is extremely important because it promotes outstanding productivity and simultaneously controls the environmental harm (Ahmad & Dar, 2020) (Fig. 2).

The integration of soil microbiome studies with precision agriculture can contribute to the development of models that describe soil-plant-microbe interactions

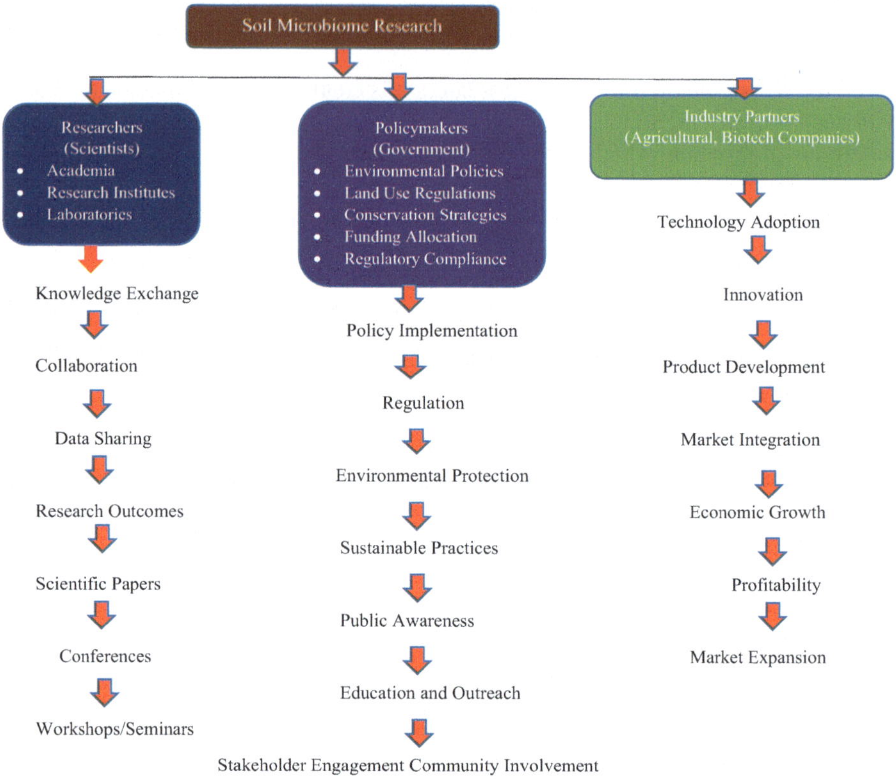

Fig. 2 Interconnected stakeholders in soil microbiome research: collaboration, regulation, and innovation

and that help in the improvement of integrated agricultural management strategies for environmental sustenance. Such tools as sensors, remote monitoring, and data analytics support timely measurements of the soil microbiome growth and crop health, and allow for more rational decisions concerning fertilization irrigation and pest control (Akanmu et al., 2021). This way farmers can manage the soil microbiome through the use of the soil microbiome data integrated into the precision agriculture platform. It helps them make soil healthier, better cycle of nutrients, and yield more crops without chemicals and environmental damage (Pahalvi et al., 2021). Future research directions could be to design bioinformatics tools to help farmers gain from the precision agriculture decision support system that uses soil microbiome data, to identify soil health and fertility indicators, and to evaluate the feasibility and commercial aspects of precision agriculture practices.

5.3 *Potential of Synthetic Biology and Microbial Engineering in Soil Microbiomes*

Synthetic biology and microbial engineering are new technologies providing solutions for the purposeful attempt to transform soil microbiomes to get certain outcomes like efficient farming, environmental clean-up, and restoration of ecosystem functions (Fazeli-Nasab et al., 2022). Researchers can engineer microbial populations with the ability to perform functions like nitrogen fixation, phosphorus solubilization, or promotion of plant growth which will further improve the fertility of the land, crop productivity, and overall resilience of the ecosystem (Hakim et al., 2021). The present status of synthetic biology in terms of tools and techniques emphasizes genome editing approaches, CRISPR-Cas systems, and artificial microbial community design which all have one objective; to have control over microbial traits and interactions (Lv et al., 2022). Using genetic engineering techniques to insert beneficial genes into soil microorganisms or by designing synthetic microbial communities with complementary complicated functions, researchers can produce site-specific solutions that cover agriculture and environmental problems (Bano et al., 2021).

6 Challenges and Opportunities

Unraveling the problematicities linked with microorganisms inhabiting soil and green advancement calls involves both the pitfalls and the advantages. This section explores the main bottlenecks and solutions in promoting research and application in the discipline, including several aspects: the studies of the microbiome functionality and interactions, the scalability of green technologies, and the concerns about regulations and public acceptance.

6.1 *Understanding Microbial Functions and Interactions*

The key obstacle in the process of understanding how soil microbiology works is connecting the functions of microbial communities in the ecosystem (Philippot et al., 2021). The large body of soil microorganisms, as well as their dynamic response to natural stimuli, present a formidable puzzle to unpick the roles and relationships of microbes (Rodríguez-Frías et al., 2021). Recently biotechnological and bioinformatic tools have enabled in defining bacterial taxa and metabolic pathways at the microbial level but connecting the data to ecosystem-level processes is like climbing the Himalayas (Bahram et al., 2018). Multi-omics (genomic, epigenetic, metabolome, proteome, etc.) approaches, with their combination with ecological modeling and systems biology, seem to be the instruments for the comprehension of microbial communities' role in soil processes (Bahram et al., 2018). Tremendous prospects exist to utilize the achievements of novel technologies and innovative interactions among the fields in decoding the complexity of the soil microbiome accompanied by the determination to increase the yields in sustainable agriculture and environmental management.

6.2 Scaling Up Green Technology Innovations

Although the green innovations that are demonstrated as enhancing sustainability in agriculture and nature management come with promising outcomes, the scaling up of these innovations is a big challenge from laboratory research to the practical world (Wigboldus et al., 2016). Social impact mechanisms such as economic imperfections (cost-effectiveness, technical feasibility), and socio-economic barriers may hinder green technological spread and diffusion on a large scale (Hochman et al., 2017).

More so, the distinctiveness of soil environments and the multiple farming systems require customizable solutions filled with the features of local area and farmer thinking (Lacombe et al., 2018; Aransiola et al., 2021). Collaborative initiatives between researchers, policymakers, producers, ropper operators, and the farming community are pivotal to developing a framework that will contribute to overcoming these challenges and adopting green technology innovations (Benito-Hernández et al., 2023). The study utilizes public-private partnerships, legislation to promote sustained practices, and the building of the capacity of this technology in agriculture and environmental management.

6.3 Regulatory and Societal Considerations

The use of Greentech innovation for soil microbiome research and management of the environment is subject to regulatory regimes, ethics, and the socio-cultural reality (Bradu et al., 2023). The lurking of regulatory uncertainties, intellectual property rights, and biosafety problems could be faced by green technologies researchers and developers and hinder them from bringing forward their commercialization (Trump et al., 2023). Also, social acceptance and stakeholder engagement are vital for fostering the development of green technologies. Without the endorsement of the relevant stakeholders, green technology innovations may not be successfully established (Baah et al., 2021). It is public risk perceptions, ethical questions, and socioeconomic impacts that have a considerable impact on forming policy and promoting the establishment of relevant technology (Wachinger et al., 2013). Among opportunities, we mention the possibility for the creation of different types of dialogs (multi-stakeholder, science-communication-oriented, public-engagement, etc.), proper integration of ethical and social aspects in decision-making, and consideration of sustainability aspects (Ziolo et al., 2019). A sound policy and

Table 1 Comparison of regulatory frameworks for green technology innovation

Regulatory aspect	United States	European Union	China
Intellectual property	Patent-centric approach	Strong IP protection	Developing IP laws
Environmental standards	EPA regulations	EU directives	National standards
Technology certification	EPA ENERGY STAR	CE mark	China compulsory cert.
Funding mechanisms	Grants and tax incentives	Horizon Europe program	Government subsidies
Market access	Open market access	Harmonized standards	Domestic preference
Research collaboration	Public-private partnerships	Horizon Europe projects	Government-led projects
Environmental impact	NEPA requirements	EIA directive	EIA regulations

societal frameworks should be instituted on green technology progress. This should reinforce the implementation and application of green innovations in soil microbiome research and soil environmental management (Table 1).

7 Conclusion

Throughout this study, we have examined the intricate relationships between soil microbiomes, green technology innovation, and environmental sustainability. We have elucidated the composition, functions, and ecological roles of soil microorganisms, highlighting their significance in nutrient cycling, soil health, and ecosystem resilience. Furthermore, we have explored cutting-edge research methodologies and applications, from omics technologies to synthetic biology, that are revolutionizing our understanding of soil microbiomes and their potential for addressing global challenges. Key findings underscore the importance of interdisciplinary collaboration, technological innovation, and policy support in harnessing the full potential of soil microbiomes and green technology innovation for sustainable agriculture and environmental management. Integrating scientific knowledge with practical solutions can enhance soil health, mitigate environmental degradation, and promote resilient and productive ecosystems. The insights gained from this study have significant implications for future research and innovation in soil microbiomes and green technology. Advances in understanding microbial functions and interactions, coupled with technological innovations, offer unprecedented opportunities for developing tailored solutions to address pressing agricultural and environmental challenges. Future research directions should focus on bridging the gap between fundamental research and on-the-ground applications, leveraging emerging technologies to optimize resource use efficiency, minimize environmental impacts, and enhance ecosystem services. Moreover, fostering collaborative networks, promoting knowledge exchange, and investing in capacity-building initiatives are essential for nurturing a vibrant innovation and entrepreneurship ecosystem in soil microbiome research and green technology development. By fostering a culture of innovation and entrepreneurship, we can accelerate the translation of scientific discoveries into tangible solutions that benefit society and the environment. As we move forward, it is imperative to translate knowledge into action and embrace sustainable practices and technology adoption in agriculture and environmental management. Stakeholders across sectors must collaborate to develop and implement evidence-based strategies that promote soil health, conserve biodiversity, and enhance ecosystem resilience. Governments, policymakers, industry leaders, researchers, and civil society organizations all have a role to play in driving this transformative agenda forward. By adopting sustainable practices, investing in green technology innovation, and promoting responsible stewardship of natural resources, we can create a more resilient, equitable, and prosperous future for generations to come. The time for action is now; together, we can build a better tomorrow for our planet and its inhabitants.

References

Abdulsalam, M., KalandarAmeer, S., Balakrishnan, S., & Pal, S. K. (2023). Maximizing sustainability: Leveraging indigenous fungal cellulases for sugarcane bagasse bioethanol production and agricultural waste management in Kano. *Journal of Advanced Zoology, 44*(S2), 2379–2393.

Abdulsalam, M., Elelu, S. A., Innocent, M. O., Ibrahim, G. O., Livinus, M. U., Lateefat, S. O., & Muhammad, A. S. (2024a). Women empowerment in environmental conservation. In *Prospects for soil regeneration and its impact on environmental protection* (pp. 351–370). Cham.

Abdulsalam, M., Ubah, F. Z. Y., Ahmed, H. U., Tafida, U. A., & Nasir, A. W. (2024b). Deciphering the genetic code: Mechanisms, evolution, and implications for biotechnology. *World Journal of Advanced Research and Reviews, 21*(1), 858–868.

Abioye, O. P., Ijah, U. J. J., Aransiola, S. A., Auta, S. H., & Ojeba, M. I. (2021). Bioremediation of toxic pesticides in soil using microbial products. In R. Prasad, S. C. Nayak, R. N. Kharwar, & N. K. Dubey (Eds.), *Mycoremediation and environmental sustainability. Fungal biology*. Springer. https://doi.org/10.1007/978-3-030-54422-5_1

Adeola, A. O., Akingboye, A. S., Ore, O. T., Oluwajana, O. A., Adewole, A. H., Olawade, D. B., & Ogunyele, A. C. (2022). Crude oil exploration in Africa: Socio-economic implications, environmental impacts, and mitigation strategies. *Environment Systems and Decisions, 42*(1), 26–50.

Ahmad, S. F., & Dar, A. H. (2020). Precision farming for resource use efficiency. In *Resources Use Efficiency in Agriculture* (pp. 109–135). Springer.

Akanmu, A. O., Babalola, O. O., Venturi, V., Ayilara, M. S., Adeleke, B. S., Amoo, A. E., & Glick, B. R. (2021). Plant disease management: Leveraging on the plant-microbe-soil interface in the biorational use of organic amendments. *Frontiers in Plant Science, 12*, 700507.

Aransiola, S. A., Victor-Ekwebelem, M. O., Ikhumetse, A. A., & Abioye, O. P. (2021). Challenges and future prospects of biotechnology. In N. R. Maddela & L. C. García (Eds.), *Innovations in biotechnology for a sustainable future*. Springer Nature Switzerland AG. https://doi.org/10.1007/978-3-030-80108-3_20

Baah, C., Opoku-Agyeman, D., Acquah, I. S. K., Agyabeng-Mensah, Y., Afum, E., Faibil, D., & Abdoulaye, F. A. M. (2021). Examining the correlations between stakeholder pressures, green production practices, firm reputation, environmental and financial performance: Evidence from manufacturing SMEs. *Sustainable Production and Consumption, 27*, 100–114.

Bahram, M., Hildebrand, F., Forslund, S. K., Anderson, J. L., Soudzilovskaia, N. A., Bodegom, P. M., & Bork, P. (2018). Structure and function of the global topsoil microbiome. *Nature, 560*(7717), 233–237.

Bandh, S. A., Parray, J. A., & Shameem, N. (Eds.). (2022). *Climate change and microbial diversity: Advances and challenges*. Apple Academic Press.

Bano, S., Wu, X., & Zhang, X. (2021). Towards sustainable agriculture: Rhizosphere microbiome engineering. *Applied Microbiology and Biotechnology, 105*(19), 7141–7160.

Benito-Hernández, S., López-Cózar-Navarro, C., & Priede-Bergamini, T. (2023). The influence of government support over environmental protection investment on SMEs: R&D collaboration and financial aspects. *Business Ethics, the Environment & Responsibility, 32*(2), 836–846.

Bhaduri, D., Sihi, D., Bhowmik, A., & Dari, B. (2022). A review on effective soil health bio-indicators for ecosystem restoration and sustainability. *Frontiers in Microbiology, 13*, 938481.

Bhattacharyya, S. S., Ros, G. H., Furtak, K., Iqbal, H. M., & Parra-Saldívar, R. (2022). Soil carbon sequestration–An interplay between soil microbial community and soil organic matter dynamics. *Science of the Total Environment, 815*, 152928.

Bradu, P., Biswas, A., Nair, C., Sreevalsakumar, S., Patil, M., Kannampuzha, S., & Gopalakrishnan, A. V. (2023). Recent advances in green technology and industrial revolution 4.0 for a sustainable future. *Environmental Science and Pollution Research, 30*(60), 124488–124519.

Calabrese, F. M., Ameur, H., Nikoloudaki, O., Celano, G., Vacca, M., Junior, W. J., et al. (2022). Metabolic framework of spontaneous and synthetic sourdough metacommunities to reveal microbial players responsible for resilience and performance. *Microbiome, 10*(1), 148.

Chataut, G., Bhatta, B., Joshi, D., Subedi, K., & Kafle, K. (2023). Greenhouse gases emission from agricultural soil: A review. *Journal of Agriculture and Food Research, 11*, 100533.

Chaudhary, P., Ahamad, L., Chaudhary, A., Kumar, G., Chen, W. J., & Chen, S. (2023). Nanoparticle-mediated bioremediation is a powerful weapon in the removal of environmental pollutants. *Journal of Environmental Chemical Engineering, 11*(2), 109591.

Fadiji, A. E., & Babalola, O. O. (2020). Metagenomics methods for the study of plant-associated microbial communities: A review. *Journal of Microbiological Methods, 170*, 105860.

Fazeli-Nasab, B., Mahdinezhad, N., & Parray, J. A. (2022). Metagenomics and microbiome engineering: Identification of core microbiome and improvement of rhizosphere. In *Core microbiome: Improving crop quality and productivity* (pp. 58–100). Wiley.

Geisen, S., Briones, M. J., Gan, H., Behan-Pelletier, V. M., Friman, V. P., de Groot, G. A., & Wall, D. H. (2019). A methodological framework to embrace soil biodiversity. *Soil Biology and Biochemistry, 136*, 107536.

Haj-Amor, Z., Araya, T., Kim, D. G., Bouri, S., Lee, J., Ghiloufi, W., & Lal, R. (2022). Soil salinity and its associated effects on soil microorganisms, greenhouse gas emissions, crop yield, biodiversity, and desertification: A review. *Science of the Total Environment, 843*, 156946.

Hakim, S., Naqqash, T., Nawaz, M. S., Laraib, I., Siddique, M. J., Zia, R., & Imran, A. (2021). Rhizosphere engineering with plant growth-promoting microorganisms for agriculture and ecological sustainability. *Frontiers in Sustainable Food Systems, 5*, 617157.

Hartmann, M., & Six, J. (2023). Soil structure and microbiome functions in agroecosystems. *Nature Reviews Earth & Environment, 4*(1), 4–18.

Haughton, G. (2021). Environmental justice and the sustainable city. In *The Earthscan reader in sustainable cities* (pp. 62–79). Routledge.

Hochman, Z., Gobbett, D. L., & Horan, H. (2017). Climate trends account for stalled wheat yields in Australia since 1990. *Global Change Biology, 23*(5), 2071–2081.

Islam, W., Noman, A., Naveed, H., Huang, Z., & Chen, H. Y. (2020). Role of environmental factors in shaping the soil microbiome. *Environmental Science and Pollution Research, 27*, 41225–41247.

Jaffe, A. B., Newell, R. G., & Stavins, R. N. (2005). A tale of two market failures: Technology and environmental policy. *Ecological Economics, 54*(2–3), 164–174.

Jansson, J. K., & Hofmockel, K. S. (2020). Soil microbiomes and climate change. *Nature Reviews Microbiology, 18*(1), 35–46.

Ji, C., Ge, Y., & Li, K. (2024). Interactions between halotolerant nitrogen-fixing bacteria and arbuscular mycorrhizal fungi under saline stress. *Frontiers in Microbiology, 15*, 1288865.

Kumar, V., Singh, K., Shah, M. P., Singh, A. K., Kumar, A., & Kumar, Y. (2021). Application of omics technologies for microbial community structure and function analysis in the contaminated environment. In *Wastewater treatment* (pp. 1–40). Elsevier.

Kupolati, W. K., Sadiku, E. R., Frattari, A., Adeboje, A. O., Kambole, C., Mojapelo, K. S., & Berkers, R. F. A. (2019). Biopolymers and nanocomposites in civil engineering applications. In *Green biopolymers and their nanocomposites* (pp. 343–378). Springer.

Lacombe, C., Couix, N., & Hazard, L. (2018). Designing agroecological farming systems with farmers: A review. *Agricultural Systems, 165*, 208–220.

Lawal, T. O., Abdulsalam, M., Aliyu, M., & Sundararajan, S. (2023). Economic and environmental implications of sustainable agricultural practices in arid regions: A cross-disciplinary analysis of plant science, management, and economics. *International Journal, 10*(3), 3100–3114.

Li, Y., Chen, J., Feng, H., & Siddique, K. H. (2022). Plastic mulching significantly improves soil enzyme and microbial activities without mitigating gaseous N emissions in winter wheat-summer maize rotations. *Field Crops Research, 286*, 108630.

Livinus, M. U., Bala, S. Z., Abdulsalam, M., Innocent, M. O., Hassan, M., & Kini, P. (2024). Natural occurrences of soil dilapidation. In *Prospects for soil regeneration and its impact on environmental protection* (pp. 205–223). Cham.

Lv, X., Hueso-Gil, A., Bi, X., Wu, Y., Liu, Y., Liu, L., & Ledesma-Amaro, R. (2022). New synthetic biology tools for metabolic control. *Current Opinion in Biotechnology, 76*, 102724.

Mona, S., Malyan, S. K., Saini, N., Deepak, B., Pugazhendhi, A., & Kumar, S. S. (2021). Towards sustainable agriculture with carbon sequestration, and greenhouse gas mitigation using algal biochar. *Chemosphere, 275*, 129856.

Mukhtar, H., Wunderlich, R. F., Muzaffar, A., Ansari, A., Shipin, O. V., Cao, T. N. D., & Lin, Y. P. (2023). Soil microbiome feedback to climate change and options for mitigation. *Science of the Total Environment, 882*, 163412.

Musa, I. O., Isibor, P. O., Samuel, J. O., Mustapha, A., Mustapha, A., Akande, S., & Adeniji, H. (2024a). Introduction to Nanotoxicology. In *Environmental Nanotoxicology: Combatting the minute contaminants* (pp. 1–22). Cham.

Musa, I. O., Samuel, J. O., Adams, M., Abdulsalam, M., Nathaniel, V., Maude, A. M., & Tiamiyu, A. G. T. (2024b). Soil erosion, mineral depletion and regeneration. In *Prospects for soil regeneration and its impact on environmental protection* (pp. 159–172). Cham.

Ortiz, D. I., Piche-Ovares, M., Romero-Vega, L. M., Wagman, J., & Troyo, A. (2021). The impact of deforestation, urbanization, and changing land use patterns on the ecology of mosquito and tick-borne diseases in Central America. *Insects, 13*(1), 20.

Pahalvi, H. N., Rafiya, L., Rashid, S., Nisar, B., & Kamili, A. N. (2021). Chemical fertilizers and their impact on soil health. In *Microbiota and Biofertilizers, Vol. 2 Ecofriendly tools for reclamation of degraded soil environs* (pp. 1–20). Springer.

Philippot, L., Griffiths, B. S., & Langenheder, S. (2021). Microbial community resilience across ecosystems and multiple disturbances. *Microbiology and Molecular Biology Reviews, 85*(2), 10–1128.

Philippot, L., Chenu, C., Kappler, A., Rillig, M. C., & Fierer, N. (2023). The interplay between microbial communities and soil properties. *Nature Reviews Microbiology*, 1–14.

Pretty, J., & Bharucha, Z. P. (2014). Sustainable intensification in agricultural systems. *Annals of Botany, 114*(8), 1571–1596.

Rodríguez-Frías, F., Quer, J., Tabernero, D., Cortese, M. F., Garcia-Garcia, S., Rando-Segura, A., & Pumarola, T. (2021). Microorganisms as shapers of human civilization, from pandemics to even our genomes: Villains or friends? A Historical Approach. *Microorganisms, 9*(12), 2518.

Sandrini, M., Nerva, L., Sillo, F., Balestrini, R., Chitarra, W., & Zampieri, E. (2022). Abiotic stress and belowground microbiome: The potential of omics approaches. *International Journal of Molecular Sciences, 23*(3), 1091.

Saravanan, A., Kumar, P. S., Vo, D. V. N., Jeevanantham, S., Karishma, S., & Yaashikaa, P. R. (2021). A review on catalytic-enzyme degradation of toxic environmental pollutants: Microbial enzymes. *Journal of Hazardous Materials, 419*, 126451.

Sharma, P., Sangwan, S., Kaur, H., Patra, A., Anamika, & Mehta, S. (2023). Diversity and evolution of nitrogen-fixing bacteria. In *Sustainable agriculture reviews 60: Microbial processes in agriculture* (pp. 95–120). Springer.

Sokol, N. W., Slessarev, E., Marschmann, G. L., Nicolas, A., Blazewicz, S. J., Brodie, E. L., & Pett-Ridge, J. (2022). Life and death in the soil microbiome: How ecological processes influence biogeochemistry. *Nature Reviews Microbiology, 20*(7), 415–430.

Stefanis, C., Alexopoulos, A., Voidarou, C., Vavias, S., & Bezirtzoglou, E. (2013). Principal methods for isolation and identification of soil microbial communities. *Folia Microbiologica, 58*, 61–68.

Suman, J., Rakshit, A., Ogireddy, S. D., Singh, S., Gupta, C., & Chandrakala, J. (2022). Microbiome is a key player in sustainable agriculture and human health. *Frontiers in Soil Science, 2*, 821589.

Tahat, M. M., Alananbeh, K. M., Othman, Y. A., & Leskovar, D. I. (2020). Soil health and sustainable agriculture. *Sustainability, 12*(12), 4859.

Tripathi, L. K., & Nailwal, T. K. (2020). Metagenomics: Applications of functional and structural approaches and meta-omics. In *Recent advancements in microbial diversity* (pp. 471–505). Academic.

Trump, B., Cummings, C., Klasa, K., Galaitsi, S., & Linkov, I. (2023). Governing biotechnology to provide safety and security and address ethical, legal, and social implications. *Frontiers in Genetics, 13*, 1052371.

Tyagi, J., Ahmad, S., & Malik, M. (2022). Nitrogenous fertilizers: Impact on environment sustainability, mitigation strategies, and challenges. *International journal of Environmental Science and Technology, 19*(11), 11649–11672.

Wachinger, G., Renn, O., Begg, C., & Kuhlicke, C. (2013). The risk perception paradox—Implications for governance and communication of natural hazards. *Risk Analysis, 33*(6), 1049–1065.

Wagg, C., Schlaeppi, K., Banerjee, S., Kuramae, E. E., & van der Heijden, M. G. (2019). Fungal-bacterial diversity and microbiome complexity predict ecosystem functioning. *Nature Communications, 10*(1), 4841.

Waldrop, M. P., Chabot, C. L., Liebner, S., Holm, S., Snyder, M. W., Dillon, M., & Mackelprang, R. (2023). Permafrost microbial communities and functional genes are structured by latitudinal and soil geochemical gradients. *The ISME Journal, 17*(8), 1224–1235.

Wigboldus, S., Klerkx, L., Leeuwis, C., Schut, M., Muilerman, S., & Jochemsen, H. (2016). Systemic perspectives on scaling agricultural innovations. A review. *Agronomy for Sustainable Development, 36*, 1–20.

Yang, F., Fu, Q., & Antonietti, M. (2023). Anthropogenic, carbon-reinforced soil as a living engineered material. *Chemical Reviews, 123*(5), 2420–2435.

Ziolo, M., Filipiak, B. Z., Bąk, I., & Cheba, K. (2019). How to design more sustainable financial systems: The roles of environmental, social, and governance factors in the decision-making process. *Sustainability, 11*(20), 5604.

Index

S. A. Aransiola et al. (eds.), *Soil Microbiome in Green Technology Sustainability*, https://doi.org/10.1007/978-3-031-71844-1

D

E

F

N

O

The manufacturer's authorised representative in the EU is Springer Nature Customer Service Centre GmbH, Europaplatz 3, 69115 Heidelberg, Germany. If you have any concerns regarding our products, please contact ProductSafety@springernature.com

Printed and bound by CPI Group (UK) Ltd, Croydon, CR0 4YY
01/10/2026
02225906-0018